SIGNAL PROCESSING in RADAR SYSTEMS

SIGNAL PROCESSING in RADAR SYSTEMS

Vyacheslav Tuzlukov

CRC Press
Taylor & Francis Group
Boca Raton London New York

CRC Press is an imprint of the
Taylor & Francis Group, an **informa** business

CRC Press
Taylor & Francis Group
6000 Broken Sound Parkway NW, Suite 300
Boca Raton, FL 33487-2742

First issued in paperback 2017

Version Date: 20120629

ISBN 13: 978-1-4398-2607-2 (hbk)
ISBN 13: 978-1-138-07546-7 (pbk)

Library of Congress Cataloging-in-Publication Data

Tuzlukov, V. P. (Viacheslav Petrovich)
Signal processing in radar systems / Vyacheslav Tuzlukov.
p. cm.
Includes bibliographical references and index.
ISBN 978-1-4398-2607-2
1. Radar. 2. Signal processing I. Title.

TK6580.T87 2012
621.382'2--dc23 2012006801

Visit the Taylor & Francis Web site at
http://www.taylorandfrancis.com

and the CRC Press Web site at
http://www.crcpress.com

To the undying memory of Dr. Peter Tuzlukov,
my dear father and teacher

Contents

PART III Stochastic Processes Measuring in Radar Systems

Preface

The essential task in radar systems is to find an appropriate solution to the problems related to robust signal processing and definition of signal parameters. There are now a number of books and papers published in journals devoted to signal processing in noise in radar systems, but many key issues remain unsolved. New approaches to these problems allow us not only to summarize investigations but also to derive better quality of robust signal processing in noise in radar systems.

This book addresses the problems of robust signal processing in complex radar systems (CRSs) based on the generalized approach to signal processing in noise. The generalized approach to signal processing in noise is proposed based on a seemingly abstract idea: the introduction of an additional noise source that does not carry any information about the signal to improve the qualitative performance of CRSs. Theoretical and experimental studies lead to the conclusion that the proposed generalized approach to signal processing in noise in CRSs allows formulating a decision-making rule based on the determination of the *jointly sufficient statistics of the mean and variance* of the likelihood function. The use of classical and modern signal processing approaches allows us to define only the *sufficient statistic of the mean* of the likelihood function (or functional).

The presence of additional information about the statistical characteristics of the likelihood function leads to better qualitative performances of robust signal processing in CRSs in comparison with optimal signal processing algorithms of classical and modern theories. The generalized approach to signal processing allows us to extend the well-known boundaries of potential noise immunity set up by classical and modern signal processing theories. The use of CRSs based on the generalized approach to signal processing in noise allows us to obtain better detection performances, particularly in comparison with CRSs constructed on the basis of optimal and asymptotic optimal signal processing algorithms of classical and modern signal processing theories.

To better understand the fundamental statements and concepts of the generalized approach, the reader is invited to consult my earlier books: *Signal Processing in Noise: A New Methodology* (IEC, Minsk, 1998), *Signal Detection Theory* (Springer-Verlag, New York, 2001), *Signal Processing Noise* (CRC Press, Boca Raton, FL, 2002), and *Signal and Image Processing in Navigational Systems* (CRC Press, Boca Raton, FL, 2005).

The radar system is an important element in the field of electrical engineering. In university engineering courses, in general, the emphasis is usually on the basic tools used by the electrical engineer, such as circuit design, signals, solid state, digital processing, electronic devices, electromagnetics, automatic control, microwaves, and so on. In the real world of electrical engineering practice, however, these are only techniques, piece parts, or subsystems that make up some type of system employed for a useful purpose.

There are various aspects to radar system design. However, before a new radar system can be manufactured, a *conceptual design* has to be made to guide the actual development, taking into consideration the requirements of the radar system that must be customer- and user-friendly. Conceptual design involves identifying the characteristics of the radar system in accordance with the radar equation and related equations and the general characteristics of the subsystems such as transmitter, antenna, receiver, signal processing, etc., that might be used. A conceptual design cannot be formed without a systems approach. Another important procedure is to define the structure of computer subsystems used in the radar for the purpose of implementing modern robust signal processing algorithms.

It should be noted that there are at least two ways in which a new CRS might be produced. One method is based on exploiting the advantages of some new invention, new technique, new device, or new knowledge. The invention of the microwave magnetron early in World War II is an example.

After the magnetron appeared, radar system design was different from what it had been before. The other, and probably more common, method for conceptual radar system design is to identify the function the new radar system has to perform, examine the various approaches available to achieve the desired capability, carefully evaluate each approach, and then select the one that best meets the needs within the operational and fiscal constraints imposed. This book discusses in detail these two methods that are based on a systems approach to design radar systems.

An important task in designing CRSs is to use robust signal processing algorithms and accurate definition of signal parameters. To this end, theory and methods of experimental investigations of stochastic processes are attracted to design the CRS. The theory of statistical estimates, for example, can be used for analyzing regularities to design and construct optimal and quasi-optimal meters of statistical parameters of stochastic processes. At the same time, significant attention is paid to investigation of systematic and random errors of statistical parameter definition as a function of considered time interval and noise level.

A detailed analysis of various procedures and methods to measure and estimate the main statistical parameters of stochastic processes, such as mean (or mathematical expectation), variance, correlation (covariance) function, power spectral density, probability density functions, spikes of energy spectra, etc., is presented. Analog and discrete procedures and methods for measurements and errors, which are characteristic of these procedures and methods, are investigated. In addition, structural block diagrams of digital meters are considered. Structural block diagrams of optimal meters to define the mathematical expectation (mean), variance, and parameters of the correlation (covariance) function are discussed. The variance of estimations and biases of the earlier-mentioned parameters is measured. A procedure to measure the mathematical expectation (mean) and variance of nonstationary stochastic process under robust signal processing used in CRSs is identified. General formulas for definition of biases and variances of statistical parameter estimations are also presented for direct analytical calculation.

I would like to thank my colleagues in the field of robust signal processing in radar systems for useful discussion about the main results, in particular, Professors V. Ignatov, A. Kolyada, I. Malevich, G. Manshin, D. Johnson, B. Bogner, Yu. Sedyshev, J. Schroeder, Yu. Shinakov, A. Kara, Kyung Tae Kim, Yong Deak Kim, Yong Ki Cho, V. Kuzkin, W. Uemura, Dr. O. Drummond, and others.

I would also like to express my gratitude to my colleagues from the Department of Information Technologies and Communications, Electronics Engineering School, College of IT Engineering, Kyungpook National University (KNU), Daegu, South Korea, for useful remarks and comments and their help in completing this project.

This research was supported by the Kyungpook National University Research Fund, 2010.

A lot of credit is also due to Nora Konopka, Kari Budyk, Richard Tressider, Suganthi Thirunavukarasu, and John Gandour as well as to the entire staff at CRC Press, Taylor & Francis Group, for their encouragement and support.

Last, but definitely not least, I would like to thank my family—my lovely wife, Elena; my sons, Anton and Dima; and my dear mom, Natali—for putting up with me during the writing of the manuscript. Without their support, this book would not have been possible!

Finally, I wish to express my lifelong, heartfelt gratitude to Dr. Peter Tuzlukov, my father and teacher, who introduced me to science.

Vyacheslav Tuzlukov

MATLAB® is a registered trademark of The MathWorks, Inc. For product information, please contact:

The MathWorks, Inc.
3 Apple Hill Drive
Natick, MA, 01760-2098 USA
Tel: 508-647-7000
Fax: 508-647-7001
E-mail: info@mathworks.com
Web: www.mathworks.com

Author

Dr. Vyacheslav Tuzlukov is currently a full professor in the Department of Information Technologies and Communication, School of Electronics Engineering, College of IT Engineering, Kyungpook National University, Daegu, South Korea. His research interests include signal processing, detection and estimation in radar, wireless communications, wireless sensor networks, remote sensing, sonar, and mobile communications. He is the author of over 170 journal and conference papers and 8 books on signal processing published by Springer-Verlag and CRC Press, in particular, *Signal Detection Theory* (2001), *Signal Processing Noise* (2002), *Signal and Image Processing in Navigational Systems* (2005), and *Signal Processing in Radar Systems* (this book). He has also contributed the chapters entitled "Underwater Acoustical Signal Processing" and "Satellite Communications Systems: Applications" to *Electrical Engineering Handbook: 3rd Edition* (2005), and the chapter entitled "Generalized Approach to Signal Processing in Wireless Communications: The Main Aspects and Some Examples" to *Wireless Communications* (2012). He is a keynote speaker, chair of sessions, tutorial instructor, and plenary speaker at major international conferences on signal processing. Dr. Tuzlukov has been highly recommended by U.S. experts of Defense Research and Engineering (DDR&E) of the United States Department of Defense (U.S. DoD) for his expertise in the field of humanitarian demining and minefield-sensing technologies and was awarded the Special Prize of the U.S. DoD in 1999. His achievements have distinguished him as one of the leading experts from around the world by *Marquis Who's Who*. His name and biography have been included in the *Who's Who in the World, 2006–2012*; *Who's Who in the World*, 25th silver anniversary edition, 2008, Marquis Who's Who, Berkeley Heights, NJ; *Who's Who in Science and Engineering, 2006–2012*; *Who's Who in Science and Engineering*, 10th anniversary edition, 2006–2009, Marquis Who's Who, Berkeley Heights, NJ; *2009–2010 Princeton Premier Business Leaders and Professionals Honors Edition*, Princeton Premier, New York; *2009 Strathmore's Who's Who Edition*, Strathmore's Who's Who, New York; *2009 Presidential Who's Who Edition*, Presidential Who's Who, New York; *Who's Who among Executives and Professionals*, 2010 edition, Marquis Who's Who, Berkeley Heights, New York; and *Who's Who in Asia 2012*, 2nd edition, Marquis Who's Who, Berkeley Heights, New York.

Introduction

This book addresses robust signal processing problems in complex radar systems (CRSs) and describes their features. Both traditional problems of synthesis and analysis of the main digital signal processing operations and new problems of the robust signal processing in noise, in particular, the generalized approach to signal processing in noise under coherent filtering, are presented in this book. Problems of adaptation and control of functioning processes in CRSs are also new problems both by the problem statement and by the methods to solve these problems. Successes in automation allow us to make definite generalizations that will promote further development.

When designing CRSs from a modern perspective, it is important to support an interrelation between the individual robust signal processing algorithms and designing the multipronged attack on an all-round algorithm of CRS functioning in parallel with a choice of robust signal processing algorithms. In this regard, this book focuses on the problems related to system design. In this book, the complexity and difficulty in realization of robust signal processing algorithms are defined by specific systems approaches, allowing us to identify realistically the requirements of CRSs.

Construction of CRSs for information and control purposes is a multistage, long-term process. One of the important stages of system construction is design. That said, the actual problem lies in essentially increasing the quality of design while simultaneously reducing system construction time. To solve this problem, extensive use of science-based methods of CRS construction, taking into account features and functioning conditions, plays an important role.

An important problem is the definition of signal parameters. The process of defining signal parameters is related to theoretical and experimental investigation methods of stochastic processes. Experimental investigations of stochastic processes are applied in the following cases:

- Under analysis of signal transformations by linear and nonlinear systems when we do not have a priori information with respect to statistic of the input stochastic process and the physical source generating the input stochastic process.
- There is a need to check the correctness of theoretical approaches applied to the investigation of CRSs.
- Mathematical description of physical processes in CRSs is cumbersome and impractical.

Analysis of statistical regularities for the design of optimal and quasi-optimal measuring systems to define CRS signal parameters based on the theory of statistical estimations is presented in this book. Much attention is also paid to the analysis of systematic and stochastic errors of signal parameter definition as a function of observation time and noise power. Procedures of experimental investigations of CRS stochastic processes are much more difficult when compared to methods of experimental investigations of deterministic processes. This can be explained due to the following reasons:

- To describe the stochastic process completely, extensive evidence of various signal parameters is required.
- It is impossible, in practice, to carry out a measurement of individual signal parameters in accordance with their definition.

Detailed investigations have been carried out for various methods to measure the main statistical characteristics of stochastic processes and their estimates such as mean, variance, correlation function, power spectral density, probability density function, probability distribution function, and so on. Analysis of measuring procedures and errors of these procedures under robust signal processing as well as block diagrams of programs for digital measuring systems are presented in this book. Structures of optimal metering systems for definition of the mean, variance, and parameters of the correlation functions are designed, and the drifts and estimate variances of these characteristics are also determined. The procedure of measurement of the mathematical expectation and variance of nonstationary stochastic processes under robust signal processing is discussed in detail. The general mathematical relations for the drifts and estimate variances of the main stochastic process characteristics are presented in an appropriate form for analytical calculations in the cases of the Gaussian and Rayleigh distributions of stochastic processes.

The following principal problems of mathematical statistics are frequently encountered under investigation of stochastic processes in robust signal processing based on the generalized approach to signal processing in noise in CRSs:

- Definition of unknown probability distribution functions and probability density functions
- Definition of unknown parameters of probability distribution function and probability density function
- Test of statistical hypotheses

It must be noted that the last of these problems is a rare case compared to the first two problems.

The law of large numbers lies at the heart of experimental methods to define characteristics of stochastic processes under robust signal processing used in CRSs. According to this law, the probability of an event can be changed by the corresponding event appearance frequency, and the mathematical expectation can be changed by an average. In practice, given that a large number of tests are conducted, we can conclude that the probability of an event and its characteristics obtained in this manner are close to the true values. However, there are situations when we need to use only a limited number of tests. For these cases, we apply the same mathematical formulas as under the large number of tests. As a result, an additional problem, related to defining the estimate of characteristics obtained based on testing and comparing this estimate with the potentially achieved accuracy of measuring devices, arises.

Definition of extreme accuracy to estimate stochastic process parameters for our purposes is based on methods and procedures of statistical decision-making theory. This theory is developed to construct and design optimal measuring devices for deterministic and quasideterministic signals in noise and to analyze signal parameters and their estimations [1–5]. Investigations devoted to the methods of experimental study of stochastic processes and accurate definitions of estimations of statistical parameters are also widely covered in the literature [6–21].

In this book, we attempt to analyze the measuring methods of the main parameters of stochastic processes using the unified methodological approach, taking into consideration the theory of statistical estimates. The stationary ergodic stochastic processes and analog procedures to measure their parameters, in general, are investigated owing to the highest measuring accuracy. Specificity of digital measuring technique lies in analog-to-digital transform of signals under robust signal processing in CRSs and in the use of radar computer subsystems [22].

This book provides a definition of and investigates potential accuracy measurement of the probability distribution and probability density functions, correlation and covariance functions, mathematical expectation, variance, spikes of energy spectrum as a function of observation time, correlation interval of considered stochastic processes, and signal-to-noise ratio. In doing so, the bias, variance, and correlation function of estimations widely used in the theory of statistical estimates and mathematical statistics are employed to measure the accuracy of stochastic processes. To obtain more simple and illustrative examples, we use the approximated solutions acceptable in practice.

This book summarizes the investigations carried out by the author over the last 30 years. It consists of three parts. Part I discusses the main design principles of the modern robust digital signal processing algorithms used in CRSs. Special attention is paid to the generalized approach of signal processing in noise. Part II covers the main principles of computer system design for modern robust digital signal processing algorithms. Some examples of designing actual CRSs are discussed in this part. Part III deals with experimental measurements of the main statistical parameters of stochastic processes and definitions of their estimations. Important estimates of statistical parameters of stochastic processes, such as mathematical expectation, variance, correlation function, probability density function, probability distribution function, and other frequency–time parameters, are defined by experimental investigations.

The book consists of 15 chapters. Chapter 1 discusses the principles of systems approach to design CRSs. It focuses on the design methodology of and main requirements of CRSs. Problems in the system design of complex automated radar systems are covered. Representation of signal processing subsystems as an object of design is also investigated.

Chapter 2 deals with signal processing by the digital generalized detector in CRSs based on the digital signal processing procedures. The main principles of analog-to-digital conversion of signals are discussed. A comparative analysis of the generalized approach to signal processing in noise and matched filtering for digital signal processing procedures is done and the key results are analyzed. In addition, comparison between the digital matched filtering and digital generalized detector constructed for coherent pulse signal is presented. Theoretical study is strengthened by computer modeling results that are discussed.

Chapter 3 presents robust digital interperiod signal processing algorithms. It investigates robust digital signal processing algorithms for selection of moving targets. Digital generalized detectors of target return signals, for cases when the noise is with known and unknown statistical parameters, are discussed. Comparative analysis with robust digital generalized detectors is carried out. Digital measuring systems of signal parameters are investigated. Complex interperiod signal processing algorithms, including robust digital generalized signal processing algorithms are also investigated and a comparison is made. Theoretical study is strengthened by computer modeling results that are discussed.

Chapter 4 explores problems of robust signal detection algorithms based on the generalized approach to signal processing in noise and trajectory tracking of targets on digital measurements carried out by measuring subsystems. The principal stages and secondary operations of robust signal processing are discussed. Trajectory detection and tracking of targets through surveillance of CRS data are investigated and discussed. Theoretical study is strengthened by computer modeling results that are discussed.

Chapter 5 provides a definition of filtering algorithms based on the generalized approach to signal processing in noise for robust signal processing and extrapolation of target trajectory parameters using the measured data obtained by CRSs. In this chapter, the initial premises to determine the estimations and errors of trajectory parameters are defined. Investigation of the filtered input stochastic process at the front end of the generalized receiver is carried out. A statistical approach to define random unknown signal parameters by filtering technique is discussed. Algorithms of linear filtering and extrapolation under fixed sample size are defined and investigated. Recurrent filtering algorithms for the definition of trajectory parameters of the nondistorted polynomial trajectory are presented and discussed. Adaptive filtering, which allows us to define the trajectory parameters of maneuvering targets, is investigated in detail. Logical block diagrams of all-round algorithms for secondary signal processing are also discussed. Theoretical study is strengthened by computer modeling results that are discussed.

Chapter 6 deals with the principles of design and construction of control algorithms by CRS functioning in a dynamic regime. Organization principles and structural block diagrams of CRS control circuits are discussed. Principles of direct control of parameters and, in particular, device parameters, are formulated. Procedures of radar scanning control under new target searching are constructed and discussed. The principles of resource management under target tracking are defined. Distribution of energy resources under overlapping of target searching and tracking operations is described. Theoretical study is strengthened by computer modeling results that are discussed.

Chapter 7 deals with the principles for designing all-round algorithms of CRS computer subsystems. The methods of assignment of all-round algorithms for computer subsystems are defined. An estimation of work content for the realization of all-round algorithms in computer subsystems is carried out. Principles of parallelization of computational processes by computer subsystems are discussed. Theoretical study is strengthened by computer modeling results that are discussed.

Chapter 8 focuses on the designing principles of CRS computer subsystems employing robust signal processing algorithms based on the generalized approach to signal processing in noise. The structure and technical requirements and parameters of computer subsystems are defined. The technical requirements concerning an effective processing speed are validated, taking into consideration the memory size and memory structure of computer subsystems. The technical characteristics of central computer subsystem microprocessors are defined. The structure and elements of computer subsystems are also discussed. The requirements and structure of central high-performance computer systems are proposed. Programmable microprocessors for robust signal preprocessing are designed. Theoretical study is strengthened by computer modeling results that are discussed.

Chapter 9 looks into an example to design CRS digital signal processing subsystems. The main statements and initial conditions are defined. The structure of designing computer subsystems is developed and validated. The microprocessor structure for robust coherent and incoherent signal preprocessing is also developed and validated. The requirements for microprocessor structure and characteristics under secondary robust signal processing are formulated. An example of a computer subsystem for robust signal processing of target return signals is discussed. Theoretical study is strengthened by computer modeling results that are discussed.

Chapter 10 is devoted to the analysis of digital signal processing subsystem samples. One of the variants of digital signal processing subsystems is proposed and analyzed. Analysis of radar measuring devices of the kinds "$n - 1 - 1$," "$n - n - 1$," and "$n - m - 1$" is carried out, and the results are discussed. A comparative analysis of the proposed and considered versions of the target tracking subsystem design is carried out, and the results are discussed. Theoretical study is strengthened by computer modeling results that are discussed.

Chapter 11 focuses on the theory and experimental measurement of statistical estimation of stochastic process parameters. Basic definitions are presented, point estimations are defined, and their main features are discussed. Effective estimations, the loss function and average risk, and Bayesian estimations for various loss functions are also defined and discussed. Theoretical study is strengthened by computer modeling results that are discussed.

Chapter 12 explores the mathematical expectation estimation of stochastic processes defined by experimental tests. The conditional functional of probability density function with estimates at the mathematical expectation is defined by experimental study. The maximal mathematical expectation likelihood estimate and the Bayesian estimate of the quadrature loss function mathematical expectation are defined experimentally. Applied approaches to estimate the mathematical expectation by tests are discussed. The mathematical expectation estimate under analysis of parameters of stochastic processes at discrete time instants is defined experimentally. The estimate of the mathematical expectation under amplitude of stochastic processes is defined experimentally. The optimal estimate of the Gaussian stochastic process mathematical expectation varying as a function in time is defined experimentally. The optimal estimate of the mathematical expectation varying as a function in time under averaging in time the stochastic process is defined by experiment at robust signal processing. The estimate of stochastic process mathematical expectation is defined experimentally by interactive approaches. The estimate of stochastic process mathematical expectation with unknown time is defined by a test. Theoretical study is strengthened by computer modeling results that are discussed.

Chapter 13 deals with the estimate of stochastic process variance, which is defined by experimental investigation. The optimal estimate of Gaussian stochastic process variance is defined by a test. The estimate of stochastic process variance at averaging in time is defined experimentally. Errors caused by difference of transducer performance from the quadratic function and limitations in instantaneous values are defined experimentally. The estimate of stochastic process variance varying as a function of time is defined experimentally. Measurements of stochastic process variance under stimulus of the noise are carried out by tests. Theoretical study is strengthened by computer modeling results that are discussed.

Chapter 14 deals with the estimate of the probability distribution function and probability density function of stochastic process defined by experimental investigation. The main features of estimates of the probability distribution function and probability density function of stochastic process are defined by test. The major characteristics of the probability distribution function and probability density function estimate of stochastic process as well as the estimation variance of the Gaussian and Rayleigh probability distribution functions are also defined experimentally. Estimation of the probability density function of stochastic process based on estimations of expansion coefficients is defined by a test. The designing principles of measuring devices of the probability distribution function and probability density function of stochastic process are discussed and confirmed by experimental investigations. Theoretical study is strengthened by computer modeling results that are discussed.

Chapter 15 focuses on frequency–time parameter estimations of stochastic process defined by experimental investigation. Estimation of the correlation function of stochastic process is defined by experimental investigation. Estimation of the correlation function of stochastic process based on series expansion is defined by experiment. Estimation of the optimal correlation function parameter of Gaussian stochastic process is defined experimentally. Methods of correlation function definition based on other principles of estimations are defined experimentally. Estimations of the power spectral density and of spike parameters of stochastic process are defined experimentally. The average quadratic frequency estimation of power spectral density of stochastic process and measurements of the correlation functions and power spectral densities of stochastic process by digital signal processing approaches are presented experimentally. Theoretical study is strengthened by computer modeling results that are discussed.

This book presents the different principles of optimization in the structure of computer subsystems in the design and construction of CRSs, taking into account all peculiarities of robust signal processing of target return signals and control algorithms, by which we are able to define the parameters of the target and target trajectory. Additionally, it describes numerous procedures and methods to measure the principal statistical parameters of target and target trajectory and to define their estimates under signal and image processing.

REFERENCES

1. Amiantov, I. 1971. *Selected Problems of Statistical Theory of Communications*. Moscow, Russia: Soviet Radio.
2. Van Trees, H. 2003. *Detection, Estimation, and Modulation Theory. Part I*. New York: John Wiley & Sons, Inc.
3. Tihonov, V. 1983. *Optimal Signal Reception*. Moscow, Russia: Radio I Svyaz.
4. Falkovich, S. and A. Homyakov. 1981. *Statistical Theory of Measuring Systems*. Moscow, Russia: Radio I Svyaz.
5. Skolnik, M. 2002. *Introduction to Radar Systems*. 3rd edn. New York: McGraw-Hill, Inc.
6. Mirskiy, G. 1971. *Definition of Stochastic Process Parameters by Measuring*. Moscow, Russia: Energy.
7. Tzvetkov, A. 1979. *Foundations of Statistical Measuring Theory*. Moscow, Russia: Energy.
8. Tihonov, V. 1970. *Spikes of Statistical Processes*. Moscow, Russia: Science.
9. Mirskiy, G. 1982. *Parameters of Statistical Correlation and Their Measurements*. Moscow, Russia: Energoatomizdat.
10. Widrow, B. and S. Stearns. 1985. *Adaptive Signal Processing*. Upper Saddle River, NJ: Prentice-Hall, Inc.
11. Lacomme, P., Hardange, J., Marchais, J., and E. Normant. 2001. *Air and Spaceborn Radar Systems: An Introduction*. New York: William Andrew Publishing.
12. Stimson, G. 1998. *Introduction to Airborne Radar*. Raleigh, NC: SciTech Publishing, Inc.
13. Morris, G. and L. Harkness. 1996. *Airborne Pulsed Doppler Radar*. 2nd edn. Norwood, MA: Artech House, Inc.

14. Nitzberg, R. 1999. *Radar Signal Processing and Adaptive Systems*. Norwood, MA: Artech House, Inc.
15. Mitchell, R. 1976. *Radar Signal Simulation*. Norwood, MA: Artech House, Inc.
16. Nathanson, F. 1991. *Radar Design Principles*. 2nd edn. New York: McGraw-Hill, Inc.
17. Tsui, J. 2004. *Digital Techniques for Wideband Receivers*. 2nd edn. Raleigh, NC: SciTech Publishing, Inc.
18. Gray, M., Hutchinson, F., Ridgley, D., Fruge, F., and D. Cooke. 1969. Stability measurement problems and techniques for operational airborne pulse Doppler radar. *IEEE Transactions on Aerospace and Electronic System*, AES-5: 632–637.
19. Scheer, J. and J. Kurtz. 1993. *Coherent Radar Performance Estimation*. Norwood, MA: Artech House, Inc.
20. Guerci, J. 2003. *Space-Time Adaptive Processing for Radar*. Norwood, MA: Artech House, Inc.
21. Tait, R. 2005. *Introduction to Radar Target Recognition*. Cornwall, U.K.: IEE.
22. Rabiner, L. and B. Gold. 1992. *Theory and Application of Digital Signal Processing*. Upper Saddle River, NJ: Prentice-Hall, Inc.

Part I

Design of Radar Digital Signal Processing and Control Algorithms

1 Principles of Systems Approach to Design Complex Radar Systems

1.1 METHODOLOGY OF SYSTEMS APPROACH

The design and construction of any complex information and control systems, including complex radar systems (CRSs), involve a long-term multistage process. The most important stage of CRS development is the design. Therefore, an essential development carried out simultaneously with a reduction in lead time is an important issue of the day. The solution of this problem is based on designing and widespread occurrence of science-based methods of system development taking into consideration structural features of systems and conditions of functioning that are founded on widely used computer subsystems.

The essential methodological principle of this approach is systems engineering. In systems engineering, we understand the term *designing* as a stage of system cycle from a compilation of requirements specification for development of complex information and control systems, to prototype production and carrying out a comprehensive test operation. The designing process is divided into two sufficiently pronounced stages:

- *System designing*: to select and organize functional operations and complex information and control systems
- *Engineering designing*: to select and develop elements of complex information and control systems

Under system designing, an object is considered as a system intended to achieve definite goals at the expense of controlled interaction subsystems, mainly. A conception of complex information and control system integrity, makes specific owing to an idea of backbone communications, for example, structural and control communications. At the stage of system designing, the most important issue is a structure or architecture of the future complex information and control system, that is, a fixed totality of elements and communications between them.

Characteristics of such complex system structures are as follows:

- *Autonomy of individual controlled subsystems*: Each subsystem controls a limited number of sub subsystems.
- *Subsystems are controlled under incomplete information*: The high-level subsystem cannot know problems and restrictions for the low-level subsystems.
- Information packing (generalization) under hierarchy motion up.
- Presence of particular problems to control each subsystem and a general problem for the system as a whole.
- Interaction between subsystems due to the presence of total restrictions.

Investigation of possible variants of a complex radar system structure allows us to solve some problems concerning the architecture of the developed CRS while putting aside for the time being the concrete element base that is used in designing the system. In doing so, we take into account the fact that structural regularities are stable. Selection and comparison of structural variants is not a prime problem. However, this problem is important because an unsuccessful choice of the CRS structure may bring to one's grave the results of the next stages of development.

Any CRS cannot be imagined without the so-called environment. Separation between the system and the environment is not well defined and can be realized in a large number of ways. The main problem is to define an optimal boundary between the system and the environment. At the same time, there is a need to take into consideration all factors affecting the system or the effects of system operation. In the case of information systems, including CRSs that operate in conflict situations, the most essential external factors are as follows:

- *Environment*: the weather, atmospheric precipitation, underlying terrain, etc.
- *Facilities* of counteracting forces (enemy)
- *Level of development* of the element base and modern technologies
- *Economic factors*: the facilities, timing of orders, time for completion of system designing, and so on
- *Human element*: the team with a good understanding and knowledge of how to perform a high-quality job

Based on a methodology viewpoint, we can emphasize the following aspects of the systems approach to design any complex information and control system:

1. The complex hierarchical system (the CRS) can be divided into a set of subsystems, and it is possible to design each individual subsystem. For this case, an optimization of subsystems does not solve the problem of optimization of the complex hierarchical system as a whole. Designing of the radar system as an integrated object with a predefined mission is related to the trade-off decisions ensuring a maximal efficiency thanks to the decrease in efficiency of some individual subsystems.
2. All alternative variants of CRS structure must be considered and analyzed at the initial designing, and a structure that satisfies all quality standards must be chosen. Now, a choice of the system structure variants, which must be optimized, is carried out by heuristic methods based on experience, intuition, creativeness, and ingenuity of engineers. Evidently, heuristic elements in the designing of radar systems are inevitable in the future.
3. A choice of the favorite variant of the complex radar system structure depends on the possibility to estimate the effectiveness of each alternative structure and the finances that are necessary to realize this alternative structure. For this purpose, there is a need to use quantitative measures of quality, namely QoS (quality of service). In design problems, the QoS criteria are also called the objective functions of optimal design. The CRS is considered effective if the following main requirements are satisfied:
   a. Under given conditions of operation, the CRS solves the assigned tasks completely and at a stated time—this is its technical efficiency.
   b. The cost of the problem solved by the CRS is not less than the cost of manufacture and the cost of maintenance during its operation.

   The criterion satisfying aforementioned requirements can be presented in the following form:

$$\mathcal{T} = \mathcal{G} - \mathcal{W}, \tag{1.1}$$

where

$\mathcal{G}$ is the positive effect following the use of the CRS for definite purposes

$\mathcal{W}$ is the designing, development, and exploitation charges of the CRS

Choice of the performance criterion is the exterior problem, which must be solved based on an analysis of high-level information and control system purposes in comparison with the considered system. Naturally, the designed CRS is a constituent of information and control system.

4. The problem of searching for the best structure of the CRS must be solved by employing a computer-aided design system and using the optimization problems.
5. The CRS model is a physical or abstract model, which can sufficiently represent some aspects of the system operation. Adequacy assumes a reproduction by the system model all features with a sufficient completeness, which are essential to reach the end purpose of a given investigation. In the design of complex radar systems we widely use the following:
   a. *Mathematical models*: a representation of the system operations using a language of mathematical relations and definitions.
   b. *Simulation models*: a reproduction of the system operation by other computer subsystems.
   c. *Modeling*: a process of representation of the investigated system by an adequate model with subsequent test operation to obtain information about system functioning.
6. Preliminary structure of the complex radar system seems uncertain at the initial moment. Solutions made at the beginning of system designing are approximated. Solutions are made clear with accumulation of knowledge. Consequently, the designing process is an iterative process, at each stage of which we search for a solution that is perfect in comparison with the previous one. The iterative character of a designed problem solution is a principal difference between the systems approach and traditional and ordinary approaches under system synthesis and designing.

Thus, the most characteristic feature of the systems approach to design CRSs is a decision searching by iterative optimization based on computer-aided designing systems. Figure 1.1 represents a

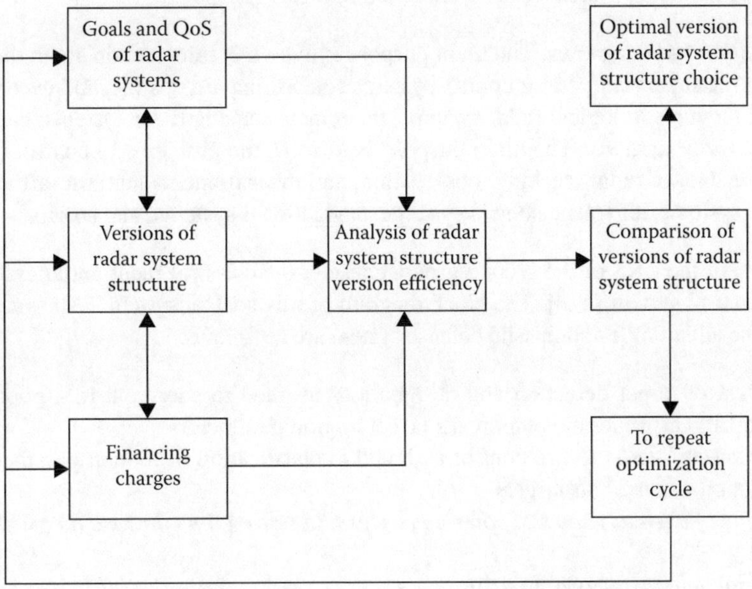

FIGURE 1.1 Block diagram of choice in optimal decision regarding the CRS structure.

block diagram to choose the optimal structure. In accordance with this diagram, the main operations of the system designing are carried out in the following order:

- Definition and generation of the end goals, restrictions for number of goals, and problems that must be solved by the system; choice and justification of the system QoS (the efficiency criterion)
- Generation of all possible alternative variants to design the system, including impossible versions, at the first sight
- Definition of investments to realize each alternative version of the system structure
- Designing the models chosen to optimize the alternatives and their software implementation, definition of QoS, and costs of the alternative system structures using the definite model
- Comparison of the alternatives and the decision making: either to recommend one or several versions of the radar system structure for further designing or to repeat the whole cycle of optimization process changing a set of initial statements and defining more exactly the QoS (the criterion of effectiveness)

In conclusion, various mathematical methods and procedures are required at the stage of system designing: the theory of probabilities and mathematical statistics; the theory of linear, nonlinear, and dynamic programming; the theory of modeling; etc.

1.2 MAIN REQUIREMENTS OF COMPLEX RADAR SYSTEMS

There are various aspects to complex radar system design. Before a new CRS that has not existed previously can be manufactured, a conceptual design has to be performed to guide the actual development. A conceptual design is based on the requirements for the system that will satisfy the customer or user. The result of the conceptual design effort is to provide a list of the radar characteristics as found in the general characteristics of the subsystems that might be employed, namely, transmitter, antenna, receiver, signal processing, and so forth.

Automated CRSs are widely used to solve, for instance, the following problems: air-traffic control, military fighter/attack, ballistic missile defense, battlefield surveillance, navigation, target tracking and control, and so on. In accordance with purposes and the nature of problems that we try to solve, these automated systems can be classified into two groups [1]:

- *Information radar systems*: The main purpose is to collect information about the searched objects (the supervisory radar control systems concerning air, cosmic, and over-water conditions; the meteorological radar systems; the remote sensing radar systems; etc.).
- *Control radar systems*: The main purpose is to solve the problems to control the objects using the data of radar tracking, observation, and measurements (antiaircraft and missile defense systems, air traffic control systems, navigational systems, and so on).

As an example of the CRS of the second group, consider the classical radar antiaircraft and missile defense and control system [2–4]. The block diagram of this control system is shown in Figure 1.2. Elements of the antiaircraft and missile defense system are as follows:

- *Subsystem* of target detection and designation assigned to carry out in a good time the detection and estimation of enemy air target motion parameters
- *One or several CRSs* to fire control assigned to clarification of motion parameters of the annihilation-oriented air targets
- *Definition of current values of pointing angles and setting time for time fuse with required accuracy*
- *Missile launchers* assigned to firing
- *Facilities* to transmit the information-bearing signals between elementary blocks and subsystems of the control system

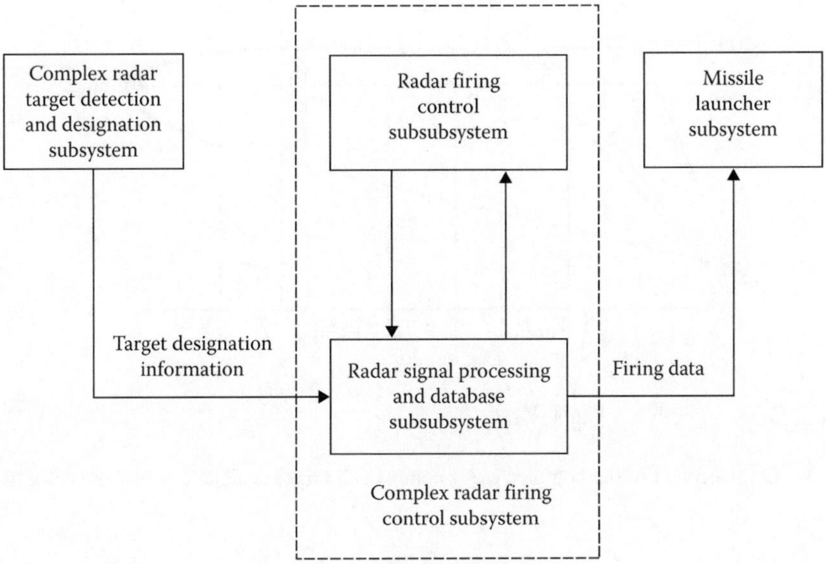

FIGURE 1.2 Radar control system of antiaircraft and missile complex.

The most general QoS (the criterion of effectiveness) of the considered control system, which is generated based on performance purposes, is the so-called averted harm given by [5]

$$\mathcal{D} = \mathcal{Q}_{\text{ob}} \prod_{j=1}^{N} \left[1 - \mathcal{D}_{\text{ob}_j} \prod_{i=1}^{L_j} \left(1 - P_{ij}\right) \right], \tag{1.2}$$

where

$\mathcal{Q}_{\text{ob}}$ is the importance of the defended objective
$\mathcal{D}_{\text{ob}_j}$ is the relative damage of the objective caused by the jth armed attack facility under absence of defense
P_{ij} is the probability of the jth armed attack facility damage by the ith defended facility (e.g., missile)
N is the number of the armed attack facilities
L_j is the number of missiles assigned to attack the armed attack facility, where $\sum_{j=1}^{N} L_j = L_0$
L_0 is the missile resource

To ensure the maximal value of damage prevention we must try to increase the probability of target destruction in accordance with Equation 1.2:

$$P_{ij} = P_{\text{td}_j} P_{2j} P_{3_{ij}}, \tag{1.3}$$

where

P_{td_j} is the probability of success under the target designation with respect to the jth target by the complex radar detection and target designation system
P_{2j} is the probability of success under the parameter clarification of the jth target and definition of firing data by the complex radar control firing system at the condition that the problem of target designation has been solved successfully
$P_{3_{ij}}$ is the probability of destruction of the jth target at the condition that the problems of target designation and control have been solved successfully

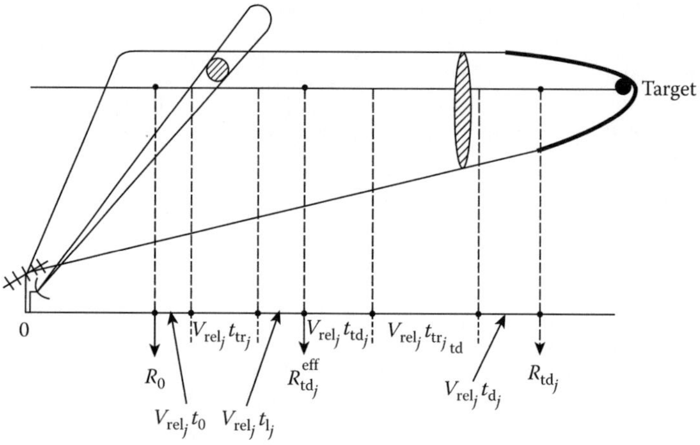

FIGURE 1.3 Emplacement of the complex radar control and target designation systems. (See the notations in text.)

Using Equation 1.3, we can define the QoS of the complex radar control and target designation systems and justify limitations for this quality performance. Consider the case when these systems are placed on the same emplacement (see Figure 1.3). The conditional probability of target destruction by missile can be presented in the following form [6]:

$$P_{\text{suc}}^{\text{dest}} = \left(1 + \frac{\sigma_{\text{miss}}^2}{R_{\text{eff}}^2}\right)^{-1}, \tag{1.4}$$

where

σ_{miss}^2 is the variance of miss

R_{eff} is the effective radius of target destruction, that is, the radius of sphere, within the limits of which a missile hits and destructs the searched target with the given probability

Based on Equation 1.4, we see that the probability of target destruction $P_{\text{suc}}^{\text{dest}}$ is inversely proportional to the variance of miss σ_{miss}^2 to deliver a missile to target area, which is characterized by the following constituents:

- The error variance to hit the target by missile σ_{hit}^2
- The error variance of antitarget guidance by launcher $\sigma_{\text{launcher}}^2$
- The error variance of missile flight path $\sigma_{\text{missile}}^2$

Since these constituents are considered as independent and uncorrelated, the variance of total error with respect to each coordinate from the set

$$\theta = \{R, \beta, \varepsilon\} \tag{1.5}$$

is defined in the following form:

$$\sigma_{\text{miss}}^2 = \sigma_{\text{hit}_\theta}^2 + \sigma_{\text{launcher}_\theta}^2 + \sigma_{\text{missile}_\theta}^2. \tag{1.6}$$

Only the first term in Equation 1.6, the error variance to hit the target by missile σ^2_{hit}, depends on the target designation accuracy given by the control system. Under the fixed values of the second and third terms in Equation 1.6, the error variance of antitarget guidance by launcher $\sigma^2_{launcher}$ and the error variance of missile flight path $\sigma^2_{missile}$ and given before restrictions to the variance of miss σ^2_{miss}, we are able to define requirements for the control system in accuracy of the observed target coordinates.

The effective radar range of the control system can be determined in the following form (see Figure 1.3):

$$R^{eff}_{td_j} = R_0 + \left(t_0 + t_{tr_j} + t_{1_j}\right)V_{rel_j}, \tag{1.7}$$

where

R_0 is the far-field region boundary of destruction area

V_{rel_j} is the relative velocity of the jth target motion with respect to the defended object

t_0 is the flight time of missile for the range R_0

t_{1_j} is the time of lock-in of the jth target to track by the firing control system

t_{tr_j} is the tracking time of the jth target to track by the system of firing control, that is, the time from the lock-in instant to the instant of reaching the required accuracy of the target tracking

Thus, to solve the assigned task successfully, the firing control system has to

- Possess the radar range ensuring the target destruction on the far-field region boundary of the destruction area
- Ensure the smoothed coordinates of the target with accuracy that is sufficient to reach the searched target

In accordance with a character of assigned tasks, the firing control system must possess the pencil-beam pattern of antenna and restricted zone of searching. For this reason, the main assigned task of the target designation system is to provide the coordinates and motion parameters of targets on the target designation line with accuracy allowing the firing control system to accomplish the target lock-in based on the target designation data without any additional searching or, at least, to limit the additional zone searching to a minimum.

Errors in the target designation are defined by the errors of coordinate measurements carried out by the target designation system, total time of smoothing the target coordinates and parameters, time of transmission, receiving, and processing of target designation commands. Let the target designation area be given by the following coordinates ΔR_{td}, $\Delta\beta_{td}$, and $\Delta\varepsilon_{td}$ in the spherical coordinate system. Then the probability to lock-in the target by the control firing system using a single target designation from the CRS of target designation, that is, the probability of success under target designation can be determined in the following form:

$$P_{td} = \int_{-0.5\Delta R_{td}}^{0.5\Delta R_{td}} \int_{-0.5\Delta\beta_{td}}^{0.5\Delta\beta_{td}} \int_{-0.5\Delta\varepsilon_{td}}^{0.5\Delta\varepsilon_{td}} f(\Delta R, \Delta\beta, \Delta\varepsilon)\, d\Delta R\, d\Delta\beta\, d\Delta\varepsilon, \tag{1.8}$$

where $f(\Delta R, \Delta\beta, \Delta\varepsilon)$ is the probability density function of target coordinate deviation from the target designation area center.

In the case of systematic bias absence of the target designation and the Gaussian normal distribution of random errors with the variances $\sigma^2_{R_{td}}$, $\sigma^2_{\beta_{td}}$, and $\sigma^2_{\varepsilon_{td}}$, the probability of success under the target designation takes the following form:

$$P_{td} = \Phi_0\left(\frac{\Delta R_{td}}{\sigma_{R_{td}}}\right)\Phi_0\left(\frac{\Delta\beta_{td}}{\sigma_{\beta_{td}}}\right)\Phi_0\left(\frac{\Delta\varepsilon_{td}}{\sigma_{\varepsilon_{td}}}\right), \tag{1.9}$$

where

$$\Phi_0(x) = \frac{2}{\sqrt{2\pi}}\int\limits_0^x \exp\left(-0.5t^2\right)dt \tag{1.10}$$

is the integral of probability. From Equation 1.9 it follows that if the required probability of target designation and the coordinates of target designation area are given before, we are able to formulate the requirements to allowable values of the target designation error variances $\sigma^2_{R_{td}}$, $\sigma^2_{\beta_{td}}$, and $\sigma^2_{\varepsilon_{td}}$.

The aforementioned statements are correct in the case of a single target designation. If there is a possibility to renew and transmit information about the target designation k times ($k > 1$), then the probability of success under the target designation is given by

$$P_{td} = 1 - \left(1 - P_{td_1}\right)\left(1 - P_{td_2}\right)\cdots\left(1 - P_{td_k}\right). \tag{1.11}$$

The repeated target designation data lead to an increase in time of target designation transmission that requires, finally, increasing the radar range of the system. The required radar range is determined by the following formula (the case of a single target designation):

$$R_{td_j} = R^{\text{eff}}_{td_j} + V_{rel_j}\left(t_{d_j} + t_{tr_{jtd}} + t_{td_j}\right), \tag{1.12}$$

where

t_{d_j} is the time required to detect the jth target

$t_{tr_{jtd}}$ is the time of the jth target tracking by the target designation system to ensure the given accuracy of coordinate and parameter definition and estimation at the extrapolated point for t_{td_j}

t_{td_j} is the time required to transmit information about the target designation to the firing control system

From Equations 1.9 and 1.12 it follows that the target designation system has to

1. Possess circular radar scanning
2. Ensure the radar range required to guarantee the target designation on the reassigned target line
3. Provide an accuracy of definition and estimation of the searched target coordinates at the predicted point of target designation, which is sufficient to lock-in the target by the control system without additional searching of the target

Thus, reasoning from the considered QoS (the criterion of effectiveness) of the antiaircraft defense system, QoS indices of radar subsystems included in the antiaircraft defense system are

1. *Radar scanning configuration of the CRS*: Requirements and configurations of radar scanning of the target designation system and the control system are different. The radar scanning of the control system in angular coordinates is restricted and a way to scan it is specific (spiral, bitmapped, etc.). The radar scanning of the target designation system is circular, as a rule, or sectored and limited on vertical plane by the special cosecant-squared directional diagram shape.
2. *Radar range* $R_{\text{td}}\left(R_{\text{td}_{\text{eff}}}\right)$ is the distance, within the limits of which the task performance of each CRS providing information to the missile firing control subsystem is ensured.
3. *Accuracy of information* at the given radar range, which is characterized by the covariance matrix of error.
4. *QoS index* characterizing influence of the external and internal noise and interferences on the considered CRS and that can be defined by numbers of detected decoy targets, which are tracked by the system over a definite period of time.

QoS indices 2 and 3 are related to each other, since the accuracy of obtaining information about the target depends on the distance between the CRS and the searched target. Moreover, there is a need to take into consideration the statistical nature of the QoS 2, 3, and 4, a relationship between them and the probability of target detection, the probability of false alarm, and the accuracy of coordinate measurements by the radar system. Since the probability of target detection, the probability of false alarm, and the accuracy of coordinate measurements depend on technical parameters of the radar systems also, there is a function between the earlier-mentioned statistical performance and the power, duration, bandwidth of scanning signals, dimensions, and type of transmitting and receiving antenna. These parameters of the radar systems must be defined during system designing.

Consider Equation 1.3 again. Taking into consideration the obtained formulas and relations, we can state that the probability of target destruction can be determined and also used to determine the averted harm given by Equation 1.2 under the known QoS of the target designation system, the control system, and the missile launcher. Consequently, the averted harm criterion (the QoS) can be defined, and it is not changed during the system designing. However, this criterion (the QoS) is related to technical parameters of the designed systems by composite and multiple-valued function and cannot be used in practice to evaluate and compare the solutions of designing. Meanwhile, according to the systems approach under the radar system designing there is a need to take into account the QoS criteria possessing a physical sense, the ability to be determined, and associated with technical parameters of the designed CRS. The criterion, which we have just considered and analyzed, does not satisfy these conditions.

Under designing the CRSs, it is difficult to define a general criterion (QoS) satisfying the aforementioned requirements owing to the complicated mathematical model of target searching by the system. For this reason, it is worthwhile to introduce an intermediate QoS instead of the general criterion (QoS) with the purpose to relate the main parameters of the radar systems and signal processing subsystems that are designed.

As a basis for the generalized criterion, we can consider the signal-to-noise ratio (SNR) given by

$$q^2 = \frac{2E_{\text{s}}}{N_{N+I}}, \tag{1.13}$$

where

E_{s} is the energy of the received signal

N_{N+I} is the power spectral density of the total noise and interferences

In accordance with the general radar formula for the case of matched signal processing in free space in set noise, we can write [7–11]

$$q^2 = \frac{2P_t^{av} t_0 G_t G_r \lambda^2 S_t^{ef}}{(4\pi)^3 R_t^4 k T_0 N_0 \mathcal{L}}, \tag{1.14}$$

where
P_t^{av} is the transmit power
t_0 is the observation time
G_t is the transmitting antenna gain
G_r is the receiving antenna gain
λ is the wavelength
S_t^{ef} is the effective target reflective surface
R_t is the distance to the target
$k = 1.38 \times 10^{-23}$ W/Hz is the Boltzmann constant
T_0 is the absolute temperature of signal source
N_0 is the power spectral density of set noise
$\mathcal{L}$ is the total loss factor

In the case of pulsed radar we can write

$$P_t^{av} = P_p \tau_s F, \tag{1.15}$$

where
P_p is the transmit pulse power
τ_s is the duration of scanning signal
F is the repetition frequency of scanning signals

Under conditions of conflict radar in practice, interferences generated by the enemy are the main noise. The power spectral density of deliberate interference is determined by [12,13]

$$N_d = \frac{\alpha P_{dI}}{4\pi R_{dI}^2 \Delta f_{dI}}, \tag{1.16}$$

where
P_{dI} is the power of noise source
α is the coefficient depending on the direction to the noise source and the performance of the noise source directional diagram and CRS directional diagram
R_{dI} is the distance between the CRS and the noise source (the generator of deliberate interference)
Δf_{dI} is the noise bandwidth

From Equations 1.14 through 1.16 it follows that SNR depends on the main parameters of the CRSs, environment, and target. On the other hand, the main QoS indices of radar signal processing are also expressed by SNR. For instance, the probability of signal detection under the Rayleigh distribution of amplitude and uniform distribution of phase of the signal can be written in the following form:

$$P_D = \exp\left[-\frac{\gamma_{rel}^2}{2(1+0.5q^2)}\right], \tag{1.17}$$

where γ_{rel} is the relative threshold. The root-mean-square value of potential error under delay measurement of the unmodulated in frequency bell-shaped pulse, which can be given by

$$x(t) = -\frac{\pi t^2}{\tau_s^2}, \tag{1.18}$$

where τ_s is the duration of scanning bell-shaped pulse, at the level 0.46 has the following form:

$$\sigma_\tau = \frac{\tau_s}{q\sqrt{\pi}}. \tag{1.19}$$

The root-mean-square value of potential error under measurement of the Doppler frequency of the coherent unmodulated in frequency and bell-shaped pulse with duration τ_s at the level 0.46 takes the following form:

$$\sigma_{f_D} = \frac{1}{q\tau_s\sqrt{\pi}}. \tag{1.20}$$

The root-mean-square value of potential error under measurement of the angular coordinates is determined by

$$\sigma_\theta = \frac{1}{ql_{eff}}, \tag{1.21}$$

where l_{eff} is the effective length of antenna aperture normalized with respect to the wavelength and is defined as

$$l_{eff} = \chi\frac{d}{\lambda}, \tag{1.22}$$

where
- d is the length of antenna aperture
- χ is the *constant*

Analogous relations take place for more complex models of signal processing techniques.

Thus, the SNR is a generalized parameter that can be used as QoS (the criterion of effectiveness) in the design of the complex radar system and radar signal processing methods.

1.3 PROBLEMS OF SYSTEM DESIGN FOR AUTOMATED COMPLEX RADAR SYSTEMS

The CRS contains a great number of interdependent elements and blocks and belongs to the complex system class. As mentioned earlier, the first step in the design of the system is a definition of functional purposes in the higher order system. In Section 1.2, we have introduced the complex radar target designation and control firing systems. Henceforth, we are limited by consideration of system design problems of the CRS carrying out operations of searching, detection, and tracking a set of targets within the limits of radar coverage and providing information with the required QoS on the reassigned line.

The main task of the system design is the choice and justification of a structural block diagram of the system. For this task, a designing process is based on existing experience of construction

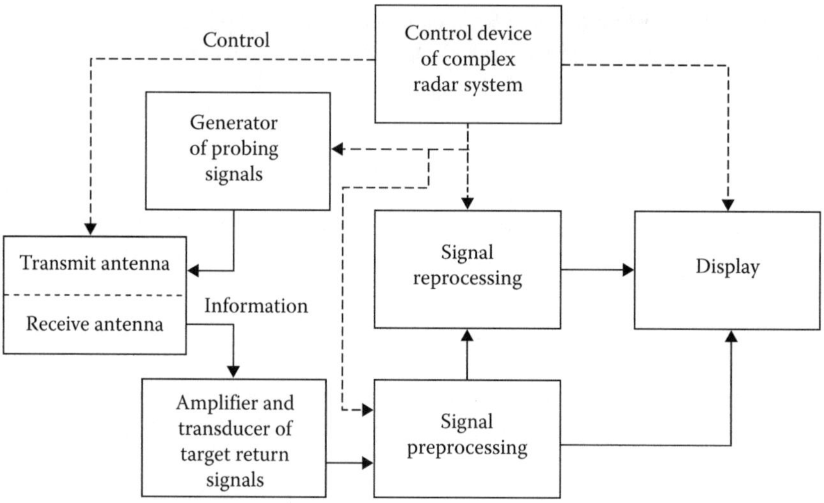

FIGURE 1.4 Structural block diagram of an automated complex radar system.

of the system with analogous application. In doing so, we should take into consideration the structural stability of the system to changes in functional purposes and initial premises of design, including motivations stimulating a new design process. In this case, we need to design and construct only a small number of elements and blocks in future.

The structural block diagram of the complex automated radar system is presented in Figure 1.4. This diagram consists of the following blocks [14,15]:

- Transmitting and receiving antennae or matched transmitting–receiving antenna
- Generator, amplifier, and guidance of scanning signals
- Amplifier and transformer of receiving signals
- *Preprocessing of the target return signals*—the receiver processing the target return signals: the filtering, accumulation, detection, and parameter estimation
- *Reprocessing of the target return signals*—a definition of target trajectory parameters
- *Computer subsystem to control the CRS*—the synchronization and adaptation to a changed environment
- Data displaying for user

Each listed block is a complex system both by elements and by structure. Each block is an objective of design on the next step of detailed structuring. Figure 1.4 shows us that the optimal design of the CRS, as a whole, is an unsatisfiable problem. In this case, according to the systems approach, the designed system is divided into individual blocks. Under partitioning, we consider a decision rule implicating to establish a single-valued function between a set QoS, as a whole, and individual blocks of the system. We take into consideration dynamical and structural functions. Under such an approach, the design process of the CRS can be divided into the following components satisfying the aforementioned requirements in general:

- Definition of energy parameters of the system and designing the generator, amplifier, and guidance of scanning signals
- Designing of devices and computer subsystems for signal processing to get information about the target from a set of radar signals in natural and artificial noise, including the target return signals
- Designing a subsystem to control the system ensuring a stability of all functions in complex and rapidly changed situations

Energy parameters of the complex radar system are the following:

- The power of scanning pulse P_{scan}
- The duration of pulse τ_{s}
- The transmitting antenna gain G_{t}
- The receiving antenna gain G_{r}
- The effective area of the receiving antenna S_{eff}

Choice of these parameters is accomplished in accordance with the end use of the CRS, the level of development of the corresponding element base, the technology of production, the technology and procedures of adjustment, and taking into account the allowable charges on production manufacturing and operating costs. As a rule, under designing the system, a definition of scanning pulse parameters, ways of scanning pulse generation, and emission is basic, and results of definition are initial data to design the receiving path and target return signal processing algorithms.

In designing the receiving path and the target return signal processing algorithms, the energy parameters of the CRS are considered as the fixed and external parameters. Thus, attention is focused on solution of problems in accordance with which there is a need to define the target return signal processing algorithms ensuring a maximal effect that can be characterized by probability performance and accuracy of definition of the target return signal parameters required by the user. Ultimately, the solution of this problem is reduced to a definition of the target return signal processing algorithms and a choice of computer subsystems for signal processing tasks at all stages, from preamplification and signal conditioning to data preparation and radar information output to the user. To solve these problems, the target return signal processing algorithms well developed in the statistical signal processing area are used, which are invariant to methods of transmission and receiving the signals. Consequently, the receiving path and the target return signal processing algorithms can be developed individually from other blocks and components of the CRSs. Henceforward, a totality of signal processing algorithms and receivers and/or detectors, which are needed to design, is called *radar signal processing system*.

Considering the *radar signal processing system* as an autonomous subsystem of the complex radar system, we can define and solve the design problems of the system as applied to its classes and groups, not only the individual CRS, based on functional purposes in high-order systems. The control system of the CRS is, per se, the autonomous system, but by the nature of solving problems, it is the system of higher order in comparison with the considered CRS. Naturally, designing of the control system can be considered as the solo problem within the limits of requirement specification presented for the CRS as a whole.

Thus, the design of the CRS is divided into three individual tasks, which are solved independently of each other but satisfying the conditions of continuous interaction and adjustment of parameters to guarantee a fulfillment and reaching the functional purposes of the systems. Problems of energy parameter definition of the systems are outside the scope of this book.

1.4 RADAR SIGNAL PROCESSING SYSTEM AS AN OBJECT OF DESIGN

The systems approach to design assumes the availability of some basic mathematical models and structures of the radar signal processing systems, which must be put into the basis of new development. Under designing the radar signal processing system, the well-known structure of the receiving path of the CRS is considered as the basic. Optimal signal processing algorithms obtained from the statistical radar theory are used as the basic mathematical models. In accordance with the

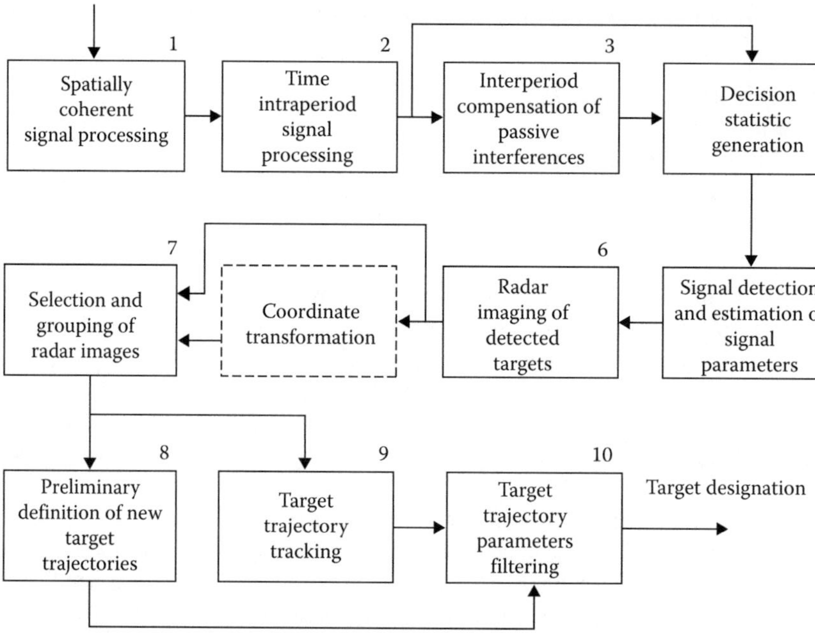

FIGURE 1.5 Block diagram of operations carried out by the optimal receiver.

conclusions of statistical radar theory, the optimal receiver must carry out the following operations (see Figure 1.5) [16–34]:

1. Spatial signal processing of the coherent target return signals using the multielement array placed in one or several signal reception points
2. Time intraperiod signal processing of the coherent target return signals including a nonlinear signal processing, namely, limitation, taking the logarithm, etc., and matched filtering or correlation signal processing
3. Interperiod compensation of correlated noise and interferences caused by reflection from objects on the Earth's surface, hydrometeors, and man-made reflectors (artificial passive interferences)
4. Accumulation of the target return signals and forming some statistics about incoming signals (the decision statistics), based on which we make a decision about the target detection and estimate the target return signal parameters
5. Comparison of decision statistics with threshold and realization of signal detection algorithms and estimation of signal parameters
6. Radar imaging of detected targets
7. Selection and grouping of new radar images by tracking target trajectory tackle and new incoming data for target tracking
8. Preliminary definition of parameters of new target trajectories
9. New radar data binding to trajectories of the tracking target
10. Filtering of the tracking target trajectory parameters during solution of the tracking target trajectory problems

Operations 1 and 2 are the stage of the intraperiod spatial signal processing for coherent single-short pulse. Operations 3–6 are the stage of the interperiod signal processing for a set of target return signals reflected by each target under an antenna beam regular scanning in the radar coverage or at multiple scanning of each direction in the radar coverage. Operations 7–10 are the stage of the surveillance radar signal processing concerning information about trajectories of tracking targets. Thus, there is a consecutive sequence in the radar signal processing by stages. Each stage possesses its own real timescale of signal processing that allows us to carry out an autonomous realization of these stages.

The radar signal processing systems are divided into three classes by the method of implementation: (a) analogous, (b) digital, and (c) analog–digital. Now, the radar signal processing systems of the third class are widely used. However, digital signal processing plays a leading role in exploited and designed CRSs owing to flexibility and universality. We can see a tendency of extension to use the digital signal processing techniques in designing CRSs owing to substitution of analog signal processing by digital one. Successes in digital signal processing, which have been achieved now, allow us to use digital signal processing techniques not only in time signal processing of coherent signals but also in spatial signal processing of coherent signals. Henceforth, we consider digital signal processing techniques starting from the matched filtering of coherent signals.

Solving the tasks of designing radar signal processing subsystems under the fixed energy parameters of the CRS, attention is paid to optimization of receiving path in natural and artificial noise and interferences. All tasks of optimal signal processing are solved by methods and procedures of the theory of statistical decisions. Because of this, the QoS indices of radar signal processing subsystems are imported from the theory of statistical decisions. For some cases and examples, these QoS indices acquired characteristic features of a radar.

Independent of any application area of CRSs, the main QoS indices of radar signal processing are

1. *Space–time signal processing*: the coefficient of energy use given by

$$k_{\text{use}} = \frac{q^2}{q_0^2}, \tag{1.23}$$

where

q^2 is the signal-to-interference-plus-noise ratio (*SINR*)

q_0^2 is the SNR [35]

2. *Radar signal preprocessing*:
   a. The probability of signal detection P_{D}
   b. The probability of false alarm P_{F}
   c. The accuracy of definition of the target coordinates and their estimation, which is characterized by the covariance matrix of estimation errors K_{mes}, in a general case, and by the variance of estimation error σ^2_{mes}
3. *Radar signal reprocessing*:
   a. The probability of target trajectory detection P_{D}^{tr}
   b. The probability of target trajectory false alarm P_{F}^{tr}
   c. The accuracy of definition of the target trajectory parameters, which is characterized by the covariance matrix of estimation errors of the target trajectory parameters $K_{\text{mes}}^{\text{tr}}$
   d. The probability of breaking up in the target tracking P_{br}

The earlier-listed QoS indices are associated directly and unambiguously with QoS indices of the CRS, as a whole (see Section 1.3). We can observe a direct relation between them: the higher the QoS indices of radar signal processing subsystems, the higher the QoS indices of the system.

Under implementation of digital signal processing, there is a need to take into consideration some limitations in speed of corresponding devices. For this reason, an important and essential QoS index under digital signal processing is the digital signal processing effort defined by the number of operations at a single realization of the radar signal processing algorithm. Data throughput is an important

QoS index under digital signal processing, which can be defined, for example, as the number of target processing by the system simultaneously. Of course, other approaches and methods to estimate the data throughput of radar digital signal processing subsystems are possible [36,37].

In designing CRSs, an initial optimization of signal processing algorithms is carried out by criteria imported from the theory of statistical decisions. For instance, the Neyman–Pearson criterion is the basic criterion in the signal detection theory. The essence of this criterion is the following: we obtain the maximal probability of detection of target return signals (target trajectories) P_D^{op} under some limitations in choice of the false alarm probability P_F:

$$P_D^{op} = \max_{\{v\}} P_D^v \quad \text{and} \quad P_F \le P_F^{ad}, \tag{1.24}$$

where

$\{v\}$ is the set of possible decision-making rules of detection
v_{opt} from the set $\{v\}$ corresponds to P_D^{max}
P_F^{ad} is the permissible probability of false alarm

Generally, estimations of signal parameters are optimized by the criterion of minimizing the average risk:

$$\mathcal{R}^{av}(a, \hat{a}_{op}) = \min_{\{\hat{a}\}} \mathcal{R}^{av}(a, \hat{a}), \tag{1.25}$$

where

$\mathcal{R}^{av}(\cdots)$ is the average risk
a is the real value of the signal parameter
$\hat{a}$ is the estimation of real value of the signal parameter

Operations and functions, which are not related to detection of the target return signals and definition and estimation of target return signal parameters, are optimized by criteria corresponding to maximal effect or maximal value of the corresponding QoS in applying restrictions on the energy supply and hardware implementation.

To optimize the system as a whole, the QoS must include all the main indices, that is, it must be a vector in mathematical form. If m subsystems of the CRS are characterized by individual QoS, for example, $q_1, q_2, \ldots, q_m$, then the system, as a whole, is characterized by the vector $\mathbf{Q} = (q_1, q_2, \ldots, q_m)$. The goal of vector optimization is to choose a CRS that possesses the best-case value of the vector $\mathbf{Q}$. At the same time, we assume that the appropriate QoS vector $\mathbf{Q}$ is given already. The theory of vector optimization of a CRS is far from completion. Simple methods that are used allow us to reduce the vector analysis directly or marginally to a scalar one. A simple procedure is used in this book. We consider all individual QoS $q_1, q_2, \ldots, q_m$ except the only one, for example, q_1, which is the most essential, as there are restrictions with further conditional optimization of the system by this QoS.

The designing of a radar digital signal processing system is divided into two stages: the designing and construction of the digital signal processing algorithms and the designing of the computer subsystems. The designing of the digital signal processing algorithms employed by the CRS is initiated from making clear the main goal of digital signal processing algorithm construction; how the main functions can be generated into the CRS; definition of basic restrictions; QoS; and designing of the objective function. A sequence of the digital signal processing algorithm design is given as follows:

- Definition of purpose and main functions of the digital signal processing algorithm.
- Designing and construction of logical block diagrams of the digital signal processing algorithms; there is a need to propose several variants.

- Off-line testing and processing of the individual digital signal processing algorithms or logical blocks.
- Simulation and definition of workability and QoS of the digital signal processing algorithm.
- Optimization and construction of the complex digital signal processing algorithm employed by the CRS, which operates in real time. The optimization is carried out by the discrete choice method of trade-off variant from a set of possible ones, which are digital signal processing algorithms corresponding to the given stage of operations. This approach allows us to combine the heuristic procedures and methods based on design-automation systems and optimization of the digital signal processing algorithms.

On the basis of designing and debugging results of a complex digital signal processing algorithm, we are able to define the main parameters of computer subsystems and to state the basic requirements to these parameters with the purpose of realizing all steps of signal processing in the complex radar systems.

Designing of special-purpose computer subsystems (SPCSs) starts from definition of the main parameters and relationships with components of SPCS structure. After definition of functions between parameters of an SPCS as a whole and parameters of structural components of an SPCS, we can formulate the requirements to each elementary structural block of SPCS based on the general requirements to computer subsystems and, by this way, to define the requirements specification to design the CRS as a whole. Relationships between the main parameters and elementary structural blocks of an SPCS are called the parametric balance. The most widely used system balances are the time balance, error balance, memory size balance, reliability balance, balance of costs, and so on [38,39].

Generally, in designing the SPCS structure, the solution can be found using the following variant of the criterion "effectiveness—cost":

- Providing the minimum time to realize the complex digital signal processing algorithm under given restrictions on the equipment investments
- Providing the minimal equipment investments under given time of realization of the complex digital signal processing algorithm

In designing CRSs, the second variant is preferable. Ultimately, the designing of an SPCS reduces to definition of the computer subsystems number for different functional purposes and conservation of algorithms to interact between computer subsystems.

In conclusion, we consider an example of systematic sequence under the system design of complex digital signal processing algorithms and computer subsystems for radar signal processing systems (see Figure 1.6). Before designing a CRS, there is a need to define the requirements specification, in which the main purposes and requirements, structural block diagrams of subsystems, and the main restrictions and requirements on the parameters at the system output are described.

Consider the main stages of the CRS design, which are presented in Figure 1.6:

- The first stage (block 1) is a formulation of the optimal designing problem, definition of external and internal system parameters and relationships between them, and choice and justification of the objective function of optimal designing. The result is a formalization of the requirements specification.
- The second stage is a decomposition of the general problem of system designing (block 2) on a set of simple tasks of subsystem designing and corresponding representation of the general objective function in the form of superposition of objective functions of the optimal subsystem designing. Success in solving the problems of the first and second stages depends on the level of development of the optimal designing methods, in general, and radar signal processing systems, in particular.

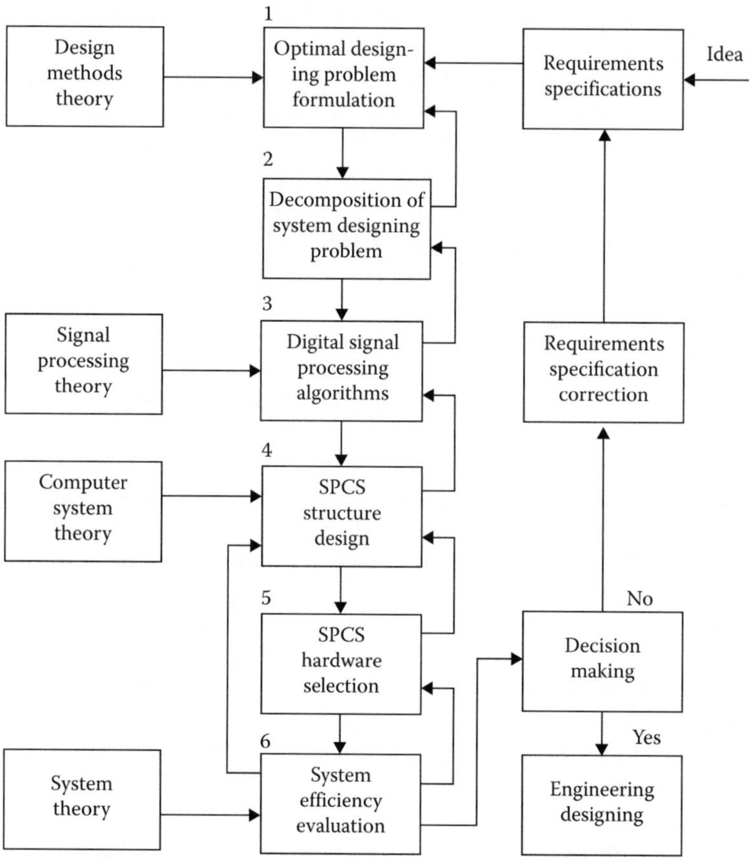

FIGURE 1.6 Stages of CRS designing of the digital signal processing algorithms and computer subsystems.

- The third stage (block 3) is the designing and investigation of digital signal processing algorithms of subsystems and the generalized algorithm of a CRS. Designing and construction of algorithms is accompanied by comprehensive analysis and testing of digital signal processing algorithms on performance and effectiveness by preliminary established criteria. The main procedure of investigations is a simulation. Success in solution of problems of the third stage depends on the level of development of the theory of signal processing in radar systems.
- The fourth and fifth stages (blocks 4 and 5) are the synthesis and selection of equipment and hardware of SPCSs to realize the digital signal processing algorithms in the CRS. Realization of these stages is carried out interacting with previous stages of digital signal processing algorithm design with the purpose to reach the optimal agreement between digital signal processing algorithms and computing facilities by the QoS criterion called "efficiency–cost." Success in problem solving of the fourth and fifth stages depends on the level of development of the theory of computer systems.
- The final stage (block 6) of system designing is the evaluation of the effectiveness of the constructed generalized digital signal processing algorithm and the computer system as a whole by the general criterion given in the requirements specification or chosen at the first stage of system designing. Completeness and reliability of this stage depend on the level of development of the theory of operation systems. Results of efficiency evaluation are used to make a decision to finish the system designing stage and complementation to technical

designing stage if the requirements specification and required QoS by efficiency criterion are satisfied. In the opposite case, there is a need to change the requirements specification and to repeat all stages of system design again.

Henceforward, in this book, we follow the considered sequence of system designing stages with attention to construction of the digital signal processing algorithms and choice of computer systems for their realization in CRSs.

1.5 SUMMARY AND DISCUSSION

The basic aspects of systems approach to design CRSs discussed in this chapter are the following.

The systems approach is considered to design CRSs. The designing process is divided into two sufficiently pronounced stages—system designing and engineering designing. A conception of complex information and control system integrity, including a CRS, makes specific owing to the idea of backbone communications, for example, structural and control communications. At the stage of system designing, the main factor to consider is the structure or architecture of the future complex information and control system, including a CRS. At the stage of engineering designing, the important issue is the choice and development of a fixed totality of CRS elements and communications between them.

Any CRS cannot be imagined without the environment. The main problem is to define an optimal boundary between the system and the environment. The environment, facilities of counteracting forces, level of development of the element base and technologies, economic factors, and human element are the factors affecting the operation of a system. Based on the methodology viewpoint, the definite aspects of the systems approach were defined to design any complex information and control system, including a CRS.

In the design of such systems, we widely use mathematical models, simulation models, and modeling as a process of representation of the CRS by an adequate model with subsequent test operation to obtain information about its functioning. The most characteristic feature of the systems approach is to search for a decision by iterative optimization based on computer-aided designing systems. The main operations of the system design are the definition and generation of the end goals; generation of all possible alternative variants; definition of investments to realize each alternative version of the system structure; designing the models chosen to optimize alternatives and their software implementation; and comparison of alternatives and the decision making.

Automated CRSs are widely used to solve different problems, namely, air-traffic control, military fighter/attack, ballistic missile defense, battlefield surveillance, navigation, target tracking and control, and so on. Based on the purposes and nature of problems, automated CRSs can be classified into two groups: *information radar systems* and *control radar systems*.

As an example, the main requirements and QoS (the criterion of effectiveness) of classical radar antiaircraft and missile defense and control systems are discussed. Reasoning from the considered QoS (the criterion of effectiveness) of the antiaircraft defense system, QoS indices of radar subsystems included in the antiaircraft defense system are defined. According to the systems approach in CRS designing, there is a need to take into account the QoS criteria possessing a physical sense, the ability to be determined, and associated with technical parameters of the designed system. In designing CRSs, it is difficult to define a general criterion (QoS) satisfying the requirements mentioned earlier owing to a complicated mathematical model to search for a target. For this reason, it is worth introducing an intermediate QoS instead of the general criterion (QoS) with the purpose of relating the main parameters of the systems and signal processing subsystems that are designed. It has been defined that the SNR is a generalized parameter that can be used as QoS (the criterion of effectiveness) in the design of the CRS and radar signal processing methods.

The main task of the system design is the choice and justification of a structural block diagram of the CRS. In this task, the designing process is based on the existing experience of construction of the system with analogous application. In doing so, we should take into consideration the structural stability of the system to changes in functional purposes and initial premises of design, including motivations stimulating a new design process. In this case, we need to design and construct only a small number of elements and blocks in the future.

The designing problem of the CRS is divided into three individual tasks, which are solved independently of each other but satisfy the conditions of continuous interaction and adjustment of parameters to guarantee fulfillment and enable achieving the functional goals of the systems. They are as follows:

1. Definition of the target return signal processing algorithms to define the parameters of the target return signal required by the user
2. Choice of computer subsystems for signal processing tasks at all stages, from preamplification and signal conditioning to data preparation and radar information output to the user
3. Definition of energy parameters of the CRSs, which is outside the scope of this book

Optimal signal processing algorithms obtained from the statistical radar theory are used as the basic mathematical models. There is a consecutive sequence in radar signal processing by stages. Each stage possesses its own real time scale of signal processing that allows us to carry out an autonomous realization of these stages. Solving the tasks of radar signal processing subsystem designing under the fixed energy parameters of the system, the main focus is on optimization of the receiving path in natural and artificial noise and interferences. All tasks of optimal signal processing are solved by methods and procedures of the theory of statistical decisions.

Independent of the application area of CRSs, the main QoS indices of radar signal processing are (a) *space–time signal processing*: the coefficient of energy use; (b) *radar signal preprocessing*: the probability of signal detection; the probability of false alarm; and the accuracy of definition of the target coordinates and their estimation, which is characterized by the covariance matrix of estimation errors, in a general case, and by the variance of estimation error; (c) *radar signal reprocessing*: the probability of target trajectory detection; the probability of target trajectory false alarm; the accuracy of definition of the target trajectory parameters, which is characterized by the covariance matrix of estimation errors of the target trajectory parameters; and the probability of breaking up in the target tracking. These QoS indices are associated directly and unambiguously with the QoS indices of the complex radar system, as a whole. We can observe a direct relation between them: the higher the QoS indices of radar signal processing subsystems, the higher the QoS indices of the complex radar system, as a whole.

The designing of a radar digital signal processing subsystem is divided into two stages: designing and construction of the digital signal processing algorithms and designing of the computer subsystems. The designing of the digital signal processing algorithms employed by the complex radar system is initiated from making clear the main goal of digital signal processing algorithm construction; how the main functions can be generated into the complex radar system; definition of basic restrictions; QoS; and designing of the objective function. Designing of computer subsystems with special purpose, which are called SPCS, starts from definition of the main parameters and relationships with components of SPCS structure. After definition of functions between parameters of SPCS as a whole and parameters of structural components of SPCS, we can formulate the requirements of each elementary structural block of SPCS based on the general requirements to computer subsystems and, in this way, define the requirements specification in designing the complex radar system as a whole. Generally, under designing the SPCS structure the solution can be found using the following variant of the criterion "effectiveness–cost."

REFERENCES

1. Skolnik, M.I. 2008. *Radar Handbook*. 3rd edn. New York: McGraw-Hill, Inc.
2. Tzvetkov, A. 1971. *Principles of Quantitative Ratings of Complex Radar System Efficiency*. Moscow, Russia: Soviet Radio.
3. Hovanessian, S. 1984. *Radar System Design and Analysis*. Norwood, MA: Artech House, Inc.
4. Meyer, D. and H. Mayer. 1973. *Radar Target Detection: Handbook of Theory and Practice*. New York: Academic Press.
5. Drujinin, B. and D. Kontorov. 1976. *Problems of Military Systems Engineering*. Moscow, Russia: Military Press.
6. Gutkin, L.S., Pestryakov, V.V., and V.H. Tipugin. 1970. *Radio Control*. Moscow, Russia: Soviet Radio.
7. Skolnik, M.I. 2001. *Introduction to Radar Systems*. New York: McGraw-Hill, Inc.
8. Lacomme, P., Hardange, J.-P., Marchais, J.-C., and E. Normant. 2001. *Air and Spaceborne Radar Systems: An Introduction*. New York: William Andrew Publishing.
9. Tait, P. 2005. *Introduction to Radar Target Recognition*. Cornwall, U.K.: IEE Press.
10. Levanon, N. and E. Mozeson. 2004. *Radar Signals*. New York: IEEE Press, John Wiley & Sons, Inc.
11. Barton, D.K. 2005. *Modern Radar System Analysis and Modeling*. Canton, MA: Artech House, Inc.
12. Stimson, G.W. 1998. *Introduction to Airborne Radar*. 2nd edn. Raleigh, NC: SciTech Publishing, Inc.
13. Nitzberg, R. 1999. *Radar Signal Processing and Adaptive Systems*. Norwood, MA: Artech House, Inc.
14. Streetly, M. 2000. *Radar and Electronic Warfare Systems*. 11th edn. Surrey, U.K.: James Information Group.
15. Guerci, J.R. 2003. *Space-Time Adaptive Processing for Radar*. Norwood, MA: Artech House, Inc.
16. Barkat, M. 2005. *Signal Detection and Estimation*. 2nd Edn., Norwood, MA: Artech House, Inc.
17. DiFranco, J.V. and W.L. Rubin. 1980. *Radar Detection*. Norwood, MA: Artech House, Inc.
18. Edde, B. 1993. *Radar Principles, Technology, Application*. Englewood Cliffs, NJ: Prentice Hall, Inc.
19. Kay, S.M. 1993. *Fundamentals of Statistical Signal Processing—Estimation Theory*. Vol. I. Englewood Cliffs, NJ: Prentice Hall, Inc.
20. Kay, S.M. 1998. *Fundamentals of Statistical Signal Processing—Detection Theory*. Vol. II. Englewood Cliffs, NJ: Prentice Hall, Inc.
21. Knott, E.F., Shaeffer, J.F., and M.T. Tuley. 1993. *Radar Cross Section*. 2nd edn. Norwood, MA: Artech House, Inc.
22. Mahafza, B.R. 1998. *Introduction to Radar Analysis*. Boca Raton, FL: CRC Press, Taylor & Francis Group.
23. Mahafza, B.R. 2000. *Radar System Analysis and Design Using MATLAB*. Boca Raton, FL: CRC Press.
24. Nathanson, F.E. 1991. *Radar Design Principles*. 3rd edn. New York: McGraw Hill, Inc.
25. Peebles, Jr., P.Z. 1998. *Radar Principles*. New York: John Wiley & Sons, Inc.
26. Richards, M.A. 2005. *Fundamentals of Radar Signal Processing*. Englewood Cliffs, NJ: Prentice Hall, Inc.
27. Brookner, E. 1988. *Aspects of Modern Radar*. Boston, MA: Artech House, Inc.
28. Franceschetti, G. and R. Lanari. 1999. *Synthetic Aperture Radar Processing*. Boca Raton, FL: CRC Press.
29. Johnson, D.H. and D.E. Dudgeon. 1993. *Array Signal Processing: Concepts and Techniques*. Englewood Cliffs, NJ: Prentice Hall, Inc.
30. Klemm, R. 1998. *Space-Time Adaptive Signal Processing: Principles and Applications*. London, U.K.: INSPEC/IEEE.
31. Sullivan, R.J. 2000. *Microwave Radar: Imaging and Advanced Concepts*. Boston, MA: Artech House, Inc.
32. Tuzlukov, V.P. 2001. *Signal Detection Theory*. New York: Springer-Verlag.
33. Tuzlukov, V.P. 2002. *Signal Processing Noise*. Boca Raton, FL: CRC Press.
34. Tuzlukov, V.P. 2004. *Signal and Image Processing in Navigational Systems*. Boca Raton, FL: CRC Press.
35. Van Trees, H.L. 2002. *Optimum Array Processing: Part IV of Detection, Estimation, and Modulation Theory*. New York: John Wiley & Sons, Inc.
36. Tsui, J.B. 2004. *Digital Techniques for Wideband Receivers*. 2nd edn. Raleigh, NC: SciTech Publishing, Inc.
37. Bogler, P.L. 1990. *Radar Principles with Applications to Tracking Systems*. New York: John Wiley & Sons, Inc.
38. Cumming, I.G. and F.N. Wong. 2005. *Digital Signal Processing of Synthetic Aperture Radar Data*. Norwood, MA: Artech House, Inc.
39. Long, M.W. 2001. *Radar Reflectivity of Land and Sea*. 3rd edn. Boston, MA: Artech House, Inc.

2 Signal Processing by Digital Generalized Detector in Complex Radar Systems

2.1 ANALOG-TO-DIGITAL SIGNAL CONVERSION: MAIN PRINCIPLES

For digital signal processing of target return signals in complex radar systems (CRSs), there is a need to carry out an analog-to-digital signal conversion. This transformation is produced by two stages. In the course of the first stage, there is a need to sample the target return signal—the continuous target return signal $x(t)$ is sampled instantaneously and at a uniform rate, once every T_s seconds. In the course of the second stage—quantization—the sampled target return signal $\{x(nT_s)\}$ is converted into a sequence of binary coded words. The sampling and quantization functions are realized by analog-to-digital converters.

The design process of analog-to-digital converters involves particularly the task of choosing a sampling interval value and number of signal quantization levels of the sampled target return signal. Simultaneously, we should take into account the problems of designing and realization both for converters and for digital receivers processing the target return signals. In the present section, we consider and discuss the main aspects and foundations of analog-to-digital signal conversion and the main principles underlying the design of analog-to-digital converters.

2.1.1 Sampling Process

In general, a sampling process of the continuous function $x(t)$ representing a target return signal is to measure its values at the time instants spaced T_s seconds apart. Consequently, we obtain an infinite sequence of samples spaced T_s seconds apart and denoted by $\{x(nT_s)\}$, where n takes on all possible integer values, both positive and negative. We refer to T_s as the *sampling period* or *sampling interval,* and to its reciprocal, $f_s = 1/T_s$, as the *sampling rate.* This ideal form of sampling is called an *instantaneous sampling.* As a rule, a value of T_s is *constant.*

A sampling device can be considered as a make circuit with the time τ and sampling period T_s. Time diagram of conversion of the continuous function $x(t)$ into a sequence of instantaneous ($\tau \to 0$) readings $\{x(nT_s)\}$ spaced T_s seconds apart is shown in Figure 2.1. Let $x_\delta(t)$ denote the signal obtained by individually weighting the elements of a periodic sequence of Dirac delta functions spaced T_s seconds apart by the sequence of numbers $\{x(nT_s)\}$ [1]:

$$x_\delta(t) = \sum_{n=-\infty}^{\infty} x(nT_s)\delta(t - nT_s). \tag{2.1}$$

We refer to $x_\delta(t)$ as the instantaneously (ideal) sampled target return signal. The term $\delta(t - nT_s)$ represents a delta function positioned at time $t = nT_s$. From the definition of the delta function [2], recall that such an idealized function has unit area. We may, therefore, view the multiplying factor $x(nT_s)$ in (2.1) as a "mass" assigned to the delta function $\delta(t - nT_s)$. A delta function weighted in this manner is closely approximated by a rectangular pulse of duration τ and amplitude $x(nT_s)/\tau$; the smaller we make τ, the better the approximation will be.

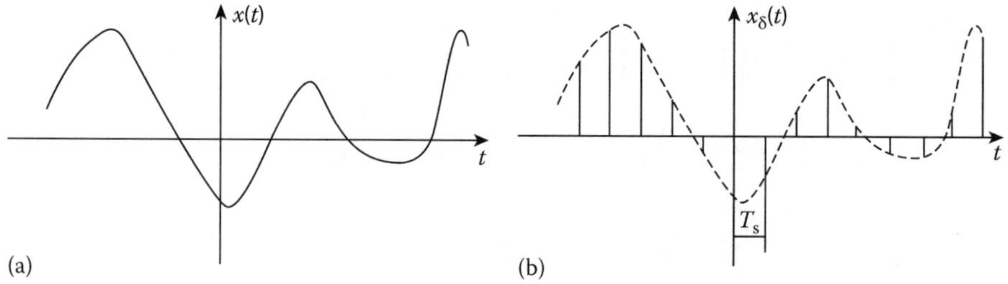

FIGURE 2.1 Illustration of the sampling process: (a) Analog waveform $x(t)$ and (b) instantaneously sampled representation of $x(t)$.

The instantaneously sampled target return signal $x_\delta(t)$ has a mathematical form similar to that of the Fourier transform of a periodic signal. This is readily established by comparing (2.1) for $x_\delta(t)$ with the Fourier transform of a periodic signal given by

$$\sum_{m=-\infty}^{\infty} x(t - mT_s) \leftrightarrow f_s \sum_{n=-\infty}^{\infty} G(nf_s)\delta(f - nf_s), \tag{2.2}$$

where $G(nf_s)$ is the Fourier transform of $x(t)$, evaluated at the frequency $f = nf_s$. This correspondence suggests that we may determine the Fourier transform of the sampled target return signal $x_\delta(t)$ by invoking the duality property of the Fourier transform [3], the essence of which is the following:

$$\text{If } x(t) \leftrightarrow G(f), \quad \text{then } G(t) \leftrightarrow g(-f). \tag{2.3}$$

Representation of the continuous function $x(t)$ in the form of the sequence $\{x(nT_s)\}$ is possible only under the well-known restrictions. One restriction in kind is a requirement of spectrum limitation of the sampled target return signal $x_\delta(t)$. By Kotelnikov's theorem concerning the signal space concept [4], the continuous target return signal $x(t)$ possessing a limited spectrum is completely defined by countable set of samples spaced $T_s \leq 1/(2f_{\max})$ seconds apart, where $f_{\max}$ is the cutoff frequency of target return signal spectrum.

Under digital signal processing in CRSs, a stochastic process at the analog receiver output is an object to sample. This output process is a narrowband process at the condition $\Delta f_s / f_c \ll 1$, where Δf_s is the signal spectrum bandwidth and f_c is the carrier frequency. This condition allows us to apply an envelope procedure to represent the narrowband signal in the following form:

$$x(t) = X(t)\cos[2\pi f_c t + \varphi(t)], \tag{2.4}$$

where

$X(t)$ is the low-frequency signal (envelope)

$\varphi(t)$ is the phase modulation law of the narrowband signal—slowly varied function in comparison with $2\pi f_c t$

Since information about the target is extracted from the envelope $X(t)$ and phase $\varphi(t)$ of the target return signal and not from the carrier frequency f_c, and, moreover, $X(t)$ and $\varphi(t)$ are the slowly varied functions in time, there is a need to convert the target return signal given by (2.4) in such a way that sampling intervals would be defined by the real signal spectrum bandwidth rather then the carrier frequency f_c. In the case of the narrowband radio signal, we can write

$$x(t) = \text{Re}\left[\dot{X}(t)\exp(j2\pi f_c t)\right], \tag{2.5}$$

where Re[·] is the real part of the complex narrowband signal and

$$\dot{X}(t) = X(t)\exp[j\varphi(t)] \tag{2.6}$$

is the complex envelope of the narrowband signal. This complex envelope of the narrowband signal can be also presented in the following form:

$$\dot{X}(t) = X(t)\cos\varphi(t) - jX(t)\sin\varphi(t) = x_{\text{I}}(t) - jx_{\text{Q}}(t), \tag{2.7}$$

where $x_I(t)$ and $x_Q(t)$ are, respectively, the in-phase and quadrature components of the narrowband target return signal $x(t)$. Moreover,

$$X(t) = \sqrt{x_{\text{I}}^2(t) + x_{\text{Q}}^2(t)}, \quad X(t) > 0, \tag{2.8}$$

$$\varphi(t) = \operatorname{arctg}\frac{x_{\text{I}}(t)}{x_{\text{Q}}(t)}, \quad -\pi \le \varphi(t) \le \pi. \tag{2.9}$$

The in-phase $x_I(t)$ and quadrature $x_Q(t)$ components can be obtained by product between the target return signal $x(t)$ and two orthogonal signals with the frequency f_c forming at the local oscillator output. The corresponding device is called the *phase detector*. The flowchart of a simple phase detector is presented in Figure 2.2, where there are multipliers followed by the low-pass filters suppressing all high-frequency harmonics. Thus, the low-pass filters pass only the low-frequency in-phase $x_I(t)$ and $x_Q(t)$ quadrature components that must be sampled by the analog-to-digital converter.

The complex envelope of the narrowband radio signal can be presented either by the envelope and phase, which are the functions of time, or by the in-phase and quadrature components. In accordance with this statement, under the narrowband radio signal sampling there is a need to use two samples: either the envelope amplitude and phase or the in-phase and quadrature components of complex amplitude of the narrowband radio signal. Thus, we deal with two-dimensional signal sampling.

The sampling theorem for the two-dimensional signal may be presented in the following form [5]

$$T_{s1} \le \frac{1}{f_{1\max}}, \quad T_{s2} \le \frac{1}{f_{2\max}}, \tag{2.10}$$

where $f_{1\max}$ and $f_{2\max}$ are the highest frequencies in spectra of the first and second components of the narrowband signal. It is important to note that if we represent the narrowband signal using the envelope amplitude and phase, the frequencies then differ: namely, $f_{1\max}$ is the maximal frequency of the amplitude-modulated narrowband signal spectrum; $f_{2\max}$ is the maximal frequency of the

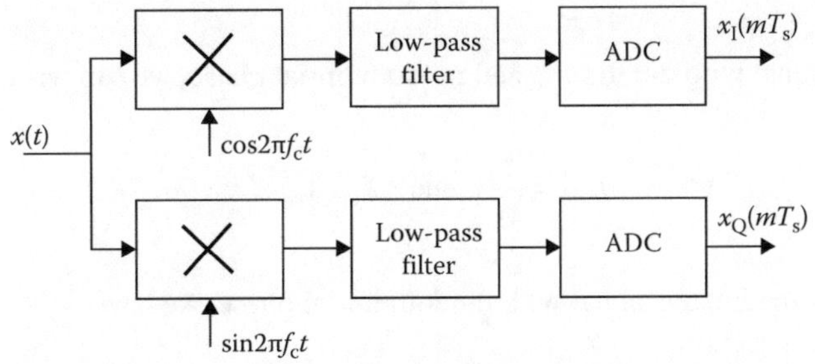

FIGURE 2.2 Example of phase detector: ADC—the analog-to-digital converter.

phase-modulated narrowband signal spectrum. For example, in the case of a completely known deterministic signal, the spectral bandwidth of the envelope is defined from the following condition [6,7]:

$$f_{1_{\max}} \times \tau_0 = 1 \quad \text{or} \quad f_{1_{\max}} = \frac{1}{\tau_0}, \tag{2.11}$$

where τ_0 is the duration of the sampled signal. Consequently, a maximal sampling period of amplitude envelope is limited by the condition $T_s \leq \tau_0$. In doing so, an initial phase of the envelope is known. By this reason, it becomes evident that there is no need to sample it. In the case of a wideband signal of the same duration τ_0 (the case of chirp modulation), the spectrum bandwidth of the modulated signal is close to double the amount of frequency deviation [8]:

$$f_{2_{\max}} \approx 2\Delta F_0. \tag{2.12}$$

Consequently, for one-valued representation of the frequency-modulated signal with constant amplitude, there is a need to set the phase

$$\varphi(t) = \int \Delta F_0 \, dt \tag{2.13}$$

at readings spaced as

$$T_{s_\varphi} \leq \frac{1}{2\Delta F_0}. \tag{2.14}$$

Number of phase counts of the signal with duration τ_0 is given by

$$N_\varphi = 2\Delta F_0 \tau_0. \tag{2.15}$$

Under the representation of complex envelope of the radio signal in the in-phase $x_I(t)$ and quadrature $x_Q(t)$ components, the maximal frequencies $f_{1_{\max}}$ and $f_{2_{\max}}$ are the same, that is,

$$f_{1_{\max}} = f_{2_{\max}} = f_{\max}. \tag{2.16}$$

Consequently, the sampling must be carried out simultaneously over every equal sampling interval:

$$T_{s_I} \leq \frac{1}{f_{\max}} \quad \text{and} \quad T_{s_Q} \leq \frac{1}{f_{\max}}. \tag{2.17}$$

In the case of the signal with duration τ_0 and random initial phase, we can write

$$f_{\max} = \frac{1}{\tau_0} \quad \text{and} \quad T_{s_I}, T_{s_Q} \leq \tau_0. \tag{2.18}$$

In the case of chirp-modulated signal with random initial phase, we have

$$f_{\max} \approx \Delta F_0. \tag{2.19}$$

Consequently,

$$T_{s_I}, T_{s_Q} \le \frac{1}{f_{max}} \le \frac{1}{\Delta F_0} \tag{2.20}$$

and the number of paired samples for the signal with duration τ_0 is determined as

$$N_I, N_Q = \Delta F_0 \tau_0. \tag{2.21}$$

In the case of the phase-manipulated pulse signal, the number of paired samples must not be less than the number of elements in code chain. If τ_0 is the duration of elementary signal, then the sampling intervals for the in-phase and quadrature components of the radio signal are given by

$$T_{s_I}, T_{s_Q} \le \tau_0. \tag{2.22}$$

2.1.2 Quantization and Signal Sampling Conversion

Under digital signal processing of target return signals in CRSs, there is a need to carry out a quantization of sampled values of complex envelope and phase or in-phase and quadrature components of radio signals in addition to sampling. Devices that carry out this function are called the *quantizers*.

The amplitude characteristic of alternating-sign sample quantizer with a fixed quantization step is presented in Figure 2.3. Here $X_1, X_2, \ldots, X_i, X_{i+1}, \ldots, X_m$ are the decision values; X_0 is the limiting level of signal; Δx is the quantization step; and $x'_1, x'_2, \ldots, x'_m$ are the sampled data of output signal covering the following center of range:

$$x'_i = \frac{X_i + X_{i+1}}{2}. \tag{2.23}$$

Under quantization of the in-phase and quadrature components of radio signal complex amplitude, the quantization step is chosen, as a rule, based on the condition

$$\Delta x = X_{min} \le \sigma_0, \tag{2.24}$$

where σ_0^2 is the variance of receiver noise.

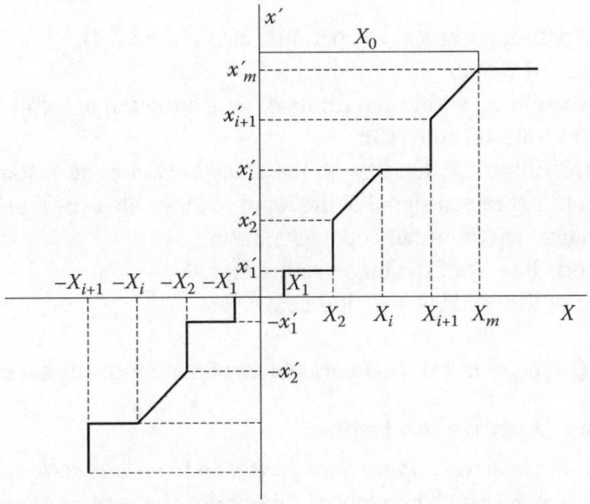

FIGURE 2.3 Alternating—sign sample quantizer with fixed quantization step.

The number of quantization levels is determined by

$$N_q = \frac{X_{max} - X_{min}}{\Delta x} = d_r - 1, \tag{2.25}$$

where d_r is the dynamic range of analog part of receiver. The number of bit site to represent the sampled target return signal is given by

$$n_b = \mathscr{F}[\log_2(N_q + 1)] = \mathscr{F}[\log_2 d_r], \tag{2.26}$$

where $\mathscr{F}[z]$ is the nearest integer NLE z. To characterize the analog-to-digital converter, we often use the following formula defining a dynamic range in dB for quantized sequence of samples for one bit of conversion [9]:

$$v = \frac{20\lg d_r}{n_b} = \frac{20\lg d_r}{\mathscr{F}[\log_2 d_r]} \approx 6\frac{\text{dB}}{\text{bit}}. \tag{2.27}$$

Under detection of the target return signal, definition of its parameters, and noise compensation by digital signal processing subsystems in CRSs, there is a need to use a capacity digit, for example, $N_b = 6 \div 8$, analog-to-digital conversion of the sampled target return signal. Capacity digit quantization at a high sampling rate f_s is a very difficult technical problem. Moreover, an increase in the sampling rate f_s and capacity digit quantization leads to overdesign of digital signal processing subsystems in complex radar systems. Because of this, we use binary quantizers and binary detectors for target return signals, in addition to capacity digit quantization [10,11]. Binary detectors are very simple under realization by digital signal processing techniques.

2.1.3 Analog-to-Digital Conversion: Design Principles and Main Parameters

Manifold types of analog-to-digital conversion are employed by digital signal processing subsystems in CRSs, for instance, the analog-to-digital conversion of voltage/current, time intervals, phase, frequency, and angular displacement. The principal structure of majority of analog-to-digital converters is the same: There are sampling block, quantization block, and data encoding.

The main engineering data are as follows:

1. Time parameters defining a speed of operation (see Figure 2.4):
   a. Sampling interval T_s
   b. Time of conversion T_c, within the limits of which the target return signal is processed by the analog-to-digital converter
   c. Time of conversion cycle T_{cc}, that is, the delay between the instant of appearance of the incoming target return signal at the input of the analog-to-digital converter and the instant of occurrence of output code generation
2. The number of code bits N_b of the target return signal
3. The element base of the analog-to-digital converter

Consider in detail the Q-factors or external parameters of analog-to-digital converter.

2.1.3.1 Sampling and Quantization Errors

There are two types of these errors: *dynamical errors* and *statical errors*. The dynamical errors are the discrete transform errors. The statical errors are the unit sample errors. The dynamical errors depend on the target return signal nature and time performance of the analog-to-digital

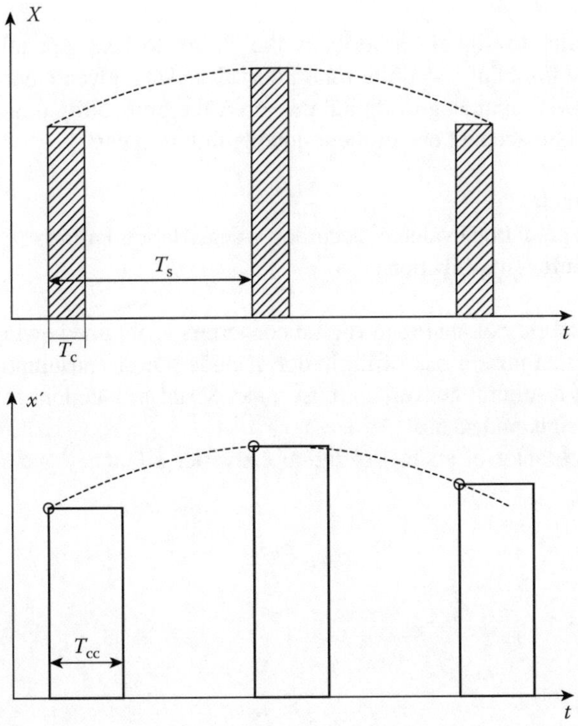

FIGURE 2.4 Time parameters of ADC.

converter. The main constituent of this kind of error is an inaccuracy caused by variations in the target return signal parameters at the analog-to-digital converter input. For example, if variations in the amplitude of the target return signal within the limits of T_{cc} are less or comparable with the quantization step Δx, then

$$T_{cc} \leq \frac{\Delta x}{V_{x_{max}}}, \tag{2.28}$$

where $V_{x_{max}}$ is the maximal speed of variations of the sampled and quantized target return signal. Taking into consideration that $T_{cc_{max}} = T_s$, we can define the sampling interval T_s based on (2.28).

The unit sample error is caused, first, by quantization error with variance

$$\sigma_q^2 = \frac{\Delta x^2}{12} \tag{2.29}$$

and, second, by deviation of actual quantization performance from ideal one (see Figure 2.3). If this deviation at the i-th quantization step is equal to ξ_i and the input target return signal possesses a uniform distribution of instantaneous values within the limits of operating range, then the variance of statical quantization error can be determined by

$$\sigma_{q\Sigma}^2 = \frac{\Delta x^2}{12} + \frac{1}{N_q}\sum_{i=1}^{N_q} \xi_i^2. \tag{2.30}$$

2.1.3.2 Reliability

Reliability of the analog-to-digital converter is the ability to keep operation accuracy at given boundaries within the limits of a definite time interval under a given environment. As a rule, to evaluate the reliability of an analog-to-digital converter the probability of no-failure operation for time t is used, taking into account one of the following failure types:

- The outage—glitch
- Errors exceed a predefined value of accuracy—degradation failure
- Intermittent failure—malfunctions

An increase in the reliability of analog-to-digital converters is obtained owing to redundancy.

Other parameters that pose a risk of limitation include power consumption, weight, and other dimensions of analog-to-digital converters, cost under serial production, manufacturability, time that is required for design, and so on.

As a generalized Q-factor of analog-to-digital converter efficiency, we consider the following ratio:

$$\mathcal{G} = \frac{n_{\text{b}} f_{\text{s}}}{Q}, \tag{2.31}$$

where

$$n_{\text{b}} f_{\text{s}} = \frac{n_{\text{b}}}{T_{\text{s}}} \tag{2.32}$$

is the data throughput of the analog-to-digital converter and Q is the size of enginery. At the digital signal processing of wideband radio signals in CRSs, rigid requirements in data throughput and reliability are applied to analog-to-digital converters. To design and produce such converters, the parallel Fourier transforms for target return signals are used.

2.2 DIGITAL GENERALIZED DETECTOR FOR COHERENT IMPULSE SIGNALS

2.2.1 Matched Filter

Recall briefly the main statements of classical detection theory. In accordance with a general theory of radar signal processing, the signal processing in time of stochastic process $x(t)$ representing an additive mixture of the signal $s(t)$ and stationary white Gaussian noise $w(t)$ is reduced to calculation of the correlation integral, which in the case of a scalar real signal $s(t, \alpha)$ with known parameter α can be presented in the following form:

$$T(\alpha) = \int_{-\infty}^{\infty} s^*(t, \alpha) x(t) dt, \tag{2.33}$$

where $s^*(t, \alpha)$ is the expected signal model generated by a local oscillator in the receiver or detector. If α is the delay of the expected signal with respect to the incoming initial input realization $x(t)$, then the correlation integral is determined as

$$T(\alpha) = \int_{-\infty}^{\infty} s^*(t - \alpha) x(t) dt. \tag{2.34}$$

Equation 2.34 is analogous to the convolution integral describing a process at the linear system output with impulse response $h(t)$ if the stochastic process $x(s)$ comes in at the linear system input:

$$Z(t) = \int_{-\infty}^{\infty} h(t-s)x(s)ds. \tag{2.35}$$

This analogy allows us to use the linear filter to calculate the correlation integral, the impulse response of which is matched with the expected signal $s(t)$. Matching is reduced to choice of a corresponding linear filter impulse response satisfying the following condition:

$$T(t_0 + \alpha) = Z(\alpha). \tag{2.36}$$

For the considered case of detection problem the impulse response of the matched linear filter must be mirrored with respect to the expected signal

$$h(t) = as(t_0 - t), \tag{2.37}$$

where

t_0 is the delay of signal peak at the matched filter output, which in the case of the pulse signal must be $t_0 \geq \tau_0$

a is the fixed scale factor

If the process $x(t) = s(t, \alpha) + w(t)$ comes in at the matched filter input, then according to (2.36) the process forming at the matched filter output at the instant $t_0 = \tau_0$ is defined in line with the following formula:

$$Z(t) = a \int_{t-\tau_0}^{\infty} x(u)s^*(\tau_0 - t + u)du. \tag{2.38}$$

In particular, when $w(t) = 0$, we obtain

$$Z(t) = a \int_{t-\tau_0}^{\infty} s(u)s^*(\tau_0 - t + u)du = aR_{ss*}(\tau_0 - t), \tag{2.39}$$

where $R_{ss*}(\tau_0 - t)$ is the autocorrelation function of expected signal $s(t, \alpha)$.

As it follows from (2.38) and (2.39), the signal at the matched filter output coincides with the mutual correlation function of the signal model and expected signal accurate within the fixed factor. When the white noise is absent, that is, $w(t) = 0$, the output signal coincides with the same accuracy with the autocorrelation function $R_{ss*}(\tau_0 - t)$ of expected signal $s(t, \alpha)$ at the time instant $(\tau_0 - t)$. Signal-to-noise ratio (SNR) by energy at the matched filter output is given by

$$\text{SNR} = \frac{2E_s}{\mathcal{N}_0}, \tag{2.40}$$

where $0.5\mathcal{N}_0$ is the two-sided power spectral density of white noise. The Neyman–Pearson detector brings us the analogous results [12]. Thus, the matched filter allows us to obtain the maximal SNR at the output within the limits of classical signal detection theory. Realization of analog matched filters in practice is very difficult, especially in the case of wideband signals. Moreover, it is impossible to carry out a parameter tuning for analog matched filters. For this reason, digital matched filters are widely used.

2.2.2 Generalized Detector

Recall the main functioning principles of the generalized detector (GD) constructed based on the generalized approach to signal processing in noise [13–17]. The GD is a composition of the linear systems, Neyman–Pearson receiver, and energy detector. A flowchart of a GD explaining the main functioning principles is shown in Figure 2.5. Here, we use the following notations: the model signal generator or local oscillator (MSG), the preliminary linear system or filter (PF), and the additional linear system of filter (AF).

Consider briefly the main statements regarding AF and PF. There are two linear systems at the GD front end that can be presented, for example, as bandpass filters, namely, the PF with the impulse response $h_{\text{PF}}(\tau)$ and the AF with the impulse response $h_{\text{AF}}(\tau)$. For simplicity of analysis, we consider that these filters have the same values for amplitude–frequency responses and bandwidths. Moreover, a resonant frequency of the AF is detuned relative to a resonant frequency of the PF on such a value that the incoming signal cannot pass through the AF. Thus, the received signal and noise can appear at the PF output and *the only noise* appears at the AF output (see Figure 2.5).

It is a well-known fact that if a value of detuning between the AF and PF resonant frequencies is more than $4 \div 5\Delta f_s$, where Δf_s is the signal bandwidth, the processes forming at the AF and PF outputs can be considered as independent and uncorrelated processes. In practice, the coefficient of

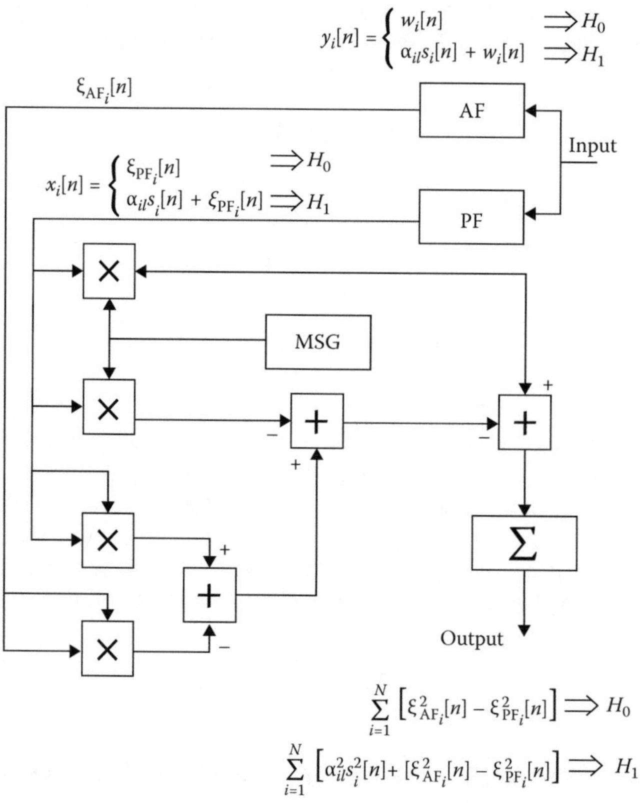

FIGURE 2.5 Generalized detector.

correlation is not more than 0.05. In the case of signal absence in the input process, the statistical parameters at the AF and PF outputs will be the same under the condition that these filters have the same amplitude–frequency responses and bandwidths by value, because the same noise is coming in at the AF and PF inputs. We may think that the AF and the PF do not change the statistical parameters of input process, since they are the linear front-end systems of a GD. For this reason, the AF can be considered as a generator of reference sample with a priori information *a "no" signal is obtained in the additional reference noise* forming at the AF output.

There is a need to make some comments regarding the noise forming at the PF and AF outputs. If the white Gaussian noise with zero mean and finite variance σ_n^2 comes in at the AF and PF inputs, the linear front-end system of the GD, the noise forming at the AF and PF outputs is Gaussian, too, because AF and PF are the linear systems and, in general, the noise takes the following form:

$$\xi_{\mathrm{PF}}(t)=\int_{-\infty}^{\infty} h_{\mathrm{PF}}(\tau)w(t-\tau)d\tau \quad \text{and} \quad \xi_{\mathrm{AF}}(t)=\int_{-\infty}^{\infty} h_{\mathrm{AF}}(\tau)w(t-\tau)d\tau, \tag{2.41}$$

where $\xi_{\mathrm{PF}}(t)$ and $\xi_{\mathrm{AF}}(t)$ are the narrowband Gaussian noise. If, for example, the additive white Gaussian noise with zero mean and two-sided power spectral density $0.5\mathcal{N}_0$ is coming in at the AF and PF inputs, then the noise forming at the AF and PF outputs is Gaussian with zero mean and variance given by [14, pp. 264–269]

$$\sigma_n^2=\frac{\mathcal{N}_0\,\omega_0^2}{8\Delta_{\mathrm{F}}}, \tag{2.42}$$

where if the AF or the PF is the RLC oscillator circuit, then the AF or the PF bandwidth Δ_{F} and resonance frequency ω_0 are defined in the following manner:

$$\Delta_{\mathrm{F}}=\pi\beta, \quad \omega_0=\frac{1}{\sqrt{LC}}, \quad \text{where } \beta=\frac{R}{2L}. \tag{2.43}$$

The main functioning condition of a GD is the *equality over the whole range of parameters* between the expected signal $s(t, \alpha)$ and the model signal forming at the MSG or local oscillator output $s^*(t-\tau_0, a)$. How we can satisfy this condition in practice is discussed in detail in Refs. [14, pp. 669–695,17]. More detailed discussion about choosing between the PF and the AF and their amplitude–frequency responses is given also in Refs. [15,16].

According to Figure 2.5 and the main functioning principle of a GD, the process forming at the GD output takes the following form:

$$Z_{\mathrm{GD}}^{\mathrm{out}}(t)=aR_{ss*}(\tau_0-t)+\xi_{\mathrm{AF}}^2(t)-\xi_{\mathrm{PF}}^2(t). \tag{2.44}$$

From (2.44) we see that the signal at the GD output coincides with the mutual correlation function of the signal model and expected signal accurate within the fixed factor. In a statistical sense, the background noise $\xi_{\mathrm{AF}}^2(t)-\xi_{\mathrm{PF}}^2(t)$ forming at the GD output tends to approach zero when the number of samples or the time interval of observation tends to approach infinity [15,17]. SNR by energy at the GD output is given by [14]

$$\mathrm{SNR}=\frac{E_{\mathrm{s}}}{\sqrt{4\sigma_n^4}}=\frac{E_{\mathrm{s}}}{2\sigma_n^2}, \tag{2.45}$$

where σ_n^2 is defined by (2.42).

2.2.3 Digital Generalized Detector

Now consider briefly the main principles of designing and construction of the digital GD (DGD). The DGD flowchart is represented in Figure 2.6. We see that processes at the outputs of MSG, PF, and AF are sampled and quantized, which is equivalent to passing these processes through digital filters. The model signal forming at the MSG output after sampling and quantization can be presented in the following form:

$$s^*(lT_s) = aT_s s^*[(n_0 - l)T_s], \tag{2.46}$$

where $n_0 = \tau_0/T_s$ is the number of discrete elements of the model signal. For simplicity, we can assume that $a = T_s^{-1}$. Then

$$s^*(lT_s) = s^*[(n_0 - l)T_s], \tag{2.47}$$

where $l = (0, 1,\ldots, n_0 - 1)$.

If the main functioning condition of GD, that is, an equality over the whole range of parameters between the expected signal $s(t, \alpha)$ and the model signal forming at the MSG output $s^*(\tau_0 - t + u)$ is satisfied, the process at the DGD output, when the additive mixture of the signal $s(t)$ and stationary white Gaussian noise $w(t)$ comes in at the input, can be represented in the following form [18–21]:

$$Z_{\text{DGD}}^{\text{out}}(kT_s) = 2\sum_{l=0}^{n_0-1} C(l)x[(k-l)T_s]s^*[(n_0-l)T_s] - \sum_{l=0}^{n_0-1} C(l)x^2[(k-l)T_s] + \sum_{l=0}^{n_0-1} C(l)\xi_{\text{AF}}^2[(k-l)T_s], \tag{2.48}$$

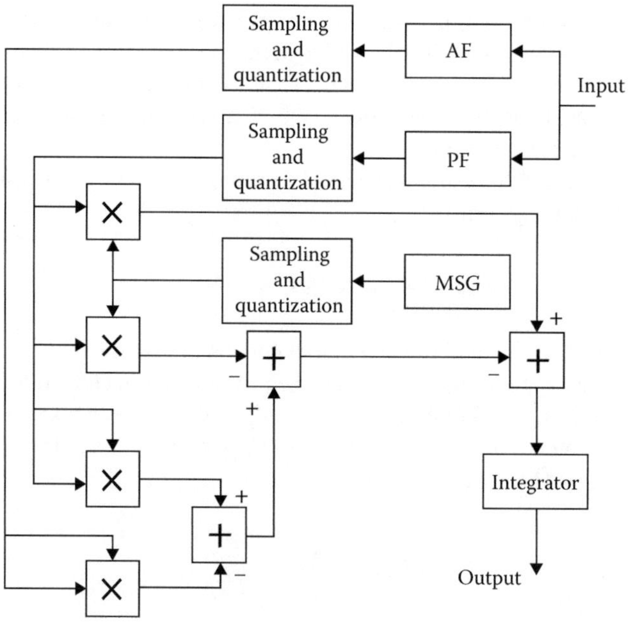

FIGURE 2.6 Digital generalized detector.

where

$$\begin{cases} x[(k-l)T_s] = x(t)\big|_{t=\tau_0}; \quad s*[(n_0-l)T_s] = s*(t)\big|_{t=\tau_0}; \quad s[(n_0-l)T_s] = s(t)\big|_{t=\tau_0}; \\ \xi_{PF}[(k-l)T_s] = \xi_{PF}(t)\big|_{t=\tau_0}; \quad \xi_{AF}[(k-l)T_s] = \xi_{AF}(t)\big|_{t=\tau_0} \end{cases} \tag{2.49}$$

and

$$C(l) = T_s C_0(l) \tag{2.50}$$

are the coefficients determined by numerical integration using the technique of rectangles (any technique may be used):

$$C_0(l) = 1, 1, \ldots, 1, 0. \tag{2.51}$$

If in (2.48) we replace $n_0 - l$ by i in the first term, after elementary mathematical transformations of the first term with the second term we obtain

$$R_{ss*}(kT_s) = \sum_i s*(iT_s)s[(k-(n_0-i))T_s]. \tag{2.52}$$

Equation 2.52 represents, by analogy with (2.39), the autocorrelation function of the expected signal $s(t, \alpha)$. The autocorrelation function (2.52) of the signal at the DGD output is a periodic function by frequency. At low values of f_s, cross-sections of the autocorrelation function (2.52) may overlap, which leads to distortions of the process forming at the DGD output. However, if f_s were chosen in agreement with the sampling theorem, these distortions would be negligible.

As it follows from (2.52), when the expected signal $s(iT_s)$ comes in at the DGD input, the signal at the DGD output is matched with the autocorrelation function (2.52) accurate within the cofactor aT_s. Since the autocorrelation function (2.52) is symmetric with respect to its maximum $R_{ss*}(0)$, the data samples of the sequence $\{Z_{GD}^{out}(kT_s)\}$ at the DGD output will at first increase and after reaching the upper limit at $kT_s = n_0 T_s$, that is, the maximal value of the autocorrelation function (2.52), decrease to zero within the limits of the time interval between $n_0 T_s$ and $2n_0 T_s$. The envelope of the data samples of the sequence $\{Z_{GD}^{out}(kT_s)\}$ at the DGD output coincides with an envelope of the autocorrelation function $R_{ss*}(kT_s)$ given by (2.52). This peculiarity of data samples of the sequence $\{Z_{GD}^{out}(kT_s)\}$ at the DGD output agrees with that of analog GD.

The limiting value of the DGD output signal energy is equal to the energy of expected signal sequence $s(kT_s)$ and is reached at the finite signal bandwidth and $T_s \le (2\Delta f_s)^{-1}$, where Δf_s is the signal bandwidth. If the signal spectrum is infinite and we must take into account some effective signal bandwidth Δf_s^{eff} under sampling, there are energy losses owing to superposition of unaccounted power spectral density tails under their mutual shift on k/T_s. These energy losses may be taken into consideration introducing an additional noise within the power spectral density $0.5\mathcal{N}_0'$. The receiver noise of DGD is the stationary random sequence with the power spectral density $0.5\mathcal{N}_0''$ uniform within the limits of the bandwidth $-(2T_s)^{-1} \le \Delta f < (2T_s)^{-1}$ and depends on T_s.

It is well known that the sampling and quantization techniques are used to digitize analog signals. For example, the sampling technique is used to discretize a signal within the limits of the time interval, and the quantization technique is used to discretize a signal by amplitude within the limits of the sampling interval. For this reason, there is a need to distinguish between the errors caused by these two digitizing techniques, which allows us to obtain a high accuracy of receiver performance.

Amplitude quantization and sampling can be considered as additional noises $\zeta_1(kT_s)$ and $\zeta_2(kT_s)$ with zero mean and finite variances $\sigma_{\zeta_1}^2$ and $\sigma_{\zeta_2}^2$, respectively. If the relationship between the chosen amplitude quantization step Δx and the mean square deviation σ' of process at sampling, and quantization block output is determined by $\Delta x < \sigma'$, then the absolute value of mutual correlation function between the amplitude quantization error and the signal is approximately 10^{-9} with respect to the values of the autocorrelation function of the signal. Therefore, it is reasonable to neglect this mutual correlation function. As a first approximation, it is reasonable to assume that the noises $\zeta_1(kT_s)$ and $\zeta_2(kT_s)$ are Gaussian.

We may suggest that the additive component $\zeta_\Sigma(kT_s)$ can be presented in the form of summary uncorrelated interferences: the interference caused by quantization $\zeta_1(kT_s)$, normal Gaussian with zero mean and the finite variance $\sigma_{\zeta_1}^2$, and the interference caused by sampling $\zeta_2(kT_s)$, normal Gaussian with zero mean, and the finite variance $\sigma_{\zeta_2}^2$. Thus, the summary additive interference $\zeta_\Sigma(kT_s)$ can be presented in the following form:

$$\zeta_\Sigma(kT_s) = \zeta_1(kT_s) + \zeta_2(kT_s) \tag{2.53}$$

and is the normal Gaussian with zero mean and the finite variance given by

$$\sigma_{\zeta_\Sigma}^2 = \sigma_{\zeta_1}^2 + \sigma_{\zeta_2}^2 \tag{2.54}$$

This is a direct consequence of Bussgang's theorem [22].

Taking into consideration the aforementioned statements, the total background noise at the DGD output is defined by the receiver noise and interferences caused by quantization and sampling:

$$\text{the total background noise} = \xi_{\text{AF}_\Sigma}^2(kT_s) - \xi_{\text{PF}_\Sigma}^2(kT_s), \tag{2.55}$$

where

$$\xi_{\text{PF}_\Sigma}(kT_s) = \xi_{\text{PF}}(kT_s) + \zeta_\Sigma(kT_s) = \xi_{\text{PF}}(kT_s) + \zeta_1(kT_s) + \zeta_2(kT_s) \tag{2.56}$$

is the noise forming at the PF output of DGD input linear system consisting of the normal Gaussian noise $\xi_{\text{PF}}(kT_s)$ with zero mean and the variance σ_n^2; the interference $\zeta_1(kT_s)$ with zero mean and the variance $\sigma_{\zeta_1}^2$, which is caused by quantization; and the interference $\zeta_2(kT_s)$ with zero mean and the variance $\sigma_{\zeta_2}^2$, which is caused by sampling. Noise $\xi_{\text{PF}}(kT_s)$ and interferences $\zeta_1(kT_s)$ and $\zeta_2(kT_s)$ do not correlate with each other.

$$\xi_{\text{AF}_\Sigma}(kT_s) = \xi_{\text{AF}}(kT_s) + \zeta_\Sigma(kT_s) = \xi_{\text{AF}}(kT_s) + \zeta_1(kT_s) + \zeta_2(kT_s) \tag{2.57}$$

is the noise forming at the AF output of DGD input linear system (additional or reference noise) [14,17–21], consisting of the normal Gaussian noise $\xi_{\text{AF}}(kT_s)$ with zero mean and the variance σ_n^2; the interference $\zeta_1(kT_s)$ with zero mean and the variance $\sigma_{\zeta_1}^2$, which is caused by quantization; and the interference $\zeta_2(kT_s)$ with zero mean and the variance $\sigma_{\zeta_2}^2$, which is caused by sampling. The noise $\xi_{\text{AF}}(kT_s)$ and the interferences $\zeta_1(kT_s)$ and $\zeta_2(kT_s)$ do not correlate to each other.

The probability density function of the total background noise forming at the DGD output is symmetric with respect to zero because the means of noises $\xi_{\text{PF}}(kT_s)$ and $\xi_{\text{AF}}(kT_s)$ and interferences $\zeta_1(kT_s)$ and $\zeta_2(kT_s)$ are equal to zero owing to the initial conditions. The probability density function of the total background noise forming at the DGD output is discussed in detail in Refs. [14, pp. 250–263,15].

Because the noise $\xi_{PF}(kT_s)$ and $\xi_{AF}(kT_s)$ and the interferences $\zeta_1(kT_s)$ and $\zeta_2(kT_s)$ do not correlate with each other, the variance of the total background noise and the interferences forming at the DGD output can be determined in the following form [18–21]:

$$\sigma^2_{\xi^2_{AF\Sigma}-\xi^2_{PF\Sigma}} = 4\sigma_n^4 + 4\sigma_{\zeta\Sigma}^4. \tag{2.58}$$

SNR by energy at the DGD output is given by the following [14, pp. 504–508, 18–21]:

$$\text{SNR} = \frac{E_s}{\sqrt{\sigma^2_{\xi^2_{AF\Sigma}-\xi^2_{PF\Sigma}}}} = \frac{E_s}{\sqrt{4\sigma_n^4 + 4\sigma_{\zeta\Sigma}^4}} = \frac{E_s}{2\sqrt{\sigma_n^4 + \sigma_{\zeta\Sigma}^4}}, \tag{2.59}$$

where σ_n^2 is defined by (2.42) and $\sigma^2_{\zeta\Sigma}$ is given by (2.54).

Thus, the losses caused by the sampling period T_s are possible in digital signal processing subsystems employed by complex radar systems.

2.3 CONVOLUTION IN TIME DOMAIN

Target return signals employed by CRSs typically have a narrowband. It is for this reason the DGD must use two channels for signal processing: *in-phase* and *quadrature* channels. The narrowband target return signals coming in at the input of DGD linear systems can be presented by the in-phase $x_I[k]$ and quadrature $x_Q[k]$ constituents at discrete sampling instants kT_s. In this case, the complex envelope of the input signal can be presented in the following form:

$$\dot{X}[k] = x_I[k] - jx_Q[k]. \tag{2.60}$$

By analogy with (2.60), the complex envelope at the output of the MSG can be presented in the following form:

$$\dot{S}^*[k] = S_I^*[k] + jS_Q^*[k]. \tag{2.61}$$

The process at the DGD output accurate within the factor $0.5T_s$ can be determined in the following form [19]:

$$Z_{DGD}^{out}[k] = 2\dot{X}[k]\dot{S}^*[k] - \dot{X}^2[k] + \xi_{AF}^2[k], \tag{2.62}$$

where $\xi_{AF}[k]$ is the noise forming at the AF output (the input linear filter of DGD). We can discard the third term in (2.62) and will consider it in the end result. Based on (2.61) and (2.62), the process at the DGD output can be written as follows:

$$\begin{aligned} Z_{DGD}^{out}[k] &= 2\left\{\left[x_I[k] - jx_Q[k]\right]\left[S_I^*[k] + jS_Q^*[k]\right]\right\} - \left\{x_I[k] - jx_Q[k]\right\}^2 \\ &= \sum_{l=0}^{n_0-1}\left\{2\left\{\left[x_I[k-l] - jx_Q[k-l]\right]\left[S_I^*[n_0-l] + jS_Q^*[n_0-l]\right]\right\} - \left\{x_I[k-l] - jx_Q[k-l]\right\}^2\right\}. \end{aligned} \tag{2.63a}$$

Replacing $n_0 - l$ by i in (2.63a), we obtain

$$Z_{\text{DGD}}^{\text{out}}[k] = \sum_{l=0}^{n_0-1}\Big\{2\big\{\{x_{\text{I}}[k-(n_0-i)] - jx_{\text{Q}}[k-(n_0-i)]\}\{S_{\text{I}}^{*}[i] + jS_{\text{Q}}^{*}[i]\}\big\}$$

$$-\{x_{\text{I}}[k-(n_0-i)] - jx_{\text{Q}}[k-(n_0-i)]\}^{2}\Big\}. \tag{2.63b}$$

According to the analysis carried out in [15, pp. 269–282] and following the main functioning DGD condition, that is, in the considered case

$$S_{\text{I}}^{*}(i) = S_{\text{I}}[k-(n_0-i)] \quad \text{and} \quad S_{\text{Q}}^{*}(i) = S_{\text{Q}}[k-(n_0-i)], \tag{2.64}$$

the in-phase and quadrature constituents of the process at the DGD output can be presented in the following form:

$$Z_{\text{DGD}_{\text{I}}}^{\text{out}} = Z_{\text{DGD}_{\text{II}}}^{\text{out}} + ZD_{\text{GD}_{\text{QQ}}}^{\text{out}} = S_{\text{I}}^{*}(i)S_{\text{I}}[k-(n_0-i)] - \xi_{\text{I}}^{2}[k-(n_0-i)]$$

$$+ S_{\text{Q}}^{*}(i)S_{\text{Q}}[k-(n_0-i)] + \xi_{\text{Q}}^{2}[k-(n_0-i)], \tag{2.65}$$

$$Z_{\text{DGD}_{\text{Q}}}^{\text{out}} = Z_{\text{DGD}_{\text{IQ}}}^{\text{out}} + Z_{\text{DGD}_{\text{QI}}}^{\text{out}} = -4S_{\text{I}}^{*}(i)S_{\text{Q}}[k-(n_0-i)] - 4S_{\text{I}}^{*}(i)\xi_{\text{Q}}[k-(n_0-i)]$$

$$+ 4S_{\text{Q}}^{*}(i)S_{\text{I}}[k-(n_0-i)] + 4S_{\text{Q}}^{*}(i)\xi_{\text{I}}[k-(n_0-i)]$$

$$+ 2\xi_{\text{Q}}[k-(n_0-i)]\xi_{\text{I}}[k-(n_0-i)], \tag{2.66}$$

where the factor 2 in the second line is caused by the presence of amplifier (see Ref. [15, pp. 269–282]). Moreover, the corresponding terms in the first and second lines of (2.66) are compensated in the statistical sense. As a result, the quadrature constituent of the process at the DGD output is caused by the autocorrelation function of the in-phase and quadrature constituents of the narrowband noise. Thus, we can write

$$R_{Z_{\text{DGD}_{\text{Q}}}^{\text{out}}}(\tau) = \sigma_{\xi_{\text{AF}\Sigma}^{2}-\xi_{\text{PF}\Sigma}^{2}}^{2}\Delta_{\text{F}}\,\text{sinc}(\Delta_{\text{F}}\tau), \tag{2.67}$$

where the variance of the total background noise at the DGD output $\sigma_{\xi_{\text{AF}\Sigma}^{2}-\xi_{\text{PF}\Sigma}^{2}}^{2}$ is given by (2.58); the DGD input linear system (PF and/or AF) bandwidth is defined by (2.43); and sinc(x) is the sinc function [1].

Further specification of digital signal processing algorithms is defined by type of convolved signals. For example, in the case of chirp modulation of the signal with rectangular envelope

$$S(t) = \sin\left[2\pi f_{\text{c}}t + \gamma t^{2}\right], \tag{2.68}$$

where $0 < t \le \tau_0$ and $\gamma = \pi\Delta F_0/\tau_0 = \text{const}$; ΔF_0 is the target return signal frequency deviation. The complex envelope can be presented in the following form:

$$\dot{S}(t) = \sin\gamma t^{2} - j\cos\gamma t^{2}. \tag{2.69}$$

Consequently, the in-phase and quadrature constituents of the target return signal at discrete instants $[k-(n_0-i)]T_s$ can be presented in the following form:

$$\begin{cases} x_I[k-(n_0-i)] = \sin\gamma[k-(n_0-i)]^2 + \xi_I[k]; \\ x_Q[k-(n_0-i)] = \cos\gamma[k-(n_0-i)]^2 + \xi_Q[k], \end{cases} \quad (2.70)$$

where $\xi_I[k]$ and $\xi_Q[k]$ are the constituents of the narrowband noise forming at the PF (the DGD input linear system) output.

In this case, the complex envelope of the model signal forming at the MSG output takes the following form:

$$\dot{S}^*(\tau_0 - t) = \sin[\gamma(\tau_0 - t)^2] + j\cos[\gamma(\tau_0 - t)^2], \quad (2.71)$$

and the in-phase and quadrature constituents at discrete instants iT_s are given by

$$S_I^*[i] = \sin\gamma[i]^2 \quad \text{and} \quad S_Q^*[i] = \cos\gamma[i]^2. \quad (2.72)$$

The flowchart of the generalized signal processing algorithm given by (2.62) is shown in Figure 2.7. There are eight convolving blocks and six summators to calculate all in-phase and quadrature constituents of the process forming at the DGD output. Each in-phase and quadrature components can be presented in the following form taking into account (2.68) through (2.72):

$$Z_{DGD_{II}}^{out} = 2\sum_{i=1}^{n_0} \sin\gamma[i]^2 x_I[k-(n_0-i)] = 2\sum_{i=1}^{n_0} \sin\gamma[i]^2 \sin\gamma[k-(n_0-i)]^2$$
$$+2\sum_{i=1}^{n_0} \sin\gamma[i]^2 \xi_I[k-(n_0-i)]^2; \quad (2.73)$$

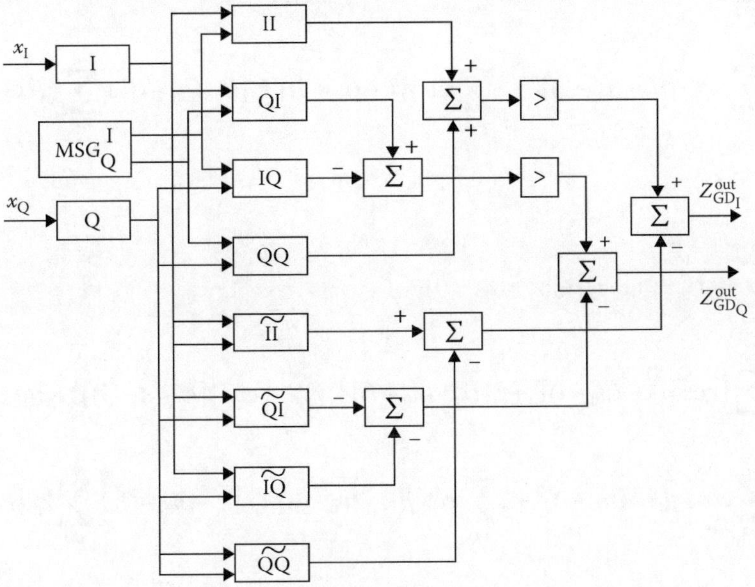

FIGURE 2.7 Convolution in time using the digital generalized detector.

$$Z^{\text{out}}_{\text{DGD}_{\text{QQ}}} = 2\sum_{i=1}^{n_0} \cos\gamma[i]^2 x_{\text{Q}}[k-(n_0-i)] = 2\sum_{i=1}^{n_0} \cos\gamma[i]^2 \cos\gamma[k-(n_0-i)]^2 + 2\sum_{i=1}^{n_0} \cos\gamma[i]^2 \xi_{\text{Q}}[k-(n_0-i)]^2; \quad (2.74)$$

$$Z^{\text{out}}_{\text{DGD}_{\text{QI}}} = 2\sum_{i=1}^{n_0} \cos\gamma[i]^2 x_{\text{I}}[k-(n_0-i)] = 2\sum_{i=1}^{n_0} \cos\gamma[i]^2 \sin\gamma[k-(n_0-i)]^2 + 2\sum_{i=1}^{n_0} \cos\gamma[i]^2 \xi_{\text{I}}[k-(n_0-i)]^2; \quad (2.75)$$

$$Z^{\text{out}}_{\text{DGD}_{\text{IQ}}} = 2\sum_{i=1}^{n_0} \sin\gamma[i]^2 x_{\text{Q}}[k-(n_0-i)] = 2\sum_{i=1}^{n_0} \sin\gamma[i]^2 \cos\gamma[k-(n_0-i)]^2 + 2\sum_{i=1}^{n_0} \sin\gamma[i]^2 \xi_{\text{Q}}[k-(n_0-i)]^2; \quad (2.76)$$

$$\begin{aligned}\tilde{Z}^{\text{out}}_{\text{DGD}_{\text{II}}} &= \sum_{i=1}^{n_0} x_{\text{I}}[k-(n_0-i)]x_{\text{I}}[k-(n_0-i)] \\ &= \sum_{i=1}^{n_0}\left\{\sin\gamma[k-(n_0-i)]^2 + \xi_{\text{I}}[k-(n_0-i)]\right\} \times \sum_{i=1}^{n_0}\left\{\sin\gamma[k-(n_0-i)]^2 + \xi_{\text{I}}[k-(n_0-i)]\right\} \\ &= \sum_{i=1}^{n_0} \sin^2\gamma[k-(n_0-i)]^2 + 2\sum_{i=1}^{n_0} \sin\gamma[k-(n_0-i)]^2 \xi_{\text{I}}[k-(n_0-i)] + \sum_{i=1}^{n_0} \xi_{\text{I}}^2[k-(n_0-i)];\end{aligned} \quad (2.77)$$

$$\begin{aligned}\tilde{Z}^{\text{out}}_{\text{DGD}_{\text{QQ}}} &= \sum_{i=1}^{n_0} x_{\text{Q}}[k-(n_0-i)]x_{\text{Q}}[k-(n_0-i)] \\ &= \sum_{i=1}^{n_0}\left\{\cos\gamma[k-(n_0-i)]^2 + \xi_{\text{Q}}[k-(n_0-i)]\right\} \times \sum_{i=1}^{n_0}\left\{\cos\gamma[k-(n_0-i)]^2 + \xi_{\text{Q}}[k-(n_0-i)]\right\} \\ &= \sum_{i=1}^{n_0} \cos^2\gamma[k-(n_0-i)]^2 + 2\sum_{i=1}^{n_0} \cos\gamma[k-(n_0-i)]^2 \xi_{\text{Q}}[k-(n_0-i)] + \sum_{i=1}^{n_0} \xi_{\text{Q}}^2[k-(n_0-i)];\end{aligned} \quad (2.78)$$

$$\tilde{Z}^{\text{out}}_{\text{DGD}_{\text{QI}}} = \sum_{i=1}^{n_0} x_{\text{Q}}[k-(n_0-i)]x_{\text{I}}[k-(n_0-i)]$$

$$= \sum_{i=1}^{n_0}\left\{\cos\gamma[k-(n_0-i)]^2 + \xi_{\text{Q}}[k-(n_0-i)]\right\} \times \sum_{i=1}^{n_0}\left\{\sin\gamma[k-(n_0-i)]^2 + \xi_{\text{I}}[k-(n_0-i)]\right\}$$

$$= \sum_{i=1}^{n_0}\cos\gamma[k-(n_0-i)]^2\sin\gamma[k-(n_0-i)]^2 + \sum_{i=1}^{n_0}\cos\gamma[k-(n_0-i)]^2\xi_{\text{I}}[k-(n_0-i)]$$

$$+\sum_{i=1}^{n_0}\sin\gamma[k-(n_0-i)]^2\xi_{\text{Q}}[k-(n_0-i)] + \sum_{i=1}^{n_0}\xi_{\text{Q}}[k-(n_0-i)]\xi_{\text{I}}[k-(n_0-i)]; \tag{2.79}$$

$$\tilde{Z}^{\text{out}}_{\text{DGD}_{\text{IQ}}} = \sum_{i=1}^{n_0} x_{\text{I}}[k-(n_0-i)]x_{\text{Q}}[k-(n_0-i)]$$

$$= \sum_{i=1}^{n_0}\left\{\sin\gamma[k-(n_0-i)]^2 + \xi_{\text{I}}[k-(n_0-i)]\right\} \times \sum_{i=1}^{n_0}\left\{\cos\gamma[k-(n_0-i)]^2 + \xi_{\text{Q}}[k-(n_0-i)]\right\}$$

$$= \sum_{i=1}^{n_0}\sin\gamma[k-(n_0-i)]^2\cos\gamma[k-(n_0-i)]^2 + \sum_{i=1}^{n_0}\sin\gamma[k-(n_0-i)]^2\xi_{\text{Q}}[k-(n_0-i)]$$

$$+\sum_{i=1}^{n_0}\cos\gamma[k-(n_0-i)]^2\xi_{\text{I}}[k-(n_0-i)] + \sum_{i=1}^{n_0}\xi_{\text{I}}[k-(n_0-i)]\xi_{\text{Q}}[k-(n_0-i)]. \tag{2.80}$$

In the case of the phase-code-manipulated signal with duration $\tau_{\Sigma_0} = N_e\tau_0$, where N_e is the number of elementary signals and τ_0 is the duration of elementary signal, the complex amplitude envelope can be presented in the following form:

$$\dot{S}(t) = \sum_{i=1}^{N_e}\dot{S}_i(t), \quad \text{where} \quad \dot{S}_i(t) = \exp(j\theta_i). \tag{2.81}$$

In the case of binary signal, we have $\theta_i = [0, \pi]$ and

$$\dot{S}_i(t) = \varsigma[i] = \pm 1. \tag{2.82}$$

Consequently, the discrete impulse response at the MSG output matched with the phase-code-manipulated target return signal is given by

$$S^*[i] = \varsigma[N_e - i] \tag{2.83}$$

and the in-phase and quadrature constituents of process forming at the DGD output take the following form:

$$\begin{cases} Z_{\mathrm{DGD_I}}^{\mathrm{out}\ 2}[k] = \sum_{i=1}^{N_e}\{\varsigma[N_e - i]\varsigma[k-(N_e-i)]\}^2 + \sum_{i=1}^{N_e}\{\xi_{\mathrm{AF_I}}^2[k-(N_e-i)] - \xi_{\mathrm{PF_I}}^2[k-(N_e-i)]\}^2; \\ Z_{\mathrm{DGD_Q}}^{\mathrm{out}\ 2}[k] = \sum_{i=1}^{N_e}\{\varsigma[N_e - i]\varsigma[k-(N_e-i)]\}^2 + \sum_{i=1}^{N_e}\{\xi_{\mathrm{AF_Q}}^2[k-(N_e-i)] - \xi_{\mathrm{PF_Q}}^2[k-(N_e-i)]\}^2, \end{cases} \tag{2.84}$$

where the complex amplitude envelope of the DGD output process is defined as follows:

$$Z_{\mathrm{DGD}}^{\mathrm{out}\ 2}[k] = \sqrt{Z_{\mathrm{GD_I}}^{\mathrm{out}\ 2}[k] + Z_{\mathrm{GD_Q}}^{\mathrm{out}\ 2}[k]}. \tag{2.85}$$

Thus, in the case of the phase-code-manipulated target return signal, the DGD employed by digital signal processing subsystem in complex radar system uses only four convolving blocks. Moreover, a convolution within the limits of each cycle equal to the elementary signal duration τ_0 is a summation of amplitude samples of the in-phase and quadrature constituents of the target return signal at instants with signs defined by values $S^*[i] = \pm 1$ in accordance with the given code of the phase-manipulated operation. Practical realization of this principle is not difficult.

Now consider the flowchart shown in Figure 2.7 and estimate a functional ability of DGD with convolution blocks in time domain to compress the chirp-modulated target return signal. As we can see from Figure 2.7, we use eight convolving blocks. Each convolving block must carry out n_0 multiplications and $n_0 - 1$ additions of N_b numbers in the course of calculation of the k-th signal value, that is, within the limits of the sampling interval T_s. Estimate a required processing speed of the convolving block, taking into account only multiplications for the widely used case of signal processing of the chirp-modulated target return signals at the sampling rate $f_s = \Delta F_0$ and the target return signal duration $\tau_0 = n_0 T_s$, namely, $V_{\mathrm{req}} = n_0 \Delta F_0$. For example, when $n_0 = 100$ and $\Delta F_0 = 5 \times 10^6$ Hz, the required processing speed must be about $V_{\mathrm{req}} = 5 \times 10^8$ multiplications per second.

Consequently, in the considered case, a direct realization of convolution operations by serial digital signal processing techniques is not possible. There is a need to apply specific procedures of calculations under digital signal processing. First of all, we may use a principle of parallelism that is characteristic of convolution problems. This principle allows us to calculate n_0 of pairwise multiplications $S^*[i] \times x[k - (n_0 - i)]$, where $i = 1, \ldots, n_0$, simultaneously using n_0 parallel multipliers with subsequent summation of partial multiplications (see Figure 2.8). In this case, each multiplier must possess the processing speed $V_{\mathrm{req}} = 5 \times 10^6$ multiplications per second, that is, one multiplication operation per 40 ns. This processing speed can be easily provided by employing very large-scale integration (VLSI) circuits.

Another example of accelerated convolution operation is the implementation of a specifically designed processor that uses the read-only memory (ROM) block to store calculations of bitwise multiplications made before. Factor codes are used as result addresses of these bitwise multiplications [22,23]. Consider in detail the convolution principle carried out by a specific processor with ROM. For this case, we can present the process at the DGD output as a convolution operation in the following form to simplify calculations:

$$Z_{\mathrm{DGD}}^{\mathrm{out}} = 2\sum_{i=1}^{N} S_i^* X_i - \sum_{i=1}^{N} X_i^2 + \sum_{i=1}^{N} \xi_{\mathrm{AF}_i}^2, \tag{2.86}$$

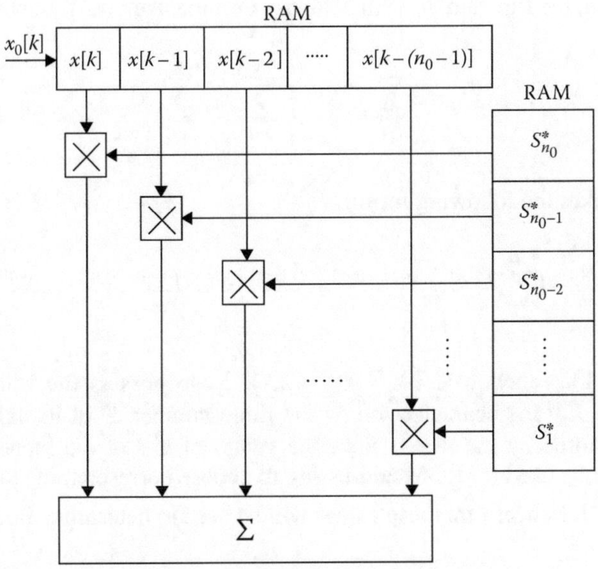

FIGURE 2.8 Principle of parallelism: RAM—the random access memory.

where

S_i^* can be considered as the weight factors of the model signal at the MSG output
X_i is the sample of the input target return signal
N is the sample size

We can assume that the input target return signals are scaled and as a sequence, $|X_i| < 1$. In addition, the input target return signals are presented in the form of n-bit additional comma-fixed code. Then (2.86) can be presented in the following form:

$$Z_{\text{DGD}}^{\text{out}} = 2\sum_{i=1}^{N} S_i^* \left(\sum_{i=1}^{n_i} X_i^{(k)} 2^{-k} - X_i^{(0)}\right) - \left(\sum_{k=1}^{n_i} X_i^{(k)} 2^{-k} - X_i^{(0)}\right)\left(\sum_{i=1}^{n_i} X_i^{(k)} 2^{-k} - X_i^{(0)}\right)$$
$$+ \left(\sum_{k=1}^{n_i} \xi_{\text{AF}_i}^{(k)} 2^{-k} - \xi_{\text{AF}_i}^{(0)}\right)\left(\sum_{k=1}^{n_i} \xi_{\text{AF}_i}^{(k)} 2^{-k} - \xi_{\text{AF}_i}^{(0)}\right), \tag{2.87}$$

where $X_i^{(k)}$ and $\xi_i^{(k)}$ are the values (0 or 1) of k-th bit of i-th sample of the input signal.

The DGD output process given by (2.87) can also be presented in the following form:

$$Z_{\text{DGD}}^{\text{out}} = \sum_{k=1}^{n-1} 2^{-k} \left[2\left(\sum_{i=1}^{N} S_i^* X_i^{(k)} - \sum_{i=1}^{N} S_i^* X_i^{(0)}\right) - \left(\sum_{i=1}^{N} X_i^{(k)} - \sum_{i=1}^{N} X_i^{(0)}\right)\left(\sum_{i=1}^{N} X_i^{(k)} - \sum_{i=1}^{N} X_i^{(0)}\right)\right.$$
$$\left. + \left(\sum_{i=1}^{N} \xi_{\text{AF}_i}^{(k)} - \sum_{i=1}^{N} \xi_{\text{AF}_i}^{(0)}\right)\left(\sum_{i=1}^{N} \xi_{\text{AF}_i}^{(k)} - \sum_{i=1}^{N} \xi_{\text{AF}_i}^{(0)}\right)\right]. \tag{2.88}$$

Now we can introduce the function $\mathcal{G}_k$ with N binary arguments in the following form:

$$\mathcal{G}_k\left(X_1^{(k)}, X_2^{(k)}, \ldots, X_N^{(k)}\right) = 2\sum_{i=1}^{N} S_i^* X_i^{(k)} - \sum_{i=1}^{N} X_i^{(k)} \sum_{i=1}^{N} X_i^{(k)} + \sum_{i=1}^{N} \xi_{\text{AF}_i}^{(k)} \sum_{i=1}^{N} \xi_{\text{AF}_i}^{(k)}. \tag{2.89}$$

In this case, (2.88) takes the following form:

$$Z_{\text{DGD}}^{\text{out}} = \sum_{k=1}^{n-1} 2^{-k} \mathcal{G}_k\left(X_1^{(k)}, X_2^{(k)}, \ldots, X_N^{(k)}\right) - \mathcal{G}_0\left(X_1^{(0)}, X_2^{(0)}, \ldots, X_N^{(0)}\right). \tag{2.90}$$

Since arguments of the function $\mathcal{G}_k\left(X_1^{(k)}, X_2^{(k)}, \ldots, X_N^{(k)}\right)$ can possess the value 0 or 1, the function $\mathcal{G}_k\left(X_1^{(k)}, X_2^{(k)}, \ldots, X_N^{(k)}\right)$ is characterized by the finite number 2^N of its values that can be calculated before and stored by the ROM. Now, the values of bits of the input target return signal $X_1^{(k)}, X_2^{(k)}, \ldots, X_N^{(k)}$ can be used for ROM addressing to choose corresponding values of the function $\mathcal{G}_k\left(X_1^{(k)}, X_2^{(k)}, \ldots, X_N^{(k)}\right)$. Henceforth, these values will be used to determine the DGD output process $Z_{\text{DGD}}^{\text{out}}$ by (2.90).

Thus, a convolution of the DGD output process $Z_{\text{DGD}}^{\text{out}}$ can be obtained by n addressing to ROM, $n - 1$ summations, and the only subtraction ($k = 0$). In doing so, a number of operations occur independent of the sample size N and are determined by quantization bits of the input target return signals. A very simple block diagram of a specifically designed processor that produces a convolution in accordance with (2.90) is presented in Figure 2.9.

Pulse packet of the target return signals is shifted by turns at the *shift registers SR*1,..., *SRN* starting from the low-order bit. At first, the values $X_i^{(n-1)}$ at the output of each shift register are used for addressing the corresponding value of the function $\mathcal{G}_{n-1}\left(X_1^{(k)}, X_2^{(k)}, \ldots, X_N^{(k)}\right)$ from ROM. This value is loaded in the register $R1$ and added to the register $R2$ content (zero content at the first step). The obtained result is recorded by the register $R3$. At the next cycle, the value of the function $\mathcal{G}_{n-2}\left(X_1^{(k)}, X_2^{(k)}, \ldots, X_N^{(k)}\right)$ is chosen and the content of the register $R3$ (the previous sum) is recorded by the register $R2$, with the right shift on one bit that corresponds to multiplication on the factor 0.5. The content of the register $R1$, that is, the value of the function $\mathcal{G}_{n-2}\left(X_1^{(k)}, X_2^{(k)}, \ldots, X_N^{(k)}\right)$, is added to the content of the register $R2$ that is a value of the function $0.5\mathcal{G}_{n-1}\left(X_1^{(k)}, X_2^{(k)}, \ldots, X_N^{(k)}\right)$. As a result,

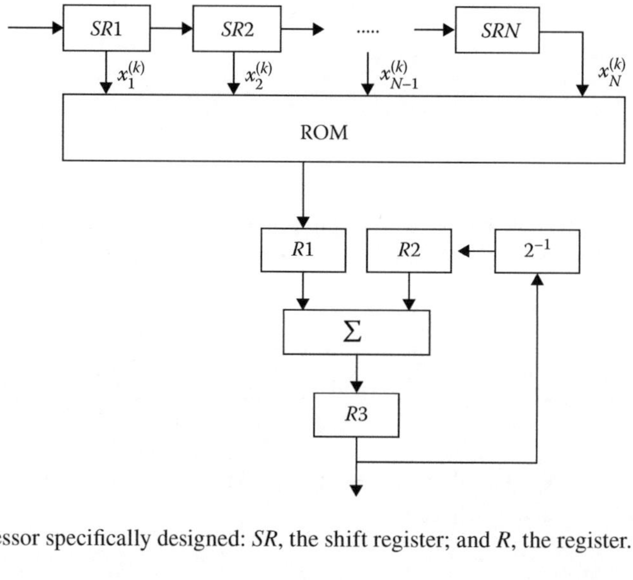

FIGURE 2.9 Processor specifically designed: *SR*, the shift register; and *R*, the register.

we obtain a regular particular result. This operation is repeated n times, and in doing so, at the last step, the function $\mathcal{G}_0(X_1^{(0)}, X_2^{(0)}, \ldots, X_N^{(0)})$ is subtracted from the accumulated sum and the final result of convolution given by (2.90) is the content of the register $R3$ after n cycles.

As we can see, a realization of the considered accelerated convolution is not difficult for low values of the sample size N, for example, $N = 10\ldots12$. With increasing N, the required ROM size becomes very large, for example, at $N = 15$, $Q_{\text{ROM}} = 32798$ words, and an access time is increased essentially. We can decrease the ROM size by partitioning a calculation process using a set of steps with later summing up results. Assume $N = LM$, where M is the number of functions $\mathcal{G}_k(X_1^{(k)}, X_2^{(k)}, \ldots, X_N^{(k)})$ and L is the number of binary arguments of these functions, then, in this case, (2.86) can be written in the following form:

$$Z_{\text{DGD}}^{\text{out}} = \sum_{i=1}^{L} \left(2S_i^* X_i - X_i^2 + \xi_{\text{AF}_i}^2 \right) + \sum_{i=L+1}^{2L} \left(2S_i^* X_i - X_i^2 + \xi_{\text{AF}_i}^2 \right) X_i^2 + \cdots + \sum_{i=(M-1)(L+1)}^{N} \left(2S_i^* X_i - X_i^2 + \xi_{\text{AF}_i}^2 \right). \tag{2.91}$$

Each individual sum can be determined by the earlier-described procedure. For this case, the ROM size is determined as $Q_{\text{ROM}} = 2^L M$ instead of $2^N = 2^M \times 2^L$ without the partition of calculation process.

2.4 CONVOLUTION IN FREQUENCY DOMAIN

Now, consider the features of discrete convolution in frequency domain. In accordance with the theory of discrete presentation of continuous functions limited by time or frequency, the function $X(t)$ that can be presented by a sequence of readings $\{X[i]\}$, $i = 0, 1, 2, \ldots, n-1$ is transformed in frequency domain by discrete Fourier transform (DFT) that for each $k = 0, 1, 2, \ldots, n-1$ takes the following form [24]:

$$\mathbf{F}_X(k) = \sum_{i=0}^{n-1} X[i] \exp\left\{ -\frac{j2\pi ik}{n} \right\} = \sum_{i=0}^{n-1} X[i] W_n^{ik}, \tag{2.92}$$

where

$$W_n = \exp\left\{ -\frac{j2\pi}{n} \right\}, \tag{2.93}$$

and vice versa, any function presented by the limited discrete spectrum $\{\mathbf{F}_X[k]\}$, $k = 0, 1, 2, \ldots, n-1$, can be reconstructed in the time domain using the inverse discrete Fourier transform (IDFT):

$$X[i] = \frac{1}{n} \sum_{k=0}^{n-1} \mathbf{F}_X[k] \exp\left\{ \frac{j2\pi ki}{n} \right\} = \frac{1}{n} \sum_{k=0}^{n-1} \mathbf{F}_X[k] W_n^{-ik}. \tag{2.94}$$

Note that the number of discrete elements of the function $X(t)$ is the same for its presentation both in the time domain and in the frequency domain.

Convolution of sequences in the frequency domain is reduced to product of DFT results. For this purpose, there is a need to realize two direct Fourier transforms, namely, to convolve a sequence of readings of the function $\{X[i]\}$, $i = 0, 1, 2, \ldots, n-1$ and a sequence of readings of the impulse response of filters used by DGD. If after convolution it is necessary to make transformations into the time domain, there is a need to carry out IDFT for sequence of spectral components $\{\mathbf{F}_X[k]\}$, $k = 0, 1, 2, \ldots, n-1$.

For complex functions (signals) an algorithm of the operation *DFT-Convolution-IDFT* takes the following form:

$$1.\quad \dot{\mathbf{F}}_H[k]=\sum_{i=0}^{n-1}\dot{\mathbf{H}}[i]W_n^{ik},\quad k=0,1,2,\ldots,n-1, \tag{2.95}$$

where $\dot{\mathbf{H}}[i]$ is a sequence of readings of the complex impulse response of the convolving filter:

$$2.\quad \dot{\mathbf{F}}_X[k]=\sum_{i=0}^{l-1}\dot{\mathbf{X}}[i]W_l^{ik}, \tag{2.96}$$

where $\dot{\mathbf{X}}[i]$ is a sequence of complex readings of the input (convolved) function (the target return signal).

$$3.\quad \dot{\mathbf{F}}_{Z_{\mathrm{DGD}}^{\mathrm{out}}}[k]=\dot{\mathbf{F}}_H[k]\dot{\mathbf{F}}_X[k],\quad k=0,1,\ldots,l+n-1. \tag{2.97}$$

$$4.\quad \dot{\mathbf{Z}}_{\mathrm{DGD}}^{\mathrm{out}}[i]=\frac{1}{l+n}\sum_{k=0}^{l+n-1}\dot{\mathbf{F}}_{Z_{\mathrm{DGD}}^{\mathrm{out}}}[k]W_{l+n-1}^{-ik},\quad i=0,1,\ldots,l+n-1. \tag{2.98}$$

Principal peculiarity of the considered algorithm is a group technology type flow procedure if the width of an input data array is higher n, that is, $l\geq n$. The resulting convolution width is $l+n-1$. Under solution of detection problems by DGD, we assume that the impulse response of all filters used by DGD is not variable, at least for probing signals of the same kind. Therefore, the DFT of impulse response of all filters used by DGD is carried out in advance and stored in the memory device of the corresponding computer. In the course of convolution, there is a need to accomplish one DFT and one IDFT. It must be emphasized that under radar detection and signal processing problems, a convolved sequence width L corresponds to radar range sweep length that is much more in comparison with the width of convolving sequence equal to the width of an impulse response n_{im} of filters used by DGD. In accordance with Section 2.1, $n_{\mathrm{im}}=n_0$, where n_0 is the number of discrete elements of the expected target return signal. Synchronous convolution of such sequences is a very cumbersome process. Because of this, as a rule, the input sequence is divided on blocks with the width l. Each element of the p-th block is generated from a general sequence $\{\dot{X}[i]\}$, $i=0,1,2,\ldots,L$ following the law

$$\dot{\mathbf{Z}}_p[i]=\dot{\mathbf{Z}}[i+pl],\quad n=0,1,2,\ldots,\mathrm{In}\left[\frac{L}{l}\right], \tag{2.99}$$

where In[L/l] is the integer part of the ratio in brackets.

For each input data block of the width l the $(l+n-1)$-point DFT is determined. For convolving sequence of impulse response of filters used by DGD the components of the $(l+n-1)$-point DFT must be determined in advance and stored in a memory device. Convolution in frequency domain for each block is obtained by multiplication between the DFT of convolved and convolving sequences at $(l+n-1)$ points. To determine the convolution in time domain the IDFT is accomplished. The widths of obtained sequence $\dot{\mathbf{Z}}_{\mathrm{DGD}_p}^{\mathrm{out}}[i]$ are equal to $(l+n-1)$ and the neighboring sequences $\dot{\mathbf{Z}}_{\mathrm{DGD}_p}^{\mathrm{out}}[i]$ and $\dot{\mathbf{Z}}_{\mathrm{DGD}_{p+1}}^{\mathrm{out}}[i]$ are overlapped at the $n-1$ points. Thus, only the l sequence values will be true. Furthermore, the sum of overlapping partial sequences is used to obtain the correct calculation results for $\dot{Z}_{\text{DGD}}^{\text{out}}[i]$ in all points. In the course of design process, the problem of selecting the optimal value l at a fixed n by criterion of minimum convolution time arises. At low values of $n_{\text{im}} \leq 100$ the condition $l_{\text{opt}} \approx 5n_{\text{im}}$ is satisfied [11,21].

Consider now the problem of work content to realize the DGD in frequency domain. DFT or IDFT requires $(l + n - 1)^2$ operations of multiplication and $(l + n - 1)$ operations of summation for complex values to obtain the $(l + n - 1)$ frequency (time) samples. The total number of required operations taking into account transformations from the frequency domain to the time domain after convolution consists of $2(l + n - 1)^2 + (l + n - 1)$ multiplications and $2(l + n - 1)$ summations for complex values. In doing so, we obtain the sample of the output data with the width l. To obtain the same length for output data in the time domain, l^2 operations of multiplications and $l - 1$ operations of summations are required for complex values. Consequently, a convolution in the frequency domain is more time-consuming compared to a convolution in the time domain, approximately eight times if $l_{\text{opt}} \approx n_{\text{im}}$. Thus, there is no purpose to implement a convolution in the discussed form for DGD.

We can essentially decrease the number of operations at convolution in the frequency domain by employing the specific DFT algorithms that are called the fast FT (FFT) [25]. Now, consider the design principles of the FFT algorithm with time decimation by modulus 2 of real sequence. Let the sequence $\{X[i]\}$ that is processed by DFT have the width M corresponding to the integer power of the number 2, that is, $M = 2^m$. This initial sequence can be divided into two parts in accordance with the following rule:

$$X_{\text{even}}[i] = X[2i] \quad \text{and} \quad X_{\text{odd}}[i] = X[2i+1]\,, \quad i = 0,1,\ldots,0.5M. \tag{2.100}$$

The sequence $X_{\text{even}}[i]$ consists of elements of the initial sequence with even numbers, and the sequence $X_{\text{odd}}[i]$ consists of elements of the initial sequence with odd numbers. The width of each sequence is equal to $0.5M$. The sequences obtained in the issue of expansion are expanded again in two parts, while $0.5M$ two-point sequences are delivered. The number of expansion steps is equal to $m = \log_2 M$.

The essence of the FFT algorithm with time decimation by modulus 2 is as follows. The DFT sequences with the width $l > 2$ are calculated by a combination of DFT of two sequences with the width equal to $0.5\ l$. In accordance with this fact, at first, the $0.5M$-point DFT of two sequences is carried out under processing of the M-point sequence by FFT. Then the obtained transformations are united for the purpose of creating the $0.25M$ four-point sequences, $0.125M$ eight-point sequences, and so on, while a transformation of the width M will be obtained after m steps. FFT determination is carried out by the following formulas:

$$\mathbf{F}[k] = \mathbf{F}_{\text{even}}[k] + \mathbf{F}_{\text{odd}}[k]W_M^k, \quad k = 0,1,\ldots,0.5M - 1, \tag{2.101}$$

$$\mathbf{F}[k + 0.5M] = \mathbf{F}_{\text{even}}[k] - \mathbf{F}_{\text{odd}}[k]W_M^k, \quad k = 0,1,\ldots,0.5M - 1, \tag{2.102}$$

where

$$\mathbf{F}_{\text{even}}[k] = \sum_{i=0}^{0.5M-1} X_{\text{even}}[i]W_{0.5M}^{ik} \tag{2.103}$$

and

$$\mathbf{F}_{\text{odd}}[k] = \sum_{i=0}^{0.5M-1} X_{\text{odd}}[i]W_{0.5M}^{ik} \tag{2.104}$$

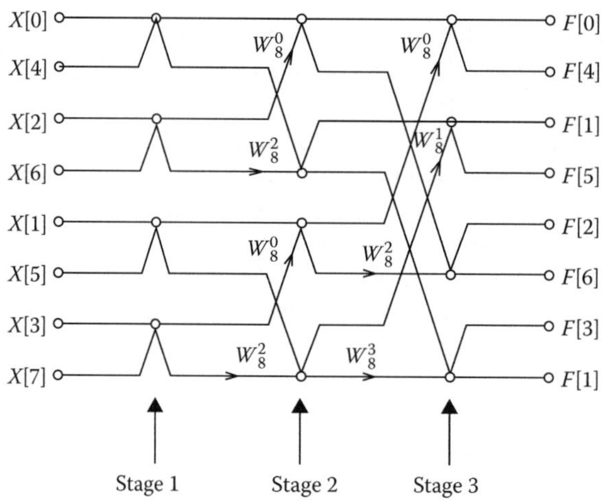

FIGURE 2.10 The directed graph for 8-point FFT.

are DFT for even and odd sequences, correspondingly; $W_M^k = (W_M)^k$ is the k-th power of the factor W_M, which is called rotating factor.

Directed graphs are used to represent graphically the FFT algorithm. The following notations are implemented under graphical representations: The point (or circle) denotes the summation–subtraction operation, the final sum appearing at the top output branch and the final subtraction appearing at the bottom output branch; the arrow on the branch denotes the multiplication by constant written over the arrow; if the arrow is absent the constant factor is equal to unit. The directed graph for 8-point FFT with time decimation by modulus 2 is shown in Figure 2.10. An order of input data assignment is obtained using a procedure of binary inversion of the numbers 0, 1, 2, 7. Such procedure makes a representation simpler and allows us to obtain the input sequence in the natural order—**F**[0], **F**[1], **F**[2], **F**[3] at the top output branches and **F**[4], **F**[5], **F**[6], **F**[7] at the bottom output branches. As we can see from Figure 2.10, the 8-point FFT is accomplished over three steps.

At the first step, the four 2-point DFTs are implemented and the condition $W_2 = \exp\{-j\pi\} = -1$ is taken into consideration. By this reason, the multiplication operations are absent, and in accordance with (2.98), we can write

$$\mathbf{F}[0] = \mathbf{F}_{\text{even}}[0] + \mathbf{F}_{\text{odd}}[0] \quad \text{and} \quad \mathbf{F}[1] = \mathbf{F}_{\text{even}}[0] - \mathbf{F}_{\text{odd}}[0]. \tag{2.105}$$

At the second step, two pairs of 2-point FFTs are combined with two 4-point FFTs according to (2.101). At the third step, two 4-point FFTs are transformed into the 8-point FFT. In general, the number of steps is defined as $m = \log_2 M$. At each step, excluding the first step, one half of M multiplications and M summations for complex values are processed. For this reason, to determinate M-point FFT we need $(M/2)\log_2 M$ multiplication operations and $M\log_2 M$ summation operations for complex values.

Previously it was shown that to determine the DFT we need M^2 multiplications and M summations for the complex values. The advantage in the number of multiplication operations under realization of FFT in comparison with the direct DFT is

$$\nu = \frac{M^2}{(M/2)\log_2 M} = \frac{2M}{\log_2 M}. \tag{2.106}$$

For example, $\nu \approx 200$ at $M = 1024$, $\nu \approx 21$ at $M = 128$.

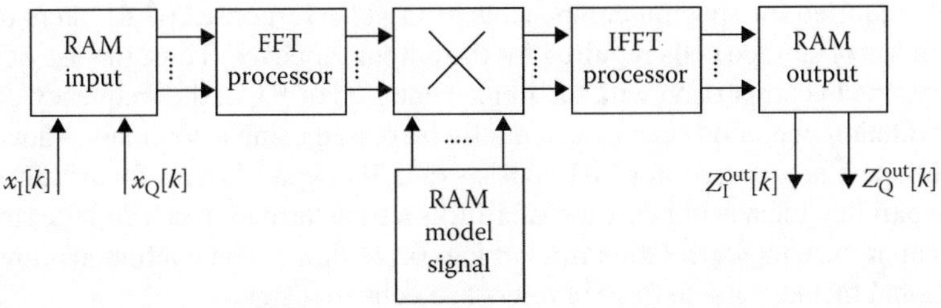

FIGURE 2.11 The FFT-specific processor.

Now, return to the problem of signal processing in frequency domain for DGD taking into account the FFT implementation. A flowchart of the corresponding FFT-specific processor is shown in Figure 2.11. The FFT-specific processor consists of the input memory device, FFT processor, IFFT processor, ROM device, multiplier, and output memory device. As can be seen from Figure 2.11, the in-phase and quadrature constituents of the target return signal come in at the input memory device simultaneously. The in-phase and quadrature constituents of the target return signal form the complex signal subjected to the transformation *FFT-Multiplication-IFFT*. Because of this, the input and output memory devices and all intermediate registers must possess a double bit width.

For signal processing by the DGD there is a need to carry out the FFT, IFFT, and multiplication between $2 \times M$-point complex values. Furthermore, we assume that $M = l + n - 1$. Since the main operation time of the considered specific processor is spent by multiplications between complex and complex conjugate values, the number of convolution operations can be presented in the following form:

$$N_{sp} = 2[0.5M \log_2 M] + m = M[1 + \log_2 M]. \tag{2.107}$$

The number of convolution operations for the in-phase and quadrature constituents of one sequence of the output signal is determined by

$$\frac{N_{sp}}{l} = \frac{M[l + \log_2 M]}{l}. \tag{2.108}$$

Under direct calculations in time domain, the number of multiplications between the complex values for a single sample is equal to n at $l = n$. Consequently, a gain in the number of multiplication operations for complex and complex conjugate values implementing the direct FFT can be determined in the following form:

$$g_{FFT} = \frac{n^2}{M[1 + \log_2 M]} = \frac{n}{2[1 + \log_2 2n]}. \tag{2.109}$$

Calculations made by this formula show that the gain $g_{FFT} > 1$ can be obtained only for the case $l \geq 12$. Thus, at $l = 2048$ we obtain $g_{FFT} = 85$. It can be seen that a greater gain in the number of multiplication operations between the complex and complex conjugate values can be obtained only for high-width sequences.

Now, let us evaluate the required memory size to realize the DGD in radar signal processing subsystem under the implementation of the FFT. Using the same memory storage cells for FFT and IFFT and the buffer to store the output signal (see Figure 2.11), the memory size required for functional DGD purposes is $Q = M_1 + M_2 + M_3 + l$, where M_1 is the number of storage cells required for input sequence sells, M_2 is the number of storage cells required for FFT and IFFT, M_3 is the number of storage cells required for spectral components of impulse responses of all filters used by DGD, and l is the number of storage cells required for the output sequence. Thus, the use of FFT requires higher memory size in comparison with the implementation of FT in the frequency domain. A frequency of convolution operations can be essentially increased using a continuous-flow FFT system. In this case, the specific processor for FFT consists of $0.5M\log_2 M$ devices for arithmetic operations functioning in parallel. Each arithmetic device fulfills a transformation at a definite stage of FFT. In doing so, we can reduce the calculation time in $\log_2 0.5M$ times. The continuous-flow FFT system requires additional memory in the form of interstage delay registers.

2.5 EXAMPLES OF SOME DGD TYPES

At the initial stage of designing, first of all, there is a need to define the following parameters of digital signal processing by DGD: the sampling rate f_s and the width of digit capacity of the target return signal amplitude samples N_b. DGD performance and requirements of digital signal processing devices used by DGD depend strongly on a choice of these parameters. Corresponding practical recommendations can be delivered by analyzing different types of DGD designing and construction by computer simulation. In what follows, we present some of the DGD simulation results for chirp-modulated target return signals with the frequency deviation ΔF_0 equal to 5 MHz.

With the sampling theorem viewpoint, the sampling period T_s of chirp-modulated signals is considered like the limiting case if the following condition $T_{s_{max}} = 1/\Delta F_0$ is satisfied for the in-phase and quadrature channels. The curve 1 in Figure 2.12 corresponds to the DGD output signal at various values T_s in neighboring area relatively to the peak of compressed signal. At $T_s = 1/\Delta F_0$ the DGD output signal is represented by a single point in the main lobe area (the point a in Figure 2.12). The main lobe area width near its base is equal to $2/\Delta F_0$.

In general, the DGD output signal has big side lobes with amplitudes for about one-half amplitude of the main lobe, which is undesirable. At $T_s = 1/2\Delta F_0$, in other words, when the sampling rate is two times higher than the limiting sampling rate $f_{s_{max}}$, two readings are within the limits of the main peak with the width equal to $1/\Delta F_0$. At $T_s = 1/5\Delta F_0$, the DGD output signal is close to the analog GD output signal. As evident from the simulation results discussed, the sampling rate must be at

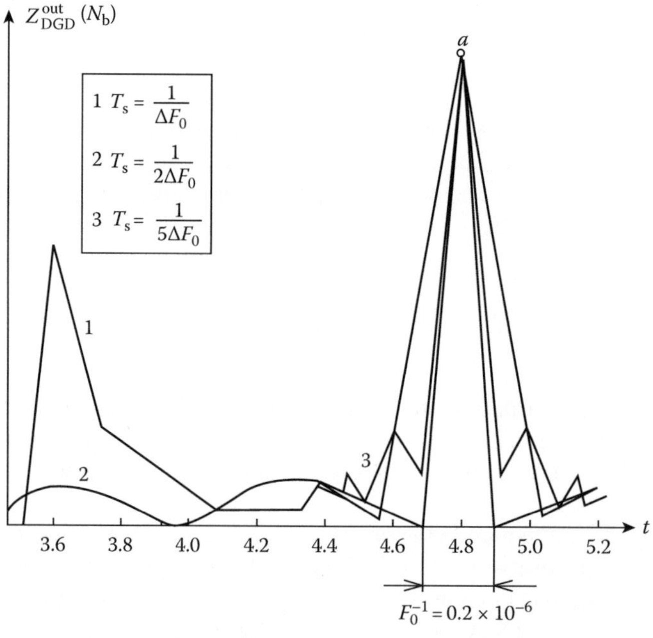

FIGURE 2.12 The DGD output signal amplitude versus the sampling period T_s.

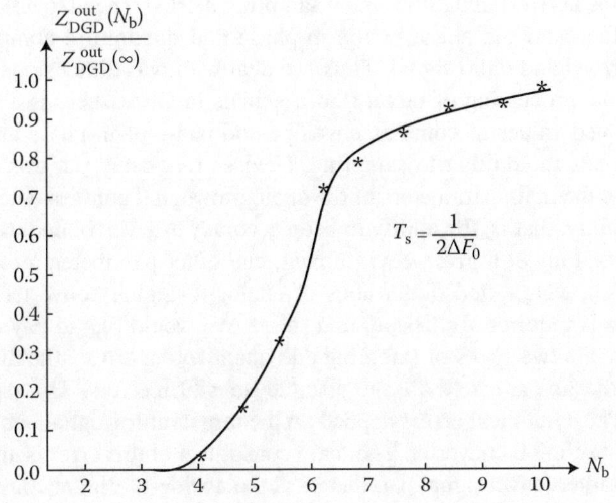

FIGURE 2.13 The DGD output signal amplitude ratio $Z_{DGD}^{out}(N_b)/Z_{DGD}^{out}(\infty)$ as a function of the digit capacity N_b of input signal.

least two times higher than the limiting sampling rate $f_{s_{max}}$. There is a need to take into account the fact that an increase in the sampling rate f_s causes a drastic increase in DGD design and realization complexity, that is, the technical specifications to required performance and memory size of digital signal processing equipment become rigid. For this reason, in the course of designing the DGD, it is recommended to choose the sampling rate f_s within the limits $2 \div 3\Delta F_0$.

The number of bits of the sampled target return signal or digit capacity sampling width plays a very important role in the definition of the DGD performance. Computer simulation results representing a DGD output signal amplitude ratio $Z_{DGD}^{out}(N_b)/Z_{DGD}^{out}(\infty)$ as a function of the digit capacity of the target return signal amplitude samples N_b are shown in Figure 2.13. Here $Z_{DGD}^{out}(N_b)$ denotes the DGD output signal at limited digit capacity N_b and $Z_{DGD}^{out}(\infty)$ denotes the GD output signal without sampling ($N_b \to \infty$). As we can see from Figure 2.13, at $N_b \geq 10\ldots12$ a dependence of the DGD output signal amplitude peak on the digit capacity N_b is weak. Thus, this digit capacity N_b can be recommended under the DGD designing in digital signal processing subsystems of complex radar systems.

2.6 SUMMARY AND DISCUSSION

The discussion in this chapter allows us to draw the following conclusions.

Under the designing of analog-to-digital converters, the task of choosing a sampling interval value and the number of signal quantization levels of the sampled target return arises. Simultaneously, we should take into account the problems of designing and realization both for converters and for digital receivers processing the target return signals.

A sampling device can be considered as a make circuit with the time τ and sampling period T_s. The instantaneously sampled target return signal $x_\delta(t)$ has a mathematical form similar to that of the Fourier transform of a periodic signal. Representation of the continuous function $x(t)$ in the form of the sequence $\{x(nT_s)\}$ is possible only under well-known restrictions. One restriction in kind is a requirement of spectrum limitation of the sampled target return signal $x_\delta(t)$. By Kotelnikov's theorem concerning the signal space concept [4], the continuous target return signal $x(t)$ possessing a limited spectrum is completely defined by a countable set of samples spaced $T_s \leq 1/(2f_{max})$ seconds apart, where f_{max} is the cutoff frequency of target return signal spectrum.

The complex envelope of a narrowband radio signal can be presented either by the envelope and the phase, which are the functions of time, or by the in-phase and quadrature components.

Accordingly, under the narrowband radio signal sampling there is a need to use two samples: either the envelope amplitude and the phase or the in-phase and quadrature components of complex amplitude of the narrowband radio signal. Thus, we deal with two-dimensional signal sampling.

Under digital signal processing of target return signals in CRSs there is a need to carry out a quantization of sampled values of complex envelope and phase or in-phase and quadrature components of radio signals in addition to sampling. Devices that carry out this function are called quantizers. We define the main parameters of the analog-to-digital conversion: sampling and quantization errors, reliability, that is, the ability to keep accuracy at given boundaries within the limits of definite time interval under a given environment, and other parameters producing a limitation in power consumption, weight and dimensions of analog-to-digital converters, cost under serial production, time that is required for design, and so on. We would like to pay attention to the following fact that there are two types of sampling and quantization errors: the dynamical errors and statical errors. The dynamical errors are the discrete transform errors. The statical errors are the unit sample errors. The dynamical errors depend on the target return signal nature and time performance of the analog-to-digital converter. The main constituent of this error is an inaccuracy caused by variations in the target return signal parameters at the analog-to-digital converter input.

The sampling and quantization techniques are used to digitize analog signals. For example, the sampling technique is used to discretize a signal within the limits of the time interval, and the quantization technique is used to discretize a signal by amplitude within the limits of the sampling interval. For this reason, there is a need to distinguish errors caused by these two digitizing techniques that allow us to obtain a high accuracy of receiver performance. Amplitude quantization and sampling can be considered as the additional noise $\zeta_1(kT_s)$ and $\zeta_2(kT_s)$ with zero mean and the finite variances $\sigma^2_{\zeta_1}$ and $\sigma^2_{\zeta_2}$, respectively. If the relationship between the chosen amplitude quantization step Δx and the mean square deviation σ' of process at sampling, and quantization block output is determined by $\Delta x < \sigma'$, then the absolute value of mutual correlation function between the amplitude quantization error and the signal is approximately equal to 10^{-9} with respect to the values of the autocorrelation function of the signal. Therefore, it is reasonable to neglect this mutual correlation function. As a first approximation, it is reasonable to assume that the noise $\zeta_1(kT_s)$ and $\zeta_2(kT_s)$ are Gaussian.

In the case of the phase-code-manipulated target return signal, the DGD employed by a digital signal processing subsystem in a CRS uses only four convolving blocks. Moreover, a convolution within the limits of each cycle equal to the elementary signal duration τ_0 is a summation of amplitude samples of the in-phase and quadrature constituents of the target return signal at instants with signs defined by values $S^*[i] = \pm 1$ in accordance with the given code of the phase-manipulated operation. Practical realization of this principle is not difficult. In the considered case, a direct realization of convolution operations by serial digital signal processing techniques is not possible. There is a need to apply specific procedures of calculations under digital signal processing. First of all, we may use a principle of parallelism that is characteristic of convolution problems. This principle allows us to calculate n_0 of pairwise multiplications $S^*[i] \times x[k-(n_0-i)]$, where $i = 1,\ldots, n_0$, simultaneously using n_0 parallel multipliers with subsequent summation of partial multiplications (see Figure 2.8). In this case, each multiplier must possess the processing speed $V_{\text{req}} = 5 \times 10^6$ multiplications per second, that is, one multiplication operation per 40 ns. This processing speed can be easily provided by employing VLSI circuits. Another example of accelerated convolution operation is the implementation of specifically designed processor that uses the ROM blocks to store calculations of bitwise multiplications made before. Factor codes are used as result addresses of these bitwise multiplications.

Convolution of sequences in the frequency domain is reduced to a product of DFT results. For this purpose, there is a need to realize two DFTs, namely, to convolve a sequence of readings of the function $\{X[i]\}$, $i = = 0, 1, 2,\ldots, n-1$ and a sequence of readings of the impulse response of filters used by DGD. If after convolution it is necessary to make transformations into the time domain, there is a need to carry out IDFT for a sequence of spectral components $\{\mathbf{F}_X[k]\}$, $k = 0, 1, 2,\ldots, n-1$.

The principal peculiarity of the considered algorithm of the operation *DFT-Convolution-IDFT* is a group technology type flow procedure if a width of input data array is higher n, that is, $l \geq n$. The resulting convolution width is $l + n - 1$. In detection problems solved by DGD, we assume that the impulse response of all filters used by DGD is not variable, at least for probing signals of the same kind. Therefore, the DFT of impulse response of all filters used by DGD is carried out in advance and stored in the memory device of a corresponding computer. In the course of convolution, there is a need to accomplish one DFT and one IDFT. It must be emphasized that under radar detection and signal processing problems, a convolved sequence width L corresponds to radar range sweep length that is greater in comparison with the width of a convolving sequence equal to a width of impulse response n_{im} of filters used by DGD.

Evaluation of the required memory size to realize the DGD in a radar signal processing subsystem under the implementation of FFT is based on the following statements. Using the same memory storage cells for FFT and IFFT and the buffer to store the output signal (see Figure 2.11), the memory size required for functional DGD purposes is $Q = M_1 + M_2 + M_3 + l$, where M_1 is the number of storage cells required for input sequence cells; M_2 is the number of storage cells required for FFT and IFFT; M_3 is the number of storage cells required for the spectral components of impulse responses of all filters used by DGD; l is the number of storage cells required for the output sequence. Thus, the use of FFT requires higher memory size compared to the implementation of FT in the frequency domain. A frequency of convolution operations can be essentially increased using a continuous-flow FFT system. In this case, the specific processor for FFT consists of $0.5M \log_2 M$ devices for arithmetic operations functioning in parallel. Each arithmetic device fulfills a transformation at a definite stage of FFT. In doing so, we can reduce the calculation time in $\log_2 0.5M$ times. The continuous-flow FFT system requires additional memory in the form of interstage delay registers.

REFERENCES

1. Haykin, S. and M. Mocher. 2007. *Introduction to Analog and Digital Communications*. 2nd edn. New York: John Wiley & Sons, Inc.
2. Lathi, B.P. 1998. *Modern Digital and Analog Communication Systems*. 3rd edn. Oxford, U.K.: Oxford University Press.
3. Ziemer, R. and B. Tranter. 2010. *Principles of Communications: Systems, Modulation, and Noise*. 6th edn. New York: John Wiley & Sons, Inc.
4. Kotel'nikov, V.A. 1959. *The Theory of Optimum Noise Immunity*. New York: McGraw-Hill, Inc.
5. Manolakis, D.G., Ingle, V.K., and S.M. Kogon. 2005. *Statistical and Adaptive Signal Processing*. Norwood, MA: Artech House, Inc.
6. Goldsmith, A. 2005. *Wireless Communications*. Cambridge, U.K.: Cambridge University Press.
7. Kamen, E.W. and B.S. Heck. 2007. *Fundamentals of Signals and Systems*. 3rd edn. Upper Saddle River, NJ: Prentice Hall, Inc.
8. Tse, D. and P. Viswanath. 2005. *Fundamentals of Wireless Communications*. Cambridge, U.K.: Cambridge University Press.
9. Simon, M.K., Hinedi, S.M., and W.C. Lindsey. 1995. *Digital Communication Techniques: Signal Design and Detection*. 2nd edn. Upper Saddle River, NJ: Prentice Hall, Inc.
10. Richards, M.A. 2005. *Fundamentals of Radar Signal Processing*. New York: McGraw-Hill, Inc.
11. Oppenheim, A.V. and R.W. Schafer. 1989. *Discrete-Time Signal Processing*. New York: Prentice Hall, Inc.
12. Kay, S. 1998. *Fundamentals of Statistical Signal Processing: Detection Theory*. Upper Saddle River, NJ: Prentice Hall, Inc.
13. Tuzlukov, V. 1998. *Signal Processing in Noise: A New Methodology*. Minsk, Belarus: IEC.
14. Tuzlukov, V. 2001. *Signal Detection Theory*. New York: Springer-Verlag, Inc.
15. Tuzlukov, V. 2002. *Signal Processing Noise*. Boca Raton, FL: CRC Press.
16. Tuzlukov, V. 2005. *Signal and Image Processing in Navigational Systems*. Boca Raton, FL: CRC Press.
17. Tuzlukov, V. 1998. A new approach to signal detection theory. *Digital Signal Processing*, 8(3): 166–184.
18. Tuzlukov, V. 2010. Multiuser generalized detector for uniformly quantized synchronous CDMA signals in AWGN noise. *Telecommunications Review*, 20(5): 836–848.

19. Tuzlukov, V. 2011. Signal processing by generalized receiver in DS-CDMA wireless communication systems with optimal combining and partial cancellation. *EURASIP Journal on Advances in Signal Processing*, 2011, Article ID 913189: 15, DOI: 10.1155/2011/913189.
20. Tuzlukov, V. 2011. Signal processing by generalized receiver in DS-CDMA wireless communication systems with frequency-selective channels. *Circuits, Systems, and Signal Processing*, 30(6): 1197–1230.
21. Tuzlukov, V. 2011. DS-CDMA downlink systems with fading channel employing the generalized receiver. *Digital Signal Processing*, 21(6): 725–733.
22. Papoulis, A. and S.U. Pillai. 2001. *Probability Random Variables and Stochastic Processes*. 4th edn. New York: McGraw-Hill, Inc.
23. Hayes, M.H. 1996. *Statistical Digital Signal Processing and Modeling*. New York: John Wiley & Sons, Inc.
24. Proakis, J.G. and D.G. Manolakis. 1992. *Digital Signal Processing*. 2nd edn. New York: Macmillan, Inc.
25. Kammler, D.W. 2000. *A First Course in Fourier Analysis*. Upper Saddle River, NJ: Prentice Hall, Inc.

3 Digital Interperiod Signal Processing Algorithms

3.1 DIGITAL MOVING-TARGET INDICATION ALGORITHMS

The main fundamental theoretical principles of moving-target indication radar are delivered well in Ref. [1]. The performance of the moving-target indication radar can be greatly improved due primarily to four advantages:

- Increased stability of radar subsystems such as transmitters, oscillators, and receivers
- Increased dynamic range of receivers and analog-to-digital converters
- Faster and more powerful digital signal processing
- Better awareness of the limitations, and therefore requisite solutions, of the adapting moving-target indication radar systems to the environment

These four advantages can make it practical to use sophisticated techniques that were considered, and sometimes tried, many years ago but were impractical to implement. Examples of early concepts that were well ahead of the adaptive technology were the velocity-indicating coherent integrator [2] and the coherent memory filter [3,4].

Although such technological developments are able to augment the moving-target indication radar capabilities substantially, there are still no perfect solutions to all problems encountered with using moving-target indication radars, and the design of moving-target indication radar systems is still as much an art as it is a science. Examples of current problems include the fact that when receivers are built with increased dynamic range, limitations arising out of systemic instability will cause increased clutter residue (relative to system noise), leading to false detections. Clutter maps, which are used to prevent false detections from clutter residues, work quite well on fixed radar systems but are difficult to implement on, for example, shipboard radars, because as the ship moves, the aspect angle and range to each clutter patch changes, creating increased residues after the clutter map. A decrease in the resolution of the clutter map to counter the rapidly changing clutter residue will preclude much of the inter-clutter visibility, which is one of the least appreciated secrets of successful moving-target indication radar operation. The moving-target indication radar must work in the environment that contains strong fixed clutter; birds, bats, and insects; weather; automobiles; and ducting. The ducting is also referred to as anomalous propagation; it causes radar return signals from clutter on the surface of the Earth to appear at greatly extended ranges, which in turn exacerbates the problems with birds and automobiles, and can also cause the detection of fixed clutter hundreds of kilometers away.

3.1.1 PRINCIPLES OF CONSTRUCTION AND EFFICIENCY INDICES

When signal processing is carried out under conditions of correlated passive interferences, an initial optimal signal processing of pulse train from N coherent pulses is reduced to ν-fold ($\nu \leq N$) interperiod subtraction of complex amplitude envelopes of the pulse train with subsequent accumulation of uncompensated remainders. A procedure of the interperiod subtraction (equalization) of fixed-position correlated interferences (interferences generated by fixed-position correlated sources) allows us to indicate targets moving relative to radar system. As a rule, this procedure is called the *moving-target indication.*

About 25–30 years ago, the moving-target indication was carried out by analog wireless devices such as delay circuit, analog filters, and so on. Currently, digital moving-target indicators are widely used for the equalization of passive interferences. The digital moving-target indicators consist of memory devices and digital filters [5–8].

The designing of digital moving-target indicators and the evaluation of their efficiency must take into consideration the following peculiarities:

- Using the digit capacity $N_b \geq 8$ of the target return signal amplitude samples in the digital moving-target indicator, quantization errors can be considered as the white Gaussian noise that is added to the receiver noise; on account of this, a synthesis of the digital moving-target indicator is carried out using an analog prototype, that is, by the digitization of well-known analog algorithms.
- Quantization of the target return signal amplitude samples leads to additional losses under the equalization of passive interferences, compared to that observed with the use of analog moving-target indicators.

The digital moving-target indicators can be presented using both two-channel and single-channel forms. The block diagram of the two-channel digital moving-target indicator is shown in Figure 3.1. After processing by DGD, the in-phase and quadrature components of the DGD output signal come in at the inputs of two similar digital filters realizing the interperiod subtraction operation. Under single subtraction, the output signals of quadrature channels of the filter can be presented in the following form:

$$\begin{cases} Z_{\text{I}}^{\text{out}}[i] = Z_{\text{DGD}_\text{I}}^{\text{out}}[i] - Z_{\text{DGD}_\text{I}}^{\text{out}}[i-1] \\ Z_{\text{Q}}^{\text{out}}[i] = Z_{\text{DGD}_\text{Q}}^{\text{out}}[i] - Z_{\text{DGD}_\text{Q}}^{\text{out}}[i-1] \end{cases}. \tag{3.1}$$

In a similar manner, we can obtain the formulas for the filter output signals in the case of any ν-fold ($\nu \leq N$) interperiod subtraction when $\nu = 2, 3, \ldots$. There is a need to take into consideration that the operations are realized on the signals obtained within the limits of the same sampling interval under sampling in time and the index i denotes the current value of the sweep. Thus, the signals $Z_{\text{I}}^{\text{out}}[i]$ and $Z_{\text{Q}}^{\text{out}}[i]$ are either accumulated at first and united after accumulation or are united at

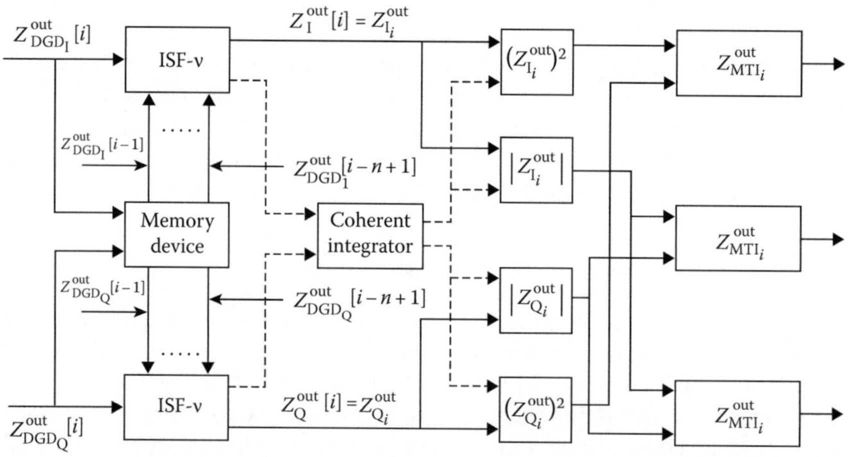

FIGURE 3.1 Flowchart of the two-channel digital moving-target indicator: ISF-ν–ν-fold interperiod subtraction filter; $Z_{\text{MTI}_1}^{\text{out}} = \sqrt{\left(Z_{\text{I}_i}^{\text{out}}\right)^2 + \left(Z_{\text{Q}_i}^{\text{out}}\right)^2}$; $Z_{\text{MTI}_2}^{\text{out}} = \left|Z_{\text{I}_i}^{\text{out}}\right| + \left|Z_{\text{Q}_i}^{\text{out}}\right|$; $Z_{\text{MTI}_3}^{\text{out}} = \left|\max\left(Z_{\text{I}_i}^{\text{out}}, Z_{\text{Q}_i}^{\text{out}}\right)\right| + 0.5\left|\min\left(Z_{\text{I}_i}^{\text{out}}, Z_{\text{Q}_i}^{\text{out}}\right)\right|$.

first and subjected to noncoherent accumulation before coming in at the input of decision circuit. Possible ways of packing of the in-phase and quadrature signal components are shown in Figure 3.1, the outputs 1, 2, and 3. The single-channel digital moving-target indicator can be obtained from a two-channel digital moving-target indicator version if only one of the channels of the two-channel digital moving-target indicator shown in Figure 3.1 is used. In this case, the block diagram of the digital moving-target indicator is significantly simplified, but additional losses in the SNR do occur.

In practice, the following indices are widely used for the evaluation of the efficacy of digital moving-target indicators:

- The coefficient of interference cancellation G_{can} that is defined as a ratio of the passive interference power at the equalizer input to the passive interference power at the equalizer output

$$G_{\text{can}} = \frac{P_{\text{pi}}^{\text{in}}}{P_{\text{pi}}^{\text{out}}} = \frac{\sigma_{\text{pi}_{\text{in}}}^2}{\sigma_{\text{pi}_{\text{out}}}^2} \quad \text{at} \quad \sigma_{\text{pi}_{\text{in}}}^2 \gg \sigma_n^2, \tag{3.2}$$

where
$P_{\text{pi}}^{\text{in}}$ is the power of the passive interference at the equalizer input
$\sigma_{\text{pi}_{\text{in}}}^2$ is the variance of the passive interference at the equalizer input
$P_{\text{pi}}^{\text{out}}$ is the power of the passive interference at the equalizer output
$\sigma_{\text{pi}_{\text{out}}}^2$ is the variance of the passive interference at the equalizer output
σ_n^2 is the receiver noise

- *Figure of merit* for the case of linear interperiod cancellation of the passive interferences can be defined in the following manner:

$$\eta = \frac{P_{\text{s}}^{\text{out}}/P_{\text{pi}}^{\text{out}}}{P_{\text{s}}^{\text{in}}/P_{\text{pi}}^{\text{in}}} = \frac{P_{\text{pi}}^{\text{in}}}{P_{\text{pi}}^{\text{out}}} \times \frac{P_{\text{s}}^{\text{out}}}{P_{\text{s}}^{\text{in}}} = G_{\text{can}} \times G_{\text{s}}, \tag{3.3}$$

where G_{s} is the coefficient characterizing a signal passing through the passive interferences equalizer. In general, when a nonlinear signal processing is used, the figure of merit η indicates to what extent the interference power at the equalizer input can be increased, without losses in detection performance.
- The coefficient of improvement is a characteristic of the response of the digital moving-target indicator on passive interferences with respect to the averaged response on target return signals:

$$G_{\text{im}} = \frac{P_{\text{s}}^{\text{out}}/P_{\text{pi}}^{\text{out}}}{\overline{\left[P_{\text{s}}^{\text{in}}/P_{\text{pi}}^{\text{in}}\right]}_{V_{\text{tg}}}}, \tag{3.4}$$

where $\overline{[P_{\text{s}}^{\text{in}}/P_{\text{pi}}^{\text{in}}]}_{V_{\text{tg}}}$ is the SNR at the equalizer input while deciding on the average values with respect to all target velocities.

The aforementioned Q-factors can be determined both for passive interferences equalizer systems using the interperiod subtraction and for equalizer systems with subsequent accumulation of residues after the passive interference cancellation.

3.1.2 Digital Rejector Filters

The main element of the digital moving-target indicator is the *digital rejector filter* that can guarantee a cancellation of correlated passive interference. In the simplest case, the digital rejector filter is constructed in the form of the filter with the ν-fold ($\nu \leq N$) interperiod subtraction that corresponds to the structure of the nonrecursive filter. The *recursive filters* are also widely used for the cancellation of passive interferences. Consider briefly the problems associated with the analysis and synthesis of rejector filters.

The signal processing algorithm of the digital nonrecursive filter takes the following form:

$$Z_{\mathrm{IQ_{nr}}}^{\mathrm{out}}(nT) = Z_{\mathrm{IQ_{nr}}}^{\mathrm{out}}[n] = \sum_{i=0}^{\nu} h_i Z_{\mathrm{IQ_{nr}}}^{\mathrm{in}}[n-i], \tag{3.5}$$

where

h_i is the weight coefficient of the filter

$Z_{\mathrm{IQ_{nr}}}^{\mathrm{in}}[n-i]$ are in-phase and quadrature components of signal at the filter input (the DGD output)

At $\nu = 2$, (3.5) defines a signal processing algorithm for the nonrecursive filter with the double interperiod subtraction. To define the coefficients of this filter we can set up the following equations:

$$\Delta Z_{\mathrm{IQ_{nr}}}^{\mathrm{out}}[n] = Z_{\mathrm{IQ_{nr}}}^{\mathrm{in}}[n] - Z_{\mathrm{IQ_{nr}}}^{\mathrm{in}}[n-1]; \tag{3.6}$$

$$\Delta Z_{\mathrm{IQ_{nr}}}^{\mathrm{out}}[n-1] = Z_{\mathrm{IQ_{nr}}}^{\mathrm{in}}[n-1] - Z_{\mathrm{IQ_{nr}}}^{\mathrm{in}}[n-2]; \tag{3.7}$$

$$Z_{\mathrm{IQ_{nr}}}^{\mathrm{out}}[n] = \Delta Z_{\mathrm{IQ_{nr}}}^{\mathrm{out}}[n] - \Delta Z_{\mathrm{IQ_{nr}}}^{\mathrm{out}}[n-1] = Z_{\mathrm{IQ_{nr}}}^{\mathrm{in}}[n] - 2Z_{\mathrm{IQ_{nr}}}^{\mathrm{in}}[n-1] + Z_{\mathrm{IQ_{nr}}}^{\mathrm{in}}[n-2]. \tag{3.8}$$

Consequently, the filter coefficients are equal to $h_0 = h_2 = 1$, $h_1 = -2$, $h_i = 0$ at $i > 2$. The flowchart of the digital filter with the interperiod subtraction at $\nu = 2$, in the case of the single quadrature channel, is shown in Figure 3.2. Similarly, we can derive the filter coefficients at $\nu > 2$. So, at $\nu = 3$ we obtain $h_0 = 1$, $h_1 = -3$, $h_2 = 3$, $h_3 = -1$, $h_i = 0$ at $i > 3$; and the signal processing algorithm takes the following form:

$$Z_{\mathrm{IQ_{nr}}}^{\mathrm{out}}[n] = Z_{\mathrm{IQ_{nr}}}^{\mathrm{in}}[n] - 3Z_{\mathrm{IQ_{nr}}}^{\mathrm{in}}[n-1] + 3Z_{\mathrm{IQ_{nr}}}^{\mathrm{in}}[n-2] - Z_{\mathrm{IQ_{nr}}}^{\mathrm{in}}[n-3]. \tag{3.9}$$

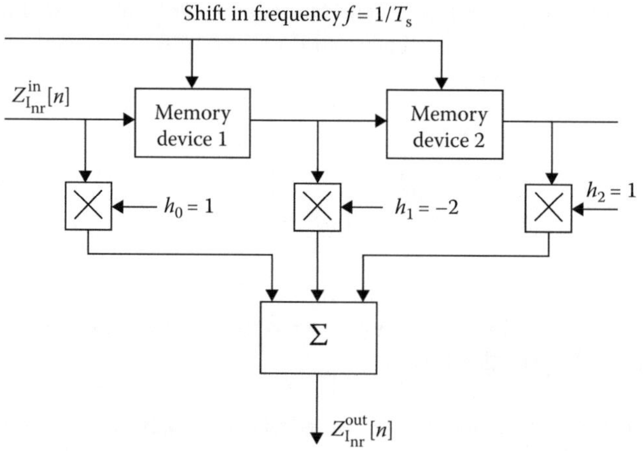

FIGURE 3.2 Flowchart of the digital filter with the interperiod subtraction at $\nu = 2$: the in-phase channel.

The digital nonrecursive filters are very simple under realization, but they have very low-angle fronts of the amplitude–frequency performance, which severely affects the effectiveness of passive interference cancellation.

In addition to the nonrecursive filters, the recursive filter can be used as the rejector filter with the following signal processing algorithm:

$$Z_{\mathrm{IQ_{nr}}}^{\mathrm{out}}[n] = \sum_{i=0}^{\mathrm{V}} a_i Z_{\mathrm{IQ_{nr}}}^{\mathrm{in}}[n-i] + \sum_{j=0}^{\mathrm{V}} b_j Z_{\mathrm{IQ_{nr}}}^{\mathrm{in}}[n-j], \tag{3.10}$$

where a_i and b_j are the coefficients of the digital recursive filter.

Methods of recursive filter synthesis are considered and discussed in numerous literatures [9–21]. Without going into detail, we note that a synthesis procedure based on a squared amplitude–frequency filter response in accordance with a technique described in Ref. [22] can be considered the most effective for the digital moving-target indicators constructed based on the digital rejector filters. As a first step, and on the basis of the given parameters of environment conditions and initial premises, an order is chosen and the amplitude–frequency characteristic of analog filter prototype is defined. After that we define an approximation of squared amplitude–frequency characteristic of digital filter to the squared amplitude–frequency characteristic of analog filter-prototype, taking into consideration the restrictions arising from structural complexity. The elliptic filter, also known as Zolotaryov–Cauer filter [23], is best suited for these requirements.

Under realization of the digital rejector filter in accordance with (3.10), it is convenient that a_i and b_j are simple binary numbers. In this case, there is no need to use ROM to store these coefficients, and the multiplication operations are changed by simple shift and summation operations. For example, in the case of the recursive elliptic filter of the second order, synthesized using the squared amplitude–frequency characteristic, we obtain the following values of coefficients [22]: $a_0 = a_2 = 1$; $a_1 = -1.9682$; $b_1 = -0.68$; and $b_2 = -0.4928$. To simplify the realization of the digital rejector filter, the coefficients a_1, b_1, and b_2 can be rounded in the following order: $a_1 = -1.875 = -2^1 + 2^{-3}$, $b_1 = -0.75 = 2^0 + 2^{-3}$, $b_2 = -0.5 = -2^{-1}$. Investigations showed that rounding off of the coefficients a_1, b_1, and b_2 can give a very small effect on the cancellation of passive interferences; however, this effect is negligible.

The digital rejector filter ensures the best cancellation of passive interferences due to an improvement in the amplitude–frequency characteristic shape, compared with the nonrecursive rejector filter of the same order. However, the degree of correlation of the passive interference remainders at the digital recursive filter output is higher compared to the correlation degree of the passive interference remainders at the digital nonrecursive filter. Moreover, the presence of positive feedback leads to an increase in the duration of transient process and corresponding losses in efficacy if the number of target return pulses in the train is less 20, that is, $N < 20$.

Now, consider some results of the comparison between the efficiency levels of the digital filter and the analog filter. At first, transform (3.10) to nonrecursive form, that is, present the signal processing algorithm of the digital recursive filter in the following form:

$$Z_{\mathrm{IQ_{nr}}}^{\mathrm{out}}[n] = \sum_{i=0}^{N-1} h_{i_{\mathrm{rf}}} Z_{\mathrm{IQ_{nr}}}^{\mathrm{in}}[n-i], \tag{3.11}$$

where

N is the number of pulses in train (sequence)

$h_{i_{\mathrm{rf}}}$ are the coefficients of digital recursive filter impulse response defined by [23,24]

$$h_{0_{\mathrm{rf}}} = 1, \quad h_{i_{\mathrm{rf}}} = a_i + \sum_{j=1}^{k} b_j h_{i-j_{\mathrm{rf}}}. \tag{3.12}$$

For example, in the case of the digital recursive second-order filter ($\nu = k = 2$) we obtain $h_{0_{\text{rf}}} = 1$, $h_{1_{\text{rf}}} = a_1 + h_{2_{\text{rf}}} = a_2 + b_1 h_{1_{\text{rf}}} + b_2$, at $i > 2$ we have $h_{i_{\text{rf}}} = b_1 h_{i-1_{\text{rf}}} + b_2 h_{i-2_{\text{rf}}}$. As follows from (3.12), the impulse response of recursive filter tends to approach infinity and the number of used coefficients is defined by the sample length of the input signal.

Efficiency of interperiod subtraction at different multiplicities of ν can be compared using the coefficient of passive interference cancellation [25]:

$$G_{\text{can}} = \frac{1}{\sum_{i=0}^{\nu} \sum_{j=0}^{\nu} b_i h_j \rho_{\text{pi}}\left[(i-j)T\right]}, \tag{3.13}$$

where

$\rho_{\text{pi}}[\cdot]$ is the normalized coefficient of interperiod correlation of the passive interferences
T is the radar bang period

At determination of G_{can} for nonrecursive filters using (3.13) we think that ν is the multiplicity of the interperiod subtraction and for the recursive filters $\nu = N_{\text{pi}} - 1$, where N_{pi} is the sample size of passive interferences.

To make calculations using (3.13) there is a need to define a model of interferences. The energy spectrum of passive interferences, that is, the signals reflected from, for example, extended territorial objects, clouds, or other types of reflectors, can be approximated by the Gaussian distribution law given by

$$f_{\text{pi}}(S_{\text{pi}}) \approx \exp\left[-2.8\left(\frac{S_{\text{pi}}}{\Delta f_{\text{pi}}}\right)^2\right], \tag{3.14}$$

where

S_{pi} is the energy spectrum of passive interferences
Δf_{pi} is the bandwidth of passive interference spectrum at the level 0.5

The normalized coefficient of correlation corresponding to this spectrum is given as

$$\rho_{\text{pi}}(iT) = \exp\left[-\frac{\pi^2(\Delta f_{\text{pi}} iT)^2}{2.8}\right]. \tag{3.15}$$

Figure 3.3 represents the coefficient of passive interference cancellation G_{can} as a function of $\Delta f_{\text{pi}} iT(i = 1)$ for various values of multiplicity of interperiod subtractions. As we can see from Figure 3.3, the use of the twofold interperiod subtraction brings a win in cancellation for about 15 dB compared to a single interperiod subtraction. In the case of the threefold interperiod subtraction, a win in cancellation in comparison with the twofold interperiod subtraction is less than 15 dB. In the case of the fourfold interperiod subtraction, a win in cancellation in comparison with the threefold interperiod subtraction is still less. The coefficient of cancellation of passive interferences for recursive filters depends on the sample size N_{pi} of the passive interference. Thus, it is of no use to compare the recursive and nonrecursive filters by the coefficient of passive interference cancellation G_{can}. Comparative efficacy of the interperiod subtraction and recursive filters can be evaluated by the figure of merit η given by (3.3), which corresponds to a win in SNR in the case of digital linear filters.

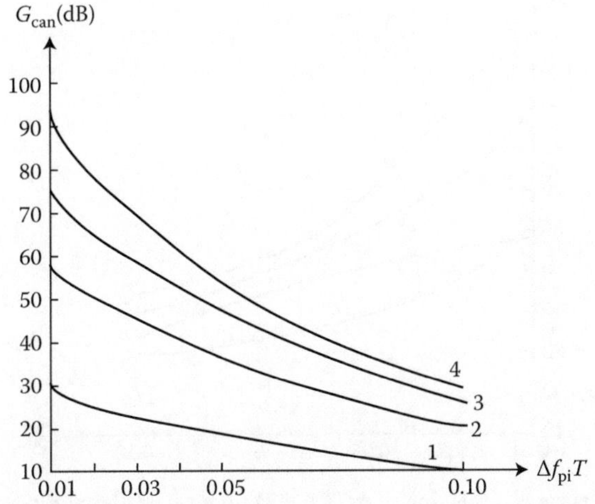

FIGURE 3.3 Coefficient of passive interference cancellation as a function of $\Delta f_{pi}T$ ($i = 1$) for various values of multiplicity of interperiod subtractions: ISF-1, ISF-2, ISF-3, and ISF-4—a single, twofold, threefold, and fourfold interperiod subtraction filters.

Assume that a passive interference source is stationary and fixed, that is, the Doppler shift is zero, $\varphi_D = 0$, and the target velocity is optimal, that is, $\varphi_D = \pi$; in such a scenario, the figure of merit η can be determined in the following form [24]:

$$\eta = \frac{\sum_{i,j=0}^{\nu_s} (-1)^{i-j} h_i h_j \rho_s\left[(i-j)T\right]}{\sum_{i,j=1}^{\nu_{in}} h_i h_j \rho_{pi}\left[(i-j)T\right]}, \tag{3.16}$$

where

ν_s is the sample size of the target return signal equal to $N_s - 1$ (N_s is the number of target return pulses in train) in the case of recursive filter and the interperiod subtraction multiplicity ν in the case of nonrecursive filter

ν_{in} is the sample size of interference determined by analogous way as for ν_s

ρ_s is the coefficient of correlation of the target return signal

ρ_{pi} is the coefficient module of passive interference interperiod correlation

There is a need to define a model of the target return signal for calculations. As a rule, we can think that the target return signal amplitude envelope is subjected to Rayleigh distribution law with the coefficient of correlation defined by

$$\rho_s(T) = \exp(-\pi \Delta f_{tg} T), \tag{3.17}$$

where Δf_{tg} is the spectrum bandwidth of the fluctuated target return signal.

The figure of merit η for a set of the digital nonrecursive filters with the twofold (the curve 1) and threefold (the curve 3) interperiod subtractions, the digital recursive filter of the second order (the curve 2), and the digital composite filter consisting of the nonrecursive filter with the single interperiod subtraction and the recursive filter of the second order connected in cascade (the curve 4) is shown in Figure 3.4 as a function of $\Delta f_{pi} iT$ ($i = 1$). As follows from Figure 3.4, the use of digital recursive filters provides a win in 10 dB in the figure of merit in comparison with the use of filters with the interperiod subtraction. The digital recursive filter order and the multiplicity of the interperiod subtraction of the digital nonrecursive filters are the same under comparison.

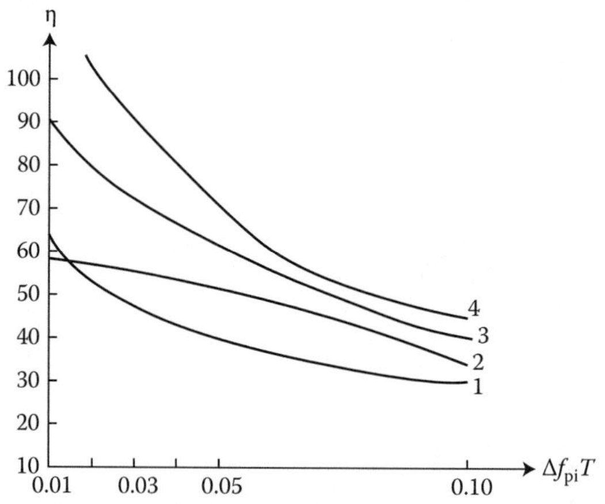

FIGURE 3.4 Figure of merit as a function of $\Delta f_{\text{pi}}T$ (i = 1) for a set of digital filters: 1—the digital nonrecursive filter with twofold interperiod subtractions; 2—the digital recursive filter of the second order; 3—the digital nonrecursive filter with the threefold interperiod subtractions; 4—the digital composite filter: the digital nonrecursive filter with the single interperiod subtraction plus the digital recursive filter of the second order.

In addition, we can see that the time of transient process in the digital recursive filters is much more in comparison with the time of transient process of the digital nonrecursive filters. Because of this, the interference sample size under the use of the digital recursive filters of the second order must be greater than 20, that is, $\nu_{\text{in}} \geq 20$, and $\nu_{\text{in}} \geq 30$, if we use the digital recursive filters of the third order. In doing so, the sample size of the target return signal ν_{s} does not affect the efficacy of the digital recursive filters of the second and third orders already if $\nu_{\text{s}} \geq 10$.

To reduce computer cost under realization of the signal processing algorithm (3.10) using the digital recursive filter, we can implement various computational processes in parallel. As an example, consider a hardware implementation by iteration network [14,26,27]. For this purpose, we can rewrite (3.10) in the following form at $\nu = k = N$

$$Z^{\text{out}}[n] = \sum_{i=0}^{N} a_i D^i \left(Z^{\text{in}}[n]\right) - \sum_{j=1}^{N} b_j D^j \left(Z^{\text{out}}[n]\right), \tag{3.18}$$

where D^i is the operator delaying the input data on i cycles. Equation 3.18 can be presented in the following detailed form:

$$Z^{\text{out}}[n] = a_0 Z^{\text{in}}[n] + D^1\left(a_1 Z^{\text{in}}[n] - b_1 Z^{\text{out}}[n]\right) + \cdots + D^i\left(a_i Z^{\text{in}}[n] - b_i Z^{\text{out}}[n]\right) + \cdots$$
$$+ D^N\left(a_N Z^{\text{in}}[n] - b_N Z^{\text{out}}[n]\right)$$

After elementary transformations we obtain

$$Z^{\text{out}}[n] = a_0 Z^{\text{in}}[n] + D^1\left\{\left(a_1 Z^{\text{in}}[n] - b_1 Z^{\text{out}}[n]\right) + D^1\left\{\left(a_2 Z^{\text{in}}[n] - b_2 Z^{\text{out}}[n]\right) + \cdots\right.\right.$$
$$\left.\left. + D^1\left\{\left(a_i Z^{\text{in}}[n] - b_i Z^{\text{out}}[n]\right) + \cdots + D^1\left(a_N Z^{\text{in}}[n] - b_N Z^{\text{out}}[n]\right)\right\} \cdots\right\}. \tag{3.19}$$

As follows from (3.19), the signal processing algorithm based on the recursive digital filter can be realized using the iteration network (see Figure 3.5) consisting of homogenous elementary blocks

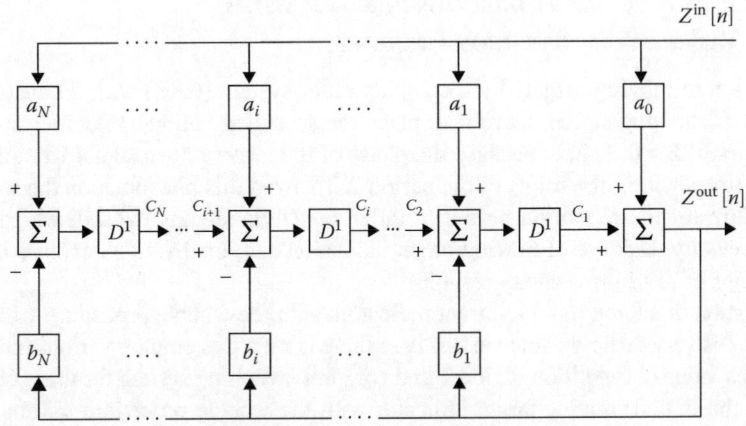

FIGURE 3.5 Digital recursive filter based on the iteration network.

with three inputs and a single output and realizing the signal processing algorithm that can be presented in the following form:

$$C_i = C_{i+1} + a_1 Z^{\text{in}}[n] - b_1 Z^{\text{out}}[n], \quad b_0 = 0. \tag{3.20}$$

Operating speed of the iteration network presented in Figure 3.5 is defined by the operating speed of a single filter cell carrying out two multiplications on constant factors and two summations. Because of this, further increase in the operating speed is made possible using specific structural designs that help with making calculations based on (3.20). When in the considered iteration network all $b_j = 0$ we can obtain a realization of the digital nonrecursive filter. Under the realization of rejector digital filters some specific losses in efficacy occur. The main sources of these losses are quantization of the input signals, rounding-off of the filter weight coefficients, and results of calculations.

As we noted in Section 2.1, under choosing the quantization step Δx based on the condition $\Delta x/\sigma_n \le 1$, where σ_n^2 is the variance of receiving path noise, a mutual correlation of quantization noise and quantized process noise is absent and the variance of the quantization noise is given by $\sigma_{\text{qn}}^2 = \Delta x^2/12$. Digital filter apparatus errors caused by round-off of the calculation results do not depend on the quantization noise and have zero mean and the variance determined by $\sigma_{\text{an}}^2 = m\delta^2/12$, where m is the number of multiplications on fractional weight coefficients in the course of single realization of signal processing algorithm by the digital filter and δ is the value of the arithmetic unit low order. Taking into consideration the error sources identified, the total error variance at the digital filter output can be presented in the following form:

$$\sigma_{\text{out}}^2 \approx \sigma_n^2 + \sigma_{\text{qn}}^2 + \sigma_{\text{an}}^2. \tag{3.21}$$

In this case, the resulting value of the figure of merit, taking into consideration the quantization and round-off errors, will be determined as

$$\eta' = \frac{\eta}{1 + \left(\left(\sigma_{\text{qn}}^2 + \sigma_{\text{an}}^2\right)/\sigma_n^2\right)}. \tag{3.22}$$

3.1.3 Digital Moving-Target Indicator in Radar System with Variable Pulse Repetition Frequency

Under indication of moving targets by a complex radar system (CRS) with a constant repetition period of radar searching signal, there take place the so-called "blind" velocities at Doppler frequencies $f_D = \pm k/T$, $k = 0, 1, 2,\ldots$ because the phase of the target return signal from moving target varies by $2k\pi$ times within the limits of the period T. To avoid this phenomenon the wobble (modulation) procedure for the repetition period of radar searching signals is used, which leads to the spreading of velocity response of moving-target indication and, finally, to a decrease in the number and depth of dips of resulting velocity response.

Implementation of analog moving-target indicators with the wobble procedure within the limits of the interval T is very difficult, since in this case there is a need to employ various individual delay circuits for each value of repetition period T and complex switching system for these circuits. Under realization of the digital moving-target indicator with the wobble procedure within the limits of the interval T, it is enough to realize only synchronization between a sample of delayed data from memory device and radar searching signal instants. In doing so, the size of memory device does not change and does not depend on the number of discrete values and the wobble procedure function within the limits of the repetition period T. The best speed performance of the digital moving-target indicators with the wobble procedure within the limits of the repetition period T can be obtained if each pulse from the target return pulse train is measured N_s, where N_s is the number of target return pulses in the train adjusted with a new individual repetition period. The individual repetition period T must vary with respect to the average value $\bar{T}$ on the multiple value of the fixed time interval $\pm\Delta T$. In this case, if N_s an odd sequence of the repetition periods T within the limits of the train takes the following form:

$$T_i = \bar{T} + i\Delta T, \quad i = 0, \pm 1, \ldots, \pm 0.5(N_s - 1). \tag{3.23}$$

The problem of design and construction of the digital moving-target indicators with the wobble procedure within the limits of the repetition period T is to make a correct selection of a value $\bar{T}$ and define the wobble function for the pulse train in sequence given by (3.23).

In the case of coherent pulse radar, we should take into consideration the following conditions choosing the values of $\bar{T}$ and ΔT. The minimal repetition period $T_{\min}$ must satisfy the condition of the unique radar range determination given before. Because of this, the following condition must be satisfied:

$$\bar{T} - 0.5(N_s - 1)\Delta T \geq T_{\min}. \tag{3.24}$$

The maximal repetition period $T_{\max}$ is defined from conditions that are not associated with the digital moving-target indicator operation. Under the given and known values $T_{\min}$ and $T_{\max}$ and symmetric disposition of the wobble interval with respect to the value $\bar{T}$, we obtain

$$\bar{T} = 0.5(T_{\min} + T_{\max}). \tag{3.25}$$

Then, knowing the value of N_s, from (3.25) we can define ΔT.

In general, the wobble procedure is defined by the criterion of figure of merit η maximization, taking into consideration the minimization of amplitude–frequency characteristic ripple in the digital filter bandwidth. As a rule, this problem is solved by simulation methods. The following wobble procedures are used:

- Linear—a consecutive increasing or decreasing T on $\pm\Delta T$ from pulse to pulse in the train
- Cross-sectional analysis, for example, following the procedure

$$\begin{cases} T_{2i} = \overline{T} + i\Delta T, & i = 0, 2, \ldots, 0.5(N_s - 1) \\ T_{2i+1} = \overline{T} - (i+1)\Delta T, & i = 1, 3, \ldots, 0.5(N_s - 2) \end{cases} \quad (3.26)$$

- Random, for example, by realization of the "bowl" model, restored from the total value of a prior given set *T*.

Calculations and simulations demonstrate that the wobble procedures for the repetition period *T* lead to a decrease in the depth of amplitude–frequency response dips of the nonrecursive and recursive digital filters. However, at the same time, a stop band of the nonrecursive and recursive digital filters is narrowed simultaneously with the frequency band extension and distortion of the interference frequency spectrum. By this reason, the effectiveness in cancellation of passive interferences is considerably reduced. Absolute losses in the figure of merit η for the recursive digital filters of the second order vary from 0.3 to 4.3 dB and from 4 to 19 dB for the recursive digital filters of the third order, in comparison with the digital moving-target indicators without wobble procedures for the repetition period *T* (at optimal velocity) [28–34].

3.1.4 Adaptation in Digital Moving-Target Indicators

In practice, spectral-correlation characteristics of passive interferences are unknown a priori and, moreover, are heterogeneous in space and not stationary in time. Naturally, the efficacy of passive interference cancellation becomes essentially inadequate. To ensure high effectiveness of the digital moving-target indication radar subsystems under conditions of a priori uncertainty and nonstationarity of passive interference parameters, the adaptive digital moving-target indication radar subsystems are employed [4]. In a general sense, the problem of adaptive moving-target indication is solved based on an implementation of the correlation feedback principle [35]. Correlation automatic equalizer of passive interferences represents a closed tracker adapting to noise and interference environment without taking into consideration the Doppler frequency. Parallel with high effectiveness, the tracker has a set of the following imperfections:

- Poor cancellation of area-extensive interference leading edge that is a consequence of high time constant value (for about 10 resolution elements) of adaptive feedback
- Decrease in passive interference cancellation efficiency over the powerful target return signal
- Very difficult realization, especially using digital signal processing

The problem of moving-target indicator adaptation can be solved using the so-called empirical Bayes approach using of which we at first, define the maximal likelihood estimation (MLE) of passive interference parameters and after that we use these parameters to determine the impulse response coefficients of rejector digital filter. In this case, we obtain an open-loop adaptation system, a transient process that is carried out within the limits of transient process of the digital filter.

The first simplest adaptation level in the open-loop adaptation system is to cancel the average Doppler frequency of passive interference caused by relocating to an interference source relative to radar system. In this case, an estimator of the average Doppler frequency of passive interference $\overline{f_{D_{pi}}}$ or the equivalent Doppler shift in phase within the limit of the period *T*, that is, $\Delta\hat{\overline{\varphi}}_{D_{pi}} = 2\pi \overline{f_{D_{pi}}} T$, must be included in the adaptive rejector digital filter. Determination of the average Doppler frequency of passive interference $\overline{f_{D_{pi}}}$ or the equivalent Doppler shift in phase within the limits of the period *T*, that is, $\Delta\hat{\overline{\varphi}}_{D_{pi}}$, must be carried out in real time. To estimate the average Doppler phase shift of passive interference within the limits of the searching period *T* implementing the MLE, the sample consisting of *k* readings of the in-phase and quadrature component pairs of passive interference is used, which are related to the adjacent elements of sampling in radar range from two adjacent searching periods. The algorithm to estimate the average Doppler phase shift of passive interference takes the following form [35]:

$$\Delta\hat{\bar{\varphi}}_{\mathrm{D_{pi}}} = \arctan\frac{\sum_{i=1}^{k}\left(Z^{\mathrm{in}}_{\mathrm{I1}_i}Z^{\mathrm{in}}_{\mathrm{Q2}_i} - Z^{\mathrm{in}}_{\mathrm{I2}_i}Z^{\mathrm{in}}_{\mathrm{Q1}_i}\right)}{\sum_{i=1}^{k}\left(Z^{\mathrm{in}}_{\mathrm{I1}_i}Z^{\mathrm{in}}_{\mathrm{I2}_i} + Z^{\mathrm{in}}_{\mathrm{Q1}_i}Z^{\mathrm{in}}_{\mathrm{Q2}_i}\right)}, \tag{3.27}$$

where $Z^{\mathrm{in}}_{\mathrm{I1}}$, $Z^{\mathrm{in}}_{\mathrm{Q1}}$ and $Z^{\mathrm{in}}_{\mathrm{I2}}$, $Z^{\mathrm{in}}_{\mathrm{Q2}}$ are the in-phase and quadrature components of the input signal within the limits of two adjacent searching periods. In (3.27) we assume that a correlation of interference in the adjacent elements of sampling in radar range is absent and the interference is a stationary process in k adjacent elements of sampling. To obtain an acceptable accuracy of computer calculation the value k must be of the order 5–10. The obtained estimations $\Delta\hat{\bar{\varphi}}_{\mathrm{D_{pi}}}$ are used to make corrections in the delayed in-phase and quadrature components corresponding to the rotation of complex amplitude envelope sum vector given by

$$\dot{\mathbf{Z}}^{\mathrm{in}}[n-1] = Z^{\mathrm{in}}_{\mathrm{I}}[n-1] + jZ^{\mathrm{in}}_{\mathrm{Q}}[n-1] \tag{3.28}$$

on the angle $\Delta\hat{\bar{\varphi}}_{\mathrm{D_{pi}}}$. Determination of the corrected in-phase and quadrature components of the vector given by (3.28) is carried out in accordance with the following formulas:

$$\begin{cases} Z'^{\mathrm{in}}_{\mathrm{I}}[n-1] = Z^{\mathrm{in}}_{\mathrm{I}}[n-1]\cos\Delta\hat{\bar{\varphi}}_{\mathrm{D_{pi}}} - Z^{\mathrm{in}}_{\mathrm{Q}}[n-1]\sin\Delta\hat{\bar{\varphi}}_{\mathrm{D_{pi}}}, \\ Z'^{\mathrm{in}}_{\mathrm{Q}}[n-1] = Z^{\mathrm{in}}_{\mathrm{I}}[n-1]\sin\Delta\hat{\bar{\varphi}}_{\mathrm{D_{pi}}} + Z^{\mathrm{in}}_{\mathrm{Q}}[n-1]\cos\Delta\hat{\bar{\varphi}}_{\mathrm{D_{pi}}}. \end{cases} \tag{3.29}$$

Flowchart of digital signal processing algorithm used by the digital moving-target indicator under adaptation to the displacement of passive interference source is shown in Figure 3.6. The block diagram consists of memory devices to store the four $(k-1)$-fold samples of input signals for each element of radar range along with normal elements; digital signal processing circuits to determine $\Delta\hat{\bar{\varphi}}_{\mathrm{D_{pi}}}$, $\cos\Delta\hat{\bar{\varphi}}_{\mathrm{D_{pi}}}$, $\sin\Delta\hat{\bar{\varphi}}_{\mathrm{D_{pi}}}$, $Z'^{\mathrm{in}}_{\mathrm{I}}[n-1]$, $Z'^{\mathrm{in}}_{\mathrm{Q}}[n-1]$; and the single-order digital filter with the interperiod subtraction. This block diagram can be used for digital filters of other types. For this purpose,

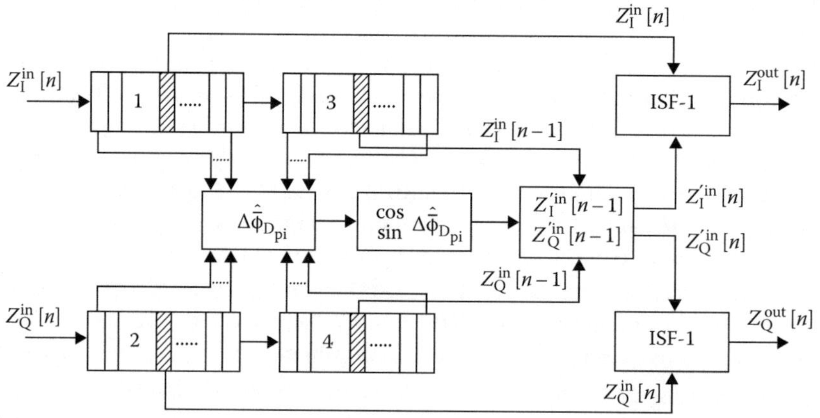

FIGURE 3.6 Block diagram of digital signal processing algorithm used by the digital moving-target indicator under adaptation to target displacement: memory devices 1, 2, 3, and 4.

there is no need to include new elements in the block diagram, but this certainly adds to the computational costs. More complete adaptation to the correlation features of passive interferences involves an estimation of the coefficient module ρ_{pi} of passive interference interperiod correlation along with the estimation $\Delta\hat{\varphi}_{D_{pi}}$. However, as calculations showed when the passive interference spectrum has a form of a single hump and the digital filter with interperiod subtraction possesses a multiplicity, the approximation of the coefficient module ρ_{pi} of passive interference interperiod correlation by unit leads to negligible losses.

3.2 DGD FOR COHERENT IMPULSE SIGNALS WITH KNOWN PARAMETERS

3.2.1 Initial Conditions

In digital signal processing systems, a process of accumulation and signal detection is realized, as a rule, at video frequency, after the union of in-phase and quadrature channels. Henceforth, the signal detection problems will be solved taking into consideration the following initial conditions:

1. A single input signal can be presented in the following form:

$$X_i = X(t_i) = S(t_i, \alpha) + \mathcal{N}(t_i), \tag{3.30}$$

where $S(t_i, \alpha)$ is the target return signal (information signal, i.e., the signal containing an information about parameters of the target), α is the function of time and signals parameters, and $\mathcal{N}(t_i)$ is the noise. The signal parameters are the delay t_d and direction of arrival θ. The total target return signal is a sequence of periodically repeated pulses (the pulse train). At uniform rotation of radar antenna in the searching plane, the pulse train is modulated by directional diagram envelope of the radar antenna. The number of pulses in train is given as

$$N_p = \frac{\varphi_{dd} f}{V_A}, \tag{3.31}$$

where

φ_{dd} is the width of radar antenna directional diagram in the searching plane at the given power P level

f is the pulse repetition frequency of searching signals

V_A is the scanning speed of antenna beam

Under discrete scanning in radar systems with phase array the pulse train envelope of the target return signals has a square shape and the number of pulses in train is defined based on the predetermined probability of detection at the searching target zone edge with minimal effective scattering surface. As for the statistical characteristics of the target return signal (information signal), as usual, we consider two cases:

a. The train of nonfluctuated pulses
b. The train of independently fluctuating pulses obeying to the Rayleigh pdf with zero mean and the variance σ_S^2, that is,

$$p(S_i) = \frac{S_i}{\sigma_S^2} \exp\left\{-\frac{S_i^2}{2\sigma_S^2}\right\}. \tag{3.32}$$

2. Under synthesis of signal detection algorithms and evaluation of target return signal parameters we use, as a rule, a noise model in the form of Gaussian random process with zero mean and the variance σ_n^2. When the correlated time interferences (namely, passive interferences) are absent, noise samples modeled by the Gaussian process do not have an interperiod correlation. When the passive interferences or their remainders are present after cancellation by the digital moving-target indicators, a sequence of passive interference samples are approximated by Markov chain. To describe statistically the Markov chain we, in addition to the variance, should know the coefficient module ρ_{pi} of passive interference interperiod correlation. The interperiod correlation coefficient of the uncorrelated noise with the variance σ_n^2 and correlated passive interference with the variance σ_{pi}^2 can be presented in the following form:

$$\rho_{ij} = \frac{\sigma_{\text{pi}}^2 \rho_{\text{pi}_{ij}} + \sigma_n^2 \delta_{ij}}{\sigma_\Sigma^2}, \tag{3.33}$$

where

$$\sigma_\Sigma^2 = \sigma_{\text{pi}}^2 + \sigma_n^2; \quad \delta_{ij} = 1 \text{ at } i = j \quad \text{and} \quad \delta_{ij} = 0 \text{ at } i \neq j. \tag{3.34}$$

As an example of additional non-Gaussian noise, we can consider a random pulse interference generated by other sources of radiation. This interference is characterized by the off-duty ratio $Q_{\text{in}}^{\text{pulse}}$ and amplitude $Z_{\text{in}}^{\text{pulse}}$ that are random variables. Analysis of the stimulus of random pulse interference under signal processing is carried out, as a rule, by simulation methods.

3. Samples of the input signal X_i in the case of absence of target reflecting surface fluctuations are subjected to the general Rayleigh pdf (the hypothesis $\mathcal{H}_1$—a "yes" signal)

$$p_{SN}^{\mathcal{H}_1}(X_i) = \frac{X_i}{\sigma_\Sigma^2} \exp\left\{-\frac{X_i^2 + S_i^2}{2\sigma_\Sigma^2}\right\} I_0\left(\frac{X_i S_i}{\sigma_\Sigma^2}\right), \quad X_i > 0, \tag{3.35}$$

where $I_0(x)$ is the zero-order Bessel function of the first kind. In the case of presence of target reflecting surface fluctuations, we can write

$$p_{SN}^{\prime\mathcal{H}_1}(X_i) = \frac{X_i}{\sigma_\Sigma^2 + \sigma_{S_i}^2} \exp\left\{-\frac{X_i^2}{2\left(\sigma_\Sigma^2 + \sigma_{S_i}^2\right)}\right\}, \tag{3.36}$$

where $\sigma_{S_i}^2$ is the variance of target return signal amplitude. Introduce the following notations: $x_i = X_i/\sigma_\Sigma$ is the relative envelope amplitude; $q_i = S_i/\sigma_\Sigma$ is the SNR by voltage; $k_i^2 = \sigma_{S_i}^2/\sigma_\Sigma^2$ is the ratio of the target return signal amplitude variance to the interference amplitude variance. Using these notations, we can rewrite (3.35) and (3.36) in the following form:

$$p_{SN}^{\mathcal{H}_1}(x_i) = x_i \exp\left\{-\frac{x_i^2 + q_i^2}{2}\right\} I_0(x_i, q_i), \tag{3.37}$$

$$p_{SN}^{\prime\mathcal{H}_1}(x_i) = \frac{x_i}{1 + k_i^2} \exp\left\{-\frac{x_i^2}{2\left(1 + k_i^2\right)}\right\}. \tag{3.38}$$

If the target return signal is absent in the input signal, the pdf is the same for the considered cases, that is,

$$p_N^{\mathcal{H}_0}(x_i) = x_i \exp\left\{-0.5x_i^2\right\}. \tag{3.39}$$

4. The joint pdf of pulse train from uncorrelated normalized samples in the case of absence of target reflecting surface fluctuations (uniform radar antenna scanning) takes the following form:

$$p_{SN}^{\mathcal{H}_1}(x_1, x_2, \ldots, x_N) = p_{SN}^{\mathcal{H}_1}\{x\}_N = \prod_{i=1}^{N}\left\{x_i \exp\left\{-\frac{x_i^2+q_i^2}{2}\right\} I_0(x_i, q_i)\right\}, \tag{3.40}$$

where
$q_i = q_0 g_i$, g_i are the weight coefficients depending on the radar antenna directional diagram shape
q_0 is the SNR at the maximum power of radar antenna directional diagram

Similarly, for the pulse train from N samples in the case of Rayleigh fluctuations of the target return signal (the target reflecting surface fluctuations), we can write

$$p_{SN}^{\mathcal{H}_1}\{x\}\left\{\frac{x_i}{1+k_i^2}\exp\left\{-\frac{x_i^2}{2\left(1+k_i^2\right)}\right\}\right\} = \prod_{i=1}^{N}\left\{\frac{x_i}{1+k_i^2}\exp\left\{-\frac{x_i^2}{2\left(1+k_i^2\right)}\right\}\right\}. \tag{3.41}$$

In the case of discrete radar antenna scanning, (3.40) and (3.41) have the same form under the conditions

$$x_1 = x_2 = \cdots = x_N, \quad a_i = a_0, \quad k_1 = k_0, \quad g_i = 1. \tag{3.42}$$

5. Digital signal detection algorithm is considered for two versions:
   a. The target return signal is quantized by amplitude in such manner that a low-order bit value does not exceed the root-mean-square value σ_n of receiver noise. In this case, a stimulus tracking of quantization by amplitude is added up to an addition of independent quantization noise to the input noise and a synthesis of digital signal processing algorithms is reduced to digital realization of optimal analog signal processing algorithms [36].
   b. The target return signal is quantized on two levels—binary quantization. In this case, there is a need to carry out a direct synthesis of signal processing algorithms and decision-making networks to process digital binary quantized signals. The required probability of signal detection P_{D} and the probability of false alarm P_{F} take the following form [37–40]:

$$P_{\mathrm{D}}\{d_i\} = \prod_{i=1}^{N} P_{SN_i}^{d_i} b_{SN_i}^{(1-d_i)}; \tag{3.43}$$

$$P_{\mathrm{F}}\{d_i\} = \prod_{i=1}^{N} P_{N_i}^{d_i} b_{N_i}^{(1-d_i)}; \tag{3.44}$$

where

$P_{SN_i}^{d_i}$ is the probability of detection

$P_{N_i}^{d_i}$ is the probability of false alarm of the i-th target return signal from the pulse train that can be presented in the following form:

$$P_{SN_i} = \int_{c_0}^{\infty} p_{SN}^{\mathcal{H}_1}(x_i)dx_i, \quad b_{SN_i} = 1 - P_{SN_i}; \tag{3.45}$$

$$P_{N_i} = \int_{c_0}^{\infty} p_{N}^{\mathcal{H}_0}(x_i)dx_i, \quad b_{N_i} = 1 - P_{N_i}; \tag{3.46}$$

and

$$d_i = \begin{cases} 1, & \text{if} \quad x_i \ge c_0 \\ 0, & \text{if} \quad x_i < c_0 \end{cases}, \tag{3.47}$$

where c_0 is the normalized threshold for binary signal quantization by amplitude.

3.2.2 DGD for Target Return Pulse Train

DGD is discussed in detail for a variety of applications in Refs. [39,40,42–69]. In this section we present a brief analysis of the digital signal processing algorithms based on the generalized approach to signal processing in noise. Our main goal in this section is to compare the work content of DGD realization with other digital signal processing algorithms. First, consider the case when the parameters of target return signals—the pulse train from N nonfluctuating pulses with the additive receiver noise with known statistical characteristics—are known.

Based on the theoretical analysis carried out in Refs. [41,44,45], we can write the likelihood ratio for the generalized signal processing algorithm in the following form:

$$\mathcal{L}_g = \frac{p_{SN}^{\mathcal{H}_1}\{x_i\}}{p_{N}^{\mathcal{H}_0}\{\tilde{x}_i\}} = \prod_{i=1}^{N} \exp\left[-0.5q_i^2\right] I_0\left[2x_i q_i - x_i^2 + \tilde{x}_i^2\right], \tag{3.48}$$

where $\tilde{x}_i$ is the reference noise with a priori information "a no signal" and the same statistical characteristics as the noise at the receiver input of radar system. Consequently, the generalized signal detection algorithm takes the following form:

$$\prod_{i=1}^{N} \exp\left[-0.5q_i^2\right] I_0\left[2x_i q_i - x_i^2 + \tilde{x}_i^2\right] \ge K_g. \tag{3.49}$$

Taking logarithm and making some mathematical transformations with respect to (3.49), we can write

$$\sum_{i=1}^{N} \ln I_0\left[2x_i q_i - x_i^2 + \tilde{x}_i^2\right] \ge \ln K_g + \sum_{i=1}^{N} 0.5q_i^2. \tag{3.50}$$

Further mathematical transformations of (3.50) are associated with an approximation of the function $\ln I_0(x)$. In the case of weak signals, that is, $q_i \ll 1$, we use the following approximation

$$\ln I_0\left[2x_i q_i - x_i^2 + \tilde{x}_i^2\right] \approx \frac{1}{2} x_i q_i^2 - x_i^2 + \tilde{x}_i^2. \tag{3.51}$$

Consequently, taking into consideration that $q_i = q_0 g_i$ and g_i are the weight coefficients depending on the radar antenna directional diagram shape, the detection algorithm of weak target return pulse train will have the following form:

$$\sum_{i=1}^{N}\left[2g_i^2 x_i^2 - x_i^4 + \tilde{x}_i^4\right] \geq K'_g, \quad \text{where} \quad K'_g = \frac{\ln K_g}{q_0^2} + 0.5q_0^2 \sum_{i=1}^{N} g_i^2. \tag{3.52}$$

Now, consider the case of powerful signals, that is, $q_i \gg 1$. In this case, we use the following approximation:

$$\ln I_0\left[2x_i q_i - x_i^2 + \tilde{x}_i^2\right] \approx 2x_i q_i - x_i^2 + \tilde{x}_i^2. \tag{3.53}$$

Consequently, taking into consideration that $q_i = q_0 g_i$ and g_i are the weight coefficients depending on the radar antenna directional diagram shape, the detection algorithm of powerful target return pulse train will have the following form:

$$\sum_{i=1}^{N}\left[2g_i x_i - x_i^2 + \tilde{x}_i^2\right] \geq K'_g, \quad \text{where} \quad K'_g = \frac{\ln K_g}{2q_0} + 0.5q_0 \sum_{i=1}^{N} g_i^2, \tag{3.54}$$

Thus, in the case of the target return pulse train with completely known parameters and modulated by the radar antenna directional diagram, the generalized signal detection algorithm comes to weight summation of normalized samples of the target return pulse train and implementation of energy detector at the output of quadrature or linear detector within the limits of target return pulse train bandwidth and comparison of accumulated statistic with the threshold K'_g.

In practice, when the real radar system is employed, the target return pulse train contains the unknown parameters, such as SNR, that is, q_0, at the maximum point of radar antenna directional diagram, the delay t_d of the target return pulse train relative to searching/scanning signal, and angular deflection of the target return pulse train center θ in the scanning plane with respect to the fixed direction θ_0. Because of this, to process information completely inside the radar coverage a realization of generalized signal detection algorithm should be organized within the limits of each interval of time sampling or radar range discretization. Accumulation of processed signals must be carried out within the limits of "tracking/moving window." The length of "tracking/moving window" must be equal to the number of target return pulses in train. In this case, at the low SNR (for rth sampling or discretization interval) the generalized detection algorithm of nonfluctuating target return pulse train (there are no fluctuations of the target reflecting surface) can be presented in the following form:

$$\ln\left\{\mathcal{L}_{g\mu}^{(r)}\right\} = \sum_{i=0}^{N-1}\left[2g_i^2\left(x_{\mu-i}^{(r)}\right)^2 - \left(x_{\mu-i}^{(r)}\right)^4 + \left(\tilde{x}_{\mu-i}^{(r)}\right)^4\right] \geq K'_g, \quad r = 1,2,\ldots,M, \quad M = \frac{T}{T_s}, \quad \mu \geq N; \tag{3.55}$$

The generalized detection algorithm of fluctuating target return pulse train (there are fluctuations of the target reflecting surface) can be presented in the following form (at $q_0 \ll 1$):

$$\ln\left\{\mathcal{L}_{g\mu}^{(r)}\right\} = \sum_{i=0}^{N-1}\left[\frac{2g_i^2k_i^2}{1+g_i^2k_i^2}\left(x_{\mu-i}^{(r)}\right)^2 - \left(x_{\mu-i}^{(r)}\right)^4 + \left(\tilde{x}_{\mu-i}^{(r)}\right)^4\right] \geq K_g'. \tag{3.56}$$

In the case of powerful signal, the generalized signal detection algorithm within the limits of the "tracking/moving window" is obtained by analogous way. Thus, the DGD of target return pulse train with unknown parameters represents a composition of the "tracking/moving window" summator and energy detector with the threshold network and signal generator indicating a detection of the target return pulse train.

3.2.3 DGD for Binary Quantized Target Return Pulse Train

Now, consider the case when the input sequences are quantized on two levels by amplitude. The generalized detection algorithm of binary quantized target return pulse train is obtained from a synthesis of the likelihood ratio comparing it with the threshold. In doing so, we use the probability of signal detection P_D and the probability of false alarm P_F given by (3.43) and (3.44). The obtained generalized signal detection algorithm has the following form in its final version:

$$\ln\left\{\mathcal{L}_{g\mu}^{(r)}\right\} = \sum_{j=0}^{N-1}\chi_j d_{\mu-j}^{(r)} \geq K_g', \tag{3.57}$$

where the weight coefficients and the threshold are given by

$$\chi_j = \ln\frac{P_{SN}b_N}{P_N b_{SN_j}}; \tag{3.58}$$

$$K_g' = \ln K_g - \sum_{j=0}^{N-1}\frac{b_{SN_j}}{b_N}; \tag{3.59}$$

and

$$P_{SN_j} = \int_{c_0}^{\infty} x_j \exp\left\{-\frac{x_j^2+q_j^2}{2}\right\} I_0(x_j, q_j)d(x_j); \quad b_{SN_j} = 1 - P_{SN_j} \tag{3.60}$$

is the probability to get the unit on the jth position of the target return pulse train;

$$P_N = \int_{c_0}^{\infty} x_j \exp\left\{-\frac{x_j^2}{2}\right\} d(x_j); \quad b_N = 1 - P_N \tag{3.61}$$

is the probability to get the unit in noise region (a "no" signal);

$$d_{\mu-j} = \begin{cases} 1, & \text{if } z_{\mu-j}^{\text{in}} \geq c_0 \\ 0, & \text{if } z_{\mu-j}^{\text{in}} < c_0 \end{cases}, \tag{3.62}$$

where c_0 is the normalized threshold for binary signal quantization by amplitude.

Thus, the generalized signal detection algorithm for the binary quantized target return pulse signals comes to summation of the weight coefficients χ_j corresponding to positions of the target return pulse train where $d_{\mu-j}^{(r)} = 1$. The generalized signal detection algorithm of nonmodulated target return pulse train (in the case of scanning with the fixed antenna) takes the following form:

$$\ln\left\{\mathcal{L}_{g_\mu}\right\} = \sum_{j=0}^{N-1} d_{\mu-j} \geq K_g''. \tag{3.63}$$

In other words, the generalized signal detection algorithm is reduced to accumulation of units within the limits of the target return pulse train length (within the limits of length of the "moving/tracking window" with the predetermined sampling increment) and comparison of the obtained end sum with the threshold.

3.2.4 DGD Based on Methods of Sequential Analysis

The use of methods of the sequential analysis takes a very important place in signal detection theory. Detectors constructed based on methods of the sequential analysis allow us to determine the logarithm of likelihood ratio by the following recurrence formula [70–75]

$$\ln\left\{\mathcal{L}_{g_\mu}\right\} = \ln\left\{\mathcal{L}_{g_{\mu-1}}\right\} + \ln\left\{\Delta\mathcal{L}_{g_\mu}\right\}, \tag{3.64}$$

where
$\mathcal{L}_{g_{\mu-1}}$ is the accumulated likelihood ratio over $\mu - 1$ steps
$\Delta\mathcal{L}_{g_\mu}$ is the likelihood ratio increment at the μ-th step of sequential analysis

The accumulated step-to-step statistic $\ln\left\{\mathcal{L}_{g_\mu}\right\}$ is compared with the upper $\ln A$ and lower $\ln B$ thresholds

$$\ln A = \ln\frac{P_D}{P_F} \quad \text{and} \quad \ln B = \ln\frac{1-P_D}{1-P_F}, \tag{3.65}$$

where
P_D is the predetermined probability of detection
P_F is the predetermined probability of false alarm, correspondingly

If following a comparison we have

$$\ln\left\{\mathcal{L}_{g_\mu}\right\} \geq \ln A, \tag{3.66}$$

the decision about signal detection is accepted and we stop analysis. If following a comparison we obtain

$$\ln\left\{\mathcal{L}_{g_\mu}\right\} \leq \ln B. \tag{3.67}$$

Then the decision about a "no" signal is accepted and we also stop analysis. If the following condition is satisfied

$$\ln B < \ln\left\{\mathcal{L}_{g_\mu}\right\} < \ln A, \tag{3.68}$$

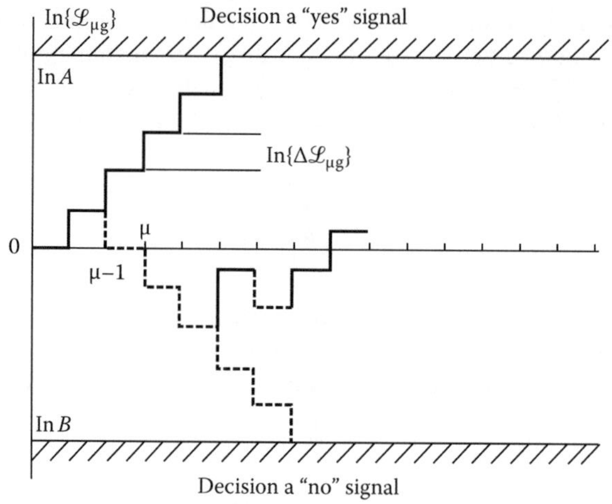

FIGURE 3.7 Accumulation of statistic and decision-making procedure under the sequential analysis.

we continue an analysis, that is, we observe a new sample and determine the likelihood increment. The process of accumulation of $\ln\left\{\mathcal{L}_{g_\mu}\right\}$ and making a decision is presented in Figure 3.7.

Logarithm of likelihood increment is determined by the following formula:

$$\ln\left\{\Delta\mathcal{L}_{g_\mu}\right\} = \ln\frac{p_{SN}(x_\mu)}{p_N(\tilde{x}_\mu)}. \tag{3.69}$$

In doing so, in the case of nonfluctuating target return pulse train model (there are no fluctuations of the target reflecting surface), we use the following logarithm of likelihood increment:

$$\ln\left\{\Delta\mathcal{L}_{g_\mu}\right\} = \ln I_0(x_\mu, q_\mu) - \frac{q_\mu^2}{2}. \tag{3.70}$$

In the case of rapidly fluctuating target return pulse train model (there are fluctuations of the target reflecting surface), we use the following logarithm of the likelihood increment:

$$\ln\left\{\Delta\mathcal{L}_{g_\mu}\right\} = \frac{k_\mu^2 x_\mu^2}{1-k_\mu^2} - \ln\left(1+k_\mu^2\right), \tag{3.71}$$

where the parameter k is given by (3.42).

Thus, to realize a procedure of sequential signal detection there is a need to first define the expected SNR by voltage, for example, in the case of nonfluctuating target return pulse train model (there are no fluctuations of the target reflecting surface), and by power in the case of rapidly fluctuating target return pulse train model (there are fluctuations of the target reflecting surface). The main characteristic of sequential analysis procedure is the average number of steps $\bar{n}$ to make a final decision a "yes" or a "no" signal in the input process. Consider how the average number of steps depends on a "yes" signal and threshold under decision making.

In the case of a "no" signal in the input process, the procedure of sequential analysis is finished by the likelihood ratio logarithm $\ln\left\{\mathcal{L}_{g_\mu}\right\}$ overrunning the lower threshold $\ln B$. Value of the lower threshold $\ln B$ does not depend on the predetermined probability of false alarm P_F since the probability of false alarm P_F varies within the limits of $10^{-3} \div 10^{-11}$ and is only defined by an acceptable value of the probability of miss, that is, $P_M = 1 - P_D$, where P_D is the probability of signal detection. By this reason, the average cycle of sequential analysis procedure, in the case of a "no" signal in the input process, depends only on the predetermined probability of signal detection P_D and increases with increasing in P_D.

In the case of a "yes" signal at the DGD input, the procedure of sequential analysis is finished, as a rule, by overrunning the upper threshold ln A. A value of the upper threshold ln A is defined by the predetermined probability false alarm P_F. Consequently, the average cycle of analysis in the case of a "yes" signal at the DGD input depends on the predetermined probability of false alarm P_F and SNR.

An advantage of the DGD based on the sequential analysis in comparison with the well-known Neyman–Pearson detector consists of the average number of target return signal samples $\overline{n}$, which is necessary to make a decision based on the predetermined probability of false alarm P_F and probability of signal detection P_D is less under sequential analysis procedure than the fixed number of samples N required for the Neyman–Pearson detector. Consequently, it is possible to reduce the computer cost and power energy of a CRS under the detection of target. This possibility can be realized by a radar system with programmable radar antenna scanning when the radar antenna beam can be delayed in scanning direction until the end decision making. However, this advantage has been proved rigorously only in the case of a single channel, that is, under signal detection within the limits of a single radar range resolution interval. In the time of radar signal processing, there is a need to make a decision taking into consideration all elements of radar range resolution simultaneously (multichannel case). In this case, the effectiveness of sequential analysis is defined by the average number of searching radar signals required to make the end decision in all elements of radar range resolution with the predetermined probability of false alarm P_F and probability of signal detection P_D. This number is determined by the following formula:

$$\overline{n} = \max_{k=1,m} \overline{n}_k, \tag{3.72}$$

where

m is the number of analyzed resolution elements in radar range

$\overline{n}_k$ is the average cycle of sequential analysis at the kth radar range resolution element

Optimality of sequential procedure in the multichannel radar systems is not proved so far. There are no analytical and theoretical methods and procedures to determine the efficacy of detectors constructed based on the sequential analysis and employed by multichannel radar systems. This is also particularly true for the DGD. For this reason, an analysis of the multichannel radar systems implementing the sequential analysis procedures is carried out using simulation techniques in each particular case. In what follows, we present some results of this analysis.

In the time of sequential analysis in the multichannel radar system there are two possible types of procedures:

- Decision making is an independent procedure: In this case, a test for the given channel is finished after reaching one of the thresholds—the upper threshold or the lower threshold.
- Simultaneous decision making: This decision making is made if all values of partial likelihood ratios overrun one of the thresholds—the upper threshold or the lower threshold. In this case, multiple crossings of the thresholds are possible.

The average number of tests both for the first case and for the second cases is defined by (3.72). However, the procedure of the second kind is more cost-effective in terms of the required average number of tests since this procedure allows us to level out the lower threshold and, consequently, decrease an uncertainty zone between the upper and lower thresholds. This statement is illustrated in Figure 3.8, where the lower threshold values are demonstrated to make a decision at the

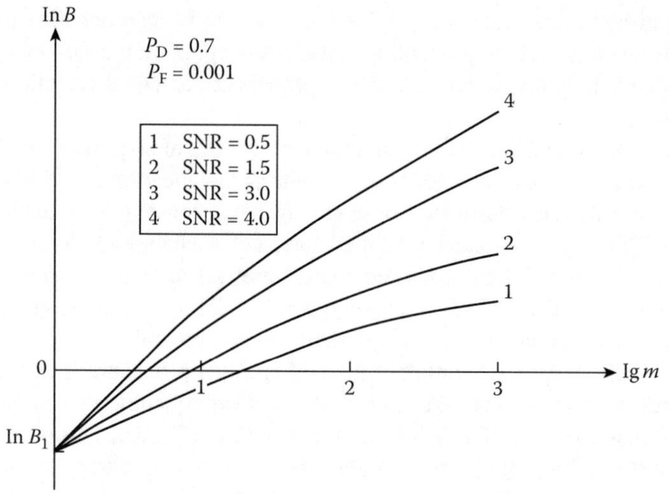

FIGURE 3.8 Simultaneous decision making for multiplicity channel radar systems.

predetermined probability of false alarm $P_F = 10^{-3}$ and the probability of signal detection $P_D = 0.7$ for various values of SNR under various ratios of the rapidly fluctuating target return pulse train model (there are fluctuations of the target reflecting surface) to the noise by power as a function of the number m (the number of analyzed resolution elements in radar range). As we can see from Figure 3.8, the value of the lower threshold increases with a corresponding increase in the SNR and m. In this case, at fixed values of the lower threshold, the probability of missing P_M of the target return signal decreases inversely to the number m of analyzed resolution elements in radar range; however, in the case of sequential analysis procedure of the first kind, the probability of missing P_M of the target return signal is independent of the number m of analyzed resolution elements in radar range.

In the case of a "no" signal at the DGD input, the average number of tests at sequential analysis procedure in a single receiver antenna radar system depends only on the predetermined probability of signal detection P_D. In the multichannel radar system, there is a dependence of the average number of tests at sequential analysis procedure both on the predetermined probability of signal detection P_D and on the number m of analyzed resolution elements in the radar range. Dependences between the average number of tests at sequential analysis procedure and the predetermined probability of signal detection P_D, the probability of false alarm P_F, and $(\text{SNR})^2 = 1.5$ obtained by simulation are presented in Figure 3.9. The average number of tests for a Neyman–Pearson detector are denoted by horizontal lines for each pair of the predetermined probability of signal detection P_D and the probability of false

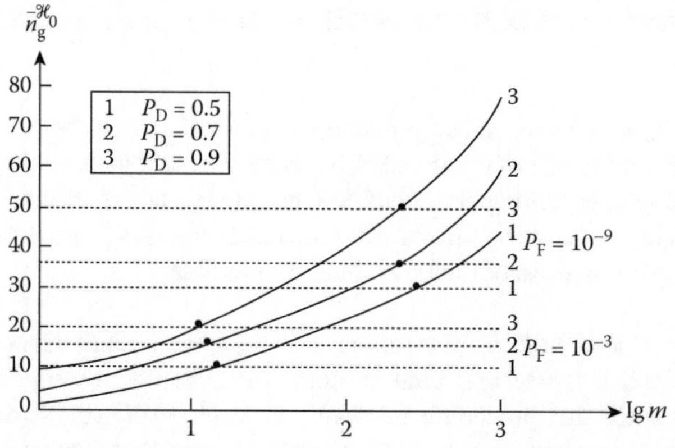

FIGURE 3.9 Average numbers of tests at the sequential analysis versus the number of resolution elements.

alarm P_F. The points of intersections of corresponding curves allow us to define the number of analyzed resolution elements in radar range at which the average number of tests for a Neyman–Pearson detector and the average number of tests for the DGD based on sequential analysis procedure is the same. Evidently, the DGD based on sequential analysis procedure is more effective in comparison with the Neyman–Pearson detector if the following condition in the case of a "no" signal

$$\bar{n}_{\text{seq}}^{g\mathcal{H}_0} < \bar{n}_{\text{N-P}}^{\mathcal{H}_0} \tag{3.73}$$

is satisfied. As follows from Figure 3.9, in the case of a "no" signal at the DGD input the efficacy of the DGD based on sequential analysis procedure is higher, and the predetermined probability of false alarm P_F is lower. For a fixed value of the probability of false alarm P_F the efficacy of the DGD based on sequential analysis procedure for m receiver antennas increases concomitant to decrease in the probability of signal detection P_D.

In the multichannel radar systems, the average number of tests at sequential analysis procedure in the direction of target presence becomes commensurable with the average number of tests at sequential analysis procedure in the direction of target absence even at small numbers of resolution elements. Consequently, the average number of tests at sequential analysis procedure before a decision making is mainly defined by the average number of tests at sequential analysis procedure in the direction of target absence. In the multichannel radar systems, the average number of tests at sequential analysis procedure can be prolonged to such limits that are not acceptable by energy or other tactical considerations. The only way to avoid this prolonged delay of antenna beam in scanning direction is to introduce a truncation procedure of sequential analysis at the n_{tr}-th step and to make a decision about exceeding some fixed threshold C. In this case, the probability of error at the truncation procedure can be presented in the following form:

$$P_{\text{er}}^{\text{tr}} = P_{\text{er}}(\bar{n} < n_{\text{tr}}) + P_{\text{er}}(C)[1 - P_{\text{er}}(\bar{n} < n_{\text{tr}})], \tag{3.74}$$

where

$P_{er}(\bar{n} < n_{tr})$ is the probability of wrong decision at sequential analysis procedure before truncation
$P_{er}(C)$ is the probability of wrong decision at sequential analysis procedure under comparison of decision statistic with the threshold C

Obviously, we can choose such values of n_{tr} and C that additional errors caused by truncation will be satisfied to the following conditions:

$$P_F(C) \le P_F^{\text{add er}} - P_F(\bar{n} < n_{\text{tr}}), \tag{3.75}$$

$$P_M(C) \le P_M^{\text{add er}} - P_M(\bar{n} < n_{\text{tr}}). \tag{3.76}$$

Algorithm to choose n_{tr} and C satisfying to the conditions (3.75) and (3.76) comes to the following procedure: At $n = n_{tr_1}$, we can choose C such that the first condition given by (3.75) is satisfied. If, at the same time, the second condition given by (3.76) is also satisfied, we stop the procedure of choice. If the (3.76) is not satisfied, we select $n_{tr_2} = n_{tr_1} + 1$ and test conditions (3.75) and (3.76) again. As was proved by the theory of sequential analysis [25,71], there are the required n_{tr} and C. Introduction of truncation does not effect essentially the average number of tests at sequential analysis procedure in the direction of target absence and slightly reduces the average number of tests at sequential analysis procedure in the direction of target presence.

At binary quantization of the target return pulse train samples, the generalized detection algorithm based on the sequential procedure becomes essentially simple, but the efficacy of detection procedure is decreased. In this case, additional losses in the average number of tests at the sequential analysis procedure can be 15% ÷ 30%. Losses caused by quantization can be reduced to 2%/5% even at 7–8 quantization levels. This fact gives us a reason to design and construct the DGDs based on the sequential analysis procedure with small digit capacity of the target return pulse train samples.

Comparative analysis of effectiveness of the DGD based on the sequential analysis procedure and a detector with the fixed sample volume, for example, the Neyman–Pearson detector, gives rise to a consideration of two-stage signal detection procedure. The two-stage signal detection procedure consists of the following. At first, the target return pulse train sample with the fixed volume n_1 is subjected to the generalized signal processing algorithm with the predetermined probability of signal detection P_{D_1} and the probability of false alarm P_{F_1}. If the decision a "no" signal is made, the procedure is stopped. Otherwise, an additional procedure is carried out with respect to "suspect data," and we use for this purpose n_2 samples of the target return pulse train and obtain the probability of signal detection P_{D_2} and the probability of false alarm P_{F_2}. In doing so, using results of the two stages, we obtain

$$P_D = P_{D_1}P_{D_2} \quad \text{and} \quad P_F = P_{F_1}P_{F_2}. \tag{3.77}$$

To solve the problem of optimization for the two-stage procedure we can choose such values of n_1 and n_2 that the average number of tests at sequential analysis procedure would be minimal. In doing so,

$$\begin{cases} \bar{n}^{\mathcal{H}_1} = n_1 + P_D\bar{n}_2^{\mathcal{H}_1}; \\ \bar{n}^{\mathcal{H}_0} = n_1 + P'_{sc}\bar{n}_2^{\mathcal{H}_0}, \end{cases} \tag{3.78}$$

where at $P_{F_1} \ll 1$ the probability of duplicate scanning in the a "no" signal direction is defined in the following form:

$$P'_{sc} = \left(1 - P_{F_1}\right)^{m_1} \approx m_1 P_{F_1}; \tag{3.79}$$

where

P_{F_1} is the probability of false alarm at the first stage

$\bar{n}^{\mathcal{H}_1}$ is the average number of tests at sequential analysis procedure in the direction of target presence

$\bar{n}^{\mathcal{H}_0}$ is the average number of tests at sequential analysis procedure in the direction of target absence

$\bar{n}_2^{\mathcal{H}_1}$ is the average number of tests at sequential analysis procedure in the direction of target presence at the second stage

$\bar{n}_2^{\mathcal{H}_0}$ is the average number of tests at sequential analysis procedure in the direction of target absence at the second stage

m_1 is the number of analyzed resolution elements in radar range at the first stage.

Two-stage DGD based on the sequential analysis procedure allows us to obtain an advantage from 25% to 40% in the average number of tests at sequential analysis procedure in comparison with the DGD or Neyman–Pearson detector with the fixed number of tests subjected to the predetermined probability of signal detection P_D and the probability of false alarm P_F. Under target detection at high values of the probability of false alarm P_F, for example, $P_F = 10^3$, and at large number of resolution elements, the two-stage DGD can be more effective in comparison with the DGD based on the sequential analysis procedure.

3.2.5 Software DGD for Binary Quantized Target Return Pulse Train

In practice, while designing the detectors for binary quantized target return pulse train, many heuristic methods are used that make a decision about signal detection based on the definition of a unit density within the limits of each sampling interval of the target return pulse train at the envelope detector output. The best-known digital signal processing algorithms from this class of detectors are software algorithms fixing an initial instant of target return pulse train appearance by the presence of l units on m adjacent positions, where $l \leq m$, $m \leq 5\ldots10$. This is the so-called criterion "l from m" or "l/m." Criterion of the initial instant fixation of target return pulse train appearance is, at the same time, a criterion of target return pulse train detection. To eliminate ambiguity, a criterion of target return pulse train end is defined under an angular coordinate reading. As a rule, the target return pulse train end is fixed by presence of series from k skipping (zeros) one by one ($k = 2\ldots3$). To determine positions between the origin and the end of target return pulse train the binary scalers are employed.

Each of the considered operations can be realized by the digital finite-state automaton. Composition of these digital finite-state automata gives us an example of the software detector block diagram presented in Figure 3.10, where $A1$ is the digital finite-state machine realizing the detection criterion "l/m"; $A2$ is the digital finite-state automaton realizing the criterion of target return pulse train end definition based on series from k zeros; and $A3$ is the digital finite-state automaton (counter) to determine the number of positions from the instant of target return pulse train detection to the instant of erasing the stored information. Using the well-known methods of abstract digital automata composition [76,77], we are able to obtain the transition matrix and graph of associated automaton realizing a program "$l/m - k$" for each data pattern l, m, and k. This graph is a starting point to design tools to realize the required digital signal detection algorithm, and the transition matrix allows us to determine rigorously the quality characteristics of software detectors.

Digital storage devices of binary quantized target return pulse train are considered as other groups of detection algorithms for software detectors. The origin of target return pulse train is fixed by the digital storage device by the first unit obtained after counter storage erasing. The target return pulse train end is fixed when a series of k zeros appears, that is, the same principle as for the software detectors. The signal detection is fixed at the instant of counter erasing if the number of storage units is equal or greater than the threshold N_{thr}. Storage devices of other types are designed in such a way that the counter stores the number of positions (cycles) within the limits of the target return pulse train bandwidth, not the number of units between series from k and more skipping one by one. In doing so, skipping of units at intrinsic positions of target return pulse train is restored. These storage devices are very convenient to realize a digital signal processing algorithm to define the angle coordinate reading of target return pulse train.

Analysis of quality characteristics of the digital software detectors and storage devices can be carried out both by analytic–theoretical method and by simulation. As an example, Figure 3.11 represents the detection performance of software detectors "3/3 – 2," "3/4 – 2," and "4/5 – 2" and digital storage device of units at the predetermined probability of false alarm $P_{\text{F}} = 10^{-4}$ for the target return pulse train consisting of 15 pulses modulated by the radar antenna directional diagram envelope of the following form:

$$g(x) = \left|\frac{\sin x}{x}\right|. \tag{3.80}$$

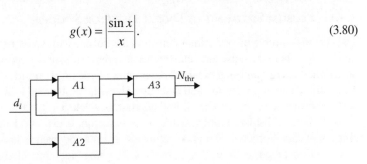

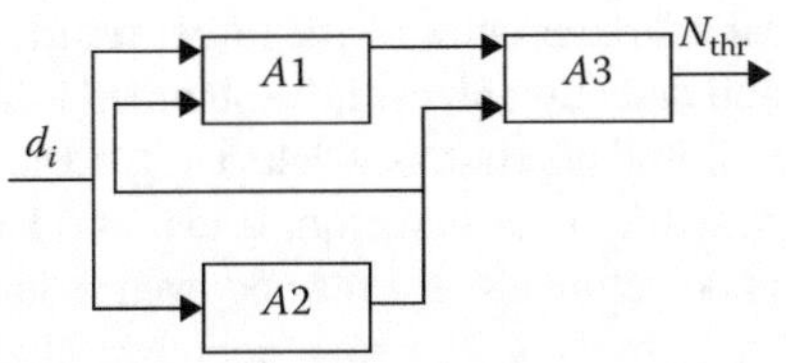

FIGURE 3.10 Software detector block diagram.

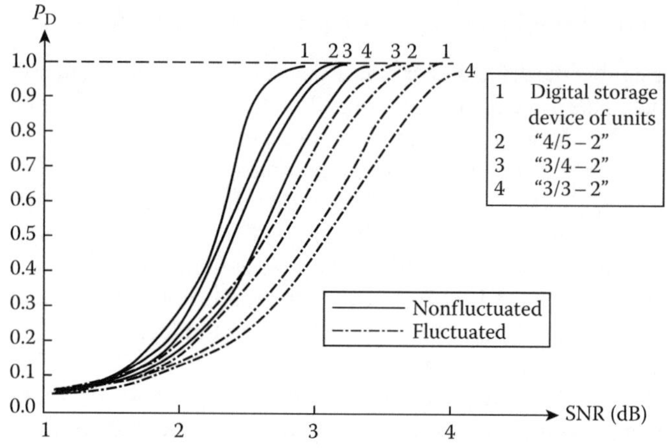

FIGURE 3.11 Detection performance of software detector.

Thresholds of binary quantization are determined by procedures discussed in Ref. [77,78]. The storage device detection threshold (the counter threshold) N_{thr} is determined in the following form:

$$N_{\text{thr}} = \text{entier}\left\{1.5\sqrt{\sum_{i=1}^{N} g_i}\right\}, \quad i = \{1, 2, \ldots, 15\}. \tag{3.81}$$

Simulation brings the following results:

- In the case of absence of target reflecting surface fluctuations under detection of the target return pulse train, the storage devices are more effective in comparison with the software detectors and vice versa; in the case of presence of target reflecting surface fluctuations under detection of the target return pulse train, the software detectors with the detection criterion "l/m" or "$l < m$" are more effective in comparison with the storage devices.
- Threshold losses are low when the target return pulse train is detected by "tracking/moving window" using both the software detectors and the storage devices. Digital realization of these detectors is sufficiently simple, which makes it advantageous to use these detectors when requirements for simplicity of realization prevail over requirements in minimization of energy losses.
- It is appropriate to employ the software detectors on the first stage under the two-stage detection procedure.

3.3 DGD FOR COHERENT IMPULSE SIGNALS WITH UNKNOWN PARAMETERS

3.3.1 Problem Statements of Digital Detector Synthesis

The synthesis problems of optimal detectors are solved with an assumption that there is a priori all information about the noise and interference and their statistical parameters and features under a predetermined energy and statistical characteristics of the target return signal (the information signal). The optimal signal processing and detection algorithms obtained under these conditions possess the best performance for established initial conditions. Changes in noise environment or deviation from characteristics considered under synthesis of detectors leads, as a rule, to a drastic decrease in the efficacy of signal processing and detection algorithms and even to losses in capacity for work.

Detection of target return signals by CRSs in practice is characterized by ambiguity, to a greater or lesser extent, in energy and statistical parameters of signals and noise. In addition, these characteristics are varied in time or are functions of time. In real practice, noise jumping can reach several tens of dB. For example, jumping on $2 \div 3$ dB only under the fixed detection threshold reduces to changes in the probability of false alarm P_F on four orders by magnitude [78]. Thus, a synthesis of robust signal processing and detection algorithms possessing a sufficiently stable quality of service (QoS) under changes in environment conditions is a special interest. As a rule, there is a need to ensure the stability of important QoS only or the probability of false alarm P_F only. In this case, we must solve the problem of constant false alarm rate or, in other words, the CFAR problem.

Depending on a priori information about the target return signals and noise, we distinguish the parametric and nonparametric uncertainty. For the parametric uncertainty, the pdf $p_{XN}^{\mathcal{H}_1}\{x_i\}$ under the hypothesis $\mathcal{H}_1$ (a "yes" signal) and the pdf $p_N^{\mathcal{H}_0}\{x_i\}$ under the hypothesis $\mathcal{H}_0$ (a "no" signal) are considered as known pdf, and only some parameters of these pdf are considered as unknown. Changes in environment and external operation conditions are changes in the noise statistical characteristics such as the mean, variance, covariance function, and so on. The robust signal detection algorithms ensuring CFAR are considered, in this case, as adaptive algorithms allowing us to obtain estimations of unknown statistical characteristics of the noise and to employ these estimations to normalize the input signal or to control the threshold. In the second case, the target return signal and noise pdf shape is, as a rule, unknown both at the hypothesis $\mathcal{H}_1$ (a "yes" signal) and at the hypothesis $\mathcal{H}_0$ (a "no" signal). In this case, a synthesis of robust signal detection algorithms is carried out based on methods of inspection of nonparametric statistical hypotheses. The obtained nonparametric signal detection algorithms bring about the probability of false alarm P_F that is independent (or invariant) of the noise envelope pdf $p_N^{\mathcal{H}_0}\{x_i\}$. However, a statistical independence of the target return signal sample values is the indispensable condition of invariance of the nonparametric signal detection algorithms. If there is a correlation between sample values of the target return signal, the mixed signal detection algorithms using parametric and nonparametric statistic are implemented.

The robust signal detection algorithms are able to ensure the best detection performance under the presence of some information about the noise pdf in comparison with the invariant signal detection algorithms and the best stability in comparison with the optimal signal detection algorithms if, in reality, the noise pdf differs from the adopted noise pdf model that is used under synthesis of signal detection algorithms. Ambiguity of the noise statistical characteristics can be given, for example, in the following form for the case of one-dimensional sample:

$$p_N^{\mathcal{H}_0}\{x_i\} = (1-\varepsilon)\hat{p}_N^{\mathcal{H}_0}\{x_i\} + \varepsilon\tilde{p}_N^{\mathcal{H}_0}\{x_i\}, \tag{3.82}$$

where

$\varepsilon > 0$ is the infinitesimal real number
$\hat{p}_N^{\mathcal{H}_0}\{x_i\}$ is the known pdf
$\tilde{p}_N^{\mathcal{H}_0}\{x_i\}$ is the unknown pdf from the given pdf class

When $\hat{p}_N^{\mathcal{H}_0}\{x_i\}$ is the normal Gaussian pdf and the noise samples are uniform, the DGD constructed based on the robust signal detection algorithm must accumulate statistic at the output that is the nonlinear function of Ref. [44], that is,

$$Z_{\text{DGD}}^{\text{out}}(z) = \begin{cases} x_0, & x > x_0, \\ x, & -x_0 < x < x_0, \\ -x_0, & x \le -x_0. \end{cases} \tag{3.83}$$

In the case of nonstationary noise (pulse noise, noise "edge"), it is recommended to employ a gate-based signal processing of the target return signal sample in the process of the robust DGD threshold computation. One of the simplest examples of such robust signal detection algorithm can be presented in the following form when N is even:

$$Z_{\text{DGD}}^{\text{out}}(x) > C\max\left[\frac{2}{N}\sum_{i=1}^{0.5N} x_i, \frac{2}{N}\sum_{i=0.5N+1}^{N} x_i\right], \tag{3.84}$$

where
N is the sample size used under threshold computation
C is the constant factor

Efficacy of signal detection algorithms in noise with unknown parameters in comparison with the optimal signal detection algorithms (in the case of known noise parameters) is evaluated by the required increase in the threshold value of SNR to obtain the same QoS. The SNR losses are defined in the following form:

$$L = 10\lg\frac{q_1^2}{q_0^2}, \tag{3.85}$$

where
q_0^2 is the threshold SNR ensuring the predetermined probability of detection P_{D} at the some fixed probability of false alarm P_{F} for the optimal signal detection algorithm
q_1^2 is the threshold SNR ensuring the same performance—the probability of detection P_{D} and the probability of false alarm P_{F} for the signal detection algorithm in noise with unknown parameters

To compare the relative efficacy of detectors we can employ the so-called coefficient of asymptotic relative efficiency:

$$\mathcal{F}(A_1, A_2, P_{\text{F}}, P_{\text{D}}) = \lim_{N_1, N_2 \to \infty}\frac{N_1}{N_2}, \tag{3.86}$$

where N_1 and N_2 are the sample sizes required for the signal detection algorithms A_1 and A_2 of two detectors to obtain the same probability of detection P_{D} at the predetermined probability of false alarm P_{F}. In doing so, it is assumed that the total signal energy is independent of the sample size. When $\mathcal{F}(\cdot) > 1$, the signal detection algorithm A_1 is more effective in comparison with the signal detection algorithm A_2.

3.3.2 Adaptive DGD

To overcome the parametric uncertainty we use estimations of unknown parameters of the target return signal and noise and their pdfs, which are defined by observations [21,27,80–88]. Then these estimations are used under solving the signal detection problems instead of unknown real parameters of the signal and noise. The signal detection algorithms that use the pdfs and their parameters or any other statistical characteristics of signals at the detector input, which are obtained based on estimations, are called the *adaptive signal detection algorithms.*

When we have information about the presence of an unknown signal parameter θ, the conditional likelihood ratio can be presented in the following form:

$$\mathcal{L}(x|\theta) = \frac{p_{XN}^{\mathcal{H}_1}\{x|\theta\}}{p_N^{\mathcal{H}_0}\{x|\theta\}}. \quad (3.87)$$

If we are able to define the estimation $\hat{\theta}$ of the signal parameter θ using any statistical procedure, we can determine the likelihood ratio and carry out a synthesis of optimal signal detection algorithm based on the determined likelihood ratio. The estimation of the unknown signal and/or noise parameter is defined by solving the following differential equation:

$$\left.\frac{dp(x|\theta)}{d\theta}\right|_{\theta=\hat{\theta}} = 0. \quad (3.88)$$

Thus, an essence of adaptation approach is the following. At first, using the limited sample size of input process we define an estimation of the maximal likelihood ratio for unknown parameters of pdf. Then we solve the problem of optimal signal detection at the fixed values of these parameters, that is, $\theta = \hat{\theta}$. Effectiveness of this approach depends on the estimation quality of unknown pdf parameters of signal and noise, which is defined by sample size used to obtain estimations (the training sample).

The main adaptation problem is a stability of false alarm level. For this reason, the adaptive DGD (see Figure 3.12) must have the network calculating an estimation of the current noise parameters (the parameters of the noise pdf). These estimated values of parameters of the noise pdf are used later in the decision statistic generation network, at the output of which we observe the decision statistic given by

$$Z_{\text{DGD}}^{\text{out}}(x) = f\left\{\sum_{i=1}^{N} \frac{x_i}{\sigma_{\Sigma_i}}\right\}, \quad (3.89)$$

to normalize the target return signals and noise and, also, to determine the adaptive detection threshold after some functional transformations, where $\sigma_{\Sigma_i}^2$ is the total variance.

However, because the sample size used to determine the parameters of the noise pdf and to define the nonstationarity of continuous noise and interference (amplitude jumping of the continuous noise and interference) is limited by m and owing to stimulus of nonstationary noise (e.g., chaotic pulses), there are significant deviations in the probability of false alarm P_{F} from the predetermined value. Moreover, it is characteristic of the fact that these deviations in the probability of false alarm P_{F} are not controlled by the considered adaptive DGD and are not used during an adaptation procedure.

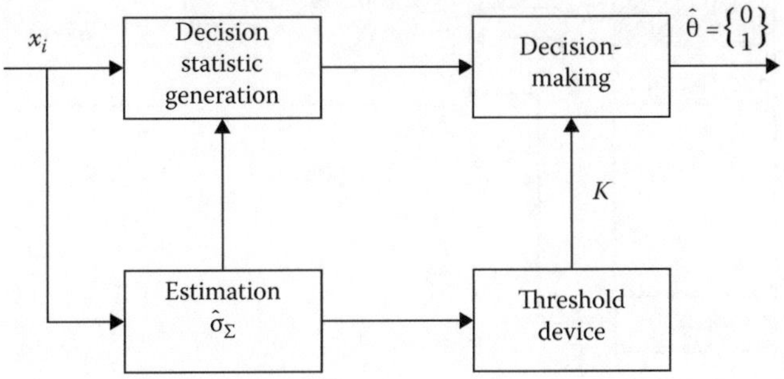

FIGURE 3.12 Adaptive DGD flow chart.

There are specific procedures and methods to reduce an impact of the nonstationarity of continuous noise and interference and chaotic pulse noise and interference in the adaptive DGD. Discuss some of them.

- To determine, for example, the total variance $\sigma_\Sigma^2 = \sigma_n^2 + \sigma_{\text{in}}^2$ of the noise and interference, assuming that the noise and interference are independent of each other, the noise and interference sampling is carried out under each scanning and in immediate proximity with a resolution element by radar range investigated to give an answer a "yes" or a "no" target. In other words, we employ m time sampling intervals neighboring with the studied resolution element, which are considered as interference. At the same time, we assume that noise samples are uncorrelated and possess a definite stationary by power interval (the quasistationary interval). To reduce an effect of jumping in noise amplitude on the shift of estimation of the variance σ_Σ^2, for example, to test a "yes" or a "no" target, we select the resolution element (elementary cell) that is in the center of $m + 1$ adjacent cells or resolution elements. The procedure to estimate the total variance σ_Σ^2 and to normalize the voltage of a "yes" signal elementary cell is explained by Figure 3.13.
- To reduce a sensitivity of the total variance σ_Σ^2 to stimulus of powerful interference, for example, chaotic pulse interference, we use a preliminary comparison of the target return signal samples obtained in two adjacent elements of sampling in radar range with follow-up restriction to the threshold used to detect chaotic pulse interference (the method of contrasts) [89]. In accordance with this method, the sampling value x_i comes in at the input of total noise variance estimation network if there is no exceeding of the threshold; that is, the condition $x_i < Cx_{i-1}$ $0 < C < 1$, is satisfied. If this condition is not satisfied, then a previous result is checked and at $x_{i-1} < Cx_{i-2}$ the value x_i is excluded and the sample size of training sample is decreased on unit. When $x_{i-1} > Cx_{i-2}$ the sampled value is replaced by the threshold value Cx_{i-1}. The constant factor C is selected in such a way that it will be possible to conform the tolerable losses in the case of interference absence to the required accuracy of estimation of the noise variance in the given range of the off-duty ratio and energy of chaotic pulse interference. Choice of the value C is made, as a rule, based on simulation.

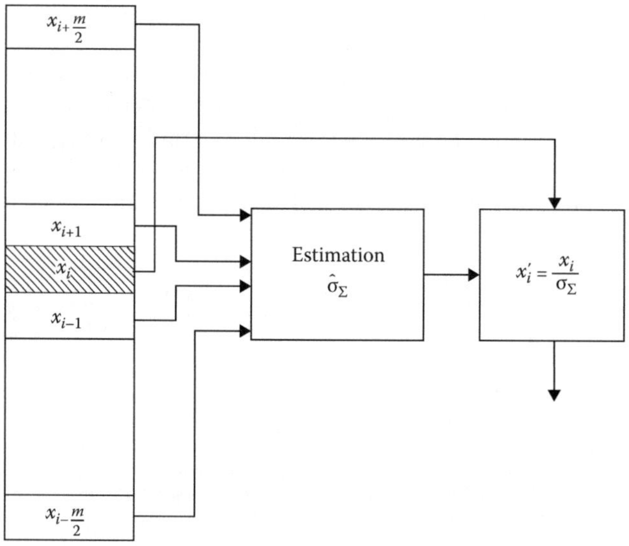

FIGURE 3.13 Procedure of estimation definition of the total variance σ_Σ^2.

- If in the course of training sample analysis there is a possibility not only to detect the chaotic pulse noise but also to measure and define their amplitude, then we can define the value of normalized factor in the adaptive DGD in the following manner. For instance, let l samples of noise from the sample size m are affected by the chaotic pulse noise. Then the estimation of the continuous noise variance (under the normal Gaussian pdf with zero mean) takes the following form:

$$\hat{\sigma}_{\Sigma}^2 = \sum_{j=1}^{m-l} \frac{x_j^2}{m-l}, \tag{3.90}$$

and the estimation of the chaotic pulse interference variance is represented in the following form:

$$\hat{\sigma}_{\text{chp}}^2 = \frac{1}{l} \sum_{j=1}^{l} \left(A_j^{\text{chp}}\right)^2, \tag{3.91}$$

where A_j^{chp} is the amplitude of chaotic pulse interference. In the considered case, the target return signal samples must be normalized by the weight

$$w = \frac{1}{1+\delta}, \quad \text{where} \quad \delta = \frac{\hat{\sigma}_{\text{chp}}^2}{\hat{\sigma}_{\Sigma}^2}. \tag{3.92}$$

This method is effective when the chaotic pulse noise sample is no more than 25%–30% of the training sample and the sample size of the chaotic pulse noise sample is defined by $m \geq 15 \ldots 20$.

The considered methods and procedures of adaptation to the noise and interference have a general disadvantage—the number of false signals at the detector output is not registered by any way. Thus, variations in this number of false signals are not detected and the detector is not controlled. In other words, there is no feedback between the detector and the number of false signals. By this reason, the hardware designed and constructed using the principle of closed loop system or open-loop one to control the threshold and decision function and ensuring the CFAR along with normalization is employed in radar systems with the automatic signal processing and signal detection subsystems where CFAR stabilization is very important.

3.3.3 Nonparametric DGD

Under the nonparametric ambiguity a shape of the pdf $p(x)$ is unknown both in the case of a "yes" signal and in the case of a "no" signal. At this condition, the nonparametric methods of the theory of statistical decisions are employed. This approach allows us to synthesize and design the signal detection algorithms with the predetermined probability of false alarm P_{F} independent of the pdf $p(x)$ shape, that is, with the CFAR in a wide class of unknown pdf $p(x)$ of input signals. Since CFAR is essential for CRS digital signal processing and detection subsystems, there is a great interest to investigate all possible ways to realize the nonparametric signal detection algorithms.

Note that direct values of sample readings of the target return signal are not used by the nonparametric digital detectors. Reciprocal order of the direct values of sample readings of the target return signal is used by the nonparametric digital detectors. This reciprocal order is characterized by the vectors of "sign" and "rank." By this reason, an initial operation of digital detectors synthesized in accordance with the nonparametric signal detection algorithms is a transformation of the input signal sequence $\{x_1, x_2,\ldots, x_N\}$ into the sequences of signs $\{\text{sgn } x_1, \text{sgn } x_2,\ldots, \text{sgn } x_N\}$ or ranks $\{\text{rank } x_1, \text{rank } x_2,\ldots, \text{rank } x_N\}$. In doing so, a statistical independence of the input signal sampling readings is the requirement for nonparametric transformation in the classical signal detection problem, that is,

$$p(x_1, x_2,\ldots, x_N) = \prod_{i=1}^{N} p(x_i). \tag{3.93}$$

Principles of designing and constructions of the *sign* and *rank* nonparametric detectors are discussed in the following.

3.3.3.1 Sign-Nonparametric DGD

When the input signal or the target return signal is the bipolar pulse, the sample of signs $\{\text{sgn } x_1, \text{sgn } x_2,\ldots, \text{sgn } x_N\}$ is formed in the following rule:

$$\text{sgn } x_i = \frac{x_i}{|x_i|}. \tag{3.94}$$

Elements of this sample have only two possible values: +1 if $x_i \geq 0$ and −1 if $x_i < 0$. For the stationary additive noise with the symmetrical pdf with respect to zero, the number of positive and negative signs in the independent noise sample will be the same as $N \to \infty$. When the positive signal appears the probability of the presence of positive signs in the input signal sample becomes higher in comparison with the probability of presence of the negative signs. This phenomenon allows us to detect the target return signal.

The flowchart presented in Figure 3.14 is employed to obtain the sign sample at the envelope detector output where an integration of in-phase and quadrature components is carried out. The input signals come in at the sign former input by two channels with the delay T_s (the sampling

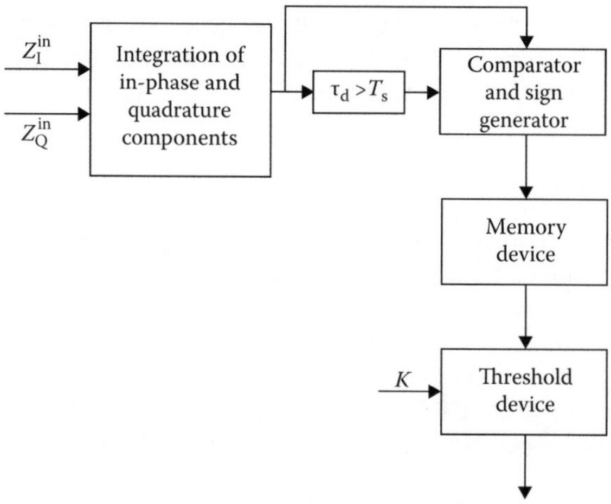

FIGURE 3.14 Flowchart of the sign-nonparametric DGD.

interval of the input signal) at one of them. Delayed and not delayed signals are compared using a comparator. The signals at the comparator output

$$\Delta x[kT_s] = x[kT_s] - x[(k-1)T_s] \tag{3.95}$$

are transformed into signs by the following law:

$$\operatorname{sgn}\{\Delta x[kT_s]\} = \begin{cases} 1, & \text{if } \Delta x[kT_s] \geq 0, \\ -1, & \text{if } \Delta x[kT_s] < 0. \end{cases} \tag{3.96}$$

For each kth sampling interval (the range ring) by a set of sign sampling

$$\operatorname{sgn}(ik) = \operatorname{sgn}\{\Delta x[kT_s]\}, \quad i = 1,2,\ldots,N, \tag{3.97}$$

the signal detection algorithm is realized by the N adjacent scanning periods based on the linear sign statistic

$$\sum_{i=1}^{N} \operatorname{sgn}(ik) \geq K, \tag{3.98}$$

where K is the detection threshold defined by the predetermined and acceptable probability of false alarm P_F.

The discussed signal detection algorithm implements the method of *one-sided contrast* [89]. The high value of the contrast function between samples of the target return signal mixed with noise and the noise only is the main precondition under employment of the sign nonparametric generalized signal detection algorithm. There are many modified sign nonparametric signal detection algorithms [70]. QoS of the sign nonparametric signal detection algorithms is characterized by coefficients of asymptotic and relative effectiveness. It is well known that in the case of nonfluctuating target return signal, the coefficient of asymptotic and relative effectiveness of the sign nonparametric generalized signal detection algorithm with respect to the optimal generalized signal detection algorithm is equal to $2/\pi = 0.65$ under the normal Gaussian pdf of the noise; that is, we see that the use of the sign nonparametric generalized signal detection algorithm has a loss rate of 35% in comparison with the optimal generalized signal detection algorithm. Moreover, when the noise is subjected to non-Gaussian pdf, the effectiveness of the sign nonparametric generalized signal detection algorithm can be higher in comparison with the optimal generalized signal detection algorithm.

3.3.3.2 Rank-Nonparametric DGD

To ensure the CFAR under the arbitrary noise pdf, the rank nonparametric detectors are employed. The rank-nonparametric detectors use rank information contained in the sampled sequence of input signals to make a decision. At the same time, the condition of independence of elements of the ranked sample is the indispensable condition also for the sign nonparametric detectors. In practice, under detection of radar signals, when the number of resolution elements (the number of channels) in radar range, in which a "no" signal exists, is much more than the number of resolution signal elements in radar range, in which a "yes" signal exists, we employ the contrast method [89], an essence of which is the following. Each ranked reading s_i, ($i = 1, 2,\ldots, N$), which is considered as the sample reading of the target return signal, is compared with a set of reference (noise) sample readings

$w_{i1}, w_{i2},\ldots, w_{im}$ taken from the adjacent resolution elements in the radar range. As a result, we determine the reading rank s_i in the following form:

$$r_i = \text{rank}\, s_i = \sum_{i=1}^{m} X_{ij}, \tag{3.99}$$

where

$$X_{ij} = \begin{cases} 1, & \text{if} \quad s_i - w_{ij} > 0, \\ 0, & \text{if} \quad s_i - w_{ij} \le 0. \end{cases} \tag{3.100}$$

The ranked and reference samples as well as the results of rank determination can be presented in the following form:

$$\begin{Vmatrix} s_1 & w_{11} & w_{12} \cdots\cdots & w_{1m} \\ s_2 & w_{21} & w_{22} \cdots\cdots & w_{2m} \\ \cdots & \cdots & \cdots & \cdots \\ s_N & w_{N1} & w_{N2} \cdots\cdots & w_{Nm} \end{Vmatrix} \Rightarrow \begin{Vmatrix} r_1 \\ r_2 \\ \vdots \\ r_N \end{Vmatrix} = R_N, \quad r_i = 1,2,\ldots, m. \tag{3.101}$$

Further signal processing is to store the rank statistic and compare with the threshold:

$$\sum_{i=1}^{N} Z_{\text{DGD}}^{\text{out}}(r_i) \ge K_{\text{g}}, \quad i = 1,2,\ldots,N, \tag{3.102}$$

where

$Z_{\text{DGD}}^{\text{out}}(r_i)$ is the known rank function
K_{g} is the threshold defined based on the acceptable value of the probability of false alarm P_{F}

The simplest rank statistic is the Wilcoxon statistic defined by summation of ranks. By Wilcoxon criterion, the decision a "yes" signal is made in accordance with the following signal detection algorithm:

$$\sum_{i=1}^{N} r_i \ge K_{\text{g}}, \quad i = 1,2,\ldots,N. \tag{3.103}$$

Relative effectiveness of the rank DGD is higher in comparison with the sign DGD. In the case of the nonfluctuating target return signal and normal Gaussian noise pdf, the relative effectiveness is estimated by $3/\pi \approx 0.995$. Thus, the rank-generalized signal detection algorithms are practically effective like the optimal generalized signal detection algorithm. A higher level of efficacy of the rank-generalized signal detection algorithms in comparison with the sign generalized signal detection algorithms is obtained owing to the complexity of signal detection algorithm, since ranking of single sample element requires $m + 1$ summations (subtractions) instead of a single summation (subtraction) per one sample element in the sign signal detection algorithm. Under sequential (moving) rank computation for all resolution elements in radar range, $m + 1$ summations (subtractions) must be made within the limits of the sampling interval T_{s}.

In conclusion, there is a need to note that the rank DGD ensures CFAR when the reference sample is uniform, that is, if the noise is a stationary process within the limits of reference sample interval. Reference sample in homogeneity destabilizes the probability of false alarm P_F. To reduce this effect there is a need to take appropriate measures, one of which is a rational choice of reference sample arrangement relative to the ranked reading.

3.3.4 Adaptive-Nonparametric DGD

The nonparametric DGDs do not ensure the CFAR when the correlated noise comes in at the DGD input. For example, in the case of sign-nonparametric DGDs, an increase in the correlation coefficient of input signals from 0 to 0.5 leads to an increase in the probability of false alarm P_F on $3 \div 4$ orders. Analogous and even greater the probability of false alarm P_F instabilities takes place in the nonparametric DGDs of other types.

One way to stabilize the probability of false alarm P_F at the nonparametric DGD output under the correlated noise conditions is an *adaptive threshold tuning* subjected to correlation features of the noise [45]. The detector designed and constructed in such a way is called the *adaptive-nonparametric* DGD. Initial nonparametric signal detection algorithm forming a basis of the adaptive-nonparametric signal detection algorithm must ensure the CFAR when the variance or the noise pdf is varied. The threshold tuning must set off the effect of nonstability of the probability of false alarm P_F when the noise correlation function is varied.

The block diagram of the adaptive-nonparametric DGD is shown in Figure 3.15. The nonparametric statistic computer implements a basic function of the initial nonparametric signal detection algorithm:

$$Z_{\text{DGD}}^{\text{out}}(x_i) = \sum_{i=1}^{N} \varsigma_i. \tag{3.104}$$

If, for example, the initial nonparametric signal detection algorithm is the sign nonparametric signal detection algorithm, then

$$\varsigma_i = \operatorname{sgn} x_i = \begin{cases} 1, & \text{if} \quad x_i \geq 0, \\ -1, & \text{if} \quad x_i < 0. \end{cases} \tag{3.105}$$

To estimate the correlation function of the noise $R_n[k]$ we can use the unclassified sample from the main analyzed sequence $\{x_i\}$, where $i = 1, 2, \ldots, N$, and the training reference sample $\{\eta_i\}$. In the

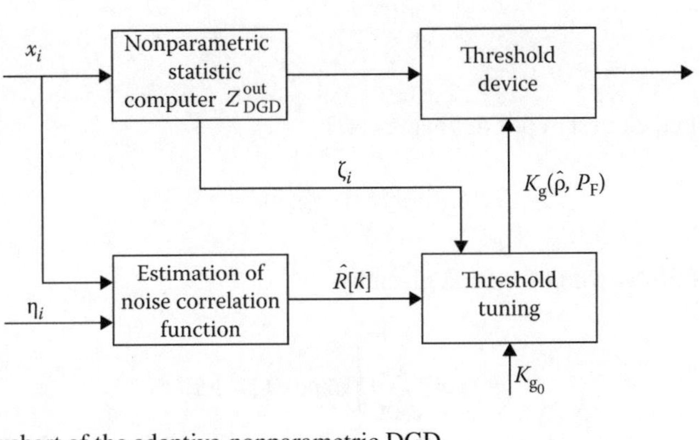

FIGURE 3.15 Flowchart of the adaptive-nonparametric DGD.

case of the use of unclassified sample, the estimate of noise correlation function is determined in the following form:

$$\hat{R}[k]=\frac{1}{N-k}\sum_{i=1}^{N-k}x_i x_{i+k}-\left(\frac{1}{N}\sum_{i=1}^{N}x_i\right)^2, \quad k=0,1,\ldots,N-1. \tag{3.106}$$

When we know a shape of the noise correlation function $R_n[k]$, for example, the exponential function, that is, $R_n[k] = (R_n[1])^k$ or Gaussian, $R_n[k]=(R_n[1])^{k^2}$, then in the case of automatic threshold tuning it is sufficiently to estimate the coefficient of noise interperiod correlation

$$\hat{\rho}=\frac{\hat{R}[1]}{\hat{R}[0]}, \tag{3.107}$$

since all other $\hat{R}[k]$ are related unambiguously with this coefficient. Assuming that the detection threshold K_g depends on the predetermined probability of false alarm P_F and the coefficient of noise interperiod correlation, the adaptive-nonparametric generalized signal detection algorithm takes the following form:

$$Z_{\text{DGD}}^{\text{out}}(x_i)=\sum_{i=1}^{N}\varsigma_i \geq K_g(\hat{\rho},P_F). \tag{3.108}$$

Determination of the threshold becomes simple if a sequence of values ς_i defined by the sequence of values $\{x_i\}$ satisfies the conditions of the central limit theorem for dependent sample readings [90]. Then the pdf of statistic $Z_{\text{DGD}}^{\text{out}}(x_i)$ forming at the adaptive-nonparametric DGD output is subjected to the asymptotic normal Gaussian pdf. In this case, the probability of false alarm P_F can be defined using the following expressions

$$P_F=2\Phi\left\{\frac{E_Z[\hat{\rho}]-K_g(\hat{\rho},P_F)}{\sigma_Z(\hat{\rho})}\right\}, \tag{3.109}$$

where

$$\Phi(x)=\frac{1}{\sqrt{2\pi}}\int_{-\infty}^{x}\exp(-0.5t^2)dt \tag{3.110}$$

is the standard normal Gaussian pdf defined as [41]

$$\Phi(x)=1-Q(x), \tag{3.111}$$

where $Q(x)$ is the well-known Q-function given by

$$Q(x)=\frac{1}{\sqrt{2\pi}}\int_{x}^{\infty}\exp(-0.5t^2)dt. \tag{3.112}$$

Reference to (3.109) through (3.112) allows us to obtain the algorithm to control and adjust the threshold in the following form:

$$K_g(\hat{\rho}, P_F) = E_Z[\hat{\rho}] - \Phi^{-1}[0.5P_F\sigma_Z(\hat{\rho})], \tag{3.113}$$

where

$E_Z[\hat{\rho}]$ is the mean of decision statistic when the input process samples correlate with each other and this mean is equal to the mean of the decision statistic of the initial nonparametric generalized signal detection algorithm

$\sigma_Z^2(\hat{\rho})$ is the variance of decision statistic when the input process samples correlate with each other

The main problem under synthesis of the adaptive-nonparametric generalized signal detection algorithms is to define the variance of decision statistic at the adaptive-nonparametric DGD output, which depends on input process stationary and the way forming the sequence of values ς_i.

The adaptive-nonparametric generalized signal detection algorithms with the adaptive tuning threshold based on the sign and rank criteria possess a satisfactory stability to changes in the noise correlation function at the adaptive-nonparametric DGD input. Simulation results show us that with increase in the coefficient of noise interperiod correlation from 0 to 0.5 the probability of false alarm P_F at the adaptive-nonparametric DGD output increases by $2 \div 5$ times, but in the case of the initial sign nonparametric DGD the probability of false alarm P_F increases by $100 \div 300$ times. The adaptive-nonparametric generalized rank signal detection algorithms have analogous characteristics on the stability of the probability of false alarm P_F.

In conclusion, we would like to note that the digital adaptive asymptotic optimal generalized signal detection algorithms and the generalized signal detection algorithms using the similarity and invariance principles can be employed in addition to the adaptive-nonparametric generalized signal detection algorithms for signals with the correlated and uncorrelated noise with arbitrary pdf. Each signal detection algorithm has its own peculiarities defining a practicability to use these signal detection algorithms under specific conditions. It is impossible to design the signal detection algorithms that will have the same efficacy when the input signals are not controlled.

3.4 DIGITAL MEASURERS OF TARGET RETURN SIGNAL PARAMETERS

Estimation of target return signal parameters comprising information about the coordinates and target characteristics is the main operation of radar signal preprocessing. Estimation of parameters starts after making the decision a "yes" signal; that is, the target has been detected in the direction of radar antenna scanning at the definite distance. At this time, the target detection is associated with rough calculation of target coordinates, for example, the azimuth accurate with the radar antenna directional diagram width and the target range accurate with dimension of the resolution element in radar range and so on. The main task of digital measurer is to obtain more specific information about primary data of estimated target return signal parameters to the predetermined values of target return signal parameters.

Henceforth, we assume that the totality of target return signals, which are used to solve the problem of definition of the target return signal parameters, is within the limits of "moving/tracking window" and dimensions of this "moving/tracking window" correspond to strob bandwidth by radar range and the width of radar antenna directional diagram by angular coordinates. All initial conditions about statistic of the target return signals used under the operation of digital signal processing algorithms of radar signals are kept. We consider uniform estimations of the main no-energy parameters of radar signals, namely, the estimations of angular coordinates, Doppler frequency shift (the radial velocity), and time delay. Quality indices of one-dimensional measurements are the variance of errors σ_θ^2, where $\theta = \{\beta, \varepsilon, f_D, t_d\}$, and the work content of the corresponding digital signal processing and detection algorithms.

3.4.1 Digital Measurer of Target Range

Definition of target range by CRSs is carried out as a result of measurements of the time delay t_d of the target return signal relative to the searching signal in accordance with the formula

$$t_d = \frac{2r_{tg}}{c}, \tag{3.114}$$

where c is the velocity of light propagation. In surveillance radar systems with uniform circular (sector) scanning by radar antenna or phased-array radar antenna implemented to define the coordinates a lot of targets, the target range is estimated by readings of the scaled pulses from the instant to send the searching radar signal Δt_{d1} (the transmit antenna) to the instant to receive the target return signal Δt_{d2} (the receive antenna, see Figure 3.16). In doing so, with sufficient accuracy in practice, we can believe that there is no target displacement within the limits of receiving the target return pulse train. Consequently, the target range measured by all N pulses of the target return pulse train can be averaged in the following form:

$$\hat{r}_{tg} = \sum_{j=1}^{N} r_{tg_j}, \tag{3.115}$$

where r_{tg_j} is the target range measure by a single reading. The variance of estimation is given by

$$\sigma_{\hat{r}}^2 = \frac{1}{N}\sum_{j=1}^{N} \sigma_{r_j}^2, \tag{3.116}$$

where $\sigma_{r_j}^2$ is the variance of estimation of the target range measure by a single reading.

Error of a single reading of the time delay t_d produced by the digital measurer of target range can be presented as a sum of two terms under condition that positions of count pulses on time axis are random variables with respect to the searching signal:

$$\Delta t_d = \Delta t_{d_1} + \Delta t_{d_2}, \tag{3.117}$$

where

Δt_{d_1} is the random shift of the first count pulse with respect to the scanning signal

Δt_{d_2} is the random shift of the target return signal with respect to the last count pulse, see Figure 3.16

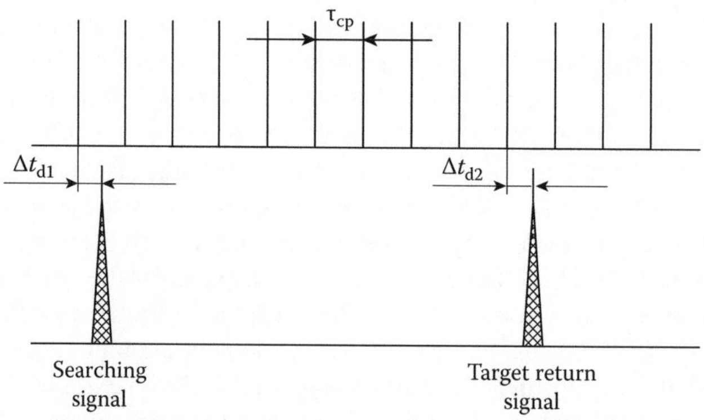

FIGURE 3.16 Definition of the range to target—a time diagram.

These errors are independent random variables uniformly distributed within the limits of the count pulse duration τ_{cp}, that is, within the limits of the interval $[-0.5\tau_{cp}, +0.5\tau_{cp}]$. By this reason in this case, the variance of error of a single reading of the time delay t_d is determined in the following form:

$$\sigma_{t_d}^2 = \frac{1}{6}\tau_{cp}^2. \tag{3.118}$$

If the initial instant of count pulses is synchronized with the searching signal, in this case, $\Delta t_{d_1} = 0$, the variance of error of a single reading of the time delay t_d is determined in the following form:

$$\sigma_{t_d}^2 = \frac{1}{12}\tau_{cp}^2. \tag{3.119}$$

The discussed flowchart of digital measurer of the target range can be realized by a specific microprocessor.

3.4.2 Algorithms of Angular Coordinate Estimation under Uniform Radar Antenna Scanning

Optimal algorithms to measure the angular coordinate are synthesized by the maximal likelihood ratio criterion, as a rule. The shape of the likelihood function depends on the statistical characteristics of the signal and noise, radar antenna directional diagram shape, and radar antenna scanning technique in the course of measuring. First, we consider the target return pulse train processing that is obtained as a result of uniform rotation of the radar antenna within the limits of radar range sampling interval.

Under multiple-level quantization of the target return signals and weight functions we obtain digital counterpart of optimal measurer of angular coordinate in the scanning plane, that is, for the two-coordinate surveillance radar system—the coordinates of target azimuth β_{tg}. The likelihood function for azimuth estimation by N normalized nonfluctuating target return pulse train in the stationary noise take the following form:

$$\mathcal{L}(x_1, x_2, \ldots, x_N | q_0, \beta_{tg}) = \prod_{i=1}^{N} p(x_i | q_i, \beta_i), \tag{3.120}$$

where for the considered case we have

$$p(x_i | q_i, \beta_i) = x_i \exp\left[-0.5\left(x_i^2 + q_i^2\right)\right] I_0(q_i, x_i). \tag{3.121}$$

In turn,

$$q_i = q_0 g(\beta_i, \beta_{tg}), \tag{3.122}$$

where q_0 is the SNR by voltage in the center of target return pulse train (see Figure 3.17a);

$$g(\beta_i, \beta_{tg}) = g\left[\frac{\beta_i - \beta_{tg}}{\varphi_0}\right] = g(\alpha) \tag{3.123}$$

is the function defining the radar antenna directional diagram envelope, for receiving and transmitting in the scanning plane; φ_0 is the one-half main beam width of the radar antenna directional diagram at zero level; and β_i is the value of azimuth angle when receiving the ith pulse of the pulse train.

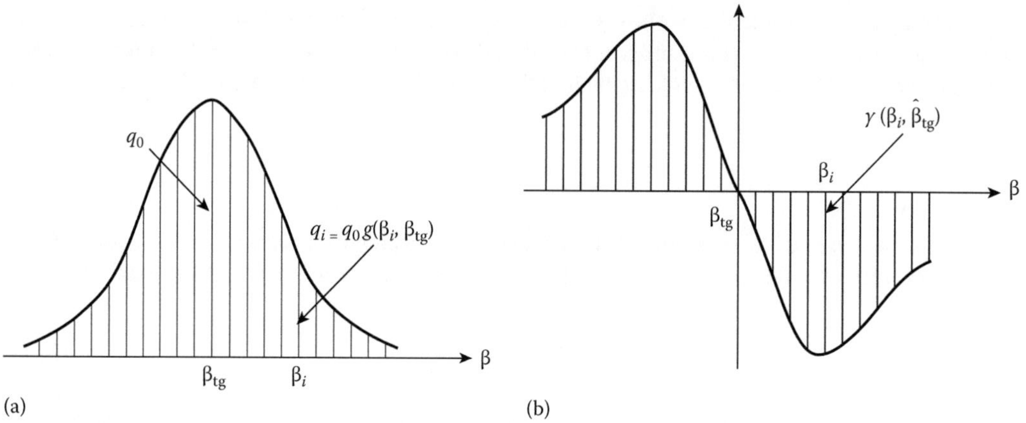

FIGURE 3.17 (a) Target return pulse train and (b) discrete weight function.

Under the fixed value of q_0, we can obtain the final expression of the likelihood function equation for estimation of the angular coordinate β_{tg} in the following form [30,90,91]:

$$\sum_{i=1}^{N} x_i \gamma(\beta_i, \hat{\beta}_{tg}) = 0, \tag{3.124}$$

where

$$\gamma(\beta_i, \hat{\beta}_{tg}) = \frac{\partial g(\beta_i, \hat{\beta}_{tg})}{\partial \hat{\beta}_{tg}}, \quad i = 1, \ldots, N \tag{3.125}$$

is the discrete weight function, that is, a sequence of the weight coefficients, to weigh the normalized amplitudes of target return pulse train (see Figure 3.17b). This weight function has a form of discriminatory characteristic, the gain slope of which is a function of the radar antenna directional diagram shape and the zero point coincides with the maximum of the radar antenna directional diagram.

Under estimation of target azimuth angle by the quick-fluctuating target return pulse train, the likelihood function equation takes the following form:

$$\sum_{i=1}^{N} x_i^2 \gamma'(\beta_i, \hat{\beta}_{tg}) = 0, \tag{3.126}$$

where

$$\gamma'(\beta_i, \hat{\beta}_{tg}) = \frac{g(\beta_i, \hat{\beta}_{tg})}{\left[1 + k_0^2 g^2(\beta_i, \hat{\beta}_{tg})\right]^2} \times \frac{\partial g(\beta_i, \hat{\beta}_{tg})}{\partial \hat{\beta}_{tg}}, \quad i = 1, \ldots, N \tag{3.127}$$

is the discrete weight function to weight the normalized amplitudes of the target return pulse train from the quick-fluctuating target surface.

Comparing (3.124) and (3.126), we see that in contrast to the case of nonfluctuating target surface to estimate the azimuth angle of target with the quick-fluctuating target surface, there is a need to sum the squared amplitudes of the target return pulse train with their weights. In this case, the weight function $\gamma'(\beta_i, \hat{\beta}_{tg})$ has a more complex form, but its character and behavior do not change. Thus, the optimal algorithm for the estimation of azimuth target angle under the uniform radar antenna scanning includes the following operations:

- The target return pulse train storage in the "moving/tracking window," the bandwidth of which corresponds to the train time
- Weighting of each target return signal amplitude in accordance with the values of corresponding weight coefficients
- Formation of half-sum of the weighted target return signal amplitudes observed by the "moving/tracking window" at the right and left sides with respect to zero value of the weight function
- Comparison of half-sums and fixation of position where the result of comparison passes through zero value

The simplest example of block diagram of realization of the target azimuth angle coordinate estimation algorithms given by (3.124) and (3.126) is presented in Figure 3.18. In accordance with this block diagram, to realize the target azimuth angle coordinate estimation algorithm there is a need to carry out $N-1$ multiplications and $N-1$ summations of multidigit binary data while receiving each target return pulse train. The problem of potential accuracy definition of the target azimuth angle coordinate estimation by the target return pulse train can be solved by analytical procedures or simulation with definite assumptions.

Under binary quantization of the target return signal amplitudes (the weight multidigit function), the likelihood function of the estimated target angular coordinate β_{tg} takes the following form:

$$\mathcal{L}(\beta_{tg}) = \prod_{i=1}^{N} P_{SN_i}^{d_i}(x_i)\,\tilde{P}_{SN_i}^{1-d_i}, \tag{3.128}$$

where

P_{SN_i} is the probability that the input signal (the target return signal) exceeds the threshold of binary quantization at the ith position of the target return pulse train

$\tilde{P}_{SN_i} = 1 - P_{SN_i}$ is the probability to get the unit on the jth position of the target return pulse train

d_i is given by (3.47)

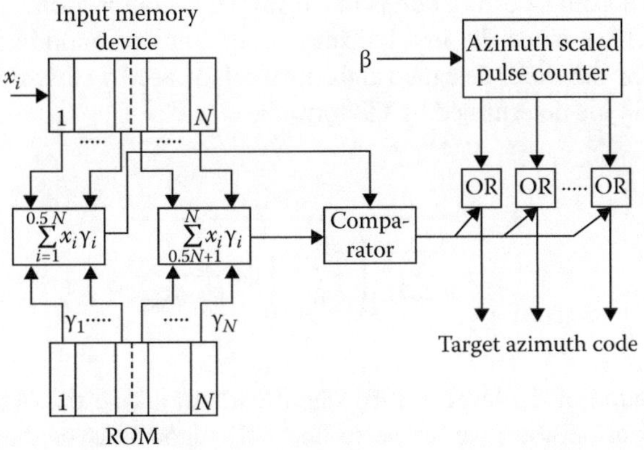

FIGURE 3.18 Block diagram of the target azimuth angle coordinate estimation algorithm given by (3.124) and (3.126).

In the case of nonfluctuating target return signals we can write

$$P_{SN_i} = \int_{c_0}^{\infty} x_i \exp\left\{-\frac{x_i^2 + q_i^2}{2}\right\} I_0(q_i, x_i) dx_i, \tag{3.129}$$

and in the case of fluctuating target return signals we have

$$P_{SN_i} = \int_{c_0}^{\infty} \frac{x_i}{1+k_i^2} \exp\left\{-\frac{x_i^2}{2\left(1+k_i^2\right)}\right\} dx_i. \tag{3.130}$$

After appropriate mathematical transformations we can write the following likelihood function equation for the considered case

$$\sum_{i=1}^{N} d_i \gamma''(\beta_i, \hat{\beta}_{tg}) = 0, \tag{3.131}$$

where

$$\gamma''(\beta_i, \hat{\beta}_{tg}) = \frac{1}{P_{SN_i} \tilde{P}_{SN_i}} \times \frac{dP_{SN_i}}{d\hat{\beta}_{tg}} \tag{3.132}$$

is the weight function of target return pulse train positions under estimation of the target azimuth angle. A type of this function is analogous to the function given by (3.125) in the case of the nonfluctuated target return signal and by (3.127) in the case of the fluctuated target return signal.

Thus, the optimal estimation of target angular coordinate comes to a sum formation of the weight coefficient values $\gamma''(\beta_i, \hat{\beta}_{tg})$ on positions where $d_i = 1$ at the right and left sides with respect to zero value of the weight function. Coordinate estimation is fixed when the sums accumulated in this manner are the same within the limits of accuracy given before. Realization of this algorithm of the target return angular coordinate estimation is much simple in comparison with the algorithm given by (3.124) through (3.127), since the multiplication operations of multidigit numbers are absent. Potential accuracy of the measurer designed and constructed based on this algorithm obtained using the Cramer–Rao equality is determined by the formula

$$\sigma_{\hat{\beta}_{tg}}^2 = \frac{1}{\sum_{i=1}^{N} \left(\frac{dP_{SN_i}}{d\hat{\beta}_{tg}}\right)^2 \times \frac{1}{P_{SN_i} \tilde{P}_{SN_i}}}. \tag{3.133}$$

Under binary quantization of the target return signal without taking the radar antenna directional diagram shape into consideration we arrive to heuristic algorithms of the target return angular coordinate estimation:

1. By a position of the first and last pulses of the target return pulse train

$$\hat{\beta}_{tg} = \{0.5[\lambda - (l-1) + \mu - k]\}\Delta_{\beta}, \tag{3.134}$$

where
λ is the position number with respect to direction selected as the origin and the first pulse of the target return pulse train is fixed on this position by the criterion of l from m (l/m)
μ is the position number and the last pulse of the target return pulse train is fixed on this position by the criterion of k one after another omission
Δ_{β} is the angular resolution of pulses in train

This algorithm provides for shift compensation of the instant of target return pulse train detection at the (l – 1)th position and a shift of detection instant of the last pulse of train on k positions.

2. By a position of the last pulse of the target return pulse train and the number of positions from the first pulse to the last pulse of the target return pulse train

$$\hat{\beta}_{tg} = [\mu - 0.5(N_p - k - 1)]\Delta_{\beta}, \tag{3.135}$$

where N_p is the number of positions appropriate to the bandwidth of the detected target return pulse train; this algorithm is realized by digital storage of binary-quantized signals.

Heuristic generalized algorithms of the target return angular coordinate estimation are the simplest in realization but lead us to 25%–30% losses in accuracy in comparison with the optimal generalized signal detection algorithms. Functional dependence of the relative variance of target return angular coordinate estimation, which is given by (3.134) and (3.135), on SNR at the center of target return pulse train at N_p = 15 and $P_F = 10^{-4}$ (see Figure 3.19) allows us to compare by accuracy the discussed target return angular coordinate estimation under the uniform radar antenna scanning.

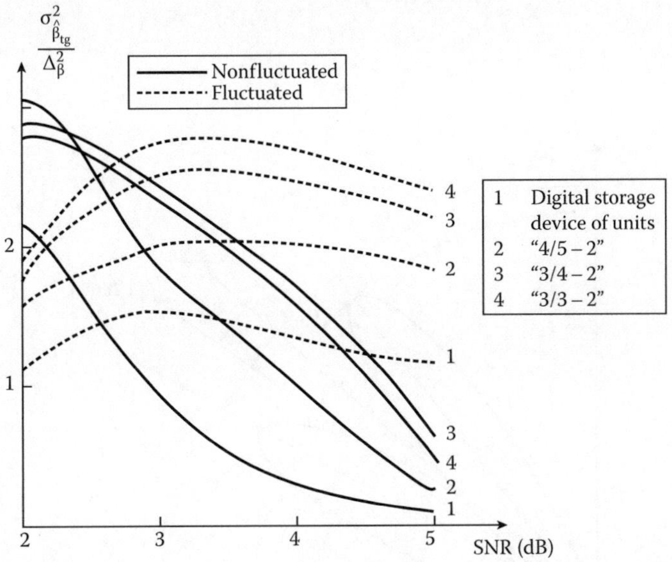

FIGURE 3.19 Relative variance of target return angular coordinate estimation as a function of SNR in the center of the target return pulse train.

3.4.3 Algorithms of Angular Coordinate Estimation under Discrete Radar Antenna Scanning

In target tracking and controlling radar systems the methods of monopulse direction finding employed by the multichannel radar systems or the discrete radar antenna scanning employed by the single channel radar systems are used for accurate angular coordinate measurement. In such cases, tracking or nontracking digital measurer can be used. Henceforth, we consider an algorithmic synthesis of the nontracking measurer of a single angular coordinate in the single channel radar system with discrete radar antenna scanning. Under measuring by the discrete scanning method the radar antenna takes two fixed positions (see Figure 3.20). In each fixed position, a direction of maximal radiation with respect to an initial direction corresponds to the angles θ_1 and $\theta_2(\theta_2 > \theta_1)$. The difference between angles θ_1 and θ_2, that is, $\Delta\theta = \theta_2 - \theta_1$ is called the angle of discrete scanning. Under angle readings relative to radar boresight a deviation of maximal radiation at scanning is $\pm\theta_0$ and a deviation of the target at scanning is θ_{tg}.

Measurement of angular coordinate in the discrete radar antenna scanning plane lies in receiving n_1 target return signals from direction θ_1 and n_2 target return signals from direction θ_2. Under target displacement θ_{tg} with respect to the radar boresight, the amplitudes of target return signals received from each direction are not the same and are equal to X_{1i} and X_{2i}, respectively. We are able to define the azimuth target angle by the amplitude ratio of these target return signals. As before, an optimal solution of the problem of target angular coordinate estimation is defined by maximal likelihood function criterion. In this case, the likelihood function equation takes the following form:

$$\left.\frac{\partial \mathcal{L}(\mathbf{x}_1,\mathbf{x}_2|\theta_{tg})}{\partial\theta_{tg}}\right|_{\theta_{tg}=\hat{\theta}_{tg}} = 0, \tag{3.136}$$

where

$$\mathbf{x}_1 = \left\| \frac{X_{11}}{\sigma_{n_1}} \; \frac{X_{12}}{\sigma_{n_1}} \cdots \frac{X_{1n_1}}{\sigma_{n_1}} \right\| = \left\| x_{11}x_{12}\cdots x_{1n_1} \right\|^T \tag{3.137}$$

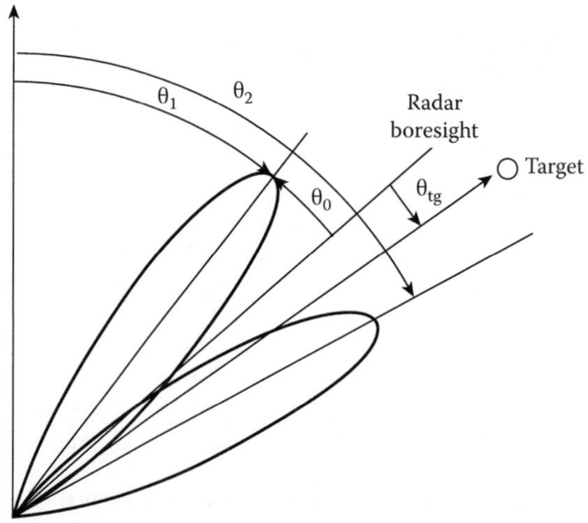

FIGURE 3.20 Fixed positions of radar antenna under discrete scanning.

and

$$\mathbf{x}_2 = \left\| \frac{X_{21}}{\sigma_{n_2}} \frac{X_{22}}{\sigma_{n_2}} \cdots \frac{X_{2n_1}}{\sigma_{n_2}} \right\| = \left\| x_{21} x_{22} \cdots x_{2n_2} \right\|^T \tag{3.138}$$

are the vectors of normalized amplitudes of target return signals received from the directions θ_1 and θ_2, respectively.

The likelihood function of sample $\mathcal{L}(\cdot)$ is differed subjecting to accepted models of the signal and noise. When the target return signal fluctuations are absent and the samples are statistically independent, the likelihood function can be presented in the following form:

$$\mathcal{L}(\mathbf{x}_1, \mathbf{x}_2 | \theta_{tg}) = \prod_{i=1}^{n_1} p_{SN}^{\mathcal{H}_1}(x_{1i}) \prod_{j=1}^{n_2} p_{SN}^{\mathcal{H}_1}(x_{2j}), \tag{3.139}$$

where

$$p_{SN}^{\mathcal{H}_1}(x_{1i}) = x_{1i} \exp\left\{ -\frac{x_{1i}^2 + q_1^2(\theta_{tg})}{2} \right\} \times I_0[x_{1i}, a_1(\theta_{tg})], \tag{3.140}$$

$$p_{SN}^{\mathcal{H}_1}(x_{2i}) = x_{2i} \exp\left\{ -\frac{x_{2i}^2 + q_2^2(\theta_{tg})}{2} \right\} \times I_0[x_{2i}, a_2(\theta_{tg})], \tag{3.141}$$

$$q_1(\theta_{tg}) = q_0 g(\theta_{tg} - \theta_0) = q_0 g(\theta_{tg} + \theta_0), \tag{3.142}$$

$$q_2(\theta_{tg}) = q_0 g(\theta_0 - \theta_{tg}), \tag{3.143}$$

where

$g(\cdot)$ is the normalized function of the radar antenna directional diagram envelope
q_0 is the SNR by voltage at the radar antenna directional diagram maximum (the same for both directions)

Substituting (3.139) in (3.136) and taking into consideration (3.140) through (3.143), in the case of powerful target return signal, we get the following result:

$$\ln I_0[x_{1i}, a_1(\theta_{tg})] \approx x_{1i} q_1(\theta_{tg}). \tag{3.144}$$

After evident mathematical transformations we obtain the final likelihood function equation in the following form:

$$\sum_{i=1}^{n_1} x_{1i} + \upsilon(\hat{\theta}_{tg}) \sum_{j=1}^{n_2} x_{2j} = q_1(\hat{\theta}_{tg}) n_1 + \upsilon(\hat{\theta}_{tg}) q_2(\hat{\theta}_{tg}) n_2, \tag{3.145}$$

where

$$\upsilon(\hat{\theta}_{tg}) = \frac{dg(\theta_0 - \hat{\theta}_{tg})}{d\hat{\theta}_{tg}} \times \left(\frac{dg(\theta_0 + \hat{\theta}_{tg})}{d\hat{\theta}_{tg}}\right)^{-1}. \quad (3.146)$$

Thus, while measuring the target angular coordinate by procedure of the discrete radar antenna scanning, the digital signal processing of target return signals is used for accumulation of the normalized target return signal amplitudes at each position of radar antenna with subsequent solution of (3.145) with respect to θ_{tg}. At this time, we assume that the signal and noise parameters characterized by q_1, σ_{n_1} and q_2, σ_{n_2} and the function describing the radar antenna directional diagram are known. Solution of (3.145) can be found, in a general case, by procedure of sequential searching, using a partition of the interval of possible target position values with respect to radar boresight equal to $2\theta_0$ on m discrete values. The number $m = 2\theta_0/\delta\theta_{tg}$, where $\delta\theta_{tg}$ is the required accuracy of the target angular coordinate estimation.

To reduce the time in finding a solution, the function $\upsilon(\theta_{tg})$ can be tabulated preliminarily with the given resolution $\delta\theta_{tg}$. Approximate solution of (3.145) can be obtained by the following way. Assume that the target is into the neighboring area that is very close to radar boresight, so we can think that as $\theta_{tg} \to 0$

$$\upsilon(\theta_{tg}) = \frac{g'(\theta_0 - \hat{\theta}_{tg})}{g'(\theta_0 + \hat{\theta}_{tg})} = -1. \quad (3.147)$$

In this case, (3.145) takes the following form:

$$\sum_{i=1}^{n_1} x_{1i} - \sum_{j=1}^{n_2} x_{2j} = q_1 n_1 + q_2 n_2. \quad (3.148)$$

Further, believing that

$$n_1 = n_2 = n, \quad \sum_{i=1}^{n} x_{1i} = \pi_1, \quad \text{and} \quad \sum_{j=1}^{n} x_{2j} = \pi_2 \quad (3.149)$$

Equation 3.148 can be transformed into the following form:

$$\pi_1 - \pi_2 = nq_0[g(\theta_0 + \hat{\theta}_{tg}) - g(\theta_0 - \hat{\theta}_{tg})]. \quad (3.150)$$

In the case of Gaussian radar antenna directional diagram, we can use the following approximation:

$$\begin{cases} g(\theta_0 - \hat{\theta}_{tg}) = \exp[-\alpha(\theta_0 - \hat{\theta}_{tg})^2] \approx 1 - \chi\hat{\theta}_{tg}, \\ g(\theta_0 + \hat{\theta}_{tg}) = \exp[-\alpha(\theta_0 + \hat{\theta}_{tg})^2] \approx 1 + \chi\hat{\theta}_{tg}, \end{cases} \quad (3.151)$$

where

$$\alpha = \frac{\beta_i - \beta_{tg}}{\varphi_0} \tag{3.152}$$

and

$$\chi = \frac{1}{g(\theta_0)}\left|\frac{dg(\theta_0 \pm \hat{\theta}_{tg})}{d\theta_{tg}}\right|_{\hat{\theta}_{tg}=\theta_0}. \tag{3.153}$$

From (3.150), taking into consideration (3.151), we obtain

$$\hat{\theta}_{tg} = -\frac{\pi_1 - \pi_2}{2nq_0\chi}. \tag{3.154}$$

3.4.4 Doppler Frequency Measurer

In complex coherent pulse radar systems, the multiple-channel filters are used to measure the Doppler frequency (see Figure 3.21a). The number n of frequency channels with the preliminary filters possessing the overlapping amplitude–frequency characteristics and DGDs has been provided (see Figure 3.21b). The required number of frequency channels is defined by the following formula:

$$n = \frac{2\Delta f_{D_{max}}}{\delta f_D}, \tag{3.155}$$

$$\Delta f_{D_{max}} = \pm\frac{2V_{tg_{max}}}{\lambda} \tag{3.156}$$

where
- $\Delta f_{D_{max}}$ is the Doppler frequency range, which is subjected to measure, defined by the well-known relationship
- λ is the position number with respect to the direction selected as the origin and the first pulse of the target return pulse train is fixed on this position by the criterion of l from $m(l/m)$
- δf_D is the CRS resolution by Doppler frequency that is characterized by a spread of signal ambiguity body section along the axis f [35]

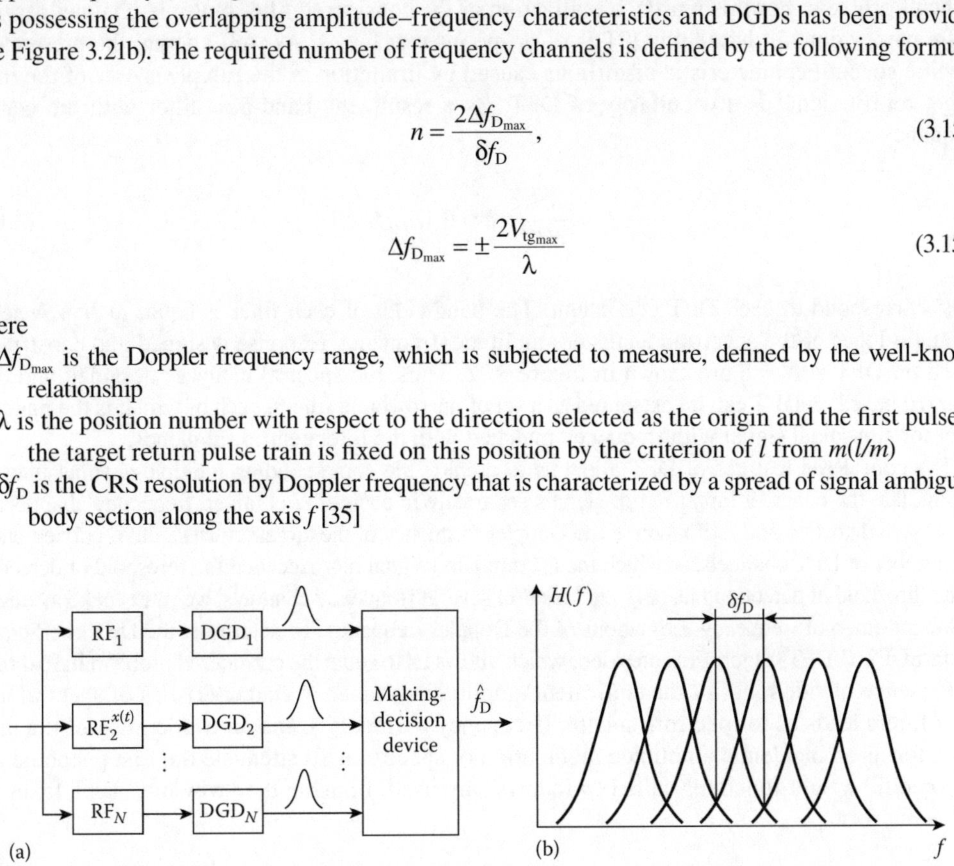

FIGURE 3.21 (a) Doppler frequency measuring by multiple channel filters and (b) amplitude–frequency responses of filters.

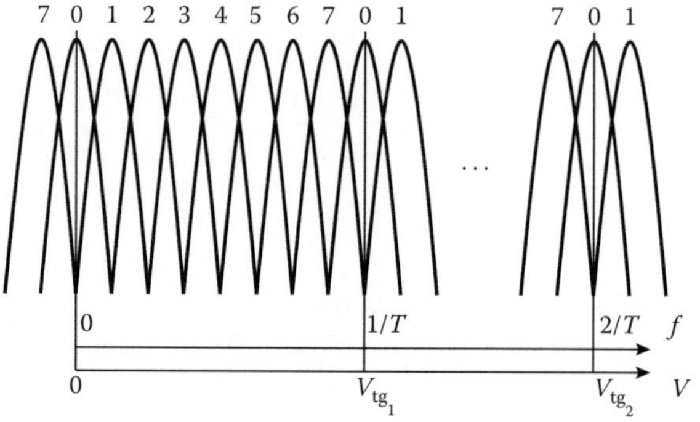

FIGURE 3.22 Amplitude–frequency responses of spectrum analyzer constructed based on DFT, $N = 0$.

Rough estimation of the Doppler frequency f_D can be defined by the channel number with the maximal amplitude of voltage at the measurer output. To increase the accuracy of definition of the Doppler frequency f_D estimation we can use the output amplitudes of three frequency channels. In this case, a position of the parabolic amplitude envelope of selected frequency channels is considered as the estimation of the Doppler frequency f_D.

Realization in the time domain of the considered filter multiple-channel network is a very difficult problem. For this reason, a realization of the considered filter multiple-channel network in frequency domain based on DFT is of prime interest for us. As well known, the presence of specific spectral characteristic distortions caused by limitation in the sample size N of the input target return signal is a peculiarity of DFT. As a result, the band-pass filter with the central frequency

$$f_k = \frac{k}{NT}, \quad k = 0,1,\ldots,N-1 \tag{3.157}$$

must correspond to each DFT coefficient. The bandwidth of each filter is equal to N^{-1}. A set of the main lobes of the spectrum analyzer amplitude–frequency response designed and constructed based on DFT at $N = 8$ are shown in Figure 3.22. Thus, the spectral analyzer designed and constructed based on DFT can be presented as a set of narrowband filters, each of which is the matched filter for sinusoidal signal with frequency matched with the filter central frequency.

The considered features of DFT allow us to employ the corresponding multiple-channel filters to accumulate the coherent target return signals separately in each of N obtained frequency channels. In this case, a detection and estimation of the Doppler frequency of the target return signal is carried out by the number of DFT channels, in which the accumulated signal has exceeded the threshold of detection. If the threshold of detection has been exceeded in several frequency channels, we must make averaging or weight union of frequency estimations of the Doppler frequency. To determine the DFT coefficients, the fast DFT (FDFT) algorithms are used, which allows us to solve the considered problem in real time.

Presence of side lobes in the equivalent amplitude–frequency characteristics of spectral analyzer filters leads us to spectrum splatter (frequency harmonics), and an overlapping of the main lobes forms an incidental amplitude modulation of spectrum. To attenuate the first phenomenon, the specific weight functions called "windows" are used. In using the "window," DFT takes the following form:

$$F_{uw}[k] = \sum_{i=0}^{N-1} w[i]u[i]\exp\left\{-\frac{j2\pi ik}{N}\right\}, \tag{3.158}$$

where

$$w[i] = w[iT], \quad i = 0,1,2,\ldots,N-1 \tag{3.159}$$

is the weight "window" function. In the case of the rectangle weight function, which corresponds practically to DFT of N input signals (the target return signals) without weighting, the side lobes of equivalent amplitude–frequency response of filters are maximal, approximately for about −13 dB. For this reason, as a rule, the "windows" with dropping weight coefficients, for example, triangle, cosine, Gauss, Dolph–Chebyshev, Hamming, and so on, are used.

Under the Doppler frequency estimation, the signal sample size N for DFT is selected to ensure the required spectral resolution

$$\delta f = \frac{\nu}{TN}, \tag{3.160}$$

where ν is the coefficient characterizing a spectral bandwidth increase for the selected "window." As a rule, the value of ν is selected equal to an equivalent noise bandwidth of the "window," that is,

$$\nu = \frac{\sum_{i=0}^{N-1} w^2[iT]}{\left(\sum_{i=0}^{N-1} w[iT]\right)^2}. \tag{3.161}$$

The incidental amplitude modulation of spectrum leads us to additional losses under signal processing of the target return pulse train, the Doppler frequency of which does not coincide with basic frequencies multiple to the frequency $1/TN$. The losses caused by the incidental amplitude modulation of spectrum are for about 1/2 dB in the case of "windows" with smoothly dropping weight coefficients. Maximal losses under DFT caused by maximal losses owing to the incidental amplitude modulation of spectrum for the given "window" and DFT losses caused by a shape of the given "window" do not exceed 3 ÷ 4 dB.

3.5 COMPLEX GENERALIZED ALGORITHMS OF DIGITAL INTERPERIOD SIGNAL PROCESSING

The complex digital generalized interperiod signal processing/detection algorithm employed by CRSs is offered as a solution to the following problems:

- To detect the target return signals
- To measure the parameters of target return signals
- To form and estimate the current coordinates, parameters, and features of targets by information contained in the target return signals
- To code the coordinates and parameters of targets and process them for follow-up processing

The complex digital generalized interperiod signal processing/detection algorithm is synthesized by the composition of particular digital interperiod signal processing/detection algorithms. There are many ways for the composition of the particular digital interperiod signal processing/detection algorithms. The number and content of construction ways of the complex digital generalized interperiod signal processing/detection algorithm depends on the operation conditions of CRSs. At the same time, there is a need to keep in mind that at any condition the CRS operates in very hard noise environment caused by deliberate interferences (active or passive) in conflict situations; reflections from ground objects (buildings, trees, vehicles, mountains, etc.) and atmospheric precipitation (rain, snow, fog, clouds); reflection from non-mechanical objects in atmosphere; and nonsynchronized (chaotic) pulse noise and interference. For this reason, the main problem that is solved in the course of designing the complex digital generalized interperiod signal processing/detection algorithm is the problem of compensation of background and scattering returns and the problem of stabilization of the probability of false alarm P_F.

While designing the complex digital generalized interperiod signal processing/detection algorithm it is appropriate to proceed from the assumption that, in each specific case of radar system operation, a definite noise environment requiring reconstruction or retuning of the considered algorithm to ensure a qualitative receiving and processing the target return signals at the fixed probability of false alarm P_F will be formed all over the radar coverage or in individual parts of radar coverage. Whereas the number of such situations is very high, we can call several most specific cases:

- Signal processing/detection in Gaussian uncorrelated (or weakly correlated) noise with unknown variance
- Signal processing/detection in Gaussian uncorrelated (or weakly correlated) noise with unknown variance plus chaotic pulse interference
- Signal processing/detection in non-Gaussian noise, for example, while using the intraperiod nonlinear signal processing/detection algorithms (limiters, logarithmic amplifiers, and so on)
- Signal processing/detection in correlated Gaussian noise with high variance by value

Under such an approach, naturally, it is required to include the adaptive generalized signal processing/detection algorithms for various noise situations and specific algorithms, which are able to recognize the type of situation and make a decision, into the complex digital generalized interperiod signal processing/detection algorithm. The algorithms recognizing specific situations can be presented in the form of individual signal processing subsystems. Finally, there is a need to provide for a high-speed switchboard in the content of the complex digital generalized interperiod signal processing/detection algorithm, which allows us to carry out all necessary switching in the course of adaptation to noise environment.

Thus, the complex digital generalized interperiod signal processing/detection algorithm for CRSs must be tuned by the totality of a given set of situations and be adaptive to noise level. The simplest version of the complex digital generalized interperiod signal processing/detection algorithm is shown in Figure 3.23. The subsystem of intraperiod signal processing, the output signals of which are the input signals for the considered complex digital generalized interperiod signal processing/detection algorithm, and the subsystem of environment analysis, the output signals of which are meant for the reconstruction of complex digital generalized interperiod signal processing/detection algorithm, are presented in Figure 3.23 by individual blocks. The following digital signal processing/detection subsystems can be produced by the considered diagram in Figure 3.23:

- *Digital signal processing and detection subsystem for slowly moving targets and targets moving with blind velocities.* The blocks 3, 5, 6, 10, and 11 form this subsystem. The block 3 carries out the generation, averaging, and storing of the noise envelope amplitudes for all resolution elements of the CRS. The block 6 generates the thresholds to detect the target return signals at the zero channel output of FDFT filter (the block 5) using the noise

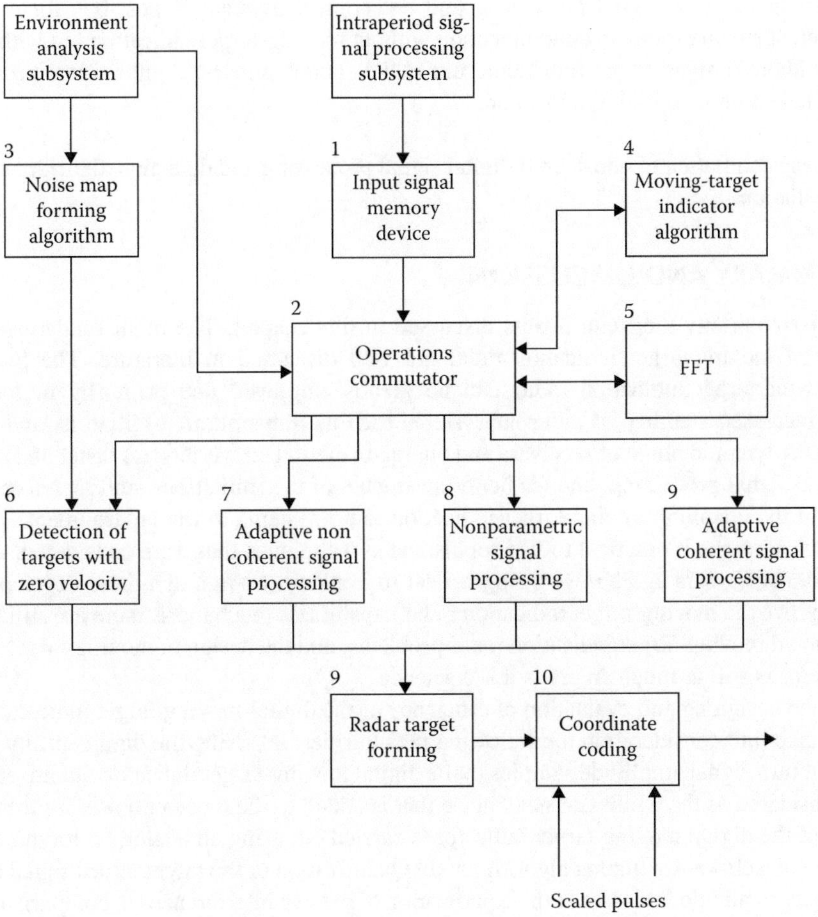

FIGURE 3.23 Simplest version of the complex digital generalized interperiod signal processing/detection algorithm.

envelope amplitudes and omitting the moving-target indicator filter. As a result, there is a greater possibility to detect the target return signals from stationary targets under the condition that the effective reflective surface of target compared to the effective specific surface of passive interference.

- *No coherent digital adaptive signal processing and detection subsystem formed by cascade connection of the blocks* 1, 7, 10, and 11. The adapter of decision-making threshold and the target return signal parameter measurer form the block 7.
- *Nonparametric digital signal processing/detection subsystem formed by the blocks* 1, 8, 10, and 11. The rank nonparametric digital signal processing/detection algorithm, the adaptive threshold of rank detector, and corresponding detectors and measurers are formed by the block 8.
- *Coherent adaptive digital signal processing and detection subsystem formed by cascade connection of blocks* 1, 4, 5, 9, 10, and 11. Moving-target indication, FDFT of noncompensated remainder samples with the purpose to accumulate target return signals and target radial velocity estimations are realized in this channel. After that, the coherent adaptive signal processing and detection algorithms and estimations of signal parameters are realized in each frequency channel. Owing to the fact of overlapping of frequency channels, the blip operations become essential.

- Forming of other signal processing and detection subsystem is possible, for example, coherent signal processing and detection without moving-target indication but with FDFT or without moving-target indication and FDFT, nonparametric digital signal processing and detection with FDFT, and so on.

Final efficacy estimation of considered digital signal processing and detection algorithms is carried out by simulation.

3.6 SUMMARY AND DISCUSSION

We summarize briefly the main results discussed in this chapter. The main fundamental theory principles of moving-target indication radar are well discussed in literature. The performance of the moving-target indication radar can be greatly improved due primarily to four advantages: (a) increased stability of radar subsystems such as transmitters, oscillators, and receivers; (b) increased dynamic range of receivers and analog-to-digital converters; (c) faster and more powerful digital signal processing; and (d) better awareness of the limitations, and therefore, requisite solutions of the adapting moving-target indication radar systems to the environment. These four advantages can make it practical to use sophisticated techniques that were considered, and sometimes tried, many years ago but were impractical to implement. Although these developments are able to improve the moving-target indication radar capabilities much more, there are still no perfect solutions to all moving-target indication radar problems, and the design of moving-target indication radar systems is still as much an art as it is a science.

Under the designing and evaluation of efficiency of the digital moving-target indicators, there is a need to take into consideration the following peculiarities: (a) Using the digit capacity $N_b \geq 8$ of the target return signal amplitude samples in the digital moving-target indicator, quantization errors can be considered as the white Gaussian noise that is added to the receiver noise; for this reason, a synthesis of the digital moving-target indicator is carried out using an analog prototype, that is, by digitization of well-known analog algorithms. (b) Quantization of the target return signal amplitude samples leads to additional losses at the equalization of passive interferences in comparison with the use of analog moving-target indicators.

In practice, under the evaluation of digital moving-target indicator efficacy the following indices are widely used: (a) the coefficient of interference cancellation G_{can} that is defined as a ratio of the input passive interference power to the power of the passive interference at the equalizer output; (b) figure of merit for the case of linear interperiod cancellation of the passive interferences as given by (3.3); in a general case, when a nonlinear signal processing is used, the figure of merit indicates to what extent we can increase an interference power at the equalizer input without decreasing in detection performance; and (c) the coefficient of improvement is a characteristic of the response of digital moving-target indicator on passive interference signals with respect to the averaged response on target return signals. The aforementioned Q-factors can be determined both for equalizer systems for passive interferences using the interperiod subtraction and for equalizer systems with subsequent accumulation of remainders after the passive interference cancellation.

The main element of the digital moving-target indicator is the digital rejector filter that can guarantee a cancellation of correlated passive interference. In the simplest case, the digital filter is constructed in the form of the filter with the ν-fold ($\nu \leq N$) interperiod subtraction that corresponds to the structure of nonrecursive filter. The recursive filters are widely used also for cancellation of passive interferences. The digital rejector filter ensures the best cancellation of passive interferences owing to an improvement of the amplitude–frequency characteristic shape in comparison with the nonrecursive rejector filter of the same order. However, there is a higher degree of the correlation of the passive interference remainders at the digital recursive filter output, in comparison with the correlation degree of the passive interference remainders at the digital nonrecursive filter. Moreover, the presence of positive feedback leads to an increase in the time of transient process and corresponding losses in efficacy if the number of target return pulses in the train is less 20, that is, $N < 20$.

Implementation of analog moving-target indicators with the wobble procedure within the limits of the interval T is very difficult since, in this case, there is a need to employ various individual delay circuits for each value of the repetition period T and complex switching system for these circuits. Under realization of the digital moving-target indicator with the wobble procedure within the limits of the interval T, it is enough to realize only the synchronization between a sample of delayed data from memory device and radar searching signal instants. In doing so, the size of memory device does not change and does not depend on the number of discrete values and the wobble procedure function within the limits of the repetition period T. The best speed performance of the digital moving-target indicators with the wobble procedure within the limits of the repetition period T can be obtained if each pulse from the target return pulse train with N_s, where N_s is the number of target return pulses in the train, would be adjusted with a new individual repetition period. The individual repetition period T must be varied with respect to the average value $\overline{T}$ on the multiple value of fixed time interval $\pm\Delta T$. The problem of design and construction of the digital moving-target indicators with the wobble procedure within the limits of the repetition period T is to make a correct selection of a value $\overline{T}$ and define the wobble function for pulse train in sequence given by (3.23). In a general case, the wobble procedure is defined by the criterion of the figure of merit η maximization taking into consideration a minimization of amplitude–frequency characteristic ripple in the digital filter bandwidth. As a rule, this problem is solved by simulation methods. Calculations and simulations demonstrate that the wobble procedures for the repetition period T lead to a decrease in the depth of amplitude–frequency response dips of the nonrecursive and recursive digital filters. However, at the same time, a stopband of the nonrecursive and recursive digital filters is narrowed simultaneously with the frequency-band extension and distortion of interference frequency spectrum. For this reason, the effectiveness needed for cancellation of passive interferences is reduced considerably. Absolute losses in the figure of merit η for the recursive digital filters of the second order vary from 0.3 to 4.3 dB and from 4 to 19 dB for the recursive digital filters of the third order in comparison with the digital moving-target indicators without wobble procedures for the repetition period T at optimal velocity.

In practice, spectral-correlation characteristics of passive interferences are unknown a priori and, moreover, heterogeneous in space and nonstationary in time. Naturally, the efficacy of passive interference cancellation is substantially reduced. To ensure high effectiveness of the digital moving-target indication radar subsystems under conditions of a priori uncertainty and nonstationarity of passive interference parameters, the adaptive digital moving-target indication radar subsystems are employed. In a general sense, the problem of adaptive moving-target indication is solved based on an implementation of the correlation feedback principle. Correlation automatic equalizer of passive interferences represents a closed tracker adapting to noise and interference environment without taking into consideration the Doppler frequency. Parallel with high effectiveness, the tracker has a set of the following imperfections: (a) poor cancellation of area-extensive interference leading edge that is a consequence of high time constant value (for about 10 resolution elements) of adaptive feedback; (b) decrease in passive interference cancellation efficiency over the powerful target return signal; and (c) very difficult realization, especially employing digital signal processing. The problem of moving-target indicator adaptation can be solved using the so-called the empirical Bayes approach using of which we at first, define the maximal likelihood estimation (MLE) of passive interference parameters and after that we use these parameters to determine the impulse response coefficients of the rejector digital filter. In this case, we obtain an open-loop adaptation system, the transient process in which is finished within the limits of the digital filter transient process.

Under the discussion of digital signal processing and detection algorithms, we consider two cases, namely, (a) the coherent impulse signals with known parameters and (b) the coherent impulse signals with unknown parameters. In digital signal processing systems, a process of accumulation and signal detection is realized, as a rule, at video frequency after union of in-phase and quadrature channels. In this chapter, we carry out analysis of signal detection problems under the following initial conditions:

1. The input signal is presented as a single signal, and the signal parameters are the delay t_{d} and direction of arrival θ; the total target return signal is a sequence of periodically repeated pulses (the pulse train). As for the statistical characteristics of the target return signal (information signal, i.e., the signal containing information about parameters of the target), we consider two cases: the train of nonfluctuated signals and the train of independently fluctuated signals obeying to the Rayleigh pdf with zero mean and the variance σ_{S}^2.
2. Under synthesis of signal detection algorithms and evaluation of target return signal parameters, we use, as a rule, a noise model in the form of Gaussian random process with zero mean and the variance σ_n^2; when the correlated in time interferences (namely, passive interferences) are absent, noise samples modeled by the Gaussian process do not have an interperiod correlation; when the passive interferences or their remainders are present after cancellation by the digital moving-target indicators, a sequence of passive interference samples are approximated by Markov chain; to describe statistically the Markov chain we, in addition to the variance, should know the coefficient module ρ_{pi} of passive interference interperiod correlation; samples of the input signal X_i in the case of absence of target reflecting surface fluctuations are subjected to the general Rayleigh pdf; in the case of presence of target reflecting surface fluctuations, we can write

$$p_{SN}^{\prime\mathcal{H}_1}(X_i) = \frac{X_i}{\sigma_\Sigma^2 + \sigma_{\text{S}_i}^2} \exp\left\{-\frac{X_i^2}{2\left(\sigma_\Sigma^2 + \sigma_{\text{S}_i}^2\right)}\right\},$$

   where $\sigma_{\text{S}_i}^2$ is the variance of target return signal amplitude.
3. A digital signal detection algorithm is considered for two versions: (a) the target return signal is quantized by amplitude in such manner that a low-order bit value does not exceed the root-mean-square value σ_n of receiver noise. In this case, a stimulus tracking of quantization by amplitude is added up to an addition of independent quantization noise to the input noise and a synthesis of digital signal processing algorithms is reduced to digital realization of optimal analog signal processing algorithms; and (b) the target return signal is quantized on two levels—binary quantization. In this case, there is a need to carry out a direct synthesis of signal processing algorithms and decision devices to process digital binary quantized signals.

Under these conditions, we consider and discuss the following detectors employed by CRSs:

1. In the case of coherent impulse signals with known parameters,
   a. DGDs for target return pulse train
   b. DGDs for binary quantized target return pulse train
   c. DGDs based on method of the sequential analysis
   d. Software DGDs for binary quantized target return pulse train
2. In the case of coherent impulse signals with unknown parameters,
   a. Adaptive DGDs
   b. Nonparametric DGDs—sign-nonparametric DGDs and rank-nonparametric DGDs
   c. Adaptive-nonparametric DGDs

Estimation of target return signal parameters comprising information about the coordinates and characteristics of target is the main operation of radar signal preprocessing. Estimation of parameters starts after making the decision a "yes" signal; that is, the target has been detected in direction of scanning at a definite distance. At this time, the target detection is associated with rough calculation of target coordinates, for example, the azimuth accurate with the radar antenna directional diagram width, and the target range accurate with the dimension of resolution element in radar range and so on. The main task of a digital measurer is to obtain more specific information about primary data of estimated target return signal parameters to the predetermined values of target return signal parameters. We investigate algorithms of angular coordinate estimation under uniform and discrete radar antenna scanning. We define the target angular coordinate estimation. In addition, we define the principles of Doppler frequency measurement.

Finally, we suggest and define a structure of the complex digital generalized interperiod signal processing/detection algorithm. While designing the complex digital generalized interperiod signal processing/detection algorithm it is appropriate to proceed from the assumption that, in each specific case of radar system operation, a definite noise environment requiring reconstruction or retuning of the considered algorithm to ensure a qualitative receiving and processing the target return signals at the fixed probability of false alarm P_{F} will be formed all over the radar coverage or in individual parts of radar coverage. Naturally, it is required to include the adaptive generalized signal processing/detection algorithms for various noise situations and specific algorithms, which are able to recognize the type of situation and make a decision, into the complex digital generalized interperiod signal processing/detection algorithm. The algorithms recognizing specific situations can be presented in the form of individual signal processing subsystems. Finally, there is a need to provide for a high-speed switchboard in the content of the complex digital generalized interperiod signal processing/detection algorithm, which allows us to carry out all necessary switching in the course of adaptation to noise environment.

REFERENCES

1. Skolnik, M. 2008. *Radar Handbook*. 3rd edn. New York: McGraw-Hills, Inc.
2. Applebaum, S. 1961. Mathematical description of VICI. Report No. AWCS-EEM-1. Syracuse, NY: General Electric Co.
3. Chow, S.M. 1967. Range and Doppler resolution of a frequency-scanned filter. *Proceedings of the IEEE*, 114(3): 321–326.
4. Shrader, W.W. 1970. MTI Radar, Chapter 17, in M.I. Skolnik, Ed. *Radar Handbook*. New York: McGraw-Hills, Inc.
5. Blackman, S. and R. Popoli. 1999. *Design and Analysis of Modern Tracking Systems*. Boston, MA: Artech House, Inc.
6. Richards, M.A. 2005. *Fundamentals of Radar Signal Processing*. New York: John Wiley & Sons, Inc.
7. Barton, D.K. and S.A. Leonov. 1997. *Radar Technology Encyclopedia*. Norwood, MA: Artech House, Inc.
8. Wehner, D.R. 1995. *High Resolution Radar*. Norwood, MA: Artech House, Inc.
9. Bozic, S.M. 1994. *Digital and Kalman Filtering*. 2nd edn. New York: Halsted Press.
10. Brown, R.G. and P.Y.C. Hwang. 1997. *Introduction to Random Signals and Applied Kalman Filtering*. 3rd edn. New York: John Wiley & Sons, Inc.
11. Diniz, P.S. 1997. *Adaptive Filtering*. Boston, MA: Kluwer Academic Publisher.
12. Figneiras-Vidal, A.R. Ed. 1996. *Digital Signal Processing in Telecommunications*. London, U.K.: Springer-Verlag.
13. Grant, P. and B. Mulgrew. 1995. Nonlinear adaptive filter: Design and application, in *Proceedings of IFAC on Adaptive Systems in Control and Signal Processing*, June 14–16, Budapest, Hungary, pp. 31–42.
14. Haykin, S. 1989. *Modern Filters*. New York: Macmillan Press, Inc.
15. Johnson, D.H. and D.E. Dudgeon. 1993. *Array Signal Processing: Concepts and Techniques*. Englewood Cliffs, NJ: Prentice Hall, Inc.
16. Lee, E.A. and D.G. Messerschmitt. 1994. *Digital Communications*. 2nd edn. Boston, MA: Kluwer Academic Publisher.
17. Mitra, S.K. 1998. *Digital Signal Processing*. New York: McGraw-Hills, Inc.

18. Regalia, P.A. 1995. *Adaptive IIR Filtering in Signal Processing and Control*. New York: Marcel Dekker, Inc.
19. Sayed, A.H. and T. Kailath. 1998. Recursive least-squares adaptive filters, in V. Madisetti, and D. Williams, Eds. *The Digital Signal Processing Handbook*. New York: CRC Press.
20. Shynk, J.J. 1989. Adaptive IIR filtering. *IEEE ASSP Magazine*, 6: 4–21.
21. Treichler, J., Johnson, C.R., and M.G. Larimore. 1987. *Theory and Design of Adaptive Filters*. New York: John Wiley & Sons, Inc.
22. Licharev, V.A. 1973. *Digital Techniques and Facilities in Radar*. Moscow, Russia: Soviet Radio.
23. Lutovac, M.D., Tosic, D.V., and B.L. Evans. 2001. *Filter Design for Signal Processing Using MATLAB and Mathematica*. Englewood Cliffs, NJ: Prentice Hall, Inc.
24. Popov, D.I. and V.A. Fedorov. 1975. Efficacy of recursive filters for moving-target indicators. *Radio Electronics*, 3: 63–68.
25. Wald, A. 1947. *Sequential Analysis*. New York: John Wiley & Sons, Inc.
26. Glushko, O.V. and L.M. Osynskiy. 1982. Microprocessor for construction of digital filtering systems. *Control Systems and Computers*, 1: 73–76.
27. Bellanger, M.G. 1987. *Adaptive Digital Filters and Signal Analysis*. New York: Marcel Dekker, Inc.
28. Purdy, R.J. et al. 2000. Radar signal processing. *Lincoln Laboratory Journal*, 12(2): 297.
29. Billingsley, J.B. 2002. *Low-Angle Radar Land Clutter—Measurements and Empirical Models*. Norwich, NY: William Andrew Publishing, Inc.
30. Barton, D.K. 2005. *Modern Radar Systems Analysis*. Norwood, MA: Artech House, Inc.
31. Skolnik, M.I. 2001. *Introduction to Radar Systems*. 3rd edn. New York: McGraw-Hills, Inc.
32. Ludloff, A. and M. Minker. 1985. Reliability of velocity measurement by MTD radar. *IEEE Transactions on Aerospace and Electronic Systems*, 21(7): 522–528.
33. Hall, T.M. and W.W. Shrader. 2007. Statistics of the clutter residue in MTI radars with IF limiting, in *IEEE Radar Conference*, April 17–20, Boston, MA, pp. 01–06.
34. Cho, J.Y.M. et al. 2003. Range-velocity ambiguity mitigation schemes for the enhanced terminal Doppler weather radar, in *31st Conference on Radar Meteorology*, August 6–12, Seattle, WA, pp. 463–466.
35. Shirman, J.D. and V.H. Manjos. 1981. *Theory and Techniques of Radar Signal Processing in Noise*. Moscow, Russia: Radio and Svyaz.
36. Popov, D.I. 1981. Synthesis of digital adaptive rejector filters. *Radiotechnika*, 10: 53–57.
37. Tuzlukov, V. 2010. Multiuser generalized detector for uniformly quantized synchronous CDMA signals in AWGN noise. *Telecommunications Review*, 20(5): 836–848.
38. Tuzlukov, V. 2011. Signal processing by generalized receiver in DS-CDMA wireless communication systems with optimal combining and partial cancellation. *EURASIP Journal on Advances in Signal Processing*, 2011, Article ID 913189: 15, DOI: 10.1155/2011/913189.
39. Tuzlukov, V. 2011. Signal processing by generalized receiver in DS-CDMA wireless communication systems with frequency-selective channels. *Circuits, Systems, and Signal Processing*, 30(6): 1197–1230.
40. Tuzlukov, V. 2011. DS-CDMA downlink systems with fading channel employing the generalized receiver. *Digital Signal Processing*, 21(6): 725–733.
41. Kay, S.M. 1998. *Statistical Signal Processing: Detection Theory*. Upper Saddle River, NJ: Prentice Hall, Inc.
42. Tuzlukov, V.P. 1998. *Signal Processing in Noise: A New Methodology*. Minsk, Belarus: IEC.
43. Tuzlukov, V.P. 1998. A new approach to signal detection theory. *Digital Signal Processing*, 8(3): 166–184.
44. Tuzlukov, V.P. 2001. *Signal Detection Theory*. New York: Springer-Verlag, Inc.
45. Tuzlukov, V.P. 2002. *Signal Processing Noise*. Boca Raton, FL: CRC Press.
46. Tuzlukov, V.P. 2005. *Signal and Image Processing in Navigational Systems*. Boca Raton, FL: CRC Press.
47. Tuzlukov, V.P. 2012. Generalized approach to signal processing in wireless communications: The main aspects and some examples. Chapter 11. In *Wireless Communications and Networks—Recent Advances*, ed. A. Eksim, 305–338. Rijeka, Croatia, IuTech.
48. Tuzlukov, V.P., Yoon, W.-S., and Y.D. Kim. 2004. Distributed signal processing with randomized data selection based on the generalized approach in wireless sensor networks. *WSEAS Transactions on Computers*, 5(3): 1635–1643.
49. Kim, J.H., Kim, J.H., Tuzlukov, V.P., Yoon, W.-S., and Y.D. Kim. 2004. FFH and MCFH spread-spectrum wireless sensor network systems based on the generalized approach to signal processing. *WSEAS Transactions on Computers*, 6(3): 1794–1801.

50. Tuzlukov, V.P., Yoon, W.-S., and Y.D. Kim. 2004. Wireless sensor networks based on the generalized approach to signal processing with fading channels and receive antenna array. *WSEAS Transactions on Circuits and Systems*, 10(3): 2149–2155.
51. Kim, J.H., Tuzlukov, V.P., Yoon, W.-S., and Y.D. Kim. 2005. Performance analysis under multiple antennas in wireless sensor networks based on the generalized approach to signal processing. *WSEAS Transactions on Communications*, 7(4): 391–395.
52. Kim, J.H., Tuzlukov, V.P., Yoon, W.-S., and Y.D. Kim. 2005. Macrodiversity in wireless sensor networks based on the generalized approach to signal processing. *WSEAS Transactions on Communications*, 8(4): 648–653.
53. Kim, J.H., Tuzlukov, V.P., Yoon, W.-S., and Y.D. Kim. 2005. Generalized detector under no orthogonal multipulse modulation in remote sensing systems. *WSEAS Transactions on Signal Processing*, 2(1): 203–208.
54. Tuzlukov, V.P. 2009. Optimal combining, partial cancellation, and channel estimation and correlation in DS-CDMA systems employing the generalized detector. *WSEAS Transactions on Communications*, 7(8): 718–733.
55. Tuzlukov, V.P., Yoon, W.-S., and Y.D. Kim. 2004. Adaptive beam-former generalized detector in wireless sensor networks, in *Proceedings of IASTED International Conference on Parallel and Distributed Computing and Networks* (*PDCN 2004*), February 17–19, Innsbruck, Austria, pp. 195–200.
56. Tuzlukov, V.P., Yoon, W.-S., and Y.D. Kim. 2004. Network assisted diversity for random access wireless sensor networks under the use of the generalized approach to signal processing, in *Proceedings of the 2nd SPIE International Symposium on Fluctuations in Noise*, May 25–28, Maspalomas, Gran Canaria, Spain, Vol. 5473, pp. 110–121.
57. Tuzlukov, V.P., Yoon, W.-S., and Y.D. Kim. 2005. MMSE multiuser generalized detector for no orthogonal multipulse modulation in wireless sensor networks, in *Proceedings of the 9th World Multiconference on Systemics, Cybernetics, and Informatics* (*WMSCI 2005*), July 10–13, Orlando, FL. (CD Proceedings)
58. Kim, J.H., Tuzlukov, V.P., Yoon, W.-S., and Y.D. Kim. 2005. Collaborative wireless sensor networks for target detection based on the generalized approach to signal processing, in *Proceedings of International Conference on Control, Automation, and Systems* (*ICCAS 2005*), June 2–5, Seoul, Korea. (CD Proceedings)
59. Tuzlukov, V.P., Chung, K.H., and Y.D. Kim. 2007. Signal detection by generalized detector in compound-Gaussian noise, in *Proceedings of the 3rd WSEAS International Conference on Remote Sensing (REMOTE'07)*, November 21–23, Venice (Venezia), Italy, pp. 1–7.
60. Tuzlukov, V.P. 2008. Selection of partial cancellation factors in DS-CDMA systems employing the generalized detector, in *Proceedings of the 12th WSEAS International Conference on Communications*, July 23–25, Heraklion, Creete, Greece. (CD Proceedings)
61. Tuzlukov, V.P. 2008. Multiuser generalized detector for uniformly quantized synchronous CDMA signals in wireless sensor networks with additive white Gaussian noise channels, in *Proceedings of the International Conference on Control, Automation, and Systems* (*ICCAS 2008*), October 14–17, Seoul, Korea. pp. 1526–1531.
62. Tuzlukov, V.P. 2009. Symbol error rate of quadrature subbranch hybrid selection/maximal-ratio combining in Rayleigh fading under employment of generalized detector, in *Recent Advances in Communications: Proceedings of the 13th WSEAS International Conference on Communications*, July 23–25, Rodos (Rhodes) Island, Greece, pp. 60–65.
63. Tuzlukov, V.P. 2009. Generalized detector with linear equalization for frequency-selective channels employing finite impulse response beamforming, in *Proceedings of the 2nd International Congress on Image and Signal Processing (CISP 2009)*, October 17–19, Tianjin, China, pp. 4382–4386.
64. Tuzlukov, V.P. 2010. Optimal waveforms for MIMO radar systems employing the generalized detector, in *Proceedings of the International Conference on Sensor Data and Information Exploitation: Automatic Target Recognition, Part of the SPIE International Symposium on Defense, Security, and Sensing*, April 5–9, Orlando, FL, Vol. 7697, pp. 76971G-1–76971G-12.
65. Tuzlukov, V.P. 2010. MIMO radar systems based on the generalized detector and space-time coding, in *Proceedings of the International Conference on Sensor Data and Information Exploitation: Automatic Target Recognition, Part of the SPIE International Symposium on Defense, Security, and Sensing,* April 5–9, Orlando, FL, Vol. 7698, pp. 769805-1–769805-12.
66. Tuzlukov, V.P. 2010. Generalized receiver under blind multiuser detection in wireless communications, in *Proceedings of IEEE International Symposium on Industrial Electronics* (*ISIE 2010*), July 4–7, Bari, Italy, pp. 3483–3488.

67. Khan, R.R. and V.P. Tuzlukov. 2010. Multiuser data fusion algorithm for estimation of a walking person position, in *Proceedings of International Conference on Control, Automation, and Systems (ICCAS 2010)*, October 27–30, Seoul, Korea. pp. 863–867.
68. Khan, R.R. and V.P. Tuzlukov. 2010. Beamforming for rejection of co-channels interference in an OFDM system, in *Proceedings of the 3rd International Congress on Image and Signal Processing (CISP 2010)*, October 16–18, Yantai, China. pp. 3318–3322.
69. Khan, R.R. and V.P. Tuzlukov. 2010. Null-steering beamforming for cancellation of co-channel interference in CDMA wireless communications systems, in *Proceedings of the 4th IEEE International Conference on Signal Processing and Communications Systems (ICSPCS'2010)*, December 13–15, Goald Coast, Queensland, Australia. (CD Proceedings).
70. Akimov, P.S., Bacut, P.A., Bogdanovich, V.A. et al. 1984. *Theory of Signal Detection*. Moscow, Russia: Radio and Svyaz.
71. Wald, A. 1950. *Statistical Decision Functions*. New York: John Wiley & Sons, Inc.
72. Fu, K. 1968. *Sequential Methods in Pattern Recognition and Machine Learning*. New York: Academic Press.
73. Ghosh, B. 1970. *Sequential Tests of Statistical Hypotheses*. Cambridge, MA: Addison-Wesley.
74. Siegmund, D. 1985. *Sequential Analysis: Tests and Confidence Intervals*. New York: Springer-Verlag.
75. Wald, A. and J. Wolfowitz. 1948. Optimum character of the sequential probability ratio test. *Annual Mathematical Statistics*, 19: 326–339.
76. Aarts, E.H.L. 1989. *Simulated Annealing and Boltzman Machines: A Stochastic Approach to Combinational Optimization and Neural Computing*. New York: John Wiley & Sons, Inc.
77. Blahut, R.E. 1987. *Principles and Practice of Information Theory*. Reading, MA: Addison-Wesley.
78. Kuzmin, S.Z. 1967. *Digital Signal Processing in Radar*. Moscow, Russia: Soviet Radio.
79. Eaves, J.L. and E.K. Reedy. 1987. *Principles of Modern Radar*. New York: Van Nostrand Reinhold.
80. Edde, B. 1995. *Radar: Principles, Technologies, Applications*. Upper Saddle River, NJ: Prentice Hall, Inc.
81. Levanon, N. 1988. *Radar Principles*. New York: John Wiley & Sons, Inc.
82. Peebles, Jr. P.Z. 1998. *Radar Principles*. New York: John Wiley & Sons, Inc.
83. Alexander, S.T. 1986. *Adaptive Signal Processing: Theory and Applications*. New York: Springer-Verlag.
84. Clarkson, P.M. 1993. *Optimal and Adaptive Signal Processing*. Boca Raton, FL: CRC Press.
85. Pitas, I. and A.N. Wenetsanopoulos. 1990. *Nonlinear Digital Filters*. Boston, MA: Kluwer Academic Publishers.
86. Sibal, L.H., Ed. 1987. *Adaptive Signal Processing*. New York: IEEE Press.
87. Widrow, B. and S. Stearns. 1985. *Adaptive Signal Processing*. Englewood Cliffs, NJ: Prentice Hall, Inc.
88. Lacomme, P., Hardange, J.-P., Marchais, J.C., and E. Normant. 2001. *Air and Spaceborne Radar Systems*. Norwich, NY: William Andrew Publishing, LLC.
89. Bakulev, P.A. and V.M. Stepin. 1986. *Methods and Hardware of Moving-Target Indication*. Moscow, Russia: Radio and Svyaz.
90. Sherman, S.M. 1986. *Monopulse Principles and Techniques*. Norwood, MA: Artech House, Inc.
91. Leonov, A.I. and K.I. Formichev. 1986. *Monopulse Radar*. Norwood, MA: Artech House, Inc.

4 Algorithms of Target Range Track Detection and Tracking

Target range tracking is accomplished by continuously measuring the time delay between the transmission of an RF pulse and the echo signal returned from the target and converting the roundtrip delay into units of distance. The range measurement is the most precise position-coordinate measurement of the radar; typically, with high SNR, it can be within a few meters at hundreds-of-kilometers range. Range tracking usually provides the major means for discriminating the desired target from other targets, although the Doppler frequency and angle discrimination are also used, by performing range gating (time gating) to eliminate the echo of other targets at different ranges from the error detector outputs. The range-tracking circuitry is also used for acquiring a desired target. Target range tracking requires not only that the time of travel of the pulse to and from the target be measured but also that the target return signal is identified as a target rather than noise and a range-time history of the target be maintained.

Although this discussion is for typical pulse-type tracking radars, the target range measurement may also be performed with continuous wave (CW) radar systems using the frequency-modulated continuous wave (FM-CW) that is typically a linear-ramp FM. The target range is determined by the range-related frequency difference between the echo-frequency ramp and the frequency of the ramp being transmitted. The performance of FM-CW radar systems, with consideration of the Doppler effect, is discussed in Ref. [1].

Acquisition: The first function of the target range tracker is acquisition of a desired target. Although this is not a tracking operation, it is a necessary first step before the target range tracking or target angle tracking (the azimuth tracking) may take place in a typical radar system. Some knowledge of target angular location (the target azimuth) is necessary for pencil-beam tracking radar systems to point their typically narrow antenna beams in the direction of the target. This information, called designation data, may be provided by surveillance radar systems or some others. It may be sufficiently accurate to place the pencil beam on the target, or it may require the tracker to scan a larger region of uncertainty. The range-tracking portion of the radar has the advantage of seeing all targets within the beam from close range out to the maximum range of the radar. It typically breaks this range into small increments, each of which may be simultaneously examined for the presence of a target. When beam scanning is necessary, the target range tracker examines the increments simultaneously for short periods, such as 0.1 s, makes its decision about the presence of a target, and allows the beam to move to a new location if a "no" target is present. This process is typically continuous for mechanical-type trackers that move the beam slowly enough that a target will remain well within the beam for the short examination period of the range increments.

Target acquisition involves consideration of the S/N threshold and integration time needed to accomplish a given probability of detection P_D with a given false alarm rate P_F similar to surveillance radar system. However, the high false alarm rates P_F, as compared with values used for surveillance radar systems, are used because the operator knows that the target is present, and operator fatigue from false alarms when waiting for a target is not involved. Optimum false alarm rates P_F are selected on the basis of performance of electronic circuits that observe each range interval to determine which interval has the target echo.

A typical technique is to set a voltage threshold sufficiently high to prevent most noise peaks from crossing the threshold but sufficiently low that a weak signal may cross. An observation is made after each transmitter pulse as to whether, in the range interval being examined, the threshold has been crossed. The integration time allows the radar to make this observation several times before deciding if there is a "yes" target. The major difference between noise and target echo is that noise spikes exceeding the threshold are random, but if a target is present, that is, a "yes" target, the threshold crossings are more regular. One typical radar system simply counts the number of threshold crossings over the integration period, and if crossings occur for more than half the number of times that the radar has transmitted, a target is indicated as being present. If the radar pulse repetition frequency is 300 Hz and the integration time is 0.1 s, the radar will observe 30 threshold crossings if there is a strong and steady target. However, because the echo from a weak target combined with noise may not always cross the threshold, a limit may be set, such as 15 crossings, that must occur during the integration period for a decision that is a "yes" target. For example, performance on a nonscintillating target has a 90% probability of detection P_D at a 2.5 dB per-pulse SNR and the false alarm probability $P_F = 10^{-5}$.

Target range tracking: Once a target is acquired in range, it is desirable to follow the target in the range coordinate to provide distance information or slant range to the target. Appropriate timing pulses provide a target range gating so the angle-tracking circuits and automatic gain control (AGC) circuits look at only the short target range interval, or time interval, when the desired echo pulse is expected. The target range tracking operation is performed by closed-loop tracking similar to the azimuth tracker. Error in centering the range gate on the target echo pulse is sensed, error voltages are generated, and circuitry is provided to respond to the error voltage by causing the gate to move in a direction to recenter on the target echo pulse.

Automatic target range tracking can generally be divided into five steps shown in Figure 4.1 and detailed here:

- Radar detection acceptance: Accepting or rejecting detections for insertion into the tracking process. The purpose of this step is to control the false track rates.
- Association of accepted detections with the existing tracks.

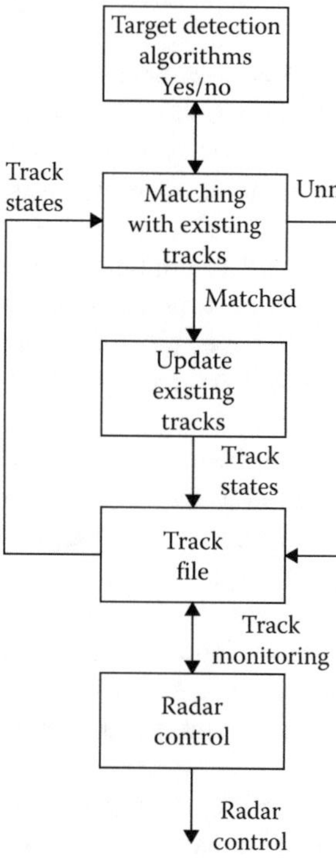

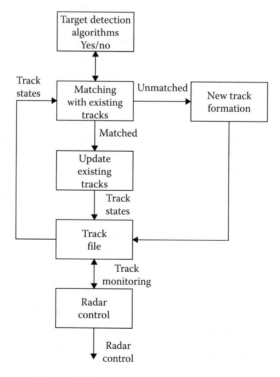

FIGURE 4.1 Target range tracking procedure.

- Updating the existing tracks with associated detections.
- New track formation using unassociated detections.
- Radar scheduling and control.

The result of the automatic target range tracking is the track file that contains a track state for each target detected by the radar system. As shown in Figure 4.1, there is a feedback loop between all these functions, so the ability to update existing tracks accurately, naturally, affects the ability to associate detections with existing tracks. Also, the ability to correctly associate detections with existing tracks affects the track's accuracy and the ability to correctly distinguish between an existing track and a new one. The detection accept/reject step makes use of feedback from the association function that measures the detection activity in different regions of the radar coverage. More stringent acceptance criteria are applied in more active regions.

When a track is established in the computer, it is assigned a track number. All parameters associated with a given track are referred to by this track number. Typical track parameters are the filtered and predicted position; velocity; acceleration (when applicable); time of last update; track quality; SNR; covariance matrices (the covariance contains the accuracy of all rank coordinates and all the statistical cross-correlations between them), if a Kalman-type filter is being used; and track history, that is, the last n detections. Tracks and detections can be accessed in various sectored, linked-list, and other data structures so that the association process can be performed efficiently [2–4]. In addition to the track file, a clutter file is maintained. A clutter number is assigned to each stationary or very slowly moving echo. All parameters associated with a clutter position are referred to by this clutter number. Again, each clutter number is assigned to a sector in azimuth for efficient association.

When the radar system has either no or limited coherent processing, not all the detections declared by the automatic detector are used in the tracking process. Rather, many of the detections (contacts) are filtered out in software using a process called activity control [5,6]. The basic idea is to use detection signal characteristics in connection with a map of the detection activity to reduce the rate of detections to one that is acceptable for forming tracks. The map is constructed by counting the unassociated detections (those that do not associate with existing tracks) at the point in the track processing shown in Figure 4.1.

4.1 MAIN STAGES AND SIGNAL REPROCESSING OPERATIONS

The signal reprocessing operations of each target are carried out in two stages [7,8]: the target range track detection and the target range tracking. Automatic target range track detection using data of 2D radar system in the Cartesian rectangular coordinates under uniform radar antenna scanning is shown in Figure 4.2. Let a *target pip* be at any point of uniform radar antenna scanning area that does not

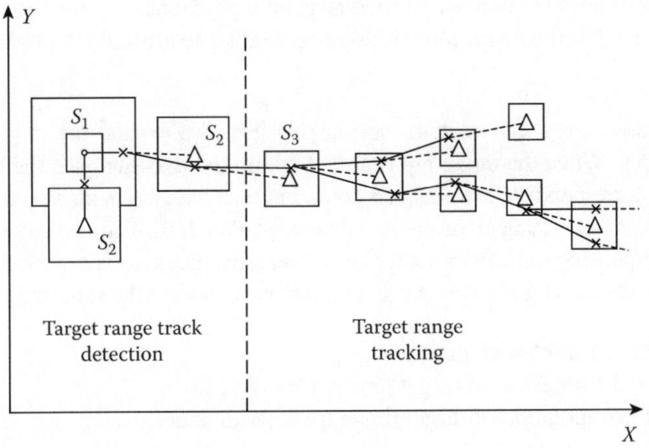

FIGURE 4.2 Target detection and target range tracking coordinate system.

correspond to existing target range tracks. This target pip is considered as a target range track datum of new target. If components by coordinate axes of minimal V_{min} and maximal V_{max} target velocities are known, then the area S_1 within the limits of which we should search the second target pip during next radar antenna scanning can be presented as an area between two rectangles. In doing so, the internal rectangle sides are defined as $2V_{X\min} T_{scan}$ and $2V_{Y\min} T_{scan}$; the external rectangle sides are defined as $2V_{X\max} T_{scan}$ and $2V_{Y\max} T_{scan}$, where T_{scan} is the radar antenna scanning period. The operation forming the area S_1 is called a gating and the area S_1 is called a gate with primary lock-in.

Several target pips may be in the gate with primary lock-in. Each target pip must be considered as one of the possible prolongations of existing target range tracks (see Figure 4.2). Using two target pips we can compute the velocity and direction of target moving and then determine a possible position of target pip for next (third) radar antenna scanning. The definition of initial parameters, namely, the velocity, target moving direction, and extrapolation of target pip position for the next radar antenna scanning, is realized by specific filtering (see Chapter 5). The extrapolated target pips are noted by triangles in Figure 4.2. Rectangle gates S_2 are formed around the extrapolated target pips. Dimensions of these rectangle gates S_2 are defined based on possible errors during extrapolation and determination of the target pip coordinates. If in the course of the third radar antenna scanning we can observe the definite target pip within the limits of the rectangle gate S_2, then we can think that this target pip belongs to the detected target range track. Taking into consideration the coordinates of this target pip we are able to obtain more specific information about target range track parameters and construct new rectangle gates. After performance of the earlier-given criterion by number of target pips that are within l consistently formed rectangle gates, we accept a detection and start a procedure of target range tracking. As follows from Figure 4.2, the detection is accepted based on three target pips following one after another—the criterion "3 from 3." Thus, in the course of target range track detection, the following operations are made:

- Gating and selection of target pip within the gate
- Checking the detection criterion
- Estimation and extrapolation of target range track parameters

Target range tracking is a sequential from measuring to measuring procedure of newly obtained target pip binding to target range track and accurate target range track determination. Under target range autotracking, the following operations are carried out:

- Accurate definition of target range track parameters in the course of newly obtained target pip binding
- Extrapolation of target range track parameters for next radar antenna scanning
- Range gating of possible newly obtained target pip positions
- Selection of the definite target pip within the gate if there are several target pips within the gate area

When there are several target pips within the tracking gate, we expand the target range track binding for each target pip. When the target pip is absent within the tracking gate, the target range track is prolongated by the corresponding extrapolated point, but the next gate is increased in area to take into consideration the errors increasing in the course of extrapolation. If there are no target pips during k radar antennas scanning following one after another, the target range tracking is canceled. Thus, at the stages of target range track detection and target range tracking, we carry out the same operations in reality:

- Radar antenna scanning area gating
- Selection and identification of target pips within the gate
- Filtering and extrapolation of target range track parameters

Consider algorithms of the first and second operations.

4.1.1 Target Pip Gating: Shape Selection and Dimensions of Gates

In accordance with the main principles of target range autotracking, a newly obtained target pip can be used to suggest a target range tracking if a target pip deviation from the gate center does not exceed a given before fixed value defined by gate dimensions or, in other words, if the following condition is satisfied:

$$\left|U_i - \hat{U}_i^{\text{center}}\right| \le 0.5\Delta U_i^{\text{gate}}, \tag{4.1}$$

where

$$U_i = \{r_i, \beta_i, \varepsilon_i\} \tag{4.2}$$

is a set of coordinates of the ith newly obtained target pip;

$$\hat{U}_i^{\text{center}} = \left\{\hat{r}_i^{\text{center}}, \hat{\beta}_i^{\text{center}}, \hat{\varepsilon}_i^{\text{center}}\right\} \tag{4.3}$$

is a set of the gate center coordinates for the ith target range track;

$$\Delta U_i^{\text{gate}} = \left\{\Delta r_i^{\text{gate}}, \Delta\beta_i^{\text{gate}}, \Delta\varepsilon_i^{\text{gate}}\right\} \tag{4.4}$$

are the gate dimensions by coordinates r, β, and ε for the ith target range track. One of the problems arising during the prolongation of the target range tracks by gating is a problem of selection of gate shape and dimensions based on known statistical characteristics of deviations of true (belonging to prolongated target range tracks) target pips from their corresponding extrapolated points. Deviation of the true target pip from the gate center is defined by total (random plus dynamic) errors of target coordinate extrapolation by previous smoothed parameters of target range track and coordinate measurement errors of newly obtained target pips. These errors are independent and identically distributed, subjected to the normal Gaussian pdf.

Let the target range track coordinate extrapolation for the next nth radar antenna scanning be carried out using the data of previous $(n-1)$th radar antennas scanning. Position of the extrapolated point is denoted by O (see Figure 4.3). We think that the origin of Cartesian coordinate system is located at the point O. The axis Y is matched with "radar system–target" direction, the axis X is a perpendicular to the "radar system–target" direction away from radar antenna rotation, and the axis Z is directed in such way that we have a right-handed coordinate system. Then the random deviations Δx, Δy, Δz of the newly obtained under nth radar antenna scanning target pip from the gate center are defined in the following form:

$$\begin{cases} \Delta x_n = \pm r\left(\Delta\beta_n^{\text{extr}} + \Delta\beta_n\right) \\ \Delta y_n = \pm\left(\Delta r_n^{\text{extr}} + \Delta r_n\right) \\ \Delta z_n = \pm r\left(\Delta\varepsilon_n^{\text{extr}} + \Delta\varepsilon_n\right) \end{cases} \tag{4.5}$$

where

Δr_n^{extr}, $\Delta\beta_n^{\text{extr}}$, and $\Delta\varepsilon_n^{\text{extr}}$ are the random errors of coordinate extrapolation in the course of the nth radar antenna scanning

Δr_n, $\Delta\beta_n$, and $\Delta\varepsilon_n$ are the random errors of coordinate measuring in the course of the nth radar antenna scanning

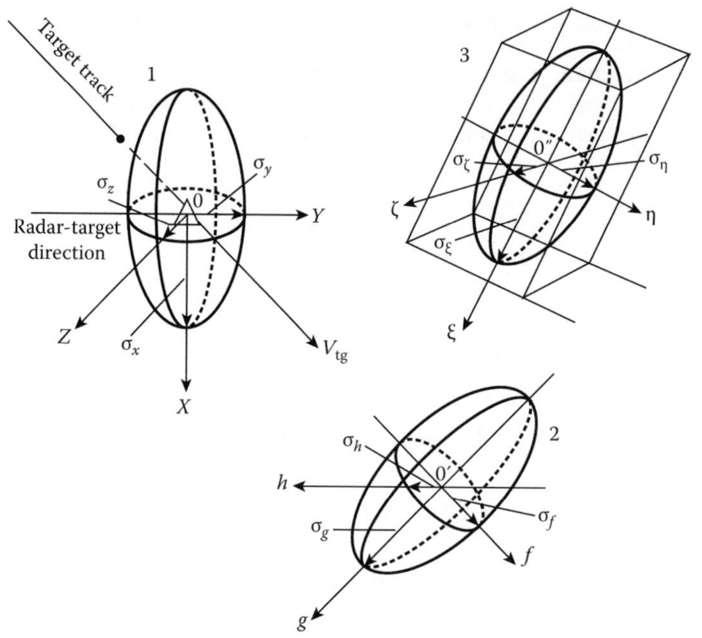

FIGURE 4.3 Extrapolation of target coordinates.

Under the definition of gate dimensions we can believe that the components Δr_n, $\Delta\beta_n$, and $\Delta\varepsilon_n$ are statistically independent, do not depend on the scanning number step n, and are subjected to the normal Gaussian pdf with zero mean and variances σ_x^2, σ_y^2, and σ_z^2, respectively. Consequently, the joint pdf can be defined in the following form:

$$f(\Delta x, \Delta y, \Delta z) = \frac{1}{\sqrt{(2\pi)^3\sigma_x^2\sigma_y^2\sigma_z^2}}\exp\left\{-\frac{1}{2}\left[\frac{(\Delta x)^2}{\sigma_x^2}+\frac{(\Delta y)^2}{\sigma_y^2}+\frac{(\Delta z)^2}{\sigma_z^2}\right]\right\}, \tag{4.6}$$

and the surface corresponding to equal pdf is defined by the following equation:

$$\frac{(\Delta x)^2}{\sigma_x^2}+\frac{(\Delta y)^2}{\sigma_y^2}+\frac{(\Delta z)^2}{\sigma_z^2} = \lambda^2, \tag{4.7}$$

where λ is the arbitrary constant. Dividing the left- and right-hand side of (4.7) by λ^2, we obtain

$$\frac{(\Delta x)^2}{\lambda^2\sigma_x^2}+\frac{(\Delta y)^2}{\lambda^2\sigma_y^2}+\frac{(\Delta z)^2}{\lambda^2\sigma_z^2} = 1. \tag{4.8}$$

Equation 4.8 is the ellipsoid equation with conjugate axles $\lambda\sigma_x$, $\lambda\sigma_y$, and $\lambda\sigma_z$. At $\lambda = 1$, we obtain the unit ellipsoid—see ellipsoid 1 in Figure 4.3.

Henceforth, we believe that dynamical errors of extrapolation caused by contingency target maneuver are also normally distributed and have independent components on the F, G, and H axes. The F axis is matched with the target velocity vector; the G axis is directed in opposition to a tangential acceleration; and the H axis supplements the system to the right-handed coordinate one. Origin of the obtained coordinate system, as well as in the case of the previous coordinate system, coincides with the extrapolated point O. For better visualization the origin is replaced at the point O' in Figure 4.3.

In 3D space the dynamic errors form an ellipsoid of equal probabilities, an equation of which takes the following form:

$$\frac{(\Delta f)^2}{\lambda^2\sigma_f^2}+\frac{(\Delta g)^2}{\lambda^2\sigma_g^2}+\frac{(\Delta h)^2}{\lambda^2\sigma_h^2}=1, \tag{4.9}$$

the ellipsoid 2 in Figure 4.3 at $\lambda = 1$. If we add ellipsoids 1 and 2, then ellipsoid 3 will be formed, and directions of conjugate axles (the directions of axes of the Cartesian coordinate system $O\eta\xi\zeta$ with respect to the axes of the Cartesian coordinate system $OXYZ$ and variances σ_η^2, σ_ξ^2, and σ_ζ^2 by these axles) are defined by summing rules in space of independent vectorial deviations, caused by random and dynamic errors. For better visualization, the origin of the coordinate system $O\eta\xi\zeta$ is replaced at the point O'' in Figure 4.3.

The joint pdf of random components $\Delta\eta$, $\Delta\xi$, and $\Delta\zeta$ takes the following form:

$$p(\Delta\eta,\Delta\xi,\Delta\zeta)=\frac{1}{\sqrt{(2\pi)^3\sigma_\eta^2\sigma_\xi^2\sigma_\zeta^2}}\exp(-0.5\lambda^2), \tag{4.10}$$

where

$$\lambda^2=\frac{(\Delta\eta)^2}{\sigma_\eta^2}+\frac{(\Delta\xi)^2}{\sigma_\xi^2}+\frac{(\Delta\zeta)^2}{\sigma_\zeta^2}. \tag{4.11}$$

Thus, a surface of equally probable deviation of true target pips from the gate center represents an ellipsoid, value and direction of conjugate axles of which relative to the direction "radar system–target" depend on coordinate measurement errors, target maneuver intensity, and vectorial direction of target moving. Under ellipsoid distribution of true target pip deviation from the gate center, it is evident that the gate must have an ellipsoid shape with the conjugate axles $\lambda\sigma_\eta$, $\lambda\sigma_\xi$, and $\lambda\sigma_\zeta$, where λ is the enlargement factor of the gate dimensions in comparison with the dimensions of unit ellipsoid to ensure the given before expectancy of hitting of true target pip within the gate.

The expectancy of random point hitting into the ellipsoid that is similar and has a similar disposition as ellipsoids of equal probabilities is defined in the following form:

$$P(\lambda)=2\left[\Phi_0(\lambda)-\frac{1}{\sqrt{2\pi}}\lambda\exp(-0.5\lambda^2)\right], \tag{4.12}$$

where $\Phi_0(\lambda)$ is the standard normal Gaussian pdf given by (3.110). At $\lambda \geq 3$, the probability $P(\lambda)$ is very close to unit. There is a need to choose just the values of λ, forming an ellipsoid gate.

In practice, forming the ellipsoidal gates is impossible both under physical and mathematical gating. By this reason, the best thing that can be done is to form a gate in the shape of parallelepiped defined around the total error ellipsoid, as shown in Figure 4.3 (3). Parallelepiped side dimensions are equal to $2\lambda\sigma_\eta$, $2\lambda\sigma_\xi$, and $2\lambda\sigma_\zeta$, and its volume is defined as

$$V_{\text{par}}=8\lambda^3\sigma_\eta\sigma_\xi\sigma_\zeta. \tag{4.13}$$

Taking into consideration the fact that the volume of total error ellipsoid is defined as

$$V_{\text{el}}=\frac{4}{3}\pi\lambda^3\sigma_\eta\sigma_\xi\sigma_\zeta, \tag{4.14}$$

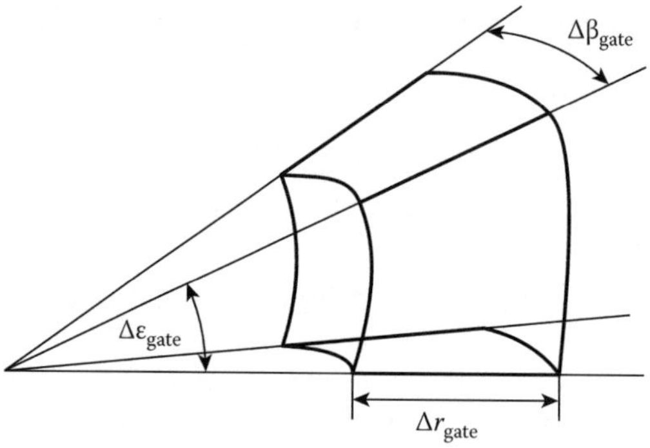

FIGURE 4.4 Definition of the simplest gate.

the gate volume is increased twice in comparison with the optimal volume. This phenomenon leads to an increase in the expectancy of hitting of false target pips within the gate or an increase in the number of target pips belonging to another target range tracks and, consequently, to deterioration of selection and resolution during gating operation.

During the processing of high number of targets in real time by the computer subsystem of a complex radar system (CRS), computation of dimensions and orientation of parallelepiped gate sides is very cumbersome. In this case, there is a need to use a simple version of gating, a sense of which is the following. The simplest gate shape is chosen for definition in that coordinate system, in which radar information is processed. In the case of spherical coordinate system, the simplest gate is given by linear dimension in target range Δr_{gate} and by two angular dimensions, namely, by the azimuth $\Delta\beta_{gate}$ and elevation $\Delta\varepsilon_{gate}$ (see Figure 4.4).

These dimensions can be preset based on maximal magnitudes of random and dynamic errors while processing all target range tracks. In short, in this case, the gate dimensions are chosen so that the ellipsoid of all possible (under all directions of target moving and flight) total deviations of true target pips from corresponding extrapolated points will be easily fit and rotated in any direction inside the gate. This is the crudest technique of gating. In conclusion, we note that all procedures, which are discussed in this subsection, to choose dimensions of 3D gate can be used in full during gating in plane for binding newly obtained target pips in 2D radar systems.

4.1.2 Algorithm of Target Pip Indication by Minimal Deviation from Gate Center

We consider a case of target pip indication under definition of target range track for a single target. At the same time, we assume that in addition to true target pips, the false target pips caused by noise and interferences passed over preprocessing filters can be present within the gate. Given the earlier-provided situation, the following varied decisions could be made:

- When there are several target pips, there is a need to continue to define the target range tracks using each target pip; in other words, we can suppose multiple target range tracks. Prolongation of target range tracks using false target pips will be canceled over several radar antennas scanning owing to absence of confirmations, but prolongation of the target range tracks using the true target pips will be continued. Such way of newly obtained target pip binding is very tedious. Moreover, when the density of false target pips is high, an avalanche-like increase in false target tracks leading to overload of memory devices in computer subsystems is possible.
- We must choose a single target pip within the gate, and we should be sure that the probability of the event that this target pip belongs to the target range tracking is the highest. Other target pips can be rejected as false. Such approach is appropriate with a view to decreasing the computation cost, but it requires a solution for the optimal target pip indication problem.

Optimization of the target pip indication process by deviation from the gate center is carried out using a criterion of maximal likelihood, on the basis of which we believe that the likelihood function is maximal only for the true target pip. Under target pip indication within 3D gate with facets that are parallel to the main axles of the total error ellipsoid (see Figure 4.3), a condition of maximal likelihood takes the following form:

$$\mathcal{L}\left(\Delta\eta_{i_{\text{true}}}, \Delta\xi_{i_{\text{true}}}, \Delta\zeta_{i_{\text{true}}}\right) = \max_{i}\left\{\mathcal{L}(\Delta\eta_i, \Delta\xi_i, \Delta\zeta_i)\right\}, \tag{4.15}$$

where i^{true} is the number of target pip considered as true, $i = 1, 2, \ldots, m$, m being the number of target pips within a gate. Condition given by (4.15) is equivalent to the condition

$$\lambda^2_{i_{\text{true}}} = \min_{i}\left[\frac{(\Delta\eta_i)^2}{\sigma^2_{\eta_i}} + \frac{(\Delta\xi_i)^2}{\sigma^2_{\xi_i}} + \frac{(\Delta\zeta_i)^2}{\sigma^2_{\zeta_i}}\right]. \tag{4.16}$$

Consequently, we should take into consideration only that target pip to prolong the target range track, the elliptical deviation of which from the gate center is minimal. A natural simplification of optimal operation discussed is indicated by the minimum of squared sums of linear deviations of the target pip from the gate center that corresponds to an assumption about the equality of the variances $\sigma^2_{\eta_i}$, $\sigma^2_{\xi_i}$, and $\sigma^2_{\zeta_i}$. If we assume that the target pip indication is carried out in a spherical coordinate system within the gate represented in Figure 4.4, then a criterion of target pip indication takes the following form:

$$\kappa^2_{i_{\text{true}}} = \min_{i}\left[\Delta r_i^2 + (r_i\Delta\beta_i)^2 + (r_i\Delta\varepsilon_i)^2\right]. \tag{4.17}$$

A quality of the target pip indication process can be evaluated by the probability of correct indication, in other words, the probability of event that the true target pip will be indicated to prolong the target range track in the next radar antennas scanning. The problem of definition of the probability of correct indication can be solved analytically if we assume that a presence of false target pips within the gate is only caused by stimulus of the noise and interference and these target pips are uniformly distributed within the limits of radar antenna scanning area. Under indication within 2D rectangle gate circumscribed around the ellipse with the parameter equal $\lambda_{\max} \geq 3$, the probability of correct indication is determined by [9,10]

$$P_{\text{ind}} = \frac{1}{1 + 2\pi\rho_S\sigma_\eta\sigma_\xi}, \tag{4.18}$$

where

ρ_S is the density of false target pips per unit square of gate

σ_η and σ_ξ are the mean square deviations of true target pips from the gate center along the axes η and ξ

In the case of target pip indication in 3D gate that can be presented in the form of parallelogram circumscribed around the total error ellipsoid (see Figure 4.3), the probability of correct indication is determined by

$$P_{\text{ind}} \approx 1 - \frac{21.33\pi\sigma_\eta\sigma_\xi\sigma_\zeta\rho_V}{\sqrt{2\pi}}, \tag{4.19}$$

where ρ_V is the density of false target pips per unit volume of gate.

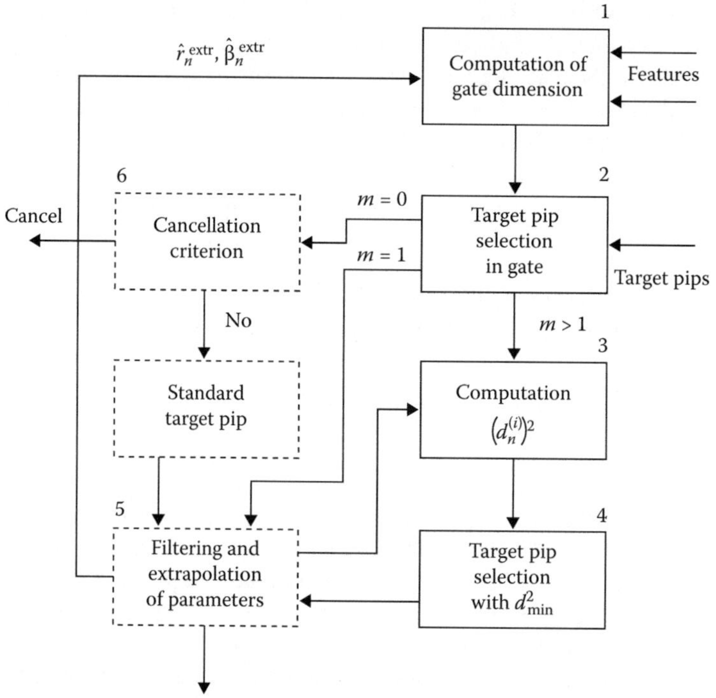

FIGURE 4.5 Algorithm of target pip indication within 2D gate.

The algorithm of target pip indication within 2D gate using a minimal linear deviation from the gate center and the polar coordinate system is represented in Figure 4.5. Blocks that are not related to the algorithm are shown by the dashed line. Sequence of operations in the considered algorithm is the following.

Step 1: Using signal processing results of the previous $(n-1)$th radar antenna scanning, we are able to define and compute the gate dimensions for the next nth radar antenna scanning. For choosing the gate dimensions, we should take into consideration the information about target maneuver and target pip misses.

Step 2: We select the target pips within the gate (block 2, Figure 4.5) using the following formulae:

$$\begin{cases} |\Delta r_n^{(i)}| = |r_n^{(i)} - \hat{r}_n^{(i)^{extr}}| \le 0.5\Delta r_n^{gate} \\ |\Delta\beta_n^{(i)}| = |\beta_n^{(i)} - \hat{\beta}_n^{(i)^{extr}}| \le 0.5\Delta\beta_n^{gate} \end{cases} \tag{4.20}$$

and determine their number. If there is a single target pip within the gate ($m = 1$), then this target pip is considered as true and comes in both at the filter input and at the input of the block of target range track parameters extrapolation (the block 5). If there are several target pips within the gate, then all of them come in at the input of computer block 3 that carries out computations of squared distances between the each target pip and the gate center by the following formula:

$$\left(d_n^{(i)}\right)^2 = \left(\Delta r_n^{(i)}\right)^2 + \left(r_n^{(i)}\Delta\beta_n^{(i)}\right)^2, \tag{4.21}$$

where $i = 1, 2, \ldots, m$, m being the number of indicated target pips within the gate.

Step 3: We compare the squared distances (the block 4) and choose the only target pip, for which

$$\left(d_n^{(i)}\right)^2 = d_{\min}^2. \tag{4.22}$$

Step 4: If there are no target pips within the gate, we check the criterion of cancellation of target range tracking (the block 6). If the criterion is satisfied, then the target range tracking is canceled. If the criterion of cancellation is not satisfied, then a command to continue to expand the target range tracking by extrapolation of target range track coordinates and target range track parameters is generated.

In conclusion, we can note that in addition to target pip deviations from the gate center, the weight features of target pips formed in the course of signal preprocessing, as some analog of SNR, can be used to indicate the target pips. In the simplest case of a binary quantized target return pulse train, we can use the number of train pulses or train bandwidth to generate the target pip weight features. Weight features of target pips can be used under target range track indication jointly with features of target pip deviations from the gate center or independently. One of the possible ways to use the target pip weight features jointly with the target pip deviations from the gate center is the following. All target pips within the gate are divided on the target pips with the weight ν_1 and target pips with the weight ν_0 depending on either the target return pulse train bandwidth exceeds the earlier-given threshold value or not, which depends on target range magnitude. If there are target pips with weight ν_1, we indicate the target pip nearest to the gate center as the true target pip from observed target pip group. If there are no target pips with weight ν_1, we indicate the target pip with weight ν_0 nearest to the gate center as the true target pip. If we can characterize the target pip weight by the number of pulses in the target return pulse train, then we are able to indicate the target range track by the maximum number of pulses. In this case, features of target pip deviations from the gate center are only used under the equality of weights of several target pips.

4.1.3 Target Pip Distribution and Binding within Overlapping Gates

In the hard noise environment, several gates formed around extrapolated points of prolonged target range tracks will overlap. Moreover, false and newly obtained target pips will appear within gates. In this case, the problem of newly obtained target pip distribution and binding to target range tracking becomes more complex. In addition, the problem of fixing a beginning of new target range track arises. First of all, this complication is associated with the necessity to use a joint target pip binding to target range track after getting newly obtained target pips liable to binding or, in other words, if these newly obtained target pips are within overlapping gates. Henceforth, there is a need to generate all possible ways or hypotheses for target pip binding and choose the most plausible version.

Consider one approach to solve the problem of target pip distribution and binding in overlapping gates under the following initial premises:

- The gates formed by each target range track have such dimensions that the probability of hitting of true target pips belonging to the given target range track is very close to unit.
- Groups of overlapping gates are selected in such way that each group can be processed individually.
- The problem is solved within 2D gates.

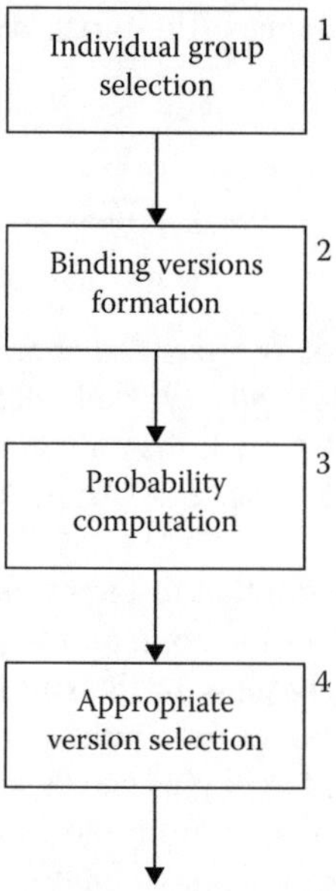

FIGURE 4.6 Target pip distribution and binding algorithms.

A logical flowchart of target pip distribution and binding algorithm within the isolated group of overlapping gates is presented in Figure 4.6. Block 1 selects the isolated groups with overlapping gates. Block 2 generates possible target pip binding versions taking into account information about the sources of newly obtained target pips forming a group. The following hypotheses are considered with respect to each target pip: The first hypothesis is that the target pip belongs to an earlier-detected target range track or to a new target range track, and the second hypothesis is that the target pip is false. Block 3 computes probabilities of all versions of the target pip binding. Block 4 defines an appropriate version of the target pip binding to begin a new target range track or to continue the earlier-obtained target range tracks.

Consider the simplest case of two overlapping gates shown in Figure 4.7a, as an example. We can see two extrapolated points of target range tracks 1 and 2 and three newly obtained target pips I, II, and III, one of which is within the overlapping gate region. Each newly obtained target pip is considered both as one belonging to the target range track 1 or 2 if this target pip is within the gate of corresponding target range track and as one belonging to a new target range track, namely, three

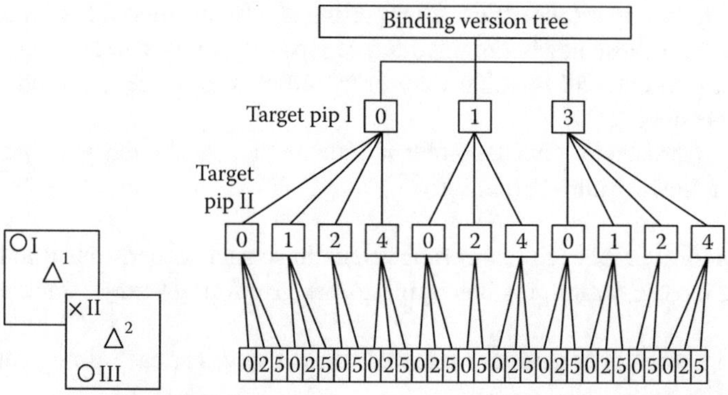

FIGURE 4.7 Possible versions of target pip binding: (a) two overlapping gates and (b) binding tree.

for target pip I, four for target pip II, and five for target pip III. In the other case, the target pip is considered false (the target range track 0). Possible versions of target pip binding forms a tree of binding hypotheses presented in Figure 4.7b. Each branch of this tree is considered as one of possible versions of newly obtained target pip distribution. For example, the first (left) branch presents a version when all newly obtained target pips are false. The last (right) branch of this tree is considered as a version when all newly obtained target pips belong to detected target range tracks with numbers 3, 4, and 5. For the considered case, the total number of versions is equal to 30.

The probabilities of events that the target pips belong to the existing target range tracks, new target range tracks, and false target range tracks are defined as follows:

1. The probability of event that the ith newly obtained target pip belongs to the observed before jth target range track, if, of course, the target pip is within the gate for this target range track, is defined by the probability of detection $P_{\text{tgp}_i}^{\text{true}}$ of a newly obtained target pip and by a distance between this newly obtained target pip and the extrapolation point, in other words, between the newly obtained target pip and the gate center of the jth target range track. The probability of target pip detection depends on the target range, the effective reflective surface of the target, and power of radar system. In the case of target range tracking, all these characteristics are known very well. For this reason, a determination of the probability of detection $P_{\text{tgp}_i}^{\text{true}}$ of a newly obtained target pip does not pose any difficulties. As before, we assume that a distance between the target pip belonging to the given target range track and the gate center along each coordinate of the Cartesian coordinate system is subjected to the normal Gaussian pdf with zero mean and the variance that can be determined in the following form:

$$\sigma_{\Sigma_{ij}}^2 = \sigma_{\text{extr}_j}^2 + \sigma_{\text{meas}_i}^2, \tag{4.23}$$

where

$\sigma_{\text{extr}_j}^2$ is the variance of errors under coordinate extrapolation of the jth target range track
$\sigma_{\text{meas}_i}^2$ is the variance of errors of coordinate measurement of the ith target pip

Hence, the probability of deviation of the ith target pip from the extrapolated point of the jth target range track, for example, by the coordinate x, is given by

$$P_{x_{ij}} = \frac{1}{\sqrt{2\pi\sigma_{\Sigma_{ij}}^2}} \exp\left[-\frac{\left(x_i - \hat{x}_{\text{extr}_i}\right)^2}{2\pi\sigma_{\Sigma_{ij}}^2}\right], \tag{4.24}$$

where

x_i is the measured target pip coordinate
$\hat{x}_{\text{extr}_i}$ is the extrapolated point coordinate

Thus, if the following condition

$$\sigma_{x_i}^2 = \sigma_{y_i}^2 \quad \text{and} \quad \left(\sigma_{x_i}^{\text{extr}}\right)^2 = \left(\sigma_{y_i}^{\text{extr}}\right)^2 \tag{4.25}$$

is satisfied, then the probability of event that the ith target pip belongs to the jth target range track is defined in the following form:

$$P_{ij}^{\text{true}} = P_{\text{tgp}_i}^{\text{true}} P_{x_{ij}} P_{y_{ij}}. \tag{4.26}$$

2. The probability of event that a newly obtained target pip is false can be determined in the following form:

$$P_i^{\text{false}} = P_{\text{tgp}\,i}^{\text{false}} L_{\text{gate}}, \tag{4.27}$$

where L_{gate} is the number of resolved elements within the gate.
3. The probability of event that the target pip belongs to a newly detected target is given by

$$P_{\text{new}_i} = P_{\text{tgp}_i}^{\text{true}} \rho_{\text{scan}} S_{\text{gate}}, \tag{4.28}$$

where

ρ_{scan} is the density of newly obtained target pips per unit area of scanning
S_{gate} is the gate area

After determination of probabilities of events that the target pips belong to the existing target range tracks, new target range tracks, and false target range tracks, we are able to determine the probabilities of all versions of the target pip binding and choose the version with the highest probability as a true version. Solution of this problem is very tedious even for the considered simplest case. This problem can be simplified if we will follow the evident practical rules:

1. Each target range track must be attached by the target pip.
2. If the target pip is not associated to the given before target range track, then a new target range track must be started independently of the probability of event that this target pip belongs to the new target range track or false target range track. In this case, we should compare only versions when there are the target pip bindings for each target range track. In the considered case, there are two such versions.
   a. The first target pip corresponds to the first target range track, and the second target pip corresponds to the second target range track.
   b. The second target pip corresponds to the first target range track, and the third target pip corresponds to the second target range track.

For this case, the third target pip at the first version and the first target pip at the second version are considered as newly obtained target pips and allow us to start a new target range track.

4.2 TARGET RANGE TRACK DETECTION USING SURVEILLANCE RADAR DATA

4.2.1 Main Operations under Target Range Track Detection

In accordance with general principles discussed in Section 4.1, the detection process of new target range tracking is started with a formation of an initial gate with primary lock-in around a single target pip. Dimensions of initial gate are selected based on the possible target moving over the radar antenna scanning period. If in the course of next radar antenna scanning period one or several target pips appear within the gate with primary lock-in, the new target range track is started using each target pip. If there are no target pips within the gate with primary lock-in, the initial target pip is canceled as false (the criterion of beginning a new target range track is "2/m") or is reserved for confirmation during the next radar antenna scanning. In doing so, dimensions of the gate with primary lock-in will be increased (fractional criteria "2/m," where $m > 2$). After target range track is begun, the target moving direction and velocity is defined. This allows us to extrapolate and gate the target position for the next radar antenna scanning. If the newly obtained target pips appear within these gates with primary lock-in, we can make a final decision about target range track detection.

Thus, the process of target range track detection is divided into two stages:

- *The first stage*: the target range track detection by criterion "2/*m*."
- *The second stage*: the confirmation of beginning the target range track, that is, the final target range track detection by criterion "*l*/*n*." In particular cases of the second stage, the target range track detection may be absent.

Algorithm of the target range track detection by criterion "2/*m*" jointly with the confirmation algorithm or final target range track detection by criterion "*l*/*n*" forms a united target range track detection algorithm by criterion "2/*m* + *l*/*n*."

The main computer subsystem operations carried out under target range track detection are as follows:

- Evaluation of the target velocity in the course of moving
- Extrapolation of coordinates
- Gating the target pips

With respect to these operations, in future we assume the following prerequisites:

- Coordinate extrapolation is carried out in accordance with a hypothesis of uniform and straight-line target moving.
- Gates at all stages of target range track detection have a shape of spherical element (see Figure 4.4). Gate dimensions Δr_{gate}, $\Delta\beta_{\text{gate}}$, $\Delta\varepsilon_{\text{gate}}$ are defined based on total errors of measurement and coordinate extrapolation at the corresponding target range track detection stage.
- Radar system range resolution by corresponding coordinates is considered as the unit volume element of gate. In this case, the gate dimensions do not depend on target range, and moreover, a distribution of false target pips within radar antenna scanning area can be considered as uniform because the probability of appearance of false target pip within each elementary gate volume is the same.

Target range track detection quality can be evaluated by the following characteristics:

- Probability of detection of true target range track
- Average number of false target range tracking per unit time
- Speed of computer subsystems used by CRSs to realize the target range track detection algorithm

Henceforth, we discuss the main results of analysis of target range track detection algorithm applying to surveillance radar systems with omnidirectional or sector radar antenna scanning.

4.2.2 Statistical Analysis of "2/*m* + 1/*n*" Algorithms under False Target Range Track Detection

First, we consider the simplest algorithm realizing the criterion "2/*m* + 1/*n*." Sequence of operations employing this algorithm under the false target range track detection can be presented using a graph with random transitions (see Figure 4.8). Beginning of algorithm functioning coincides with transition from state a_0 to state a_1 when a single target pip considered as the starting point of detected target range track appears. Henceforth, we check the appearance of newly obtained target pips within the gate V_i with primary lock-in, where $i = 1, 2,\ldots, m-1$. If even a single target pip appears within one from $(m-1)$ gates with primary lock-in, the graph passes into the state a_m corresponding to making decision about the detection of target range track beginning. Otherwise, the graph passes from the state a_{m-1} to the state a_0 (the initial point of assumed target range track is canceled as false point). After getting the second target pip by criterion "2/*m*," the second stage of target range track detection is started. At this stage, we check the presence of newly obtained target pips within gates V_{m+j}, where

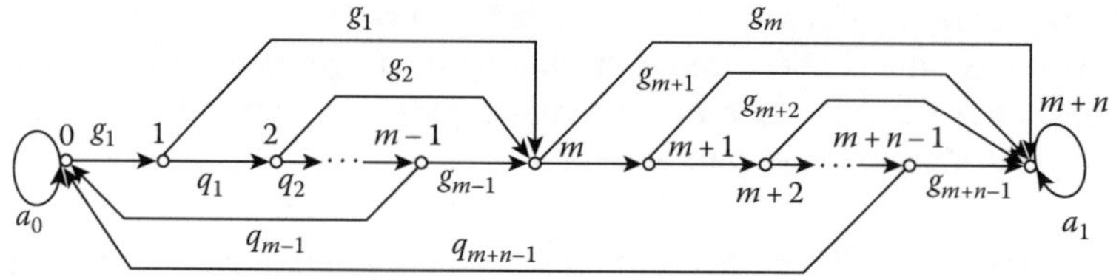

FIGURE 4.8 False target range track detection algorithm by criterion "2/*m* + 1/*n*."

$j = 1, 2, \ldots, n - 1$. Gate center position for the next radar antenna scanning is defined by coordinate extrapolation using two target pips. The dimension of gates is defined based on the measurement and coordinate extrapolation errors. If a newly obtained target pip appears within one of the n gates of the second stage, the graph passes into the state a_{m+n} that corresponds to making a final decision about the target range track detection. Otherwise, the graph passes into the state a_0 from the state a_{m+n-1} and a beginning of new target range track is canceled as false one. Volume of gates with primary lock-in defined by the number of elementary or resolved volumes is determined in the following form [11–13]:

$$V_i^{\mathrm{I}} = V_1 i^3,$$

where

$$V_1 = \frac{8 v_{r_{\max}} v_{\beta_{\max}} v_{\varepsilon_{\max}}}{\delta_r \delta_\beta \delta_\varepsilon} T_0^3. \tag{4.29}$$

The volume of gates at the second stage of target range track detection is determined by

$$V_j^{\mathrm{II}} = 8k^3 \frac{\sigma_{\Sigma_{jr}} \sigma_{\Sigma_{j\beta}} \sigma_{\Sigma_{j\varepsilon}}}{\delta_r \delta_\beta \delta_\varepsilon}, \tag{4.30}$$

where $k \approx 3$ is the coefficient of gate dimension increasing in comparison with magnitudes of the total mean square deviations of target pips from the gate center. There is a need to emphasize that the gate with definite dimensions corresponds to each graph state from 1 to $(m + n - 1)$.

To define the probability of false target range track detection, first of all, there is a need to define the conditional probability p_1 of event that a random point starting its movement from state a_1 can reach anywhere in state a_{m+n} and will stay at this state. This probability can be defined as solving a system of linear equations that can be found based on the graph presented in Figure 4.8:

$$\begin{cases} p_1 = q_{\mathrm{f}_1} p_2 + g_{\mathrm{f}_1} p_m, \\ p_2 = q_{\mathrm{f}_2} p_3 + g_{\mathrm{f}_2} p_m, \\ \ldots\ldots\ldots\ldots\ldots\ldots\ldots, \\ p_{m-1} = g_{\mathrm{f}_{m-1}} p_m, \\ p_m = q_{\mathrm{f}_m} p_{m+1} + g_{\mathrm{f}_m}, \\ p_{m+1} = q_{\mathrm{f}_{m+1}} p_{m+2} + g_{\mathrm{f}_{m+1}}, \\ \ldots\ldots\ldots\ldots\ldots\ldots\ldots, \\ p_{m+n-1} = g_{\mathrm{f}_{m+n-1}}, \end{cases} \tag{4.31}$$

where $g_{f_i}, i = 1,2,\ldots,m-1$, is the probability of appearance of the false target pips within the gates with primary lock-in:

$$q_{f_i} = 1 - g_{f_i}; \tag{4.32a}$$

$g_{f_j}, j = m, m+1,\ldots,m+n-1$, is the probability of appearance of the false target pips within the confirmation gates:

$$q_{f_j} = 1 - g_{f_j}. \tag{4.32b}$$

Solution of the linear equation system for arbitrary *m* and *n* is given by

$$p_1 = \left(1 - \prod_{i=1}^{m-1} q_{f_i}\right)\left(1 - \prod_{j=m}^{m+n-1} q_{f_j}\right). \tag{4.33}$$

Equation 4.33 allows us to define the conditional probability of detection of the false target range track. This probability can be presented in the following form:

$$p_1 = p_{beg} \times p_{conf}, \tag{4.34}$$

where
p_{beg} is the conditional probability of new target range track beginning
p_{conf} is the conditional probability of confirmation of new target range track beginning

The unconditional probability of detection of the false target range track is given by

$$P_{f_{tr}} = P_1 \times p_1, \tag{4.35}$$

where P_1 is the probability of a single false target pip appearance that can be considered as the new target range track beginning. If we consider a general case for the confirmation criterion "*l/m*," it is impossible to obtain the probability p_{conf} in a general form. For this reason, these criteria are analyzed individually.

Filtering ability of the target range track detection algorithm can be characterized by average number $\bar{N}_{f_{tr}}$ of false target range tracks that will be analyzed by radar target trackers within the radar antenna scanning period. The number $\bar{N}_{f_{tr}}$ is related to the probability p_1, which can be presented by the following formula:

$$\bar{N}_{f_{tr}} = p_1 \times \bar{N}_1, \tag{4.36}$$

where $\bar{N}_1$ is the average number of individual target pips within the radar antenna scanning that are considered as initial points of false target range tracks in steady working state of the CRS. Thus, to define $\bar{N}_{f_{tr}}$, first of all, there is a need to obtain the formula to determine the average number $\bar{N}_1$. This formula can be obtained in the case of the simplest algorithm realizing the criterion "2/*m* + 1/*n*."

The number of individual target pips becoming initial points of newly obtained false target range tracks after the $(r + 1)$th radar antenna scanning can be determined by the following formula:

$$N_1(r+1) = N_f(r+1) - \sum_{j=1}^{m+n-1} g_{f_j} N_j(r), \tag{4.37}$$

where

$j = 1, 2, \ldots, m + n - 1$

$N_f(r + 1)$ is the number of false target pips that is reprocessed by computer subsystem within the $(r + 1)$th radar antenna scanning

g_{f_j} is the probability of appearance of false target pips within the gate of volume V_j; the number of such gates is equal to $m + n - 1$

$N_j(r)$ is the number of gates with the volume V_j formed within the rth radar antenna scanning

$\sum_{j=1}^{m+n-1} g_{f_j} N_j(r)$ is the number of false target pips of the current radar antenna scanning appearing within the gates of all false target range tracks that are under detection process, but a condition of no overlapping gates must be satisfied

In turn,

$$N_j(r) = \sum_{i=0}^{r} N_1(i) P_{1j}^{(r-i)}, \tag{4.38}$$

where

$N_1(i)$ is the number of initial points of the false target range tracks formed within the ith radar antenna scanning

$P_{1j}^{(r-i)}$ is the probability of system transition (see Figure 4.8) from the initial state a_1 to the state a_j for $(r - j)$ steps

Using (4.37) we should take into consideration that

$$N_1(0) = N_f(0) \quad \text{and} \quad P_{1j}^{(0)} = \begin{cases} 1 & \text{at} \quad j = 1, \\ 0 & \text{at} \quad j > 1. \end{cases} \tag{4.39}$$

Introduce a new variable $s = r - i$ in (4.39) that corresponds to the relocation of the origin at the instant of finishing the ith radar antenna scanning. The variable s represents the number of steps (the radar antenna scanning) necessary for the transition of target range track detection algorithm graph from the initial state a_1 to the state a_j. In the case of target range track detection algorithm with criterion "$2/m + 1/n$," the maximum value of s corresponds to the maximum number of steps needed for transition from the initial state a_1 to the state a_{m+n-1} and is equal to

$$s_{\max} = m + n - 2. \tag{4.40}$$

It is easy to verify this condition if we can define the longest branch in the algorithm graph presented in Figure 4.8. Taking into consideration all transformations discussed earlier, we obtain

$$N_j(r) = \sum_{s=0}^{m+n-2} N_1(r-s) P_{1j}^{(s)}. \tag{4.41}$$

Analyzing (4.41), we see that the terms for which the following condition $P_{1j}^{(s)} \neq 0$ is satisfied are not equal to zero. Furthermore, we are going to determine the probabilities of transitions $P_{1j}^{(s)}$. System of recurrent equations to determine the probabilities $P_{1j}^{(s)}$ takes the following form:

$$\begin{cases} P_{11}^{(s)} = \begin{cases} 1 & \text{at} \quad s = 0, \\ 0 & \text{at} \quad s > 0, \end{cases} \\ P_{12}^{(s)} = P_{11}^{(s-1)} q_{f_1}, \\ P_{13}^{(s)} = P_{12}^{(s-1)} q_{f_2}, \\ \ldots\ldots\ldots\ldots\ldots\ldots, \\ P_{1m}^{(s)} = P_{11}^{(s-1)} g_{f_1} + P_{12}^{(s-1)} g_{f_2} + \cdots + P_{1,m-1}^{(s-1)} g_{f_{m-1}} = \sum_{i=1}^{m-1} P_{1i}^{(s-1)} q_{f_i}, \\ P_{1,m-1}^{(s)} = P_{1,m}^{(s-1)} q_{f_m}, \\ \ldots\ldots\ldots\ldots\ldots\ldots, \\ P_{1,m+n-1}^{(s)} = P_{1,m+n-2}^{(s-1)} q_{f_{m+n-2}}. \end{cases} \tag{4.42}$$

Taking into consideration (4.41), we can present (4.37) in the following form:

$$N_1(r+1) = N_f(r+1) - \sum_{j=1}^{m+n-1} g_{f_j} \sum_{s=0}^{m+n-2} N_1(r-s) P_{1j}^{(s)}. \tag{4.43}$$

In steady working state ($r \to \infty$) we can assume

$$N_1(r+1) = N_1(r) = N_1(r-1) = \cdots = N_1[r-(m+n-2)] = \bar{N}_1. \tag{4.44}$$

Then we obtain

$$N_f(r+1) = N_f(r) = \cdots = N_f\left[r-(m+n-2)\right] = \bar{N}_f; \tag{4.45}$$

$$\bar{N}_1 = \bar{N}_f - \bar{N}_1 \sum_{j=1}^{m+n-1} g_{f_j} \sum_{s=0}^{m+n-2} P_{1j}^{(s)} \tag{4.46}$$

and, finally we obtain

$$\bar{N}_1 = \frac{\bar{N}_f}{1 + \sum_{j=1}^{m+n-1} g_{f_j} \sum_{s=0}^{m+n-2} P_{1j}^{(s)}}. \tag{4.47}$$

Formula (4.47) combined with linear equation system (4.42) allows us to determine the average number of starting points of false target range tracks per radar antenna scanning in the steady working state of radar system. Knowing $\bar{N}_1$ and the conditional probability of lock-in p_1 and using (4.36), we can define a function between the average number of false target range tracking $\bar{N}_{f_{tr}}$ and the average number of false target pips N_f. Equation 4.47 is satisfied for target range track detection algorithms when the confirmation criterion is differed from "1/*n*." It is necessary only to make the information with respect to upper limits of summing by j and s more exact. The upper limit of summing by s is defined by the total number of gates formed in the course of realization of target range track detection algorithms. In the case of target range track detection algorithm with the confirmation criterion "*l*/*n*" ($l > 1$), this number is greater than $m + n - 1$. For this reason, in general the upper limit of summing by s will be defined by the maximum number of steps that can be made in the course of transition from state a_1 to the state that is previously relative to the accepted state. It is easy to show that this number does not depend on l if we use the criterion "*l*/*n*" and always equals $m + n - 2$. Thus, in a general form, the formula defining the number of individual target pips considered as the false target range tracks beginning under the use of criterion "2/*m* + *l*/*n*" is as follows:

$$\bar{N}_1 = \frac{\bar{N}_{\text{f}}}{1 + \sum_{j=1}^{m+\nu-1} g_{\text{f}_j} \sum_{s=0}^{m+n-2} P_{1j}^{(s)}}, \tag{4.48}$$

where ν is the number defined from target range track detection algorithm graph.

Results of determination of the average number $\bar{N}_{\text{f}_{\text{tr}}}$ of target range tracks that are needed to be tracked as a function of the average number of false target pips within the radar antenna scanning that are subjected to signal reprocessing by computer subsystems are presented in Figure 4.9, with the purpose of comparing the filtering ability. Analyzing results presented in Figure 4.9, we can conclude as follows:

- Filtering ability of target range track detection algorithms realizing the criterion "2/*m* + *l*/*n*" increases with a decrease in m and n and with an increase in l.
- Incrementing l on unit leads to a great increase in the filtering ability, in comparison with a corresponding decrease in m and n.

These features must be taken into consideration while choosing the target range track detection algorithm implemented in practice. If the computer subsystem of a CRS has limitations in the number of false target range tracks that are to be tracked, then joint choice of the average number $\bar{N}_{\text{f}_{\text{tr}}}$ and criteria of target range track detection algorithms during production tests allows us to raise a demand to the noise and interference level at the reprocessing computer subsystem input, that is, SNR.

While designing the reprocessing of the CRS computer subsystem, there is a need to have information about the average number of false target range tracks that are under detection in the steady

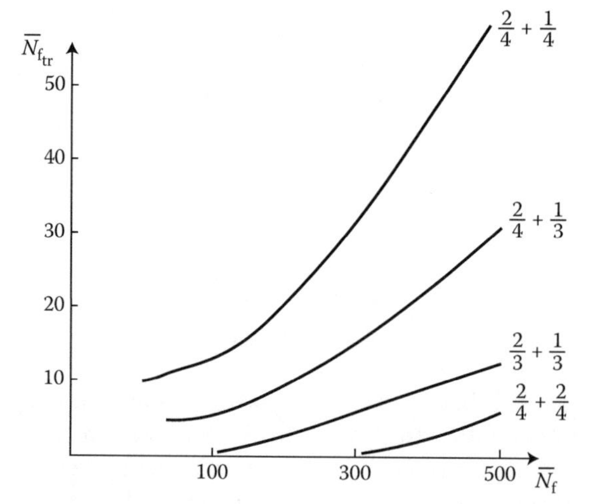

FIGURE 4.9 Comparison of filtering ability.

working state of a CRS. Denote this number by $N_{\text{ftr}}^{\text{d}}$. It is evident that all false target range tracks, with respect to which the final decision about tracking or cancellation has been not made, are under the detection process. The number of such target range tracks in the steady working state of the CRS is determined by

$$N_{\text{ftr}}^{\text{d}} = \sum_{j=1}^{m+\nu-1} \bar{N}_j, \tag{4.49}$$

where in accordance with (4.41) in the steady working state of the CRS we have

$$\bar{N}_j = \bar{N}_1 \sum_{s=0}^{m+n-2} P_{1j}^{(s)}. \tag{4.50}$$

Substituting (4.49) in (4.50), we obtain

$$N_{\text{ftr}}^{\text{d}} = \bar{N}_1 \sum_{j=1}^{m+\nu-1} \sum_{s=0}^{m+n-2} P_{1j}^{(s)}, \tag{4.51}$$

or taking into consideration (4.48), we obtain finally

$$N_{\text{ftr}}^{\text{d}} = \frac{\bar{N}_{\text{f}} \sum_{j=1}^{m+\nu-1} \sum_{s=0}^{m+n-2} P_{1j}^{(s)}}{1 + \sum_{j=1}^{m+\nu-1} g_{\text{f}_j} \sum_{s=0}^{m+n-2} P_{1j}^{(s)}}. \tag{4.52}$$

4.2.3 Statistical Analysis of "2/*m* + 1/*n*" Algorithms under True Target Range Track Detection

First, we consider the simplest algorithm realizing the criterion "2/*m* + 1/*n*." Functioning of the true target range track detection algorithm by the criterion "2/*m* + 1/*n*" is explained using Figure 4.10. A great feature of the true target range track detection algorithm graph in comparison with the false target range track detection algorithm graph shown in Figure 4.8 is the presence of nonzero probabilities of transitions to the initial state a_0 from any intermediate state a_j, where $j = 1, 2,\ldots, m + n - 1$. The probability of replacement from the intermediate state a_i, where $i = 1, 2,\ldots, m - 1$, to the initial state a_0 is equal to the probability of two simultaneous events. The first event is that the true target pip does not appear in the corresponding gate with primary lock-in; the second event is that at least one false target pip appears within the gate with primary lock-in, that is,

$$p_{i0}(r) = \left[1 - p_{\text{tr}}(r)\right] g_{\text{f}_i}, \tag{4.53}$$

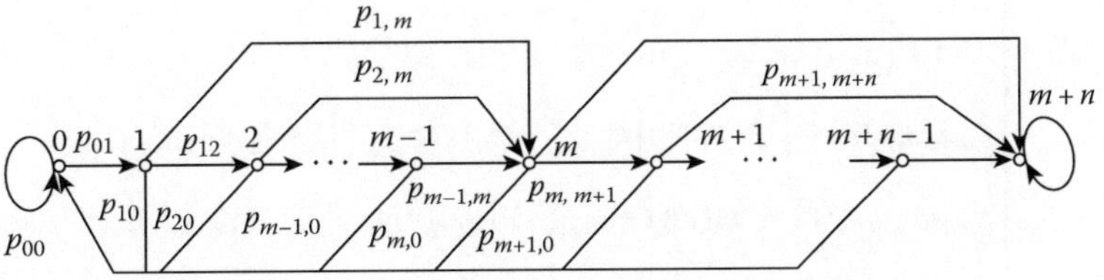

FIGURE 4.10 True target track detection algorithm graph by criterion "2/*m* + 1/*n*."

where

$p_{tr}(r)$ is the probability of true target pip detection in the course of the rth radar antenna scanning, which is independent of the gate volume V_i

g_{f_i} is the probability of detection at least one false target pip within the gate volume V_i

The probability of replacement to the initial state a_0 from any intermediate state a_j, where $j = m, m + 1,\ldots, m + n - 1$, is equal to the sum of the probabilities of two inconsistent events A_r and B_r:

$$p_{i0}(r) = P_j(A_r) + P_j(B_r), \tag{4.54}$$

where $j = m, m + 1,\ldots, m + n - 1$;

$$P_j(A_r) = \left[1 - p_{tr}(r)\right] g_{f_i}; \tag{4.55}$$

$$P_j(B_r) = p_{tr}(r) g_{f_i} \left[1 - P_{ind}(V_j)\right]; \tag{4.56}$$

$P_{ind}(V_j)$ is the probability of indication of true target pip among false target pips within the gate volume V_j.

Thus, based on the general principles of target range track detection algorithm functioning and taking into consideration (4.53) through (4.56), we can define the probabilities of transitions $p_{ij}(r)$ for the algorithm graph shown in Figure 4.10 or elements of corresponding transient probability matrix. These probabilities are determined as follows:

$$\begin{cases} p_{00}(r) = 1 - p_{tr}(r); \\ p_{01}(r) = p_{tr}(r); \\ p_{10}(r) = g_{f_i}\left[1 - p_{tr}(r)\right]; \\ p_{12}(r) = (1 - g_{f_i})\left[1 - p_{tr}(r)\right]; \\ p_{1m}(r) = p_{tr}(r); \\ \ldots\ldots\ldots\ldots\ldots\ldots \\ p_{m-1,0}(r) = 1 - p_{tr}(r); \\ p_{m-1,m}(r) = p_{tr}(r); \\ \ldots\ldots\ldots\ldots\ldots\ldots \\ p_{m0}(r) = \left[1 - p_{tr}(r)\right] g_{f_m} + p_{tr}(r) g_{f_m}\left[1 - P_{ind}(V_m)\right]; \\ p_{m,m+1}(r) = \left[1 - p_{tr}(r)\right](1 - g_{f_i}); \\ p_{m,m+n}(r) = p_{tr}(r)(1 - g_{f_i}) + p_{tr}(r) g_{f_m} P_{ind}(V_m); \\ \ldots\ldots\ldots\ldots\ldots\ldots \\ p_{m+n-1,0}(r) = \left[1 - p_{tr}(r)\right] g_{f_{m+n-1}} + p_{tr}(r) g_{f_{m+n-1}}\left[1 - P_{ind}(V_{m+n-1})\right]; \\ p_{m+n-1,m+n}(r) = p_{tr}(r)(1 - g_{f_{m+n-1}}) + p_{tr}(r) g_{f_{m+n-1}} P_{ind}(V_{m+n-1}); \\ p_{m+n,m+n}(r) = 1. \end{cases} \tag{4.57}$$

We would like to stress again that the probability of the target return signal detection for the formulas of probabilities of transition given by (4.57) is determined by the following [14–17]:

$$\mathbf{P}(r) = \left\| P_{0(r)} P_1(r) \ldots P_{m+n}(r) \right\| = \mathbf{P}(r-1)\mathbf{\Pi}(r), \tag{4.58}$$

where

$\mathbf{P}(r-1)$ is the row vector of probabilities of states at the previous $(r-1)$th radar antenna scanning step

$\mathbf{\Pi}(r)$ is the matrix of transient probabilities at the rth radar antenna scanning step

In accordance with (4.58) and taking into consideration (4.57), these equations define the elements of the transient probability matrix $\mathbf{\Pi}(r)$, and the system of recurrent equations to define components of the vector $\mathbf{P}(r)$ can be presented in the following form:

$$\begin{cases} P_0(r) = P_0(r-1)p_{00}(r) + P_1(r-1)p_{10}(r) + \cdots + P_{m+n-1}(r)p_{m+n-1,0}(r) = \sum_{j=1}^{m+n-1} P_j(r-1)p_{j0}(r); \\ P_1(r) = P_0(r-1)p_{01}(r); \\ \ldots\ldots\ldots\ldots\ldots\ldots \\ P_m(r) = P_1(r-1)p_{1,m}(r) + P_2(r-1)p_{2,m}(r) + \cdots + P_{m-1}(r-1)p_{m-1,m}(r) = \sum_{j=1}^{m+n-1} P_j(r-1)p_{j,m}(r); \\ P_{m+1}(r) = P_m(r-1)p_{m,m+1}(r); \\ \ldots\ldots\ldots\ldots\ldots\ldots \\ P_{m+n}(r) = \sum_{j=m}^{m+n-1} P_j(r-1)p_{j,m-n}(r) + P_{m+n}(r-1) = P_{\text{true}}(r). \end{cases} \tag{4.59}$$

The last line in (4.59) defines the total probability of true target range track detection at the rth radar antenna scanning increasing from scanning to scanning.

If we consider a general case of the criterion "$2/m + 1/n$," it is impossible to obtain the formula for increasing the probability of true target range track detection at an arbitrary value of l if $l > 1$. There is a need to carry out analysis for each individual target range track detection algorithm [18–20]. To compare various criteria of the type "$2/m + l/n$" used under the true target range track detection, the probability of detection as a function of normalized target range $d_r/d_{\max}$ determined based on discussed procedure is presented in Figure 4.11. In accordance with adopted notations, $d_{\max}$ is the maximal horizontal radar range; d_r is the current horizontal radar range:

$$d_r = d_{\max} - r\Delta d(T_{sc}) \tag{4.60}$$

where $d(T_{sc})$ is the variation of radar range coordinate within the limits of the radar antenna scanning period T_{sc}. The probability of target detection at the rth radar antenna scanning is determined by the formula

$$P_{tg}(r) = \exp\left[-\frac{0.68 d_r^4}{d_{\max}^4}\right]. \tag{4.61}$$

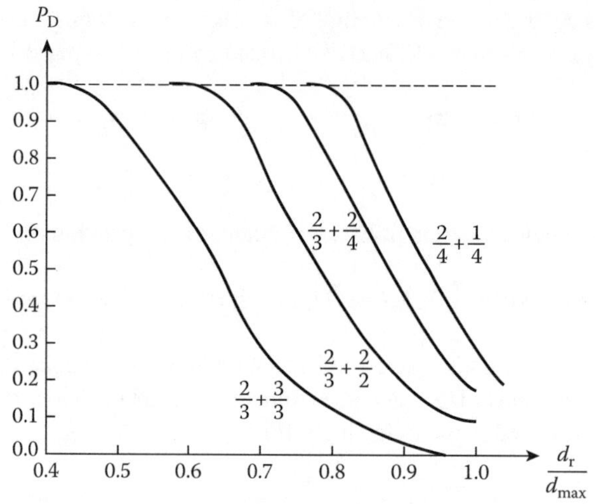

FIGURE 4.11 Probabilities of true target range track detection for various criteria "$2/m + l/n$."

The probability of true indication of target pips within confirmation gates is considered constant and equal to $P_{ind} = 0.95$. Density of false target pips per unit volume of radar antenna scanning area is equal to 10^{-4}.

Analysis and comparison of performance shown in Figure 4.11 allow us to conclude that it is worthwhile to employ the criterion "$2/m + 1/n$" based on the viewpoint of decreasing the number of steps under detection of true target range track. In doing so, small deviations in m and n do not lead to great changes in the number of steps to ensure the probability of true target range track detection close to the unit. The probability of true target range track detection equal to 0.95 using the discussed criteria is obtained for target range corresponding to 0.75 of the maximal radar system range. Implementation of confirmation criteria "l/n" ($l > 1$) leads to the prolongation of time to detect the true target range tracks.

4.3 TARGET RANGE TRACKING USING SURVEILLANCE RADAR DATA

If the earlier-given criterion of target detection is satisfied, the signal processing of the target return signal is finished at the stage of target range track detection. After that initial parameters of detected target range track are determined, and this target range track will be under *autotracking*. Henceforth, the target autotracking is considered as automatic target range track prolongation of target moving and accurate definition of prolongated target range track parameters. Thus, the terms "target autotracking" and "autotracking" (or "tracking") must be understood in the same sense. Furthermore, we prefer to use the term "target range autotracking" or "target autotracking."

4.3.1 Target Range Autotracking Algorithm

In the course of each target range tracking, two main problems are solved: gating and selection of newly obtained target pips for prolongation of target range track, estimation of target range track parameters, and definition of variation of these parameters as a function of time. In principle, solution of both problems can be realized employing the same detection algorithm. In this case, a required quality of the solved target range track estimation parameter problem must be in agreement with the customer. However, there is such a target range autotracking system in which the autotracking algorithm hunts only for target range track. To achieve a high-quality estimate of the target range track parameters, the individual target range track detection algorithm must be designed. Henceforth, we term this detection algorithm as target range track detection algorithm computations. Expediency to design the target range track detection algorithm computations arises from the following principles:

- Operations of estimation and extrapolation of target range track parameters must be carried out in the radar coordinate system to ensure the continuity of target range autotracking as soon as the input information is changed. There are no hard restrictions over the accuracy of these operations that allow us to use simple formulas for computations based on a hypothesis about straight-line target moving.
- Computations of target range track parameters must be carried out using very precise formulas, taking into consideration all available information about the target movement character (air or cosmic target, maneuvering or nonmaneuvering target, etc.) in the interests of customers. In doing so, the parameter output can be presented in other coordinate systems that are different from a radar coordinate system, for example, the Cartesian coordinate system with the center at the point gathering information. Moreover, to estimate the necessary target range track parameters, for example, a course and velocity vector modulus under aircraft autotracking, the parameters that are not associated with the target range autotracking can be selected based on the needs of the customer or with the purpose of matching them with other detection algorithms of a CRS.
- The customer is interested, first of all, in target information that is very important for a CRS, for example, the information about the type and number of aircraft following for landing stored into the automatic control system of airport. The exact target range track parameters must be determined using just these targets. Naturally, not all detected targets within the radar antenna scanning area are important and some of them are not interested in a CRS, for example, recessive targets, transiting targets, and so on. Consequently, an estimation of target range track parameters with high accuracy is necessary only for a very small part of autotracked targets. In the considered case, definition of individual target range autotracking algorithm allows us to reduce requirements to computer system speed.

Logical block diagram of the target range tracking algorithm is presented in Figure 4.12 based on the statements discussed earlier. Block 1 solves the problem of selection and indication of target pip to continue the target range tracking. The gating algorithm and indication of target pips within the

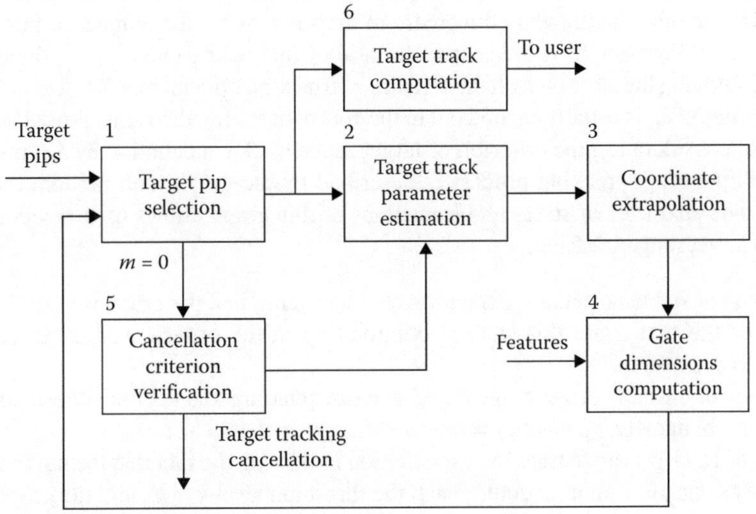

FIGURE 4.12 Block diagram of the target range tracking algorithm.

gate is designed in accordance with theoretical assumptions discussed in Section 4.1. The indicated target pip is given the number of tracked target range trajectory and it is processed by the block of target range track computations (the block 6). Simultaneously, the newly obtained target pip is used to estimate the parameters of target range track and to extrapolate the target coordinate for the next radar antenna scanning process, that is, to prepare a new gating cycle and indication. For this purpose the following operations are carried out:

- Estimation of target range track parameters under simplest conditions of target moving and coordinate measurement errors (the block 2).
- Determination of the extrapolated coordinates for the next radar antenna scanning (the block 3). Extrapolation is carried out by linear law.
- Determination of gate dimensions (the block 4). At the same time, the accurate characteristics of measured and extrapolated coordinates and information about a target pip miss within the gate are used.
- When the newly obtained target pip is absent, we check the criterion of cancellation of the target range tracking with the purpose to prolong the target range track (the block 5). If the cancellation criterion is satisfied, the target range tracking is stopped and the previous information about this target range track is removed. When the cancellation criterion is not satisfied, the coordinates of extrapolated point are used as the coordinates of newly obtained target pip and the computation process is repeated.

In general, in addition to the presence of the target pips within the gate that prolongs the target range track, we should take into consideration a set of other factors such as the target importance; target maneuverability, that is, to change a target range track during the flight; current target coordinates; direction of target moving; and length of target visibility within radar antenna scanning area and so on to make a decision about the target range tracking cancellation. However, a recordkeeping of these factors is very difficult and is not accessible forever owing to limitations in the speed of a computer system. For this reason, the main criterion while making a decision about the target range tracking cancellation is the appearance of some threshold series k_{th} of target pip misses within tracking gates. This criterion of cancellation of the target range tracking does not take into consideration individual peculiarities of each target range track and does not use information about accumulated accuracy level at the instant of appearance of the target pip miss series. A single advantage of this criterion is its simplicity.

While choosing k_{th} there is a need to proceed from the following assumptions. The greater is the value of k_{th}, the smaller is the probability to make a false decision about cancellation of the true target range track. On the other hand, with an increase in the value of k_{th}, the number of false target range tracking and its average length is increased. Because of this, while choosing k_{th} there is a need to take into consideration the statistical characteristics of true target pip misses (no detections). The final choice of the value of k_{th} is usually carried out in the course of testing the signal processing subsystem.

Taking into consideration the criterion of target range track cancellation by k_{th} misses one after another, the target range tracking process is described by the graph with random transitions (see Figure 4.13). The character of states and transitions of this graph allows us to select the following modes of the target range tracking:

- The mode of stable target range tracking characterizing that the graph is at the initial state a_{m+n}. For the first time, this state is obtained when the criterion of target range track detection is satisfied.
- The mode of unstable target range tracking corresponding to one of the intermediate states of the graph, namely, a_j, where $j = m + n - 1,\ldots, m + n + k_{\text{th}} - 1$.
- The mode of target range tracking cancellation indicating the fact that the number of target pip misses one after another could reach the threshold level $k = k_{\text{th}}$ and the graph has been passed into the state $a_{m+n+k_{\text{th}}}$.

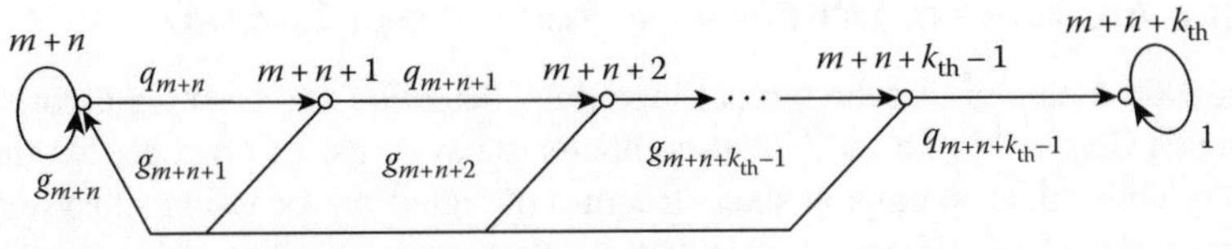

FIGURE 4.13 Target range tracking graph with random transitions.

In this case, the target range tracking algorithm graph is analogous to the graph of binary quantized target return pulse train latching algorithm [21–23]. For this reason, a procedure of analysis of these algorithms is the same.

Under statistical analysis of target range tracking algorithms, the main interest is the average time of false target range track observation and the average number of false target range tracking in the steady working state, which is associated with the average time of false target range track observation. Moreover, it is very interesting to define the probability of true target range tracking cancellation at the given probability of target pip detection within the gate. Let us define a function between the average number of false target range tracking for each radar antenna scanning period and the average number of false target range tracks in the steady working state. For this purpose, first of all, there is a need to define the probability to stop exactly the false target range tracking process at the μth step (the radar antenna scanning period) after starting the false target range tracking at the instant $\mu = 0$. In the case of cancellation criterion by k_{th} misses one after another, the probability to stop exactly the false target range tracking process at the μth step (the radar antenna scanning period) is equal to the probability of graph transition from the state a_{m+n} to the state $a_{m+n+k_{\text{th}}}$ in the course of μ steps (see Figure 4.13):

$$P_{\text{can}}(\mu) = P_{m+n+k_{\text{th}}}(\mu). \tag{4.62}$$

To determine the probability $P_{m+n+k_{\text{th}}}(\mu)$, we can use the following recurrent formulas:

$$\begin{cases} P_{m+n}(0) = 1, \\ P_{m+n}(\mu) = \sum_{j=m+n}^{m+n+k_{\text{tr}}-1} P_j(\mu-1) g_{f_i}, \\ P_{m+n+1}(\mu) = P_{m+n}(\mu-1) g_{f_{m+n}}, \\ \dots\dots\dots\dots\dots\dots\dots\dots\dots \\ P_{m+n+k_{\text{th}}}(\mu) = P_{m+n+k_{\text{th}}-1}(\mu-1) g_{f_{m+n+k_{\text{th}}-1}}. \end{cases} \tag{4.63}$$

The average length of the false target range tracks expressed as the number of radar antenna scanning periods is defined in the following form:

$$\bar{\mu} = \sum_{\mu=k_{\text{th}}}^{\infty} \mu P_{m+n+k_{\text{th}}}(\mu). \tag{4.64}$$

Furthermore, when we know the average number of the false target range tracks, the average number of the false target range tracks being under tracking is determined by

$$\bar{N}_{\text{f}_{\text{track}}}^{\text{tracking}} = \bar{N}_{\text{f}_{\text{track}}} \bar{\mu}. \tag{4.65}$$

The average number of the false target range tracks being under tracking is taken into account under the definition of computer subsystem cost and speed.

4.3.2 United Algorithm of Detection and Target Range Tracking

Until now, we have assumed that the target range track detection and tracking algorithms are realized individually, that is, by the individual computer subsystems. In practice, for the majority of cases, it is very convenient to employ such structure of signal reprocessing subsystems where the target range track detection and tracking algorithms are united and presented in the form of a single algorithm of target range track detection and tracking, and realization of this united algorithm is carried out by individual computer subsystem. Henceforth, we consider this version of structure of signal reprocessing subsystem.

If the criterion of beginning the target range track "2/*m*," the criterion of confirmation of the target range track "*l*/*n*," and the criterion of cancellation of target range tracking, for example, the criterion using k_{th} misses one after another, are given, then the united criterion of detection and tracking of the target range track can be written in the symbolic notation "$2/m + l/n - k_{th}$." The graph of the united algorithm under detection and tracking of target range track by criterion "$2/m + l/n - k_{th}$" is shown in Figure 4.14. This graph allows us to analyze the processes of detection and tracking of target range tracks as a whole instead of analysis by parts discussed earlier. Based on this graph, we can present an accurate formula defining the initial points of false target range tracks forming in the course of the steady working state (see (4.48)).

In the united algorithm realizing the criterion "$2/m + l/n - k_{th}$," the number of gates is equal to $m + \nu + k_{th} - 1$. Because of this, the upper limit of summation by j in (4.48) will be $j_{max} = m + \nu + k_{th} - 1$. The upper limit of summation by s will be defined as $s_{max} \to \infty$ since the number of steps under transition from the state a_1 to the state $a_{m+n+k_{th}-1}$ will be arbitrary large.

Thus, in the united algorithm case, the number of false target pips obtained at the initial points of new target range tracks is determined by the following formula:

$$\bar{N}_1' = \frac{\bar{N}_f}{1 + \sum_{j=1}^{m+\nu+k_{th}-1} g_{f_j} \sum_{s=0}^{\infty} P_{1j}^{(s)}}. \tag{4.66}$$

If the criterion of confirmation takes the form "1/*n*," we obtain $\nu = n$. The average number of false target range tracks for tracking is determined as

$$N_{f_{tr}}' = \bar{N}_1' p_1, \tag{4.67}$$

which is less than that of individual realization since the number of initial points is decreased. The number of false target range tracks being under tracking will also decrease correspondingly.

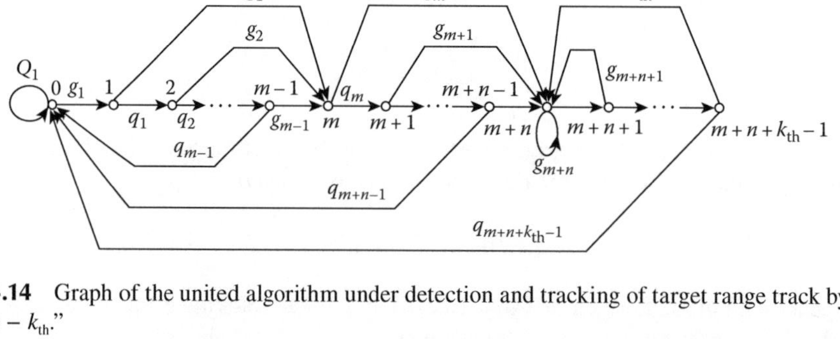

FIGURE 4.14 Graph of the united algorithm under detection and tracking of target range track by criterion "$2/m + l/n - k_{th}$."

4.4 SUMMARY AND DISCUSSION

Target range tracking is accomplished by continuously measuring the time delay between the transmission of an RF pulse and the echo signal returned from the target, and converting the roundtrip delay into units of distance. The target range measurement may also be performed with continuous wave (CW) radar systems using the frequency-modulated continuous wave (FM-CW) that is typically a linear-ramp FM. The target range is determined by the range-related frequency difference between the echo-frequency ramp and the frequency of the ramp being transmitted. The first function of the target range tracker is acquisition of a desired target. Although this is not a tracking operation, it is a necessary first step before the target range tracking or target angle tracking (the azimuth tracking) may take place in a typical radar system. The range-tracking portion of the radar has the advantage of seeing all targets within the beam from close range out to the maximum range of the radar. It typically breaks this range into small increments, each of which may be simultaneously examined for the presence of a target. When beam scanning is necessary, the target range tracker examines the increments simultaneously for short periods, such as 0.1 s, makes its decision about the presence of a target, and allows the beam to move to a new location if a "no" target is present.

Once a target is acquired in range, it is desirable to follow it in the range coordinate to provide distance information or slant range to the target. Appropriate timing pulses provide a target range gating so the angle-tracking circuits and automatic gain control (AGC) circuits look at only the short target range interval, or time interval, when the desired echo pulse is expected. The target range tracking operation is performed by closed-loop tracking similar to the azimuth tracker. Error in centering the range gate on the target echo pulse is sensed, error voltages are generated, and circuitry is provided to respond to the error voltage by causing the gate to move in a direction to recenter on the target echo pulse. Automatic target range tracking can generally be divided into five steps: (a) radar detection acceptance, accepting or rejecting detections for insertion into the tracking process—the purpose of this step is to control false track rates; (b) association of accepted detections with existing tracks; (c) updating the existing tracks with associated detections; (d) new track formation using unassociated detections; and (e) radar scheduling and control.

In the course of target range track detection the following operations are carried out: (a) gating and selection of target pip within the gate, (b) checking the detection criterion, and (c) estimation and extrapolation of target range track parameters. Target range tracking is sequential from measuring to measuring procedure of newly obtained target pip binding to target range track and accurate target range track determination. Under target range autotracking, the following operations are carried out: (a) accurate definition of target range track parameters in the course of newly obtained target pip binding, (b) extrapolation of target range track parameters for the next radar antenna scanning, (c) range gating of possible newly obtained target pip positions, and (d) selection of the definite target pip within the gate if there are several target pips within the gate area. When there are several target pips within the tracking gate, we expand the target range track binding for each target pip. When the target pip is absent within the tracking gate, the target range track is prolonged by the corresponding extrapolated point, but the next gate is increased in area to take into consideration the errors increasing in the course of extrapolation. If there are no target pips during k radar antennas scanning following one after another, the target range tracking is canceled. Thus, at various stages of target range track detection and target range tracking, we carry out the same operations in reality: (a) radar antenna scanning area gating, (b) selection and identification of target pips within the gate, and (c) filtering and extrapolation of target range track parameters.

While the CRS computer subsystem processes a high number of targets in real time, computation of dimensions and orientation of parallelepiped gate sides is very cumbersome. In this case, there is a need to use a simple version of gating, which is described as follows. The simplest gate shape is chosen for definition in that coordinate system, wherein radar information is processed. In the case of spherical coordinate system, the simplest gate is given by a linear dimension in target range Δr_{gate} and by two angular dimensions, namely, by the azimuth $\Delta\beta_{gate}$ and elevation $\Delta\varepsilon_{gate}$

(see Figure 4.4). These dimensions can be set in advance, based on the maximal magnitudes of random and dynamic errors while processing all target range tracks. In short, in this case, the gate dimensions are chosen so that the ellipsoid of all possible (under all directions of target moving and flight) total deviations of true target pips from corresponding extrapolated points will be easily fit and rotated in any direction inside the gate. This is the crudest technique of gating. We note that all procedures to choose dimensions of 3D gate can be used in full under gating in plane for binding newly obtained target pips in 2D radar systems.

The quality of target pip indication process can be evaluated by the probability of correct indication; in other words, the probability of event that the true target pip will be indicated to prolong the target range track in the next radar antennas scanning. The problem of definition of the probability of correct indication can be solved analytically if we assume that a presence of false target pips within the gate is only caused by a stimulus of the noise and interference and these target pips are uniformly distributed within the limits of radar antenna scanning area. Under indication within 2D rectangle gate circumscribed around the ellipse with the parameter equal $\lambda_{max} \geq 3$, the probability of correct indication is determined by

$$P_{ind} = \frac{1}{1 + 2\pi\rho_S\sigma_\eta\sigma_\xi},$$

where

ρ_S is the density of false target pips per unit square of gate

σ_η and σ_ξ are the mean square deviations of true target pips from the gate center along the axes η and ξ

In the case of target pip indication in 3D gate that can be presented in the form of parallelogram circumscribed around the total error ellipsoid, the probability of correct indication is determined by

$$P_{ind} \approx 1 - \frac{21.33\pi\sigma_\eta\sigma_\xi\sigma_\zeta\rho_V}{\sqrt{2\pi}},$$

where ρ_V is a density of false target pips per unit volume of gate.

We can note that in addition to target pip deviations from the gate center, the weight features of target pips formed in the course of signal preprocessing, as some analog of SNR, can be used to indicate the target pips. In the simplest case of binary quantized target return pulse train, we can use the number of train pulses or train bandwidth to generate the target pip weight features. Weight features of target pips can be used under target range track indication jointly with features of target pip deviations from the gate center or independently. One of the possible ways to use the target pip weight features jointly with the target pip deviations from the gate center is as follows. All target pips within the gate are divided with weight v_1 and with weight v_0 depending on whether the target return pulse train bandwidth exceeds the earlier-given threshold value or not, which in turn depends on target range magnitude. If there are target pips with weight v_1 we indicate the target pip that is nearest to the gate center as the true target pip from the observed target pip group. If there are no target pips with weight v_1, we indicate the target pip with weight v_0 that is nearest to the gate center as the true target pip. If we can characterize the target pip weight by the number of pulses in the target return pulse train, then we are able to indicate the target range track by the maximum number of pulses in the target return pulse train. In this case, features of target pip deviations from the gate center are only used under equality of weights of several target pips.

Under a hard noise environment, several gates formed around extrapolated points of prolonged target range tracks will be overlapped. Moreover, false and newly obtained target pips will appear within the gates. In this case, the problem of newly obtained target pip distribution and binding to target range tracking becomes more complex. In addition, the problem of fixing the beginning of a new target range track arises. First of all, this complication is associated with the necessity to use a joint target pip binding to target range track after getting newly obtained target pips liable to binding or, in other words, if these newly obtained target pips are within overlapping gates. There is a need to generate all possible ways or hypotheses for target pip binding and choose the most plausible version. After the determination of probabilities of events that the target pips belong to the existing target range tracks, new target range tracks, and false target range tracks, we are able to determine the probabilities of all versions of the target pip binding and to choose the version with the highest probability as a true version. Solution of this problem is very tedious even for the case that is considered the simplest. This problem can be simplified if the following practical rules are followed: (a) each target range track must be attached by the target pip, and (b) if the target pip is not associated to the given-before target range track, then a new target range track must be started independently of the probability of event that this target pip belongs to the new target range track or false target range track; in this case, we should compare only versions when there are target pip bindings for each target range track.

The process of target range track detection is divided into two stages: (a) *the first stage* is the target range track detection by criterion "2/*m*" and (b) *the second stage* is the confirmation of beginning the target range track, that is, the final target range track detection by criterion "*l*/*n*." In particular cases of the second stage, the target range track detection may be absent. Algorithm of the target range track detection by criterion "2/*m*" jointly with the confirmation algorithm or final target range track detection by criterion "*l*/*n*" forms a united target range track detection algorithm by criterion "2/*m* + *l*/*n*." The main computer subsystem operations carried out under target range track detection are as follows: (a) evaluation of target velocity under moving, (b) extrapolation of coordinates, and (c) gating the target pips. Target range track detection quality can be evaluated by the following characteristics: (a) the probability of detection of true target range track, (b) average number of false target range tracking per unit time, and (c) speed of computer systems used to realize the target range track detection algorithm.

Results of determination of the average number $\bar{N}_{\text{ftr}}$ of target range tracks that are needed to be tracked as a function of the average number of false target pips within the radar antenna scanning that are subjected to signal reprocessing by computer subsystems are presented in Figure 4.9, for the purpose of comparing the filtering ability. Analyzing these results, we can conclude as follows: (a) filtering ability of target range track detection algorithms, realizing that the criterion "2/*m* + *l*/*n*" increases with a decrease in *m* and *n* and with an increase in *l*, and (b) incrementing *l* on unit leads to a great increase in the filtering ability, compared to a corresponding decrease in *m* and *n*. These features must be taken into consideration while choosing the target range track detection algorithm employed in practice. If the CRS computer subsystem has limitations in the number of false target range tracks that are needed to be tracked, then a combination of the average number $\bar{N}_{\text{ftr}}$ and criteria of target range track detection algorithms during production tests allows us to raise a demand to the noise and interference level at the reprocessing computer subsystem input, that is, SNR. While designing the reprocessing computer subsystem of a CRS, information about the average number of false target range tracks that are under detection in the steady working state of a CRS is needed. Evidently, all false target range tracks, with respect to which the final decision about tracking or cancellation has not been made, are under detection process.

Analysis and comparison of performance shown in Figure 4.11 allow us to conclude that it is worthwhile to employ the criterion "2/*m* + 1/*n*" based on the viewpoint of decreasing the number of steps under detection of true target range track. In doing so, small deviations in *m* and *n* do not lead to great changes in the number of steps to ensure the probability of true target range track detection close to the unit. The probability of true target range track detection equal to 0.95 using the discussed criteria is obtained for target range corresponding to 0.75 of the maximum radar system range. Implementation of confirmation criteria "*l*/*n*" ($l > 1$) leads to the prolongation of time to detect the true target range tracks.

In the course of each target range tracking, two main problems are solved: gating and selection of newly obtained target pips for the prolongation of target range track, estimation of target range track parameters, and definition of variations in these parameters as a function of time. In principle, solutions to both problems can be obtained by employing the same detection algorithm. In this case, the required quality of a solved target range track estimation parameter problem must be in agreement with the customer requirements. However, there is a version of target range autotracking system in which the autotracking algorithm hunts only for target range track. To achieve a high-quality estimate of the target range track parameters, the individual target range track detection algorithm must be designed.

In general, in addition to the presence of the target pips within the gate to prolong the target range track, we should take into consideration a set of other factors such as the target importance; target maneuverability, that is, to change a target range track during the flight; current target coordinates; direction of target moving; and length of target visibility within radar antenna scanning area and so on to make a decision about the target range tracking cancellation. However, a recordkeeping of these factors is very difficult and is not accessible forever owing to limitations in the computer system speed. For this reason, the main criterion for decision about the target range tracking cancellation is the appearance of some threshold series k_{th} of target pip misses within tracking gates. This criterion of cancellation of the target range tracking does not take into consideration individual peculiarities of each target range track and does not use information about accumulated accuracy level at the instant of appearance of the target pip miss series. A single advantage of this criterion is its simplicity.

In practice, in the majority of cases, it is very convenient to employ a structure of signal reprocessing subsystems in which the target range track detection and tracking algorithms are united and presented in the form of a single algorithm of target range track detection and tracking, and realization of this united algorithm is carried out by the individual computer subsystem. If the criterion of beginning the target range track "$2/m$," the criterion of confirmation of the target range track "l/n," and the criterion of cancellation of target range tracking, for example, the criterion using k_{th} misses one after another are given, then the united criterion of detection and tracking of the target range track can be written in the following symbolic notation "$2/m + l/n - k_{\text{th}}$." The graph of the united algorithm under detection and tracking of target range track by criterion "$2/m + l/n - k_{\text{th}}$" is shown in Figure 4.14. It allows us to analyze the processes of detection and tracking of target range tracks as a whole instead of analysis by parts, as discussed previously. On the basis of this graph, we can present an accurate formula that defines the initial points of false target range tracks forming in the course of the steady working state (see (4.48)).

REFERENCES

1. Sherman, S.M. 1986. *Monopulse Principles and Techniques*. Norwood, MA: Artech House, Inc.
2. Trunk, G.V. 1978. Range resolution of targets using automatic detectors. *IEEE Transactions on Aerospace and Electronic Systems*, 14(9): 750–755.
3. Trunk, G.V. 1984. Range resolution of targets. *IEEE Transactions on Aerospace and Electronic Systems*, 20(11): 789–797.
4. Trunk, G.V. and M. Kim. 1994. Ambiguity resolution of multiple targets using pulse Doppler waveforms. *IEEE Transactions on Aerospace and Electronic Systems*, 30(10): 1130–1137.
5. Bath, W.G., Biddison, L.A., Haase, S.F., and E.C. Wetzlat. 1982. False alarm control in automated radar surveillance system, in *IEE International Radar Conference*, London, U.K., pp. 71–75.
6. Stuckey, W.D. 1992. Activity control principles for automatic tracking algorithms, in *IEEE Radar'92 Conference*, October 12–13, Birghton, U.K., pp. 86–89.
7. Leonov, A.I. and K.I. Formichev. 1986. *Monopulse Radar*. Norwood, MA: Artech House, Inc.
8. Barton, D.K. 1988. *Modern Radar System Analysis*. Norwood, MA: Artech House, Inc.
9. Bar-Shalom, Y. and T. Forthmann. 1988. *Tracking ad Data Association*. Orlando, FL: Academic Press.
10. Mookerjee, P. and F. Reifler. 2004. Reduced state estimator for system with parametric inputs. *IEEE Transactions on Aerospace and Electronic Systems*, 40(2): 446–461.
11. Leung, H., Hu, Z., and M. Blanchette. 1999. Evaluation of multiple radar target trackers in stressful environments. *IEEE Transactions on Aerospace and Electronic Systems*, 35(12): 663–674.

12. Blair, W.D. and Y. Bar-Shalom. 1996. Tracking maneuvering targets with multiple sensors: Does more data always mean better estimates? *IEEE Transactions on Aerospace and Electronic Systems*, 32(1): 450–456.
13. Simson, G.W. 1998. *Introduction to Airborne Radar*. 2nd edn. Mendham, NJ: SciTech Publishing.
14. Gumming, I.G. and F.N. Wong. 2005. *Digital Processing on Synthetic Aperture Radar Data*. Norwood, MA: Artech House, Inc.
15. Guerci, J.R. 2003. *Space-Time Adaptive Processing for Radar*. Norwood, MA: Artech House, Inc.
16. Levanon, N. and E. Mozeson. 2004. *Radar Signals*. New York: John Wiley & Sons, Inc.
17. Sullivan, R.J. 2000. *Microwave Radar: Imaging and Advanced Concepts*. Boston, MA: Artech House, Inc.
18. Peebles, P.Z. Jr. 1998. *Radar Principles*. New York: John Wiley & Sons, Inc.
19. Hayes, M.H. 1996. *Statistical Digital Signal Processing and Modeling*. New York: John Wiley & Sons, Inc.
20. Billingsley, J.B. 2001. *Radar Clutter*. New York: John Wiley & Sons, Inc.
21. Blackman, S. and R. Popoli. 1999. *Design and Analysis of Modern Tracking Systems*. Boston, MA: Artech House, Inc.
22. Cooperman, R. 2002. Tactical ballistic missile tracking using the interacting multiple model algorithm, in *Proceedings of the Fifth International Conference on Information Fusion*, July 8–11, Vol. 2, Annapolis, Maryland, pp. 824–831.
23. Pisacane, V.J. 2005. *Fundamentals of Space Systems*. 2nd edn. Oxford, U.K.: Oxford University Press.

5 Filtering and Extrapolation of Target Track Parameters Based on Radar Measure

A track represents the belief that a physical object or *target* is present and has actually been detected by the complex radar system (CRS). An automatic radar-tracking subsystem forms a track when enough radar detections are made in a believable enough pattern to indicate that a target is actually present (as opposed to a succession of false alarms) and when enough time has passed to allow accurate calculation of the target's kinematics state—usually the target position and velocity. Thus, a goal of tracking is to transform a time-lapse detection picture, consisting of target detections, false alarms, and clutter, into a track picture, consisting of tracks on real targets, occasional false tracks, and occasional deviations of track position from true target positions.

When a track is established in a computer subsystem, it is assigned a track number (the track file). All parameters associated with a given track are referred to by this track number. Typical track parameters are the filtered and predicted position; velocity; acceleration (when applicable); time of last update; track quality; SNR; covariance matrices—the covariance contains the accuracy of all the track coordinates and all the statistical cross-correlations between them, if a Kalman-type filter is being used; and track history, that is, the last n detections. Tracks and detections can be accessed in various sectored, linked-list, and other data structures so that the association process can be performed efficiently. In addition to the track file, a clutter file is maintained. A clutter number is assigned to each stationary or slowly moving echo. All parameters associated with a clutter point are referred to by this clutter number. Again, each clutter number is assigned to a sector in azimuth for efficient association.

There are many sources of error in radar-tracking performance. Fortunately, most are insignificant except for high-precision tracking-radar applications such as range instrumentation, where the angle precision required may be on the order 0.05 mrad (mrad, or milliradian, is 1000th of a radian, or the angle subtended by 1 m cross-range at 1000 m range). Many sources of error can be avoided or reduced by radar design or modification of the tracking geometry. Cost is a major factor in providing high-precision-tracking capability. Therefore, it is important to know how much errors can be tolerated, which sources of error affect the application, and what are the most cost-effective means to satisfy the accuracy requirements.

Because tracking radar subsystems track targets not only in angle but also in range and sometimes in Doppler, the errors in each of these target parameters must be considered in most error budgets. It is important to recognize the actual CRS information output. For a mechanically moved antenna, the angle-tracking output is usually obtained from the shaft position of the elevation and azimuth antenna axes. Absolute target location relative to each coordinates will include the accuracy of the survey of the antenna pedestal site. Phased array instrumentation radar system, such as the multiobject tracking radar (MOTR), provides electronic beam movement over a limited sector of about ±45° to approximately ±60° plus mechanical movement of the antenna to move the coverage sector [1–4].

Radar target tracking is performed by the use of the echo signal reflected from a target illuminated by the radar transmit pulse. This is called *skin tracking* to differentiate it from *beacon tracking,* where a beacon or a transponder transmits its signal to the radar and usually provides a stronger point-source signal. Because most targets, such as aircraft, are complex in shape, the total echo signal is composed of the vector sum of a group of superimposed echo signals from the individual parts of the target, such as the engines, propellers, fuselage, and wing edges. The motions of a target with respect to the CRS cause the total echo signal to change with time, resulting in random fluctuations in the radar measurements of the parameters of the target. These fluctuations caused by the target only, excluding atmospheric effects and radar receiver noise contributions, are called the *target noise*. This discussion of target noise is based largely on aircraft, but it is generally applicable to any target, including land targets of complex shape that are large with respect to a wavelength. The major difference is in the target motion, but the discussions are sufficiently general to apply to any target situation.

The echo return from a complex target differs from that of a point source by the modulations that are produced by the change in amplitude and relative phase of the returns from the individual elements. The word *modulations* is used in plural form because five types of modulation of the echo signal that are caused by a complex target affect the CRS. These are amplitude modulation, phase front modulation (glint), polarization modulation, Doppler modulation, and pulse time modulation (range glint). The basic mechanism by which the modulations are produced is the motion of the target, including yaw, pitch, and roll, which causes the change in relative range of the various individual elements with respect to the radar. Although the target motions may appear small, a change in relative range of the parts of a target of only one-half wavelength (because of the two-way radar signal path) causes a full 360° change in relative phase.

5.1 INITIAL CONDITIONS

The problem to define the estimation $\hat{\boldsymbol{\theta}}(t_i)$ of target track parameters $\boldsymbol{\theta}(t)$ by the observed coordinate sample $\mathbf{Y}(t_i)$, $i = 0, 1, \ldots, n, \ldots$ is solved in all stages of signal reprocessing in a CRS. In the course of signal reprocessing, we can distinguish the following kinds of target track parameter estimations [5,6]:

- Estimation obtained at the instant $t_j = t_n$: the last measurement, that is, target track parameter filtering
- Estimation obtained at predicted coordinates at the point $t_j > t_n$—the extrapolation of target track parameters
- Estimation obtained at the points lying within the limits of the interval of observation $0 \le t_j < t_n$—the target track parameter smoothing

In this chapter, we discuss only the first and second problems of target track parameter estimation definition. The class of considered objects is mainly limited by air targets. There is a need to take into consideration that in this chapter we discuss only the main results of filtering theory and extrapolation methods with reference to signal reprocessing problems in CRSs. We use the representation form that is understandable to designers and engineers of signal processing algorithms, and computer architecture systems and other experts exploiting CRSs for various areas of application.

5.2 PROCESS REPRESENTATION IN FILTERING SUBSYSTEMS

5.2.1 Target Track Model

In solving the filtering problems, a presentation procedure of variation of the filtered target parameters as a function of time takes a principal meaning. In our case, this corresponds to the choice of target track model. In the case of signal reprocessing problems in CRSs, taking into consideration sampling and discretization of target coordinate measuring and radio frequency interference, the target track model can be described by a system of linear difference equations that can be presented in the following vector form:

$$\boldsymbol{\theta}_{n+1} = \boldsymbol{\Phi}_n \boldsymbol{\theta}_n + \boldsymbol{\Gamma}_n \boldsymbol{\eta}_n = \boldsymbol{\theta}'_{n+1} + \boldsymbol{\Gamma}_n \boldsymbol{\eta}_n, \tag{5.1}$$

where

- $\boldsymbol{\theta}_n$ is the s-dimensional target track parameter vector at the nth step
- $\boldsymbol{\Phi}_n$ is the known $s \times s$ transfer matrix
- $\boldsymbol{\eta}_n$ is the h-dimensional vector of target track parameter disturbance
- $\boldsymbol{\Gamma}_n$ is the known $s \times h$ matrix
- $\boldsymbol{\theta}'_{n+1}$ is the deterministic (undisturbed) component of the target track parameter vector at the $(n+1)$th step

Under polynomial representation of independent target coordinates, a prediction of parameters of an undisturbed target track, for example, by target range coordinate $r(t)$, is carried out in the following manner:

$$\begin{aligned} r_{n+1} &= r_n + \dot{r}_n \tau_n + 0.5\ddot{r}_n \tau_n^2 + \cdots, \\ \dot{r}_{n+1} &= \dot{r}_n + \ddot{r}_n \tau_n + \cdots, \\ \ddot{r}_{n+1} &= \ddot{r}_n + \dddot{r}_n \tau_n + \cdots, \\ &\cdots\cdots\cdots\cdots\cdots\cdots \end{aligned} \tag{5.2}$$

where

- $\tau_n = t_{n+1} - t_n$; t_n and t_{n+1} are the readings of the function $r(t)$
- $r_n, \dot{r}_n, \ddot{r}_n, \ldots$ are the target track parameters representing the target track coordinate, velocity of target track coordinate variation, and acceleration by the target track coordinate

Using the vector-matrix representation in (5.2), we obtain

$$\boldsymbol{\theta}'_{n+1(r)} = \boldsymbol{\Phi}_n \boldsymbol{\theta}_{n(r)}, \tag{5.3}$$

where

$$\boldsymbol{\theta}_{n(r)} = \begin{Vmatrix} r_n \\ \dot{r}_n \\ \ddot{r}_n \\ \vdots \end{Vmatrix} \quad \text{and} \quad \boldsymbol{\Phi}_n = \begin{Vmatrix} 1 & \tau_n & 0.5\tau_n^2 & \cdots \\ 0 & 1 & \tau_n & \cdots \\ 0 & 0 & 1 & \cdots \\ \cdot & \cdot & \cdot & \cdot \end{Vmatrix}. \tag{5.4}$$

Formulas for undisturbed target track parameters by other independent coordinates take an analogous form.

The second term in the target track model given by (5.1) must take into consideration, first of all, the target track parameter disturbance caused by the nonuniform environment in which the target moves, atmospheric conditions, and, also, inaccuracy and inertness of control and target parameter stabilization system in the course of target moving. We may term this target track parameter disturbance as the control noise, that is, the noise generated by the target control system. As a rule, the control noise is presented as the discrete white noise with zero mean and the correlation matrix defined as follows:

$$E[\eta_i\eta_j^T] = \sigma_n^2\delta_{ij}, \tag{5.5}$$

where

- E is the mathematical expectation
- σ_n^2 is the variance of control noise
- δ_{ij} is the Kronecker symbol, that is, $\delta_{ij} = 1$, if $i = j$ and 0 if $i \neq j$

In addition to the control noise, the target track model must take into consideration specific disturbances caused by unpredictable changes in target track parameters that are generated by target in-flight maneuvering. We call these disturbances the target maneuver noise. In a general case, the target maneuver noise is neither the white noise nor the Gaussian noise. One possible example of the target maneuver noise is the pdf of target (aircraft) acceleration, the so-called *target maneuver intensity*, by one coordinate (see Figure 5.1), where P_0 is the probability of target maneuver absence and P_1 is the probability of target maneuver with the maximal acceleration $\pm g_{m_{\max}}$. The probability of any intermediate value of the target maneuver intensity or the pdf of target acceleration is given by

$$p = \frac{1-(2P_1+P_0)}{2g_{m_{\max}}}. \tag{5.6}$$

The equiprobability of intermediate values of the target maneuver intensity or the pdf of target acceleration can be founded, for example, by the fact that the projection of the target maneuver intensity by the target course (the most frequent target maneuver case) onto arbitrary direction takes a value within the limits $\pm g_{m_{\max}}$. In the case of a set of the target maneuvers in time and space, we can believe that these values are equally probable.

Since performing the target maneuver requires, as a rule, so much time (in any case, more time in comparison with the time interval τ_n defined between two target coordinate measurements), the target maneuver intensity or pdf of target acceleration is correlated, at some instant of observation, with the target maneuver intensity observed previously or in succeeding instants. By this reason, we must know the autocorrelation function of the target maneuver noise to characterize it statistically.

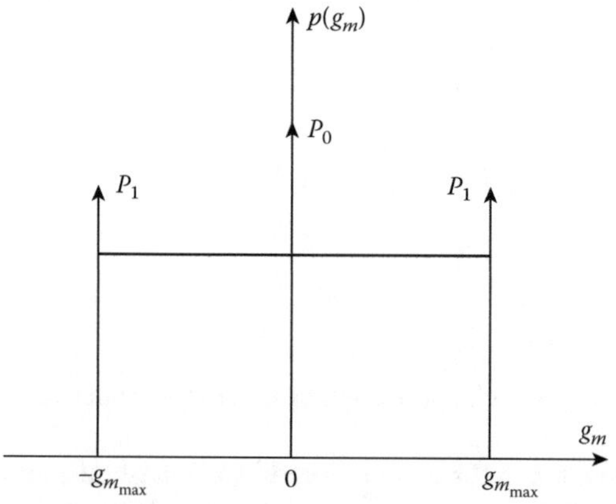

FIGURE 5.1 Example of pdf of the target maneuver intensity (acceleration).

Usually, the autocorrelation function of the target maneuver intensity can be presented in exponential form as follows:

$$E[g_m(t)g_m(t+\tau)] = R_m(\tau) = \sigma_m^2 \exp(-\alpha \mid \tau \mid), \tag{5.7}$$

where

α is the value that is inversely proportional to the average duration of target maneuver T_m, that is, $\alpha = T_m^{-1}$

σ_m^2 is the variance of the target maneuver intensity

In the case of equivalent discrete time, the autocorrelation function of the target maneuver intensity can be presented in the following form:

$$R_m(n) = \sigma_m^2 \exp\{-\alpha \mid nT_{\text{new}} \mid\} = \sigma_m^2 \rho^{|n|}, \tag{5.8}$$

where

$$\rho = \exp\{-\alpha T_{\text{new}}\}, \tag{5.9}$$

T_{new} is the period of getting new information about the target. Further values of the target maneuver intensity can be presented using the previous one, for example,

$$g_{m_{n+1}} = \rho g_{m_n} = \sqrt{\sigma_m^2(1-\rho^2)}\;\xi_n, \tag{5.10}$$

where ξ_n is the white noise with zero mean and the variance equal to unit.

In practice, in designing the CRSs, we can conditionally think that a set of targets can be divided into maneuvering and nonmaneuvering targets. The target is not considered as a maneuvering target if it moves along a straight line with the constant velocity accurate within the action of the control noise intensity. In other cases, the target is maneuvering. For example, in the case of aerodynamic objects, the nonmaneuvering target model is considered as the main model and each coordinate of this model is defined by the polynomial of the first order. However, this classification has a sense only in the case when the filtering parameters of target track are represented in the Cartesian coordinate system under signal processing in CRSs. If the filtering parameters of the target track are represented in the spherical coordinate system, then they are changed nonlinearly even when the target is moving along the straight line and uniformly. In this case, the polynomial of the second order must be applied to represent the independent target coordinates.

5.2.2 Measuring Process Model

In the solution of the filtering problems, in addition to the target track model there is a need to specify a function between the m-dimensional vector of measured coordinates $\mathbf{Y}_n$ and s-dimensional vector of estimated parameters $\boldsymbol{\theta}_n$ at the instant of nth measuring. This function, as a rule, is given by a linear equation:

$$\mathbf{Y}_n = \mathbf{H}_n\boldsymbol{\theta}_n + \Delta\mathbf{Y}_n, \tag{5.11}$$

where

$\mathbf{H}_n$ is the known $m \times s$ matrix defining a function between the observed coordinates and estimated parameters of target track

$\Delta\mathbf{Y}_n$ is the error of coordinate measuring

In the considered case, the observed coordinates are the current target coordinates in a spherical coordinate system—the target range r_n, the azimuth β_n, the elevation ε_n, or some specific coordinates for CRSs—and radar coordinates, for example, the radar range, cosine between the radar antenna array axis and direction to the target. In some CRSs, the radial velocity $\dot{r}_n$ can serve as the measured coordinate. The matrix $\mathbf{H}_n = \mathbf{H}$ takes the simplest form and consists of zeros and units when the target track parameters are estimated by the observed coordinates in the spherical coordinate system. For example, if we measure the target spherical coordinates r_n, β_n, and the parameters $\hat{r}_n, \hat{\dot{r}}, \hat{\beta}_n, \hat{\dot{\beta}}_n$ (the linear approximation) are filtered, then the matrix $\mathbf{H}_n$ takes the following form:

$$\mathbf{H}_n = \left(\begin{Vmatrix} 1 & 0 & 0 & 0 \\ 0 & 0 & 1 & 0 \end{Vmatrix} \begin{matrix} \to r_n \\ \to \beta_n \end{matrix} \right) \quad \begin{matrix} \downarrow & \downarrow & \downarrow & \downarrow \\ \hat{r}_n & \hat{\dot{r}}_n & \hat{\beta}_n & \hat{\dot{\beta}}_n \end{matrix} . \tag{5.12}$$

When the target track parameters are filtered by measured spherical coordinates in the Cartesian coordinate system, then computation of elements of the matrix $\mathbf{H}_n$ is carried out by differentiation of transformation formulas from the spherical coordinate system to the Cartesian one:

$$\mathbf{H}_n = \left(\begin{Vmatrix} \dfrac{dr_n}{dx_n} & 0 & \dfrac{dr_n}{dy_n} & 0 \\ \dfrac{d\beta_n}{dx_n} & 0 & \dfrac{d\beta}{dy_n} & 0 \end{Vmatrix} \begin{matrix} \to r_n \\ \\ \to \beta_n \end{matrix} \right) \quad \begin{matrix} \downarrow & \downarrow & \downarrow & \downarrow \\ \hat{x}_n & \hat{\dot{x}}_n & \hat{y}_n & \hat{\dot{y}}_n \end{matrix} . \tag{5.13}$$

Analogously, we can define the elements of matrix $\mathbf{H}_n$ for other compositions of measured coordinates and filtered parameters.

Errors in coordinate measuring given by the vector $\mathbf{\Delta Y}_n = \mathbf{\Delta Y}(t_n)$ in (5.11) can be considered, in a general case, as the random sequence subjected to the normal Gaussian pdf. The following initial conditions can be accepted with respect to this random sequence:

- Errors in measuring the independent observed coordinate are independent. This condition allows us to solve the filtering problems by each observed coordinate individually.
- In a general case, a set of errors under measurements of each coordinate at the instants $t_1, t_2, \ldots, t_n$ is the n-dimensional system of correlated normally distributed random variables with the $n \times n$ correlation matrix $\mathbf{R}_n$:

$$\mathbf{R}_n = \begin{Vmatrix} \sigma_1^2 & R_{12} & R_{13} & \cdots & R_{1n} \\ R_{21} & \sigma_2^2 & R_{23} & \cdots & R_{2n} \\ \cdots & \cdots & \cdots & \cdots & \cdots \\ R_{n1} & R_{n2} & R_{n3} & \cdots & \sigma_n^2 \end{Vmatrix} . \tag{5.14}$$

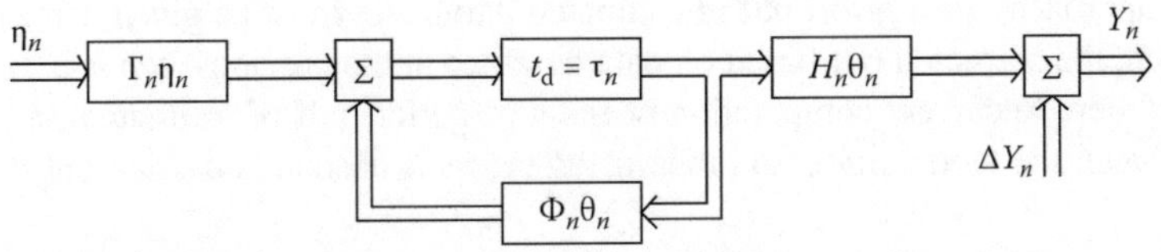

FIGURE 5.2 Flowchart of the united dynamic model.

The elements of the correlation matrix $\mathbf{R}_n$ are symmetric relative to the diagonal and equal between each other, that is, $R_{ij} = R_{ji}$. It means that $\mathbf{R}_n^T = \mathbf{R}_n$. When measurement errors are uncorrelated, all elements of the correlation matrix $\mathbf{R}_n$ excepting diagonal elements are equal to zero. That matrix is called the diagonal matrix.

In conclusion, we note that the target track model combined with the model of measuring process forms the model of united dynamic system representing a process subjected to filtering. The flowchart of the united dynamic model is shown in Figure 5.2, where the double arrows denote multidimensional or vectors relations.

5.3 STATISTICAL APPROACH TO SOLUTION OF FILTERING PROBLEMS OF STOCHASTIC (UNKNOWN) PARAMETERS

The filtering problem of stochastic (unknown) parameters is posed in the following way. Let the sequence of measured coordinate vectors $\{\mathbf{Y}\}_n = \{Y_1, Y_2, \ldots, Y_n\}$, which is statistically related to the sequence of dynamic system state vectors $\{\boldsymbol{\theta}\}_n = \{\theta_1, \theta_2, \ldots, \theta_n\}$ in accordance with (5.1) and (5.11), be observed. There is a need to define the current estimation $\hat{\boldsymbol{\theta}}_n$ of the state vector $\boldsymbol{\theta}_n$. The general approach to solve the assigned problem is given in the theory of statistical decisions. In particular, the optimal estimation of unknown vector parameter by minimal average risk criterion at the quadratic function of losses is determined from the following equation:

$$\hat{\boldsymbol{\theta}}_n = \int_{\boldsymbol{\Theta}} \boldsymbol{\theta}_n p(\boldsymbol{\theta}_n | \{\mathbf{Y}\}_n) d\boldsymbol{\theta}_n, \quad (5.15)$$

where

$p(\boldsymbol{\theta}_n | \{\mathbf{Y}\}_n)$ is the a posteriori pdf of the current value of parameter vector $\boldsymbol{\theta}_n$ by sequence of measured data $\{\mathbf{Y}\}_n$

$\boldsymbol{\Theta}$ is the space of possible values of the estimated vector parameter $\boldsymbol{\theta}_n$

If the a posteriori pdf is the unimodal function and symmetric with respect to the mode, the optimal parameter estimation is defined based on the solution of the following equation:

$$\left.\frac{\partial p(\boldsymbol{\theta}_n | \{\mathbf{Y}\}_n)}{\partial \boldsymbol{\theta}_n}\right|_{\boldsymbol{\theta}_n = \hat{\boldsymbol{\theta}}_n} = 0 \quad \text{if} \quad \frac{\partial^2 p(\boldsymbol{\theta}_n | \{\mathbf{Y}\}_n)}{\partial \boldsymbol{\theta}_n^2} < 0. \quad (5.16)$$

In this case, the optimal parameter estimation is called the optimal estimation by criterion of the maximum a posteriori pdf. Thus, in the considered case and in the cases of any other criteria to define an estimation quality, a computation of the a posteriori pdf is a sufficient procedure to define the optimal estimations.

In accordance with developed procedures and methods to carry out statistical tests in mathematical statistics, the following approaches are possible to calculate the a posteriori pdf and, consequently, to estimate the parameters: the batch method when the fixed samples are used and the recurrent algorithm consisting in sequent accurate definition of the a posteriori pdf after each new measuring.

Using the first approach, the a priori pdf of estimated parameter must be given. Under the use of the second approach, the predicted pdf based on data obtained at the previous step is used as the a priori pdf on the next step. Recurrent computation of the a posteriori pdf of estimated parameter, when a correlation between the model noise and measuring errors is absent, is carried out by the following formula [6]:

$$p(\boldsymbol{\theta}_n | \{\mathbf{Y}\}_n) = \frac{p(\mathbf{Y}_n | \boldsymbol{\theta}_n) p(\boldsymbol{\theta}_n | \{\mathbf{Y}\}_{n-1})}{\int_\Theta p(\mathbf{Y}_n | \boldsymbol{\theta}_n) p(\boldsymbol{\theta}_n | \{\mathbf{Y}\}_{n-1}) d\boldsymbol{\theta}_n}, \tag{5.17}$$

where

$p(\boldsymbol{\theta}_n | \{\mathbf{Y}\}_{n-1})$ is the pdf of predicted value of the estimated parameter $\boldsymbol{\theta}_n$ at the instant of the nth measuring by sequence data of previous $(n - 1)$ coordinate measurements

$p(\mathbf{Y}_n | \boldsymbol{\theta}_n)$ is the likelihood function of the last nth coordinate measuring

In a general case of nonlinear target track models and measuring process models, computations by the formula (5.17) are impossible, as a rule, in closed forms. Because of this, in solving filtering problems in practice, various approximations of models and statistical characteristics of CRS noise and measuring processes are used. Methods of linear filtering are widely used in practice. Models of system state and measuring of these methods are supposedly linear, and the noise is considered as the Gaussian noise. Further, we mainly consider algorithms of linear filtering.

5.4 ALGORITHMS OF LINEAR FILTERING AND EXTRAPOLATION UNDER FIXED SAMPLE SIZE OF MEASUREMENTS

Algorithms of linear filtering and target track parameter extrapolation are obtained in this section using the following initial premises:

1. The model of undisturbed target track by each of the independent coordinates is given in the form of polynomial function:

$$X(\boldsymbol{\theta}, t) = \sum_{l=0}^{s} \theta_l \frac{t^l}{l!}, \tag{5.18}$$

where the power s of this polynomial function is defined by the accepted hypothesis of target moving. In (5.18) the polynomial coefficients take a sense of coordinates, coordinate change velocity, acceleration, etc., which are the parameters of target track. The set of parameters θ_l presented in the column form forms $(s + 1)$-dimensional vector of the target track parameters

$$\boldsymbol{\theta} = \|\theta_0, \theta_1, \ldots, \theta_s\|^T. \tag{5.19}$$

We assume that this vector is not variable within the limits of measurement time.

2. Measurement results of the coordinate Y_i at discrete time instants $t_1, t_2, \ldots, t_n$ are linearly related to the parameter vector by the following equation:

$$Y_i = \sum_{l=0}^{s} \theta_l \frac{\tau_i^l}{l!} + \Delta Y_i, \quad \tau_i = t_i - t_0, \tag{5.20}$$

where ΔY_i is the measure error.

3. The conditional pdf of measure error under individual measurement takes the following form:

$$p(Y_i \mid \boldsymbol{\theta}) = \frac{1}{\sqrt{2\pi\sigma_{Y_i}^2}} \exp\left[-\frac{\left(Y_i - \sum_{l=0}^{s} \theta_l \frac{\tau_i^l}{l!}\right)^2}{2\sigma_{Y_i}^2}\right], \tag{5.21}$$

where $\sigma_{Y_i}^2$ is the measure error variance.

4. In a general case, a totality of coordinate measurement errors $\Delta Y_1, \Delta Y_2, \ldots, \Delta Y_N$ represents the N-dimensional system of correlated and normal distributed random variables and is characterized by $N \times N$-dimensional correlation matrix $\mathbf{R}_N$ (5.14). In solving the filtering problems, this matrix must be known. The conditional pdf of N-dimensional sample of correlated normally distributed random variables can be presented in the following form:

$$p(Y_1, Y_2, \ldots, Y_N \mid \boldsymbol{\theta}) = \frac{1}{(2\pi)^{N/2}\sqrt{|\mathbf{R}_N|}} \exp\left[-0.5\left(\Delta\mathbf{Y}_N^T \mathbf{R}_N^{-1} \mathbf{Y}_N\right)\right], \tag{5.22}$$

$$\Delta\mathbf{Y}_N^T = \|\Delta Y_1, \Delta Y_2, \ldots, \Delta Y_N\|, \tag{5.23}$$

$$\Delta Y_i = \left(Y_i - \sum_{l=0}^{s} \theta_l \frac{\tau_i^l}{l!}\right) = [Y_i - X(\boldsymbol{\theta}, \tau_i)], \tag{5.24}$$

where

$\mathbf{R}_N^{-1}$ is the inverse correlation matrix of measurer error

$|\mathbf{R}_N|$ is the determinant of correlation matrix

5. A priori information about the filtered parameters is absent. This corresponds to the case of parameter evaluation at the initial part of target track, that is, at the start of the target track by a set of target pips selected in a special way. The estimations obtained in this way are used at a later time as a priori data for the next stages of filtering. When a priori information is absent, the optimal filtering problems are solved using the maximum likelihood criterion. Thus, in the present section we consider the filtering algorithms and extrapolation of polynomial target track parameters using the fixed sample of measurements, which are optimal by the maximal likelihood criterion.

5.4.1 Optimal Parameter Estimation Algorithm by Maximal Likelihood Criterion for Polynomial Target Track: A General Case

The likelihood function of the vector parameter $\boldsymbol{\theta}_N$ estimated by sequent measurements $\{\mathbf{Y}_N\}$ takes the following form in vector–matrix representation:

$$\mathcal{L}(\boldsymbol{\theta}_N) = C \exp\left[-0.5\left(\Delta\mathbf{Y}_N^T \mathbf{R}_N^{-1} \Delta\mathbf{Y}_N\right)\right]. \tag{5.25}$$

This is analogous to the conditional pdf of the N-dimensional sample of the correlated normally distributed random variables. It is more convenient to use the natural logarithm of the likelihood function, that is,

$$\ln \mathcal{L}(\boldsymbol{\theta}_N) = \ln C - 0.5 \Delta\mathbf{Y}_N^T \mathbf{R}_N^{-1} \Delta\mathbf{Y}_N. \tag{5.26}$$

To define the estimations of target track parameters in accordance with the maximal likelihood method, there is a need to differentiate (5.26) with respect to vector components of estimated results at each measure point and to equate zero at $\boldsymbol{\theta}_N = \hat{\boldsymbol{\theta}}_N$. As a result, we obtain the following vector likelihood equation [7,8]:

$$\mathbf{A}_N^T \mathbf{R}_N^{-1} \left\| Y_i - \sum_{l=0}^{s} \hat{\theta}_l \frac{\tau_i^l}{l!} \right\| = 0, \tag{5.27}$$

where

$$\mathbf{A}_N^T = \left\| \begin{matrix} \frac{dX(\hat{\boldsymbol{\theta}},\tau_1)}{d\theta_0} & \frac{dX(\hat{\boldsymbol{\theta}},\tau_2)}{d\theta_0} & \cdots & \frac{dX(\hat{\boldsymbol{\theta}},\tau_N)}{d\theta_0} \\ \frac{dX(\hat{\boldsymbol{\theta}},\tau_1)}{d\theta_1} & \frac{dX(\hat{\boldsymbol{\theta}},\tau_2)}{d\theta_1} & \cdots & \frac{dX(\hat{\boldsymbol{\theta}},\tau_N)}{d\theta_1} \\ \cdots & \cdots & \cdots & \cdots \\ \frac{dX(\hat{\boldsymbol{\theta}},\tau_1)}{d\theta_s} & \frac{dX(\hat{\boldsymbol{\theta}},\tau_2)}{d\theta_s} & \cdots & \frac{dX(\hat{\boldsymbol{\theta}},\tau_N)}{d\theta_s} \end{matrix} \right\| \tag{5.28}$$

is the $(s + 1) \times N$ matrix of differential operators.

In a general case, the final solution of likelihood equation for correlated measure errors takes the following form:

$$\hat{\boldsymbol{\theta}}_N = \mathbf{B}_N^{-1} \mathbf{A}_N^T \mathbf{R}_N^{-1} \mathbf{Y}_N, \tag{5.29}$$

where

$$\mathbf{B}_N = \mathbf{A}_N^T \mathbf{R}_N^{-1} \mathbf{A}_N \tag{5.30}$$

and $\mathbf{Y}_N$ is the N-dimensional measure vector. When the measure errors are uncorrelated, then

$$\mathbf{R}_N^{-1} \mathbf{Y}_N = \mathbf{Y}'_N = \left\| \begin{matrix} w_1 Y_1 \\ w_2 Y_2 \\ \vdots \\ w_N Y_N \end{matrix} \right\|, \tag{5.31}$$

where $w_i = \sigma_{Y_i}^{-2}$ is the weight of the ith measurement. In this case, (5.29) and (5.30) take a form

$$\boldsymbol{\theta}_N = \mathbf{B}_N^{-1} \mathbf{A}_N^T \mathbf{Y}'_N, \tag{5.32}$$

which is matched with estimations obtained by the least-squares method.

Potential errors of target track parameters estimated by the considered procedure can be obtained using a linearization of the likelihood equation (5.27). The final form of the error correlation matrix of target track parameter estimations can be presented in the following form:

$$\boldsymbol{\Psi}_N = \mathbf{B}_N^{-1} = \left(\mathbf{A}_N^T \mathbf{R}_N^{-1} \mathbf{A}_N \right)^{-1}. \tag{5.33}$$

Further study in detail of (5.33) is carried out in the following for specific examples.

5.4.2 Algorithms of Optimal Estimation of Linear Target Track Parameters

Let us obtain the optimal algorithms of target track parameter estimations of the coordinate $x(t)$, which is varied linearly using discrete readings x_i, $i = 1, 2,\ldots, N$, characterized by errors $\sigma_{x_i}^2$. We consider the coordinate x_N and its increment $\Delta_1 x_N$ as the estimated parameters at the last observation point t_N. For simplicity, we believe that measurements are carried out with the period T_{eq} and the measure errors are uncorrelated from measure to measure. The coordinate is varied according to the following equation:

$$x(t_i) = x_i = x_N - (N-i)\Delta_1 x_N, \quad i = 1,2,\ldots,N, \tag{5.34}$$

where

$$\Delta_1 x_N = T_{\text{eq}}\dot{x}_N \tag{5.35}$$

is the coordinate increment. Thus, in the considered case, the vector of estimated parameters takes the following form:

$$\boldsymbol{\theta}_N = \begin{Vmatrix} \hat{\theta}_{0N} \\ \hat{\theta}_{1N} \end{Vmatrix} = \begin{Vmatrix} \hat{x}_N \\ \Delta_1\hat{x}_N \end{Vmatrix}, \tag{5.36}$$

and the transposed matrix of differential operators given by (5.28) takes a form

$$\mathbf{A}_N^T = \begin{Vmatrix} \dfrac{d\hat{x}(t_1)}{d\hat{x}_N} & \dfrac{d\hat{x}(t_2)}{d\hat{x}_N} & \cdots & \dfrac{d\hat{x}(t_N)}{d\hat{x}_N} \\ \cdots & \cdots & \cdots & \cdots \\ \dfrac{d\hat{x}(t_1)}{d\Delta_1\hat{x}_N} & \dfrac{d\hat{x}(t_2)}{d\Delta_1\hat{x}_N} & \cdots & \dfrac{d\hat{x}(t_N)}{d\Delta_1\hat{x}_N} \end{Vmatrix} = \begin{Vmatrix} 1 & 1 & \ldots & 1 \\ N-1 & N-2 & \ldots & 0 \end{Vmatrix}. \tag{5.37}$$

In the considered case, the error correlation matrix will be diagonal. By this reason, the inverse error correlation matrix can be presented in the following form:

$$\mathbf{R}_N^{-1} = \| \, w_i\delta_{ij} \, \|, \tag{5.38}$$

where $w_i = \sigma_{x_i}^{-2}$; $\delta_{ij} = 1$ if $i = j$ and $\delta_{ij} = 0$ if $i \neq j$. Substituting (5.37) and (5.38) in (5.27), we obtain the two-equation system to estimate the parameters of target linear track:

$$\begin{cases} f_N\hat{x}_N - g_N\Delta_1\hat{x}_N = \sum_{i=1}^{N} w_i x_i \\ g_N\hat{x}_N - h_N\Delta_1\hat{x}_N = \sum_{i=1}^{N} w_i (N-i) x_i \end{cases}, \tag{5.39}$$

where we have just introduced the following notations:

$$\begin{cases} f_N = \sum_{i=1}^{N} w_i; \\ g_n = \sum_{i=1}^{N} (N-i)w_i; \\ h_N = \sum_{i=1}^{N} (N-i)^2 w_i. \end{cases} \tag{5.40}$$

Solution of this equation system takes the following form:

$$\begin{cases} \hat{x}_N = \dfrac{h_N \sum_{i=1}^{N} w_i x_i - g_N \sum_{i=1}^{N} w_i(N-i)x_i}{G_N}; \\ \Delta_1\hat{x}_N = \dfrac{g_N \sum_{i=1}^{N} w_i x_i - f_N \sum_{i=1}^{N} w_i(N-i)x_i}{G_N}; \end{cases} \tag{5.41}$$

where

$$G_N = f_N h_N - g_N^2. \tag{5.42}$$

Now, assume that the measured coordinates can be considered as uniformly precise within the limits of the finite observation interval, that is, $w_1 = w_2 = \cdots = w_N = w$. In this case, we can write

$$\begin{cases} f_N = Nw; \\ g_n = \dfrac{N(N-1)}{2} w; \\ h_N = \dfrac{N(N-1)(2N-1)}{6} w. \end{cases} \tag{5.43}$$

The final formulas for estimations of linear target track parameters under uniformly precise and equally discrete readings take the following form:

$$\begin{cases} \hat{x}_N = \sum_{i=1}^{N} \eta_{\hat{x}}(i) x_i; \\ \Delta_1\hat{x}_N = \sum_{i=1}^{N} \eta_{\Delta_1\hat{x}}(i) x_i; \end{cases} \tag{5.44}$$

where

$$\begin{cases} \eta_{\hat{x}}(i) = \dfrac{2(3i - N - 1)}{N(N+1)}; \\ \\ \eta_{\Delta_1\hat{x}}(i) = \dfrac{6(2i - N - 1)}{N(N^2 - 1)} \end{cases} \tag{5.45}$$

are the weights of measurements under the coordinate estimation and the first increment, respectively. For example, at $N = 3$ we obtain the following:

$$\begin{cases} \eta_{\hat{x}}(1) = -\dfrac{1}{6}; \\ \\ \eta_{\hat{x}}(2) = \dfrac{2}{6}; \\ \\ \eta_{\hat{x}}(3) = \dfrac{5}{6}; \end{cases} \quad \text{and} \quad \begin{cases} \eta_{\Delta_1\hat{x}}(1) = -\dfrac{1}{2}; \\ \\ \eta_{\Delta_1\hat{x}}(1) = 0; \\ \\ \eta_{\Delta_1\hat{x}}(1) = \dfrac{1}{2}. \end{cases} \tag{5.46}$$

Consequently,

$$\begin{cases} \hat{x}_3 = \dfrac{1}{6}(5x_3 + 2x_2 - x_1); \\ \\ \Delta_1\hat{x}_3 = \dfrac{1}{2}(x_3 - x_1). \end{cases} \tag{5.47}$$

Note that the following condition is satisfied forever for the weight coefficients:

$$\begin{cases} \sum_{i=1}^{N} \eta_{\hat{x}}(i) = 1; \\ \\ \sum_{i=1}^{N} \eta_{\Delta_1\hat{x}}(i) = 0. \end{cases} \tag{5.48}$$

Parallel with the estimation of linear target track parameters, the error correlation matrix arising under the definition of linear target track parameter estimations has to be determined by (5.33). Under equally discrete but not uniformly precise readings, the error correlation matrix of linear target track parameter estimations is given by

$$\mathbf{\Psi} = \frac{1}{G_N}\begin{Vmatrix} h_N & g_N \\ g_N & f_N \end{Vmatrix}. \tag{5.49}$$

Under uniformly precise measures, the elements of this matrix depend only on the number of measurements:

$$\mathbf{\Psi}_N = \begin{Vmatrix} \dfrac{2(2N-1)}{N(N+1)} & \dfrac{6}{N(N+1)} \\ \\ \dfrac{6}{N(N+1)} & \dfrac{12}{N(N^2-1)} \end{Vmatrix} \sigma_x^2. \tag{5.50}$$

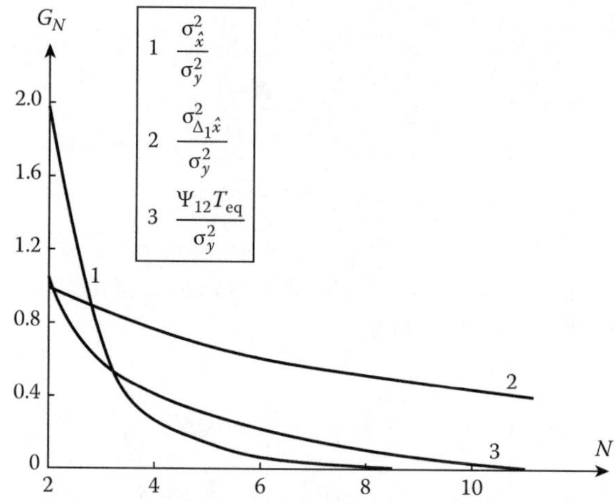

FIGURE 5.3 Coefficient of accuracy in determination of the normalized elements of correlation error matrix of target linear track parameters as a function of the number of measurements.

For example, at $N = 3$ the error correlation matrix of parameter estimation of linear target track takes the following form:

$$\mathbf{\Psi}_3 = \begin{Vmatrix} \frac{5}{6} & \frac{1}{2} \\ \frac{1}{2} & \frac{1}{2} \end{Vmatrix} \sigma_x^2. \tag{5.51}$$

Consequently, the variance of error of the smoothed coordinate estimation by three uniformly precise measurement is equal to 5/6 of the variance of individual measurement error. The variance of error of the coordinate increment estimation is only one half of the variance of error of individual measurement error owing to inaccurate target velocity estimation, and the correlation moment of relation between the errors of coordinate estimation and its increment is equal to one half of the variance of error of individual coordinate measurement. Figure 5.3 represents a coefficient of accuracy under determination of the normalized elements of error correlation matrix of linear target track parameter estimation versus the number of measurements. As follows from Figure 5.3, to obtain the required accuracy of estimations there is a need to carry out a minimum of 5–6 measurements.

Turning again to the algorithms of linear target track parameter estimations (5.43) and (5.44), we can easily see that these algorithms are the algorithms of nonrecursive filters and the weight coefficients $\eta_{\hat{x}}(i)$ and $\eta_{\Delta_1\hat{x}}(i)$ form a sequence of impulse response values of these filters. To apply filtering processing using these filters, there is a need to carry out N multiplications between the measured coordinate values at each step, that is, after each coordinate measurement, and the corresponding weight coefficients and, additionally, N summations of the obtained partial products. To store in a memory device $(N - 1)$ records of previous measurements, there is a need to use high-capacity memory. As a result, to realize such a filter, taking into account that $N > 5$, there is a need to use a high-capacity memory, and this realization is complex.

5.4.3 Algorithm of Optimal Estimation of Second-Order Polynomial Target Track Parameters

When we use the second-order polynomial to represent the target track, the coordinate $\hat{x}_N$, the first coordinate increment $\Delta_1\hat{x}_N$, and the second coordinate increment $\Delta_2\hat{x}_N$ are considered as estimated parameters. As in the previous section, we believe that measurements are equally sampled with the period T_{eq} and measure errors are uncorrelated. In this case, the coordinate is varied according to the following law:

$$x_i = x(t_i) = x_N - (N-i)\Delta_1 x_N - (N-i)^2\Delta_2 x_N, \quad i = 1,2,\ldots,N, \tag{5.52}$$

where

$$\Delta_1 x_n = T_{\text{eq}}\dot{x}_N \tag{5.53}$$

is the first increment of the coordinate x;

$$\Delta_2 x_N = 0.5T_{\text{eq}}^2\ddot{x}_N \tag{5.54}$$

is the second increment of the coordinate x; $\dot{x}_N$ is the velocity of changes of the coordinate x; $\ddot{x}_N$ is the acceleration by the coordinate x.

An order of obtaining the corresponding formulas to estimate the parameters of a second-order polynomial target track is the same as in the case of a linear target track. Omitting intermediate mathematical transformations, we can write the final formulas in the following form:

$$\hat{x}_N = \frac{1}{I_N}\left[\alpha_N\sum_{i=1}^{N} w_i x_i + \gamma_N\sum_{i=1}^{N} w_i(N-i)x_i + \delta_N\sum_{i=1}^{N} w_i(N-i)^2 x_i\right]; \tag{5.55}$$

$$\Delta_1\hat{x}_N = \frac{1}{I_N}\left[\gamma_N\sum_{i=1}^{N} w_i x_i - \xi_N\sum_{i=1}^{N} w_i(N-i)x_i + \eta_N\sum_{i=1}^{N} w_i(N-i)^2 x_i\right]; \tag{5.56}$$

$$\Delta_2\hat{x}_N = \frac{1}{I_N}\left[\delta_N\sum_{i=1}^{N} w_i x_i - \eta_N\sum_{i=1}^{N} w_i(N-i)x_i + \mu_N\sum_{i=1}^{N} w_i(N-i)^2 x_i\right]; \tag{5.57}$$

where

$$\alpha_N = h_N e_N - d_N^2; \tag{5.58}$$

$$\gamma_N = g_N e_N - h_N d_N; \tag{5.59}$$

$$\delta_N = g_N d_N - h_N^2; \tag{5.60}$$

$$\xi_N = f_N e_N - h_N^2; \tag{5.61}$$

$$\eta_N = f_N d_N - g_N h_N; \tag{5.62}$$

$$\mu_N = f_N h_N - g_N^2; \tag{5.63}$$

$$d_N = \sum_{i=1}^{N} w_i (N-i)^3; \tag{5.64}$$

$$e_N = \sum_{i=1}^{N} w_i (N-i)^4; \tag{5.65}$$

$$I_N = \left[e_N \left(f_N h_N - g_N^2 \right) + d_N (g_N h_N - f_N d_N) + h_N \left(g_N d_N - h_N^2 \right) \right]. \tag{5.66}$$

The error correlation matrix of second-order polynomial target track parameter estimation takes the following form:

$$\mathbf{\Psi}_N = \mathbf{B}_N^{-1} = \frac{1}{I_N} \begin{Vmatrix} \alpha_N & -\gamma_N & \delta_N \\ -\gamma_N & \xi_N & -\eta_N \\ \delta_N & -\eta_N & -\mu_N \end{Vmatrix}. \tag{5.67}$$

In the case of uniformly precise measurements, at $w_i = w$ from (5.58) through (5.65) it follows that

$$d_N = w_i \sum_{i=1}^{N} (N-i)^3 = \frac{N^2 (N-1)^2}{4} w; \tag{5.68}$$

$$e_N = w \sum_{i=1}^{N} (N-i)^4 = \frac{N(N-1)(2N-1)(3N^2-3N-1)}{30} w. \tag{5.69}$$

The required target track parameter estimations can be determined in the following form:

$$\hat{x}_N = \sum_{i=1}^{N} \eta_{\hat{x}}(i) x_i; \tag{5.70}$$

$$\Delta_1 \hat{x}_N = \sum_{i=1}^{N} \eta_{\Delta_1 \hat{x}}(i) x_i; \tag{5.71}$$

$$\Delta_2 \hat{x}_N = \sum_{i=1}^{N} \eta_{\Delta_2 \hat{x}}(i) x_i, \tag{5.72}$$

where $\eta_{\hat{x}}(i)$, $\eta_{\Delta_1 \hat{x}}(i)$, and $\eta_{\Delta_2 \hat{x}}(i)$ are the discrete weight coefficients of measure records under the determination of estimations of the coordinate, the first coordinate increment, and the second coordinate increment, respectively:

$$\eta_{\hat{x}}(i) = \frac{3\left[(N+1)(N+2) - 2i(4N+3) + 10i^2 \right]}{N(N+1)(N+2)}; \tag{5.73}$$

$$\eta_{\Delta_1\hat{x}}(i) = \frac{6\left[(N+1)(N+2)(6N-7) - 2i(16N^2-19) + 30i^2(N-1)\right]}{N(N^2-1)(N^2-4)}; \tag{5.74}$$

$$\eta_{\Delta_2\hat{x}}(i) = \frac{30\left[(N+1)(N+2) - 6i(N+1) + 6i^2\right]}{N(N^2-1)(N^2-4)}. \tag{5.75}$$

Formulas (5.70) through (5.75) show us that in the case of equally discrete and uniformly precise measures, the optimal estimation of the second-order polynomial target track parameters is determined by the weighted summing of measured coordinate values. The weight coefficients are the functions of the sample size N and the sequence number of sample i in the processing series. When the sample size is minimal, that is, $N = 3$, the target track parameters are determined by the following formulas:

$$\hat{x}_3 = x_3; \tag{5.76}$$

$$\Delta_1\hat{x}_3 = \dot{x}_3 T_{\text{eq}} = 0.5x_3 - 2x_2 + 1.5x_1; \tag{5.77}$$

$$\Delta_2\hat{x}_3 = 0.5\ddot{x}_3 T_{\text{eq}}^2 = 0.5(x_3 - 2x_2 + x_1). \tag{5.78}$$

In the considered case, the error correlation matrix of target track parameter estimation is produced from the matrix given by (5.67) by substituting for I_N, α_N, γ_N, δ_N, ξ_N, η_N, and μ_N of the corresponding values f_N, g_N, h_N, d_N, e_N given by (5.42), (5.68), and (5.69). As a result, the elements of the error correlation matrix of target track parameter estimation $\mathbf{\Psi}_N$ can be presented in the following form:

$$\begin{cases} \psi_{11} = \dfrac{3(3N^2-3N+2)}{N(N+1)(N+2)}\sigma_x^2; \\ \psi_{12} = \psi_{21} = -\dfrac{18(2N-1)}{N(N+1)(N+2)}\sigma_x^2; \\ \psi_{13} = \psi_{31} = \dfrac{30}{N(N+1)(N+2)}\sigma_x^2; \\ \psi_{22} = \dfrac{12(2N-1)(8N-11)}{N(N^2-4)(N^2-1)}\sigma_x^2; \\ \psi_{23} = \psi_{32} = -\dfrac{180}{N(N^2-4)(N+1)}\sigma_x^2; \\ \psi_{33} = \dfrac{180}{N(N^2-4)(N^2-1)}\sigma_x^2; \end{cases} \tag{5.79}$$

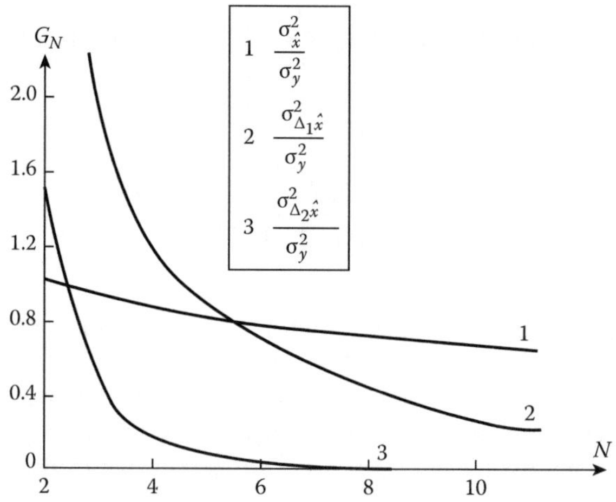

FIGURE 5.4 Coefficient of accuracy in determination of the normalized elements of error correlation matrix of second-order polynomial target track parameter estimation versus the number of measurements.

For example, at $N = 3$ we obtain

$$\boldsymbol{\Psi}_3 = \begin{Vmatrix} 1 & -\frac{3}{2} & \frac{1}{2} \\ -\frac{3}{2} & \frac{13}{2} & -3 \\ \frac{1}{2} & -3 & \frac{3}{2} \end{Vmatrix} \sigma_x^2. \tag{5.80}$$

Coefficient of accuracy under determination of the normalized elements of error correlation matrix of the second-order polynomial target track parameter estimation versus the number of measurements is shown in Figure 5.4. Comparison of the diagonal elements of the matrix (5.80) that are characteristics of accuracy under estimation of the second-order polynomial target track coordinate and its first increment with analogous elements of the error correlation matrix of linear target track parameter estimation (see Figure 5.4) shows that at low values of N, the accuracy of linear target track parameter estimation is much higher than the accuracy of the second-order polynomial target track parameter estimation. Consequently, within the limits of small observation intervals, it is worth presenting the target track as using the first-order polynomial. In this case, the high quality of cancellation of random errors of target track parameter estimation by filtering is guaranteed. Dynamic errors arising owing to mismatching between the target moving hypotheses can be neglected as a consequence of the narrow approximated part of the target track.

5.4.4 Algorithm of Extrapolation of Target Track Parameters

The extrapolation problem of target track parameters is to define the target track parameter estimations at the point that is outside the observation interval using the magnitudes of target track parameters determined during the last observation or using a set of observed coordinate values. Under polynomial representation of independent coordinates, the target track parameters extrapolated using the time τ_{ex} are defined by the following formulas:

$$\hat{x}_{ex} = \hat{x}_N + \hat{\dot{x}}_N\tau_{ex} + \hat{\ddot{x}}\frac{\tau_{ex}^2}{2} + \cdots + \hat{x}_N^{(s)}\frac{\tau_{ex}^{(s)}}{s!}, \tag{5.81}$$

$$\hat{x}_{\text{ex}} = \hat{x}_N + \hat{\dot{x}}_N \tau_{\text{ex}} + \hat{\ddot{x}} \frac{\tau_{\text{ex}}^2}{2} + \cdots + \hat{x}_N^{(s)} \frac{\tau_{\text{ex}}^{(s-1)}}{(s-1)!}, \tag{5.82}$$

$$\hat{x}_{\text{ex}}^{(s)} = \hat{x}_N^{(s)}, \tag{5.83}$$

where $\tau_{\text{ex}} = t_{\text{ex}} - t_N$ is the interval of extrapolation time. Equations 5.81 through 5.83 allow us to define the extrapolated coordinates for each specific case of target track representation. For example, in the case of linear target track at equally discrete coordinate measurement we obtain the following:

$$\hat{x}_{N+p} = \hat{x} + \hat{\dot{x}}_N \frac{\tau_{\text{ex}}}{T_0} T_{\text{eq}} = \hat{x}_N + \Delta_1 \hat{x}_N \frac{\tau_{\text{ex}}}{T_0}, \tag{5.84}$$

$$\Delta_1 \hat{x}_{N+p} = \Delta_1 \hat{x}_N. \tag{5.85}$$

Substituting in (5.84) and (5.85) the corresponding formulas for smoothed parameters, we obtain

$$x_{N+p} = \frac{1}{G_N} \left[\left(h_N + \frac{\tau_{\text{ex}}}{T_0} g_N \right) \sum_{i=1}^{N} w_i x_i - \left(g_N + \frac{\tau_{\text{ex}}}{T_0} f_N \right) \sum_{i=1}^{N} w_i (N-i) x_i \right]. \tag{5.86}$$

If, additionally, these measurements are uniformly precise, we obtain

$$\hat{x}_{N+p} = \sum_{i=1}^{N} \left[\eta_{\hat{x}}(i) + \frac{\tau_{\text{ex}}}{T_{\text{eq}}} \eta_{\Delta_1 \hat{x}}(i) \right] x_i. \tag{5.87}$$

At $\tau_{\text{ex}} = T_{\text{eq}}$, we have

$$\hat{x}_{N+1} = \sum_{i=1}^{N} \eta_{\hat{x}_{N+1}}(i) x_i, \tag{5.88}$$

where

$$\eta_{\hat{x}_{N+1}}(i) = \frac{2(3-N-2)}{N(N-1)} \tag{5.89}$$

is the weight function of measured coordinate magnitudes under the target track parameter extrapolation per one measurement period.

The error correlation matrix of linear target track parameter extrapolation takes the following form at equally discrete coordinate measurement:

$$\mathbf{\Psi}_{N+p} = \frac{1}{G_N} \left\| \begin{matrix} h_N + 2\dfrac{\tau_{\text{ex}}}{T_0} g_N + \left(\dfrac{\tau_{\text{ex}}}{T_0} \right)^2 & g_N + \dfrac{\tau_{\text{ex}}}{T_0} f_N \\ g_N + \dfrac{\tau_{\text{ex}}}{T_0} f_N & f_N \end{matrix} \right\|. \tag{5.90}$$

If, additionally, these measurements are uniformly precise, we obtain the following elements of the error correlation matrix of linear target track parameter extrapolation:

$$\psi_{11}=\frac{2\left[(N-1)(2N-1)+6(\tau_{ex}/T_{eq})(N-1)+6\left(\tau_{ex}/T_{eq}\right)^2\right]}{N(N^2-1)}\sigma_x^2; \tag{5.91}$$

$$\psi_{12}=\psi_{21}=\frac{6\left[(N-1)+(\tau_{ex}/T_{eq})\right]}{N(N^2-1)}\sigma_x^2; \tag{5.92}$$

$$\psi_{22}=\frac{12}{N(N^2-1)}\sigma_x^2. \tag{5.93}$$

If the independent target track parameter coordinate is represented by the polynomial of the second order, the target track parameter extrapolation formulas and the error correlation matrix of linear target track parameter extrapolation are obtained in an analogous way.

5.4.5 Dynamic Errors of Target Track Parameter Estimation Using Polar Coordinate System

Sometimes we use the polynomial deterministic model of target track parameter estimation to represent changes in polar coordinate r_{tg} and β_{tg} of the target track. The polynomial representation of polar coordinates does not reflect the true law of target moving and allows us only to approximate with the previously given accuracy the target movement law within the limits of finite observation interval. In the case of a uniform and straight-line target moving with arbitrary course at the fixed altitude with respect to the stationary radar system, the law of changing the polar coordinates (see Figure 5.5) is determined by the following form:

$$r_{tg}(t)=\sqrt{r_0^2+[V_{tg}(t-t_0)]^2}; \tag{5.94}$$

$$\beta_{tg}(t)=\beta_0+\arctan\left[\frac{V_{tg}(t-t_0)}{r_0}\right], \tag{5.95}$$

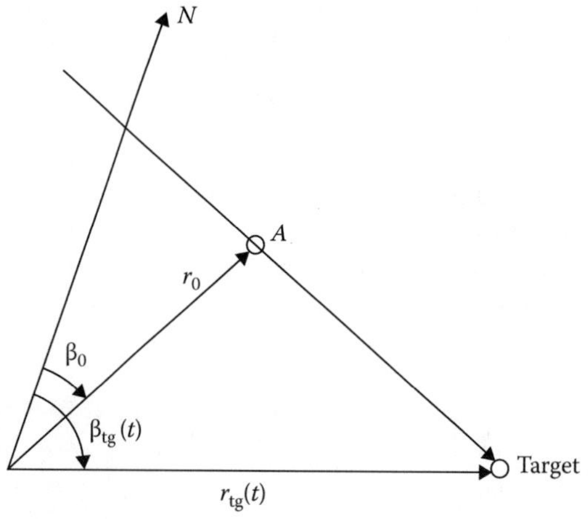

FIGURE 5.5 Target moving in polar coordinates.

where

r_0 and β_0 are the range and azimuth of the nearest to origin point on the target track (the point A on Figure 5.5)

t_0 is the time of target flight to the point A

V_{tg} is the target velocity. In the case of a target moving on circular arc within the limits of maneuver segment or mutual target moving and replacement of the CRS, we obtain the formulas that are analogous to (5.94) and (5.95) by structure but complex by essence

As follows from (5.94) and (5.95), even in the simplest cases of linear target movement and a stationary CRS, the polar coordinates are changed according to nonlinear laws. This nonlinearity becomes intensified as it is passed to complex target moving models, especially under the moving CRS. Inconsistency between the polynomial model and nonlinear character in changing the vector of estimated target track parameters leads to dynamical errors in smoothing, which can be presented in the form of differences between the true value of the vector of estimated target track parameters and the mathematical expectation of estimate of this vector, namely,

$$\Delta\boldsymbol{\theta}_g = [\boldsymbol{\theta} - E(\hat{\boldsymbol{\theta}})]. \tag{5.96}$$

To describe the dynamical errors of estimations of the target track parameters by algorithms that are synthesized by the maximum likelihood criterion and using the target track polynomial model, we employ the well-known theory of errors in the approximation of arbitrary continuously differentiable function $f(t)$ within the limits of the interval $t_N - t_0$ by the polynomial function of the first or second order using the technique of least squares. This function takes the following form:

$$\begin{cases} \mathcal{F}_1(t) = a_0 + a_1(t - t_0), \\ \mathcal{F}_2(t) = a_0 + a_1(t - t_0) + 0.5a_2(t - t_0)^2. \end{cases} \tag{5.97}$$

In the case of linear target track, the coefficients a_0 and a_1 are defined from the following system of equations:

$$\begin{cases} \int_{t_0}^{t_N} [f(t) - a_0 - a_1(t - t_0)]dt = 0, \\ \int_{t_0}^{t_N} [f(t) - a_0 - a_1(t - t_0)]t\,dt = 0. \end{cases} \tag{5.98}$$

Solution of the equation system (5.98) is found under expansion of the function $f(t)$ by Taylor series at the point $0.5(t_N + t_0)$, that is, in the middle of the approximation interval, namely,

$$f(t) = f[0.5(t_0 + t_N)] + \sum_{k=1}^{\infty} \frac{b_k}{k!}[t - 0.5(t_0 + t_N)]^k; \tag{5.99}$$

$$b_k = \left. \frac{d^k[f(t)]}{dt^k} \right|_{t=0.5(t_0+t_N)}. \tag{5.100}$$

The error of approximation at the point $t = t_N$ can be determined in the following form:

$$\Delta f(t_N) = f(t_N) - \mathcal{F}_1(t_N). \tag{5.101}$$

Investigations and computations made by this procedure give us the following results. Under linear approximation of polar coordinates, the maximum dynamic errors take the following form:

$$\begin{cases} \Delta r_{\text{d}}^{\max} \approx \dfrac{(N-1)^2 T_{\text{eq}}^2 V_{\text{tg}}^2}{12 r_{\min}}, \\ \Delta \beta_{\text{d}}^{\max} \approx \dfrac{\sqrt{3}(N-1)^2 T_{\text{eq}}^2 V_{\text{tg}}^2}{32 r_{\min}^2}; \end{cases} \tag{5.102}$$

$$\begin{cases} \Delta \dot{r}_{\text{d}}^{\max} \approx \dfrac{(N-1) T_{\text{eq}} V_{\text{tg}}^2}{2 r_{\min}}, \\ \Delta \dot{\beta}_{\text{d}}^{\max} \approx \dfrac{3\sqrt{3}(N-1) T_{\text{eq}} V_{\text{tg}}^2}{16 r_{\min}^2}; \end{cases} \tag{5.103}$$

where N is the number of coordinate readings within the limits of the observation interval $t_N - t_0$. In the case of quadratic approximation of the polar coordinates, the formulas for the maximum dynamic errors are more complex and we omit them.

Comparison of the dynamic errors of approximation of the polar coordinates by the polynomials of the first and second power at the same values of the target track parameters V_{tg}, $r_{\min}$, and T_{eq} and the sample size $(N-1)T_{\text{eq}}$ shows that the dynamic errors under linear approximation are approximately one order higher than those under quadratic approximation. Moreover, under approximation of $r_{\text{tg}}(t)$ and $\beta_{\text{tg}}(t)$ by the polynomials of the second power, the dynamical errors are small in comparison with random errors and we can neglect them. However, as shown by simulation, if the target is in an in-flight maneuver and the CRS is moving, the dynamical errors increase essentially, because nonlinearity in changing the polar coordinates increases sharply at the same time.

5.5 RECURRENT FILTERING ALGORITHMS OF UNDISTORTED POLYNOMIAL TARGET TRACK PARAMETERS

5.5.1 Optimal Filtering Algorithm Formula Flowchart

Methods of estimation of the target track parameters based on the fixed sample of measured coordinates, which are discussed in the previous sections, are used, as a rule, at the beginning of the detected target trajectory. Implementation of these methods in the course of tracking is not worthwhile owing to the complexity and insufficient accuracy defined by the small magnitude of employed measures. Accordingly, there is a need to employ the recurrent algorithms ensuring a sequential, that is, after each new coordinating measurement, adjustment of target track parameters and their filtering purposes. At the recurrent filter output, we obtain the target track parameter estimations caused by the last observation. For this reason, a process of recurrent evaluation is called further the sequential filtering, and corresponding algorithms are called the algorithms of sequential filtering of target track parameters.

In a general case, the problem of synthesis of the sequential filtering algorithm for a set or vector of target track parameters is defined in the following way. Let the model of the undistorted target track be given by the following difference equation:

$$\boldsymbol{\theta}_n = \boldsymbol{\Phi}_n \boldsymbol{\theta}_{n-1}, \tag{5.104}$$

and the observed random sequence is given by

$$\mathbf{Y}_n = \mathbf{H}_n \boldsymbol{\theta}_n + \Delta\mathbf{Y}_n, \tag{5.105}$$

where

$\boldsymbol{\theta}_n$ is the $(s + 1)$-dimensional vector of filtered target track parameters
$\mathbf{Y}_n$ is the l-dimensional vector of observed coordinates
$\Delta\mathbf{Y}_n$ is the l-dimensional vector of measurement errors

The sequence of these vectors is the uncorrelated random sequence with zero mean and known correlation matrix $\mathbf{R}_n$; $\boldsymbol{\Phi}_n$ and $\mathbf{H}_n$ are the known matrices defined in Section 4.2. Further, we consider that $\hat{\boldsymbol{\theta}}_{n-1}$ is the vector of estimations of target track parameters determined by $(n - 1)$ coordinating measurements; $\boldsymbol{\Psi}_{n-1}$ is the corresponding correlation matrix of estimation errors. There is a need to obtain the formulas for $\hat{\boldsymbol{\theta}}_n$, using for this purpose the vector $\boldsymbol{\theta}_{n-1}$ of previous estimations and results of new measurement $\mathbf{Y}_n$, and also the formula for the error correlation matrix $\boldsymbol{\Psi}_n$ based on the known matrices $\boldsymbol{\Psi}_{n-1}$ and $\mathbf{R}_n$.

In accordance with a general estimation theory, the optimal solution of the sequential filtering problem is reduced, first of all, to a definition of the a posteriori pdf of the filtered target track parameters, because this pdf possesses all information obtained from a priori sources and observation results. Differentiating the a posteriori pdf, we can obtain the optimal estimation of target track parameters, which are interesting for us, by the maximum a posteriori probability criterion. Henceforth, we consider the optimal sequential filtering, namely, in this sense.

Thus, let the estimation vector $\hat{\boldsymbol{\theta}}_{n-1}$ of the target track parameter vector $\boldsymbol{\theta}_n$ obtained by observations of the previous $(n - 1)$ coordinating measurement be given. We assume that pdf of the vector $\hat{\boldsymbol{\theta}}_{n-1}$ is the normal Gaussian with the mathematical expectation $\boldsymbol{\theta}_{n-1}$ and correlation matrix $\boldsymbol{\Psi}_{n-1}$. The vector $\hat{\boldsymbol{\theta}}_{n-1}$ is extrapolated during the next nth measurement in accordance with the equation

$$\hat{\vartheta}_{n|n-1} = \boldsymbol{\theta}_{\text{ex}_n} = \boldsymbol{\Phi}_n \boldsymbol{\theta}_{n-1}. \tag{5.106}$$

The specific form of the extrapolation matrix $\boldsymbol{\Phi}_n$ is defined by the target track model. For example, in the case of the only coordinate x_n that can be presented by the polynomial of the second order

$$\theta_{n-1} = \left\| \hat{x}_{n-1} \; \hat{\dot{x}}_{n-1} \; \hat{\ddot{x}}_{n-1} \right\|^T, \tag{5.107}$$

we have

$$\boldsymbol{\Phi}_n = \begin{Vmatrix} 1 & \tau_{\text{ex}} & 0.5\tau_{\text{ex}}^2 \\ 0 & 1 & \tau_{\text{ex}} \\ 0 & 0 & 1 \end{Vmatrix} \tag{5.108}$$

and (5.107) can be presented in the following form:

$$\hat{\boldsymbol{\theta}}_{\text{ex}_n} = \begin{Vmatrix} \hat{x}_{\text{ex}_n} \\ \hat{\dot{x}}_{\text{ex}_n} \\ \hat{\ddot{x}}_{\text{ex}_n} \end{Vmatrix} = \begin{Vmatrix} 1 & \tau_{\text{ex}} & 0.5\tau_{\text{ex}}^2 \\ 0 & 1 & \tau_{\text{ex}} \\ 0 & 0 & 1 \end{Vmatrix} \begin{Vmatrix} \hat{x}_{n-1} \\ \hat{\dot{x}}_{n-1} \\ \hat{\ddot{x}}_{n-1} \end{Vmatrix}, \tag{5.109}$$

where $\tau_{\text{ex}} = t_n - t_{n-1}$. The correlation matrix $\boldsymbol{\Psi}_{n-1}$ is also extrapolated at the instant t_n by the following formula:

$$\boldsymbol{\Psi}_{n|n-1} = \boldsymbol{\Psi}_{\text{ex}_n} = \boldsymbol{\Phi}_n \boldsymbol{\Psi}_{n-1} \boldsymbol{\Phi}_n^T. \tag{5.110}$$

Taking into consideration a linearity property of the extrapolation operator $\boldsymbol{\Phi}_n$, the pdf of the vector of extrapolated target track parameters will also be normal Gaussian:

$$p(\hat{\boldsymbol{\theta}}_{\text{ex}_n}) = C_1 \exp\left[-0.5(\hat{\boldsymbol{\theta}}_{\text{ex}_n} - \boldsymbol{\theta}_n)^T \boldsymbol{\Psi}_{\text{ex}_n}^{-1} (\hat{\boldsymbol{\theta}}_{\text{ex}_n} - \boldsymbol{\theta}_n)\right], \tag{5.111}$$

where
- $\boldsymbol{\theta}_n$ is the vector of true values of target track parameters at the instant t_n
- C_1 is the normalizing factor

The pdf given by (5.111) is the a priori pdf of the vector of estimated target track parameters before the next nth measurement. At the instant t_n, the regular measurement of target coordinates is carried out. In a general case of a 3-D CRS we obtain

$$\mathbf{Y}_n = \| r_n \;\; \beta_n \;\; \varepsilon_n \|^T. \tag{5.112}$$

It is assumed that the errors under coordinating measurements are subjected to the normal Gaussian distribution and uncorrelated between each other for neighbor radar antenna scanning. Consequently,

$$p(\mathbf{Y}_n | \boldsymbol{\theta}_n) = C_2 \exp\left[-0.5(\mathbf{Y}_n - \mathbf{H}_n\boldsymbol{\theta}_n)^T \mathbf{R}_n^{-1} (\mathbf{Y}_n - \mathbf{H}_n\boldsymbol{\theta}_n)\right], \tag{5.113}$$

where $\mathbf{R}_n^{-1}$ is the inverse correlation matrix of measure errors.

Under the assumption that there is no correlation between the measurement errors in the case of the neighbor radar antenna scanning, the a posteriori pdf for the target track parameter $\boldsymbol{\theta}_n$ after n measurements is defined using the Bayes formula

$$p(\hat{\boldsymbol{\theta}}_n | \mathbf{Y}_n) = C_3 p(\hat{\boldsymbol{\theta}}_{\text{ex}_n}) p(\mathbf{Y}_n | \boldsymbol{\theta}_n), \tag{5.114}$$

and owing to the fact that pdfs of components are the normal Gaussian, the a posteriori pdf (5.115) will also be the normal Gaussian:

$$p(\hat{\boldsymbol{\theta}}_n | \mathbf{Y}_n) = C_4 \exp\left[-0.5(\hat{\boldsymbol{\theta}}_n - \boldsymbol{\theta}_n)^T \boldsymbol{\Psi}_n^{-1} (\hat{\boldsymbol{\theta}}_n - \boldsymbol{\theta}_n)\right], \tag{5.115}$$

where
- $\hat{\boldsymbol{\theta}}_n$ is the vector of estimated target track parameters by n measurements
- $\boldsymbol{\Psi}_n$ is the error correlation matrix of estimated target track parameters

In the case of the normal Gaussian distribution, $\max p(\hat{\boldsymbol{\theta}}_n | \mathbf{Y}_n)$ corresponds to the mathematical expectation of the vector of estimated target track parameters. Consequently, the problem of target track parameter estimation by a posteriori probability maximum is reduced, in our case, to definition of the parameters $\hat{\boldsymbol{\theta}}_n$ and $\boldsymbol{\Psi}_n$ in (5.115). Using (5.111) through (5.114) for pdfs included in (5.115), we obtain after the logarithmic transformations the following formula:

$$(\hat{\boldsymbol{\theta}}_n - \boldsymbol{\theta}_n)^T \boldsymbol{\Psi}_n^{-1} (\hat{\boldsymbol{\theta}}_n - \boldsymbol{\theta}_n) = (\hat{\boldsymbol{\theta}}_{\text{ex}_n} - \boldsymbol{\theta}_n)^T \boldsymbol{\Psi}_{\text{ex}_n}^{-1} (\hat{\boldsymbol{\theta}}_{\text{ex}_n} - \boldsymbol{\theta}_n) + (\mathbf{Y}_n - \mathbf{H}_n \boldsymbol{\theta}_n)^T \mathbf{R}_n^{-1} (\mathbf{Y}_n - \mathbf{H}_n \boldsymbol{\theta}_n) + \text{const.} \tag{5.116}$$

From this equation, we can find that

$$\begin{cases} \boldsymbol{\Psi}_n^{-1} = \boldsymbol{\Psi}_{\text{ex}_n}^{-1} + \mathbf{H}_n^T \mathbf{R}_n^{-1} \mathbf{H}_n; \\ \hat{\boldsymbol{\theta}}_n = \hat{\boldsymbol{\theta}}_{\text{ex}_n} + \boldsymbol{\Psi}_n \mathbf{H}_n^T \mathbf{R}_n^{-1} (\mathbf{Y}_n - \mathbf{H}_n \hat{\boldsymbol{\theta}}_{\text{ex}_n}). \end{cases} \tag{5.117}$$

Taking into consideration (5.106) and (5.110) for $\hat{\boldsymbol{\theta}}_{\text{ex}_n}$ and $\boldsymbol{\Psi}_{\text{ex}_n}$, the main relationships of the optimal sequential filtering algorithm can be presented in the following form:

$$\begin{cases} \hat{\boldsymbol{\theta}}_{\text{ex}_n} = \boldsymbol{\Phi}_n \hat{\boldsymbol{\theta}}_{n-1}; \\ \boldsymbol{\Psi}_{\text{ex}_n} = \boldsymbol{\Phi}_n \boldsymbol{\Psi}_{n-1} \boldsymbol{\Phi}_n^T; \\ \boldsymbol{\Psi}_n^{-1} = \boldsymbol{\Psi}_{\text{ex}_n}^{-1} + \mathbf{H}_n^T \mathbf{R}_n^{-1} \mathbf{H}_n; \\ \mathbf{G}_n = \boldsymbol{\Psi}_n \mathbf{H}_n^T \mathbf{R}_n^{-1}; \\ \hat{\boldsymbol{\theta}}_n = \hat{\boldsymbol{\theta}}_{\text{ex}_n} + \mathbf{G}_n (\mathbf{Y}_n - \mathbf{H}_n \hat{\boldsymbol{\theta}}_{\text{ex}_n}). \end{cases} \tag{5.118}$$

The system of equation (5.118) represents the optimal recurrent linear filtering algorithm and is called the Kalman filtering equations [9–16]. These equations can be transformed into a more convenient form for realization:

$$\begin{cases} \hat{\boldsymbol{\theta}}_{\text{ex}_n} = \boldsymbol{\Phi}_n \hat{\boldsymbol{\theta}}_{n-1}; \\ \boldsymbol{\Psi}_{\text{ex}_n} = \boldsymbol{\Phi}_n \boldsymbol{\Psi}_{n-1} \boldsymbol{\Phi}_n^T; \\ \mathbf{G}_n = \boldsymbol{\Psi}_{\text{ex}_n} \mathbf{H}_n^T \left(\mathbf{H}_n \boldsymbol{\Psi}_{\text{ex}_n} \mathbf{H}_n^T + \mathbf{R}_n \right)^{-1}; \\ \hat{\boldsymbol{\theta}}_n = \hat{\boldsymbol{\theta}}_{\text{ex}_n} + \mathbf{G}_n (\mathbf{Y}_n - \mathbf{H}_n \hat{\boldsymbol{\theta}}_{\text{ex}_n}); \\ \boldsymbol{\Psi}_n = \boldsymbol{\Psi}_{\text{ex}_n} - \mathbf{G}_n \mathbf{H}_n \boldsymbol{\Psi}_{\text{ex}_n}. \end{cases} \tag{5.119}$$

A general filter flowchart realizing (5.119) is shown in Figure 5.6. A discrete optimal recurrent filter possesses the following properties:

- Filtering equations have recurrence relations and can be realized well by a computer system.
- Filtering equations represent simultaneously a description of procedure to realize this filter; in doing so, a part of the filter is similar to the model of target track (compare Figures 5.2 and 5.6).

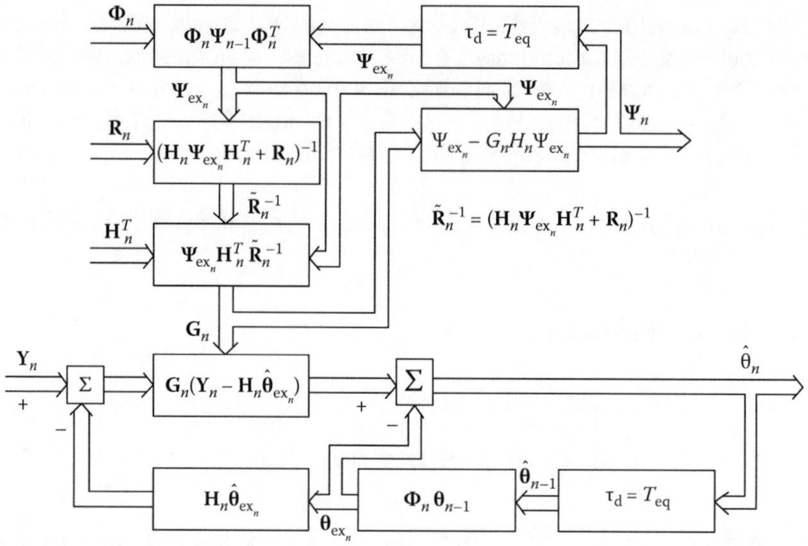

FIGURE 5.6 General Kalman filter flowchart.

- The correlation matrix of errors of target track parameter estimation Ψ_n is computed independently of measuring Y_n; consequently, if statistical characteristics of measuring errors are given, then the correlation matrix Ψ_n can be computed in advance and stored in a memory device; this essentially reduces the time of realization of target track parameter filtering.

5.5.2 Filtering of Linear Target Track Parameters

Formulas of the sequential filtering algorithm of linear target track parameters are obtained directly from (5.119). The target track coordinate and velocity of its changes at the instant of last nth measuring are considered as the filtered parameters. We assume that all measurements are equally discrete with the period T_{eq}.

1. Assume the vector of filtered target track parameters

$$\hat{\theta}_{n-1} = \begin{Vmatrix} \hat{x}_{n-1} \\ \hat{\dot{x}}_{n-1} \end{Vmatrix} \tag{5.120}$$

and the error correlation matrix of target track parameter estimations

$$\Psi_{n-1} = \frac{1}{\gamma_{n-1}} \begin{Vmatrix} h_{n-1} & \dfrac{g_{n-1}}{T_{eq}} \\ \dfrac{g_{n-1}}{T_{eq}} & \dfrac{f_{n-1}}{T_{eq}^2} \end{Vmatrix} \tag{5.121}$$

are obtained by the $(n-1)$ previous measurements of the coordinate x.

2. In accordance with the adopted target track model, an extrapolation of the target track coordinates at the instant of next measuring is carried out by the formula

$$\hat{\boldsymbol{\theta}}_{\text{ex}_n} = \begin{Vmatrix} \hat{x}_{\text{ex}_n} \\ \hat{\dot{x}}_{\text{ex}_n} \end{Vmatrix} = \begin{Vmatrix} \hat{x}_{n-1} + \hat{\dot{x}}_{n-1}T_{\text{eq}} \\ \hat{\dot{x}}_{n-1} \end{Vmatrix}. \tag{5.122}$$

3. The correlation matrix of extrapolation errors is computed by the following formula:

$$\boldsymbol{\Psi}_{\text{ex}_n} = \boldsymbol{\Phi}_n \boldsymbol{\Psi}_{n-1} \boldsymbol{\Phi}_n^T. \tag{5.123}$$

The final version after elementary mathematical transformations takes the following form:

$$\boldsymbol{\Psi}_{\text{ex}_n} = \frac{1}{\gamma_{n-1}} \begin{Vmatrix} h_{n-1} + 2g_{n-1} + f_{n-1} & \dfrac{g_{n-1} + f_{n-1}}{T_{\text{eq}}} \\ \dfrac{g_{n-1} + f_{n-1}}{T_{\text{eq}}} & \dfrac{f_{n-1}}{T_{\text{eq}}^2} \end{Vmatrix}. \tag{5.124}$$

4. After nth measuring of the coordinate x with the variance of measuring error $\sigma_{x_{\text{m}}}^2$, we can compute the error correlation matrix of target track parameter filtering:

$$\boldsymbol{\Psi}_n = \frac{1}{\gamma_n} \begin{Vmatrix} h_n & \dfrac{g_n}{T_{\text{eq}}} \\ \dfrac{g_n}{T_{\text{eq}}} & \dfrac{f_n}{T_{\text{eq}}^2} \end{Vmatrix}, \tag{5.125}$$

where

$$\begin{cases} h_n = h_{n-1} + 2g_{n-1} + f_{n-1}; \\ g_n = g_{n-1} + f_n; \\ f_n = f_{n-1} + v_n; \\ \gamma_n = \gamma_{n-1} + v_n h_n; \end{cases} \tag{5.126}$$

and $v_n = \sigma_{x_{\text{m}}}^{-2}$ is the weight of the last measuring. Formulas in (5.126) allow us to form the elements of the matrix $\boldsymbol{\Psi}_n$ directly from the elements of the matrix $\boldsymbol{\Psi}_{n-1}$ taking into consideration the weight of the last measuring.

5. The matrix coefficient of filter amplification defined as

$$\mathbf{G}_n = \boldsymbol{\Psi}_n \mathbf{H}_n^T \mathbf{R}_n^{-1} \tag{5.127}$$

in the considered case takes the following form:

$$\mathbf{G}_n = \begin{Vmatrix} A_n \\ \dfrac{B_n}{T_{\text{eq}}} \end{Vmatrix}, \tag{5.128}$$

where

$$\begin{cases} A_n = \dfrac{h_n v_n}{\gamma_n}; \\ B_n = \dfrac{g_n v_n}{\gamma_n}. \end{cases} \tag{5.129}$$

6. Taking into consideration the relations obtained, we are able to compute the estimations of linear target track parameters in the following form:

$$\hat{x}_n = \hat{x}_{\text{ex}_n} + A_n(x_{n_\text{m}} - \hat{x}_{\text{ex}_n}), \tag{5.130}$$

$$\hat{\dot{x}}_n = \hat{\dot{x}}_{n-1} + \frac{B_n}{T_\text{eq}}(x_{n_\text{m}} - \hat{x}_{\text{ex}_n}). \tag{5.131}$$

7. Under equally discrete and uniformly precise measurements of the target track coordinates, we obtain

$$\begin{cases} f_n = nv; \\ g_n = \dfrac{n(n-1)}{2}v; \\ h_n = \dfrac{n(n-1)(2n-1)}{6}v; \\ \gamma_n = \dfrac{n^2(n^2-1)}{12}v^2. \end{cases} \tag{5.132}$$

Substituting (5.132) into (5.128) and (5.129), we obtain finally

$$\begin{cases} A_n = \dfrac{2(2n-1)}{n(n+1)}; \\ B_n = \dfrac{6}{n(n+1)}. \end{cases} \tag{5.133}$$

Dependences of the coefficients A_n and B_n versus the number of observations n are presented in Figure 5.7. As we can see from Figure 5.7, with an increase in n the filter gains on the coordinate and velocity are approximated asymptotically to zero. Consequently, with an increase in n the results of last measurements at filtering the coordinate and velocity are taken into consideration with the less weight, and the filtering algorithm ceases to respond to changes in the input signal. Moreover, essential problems in the realization of the filter arise in computer systems with limited capacity of number representation. At high n, the computational errors are accumulated and are commensurable with a value of the lower order of the computer system, which leads to losses in conditionality and positive determinacy of the correlation matrices of extrapolation errors and filtering of the target track parameters. The "filter divergence" phenomenon appears

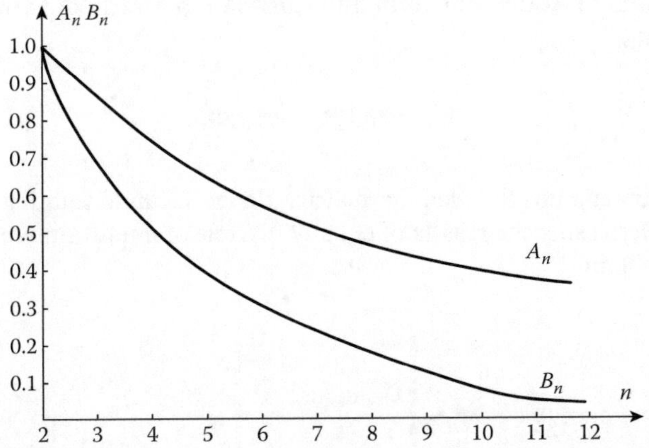

FIGURE 5.7 Coefficients A_n and B_n versus the number of observations n.

when the filtering errors increase sharply and the filter stops operating. Thus, if we do not take specific measures in correction, the optimal linear recurrent filter cannot be employed in complex automatic radar systems. Overcoming this problem is a great advance in automation of CRSs.

5.5.3 Stabilization Methods for Linear Recurrent Filters

In a general form, the problem of recurrent filter stabilization is the problem of ill-conditioned problem solution, namely, the problem in which small deviations in initial data cause arbitrarily large but finite deviations in solution. The method of stable (approximated) solution was designed for ill-conditioned problems. This method is called the regularization or smoothing method [17]. In accordance with this method, there is a need to add the matrix $\alpha\mathbf{I}$ to the matrix of measuring errors $\mathbf{R}_n$ given by (5.127), where $\mathbf{I}$ is the identity matrix, under a synthesis of the regularizing algorithm of optimal filtering of the unperturbed dynamic system parameters:

$$\mathbf{R}'_n = \mathbf{R}_n + \alpha\mathbf{I}. \tag{5.134}$$

In doing so, the regularization parameter α must satisfy the following condition:

$$\frac{\delta}{\varepsilon(\delta)} \le \alpha \le \alpha_0(\delta), \tag{5.135}$$

where

δ is the accuracy of matrix $\mathbf{R}_n$ assignment

$\varepsilon(\delta)$ and $\alpha_0(\delta)$ are the arbitrary decreasing functions tending to approach zero as $\delta \to 0$

Thus, in the given case, a general approach to obtain the stable solutions by the regularization or smoothing method is the artificial rough rounding of measuring results. However, the employment of this method in a pure form is impossible since a way to choose the regularization parameter α is generally unknown. In practice, a cancellation of divergence of the recurrent filter can be ensured by effective limitation of memory device capacity including in the recurrent filter. Consider various ways to limit a memory device capacity in the recurrent filter.

5.5.3.1 Introduction of Additional Term into Correlation Matrix of Extrapolation Errors

In this case, we obtain

$$\boldsymbol{\Psi}'_{\text{ex}_n} = \boldsymbol{\Phi}_n(\boldsymbol{\Psi}_{n-1} + \boldsymbol{\Psi}_0)\boldsymbol{\Phi}_n^T, \tag{5.136}$$

where $\boldsymbol{\Psi}_0$ is the arbitrary positive definite matrix. Under separate filtering of the polynomial target track parameters using the results of equally discrete and uniformly precise coordinating measurements, we obtain

$$\boldsymbol{\Psi}_0 = \begin{Vmatrix} c_0 & 0 & \dots & 0 \\ 0 & c_1 & \dots & 0 \\ \dots & \dots & \dots & \dots \\ 0 & 0 & \dots & c_s \end{Vmatrix} \sigma_{x_\text{m}}^2, \tag{5.137}$$

where

$\sigma_{x_\text{m}}^2$ is the variance of coordinating measuring errors
c_i is the constant coefficients, $i = 0, \dots, s$

In order for the recurrent filter to have a limited memory capacity, there is a need that the components of the vector $\mathbf{G}_n$ should be converged to the constant values $0 < \gamma_0 < 1, \dots, 0 < \gamma_s < 1$ and be in the safe operating area of the recurrent filter.

Using filtering equations, we are able to establish a relation for each specific case between the gains $\gamma_{1n}, \dots, \gamma_{sn}$ and the values $c_0, c_1, \dots, c_s$. Passing to limit as $n \to \infty$, we are able to obtain a function between the steady-state values of the recurrent filter gain $\gamma_0 = \lim_{n\to\infty} \gamma_{0n}, \dots, \gamma_s = \lim_{n\to\infty} \gamma_{sn}$ and the values $c_0, c_1, \dots, c_s$. In the case of linear target track [18], we have

$$\begin{cases} c_0 = \dfrac{12}{\left(n_{\text{ef}}^2 - 1\right)}; \\ c_1 = \dfrac{144}{\left(n_{\text{ef}}^2 - 1\right)\left(n_{\text{ef}}^2 - 4\right)}, \end{cases} \tag{5.138}$$

where n_{ef} is the fixed effective finite memory capacity of the recurrent filter. At the same time, the variances of random errors of target track parameter estimations in steady filtering coincide with analogous parameters for nonrecursive filters, namely,

$$\sigma_{\hat{x}}^2 = \frac{2(2n_{\text{ef}} - 1)}{n_{\text{ef}}(n_{\text{ef}} + 1)} \sigma_{x_\text{m}}^2; \tag{5.139}$$

$$\sigma_{\hat{\dot{x}}}^2 = \frac{12\sigma_{x_\text{m}}^2}{T_{\text{eq}}^2 n_{\text{ef}}(n_{\text{ef}} - 1)}. \tag{5.140}$$

Thus, the recurrent filter with additional term introduced into the error correlation matrix of extrapolation approximates the filter with finite memory capacity at the corresponding choice of the coefficients $c_0, c_1, \dots, c_s$.

5.5.3.2 Introduction of Artificial Aging of Measuring Errors

This operation is equivalent to a replacement of the correlation matrix $\mathbf{R}_{n-i}$ of measuring errors at the instant t_{n-i} by the matrix

$$\mathbf{R}_{n-1}^{*} = \exp[c(t_n - t_{n-i})]\mathbf{R}_{n-i}, \quad c > 0. \tag{5.141}$$

Under equally discrete measurements, we have

$$\begin{cases} t_n - t_{n-i} = iT_0; \\ \exp[c(t_n - t_{n-i})] = \exp[ciT_0] = s^i, \end{cases} \tag{5.142}$$

where $s = \exp(cT_0) > 1$. At the same time, the correlation matrix of extrapolation errors is determined in the following form:

$$\mathbf{\Psi}'_{\text{ex}_n} = \mathbf{\Phi}_n[s\mathbf{\Psi}_{n-1}]\mathbf{\Phi}_n^T, \tag{5.143}$$

In this case, under the equally discrete and uniformly precise coordinating measurements, the smoothing filter coefficients are converged to positive constants in the filter safe operating area. However, it is impossible to find the parameter s for this filter so that the variance and dynamic errors of these filters and the filter with finite memory capacity are matched.

5.5.3.3 Gain Lower Bound

In the simplest case of the equally discrete and uniformly precise coordinating measurements, the gain bound is defined directly by the formulas for $\mathbf{G}_n$ at the given effective memory capacity of the filter. Computation and simulation show that the last procedure is the best way by the criterion of realization cost and rate to define the variances of errors under the equally discrete and uniformly precise coordinating measurements from the considered procedures to limit the recurrent filter memory capacity. The first procedure, that is, the introduction of an additional term into the correlation matrix of extrapolation errors, is a little worse. The second procedure, that is, an introduction of the multiplicative term into the correlation matrix of extrapolation errors, is worse in comparison with the first and third ones by realization costs and rate of convergence of the error variance to the constant magnitude.

5.6 ADAPTIVE FILTERING ALGORITHMS OF MANEUVERING TARGET TRACK PARAMETERS

5.6.1 Principles of Designing the Filtering Algorithms of Maneuvering Target Track Parameters

Until now, in considering filtering methods and algorithms of target track parameters we assume that a model equation of target track corresponds to the true target moving. In practice, this correspondence is absent, as a rule, owing to target maneuvering. One of the requirements of successful solution of problems concerning the real target track parameter filtering is to take into consideration a possible target maneuver.

The state equation of maneuvering target takes the following form:

$$\mathbf{\theta}_n = \mathbf{\Phi}_n\mathbf{\theta}_{n-1} + \mathbf{\Gamma}_n\mathbf{g}_{\text{m}_n} + \mathbf{K}_n\mathbf{\eta}_n, \tag{5.144}$$

where

$\mathbf{\Phi}_n \mathbf{\theta}_{n-1}$ is the equation of undisturbed target track, that is, the polynomial of the first order

$\mathbf{g}_{m_n}$ is the l-dimensional vector of the disturbed target track parameters caused by the willful target maneuver

$\mathbf{\eta}_n$ is the p-dimensional vector of disturbances caused by stimulus of environment and uncertainty in control (control noise)

$\mathbf{\Gamma}_n$ and $\mathbf{K}_n$ are the known matrices

According to precision of CRS characteristics and estimated target maneuver, three approaches are possible to design the filtering algorithm of real target track parameters.

5.6.1.1 First Approach

We assume that the target has limited possibilities for maneuver; for example, there are only random unpremeditated disturbances in the target track. In this case, the second term in (5.145) is equal to zero and sampled values of the vector $\mathbf{\eta}_n$ are the sequence subjected to the normal Gaussian distribution with zero mean and the correlation matrix:

$$\mathbf{\Psi}_\eta = \begin{Vmatrix} \sigma_{\eta r}^2 & 0 & 0 \\ 0 & \sigma_{\eta\beta}^2 & 0 \\ 0 & 0 & \sigma_{\eta\varepsilon}^2 \end{Vmatrix}. \tag{5.145}$$

Nonzero elements of this matrix represent a set of a priori data about an intensity of target maneuver by each coordinate (r, β, ε). In this case, a record of target track disturbance in the filtering algorithm is reduced to filter bandwidth increase. For this purpose, a computation of the error correlation matrix of target track parameter estimations into an extrapolated point is carried out by the formula

$$\mathbf{\Psi}_{\text{ex}_n} = \mathbf{\Phi}_n \mathbf{\Psi}_{n-1} \mathbf{\Phi}_n^T + \mathbf{K}_n \mathbf{\Psi}_\eta \mathbf{K}_n^T, \tag{5.146}$$

where the $s \times l$ matrix $\mathbf{K}_n$ can be presented in the following form:

$$\mathbf{K}_n = \begin{Vmatrix} 0.5\tau_{\text{ex}_n}^2 & 0 & 0 \\ \tau_{\text{ex}_n} & 0 & 0 \\ 0 & 0.5\tau_{\text{ex}_n}^2 & 0 \\ 0 & \tau_{\text{ex}_n} & 0 \\ 0 & 0 & 0.5\tau_{\text{ex}_n}^2 \\ 0 & 0 & \tau_{\text{ex}_n} \end{Vmatrix}, \tag{5.147}$$

$\tau_{\text{ex}_n} = t_n - t_{n-1}$ is the interval of target track parameter extrapolation. Other formulas of the recurrent filtering algorithm are the same as in the case of a nonmaneuvering target.

There is a need to note that the methods discussed in the previous section to limit the memory capacity of the recurrent filter by their sense and consequences are equivalent to the considered method to record the target maneuver since a limitation in memory capacity of the recurrent filter, increasing a stability, can simultaneously decrease the sensitivity of the filter to short target maneuvering.

5.6.1.2 Second Approach

We assume that the target performs only a single premeditated maneuver of high intensity within the observation time interval. In this case, the target track can be divided into three parts: before the start of, in the course of, and after finishing the target maneuver. In accordance with this target track division, the intensity vector of a premeditated target maneuver can be presented in the following form:

$$\mathbf{g}_{\mathrm{m}}(t_i) = \begin{cases} 0 & \text{at} \quad t_i < t_{\text{start}}; \\ g_{\mathrm{m}_i} & \text{at} \quad t_{\text{start}} \le t_i \le t_{\text{finish}}; \\ 0 & \text{at} \quad t_i > t_{\text{finish}}; \end{cases} \tag{5.148}$$

where t_{start} and t_{finish} are the instants to start and finish the target maneuver. In this case, the intensity vector of target maneuver as well as the instants to start and finish the target maneuvering are subjected to statistical estimation by totality of input signals (measuring coordinates). Consequently, in the given case, the filtering problem is reduced to designing a switching filtering algorithm or switching filter with switching control based on the analysis of input signals. This algorithm concerns the class of simplest adaptive algorithms with a self-training system.

5.6.1.3 Third Approach

It is assumed that the targets subjected to tracking have good maneuvering abilities and are able to perform a set of maneuvers within the limits of observation time. These maneuvers are related to air miss relative to other targets or flight in the space given earlier. In this case, to design the formula flowchart of filtering algorithms of target track parameters, there is a need to have data about the mathematical expectation $E(g_{\mathrm{m}_n})$ and the variance σ_{g}^2 of target maneuver intensity for each target and within the limits of each interval of information updating. These data (estimations) can be obtained only based on an input information analysis, and the filtering is realized by the adaptive recurrent filter.

Henceforth, we consider the principles of designing the adaptive recurrent filter based on the Bayes approach to define the target maneuver probability [19].

5.6.2 Implementation of Mixed Coordinate Systems under Adaptive Filtering

The problems of target maneuver detection or definition of the probability to perform a maneuver by target are solved in one form or another by adaptive filtering algorithms. Detection of target maneuver is possible by deviation of target track from a straight-line trajectory by each filtered target track coordinate. However, in a spherical coordinate system, when the coordinates are measured by a CRS, the trajectory of any target, even in the event that the target is moving uniformly and in a straight line, is defined by nonlinear functions. For this reason, a detection and definition of target maneuver characteristics under filtering of target track parameters using the spherical coordinate system are impossible.

To solve the problem of target maneuver detection and based on other considerations, it is worthwhile to carry out filtering of the target track parameters using the Cartesian coordinate system, the origin of which is matched with the CRS location. This coordinate system is called the *local Cartesian coordinate system*. The formulas of coordinate transformation from the spherical coordinate system to the local Cartesian one are given by (see Figure 5.8)

$$\begin{cases} x = r\cos\varepsilon\cos\beta; \\ y = r\cos\varepsilon\sin\beta; \\ z = r\sin\varepsilon. \end{cases} \tag{5.149}$$

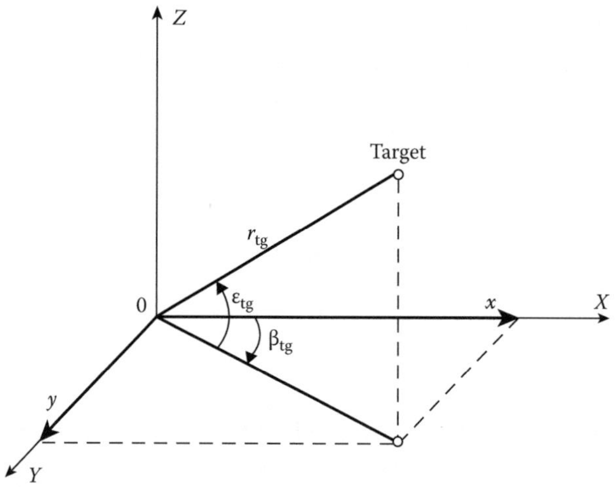

FIGURE 5.8 Transformation from the spherical coordinate system into the local Cartesian coordinate system.

Transformation to the local Cartesian coordinate system leads to an appearance of nonuniform precision and correlation between the coordinates at the filter input, which in turn leads to complexity of filter structure and additional computer cost in realization. Also, there is a need to take into consideration that other operations of signal reprocessing by a CRS, for example, target pip gating, target pip identification and so on, are realized in the simplest form in the spherical coordinate system. For this reason, the filtered target track parameters must be transformed from the local Cartesian coordinate system to the spherical one during each step of target information update.

Thus, to solve the problem of adaptive filtering of the track parameters of maneuvering targets, it is worthwhile to employ the recurrent filters, where the filtered target track parameters are represented in the Cartesian coordinate system and comparison between the measured and extrapolated coordinates is carried out in the spherical coordinate system. In this case, a detection of target maneuver or definition of the probability performance of target maneuver can be organized based on the analysis of deviation of the target track parameter estimations from magnitudes corresponding to the hypotheses of straight-line and uniform target moving.

Consider the main relationships of the recurrent filter, in which the filtered target track parameters are represented in the local Cartesian coordinate system and a comparison between the extrapolated and measured target track parameters coordinates is carried out in the spherical coordinate system. Naturally, the linear filter is considered as the basic filter, and equations defining the operation abilities of the considered linear filter are represented by the equation system given by (5.119). For simplicity, we can be restricted by the case of a two-dimensional radar measured CRS that defines the target range r_{tg} and azimuth β_{tg}. In this case, the transposed vector of target track moving parameters at the previous $(n-1)$ step takes the following form:

$$\boldsymbol{\theta}_{n-1}^{T} = \left\| \hat{x}_{n-1} \quad \hat{\dot{x}}_{n-1} \quad \hat{y}_{n-1} \quad \hat{\dot{y}}_{n-1} \right\|, \tag{5.150}$$

and the error correlation matrix of the target track parameter estimations contains 4×4 nonzero elements:

$$\boldsymbol{\Psi}_{n-1} = \begin{Vmatrix} \Psi_{11(n-1)} & \Psi_{12(n-1)} & \Psi_{13(n-1)} & \Psi_{14(n-1)} \\ \Psi_{21(n-1)} & \Psi_{22(n-1)} & \Psi_{23(n-1)} & \Psi_{24(n-1)} \\ \Psi_{31(n-1)} & \Psi_{32(n-1)} & \Psi_{33(n-1)} & \Psi_{34(n-1)} \\ \Psi_{41(n-1)} & \Psi_{42(n-1)} & \Psi_{43(n-1)} & \Psi_{44(n-1)} \end{Vmatrix}. \tag{5.151}$$

Extrapolation of target track parameters is carried out in accordance with the hypothesis about the target straight-line movement, and the correlation matrix is extrapolated by the following rule:

$$\mathbf{\Psi}_{\text{ex}_n} = \mathbf{\Phi}_n \mathbf{\Psi}_{n-1} \mathbf{\Phi}_n^T, \tag{5.152}$$

where

$$\mathbf{\Phi}_n = \mathbf{\Phi} = \begin{Vmatrix} 1 & T_0 & 0 & 0 \\ 0 & 1 & 0 & 0 \\ 0 & 0 & 1 & T_0 \\ 0 & 0 & 0 & 1 \end{Vmatrix}. \tag{5.153}$$

Extrapolated polar coordinates are determined from the Cartesian coordinates by the following formulas:

$$\hat{r}_{\text{ex}_n} = \sqrt{\hat{x}_{\text{ex}_n}^2 + \hat{y}_{\text{ex}_n}^2}, \tag{5.154}$$

$$\beta_{\text{ex}_n} = \begin{cases} A = \arctan\left|\dfrac{\hat{y}_{\text{ex}_n}}{\hat{x}_{\text{ex}_n}}\right|, & \hat{x}_{\text{ex}_n} > 0, \quad \hat{y}_{\text{ex}_n} > 0, \\ \pi - A, & \hat{x}_{\text{ex}_n} < 0, \quad \hat{y}_{\text{ex}_n} > 0, \\ \pi - A, & \hat{x}_{\text{ex}_n} < 0, \quad \hat{y}_{\text{ex}_n} < 0, \\ 2\pi - A, & \hat{x}_{\text{ex}_n} > 0, \quad \hat{y}_{\text{ex}_n} < 0. \end{cases} \tag{5.155}$$

The vector of measured target track parameter coordinates and the error correlation matrix take the following form:

$$\mathbf{Y}_n = \begin{Vmatrix} r_n \\ \beta_n \end{Vmatrix} \tag{5.156}$$

and

$$\mathbf{R}_n = \begin{Vmatrix} \sigma_{r_n}^2 & 0 \\ 0 & \sigma_{r_n}^2 \end{Vmatrix}. \tag{5.157}$$

To establish a relationship between the measured coordinates and estimated target track parameters, we use the linearized operator:

$$\mathcal{P} = \begin{Vmatrix} \dfrac{\partial r}{\partial x} & 0 & \dfrac{\partial r}{\partial y} & 0 \\ \dfrac{\partial \beta}{\partial x} & 0 & \dfrac{\partial \beta}{\partial y} & 0 \end{Vmatrix}_{x=\hat{x}_{\text{ex}_n}, y=\hat{y}_{\text{ex}_n}}. \tag{5.158}$$

This operator can be presented in the following form:

$$\mathcal{P} = \begin{Vmatrix} \mathcal{P}_r^T \\ \mathcal{P}_\beta^T \end{Vmatrix}, \tag{5.159}$$

where

$$\mathcal{P}_U^T = \begin{Vmatrix} \dfrac{\partial U}{\partial x} & 0 & \dfrac{\partial U}{\partial y} & 0 \end{Vmatrix} \quad \text{and} \quad U = \{r, \beta\}. \tag{5.160}$$

Henceforward, the error correlation matrix $\boldsymbol{\Psi}_n$ of the target track parameter estimations is determined by n measurements. The matrix filter gain is determined in the following way:

$$\mathbf{G}_n = \boldsymbol{\Psi}_n \mathcal{P}^T \mathbf{R}_n^{-1} = \boldsymbol{\Psi}_n \begin{Vmatrix} \mathcal{P}_r & \mathcal{P}_\beta \end{Vmatrix} \begin{Vmatrix} w_{r_n} & 0 \\ 0 & w_{\beta_n} \end{Vmatrix}. \tag{5.161}$$

By this reason, the vector of target track parameter estimations is determined by the following formula:

$$\hat{\boldsymbol{\theta}}_n = \hat{\boldsymbol{\theta}}_{\text{ex}_n} + \boldsymbol{\Psi}_n \begin{Vmatrix} \mathcal{P}_r & \mathcal{P}_\beta \end{Vmatrix} \begin{Vmatrix} w_{r_n} & 0 \\ 0 & w_{\beta_n} \end{Vmatrix} \begin{Vmatrix} r_n & - & \hat{r}_{\text{ex}_n} \\ \beta_n & - & \hat{\beta}_{\text{ex}_n} \end{Vmatrix}. \tag{5.162}$$

From this formula it is easy to obtain the final expression for components of the vector $\boldsymbol{\theta}_n$. Thus, in the case of $\hat{x}_n$ we obtain

$$\hat{x}_n = \hat{x}_{\text{ex}_n} + \alpha_r w_{r_n}(r_n - \hat{r}_{\text{ex}_n}) + \alpha_\beta w_{\beta_n}(\beta_n - \hat{\beta}_{\text{ex}_n}), \tag{5.163}$$

where

$$\alpha_r = \Psi_{11(n)} \frac{\partial r}{\partial x} + \Psi_{13(n)} \frac{\partial r}{\partial y}; \tag{5.164}$$

$$\alpha_\beta = \Psi_{11(n)} \frac{\partial \beta}{\partial x} + \Psi_{13(n)} \frac{\partial \beta}{\partial y}. \tag{5.165}$$

Other components are determined by analogous formulas.

From the earlier discussion and the relationships obtained, it follows that there is a statistical dependence between estimations of target track parameters by all coordinates in the considered filter. This fact poses difficulties in obtaining the target track parameter estimations and leads to tightened requirements for computer subsystems of a CRS. We can decrease the computer cost by refraining from optimal filtering of target track parameters that makes the filter simple. In particular, a primitive simplification is a rejection of joint filtering of target track parameters and passage to individual filtering of Cartesian coordinates with subsequent transformation of obtained target track parameter estimation coordinates to the polar coordinate system. The procedure of simplified filtering follows.

- Each pair of measured coordinates r_n and β_n is transformed to the Cartesian coordinate system before filtering using the following formulas:

$$x_n = r_n \cos\beta_n \quad \text{and} \quad y_n = r_n \sin\beta_n. \tag{5.166}$$

  The obtained values x_n and y_n are considered as independent measured coordinates with the variances of measurement errors:

$$\sigma_{x_n}^2 = \cos^2\beta_n\sigma_{r_n}^2 + r_n^2\sin^2\beta_n\sigma_{\beta_n}^2, \tag{5.167}$$

$$\sigma_{y_n}^2 = \sin^2\beta_n\sigma_{r_n}^2 + r_n^2\cos^2\beta_n\sigma_{\beta_n}^2. \tag{5.168}$$

- Each Cartesian coordinate is filtered independently in accordance with the adopted hypothesis of Cartesian coordinate changing. The problems of detection or definition of statistical parameters of target maneuver by each coordinate are solved simultaneously.
- Extrapolated values of Cartesian coordinates are transformed into the polar coordinate system by the formulas (5.154) and (5.155).

Comparison of accuracy characteristics of optimal and simple filtering procedures with double transformation of coordinates is carried out by simulation. For example, the dependences of root-mean-square error magnitudes of target track azimuth coordinate versus the number of measurements (the target track is indicated in the right top corner) for optimal (curve 1) and simple (curve 2) filtering algorithms at $r_{\min}$ = 10 and 20 km are shown in Figure 5.9. From Figure 5.9, we see that in the case of a simple filter, accuracy deteriorates from 5–15% against the target range, course, speed, and the number of observers. At $r_{\min}$ < 10 km, this deterioration in accuracy is for about 30%. However, the computer cost is decreased approximately by one order.

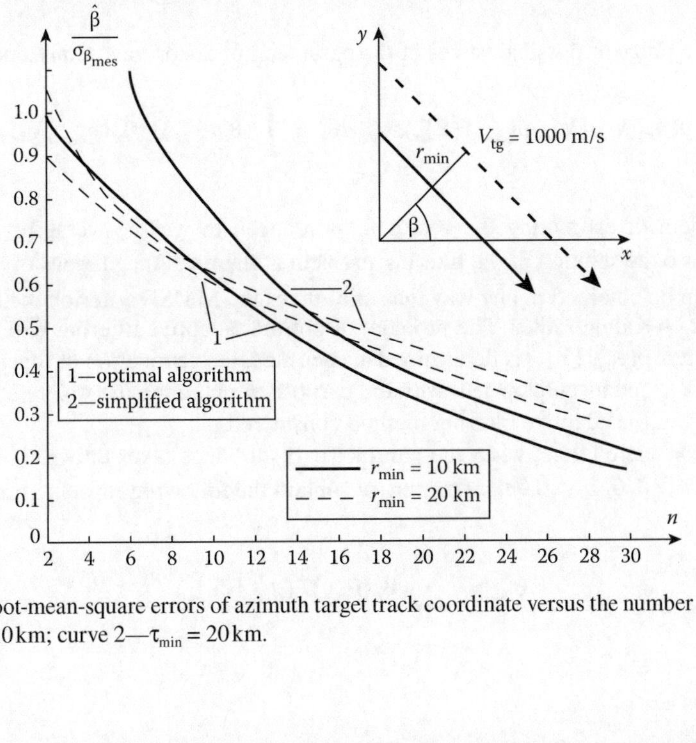

FIGURE 5.9 Root-mean-square errors of azimuth target track coordinate versus the number of measurements. Curve 1—$\tau_{\min}$ = 10 km; curve 2—$\tau_{\min}$ = 20 km.

5.6.3 Adaptive Filtering Algorithm Version Based on Bayesian Approach in Maneuvering Target

Under adaptive filtering, the linear dynamic system described by the state equation given by (5.144) is considered as the target track model. Target track distortions caused by a deliberate target maneuver are represented as a random process, the mean $E(g_{m_n})$ of which is changed step-wise taking a set of fixed magnitudes (states) within the limits of range $[-g_{m_{max}}, +g_{m_{max}}]$. Transitions of a step-wise process from the state i to the state j are carried out with the probability $P_{ij} \geq 0$ defined by a priori data about the target maneuver. The time when the process is in the state i before transition into the state j is the random variable with arbitrary pdf $p(t_i)$. The mathematical model of this process is the semi-Markov random process. Distortions of the target track caused by a deliberate target maneuver and errors of intensity estimations of deliberate target maneuver are characterized by the random component η_n in an adaptive filtering algorithm. The matrices $\mathbf{\Phi}_n$, $\mathbf{\Gamma}_n$, and $\mathbf{K}_n$ are considered as the known matrices.

Initially, we consider the Bayesian approach to design the adaptive filtering algorithm for the case of continuous distortion action g_{m_n}. As is well known, an optimal estimation of the parametric vector $\mathbf{\theta}_n$ at the quadratic loss can be defined from the following relationship:

$$\hat{\mathbf{\theta}}_n = \int\limits_{(\Theta)} \mathbf{\theta}_n p(\mathbf{\theta}_n | \{\mathbf{Y}\}_n) d\mathbf{\theta}_n, \tag{5.169}$$

where

(Θ) is the space of possible values of estimated target track parameter

$p(\mathbf{\theta}_n | \{\mathbf{Y}\}_n)$ is the a posteriori pdf of the vector $\mathbf{\theta}_n$ by data of n-dimensional sequence of measurements $\{\mathbf{Y}\}_n$

Under the presence of the distortion parameter g_m, the a posteriori pdf of the vector $\mathbf{\theta}_n$ can be written in the following form:

$$p(\mathbf{\theta}_n | \{\mathbf{Y}\}_n) = \int\limits_{(g_m)} p(\mathbf{\theta}_n | g_{m_n}, \{\mathbf{Y}\}_n) p(g_{m_n} | \{\mathbf{Y}\}_n) dg_{m_n}, \tag{5.170}$$

where (g_m) is the range of possible values of the parameter of distortions. Consequently,

$$\hat{\mathbf{\theta}}_n = \int\limits_{(\Theta)} \mathbf{\theta}_n \int\limits_{(g_m)} p(\mathbf{\theta}_n | g_{m_n}, \{\mathbf{Y}\}_n) p(g_{m_n} | \{\mathbf{Y}\}_n) dg_{m_n} d\mathbf{\theta}_n = \int\limits_{(g_m)} \hat{\mathbf{\theta}}_n(g_{m_n}) p(\hat{\mathbf{\theta}}_n | g_{m_n}, \{\mathbf{Y}\}_n) dg_{m_n}. \tag{5.171}$$

Thus, the problem of estimating the vector $\hat{\mathbf{\theta}}_n$ is reduced to weight averaging the estimations $\hat{\mathbf{\theta}}_n(g_{m_n})$, which are the solution of the filtering problem at the fixed magnitudes of g_{m_n}. The estimations $\hat{\mathbf{\theta}}_n(g_{m_n})$ can be obtained in any way that minimizes the MMSE criterion including the recurrent linear filter or Kalman filter. The problem of optimal adaptive filtering will be solved when the a posteriori pdf $p(g_{m_n} | \{\mathbf{Y}\}_n)$ is determined at each step. Determination of this pdf by sample of measurements $\{\mathbf{Y}\}_n$ and its employment with the purpose of obtaining the weight estimations is the main peculiarity of the adaptive filtering method considered.

In the case considered here, when the parametric disturbance takes only the fixed magnitudes g_{m_j}, $j = -0.5m, \ldots, -1, 0, 1, \ldots, 0.5m$, m is even, we obtain the following formula instead of (5.171):

$$\hat{\mathbf{\theta}}_{m_n} = \sum_{j=-0.5m}^{0.5m} \mathbf{\theta}_n(g_{m_{jn}}) P(g_{m_{jn}} | \{\mathbf{Y}\}_n), \tag{5.172}$$

where $P(g_{\mathrm{m}_{jn}}|\{\mathbf{Y}\}_n)$ is the a posteriori probability of the event $g_{\mathrm{m}_{jn}} = g_{\mathrm{m}_j}$ by data of n measurements $\{\mathbf{Y}\}_n$. To determine the a posteriori probability $P(g_{\mathrm{m}_{jn}}|\{\mathbf{Y}\}_n)$, we use the Bayes rule, in accordance with which (5.17) we can write

$$P(g_{\mathrm{m}_{jn}}|\{\mathbf{Y}\}_n) = P_{nj} = \frac{P(g_{\mathrm{m}_{jn}}|\{\mathbf{Y}\}_{n-1})p(\mathbf{Y}_n|g_{\mathrm{m}_{j,n-1}})}{\sum_{j=-0.5m}^{0.5m} P(g_{\mathrm{m}_{jn}}|\{\mathbf{Y}\}_{n-1})p(\mathbf{Y}_n|g_{\mathrm{m}_{j,n-1}})}. \quad (5.173)$$

In this formula, $P(g_{\mathrm{m}_{jn}}|\{\mathbf{Y}\}_{n-1})$ is the a priori probability of the parameter g_{m_j} at the nth step obtained by $(n-1)$ measurements and computed by the formula

$$P(g_{\mathrm{m}_{jn}}|\{\mathbf{Y}\}_{n-1}) = \sum_{i=-0.5m}^{0.5m} P_{ij} P(g_{\mathrm{m}_{j,n-1}}|\{\mathbf{Y}\}_{n-1}), \quad (5.174)$$

where

$$P_{ij} = P(g_{\mathrm{m}_n} = g_{\mathrm{m}_j}|g_{\mathrm{m}_{n-1}} = g_{\mathrm{m}_i}) \quad (5.175)$$

is the conditional probability of the transition of disturbance process from the state i at the $(n-1)$th step to the state j at the nth step; $p(\mathbf{Y}_n|g_{\mathrm{m}_{j,n-1}})$ is the conditional pdf of observed magnitude of the coordinate $\mathbf{Y}_n$ when the parametric disturbance at the previous $(n-1)$th step takes the magnitude g_{m_j}. This pdf can be approximated by the normal Gaussian distribution with the mean determined by

$$\hat{\mathbf{Y}}_{\mathrm{ex}_{n,j}} = \mathbf{H}_n[\mathbf{\Phi}_{n-1}\mathbf{\theta}_{n-1} + \mathbf{\Gamma}_{n-1}g_{\mathrm{m}_j}] \quad (5.176)$$

and the variance given by

$$\sigma_n^2 = \mathbf{H}_n\mathbf{\Psi}_{\mathrm{ex}_n}\mathbf{H}_n^T + \sigma_{Y_n}^2. \quad (5.177)$$

Taking into consideration (5.177) and (5.178), we obtain the a posteriori probability in the following form:

$$P_{nj} = \frac{\sum_{i=-0.5m}^{0.5m} P_{ij}P(g_{\mathrm{m}_{i,n-1}}|\{\mathbf{Y}\}_{n-1})\exp\left[-(\hat{\mathbf{Y}}_n - \hat{\mathbf{Y}}_{\mathrm{ex}_{n,j}})^2/2\sigma_n^2\right]}{\sum_{j=-0.5m}^{0.5}\sum_{i=-0.5m}^{0.5m} P_{ij}P(g_{\mathrm{m}_{i,n-1}}|\{\mathbf{Y}\}_{n-1})\exp\left[-(\mathbf{Y}_n - \mathbf{Y}_{\mathrm{ex}_{n,j}})^2/2\sigma_n^2\right]}. \quad (5.178)$$

The magnitudes P_{nj} for each j are the weight coefficients at averaging of estimations of filtered target track parameters.

Henceforth, we assume that the target track parameters are filtered individually by each Cartesian coordinate of the target. Measured values of spherical coordinates are transformed into Cartesian coordinates outside the filter. Correlation of measure errors of the Cartesian coordinates is not taken into consideration. The Cartesian coordinates fixing the intensity g_{m_x}, g_{m_y}, and g_{m_z} of deliberate target maneuver are also considered as independent between each other and $g_{\mathrm{m}_x} = \ddot{x}_{\mathrm{m}}$, $g_{\mathrm{m}_y} = \ddot{y}_{\mathrm{m}}$, and $g_{\mathrm{m}_z} = \ddot{z}_{\mathrm{m}}$. In the following, the equations of adaptive filtering algorithm applying to the Cartesian coordinate x are written in detailed form.

Step 1: Let at the $(n-1)$th step be obtained $\hat{x}_{n-1}$ and $\hat{\dot{x}}_{n-1}$—the estimations of target track parameters; the error correlation matrix of the target track parameter estimations takes the following form:

$$\boldsymbol{\Psi}_{n-1} = \begin{Vmatrix} \psi_{11,(n-1)} & \psi_{12,(n-1)} \\ \psi_{21,(n-1)} & \psi_{22,(n-1)} \end{Vmatrix}; \tag{5.179}$$

$P(\ddot{x}_{\mathrm{m}_{j,n-1}} | \{x\}_{n-1})$—the a posteriori probabilities of magnitudes of the disturbance parameter $j = -0.5m,\ldots,-1,0,+1,\ldots,0.5m$; $\sigma^2_{\ddot{x}_{n-1}}$—the variance of random component disturbed the target track.

Step 2: Extrapolation of target track parameters for each possible magnitude of $\ddot{x}_{\mathrm{m}j}$ is carried out in accordance with the following formulas:

$$\begin{cases} \hat{x}_{\mathrm{ex}_{nj}} = \hat{x}_{n-1} + \tau_{\mathrm{ex}}\hat{\dot{x}}_{n-1} + 0.5\tau^2_{\mathrm{ex}}\ddot{x}_{\mathrm{m}_j}; \\ \hat{\dot{x}}_{\mathrm{ex}_{nj}} = \hat{\dot{x}}_{n-1} + \tau_{\mathrm{ex}}\ddot{x}_{\mathrm{m}_j}. \end{cases} \tag{5.180}$$

Step 3: Elements of the error correlation matrix of extrapolation are defined by the following formulas:

$$\begin{cases} \psi_{11_{\mathrm{ex}_n}} = \psi_{11,n-1} + 2\tau_{\mathrm{ex}}\psi_{12,n-1} + \tau^2_{\mathrm{ex}}\psi_{22,n-1} + 0.25\tau^4_{\mathrm{ex}}\sigma^2_{\ddot{x}_{n-1}}; \\ \psi_{12_{\mathrm{ex}_n}} = \psi_{21_{\mathrm{ex}_n}} = \psi_{12,n-1} + \tau_{\mathrm{ex}}\psi_{22,n-1} + 0.5\tau^3_{\mathrm{ex}}\sigma^2_{\ddot{x}_{n-1}}; \\ \psi_{22,n} = \psi_{22,n-1} + \tau^2_{\mathrm{ex}}\sigma^2_{\ddot{x}_{n-1}}. \end{cases} \tag{5.181}$$

Step 4: Components of the filter gain vector at the nth step are defined in the following form:

$$\begin{cases} \mathbf{G}_{1n} = \dfrac{\psi_{11,\mathrm{ex}_n}}{\psi_{11,\mathrm{ex}_n} + \sigma^2_{x_n}} = \psi_{11,\mathrm{ex}_n} z_n^{-1} \\ \mathbf{G}_{2n} = \dfrac{\psi_{21,\mathrm{ex}_n}}{\psi_{11,\mathrm{ex}_n} + \sigma^2_{x_n}} = \psi_{21,\mathrm{ex}_n} z_n^{-1} \end{cases}; \tag{5.182}$$

where

$$z_n^{-1} = \frac{1}{\psi_{11,\mathrm{ex}_n} + \sigma^2_{x_n}}, \tag{5.183}$$

$\sigma^2_{x_n}$ is the variance of errors under measuring the coordinate x at the nth step.

Step 5: Elements of correlation matrix of errors of target track parameter estimation at the nth step are defined by the following formulas:

$$\begin{cases} \psi_{11,n} = \psi_{11,\mathrm{ex}_n} z_n^{-1}\sigma^2_{x_n}; \\ \psi_{12,n} = \psi_{21,n} = \psi_{11,\mathrm{ex}_n}\sigma^2_{x_n}; \\ \psi_{22,n} = \psi_{22,\mathrm{ex}_n} - \psi^2_{12,\mathrm{ex}_n} z_n^{-1}. \end{cases} \tag{5.184}$$

Step 6: Estimations of filtered target track parameters for each discrete value of parametric disturbance are defined in the following form:

$$\begin{cases} \hat{x}_{nj} = \hat{x}_{\text{ex}_{nj}} + \gamma_{1n}(x_n - \hat{x}_{\text{ex}_{nj}}); \\ \hat{\dot{x}}_{nj} = \hat{\dot{x}}_{\text{ex}_{nj}} + \gamma_{2n}(x_n - \hat{x}_{\text{ex}_{nj}}), \end{cases} \tag{5.185}$$

where x_n is the result of measuring the coordinate x at the nth step.

Step 7: The weight of discrete values of disturbance is defined as

$$P(\ddot{x}_{\text{m}_{jn}} | \{\mathbf{Y}\}_n) = \frac{\sum_{i=-0.5m}^{0.5m} P_{ij} P(\ddot{x}_{\text{m}_{i,n-1}} | \{\mathbf{Y}\}_{n-1}) \exp\left[-0.5(x_n - \hat{x}_{\text{ex}_{nj}})^2 z_n^{-1}\right]}{\sum_{j=-0.5m}^{0.5m} \sum_{i=-0.5m}^{0.5m} P_{ij} P(\ddot{x}_{\text{m}_{i,n-1}} | \{\mathbf{Y}\}_{n-1}) \exp\left[-0.5(x_n - \hat{x}_{\text{ex}_{nj}})^2 z_n^{-1}\right]}. \tag{5.186}$$

Step 8: The weight values of target track parameter estimations are defined as

$$\begin{cases} \hat{x}_n = \sum_{j=-0.5m}^{0.5m} \hat{x}_{n_j} P(\ddot{x}_{\text{m}_{jn}} | \{\mathbf{Y}\}_n); \\ \hat{\dot{x}}_n = \sum_{j=-0.5m}^{0.5m} \hat{\dot{x}}_{n_j} P(\ddot{x}_{\text{m}_{jn}} | \{\mathbf{Y}\}_n). \end{cases} \tag{5.187}$$

Step 9: The weight value of discrete parametric disturbance at the nth step is defined as

$$\hat{\ddot{x}}_{\text{m}_n} = \sum_j \ddot{x}_{\text{m}_j} P(\ddot{x}_{\text{m}_{jn}} | \{\mathbf{Y}\}_n). \tag{5.188}$$

Step 10: The weight variance of continuous disturbance at the nth step is defined as

$$\sigma^2_{\text{m}_n} = \sum_j (\ddot{x}_{\text{m}_j} - \hat{\ddot{x}}_{\text{m}_n})^2 P(\ddot{x}_{\text{m}_{jn}} | \{\mathbf{Y}\}_n) = \sigma^2_{\text{in}}, \tag{5.189}$$

where σ^2_{in} is the variance of inner fluctuation noise of the control subsystem.

A flowchart of the adaptive filter realizing the described system of equations is shown in Figure 5.10. The adaptive filter consists of (m + 1) Kalman filters connected in parallel, and each of them is tuned on one of possible discrete values of parametric disturbance. The resulting estimation of filtered target track parameters is defined as the weight summation of conditional estimations at the outputs of these elementary filters. The weight coefficients $P(\ddot{x}_{\text{m}_{jn}} | \{\mathbf{Y}\}_n)$ are made more exact at each step, that is, after each measurement of the coordinate x using the recurrent formula (5.187). Blocks for computation of the correlation matrix $\mathbf{\Psi}_n$ of errors of target track parameter estimation and filter gain $\mathbf{G}_n$ are shared for all elementary filters. For this reason, a complexity of realization of the considered adaptive filter takes place owing to (m + 1) multiple computations of extrapolated and smoothed magnitudes of filtered target track parameters and computation of the weights $P(\ddot{x}_{\text{m}_{jn}} | \{\mathbf{Y}\}_n)$ at each step of updating the target information.

The adaptive filter designed based on the principle of weight of partial estimations can be simplified to carrying out the weight of extrapolated values only of filtered target track parameters, and using

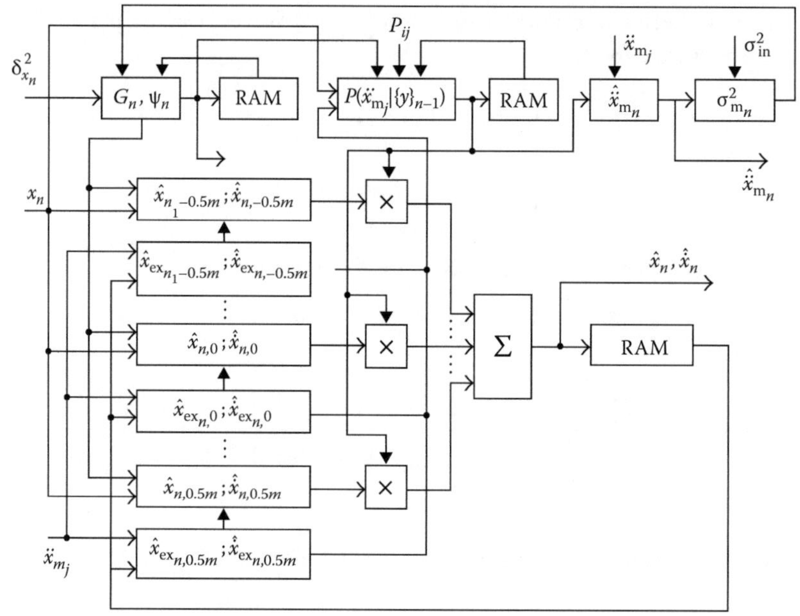

FIGURE 5.10 Adaptive filter flowchart: general case.

this weight, to compute the filtered target track parameters by the usual (undivided) filter instead of weighting the output estimations of filtered target track parameters. The equation system of the simplified adaptive filter is different from the previous one by the fact that the extrapolated values of filtered target track parameters determined by (5.180) are averaged with the following weights:

$$\begin{cases} \hat{x}_{\text{ex}_n} = \sum_j \hat{x}_{\text{ex}_{nj}} P(\ddot{x}_{\text{m}_{jn}} | \{\mathbf{Y}\}_n); \\ \hat{\dot{x}}_{\text{ex}_n} = \sum_j \hat{\dot{x}}_{\text{ex}_{nj}} P(\ddot{x}_{\text{m}_{jn}} | \{\mathbf{Y}\}_n). \end{cases} \tag{5.190}$$

After that, we can define more exactly the estimations of filtered target track parameters taking into consideration the nth coordinating measure using the well-known formulas for the Kalman filter. A flowchart of the simplified adaptive filter is shown in Figure 5.11. It consists of blocks for computation of the correlation matrix of errors $\mathbf{\Psi}_n$, the filter gain $\mathbf{G}_n$, and the probabilities $P(\ddot{x}_{\text{m}_{jn}} | \{\mathbf{Y}\}_n)$ that are shared by the filter as a whole; the block to compute the estimations $\hat{x}_n$ and $\hat{\dot{x}}_n$ of target track parameters; $(m + 1)$th blocks of target track parameter extrapolation for each fixed magnitude of acceleration $\ddot{x}_{\text{m}_{jn}}$; the weight device to compute averaged extrapolated coordinates. From Figure 5.11, it is easy to understand the interaction of all blocks. Employment of adaptive filters allows us to decrease essentially the dynamic error of filtering the target track parameters within the limits of the target maneuver range. In doing so, when there is no target maneuver, the root-mean-square magnitude of random filtering error is slightly increased, on average 10%–15%.

The relative dynamic errors of filtering on the coordinate x by the adaptive filter (solid lines) and nonadaptive filter (dashed lines) with $\sigma_{\ddot{x}}^2 = 0.5g$ for the target track shown in the right top are presented in Figure 5.12. The target makes a maneuver with accelerations $d_1 = 4g$ and $d_1 = 6g$, where g is the gravitational acceleration, moving with the constant velocity $V_{\text{tg}} = 300\,\text{m/s}$. Maneuver is observed within the limits of six scanning periods of the radar antenna. In the case of the adaptive filter, $\ddot{x}_1 = -8g, \ddot{x}_2 = 0$, and $\ddot{x}_3 = 8g$ are considered as the discrete magnitudes of acceleration. As follows from Figure 5.12, the adaptive filter allows us to decrease the dynamic error of filtering

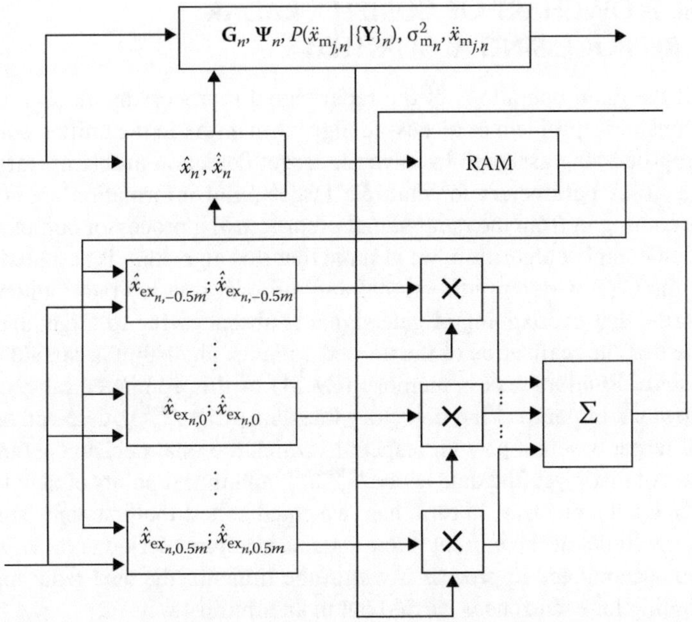

FIGURE 5.11 Flowchart of simplified adaptive filter.

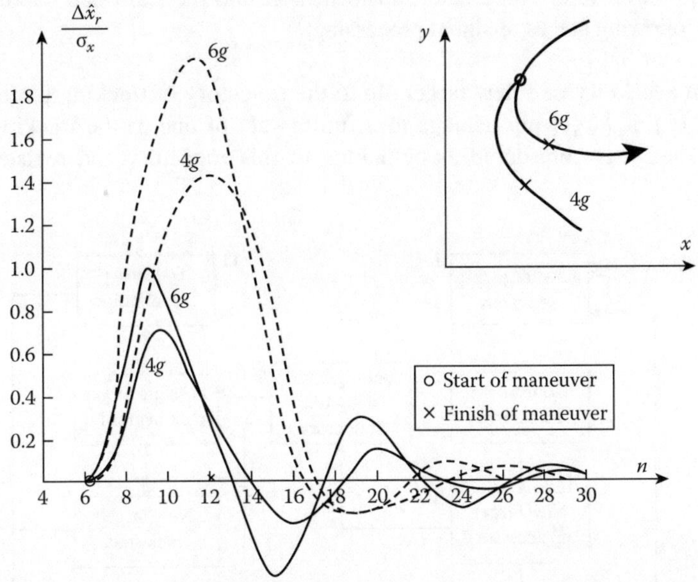

FIGURE 5.12 Dynamic errors of filtering: the adaptive filter (solid line) and nonadaptive filter (dashed line).

twice in comparison with the nonadaptive filter even in the case of low-accuracy fragmentation of the possible acceleration range. In the case considered in Figure 5.12, these errors do not exceed the variance of errors under the coordinating measurement. There is a need to take into consideration that the labor intensiveness of the considered adaptive filter realization by the number of arithmetic operations is twice higher in comparison with the labor intensiveness of the nonadaptive filter realization. With an increase in the number m of discrete values of maneuver acceleration (fragmentation with high accuracy) within the limits of the range $(-\ddot{x}_{\max}\cdots\ddot{x}_{\max})$, the labor intensiveness of the considered adaptive filter realization is essentially increased.

5.7 LOGICAL FLOWCHART OF COMPLEX RADAR SIGNAL REPROCESSING ALGORITHM

We discussed all the main operations of the radar signal reprocessing algorithms in Chapters 4 and 5. Now we consider a problem of discussed algorithm union in the unified complex algorithm of radar signal reprocessing assigned to solve the target detection problems, target tracking and filtering of target track parameters for multiple targets, and information about the contents in radar target pips coming in from the radar signal preprocessing processor output. Under consideration of the unified complex algorithm, we assume that this algorithm is assigned for radar signal reprocessing by the CRS with omnidirectional and uniform rotation radar antenna scanning. In addition, we assume that overlapping of gate signal is absent owing to target and noise environment. We assume that the realization of the unified complex algorithm is carried out by a specific computer subsystem. Random access memory (RAM) of this computer subsystem has specific data array area to store: target tracking trajectory (the data array D_{tg}^{track}); detected target tracks, that is, beginnings of target trajectories with respect to which the final decision about target tracking or cancellation is not made yet (the data array $D_{tg}^{decision}$); and initial points of new target tracks (the data array $D_{tg}^{initial}$). Each data array, in turn, has two equal zones: the first zone is to store information used within the limits of the current radar antenna scanning period and the second zone is to store information accumulated to process it within the limits of the next radar antenna scanning. Information recording for each zone is carried out in an arbitrary way.

A logical flowchart of a possible version of the unified complex radar signal reprocessing algorithm is shown in Figure 5.13. In accordance with this block diagram, each new target pip selected from the buffer after coordinate transformation into the Cartesian coordinate system is subjected to the following stages of signal processing:

- At first, an accessory of a new target pip to the trajectory of tracking target is checked (block 3). If this target pip belongs to a future gate of one of the tracking target trajectories, then it is considered as belonging to this trajectory and registered for this

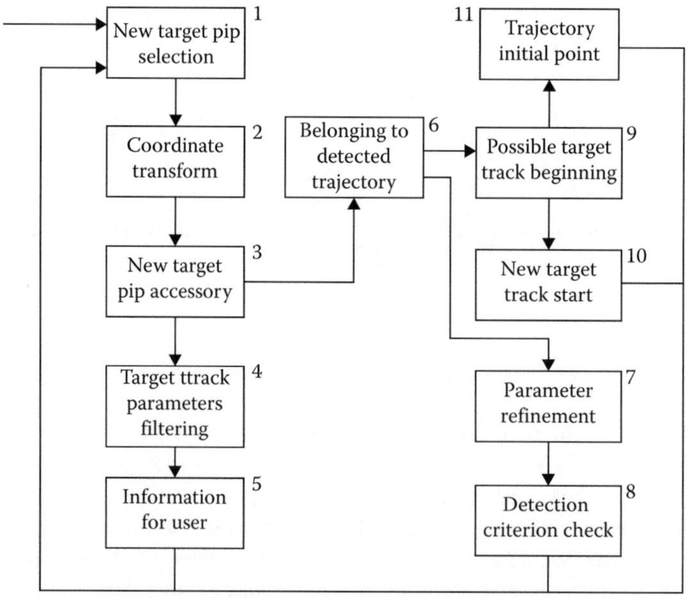

FIGURE 5.13 Logical flowchart of possible version of the united complex algorithms of radar signal reprocessing.

target track. Further, taking into consideration the coordinates of the new target pip, we make more exact the parameters of target track (block 4). For this purpose, one of the algorithms discussed in the this chapter is used. As a rule, this is the algorithm ensuring a filtering of maneuvering target track parameters, since the output information is used in the interests of the user (block 5). After completing the operations carried out by block 5, the processing of the next target pip is finished and the algorithm goes to choose a new target pip from the buffer.

- If a new target pip does not fall into the gate of any trajectory of a tracking target, in this case, this target pip is checked for a target track belonging to any detected trajectory (block 6). For this purpose, the target pip belonging to future gates formed by results of beginnings of new target tracks is checked. If the target pip belongs to the periodical gate selected from the data array $D_{\text{tg}}^{\text{decision}}$ of detected trajectories, then this target pip is considered as an extension of this target track. Taking into consideration the coordinates of the new target pip, we make more exact the parameters of the detected tracking target trajectory (block 7). After that, we check a detection criterion and the target track is recorded into the target track array and the algorithm goes to process the next target pip.
- If a new target pip belongs to no one gates of detected trajectories of tracking targets, then this target pip is checked to belong to the gates of initial lock-in formed around the individual target pips considered at the previous stages as beginnings of possible new trajectories of tracking targets (block 9). If a new target pip belongs to a regular initial lock-in gate from the data array $D_{\text{tg}}^{\text{initial}}$, then we compute initial values of target track parameters by two target pips, that is, the algorithm of beginning a new trajectory of tracking target is realized (block 10). This beginning is transferred to the data array $D_{\text{tg}}^{\text{decision}}$, and the algorithm starts to process the next target pip.
- If a new target pip does not belong to any initial lock-in gates, then it is recorded into the data array $D_{\text{tg}}^{\text{initial}}$ as a possible beginning of a new tracking target trajectory.

Thus, under an adopted structure to process new target pips (to form beginnings of tracking target trajectory) a cycle of single realization of the considered algorithm is the random variable, statistical characteristics of which are discussed in the next chapters. In conclusion, we note that the discussed flowchart of the unified complex radar signal reprocessing algorithm is not typical, that is, acceptable for all practical situations. This is only an example allowing us to make some calculations during further analysis.

5.8 SUMMARY AND DISCUSSION

There are many sources of error in radar-tracking performance. Fortunately, most are insignificant except for very high-precision tracking-radar applications such as range instrumentation, where the angle precision required may be on the order 0.05 mrad (mrad, or milliradian, is one thousandth of a radian, or the angle subtended by 1-m cross range at 1000-m range). Many sources of error can be avoided or reduced by radar design or modification of the tracking geometry. Cost is a major factor in providing high-precision tracking capability. Therefore, it is important to know how much error can be tolerated, which sources of error affect the application, and what are the most cost-effective means to satisfy the accuracy requirements. Because tracking radar subsystems track targets not only by angle but also in range and sometimes in Doppler, the errors in each of these target parameters must be considered in most error budgets. It is important to recognize the actual CRS information output. For a mechanically moved antenna, the angle-tracking output is usually obtained from the shaft position of the elevation and azimuth antenna axes. Absolute target location relative to their coordinates will include the accuracy of the survey of the antenna pedestal site.

The motions of a target with respect to the CRS causes the total echo signal to change with time, resulting in random fluctuations in the radar measurements of the parameters of the target. These fluctuations caused by the target only, excluding atmospheric effects and radar receiver noise contributions, are called the "target noise." This discussion of target noise is based largely on aircraft, but it is generally applicable to any target, including land targets of complex shape that are large with respect to a wavelength. The major difference is in the target motion, but the discussions are sufficiently general to apply to any target situation. The target track model given by (5.1) must take into consideration, first of all, the target track parameter disturbance caused by a nonuniform environment, in which the target moves, atmospheric conditions, and, also, inaccuracy and inertness of control and target parameter stabilization system in the course of target moving. We may term this target track parameter disturbance as the control noise, that is, the noise generated by the target control system. As a rule, the control noise is presented as the discrete white noise with zero mean and the variance σ_n^2. In addition to the control noise, the target track model must take into consideration specific disturbances caused by unpredictable changes in target track parameters that are generated by target in-flight maneuvering. We call these disturbances the target maneuver noise. In a general case, the target maneuver noise is neither white noise nor Gaussian noise.

In the solution of the filtering problems, additionally to the target track model there is a need to specify a function between the m-dimensional vector of measured coordinates $\mathbf{Y}_n$ and s-dimensional vector of estimated parameters $\boldsymbol{\theta}_n$ at the instant of nth measuring. This function, as a rule, is given by a linear equation. In the considered case, the observed coordinates are the current target coordinates in a spherical coordinate system, the target range r_n, the azimuth β_n, and the elevation ε_n, or some specific coordinates for the CRS and radar coordinates, for example, the radar range, cosine between the radar antenna array axis and direction to the target. In some CRSs, the radial velocity $\dot{r}_n$ can serve as the measured coordinate. We note that the target track model jointly with the model of measuring process forms the model of united dynamic system representing a process subjected to filtering.

In accordance with developed procedures and methods to carry out statistical tests in mathematical statistics, the following approaches are possible to calculate the a posteriori pdf and, consequently, to estimate the parameters: the batch method when the fixed samples are used and the recurrent algorithm consisting in sequent accurate definition of the a posteriori pdf after each new measuring. Using the first approach, the a priori pdf of the estimated parameter must be given. Under the use of the second approach, the predicted pdf based on data obtained at the previous step is used as the a priori pdf on the next step. Recurrent computation of the a posteriori pdf of estimated parameter, when a correlation between the model noise and measuring errors is absent, is carried out by the formula (5.17). In a general case of nonlinear target track models and measuring process models, computations by the formula (5.17) are impossible, as a rule, in closed forms. Because of this, under solution of filtering problems in practice, various approximations of models and statistical characteristics of CRS noise and measuring processes are used. Methods of linear filtering are widely used in practice. Models of system state and measuring of these methods are supposedly linear, and the noise is considered as the Gaussian noise.

In the course of discussion of linear filtering algorithms and extrapolation under the fixed sample size of measurements, we consider several types of target track, namely, (a) the linear target track, (b) the polynomial target track, and (c) the second-order polynomial target track. Additionally, we consider the algorithm of extrapolation of target track parameters and dynamic errors of target track parameter estimation. The likelihood function of the vector parameter $\boldsymbol{\theta}_N$ estimated by sequent measurements $\{\mathbf{Y}_N\}$ is given by (5.25). The vector likelihood equation is presented by (5.27). In a general case, the final solution of likelihood equation for correlated measure errors is defined by (5.29). Potential errors of target track parameters estimated by the polynomial target track model can be obtained using a linearization of the likelihood equation (5.27). The final form of the correlation matrix of errors of target track parameter estimations can be presented by (5.33). The variance of error of the smoothed coordinate estimation using three uniformly precise measurements is equal to 5/6 of the variance of individual measurement error. The variance of error of the coordinate increment estimation is only one half of the variance of error of individual measurement error owing to inaccurate target velocity estimation, and the correlation moment of relation between the errors of coordinate estimation and its increment is equal to one half of the variance of error of individual coordinate measurement. Figure 5.3 represents a coefficient of accuracy under determination of the normalized elements of correlation error matrix of linear target track parameter estimation versus the number of measurements. As follows from Figure 5.3, to obtain the required accuracy of estimations there is a need to carry out a minimum of 5–6 measurements. Turning to the algorithms of linear target track parameter estimations (5.43) and (5.44), we can easily see that these algorithms are the algorithms of nonrecursive filters and the weight coefficients $\eta_{\hat{x}}(i)$ and $\eta_{\Delta_1\hat{x}}(i)$ form a sequence of impulse response values of these filters. To apply filter processing using these filters, there is a need to carry out N multiplications between the measured coordinate values at each step, that is, after each coordinate measurement, and the corresponding weight coefficients and, additionally, N summations of the obtained partial products. To store in memory device $(N-1)$ records of previous measurements, there is a need to use high-capacity memory. As a result, to realize such a filter, taking into account that $N > 5$, there is a need to use high-capacity memory and this realization is complex. Formulas (5.70) through (5.75) show that in the case of equally discrete and uniformly precise measurements, the optimal estimation of target second-order polynomial track parameters is determined by the weighted summing of measured coordinate values. The weight coefficients are the functions of the sample size N and the sequence number of sample i in the processing series. The coefficient of accuracy under determination of the normalized elements of correlation error matrix of target second-order polynomial track parameter estimation versus the number of measurements is shown in Figure 5.4. Comparison of diagonal elements of the matrix (5.80) that are characteristics of accuracy under estimation of the second-order polynomial target track coordinate and the first increment with analogous elements of the correlation error matrix of target linear track parameter estimation (see Figure 5.3) shows that at low values of N the accuracy of target linear track parameter estimation is much higher in comparison with the accuracy of target second-order polynomial track parameter estimation. Consequently, within the limits of small observation intervals it is worthwhile to present the target track as using the first-order polynomial. In this case, a high quality of cancellation of random errors of target track parameter estimation by filtering is guaranteed. Dynamical errors arising owing to mismatching of target movement hypotheses can be neglected as a consequence of the narrow approximated part of the target track.

The extrapolation problem of target track parameters is to define the target track parameter estimations at the point that is outside the observation interval using the magnitudes of target track parameters determined during the last observation or using a set of observed coordinate values. Under polynomial representation of an independent coordinate, the target track parameters extrapolated using the time τ_{ex} are defined by (5.81) through (5.83), which allow us to define the extrapolated coordinates for each specific case of target track representation. The error correlation matrix of target linear track parameter extrapolation at equally discrete coordinate measurement is given by (5.90). If the independent target track parameter coordinate is represented by the polynomial of the second order, the target track parameter extrapolation formulas and the error correlation matrix of target linear track parameter extrapolation are obtained in an analogous method.

Inconsistency between the polynomial model and nonlinear character in changing the vector of estimated target track parameters leads to dynamic errors in smoothing that can be presented in the form of differences between the true value of the vector of estimated target track parameters and the mathematical expectation of the vector estimate. To describe the dynamic errors of estimations of the target track parameters by algorithms that are synthesized by the maximum likelihood criterion and using the target track polynomial model, we employ the well-known theory of errors at approximation of arbitrary continuously differentiable function $f(t)$ within the limits of the interval $t_N - t_0$ by the polynomial function of the first or second order using the technique of least squares.

Comparison of the dynamical errors of approximation of the polar coordinates by the polynomials of the first and second power at the same values of target track parameters V_{tg}, r_{min}, T_{eq} and the sample size $(N - 1)T_{eq}$ shows us that the dynamic errors under linear approximation are approximately one order higher than under quadratic approximation. Moreover, under approximation of $r_{tg}(t)$ and $\beta_{tg}(t)$ by the polynomials of the second power, the dynamic errors are small in comparison with random errors and we can neglect them. However, as shown by simulation, if the target is in an in-flight maneuver and the CRS is moving, the dynamic errors are essentially increased, because nonlinearity in changing the polar coordinates is increased sharply at the same time.

Methods of estimation of the target track parameters based on the fixed sample of measured coordinate, which are discussed in the previous sections, are used, as a rule, at the beginning of the detected target trajectory. Implementation of these methods in the course of tracking is not worthwhile owing to complexity and insufficient accuracy defined by the small magnitude of employed measures. Accordingly, there is a need to employ the recurrent algorithms ensuring a sequential, that is, after each new coordinating measurement, adjustment of target track parameters with their filtering purposes. At the recurrent filter output, we obtain the target track parameter estimations caused by the last observation. By this reason, a process of recurrent evaluation is further called sequential filtering, and corresponding algorithms are called the algorithms of sequential filtering of target track parameters.

In a discussion of the recurrent filtering algorithms, we obtain that the optimal recurrent linear filtering algorithm is defined by the Kalman filtering equations. A discrete optimal recurrent filter possesses the following properties: (a) filtering equations have recurrence relations and can be realized well by the computer system; (b) filtering equations simultaneously represent a description of procedure to realize this filter, and in doing so, a part of the filter is similar to the model of target track (compare Figures 5.2 and 5.6); (c) the error correlation matrix of target track parameter estimation $\mathbf{\Psi}_n$ is computed independently of measuring $\mathbf{Y}_n$; consequently, if statistical characteristics of measuring errors are given then the correlation matrix $\mathbf{\Psi}_n$ can be computed in advance and stored in a memory device; this essentially reduces the time of realization of target track parameter filtering.

As we can see from Figure 5.7, with an increase in n the filter gains in the coordinate and velocity are approximated asymptotically to zero. Consequently, with an increase in n the results of last measurements at filtering the coordinate and velocity are taken into consideration with less weight, and the filtering algorithm ceases to respond to changes in the input signal. Moreover, essential problems under realization of the filter arise in computer systems with a limited capacity of number representation. At high n, the computational errors are accumulated and commensurable with a value of the lower order of computer system that leads to losses in conditionality and positive determinacy of the correlation matrices of extrapolation errors and filtering of the target track parameters. The "filter divergence" phenomenon, when the filtering errors are increased sharply and the filter stops to operate, appears. Thus, if we do not take specific measures in correction, the optimal linear recurrent filter cannot be employed in complex automatic radar systems. To overcome this problem is a great advance in automation of CRSs.

In a general form, the problem of recurrent filter stabilization is the problem of ill-conditioned problem solution, namely, the problem in which small deviations in initial data cause arbitrarily large but finite deviations in solution. The method of stable (approximated) solution was designed for ill-conditioned problems. This method is called the regularization or smoothing method. In accordance with this method, there is a need to add the matrix $\alpha\mathbf{I}$ to the matrix of measuring errors $\mathbf{R}_n$ given by (5.127), where $\mathbf{I}$ is the identity matrix, under the synthesis of a regularizing algorithm of optimal filtering of the unperturbed dynamic system parameters (5.134). A general approach to obtain the stable solutions by the regularization or smoothing method is the artificial rough rounding of measuring results. However, the employment of this method in a pure form is impossible since a way to choose the regularization parameter α, is generally unknown. In practice, a cancellation of divergence of the recurrent filter can be ensured by effective limitation of memory device capacity including in the recurrent filter. In line with this, we consider several cases. For example, in the case of introduction of an additional term into the correlation matrix of extrapolation errors we obtain that the recurrent filter with an additional term approximates the filter with finite memory capacity at the corresponding choice of the coefficients $c_0, c_1, \ldots, c_s$. Under introduction of artificial aging of measuring errors and equally discrete and uniformly precise coordinating measurements, the smoothing coefficients of the filter are converged to positive constants in the filter safe operating area. However, it is impossible to find the parameter s for this filter so that the variance and dynamic errors of these filters and the filter with finite memory capacity are matched. In the case of gain lower bound for the simplest case of the equally discrete and uniformly precise coordinating measurements, the gain bound is defined directly by the formulas for $\mathbf{G}_n$ at the given effective memory capacity of the filter. Computation and simulation show that the last procedure is the best way by criterion of realization cost and rate to define the variances of errors under the equally discrete and uniformly precise coordinating measurements from considered procedures to limit the recurrent filter memory capacity. The first procedure, that is, the introduction of an additional term into the correlation matrix of extrapolation errors, is a little worse. The second procedure, that is, an introduction of the multiplicative term into the correlation matrix of extrapolation errors, is worse in comparison with the first and third ones by realization costs and rate of convergence of the error variance to the constant magnitude.

In the consideration of filtering methods and algorithms of target track parameters, we assume that the model equation of target track corresponds to the true target moving. In practice, this correspondence is absent, as a rule, owing to the target maneuvering. One requirement of successful solution of problems concerning the real target track parameter filtering is to take into consideration a possible target maneuver. The state equation of maneuvering target is given by (5.144). According to precision of CRS characteristics and estimated target maneuver, three approaches are possible for designing the filtering algorithm of real target track parameters: (a) the target has limited possibilities for maneuver, for example, there are only random unpremeditated disturbances in the target track (there is a need to note that the methods discussed in the previous section to limit a memory capacity of the recurrent filter by their sense and consequences are equivalent to the considered method to record the target maneuver since a limitation in memory capacity of the recurrent filter, increasing a stability, can simultaneously decrease the sensitivity of the filter to short target maneuvering); (b) we assume that the target performs only a single premeditated maneuver of high intensity within the observation time interval; in this case, the target track can be divided into three parts: before starting, in the course of, and after finishing the target maneuver, as the intensity vector of target maneuver at the instants of start and finish of target maneuvering are subjected to statistical estimation in this case by totality of input signals (measuring coordinates); consequently, in the given case, the filtering problem is reduced to designing a switching filtering algorithm or switching filter with switching control based on analysis of input signals; this algorithm concerns a class of simple adaptive algorithms with a self-training system; (c) it is assumed that the targets subjected to tracking have good maneuvering abilities and are able to perform a set of maneuvers within the limits of observation time; these maneuvers are related to air miss relative to other targets or flight in the given space; in this case, to design the formula flowchart of filtering algorithms of target track parameters there is a need to have data about the mathematical expectation $E(g_{\mathrm{m}_n})$ and the variance σ_{g}^2 of target maneuver intensity for each target and within the limits of each interval of information updating; these data (estimations) can be obtained only based on an input information analysis, and the filtering is realized by the adaptive recurrent filter.

The problems of target maneuver detection or definition of the probability to perform a maneuver by target are solved in one form or another by adaptive filtering algorithms. Detection of target maneuver is possible by deviation of target track from a straight-line trajectory by each filtered target track coordinate. However, in a spherical coordinate system, when the coordinates are measured by a CRS, the trajectory of any target, even in the event that the target is moving uniformly and in straight line, is defined by nonlinear functions. For this reason, a detection and definition of target maneuver characteristics under filtering of target track parameters using the spherical coordinate system are impossible. To solve the problem of target maneuver detection and based on other considerations, it is worthwhile to carry out a filtering of the target track parameters using the Cartesian coordinate system with the origin in the CRS location. This coordinate system is called the local Cartesian coordinate system. The formulas of coordinate transformation from the spherical coordinate system to the local Cartesian one are given by (5.149). Transformation to the local Cartesian coordinate system leads to an appearance of nonuniformly precise correlation between the coordinates at the filter input, which, in turn, leads to the complexity of filter structure and additional computer cost under realization. Also, there is a need to take into consideration that other operations of signal reprocessing by a CRS, for example, target pip gating, target pip identification, and so on, are realized in the simplest form in the spherical coordinate system. Therefore, the filtered target track parameters must be transformed from the local Cartesian coordinate system to the spherical one during each step of target information update. Thus, to solve the problem of adaptive filtering of the track parameters of maneuvering targets it is worth employing the recurrent filters, where the filtered target track parameters are represented in the Cartesian coordinate system and comparison between the measured and extrapolated coordinates is carried out in the spherical coordinate system. In this case, the detection of target maneuver or definition of the probability performance of target maneuver can be organized based on analysis of deviation of the target track parameter estimations from magnitudes corresponding to the hypotheses of straight-line and uniform target moving.

There is a statistical dependence between estimations of target track parameters by all coordinates in the considered recurrent filter. This fact poses difficulties in obtaining the target track parameter estimations and leads to tightened requirements for CRS computer subsystems. We can decrease the computer cost refusing from optimal filtering of target track parameters, making the filter simple. In particular, a primitive simplification is a rejection of joint filtering of target track parameters and passage to individual filtering of Cartesian coordinates with subsequent transformation of obtained target track parameter estimation coordinates to the polar coordinate system. The procedure of simplified filtering version is discussed. Comparison of accuracy characteristics of optimal and simple filtering procedures with double transformation of coordinates is carried out by simulation. For example, the dependences of root-mean-square error magnitudes of target track azimuth coordinate versus the number of measurements (the target track is indicated in the right top corner) for optimal (curve 1) and simple (curve 2) filtering algorithms at $r_{\min} = 10$ and 20 km are shown in Figure 5.9. In Figure 5.9, we see that in the case of a simple filter, accuracy deteriorates from 5–15% against target range, course, speed, and the number of observers. At $r_{\min} < 10$ km, this deterioration in accuracy is about 30%. However, the computer cost is decreased approximately one order.

Under adaptive filtering, the linear dynamic system described by the state equation given by (5.144) is considered as the target track model. Target track distortions caused by a deliberate target maneuver are represented as a random process, the mean $E(g_{\mathrm{m}_n})$ of which is changed step-wise taking a set of fixed magnitudes (states) within the limits of range $[-g_{\mathrm{m}_{\max}}, +g_{\mathrm{m}_{\max}}]$. Transitions of step-wise process from the state i to the state j are carried out with the probability $P_{ij} \geq 0$ defined by a priori data about the target maneuver. The time when the process is in the state i before transition into the state j is the random variable with arbitrary pdf $p(t_i)$. The mathematical model is the semi-Markov random process. Distortions of the target track caused by a deliberate target maneuver and errors of intensity estimations of deliberate target maneuver are characterized by the random component η_n in adaptive filtering algorithm. The matrices $\mathbf{\Phi}_n$, $\mathbf{\Gamma}_n$, and $\mathbf{K}_n$ are considered as known. Initially, we consider the Bayesian approach to design the adaptive filtering algorithm for the case of continuous distortion action g_{m_n}.

As is well known, an optimal estimation of the parametric vector $\boldsymbol{\theta}_n$ at the quadratic loss can be defined from (5.170). The problem to estimate the vector $\hat{\boldsymbol{\theta}}_n$ is reduced to weight averaging the estimations $\hat{\boldsymbol{\theta}}_n(g_{\mathrm{m}_n})$ that are the solution of filtering problem at the fixed magnitudes of g_{m_n}. The estimations $\hat{\boldsymbol{\theta}}_n(g_{\mathrm{m}_n})$ can be obtained by any way that minimizes the MMSE criterion including the recurrent linear filter or Kalman filter. The problem of optimal adaptive filtering will be solved when the a posteriori pdf $p(g_{m_n}|\{\mathbf{Y}\}_n)$ is determined at each step. Determination of this pdf by a sample of measurements $\{\mathbf{Y}\}_n$ and its employment with the purpose to obtain the weight estimations is the main peculiarity of the considered adaptive filtering method.

A flowchart of the adaptive filter realizing the described system of equations is shown in Figure 5.10. The adaptive filter consists of $(m + 1)$ Kalman filters connected in parallel and each of them is tuned on one of possible discrete values of parametric disturbance. The resulting estimation of filtered target track parameters is defined as the weight summation of conditional estimations at the outputs of these elementary filters. The weight coefficients $P(\ddot{x}_{m_{jn}}|\{\mathbf{Y}\}_n)$ are made more exact at each step, that is, after each measurement of the coordinate x using the recurrent formula (5.186). Blocks for computation of the correlation matrix $\mathbf{\Psi}_n$ of errors of target track parameter estimation and filter gain $\mathbf{G}_n$ are shared for all elementary filters. Therefore, the complexity of realization of the considered adaptive filter takes place owing to $(m + 1)$ multiple computations of extrapolated and smoothed magnitudes of filtered target track parameters and computation of the weights $P(\ddot{x}_{m_{jn}}|\{\mathbf{Y}\}_n)$ at each step of updating the target information.

The adaptive filter design based on the principle of weight of partial estimations can be simplified carrying out the weight of extrapolated values only of filtered target track parameters, and using this weight, to compute the filtered target track parameters by the usual (undivided) filter instead of weighting the output estimations of filtered target track parameters. The equation system of the simplified adaptive filter differs from the previous one by the fact that the extrapolated values of filtered target track parameters determined by (5.180) are averaged with the definite weights. After that we can define more exactly the estimations of filtered target track parameters taking into consideration the nth coordinate measure using the well-known formulas for the Kalman filter.

As follows from Figure 5.12, the adaptive filter allows us to decrease the dynamic error of filtering twice in comparison with the nonadaptive filter even in the case of low-accuracy fragmentation of possible acceleration range. In the case considered in Figure 5.12, these errors do not exceed the variance of errors under coordinating measurement. There is a need to take into consideration that the labor intensiveness of the considered adaptive filter realization by the number of arithmetic operations is twice higher in comparison with the labor intensiveness of the nonadaptive filter realization. With an increase in the number m of discrete values of maneuver acceleration (fragmentation with high accuracy) within the limits of the range $(-\ddot{x}_{\max}\ldots\ddot{x}_{\max})$, the labor intensiveness of the considered adaptive filter realization is essentially increased.

A logical flowchart of a possible version of the unified complex radar signal reprocessing algorithm is shown in Figure 5.13. In accordance with this block diagram, each new target pip selected from the buffer after coordinate transformation into the Cartesian coordinate system is subjected to the definite stages of signal processing discussed in this chapter. Under an adopted structure to process new target pips (to form beginnings of tracking target trajectories), a cycle of a single realization of the considered algorithm is the random variable, statistical characteristics of which will be discussed in the next chapters. In conclusion, we note that the discussed flowchart of the unified complex radar signal reprocessing algorithm is not typical, that is, acceptable for all practical situations. This is only an example allowing us to make some calculations during further analysis.

REFERENCES

1. Milway, W.B. 1985. Multiple targets instrumentation radars for military test and evaluation, in *Proceedings International Telemetry Council*, October 28–31, Las Vegas, Nevada, Vol. XXI, pp. 625–631.
2. Stegall, R.L. 1987. Multiple object tracking radar: System engineering considerations, in *Proceedings International Telemetry Council*, October 26–29, San Diego, California, Vol. XXIII, pp. 537–544.
3. Noblit, R.S. 1967. Reliability without redundancy from a radar monopulse receiver. *Microwave*, 12: 56–60.

4. Sakamoto, H. and P.Z. Peeblez. 1978. Monopulse radar. *IEEE Transactions on Aerospace and Electronic Systems*, 14(1): 199–208.
5. Stark, H. and J.W. Woods. 2002. *Probability and Random Processes with Applications to Signal Processing*. 3rd edn. Upper Saddle River, NJ: Prentice-Hall, Inc.
6. Thomas, M.C. and J.A. Thomas. 2006. *Elements of Information Theory*. 2nd edn. New York: John Wiley & Sons, Inc.
7. Rappaport, T.S. 2002. *Wireless Communications Principles and Practice*. Upper Saddle River, NJ: Prentice Hall, Inc.
8. Kay, S. 2006. *Intuitive Probability and Random Processes Using Matlab*. New York: Springer, Inc.
9. Castella, F.R. 1974. Analytical results for the *x*, *y* Kalman tracking filter. *IEEE Transactions on Aerospace and Electronic Systems*, 10(11): 891–894.
10. Kalman, R.E. 1960. A new approach to linear filtering and prediction problem. *Journal of Basic Engineering (ASME Transactions, Ser. D)*, 82: 35–45.
11. Kalman, R.E. and R.S. Bucy. 1961. New results in linear filtering and prediction theory. *Journal of Basic Engineering (ASME Transactions, Ser. D)*, 83: 95–107.
12. Sorenson, H. 1970. Least-squares estimation: From Gauss to Kalman. *IEEE Spectrum*, 7: 63–68.
13. Stewart, R.W. and R. Chapman. 1990. Fast stable Kalman filter algorithm utilizing the square root, in *Proceedings International Conference on Acoustics, Speech and Signal Processing*, April 3–6, Albuquerque, New Mexico, pp. 1815–1818.
14. Lin, D.W. 1984. On the digital implementation of the fast Kalman algorithm. *IEEE Transactions on Acoustics, Speech and Signal Processing*, 32: 998–1005.
15. Bozic, S.M. 1994. *Digital and Kalman Filtering*. 2nd edn. New York: Halsted Press, Inc.
16. Bellini, S. 1977. Numerical comparison of Kalman filter algorithms: Orbit determination case study. *Automatica*, 13: 23–35.
17. Tikhonov, A.N. and V. Ya. Arsenin. 1979. *Methods of Solution for Ill-Conditioned Problems*. Moscow, Russia: Science, Inc.
18. Rybova-Oreshkova, A.P. 1974. Investigations of recurrent filters with limited memory capacity. *News of Academy of Sciences of the USSR Series. Engineering Cybernetics*, 5: 173–187.
19. Sayed, A. 2003. *Fundamentals of Adaptive Filtering*. New York: Wiley Interscience/IEEE Press, Inc.

6 Principles of Control Algorithm Design for Complex Radar System Functioning at Dynamical Mode

An elementary basic flowchart presenting the subsystems usually found in a radar is shown in Figure 6.1. The *transmitter*, which is shown here as a *power amplifier*, generates a suitable waveform for the particular job the radar is to perform. It might have an average power as small as milliwatts or as large as megawatts. The average power is a far better indication of the capability of a radar's performance than its peak power. Most radars use a short-pulse waveform so that a single antenna can be used on a time-shared basis for both transmitting and receiving. The function of the *duplexer* is to allow a single antenna to be used by protecting the sensitive receiver from burning out while the transmitter is on and by directing the received echo signal to the receiver rather than to the transmitter. The *antenna* is the device that allows the transmitted energy to be propagated into space and then collects the echo energy on receiver. It is almost always a directive antenna, one that directs the radiated energy into a narrow beam to concentrate the power as well as to allow the determination of the direction of the target. An antenna that produces a narrow directive beam on transmit usually has a large area on receive to allow the collection of weak echo signals from target. The antenna not only concentrates the energy on transmit and collects the echo energy on receive but it also acts as a spatial filter to provide angle resolution and other capabilities.

The receiver amplifies the weak received signal to a level where its presence can be detected. Because noise is the ultimate limitation on the ability of a radar to make a reliable detection decision and extract information about the target, care is taken to ensure that the receiver produces very little noise of its own. At the microwave frequencies, where most radars are found, the noise that affects radar performance is usually from the first stage of the receiver, shown here in Figure 6.1 as a *low-noise amplifier*. For many radar applications where the limitation to detection is the unwanted radar echoes from the environment called the *clutter*, the receiver needs to have a large enough dynamic range so as to avoid having the clutter echoes adversely affect detection of wanted moving targets by causing the receiver to saturate. The *dynamic range* of the receiver, usually expressed in decibels, is defined as the ratio of the maximum to the minimum signal input power levels over which the receiver can operate with some specified performance. The maximum signal level might be set by the nonlinear effects of the receiver response that can be tolerated, for example, the signal power at which the receiver begins to saturate, and the minimum signal might be the minimum detectable signal. The *signal processor*, which is often in the intermediate frequency portion of the receiver, might be described as being the part of the receiver that separates the desired signal from the undesired signals that can degrade the detection process. Signal processing includes the *generalized detector* that maximizes the output signal-to-noise ratio (SNR). Signal processing also includes the Doppler processing that maximizes the signal-to-clutter ratio of a moving target when clutter is larger than the receiver noise, and it separates one moving target from other moving targets or from clutter echoes. The *detection decision* is

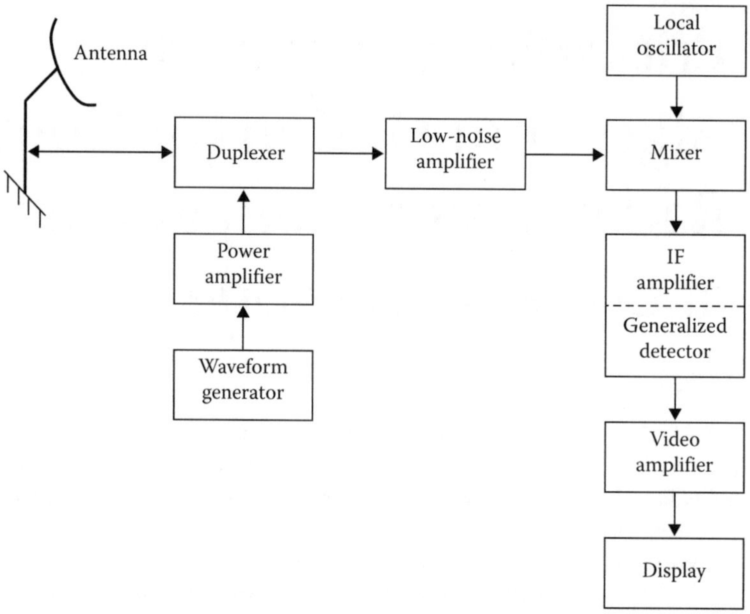

FIGURE 6.1 Flowchart of radar system.

made at the output of the receiver, so a target is declared to be present when the receiver output exceeds a predetermined threshold. If the threshold is set too low, the receiver noise can cause excessive false alarms. If the threshold is set too high, detections of some targets might be missed that would otherwise have been detected. The criterion to determine the level of the decision threshold is to set the threshold so that it produces an acceptable predetermined average rate of false alarms due to receiver noise.

After the detection decision is made, the track of a target can be determined, where a track is the locus of target locations measured over time. This is an example of *data processing*. The processed target detection information or its track might be displayed to an operator; or the detection information might be used to automatically direct a missile to a target; or the radar output might be further processed to provide other information about the nature of the target. The *radar control* ensures that the various parts of radar operate in a coordinated and cooperative manner, as, for example, providing timing signals to various parts of the radar as required.

The radar engineer has a resource *time* that allows good Doppler processing, *bandwidth* for good range resolution, *space* that allows a large antenna, and *energy* for long-range performance and accurate measurements. External factors affecting radar performance include the *target characteristics*; *external noise* that might enter via the antenna; unwanted *clutter* echoes from land, sea, birds, or rain; *interference* from other electromagnetic radiators; and propagation effects due to the Earth's surface and atmosphere. These factors are mentioned here to emphasize that they can be highly important in the design and application of radar systems.

6.1 CONFIGURATION AND FLOWCHART OF RADAR CONTROL SUBSYSTEM

The radar control subsystem is considered as a purposeful change in the structure of a complex radar system (CRS) and, as a consequence, a variation of radar subsystem parameters to reach a maximum effect under radar control. The main tasks of a complex radar control subsystem are to organize an optimal functioning of the radar system as a subsystem in the control subsystem of a higher level. A necessity to control the parameters and structure of a CRS in dynamical mode is caused by the complexity and abruptness of environment conditions, that is, man-made and/

or echo interferences, reflection from the Earth's surface, various objects, such as trees, buildings, hydrometeors, and so on, transient observations, a variety of problems that are to be solved, and limitations in energy and computational resources of the CRS. Hence, the control subsystem must be considered, first, as the means toward this end assigned for the CRS and, second, as a method to compensate for unwanted changes in environment that are interferences for its optimal functioning [1].

The main requirement to organize the control subsystem in a CRS structure is the knowledge of controlled parameters. If the number of controlled parameters is increased, the possibilities to organize the control system become high. Most control problems can be solved by multifunctional complex radar subsystems [2–7]. Therefore, in this chapter we consider mainly the problem statements and solutions of control problems in CRSs of this kind. Henceforth, all necessary signal and data processing algorithms assigned to reach the given control goal are considered as the *control subsystem* of a CRS. In doing so, the word *control* is used to designate both a process to reach the assigned task and in the sense of purposeful impact on parameters and structure of complex radar subsystems.

The first stage of designing the CRS control subsystem is a choice or formalization of goals, which must be attained in the course of controlling. In a general case, a high-level efficacy of CRS functioning is the main aim of a control subsystem. A degree of fulfillment of this aim can be evaluated by estimation of the vector $\mathbf{Y}(t)$—the output parameters of a CRS—that is used to define its efficacy index: $\mathcal{W} = \mathcal{W}[\mathbf{Y}(t)]$ [8–10]. To realize the control subsystem, there is a need to design control subsystem channels to transfer control signal $\mathbf{U}(t)$. Internal parameters and methods of limited resource consumption of CRS are the object of a control subsystem. The figure of merit of required information is a function of an environment statement $\mathbf{Z}(t)$, the set of the internal parameters $\mathbf{X}_{\mathrm{U}}(t)$ corrected by the control system, and special controlling decision $\mathbf{U}(t)$:

$$\mathbf{Y}(t) = \mathbf{F}\left[\mathbf{t}, \mathbf{Z}(t), \mathbf{X}_{\mathrm{U}}(t), \mathbf{U}(t)\right], \tag{6.1}$$

where $\mathbf{F}$ is the vector-operator of the controlled object functioning. In a general case, the goal of a control subsystem is to obtain such values of the vector $\mathbf{Y}(t) = \mathbf{Y}^{\bullet}(t)$ that ensure a condition

$$\max_{(\mathbf{Y})} \mathcal{W}\left[\mathbf{Y}(t)\right] = \mathcal{W}\left[\mathbf{Y}^{\bullet}(t)\right] = \mathcal{W}^{\bullet}(t). \tag{6.2}$$

Now, informed about the environment conditions, controlled object, and control subsystem aim, we are able to present the control signal $\mathbf{U}(t)$ in the form of an algorithm:

$$\mathbf{U}(t) = f[\mathcal{W}^{\bullet}(t), \mathbf{Z}(t), \mathbf{X}_{\mathrm{U}}(t), \mathbf{Y}(t)]. \tag{6.3}$$

Owing to the complexity of a radar system as a controlled object, the solution of the control problem, in a general form, has some difficulties. On the one hand, these difficulties are caused by the required computer costs and RAM size to solve the control problems within the limits of very short and strictly limited time intervals defined by dynamical changes of environment and CRS functioning. On the other hand, these difficulties have a principal character, since there are no methods of formalization and optimization of the control problem, which allow us to obtain a quantitative assessment as a whole for such a system as the CRS. Hence, in the course of designing the CRS control subsystem, the well-known system engineering procedure of complex system division in individual subsystems is widely used. In doing so, a choice of the structure and dimension of subsystem is carried out in accordance with universally adopted principles to save the backbone connections, and also taking into consideration the definite stages of CRS signal processing. Naturally, in division we should take into consideration possibilities to ensure the best control process for subsystems and the CRS as a whole. Division or decomposition of the CRS into subsystems assumes that each subsystem has a criterion function to control that is generated from a general objective function of the CRS.

To increase efficiency and decrease work content for the solution of control tasks, a subsystem of control algorithms is designed based on a hierarchical approach [11]. The hierarchical structure of the control subsystem is characterized by the presence of some control levels associated with each other, by which a general task is divided. In doing so, the control algorithms of high levels define and coordinate the working abilities of lower levels. Based on the aforementioned general statements, the CRS, as an object to control, can be divided into *controlled subsystems*, including the signal detection and signal processing subsystems. The controlled subsystems are as follows:

1. *Transmit and receive antenna with controlled parameters:*
   a. The number of generated beams of transmit N_{tr} and receive N_r antennas
   b. The directional diagram beam width of the transmit $\left(\theta_\beta^{tr}, \theta_\varepsilon^{tr}\right)$ and receive $\left(\theta_\beta^{r}, \theta_\varepsilon^{r}\right)$ antennas
   c. A degree of beam overlapping for the multibeam receive antenna directional diagram $(\delta\beta_r, \delta\varepsilon_r)$
   d. Function of changes in the main lobe direction of the transmit antenna directional diagram (β_0, ε_0)
   e. The number and direction of radar antenna directional diagram valleys
2. *Transmitter with controlled parameters:*
   a. The duration of searching signal τ_s
   b. The power of transmitted pulse P_p
   c. The carrier frequency f_c or sequence of carrier frequencies $\{f_{ci}\}$ under multifrequency operation
   d. The spectrum bandwidth of searching signals Δf_s
   e. The parameters and way of chirp modulation of searching signals
   f. The frequency F of searching signal repetition
3. *Receiver with controlled parameters:*
   a. The optimal structure with a viewpoint of the maximal SNR under conditions of complex noise environment
   b. Operation mode
   c. The spectrum bandwidth of input linear tract Δf_r
   d. The dimension of gated area—the number of elements in the gate and coordinates of the gate center
4. *Signal preprocessing processor of target return signals with controlled parameters:*
   a. The number of signal processing stages under detection and estimation of target return signal parameters
   b. The upper A and lower B detection thresholds under sequential detection of target return signals
   c. The truncation threshold n_{tr} under sequential analysis of target return signals
   d. The number N of accumulated pulses by detector with the fixed sample number
   e. The threshold U_0 of binary quantization
   f. The logic operations (l, m, k) under detection of binary quantized target return signals
5. *Signal reprocessing processor of target return signals and preparation of obtained information for user with controlled parameters:*
   a. The criterion to begin the target track ("$2/m + l/n$")
   b. The criterion to cancel the target range tracking k_{tr}^{can}

c. The algorithm and parameters of the linear filter used under the target range tracking
d. The way of refreshment of the tracked target
e. The target range parameters and precision characteristics transmitted to the user

The majority of controlled parameters of the CRS and signal processing subsystems depend on each other. For instance, the duration of the scanning signal τ_s, the period or frequency of repetition T or F, the power of transmitted pulse P_p, and the average power of scanning signal $\bar{P}_{scan}$ are related by the evident formula:

$$\bar{P}_{scan} = P_p\left(\frac{\tau_s}{T}\right). \tag{6.4}$$

As the average power of scanning signal $\bar{P}_{scan}$ is a constant value, then the power resources of the CRS can be controlled by the following parameters: τ_s, T, and P_p.

The radar antenna directional diagram beam width depends on the angle between the main axis of radar antenna and antenna beam:

$$\theta_b = \frac{\theta_0}{\cos(\theta_b - \theta_0)}, \tag{6.5}$$

where θ_0 is the beam width in the main axis direction of the radar antenna. The radar antenna directional diagram beam broadening under misalignment of radar antenna leads us to a decrease in the coefficient of amplification of radar antenna, and consequently, to increasing the required number N of accumulated target return signals to ensure the given SNR by power. There are other associations that must be taken into consideration under the control process. Interrelating control parameters do not allow us to organize a control process insulatedly by each of the controlled blocks or subsystems of the CRS, which indicates the impossibility of spatial division of the control subsystem. At the same time, a specificity of the CRS operation allows us to organize a control process step by step in time and design a multilevel hierarchical structure of the control subsystem (see Figure 6.2).

The first level is the control of CRS and signal processing subsystem parameters and in the course of searching for the preliminary chosen direction in radar coverage taking into consideration a radar functional operation, namely, searching, target range tracking, and so on. Objective control function at this stage is to ensure a minimal energy consumption to obtain the given effect from searching the chosen direction, namely, the given probability of detection P_D, the given accuracy to measure angle coordinates σ_θ^2, and so on. Limitations are acceptance bands of CRS technical parameters and ways to integrate them. During the first stage, a control process is carried out by direct selection of controlled parameters of the CRS and signal preprocessing subsystems in real time.

The second control level is an optimization of CRS functional modes. In the case of radar target detection and target range tracking subsystems, there is a need to optimize a searching of new targets and target range tracking functioning. A way to observe the radar coverage and ability to refresh the tracked targets are the *controlled functions*. The objective function of an optimal control process is the minimization of specific CRS power resource consumption to detect the target under scanning, refreshment, and estimation of target range track parameters under target tracking. CRS technical parameters and specifications of the high-level system are the main limitations for target range tracking. The control process at the second level is carried out by optimal scheduling of radar coverage element observation in the scanning mode and refreshes each target during the target tracking mode. Periodicity of changes in the control process at this stage is mainly defined by

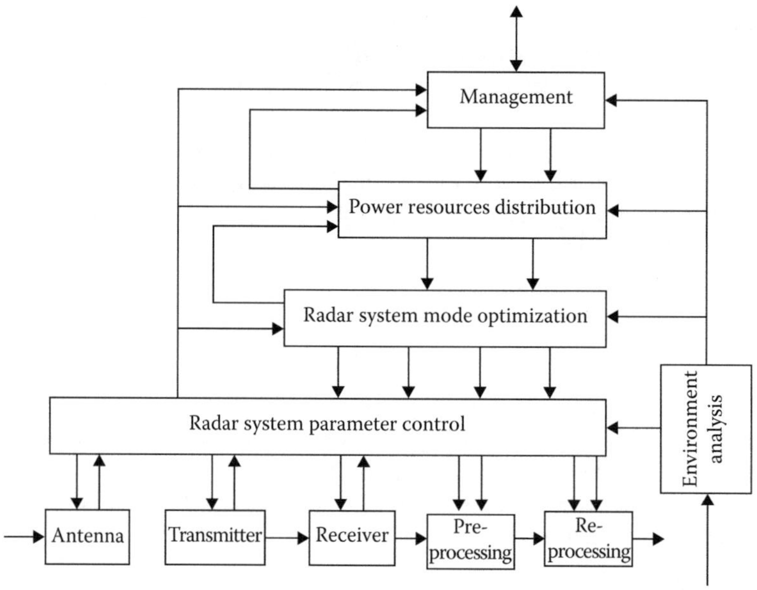

FIGURE 6.2 Hierarchical control system.

dynamics of dirty data within the limits of radar coverage—the number of targets, importance of targets, environmental noise, and so on.

The third control level is a distribution of limited CRS power resources between functional modes: the detection and target tracking modes in the case of two functional CRSs. The objective functions of optimal control at this stage are the maximization of CRS efficiency criterion for higher hierarchical level, at which the CRS operates as the detection and target tracking system, and the maximization of the processed target number taking into consideration their importance. The control process at the third level is carried out by the redistribution of power resource parts assigned for scanning and target tracking modes. The periodicity of the control process depends on the a priori and a posteriori environment data within the radar range and ability of facilities for target servicing and so on.

The fourth control level is an administration to keep the CRS under required operational conditions taking into consideration information about the operation quality of its subsystems, changed environment, information from subsystems of the same levels, and control commands from subsystems of a higher hierarchical level. At this level, a contingency administration is mainly carried out.

Thus, we obtain the multilevel hierarchical structure of a radar control subsystem, the simplest flowchart of which is shown in Figure 6.2. The hierarchy of control process is characterized by a sequential release of information from complex radar subsystem of higher to lower level subsystems. At the same time, the results of particular problem solutions at lower levels are used as the basis to make decisions at the higher level. Thus, information about the energy consumption for scanning in each direction, the number of detected targets, and the energy consumption under target tracking mode comes to the second-level subsystem as a result of the first-level task solutions. The information about the number of tracked targets $n_{\text{tg}}^{\text{track}}$, their importance, and also about a scanning interval of new target zones is transferred to the subsystem of the third level as a result of solutions of tasks at the second level. Simultaneously, the subsystems of the second and third levels of the control subsystem transfer information about an ability of facilities and quality to solve the tasks to the subsystem of the fourth level, where an administration of the CRS is carried out as a whole.

6.2 DIRECT CONTROL OF COMPLEX RADAR SUBSYSTEM PARAMETERS

6.2.1 Initial Conditions

The control process in the course of scanning the selected direction is carried out by direct choice of controlled parameters of the CRS and signal preprocessing subsystem with the purpose of adapting to dynamical environmental conditions under the given operation mode. The main modes of CRS functioning are as follows:

- Scan new targets in radar coverage.
- Refresh met target tracks or search mode of detected targets.

Other possible functioning modes can be considered as particular cases of the aforementioned modes. Additionally, an extension of the mode numbers does not change control process principles.

The external environment for a CRS is characterized by the presence of targets, noise, and different kinds of interferences such as the following:

- The inadvertent interference—radio radiation of the world network, various radio engineering systems, mobile systems, and so on
- The lumped interference—radiation, the spectrum of which is concentrated in the neighborhood of the central or resonant frequency and, with a spectral bandwidth that is larger in comparison to the spectral bandwidth of the signal at the radar receiver or detector input
- The pulse interference—a sharp short-term increase in the spectral power density of the additive noise that occurs within the spectrum bandwidth limits of the signal
- The deliberate interference—the interference generated specifically for the purpose to reduce the efficiency and noise immunity of the CRS

Henceforth, we assume that all information about target conditions in radar coverage is reconstructed by the a priori and a posteriori data in the course of CRS operation. The noise environment is evaluated by a special signal processing subsystem assigned to analyze changed conditions of the environment. We assume that analysis of additive noise parameters, namely, the type, power level, spectral power density, and so on, is carried out immediately before scanning the chosen direction. Processing and analysis of passive interferences and inclusion of protective equipment against this kind of interferences are carried out by control facilities of signal preprocessing subsystems during scanning the radar coverage.

There is a need to take into consideration that the tasks of choice and task-oriented changes of complex radar subsystem parameters must be solved within the limits of finite time intervals. This requirement is very strong. In the course of the control process, the pulsing rate defined by repetition frequency cannot be changed and must be constant. All processes of decision making should be finished between neighboring scanning. In line with this, the table methods to make a decision can be widely used based on results of noise environment analysis in the radar range.

6.2.2 Control under Directional Scan in Mode of Searched New Targets

In this case, the control process is carried out by two stages. At the first stage, the calculation and choice of scanning signal parameters are carried out. The scanning signals must correspond to the functional mode of searching targets and maximal radar range in the given direction. Additionally, the receiver performance and signal processing algorithms for signal preprocessing subsystems should be defined and computed. This stage is carried out before scanning pulsing in the given direction. Control of accumulated received signals up to making decisions about the target detection/no detection using all resolution elements in the radar range (the case of sequential analysis) or up

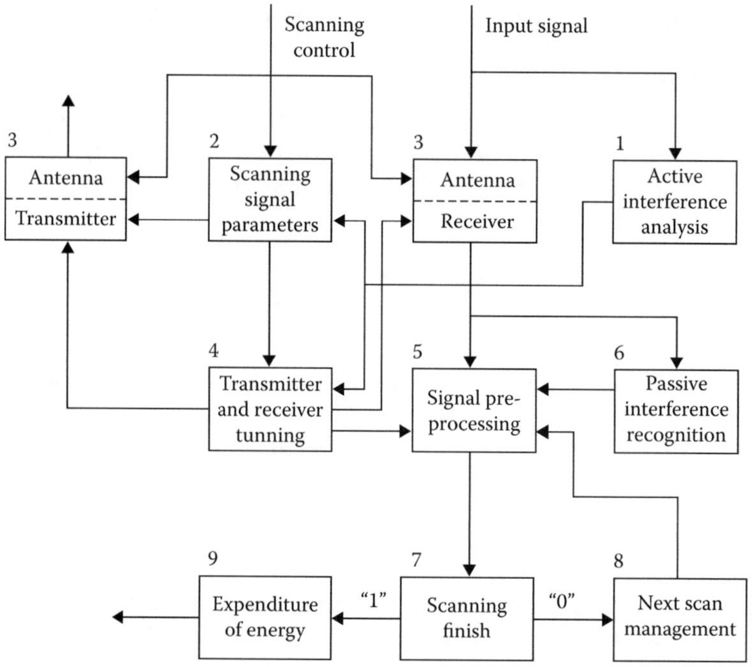

FIGURE 6.3 Logical structure of algorithm to search for a direction in the scanning mode.

to accumulation of a determined number of the target return signals using detectors with the fixed sample sizes is carried out at the second stage. In the course of accumulation of the target return signals, the initial parameter values of scanning signals are the same excepting the carrier frequency f_c. In the following, we give a description of a logical algorithmic flowchart to solve the control tasks under direction scanning using the searching mode (see Figure 6.3).

Step 1: After setting transmit and receive antenna beams along the given direction β_i, ε_i, we carry out an analysis of active interferences received along this direction (block 1). First of all, we should define the power σ_n^2 and the power spectral density $\mathcal{N}(f)$ of the noise. Evaluation of the noise power is used further to make a decision to power on that or other protection equipment from active interferences and, also, to determine predetection SNR. Estimation of power spectral density of active interferences allows us to define the "spectrum troubles" of interference within the limits of considered frequency bandwidth and select the scanning signal carrier frequency on the interval where an action of interference is weakest. In a general case, under analysis of interference there is a need to distinguish and classify the active interferences with definite integrity to create a possibility to use the obtained results to tune and learn the radar receiving channel.

Step 2: At this stage, we solve the problems of the scanning signal parameter selection (block 2). For this purpose, first of all, there is a need to determine a distance r_{ε_i} from the radar to the radar coverage bound under the given elevation angle value ε_i. Computation is carried out taking into consideration the shape of radar antenna directional diagram. For scanning and tracking radar, a vertical cross section of idealized zone of target searching is shown in Figure 6.4. All searching zone characteristics are considered unknown.

2.1 As it follows from the geometry presented in Figure 6.4, if $\varepsilon_{\max} > \varepsilon_n > \varepsilon_0$, we have

$$r_{\varepsilon_i} = \frac{R_{\text{Earth}} + H_{\max}}{\sin(0.5\pi + \varepsilon_i)} \sin\gamma_i, \tag{6.6}$$

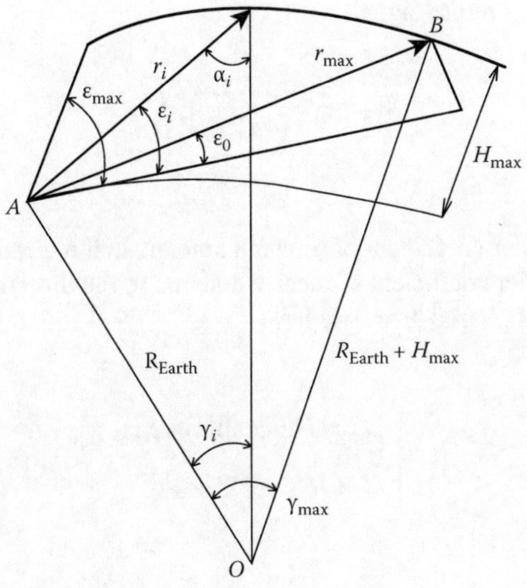

FIGURE 6.4 Vertical cross section of ideal target scanning zone.

where

$$\gamma_i = 0.5\pi - \varepsilon_i - \alpha_i \quad \text{and} \quad \gamma_i < 0.5\pi; \tag{6.7}$$

$$\sin\alpha_i = \frac{R_{Earth}\sin(0.5\pi + \varepsilon_i)}{R_{Earth} + H_{max}}, \quad \alpha_i < 0.5\pi. \tag{6.8}$$

Determination of r_{ε_i} can be made in advance for the whole range of the elevation angles $\varepsilon_0 \ldots \varepsilon_{max}$ with discrete equal, for example, to the radar antenna directional diagram width in vertical plane.

2.2 At the known range r_{ε_i}, the period of pulsing is determined from relation

$$T_i = \frac{2r_{\varepsilon_i}}{c}, \tag{6.9}$$

where $c = 3\times10^8$ m/s.

2.3 The nonmodulated scanning signal with duration of τ_s is selected under the searching mode.

2.4 The pulse power of the scanning signal is given by

$$P_{p_i} = \bar{P}_s T_i \tau_{s_i}^{-1}. \tag{6.10}$$

Computed and selected parameters of scanning signals, namely, $f_c, P_{p_i}, T_i, \tau_{s_i}$, are set in the transmitter (block 3, Figure 6.3).

Step 3: Results of scanning signal parameter computation and selection are used to define the minimal SNR on the radar coverage bound and signal preprocessing algorithm adjustments related to SNR.

3.1 The energy of target return signal

$$E_{\text{tg}} = \frac{P_{\text{p}_i}\sigma_{\text{s}_i}G_{\text{tr}_i}G_{\text{r}_i}\lambda^2 S_{\text{tg}}}{(4\pi)^3 r_{\varepsilon_i}^4 \mathcal{K}}, \tag{6.11}$$

where

G_{tr} is the amplifier coefficient of transmit antenna in the direction of ε_i, β_i

G_{r} is the amplifier coefficient of receive antenna in the direction of ε_i, β_i; in doing so, in the case of phased array, we have

$$\begin{cases} G_{\text{tr}_i} = G_{\text{tr}_0}\cos\beta_i'\cos\varepsilon_i'; \\ G_{\text{r}_i} = G_{\text{r}_0}\cos\beta_i'\cos\varepsilon_i', \end{cases} \tag{6.12}$$

where

$$\begin{cases} \beta_i' = |\beta_i - \beta_{\text{angle}}| < 0.5\pi; \\ \varepsilon_i' = |\varepsilon_i - \varepsilon_{\text{angle}}| < 0.5\pi, \end{cases} \tag{6.13}$$

G_{tr_0} and G_{r_0} are the amplifier coefficients of transmit and receive antennas on the phased array axis, respectively

β_{angle} and $\varepsilon_{\text{angle}}$ are the angular direction of phased array axis

S_{tg} is the effective reflecting surface of target that is considered as standard in the course of detection performance computation

$\mathcal{K}$ is the generalized loss factor

3.2 The total noise power spectral density at the CRS receiver input without taking into consideration the passive interferences is given by

$$\mathcal{N}_{\Sigma} = \overline{\mathcal{N}}_{\text{in}}^{\text{active}}\mu + \mathcal{N}_0, \tag{6.14}$$

where

$\mathcal{N}_{\text{in}}^{\text{active}}$ is the average power spectral density of active interferences

μ is the coefficient of active interference suppression by compensating and other protection facilities

$$\mathcal{N}_0 = 2kTB_N, \tag{6.15}$$

where

k is the Boltzmann constant

T is the absolute temperature

B_N is the receiver bandwidth

3.3 The SNR by power for the radar range r_{ε_i} is determined by

$$\text{SNR} = \frac{2E_{\text{s}}}{\mathcal{N}_{\Sigma}}. \tag{6.16}$$

Step 4: This SNR is used later to adjust a control of the CRS signal preprocessing subsystem detectors, that is, to define the required number of scans to reach the given probability of detection P_D and the probability of false alarm P_F for detectors with the fixed sample size or to define the weights of units and zeros under sequential analysis of binary quantized target return signals (block 4, Figure 6.3). There is a need to note that a realization of the a priori operations discussed earlier on selection of parameters for transmitter, receiver, and signal processing subsystems must be carried out within the limits of finite time intervals or the so-called *free running of sweep time* equal to $(T - \tau_d^{max})$. Because of this, we can employ the tables of parameters and settings corresponding to discrete values of noise characteristics, elevation angles, and other given parameters that are kept in a memory device instead of detailed computations.

Step 5: After finishing the preliminary stage, we can start the process of directional scanning and accumulation of target return signals (block 5, Figure 6.3) taking into account the results of distinguishing the passive interferences and using the protection facilities against the passive interferences (block 6, Figure 6.3). At this stage, the control process is carried out by the signal preprocessing subsystem algorithms in accordance with the operations realized by blocks 7 and 8, Figure 6.3.

After finishing the accumulation stage and making a decision about the scanned direction, the results of decision making are issued and expenditure of energy and timetable are computed to observe the scanned direction (block 9, Figure 6.3). The radar antenna beam control is assigned for an algorithm of scanning management in the searching mode or mode controller. In conclusion, we note once more that the discussed logical flowchart is the way only to assign a control process under the searching new target mode. Naturally, there are possibly other ways, but a general idea to control, namely parametric matching of CRS subsystems and signal and data processing subsystems with environmental parameters and solved problems and tasks is the same.

6.2.3 Control Process under Refreshment of Target in Target Tracing Mode

Consider a possible way to design a control algorithm to detect and measure target coordinates in the course of target tracking using information that is discretely refreshed in the CRS with a controlled radar antenna beam. At the same time, we use the following assumptions:

- Time instants of regular refreshment of target tracking are computed each time after clarification of target tracing parameters at the previous step. Time intervals between the previous and next measurements are determined based on the condition of nonrotating radar.
- Owing to possible overlapping of location intervals of the tracked targets and under stimulus of more foreground modes, delays of instruction executions, for example, the target coordinate measure, are possible. Probability and delay time are high, a loading of the CRS is high, or the number of targets in the radar coverage is high. To exclude cancellations of target tracking owing to the delays mentioned earlier, the possibility to expand a searching zone of regular renewed target pip by the angle coordinates and radar range is provided. Expanding in the searching zone by angular coordinates is carried out by assigning some additional directions of scanning around an extrapolated direction taking into consideration the target track parameters, the so-called *additional target searching*. The number of directions of the additional target searching is not high—from 3 to 5. Expanding in the searching zone by the radar range is carried out by increasing the dimensions of gate Δr_{gate}.
- Expanding in the searching zone is carried out also if there are misses of the target pips on target tracing during one or several scanning jobs.

In accordance with the assumptions discussed earlier, the control algorithm to measure the target coordinates in the course of target tracking mode is mainly reduced to carrying out the following operations (see Figure 6.5):

- Computation of delay $\Delta\tau_{s_i}$ to start a service of the selected target t_0 in comparison with the prearranged service start t_{sp_i}. This delay is compared with the allowable one computed based on the given probability of the target pip hit belonging to the given target tracing, into the gate limited by the radar antenna directional diagram beam width by the angle coordinates and bounds of the analyzed part of the radar range (blocks 1 and 2, Figure 6.5).
- If the delay $\Delta\tau_{s_i}$ is less than allowable and there is no feature of miss, that is, the probability of miss P_{miss} is equal to zero in the course of the previous session of the target coordinate renewal, a computation of narrow gate by the radar range is carried out and a location in the computed or extrapolated direction is also carried out (the blocks 3, 4, and 7).
- If the delay $\Delta\tau_{s_i}$ is greater than allowable one or $P_{miss} = 1$, there is a need to compute a wide gate by the radar range and angle coordinates of the *additional target searching* directions (blocks 5 and 6, Figure 6.5). After that, a target return signal location is provided, at the beginning, in the extrapolated direction, and after, if the target is not detected in the extrapolated direction, there is a need to provide a search in additional directions until the detection of target pip is fixed in the regular direction of the entire assigned *additional target searching* directions observed (blocks 14 and 16, Figure 6.5).
- After selection of regular target location direction, we carry out an analysis of noise environment in this direction. Henceforward, the parameters of scanning signals are calculated or selected, and tuning of the receiver is carried out (block 8, Figure 6.5), and the control process is assigned to the signal preprocessing subsystem algorithm in the target tracking mode (blocks 9 and 10, Figure 6.5).
- The amount of directional scanning in the target tracking mode is defined based on a condition to obtain the given accuracy to measure the radar range and angle target coordinates. In a hard noise environment, this number can be peak clipped based on established balance of expenditure of energy for the target searching and target tracking modes.

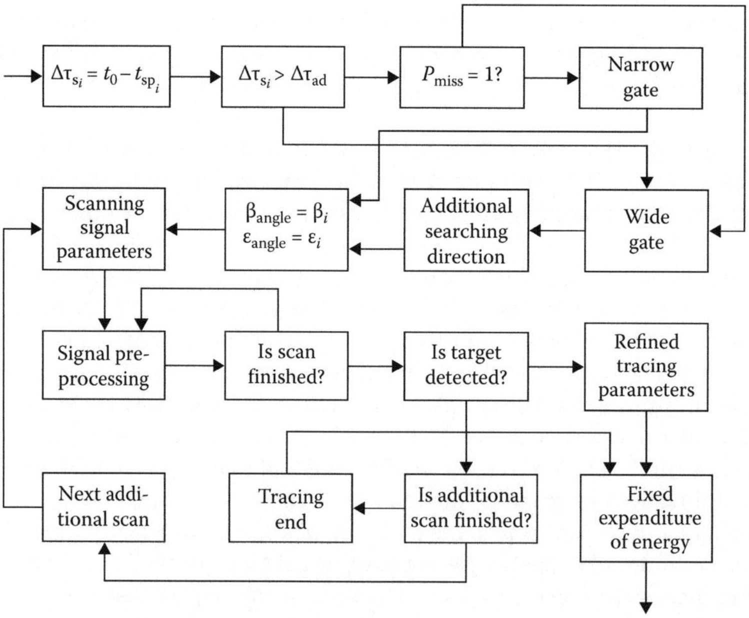

FIGURE 6.5 Flowchart of control algorithm to measure the target coordinates.

- If the target has been detected and the target coordinates have been measured with the given accuracy, then the corresponding target pip is produced to clarify the target tracing parameters. After that, the costs of refreshment about the detected target tracing are fixed and the control process is assigned to a mode administrator (blocks 11–13, Figure 6.5).
- If after observation of the entire assigned *additional target searching* directions the target has not been detected, then the algorithm of target tracing end detection is powered on, which produces a decision to continue or to stop target tracking of the considered target tracing (block 15, Figure 6.5). Naturally, in this case, the expenditure of energy is also fixed.

Thus, the considered control algorithm operates interactively in a close way with the algorithms of the main operations of signal preprocessing and reprocessing subsystems within the limits of gate coordinating these algorithms into the unique signal processing subsystem in the individual CRS target tracking mode with the control beam of radar antenna directional diagram.

6.3 SCAN CONTROL IN NEW TARGET SEARCHING MODE

6.3.1 Problem Statement and Criteria of Searching Control Optimality

The problem to control the searching mode is related to a class of observation control tasks from mathematical statistics and is given in the following form. Let the pdf $p(x|\lambda)$ of the observed random variable X be not given completely and possess some indefinite parameter ν that is selected by the statistician before a next observation of the random variable X. Thus, the following relation takes a place:

$$p(x|\lambda) = p(x|\lambda,\nu). \tag{6.17}$$

In this case, there is a need to choose the parameter ν of the pdf $p(x|\lambda,\nu)$ before observation of the random variable X. As a result, the experimentator controlling the observations affects the statistical features of observed random variables.

In radar, the control process is based on obtaining the radar coverage scanning, selection of angle direction and interval on radar range scanning, shape of radar antenna directional diagram, type of searching signal, and so on. In our case, the control process of scanning is reduced to controlling the radar antenna directional diagram position, the shape and angle dimensions of which are fixed under the condition of realization of the discussed algorithms. Under synthesis of observation control algorithms, it is natural to follow the principle of average risk minimization, that is, minimization of losses arising under the decision error. However, in this case, a choice of the corresponding function of losses is a difficult problem.

As an example, consider the following discussion on generation of the quadratic function of losses widely known from the theory of statistical decisions [12–14]. Let a set of $\{N_{\text{tg}};\theta_1,\theta_2,\ldots,\theta_N\}$ be a description of a true situation in the radar coverage, where N_{tg} is the number of targets and θ_1 are the target parameters. Let $v(t,\theta_i)$ be the loss function formed by the target with the parameter θ_i. Then $\sum_{i=1}^{N_{\text{tg}}} v(t,\theta_i)$ is the total loss formed by all targets. As a consequence of searching or experiments, we obtain the estimations in the radar coverage in the following form $\{\hat{N}_{\text{tg}};\ \hat{\theta}_1,\ \hat{\theta}_2,\ldots,\ \hat{\theta}_{\hat{N}}\}$ and the estimation of the total loss will be $\sum_{i=1}^{\hat{N}_{\text{tg}}} v(t,\ \hat{\theta}_i)$. As a figure of merit of the total loss estimation, we can consider the quadratic function of losses in the following form:

$$\int_{T_S}\left[\sum_{i=1}^{N_{\text{tg}}} v(t,\theta_i) - \sum_{i=1}^{\hat{N}_{\text{tg}}} v(t,\hat{\theta}_i)\right]^2 dt = C\{N_{\text{tg}};\theta_1,\theta_2,\ldots,\theta_{N_{\text{tg}}};\ \hat{N}_{\text{tg}};\ \hat{\theta}_1,\ \hat{\theta}_2,\ldots,\ \hat{\theta}_{\hat{N}_{\text{tg}}}\}. \tag{6.18}$$

If we are able to define the a posteriori pdfs $p(\theta_i|\mathbf{X},\mathbf{V})$, where $\mathbf{X}$ is the vector of observed data and $\mathbf{V}$ is the control vector under searching, then the average risk can be defined in the following form:

$$\mathcal{R}(\hat{\theta}\,|\,\mathbf{V}) = \int\limits_{\theta} C\{N_{\text{tg}};\theta_1,\theta_2,\ldots,\theta_{N_{\text{tg}}};\ \hat{N}_{\text{tg}};\ \hat{\theta}_1,\ \hat{\theta}_2,\ldots,\ \hat{\theta}_{\hat{N}_{\text{tg}}}\}\,p(\theta\,|\,\mathbf{X},\mathbf{V})\,d\theta. \tag{6.19}$$

In this case, a minimization of the average risk given by (6.17) must be carried out by selection of the control vector:

$$\mathbf{V}: \mathcal{R}(\hat{\theta})_{\min} = \min_{\mathbf{V}} \mathcal{R}(\hat{\theta}|\mathbf{V}). \tag{6.20}$$

It is clear that a realization of such criteria is very difficult, especially within the limited time to scan the radar coverage zone.

The information criterion of efficacy reducing to maximization of average amount of information during scanning and observation of radar coverage was suggested in Refs. [15,16]. By this criterion, the control must be organized in such a way that an increment of information would be obtained at each step of observation. In a general case, a solution by both criteria is difficult. An approximate approach is based on a set of assumptions, of which the following are important:

- The problem is solved for a single target or set of uncorrelated targets.
- The observation space consists of cells, and we can scan a single cell only within the limits of observation step (the scanning period); in this case, a control process is the integer scalar value υ coinciding with the number of observed cells.
- The υ th cell random value X observed under scanning is a scalar and subjected to the pdf $p_0(x)$ in the case of target absent and the pdf $p_1(x)$ if the target is present; These pdfs are considered as known.

6.3.2 Optimal Scanning Control under Detection of Single Target

We start an analysis of optimal control methods to scan with the simplest case, namely, the detection of a single target within the limits of radar coverage consisting of N cells. Assume that there is a single target in the radar coverage or there are no targets. If there are no targets in the radar coverage, then the target can appear during the time τ within the ith cell ($i = 1, 2, \ldots, N$) with the probability $P_\tau(i)$. If there is a single target within the limits of radar coverage, the probability of appearance of new targets is equal to zero. The target can be replaced by cells of scanning area with the probability of passing $P_\tau(i\,|\,j)$. Under a synthesis of control algorithm, we can consider the time (the number of cycles) from the beginning of scan to the instant of target detection as a figure of merit. The optimization process of the control algorithm is a minimization of this time.

The main result of synthesis shows us that the optimal control algorithm of radar coverage scanning is reduced to get the a posteriori probabilities $P_t(i)$ of target presence in the cells ($i = 1, 2,\ldots, N$) and to select the cell for next scanning ($i = \upsilon$), within the limits of which the a posteriori probabilities $P_t(i)$ of target presence are maximal, that is,

$$P_t(\upsilon) = \max_i P_t(i). \tag{6.21}$$

In the absence of the a priori data, the initial values of the probability $P_0(i)$ at $t = 0$ are chosen based on the condition that the probability of target presence in any radar coverage cell is the same. After regular scanning (e.g., the υ th cell) at the time instant t_N the a posteriori probabilities of target presence in cells are defined by the formulas:

- In the case that the radar coverage cell has been just scanned ($i = \upsilon$),

$$P_{t_N}(\upsilon) = \frac{P^*_{t_N}(\upsilon)p_1(x)}{\left[1 - P^*_{t_N}(\upsilon)\right]p_0(x) + P^*_{t_N}(\upsilon)p_1(x)}; \tag{6.22}$$

- In the case of no scanned radar coverage cells ($i \neq \upsilon$),

$$P_{t_N}(\upsilon) = \frac{P^*_{t_N}(\upsilon)p_0(x)}{\left[1 - P^*_{t_N}(\upsilon)\right]p_0(x) + P^*_{t_N}(\upsilon)p_1(x)}; \tag{6.23}$$

where $P^*_{t_N}(\upsilon)$ is the a posteriori probability of target presence in the υ th cell determined at the instant t_N.

Thus, after scanning the υ th cell, the probabilities of target presence both in the scanned cell and in the unscanned cells are changed and, for example, an increase in the probability of target presence in a single cell leads us to a decrease in the probability of target presence in other cells.

The next scanning of the ith cell will be over the time interval τ. For this reason, under computation of the a posteriori probability of target presence $P_{t_N}(\upsilon)$ at the instant $t_N + \tau$ there is a need to take into consideration the probability of passing the target in other cells, that is, the probability $P_\tau(i\,|\,j)$, and the probability of new target appearance $P_\tau(i)$ within the limits of radar coverage, if a new target is absent. Computation of the probability of target presence in the ith cell at the instant $t_N + \tau$ is carried out by the formula

$$P^*_{t_{N+\tau}}(i) = \left[1 - \sum_{i=1}^{N} P_{t_N}(t)\right]P_\tau(i) + \sum_{j=1}^{N} P_\tau(i|j)P_{t_N}(j). \tag{6.24}$$

The considered control algorithm operates in such a way that at the end of scanning the antenna beam is directed to the cell with the maximal a posteriori probability of target presence and is not able to be switched to scan other cells. The scanning stop is carried out when the given a posteriori probability of target presence is reached (the threshold).

6.3.3 Optimal Scanning Control under Detection of Unknown Number of Targets

As a model of a radar coverage situation, we consider an uncorrelated set of targets that appear and move within the limits of the radar coverage (from cell to cell) independently of each other. The probability of a presence in each cell of more than one target is equal to zero, which corresponds to a model of rarefied flow. As earlier, the probability of target appearance in the ith cell during the time τ is equal to $P_\tau(i)$, and replacement of targets from cell to cell is described by the passing probability $P_\tau(i|j)$. A procedure of optimal scanning is to choose that cell at the next step for which the a posteriori probability of target presence is maximal. However, the following fact plays a very important role in this procedure: there are several targets within the limits of the radar coverage. By this reason, there is a need to ensure a mode of sequential addressing to the different cells including the cells with the minimal probability of target presence. In accordance with this fact, a sequence of control algorithm operations must be the following:

- At the beginning of scan when the a posteriori probabilities of target presence in the cells are low, the υ th cell with the maximal probability $P_t(\upsilon) = \max_i P_t(i)$ is selected to scan.
- If after regular scanning the probability $P_t(\upsilon)$ is continuously increased, the scanning of the υ th cell is continued until this probability reaches the threshold value C_1 chosen based on a permissible value of the probability of false alarm P_F. After that, a decision about a target presence in the given cell is made and scanning of this cell is stopped.
- Scanning locator is switched on one no scanned cell with the maximal probability $P_t(i)$. In doing so, if after regular scanning the a posteriori probability of target presence in this cell is decreased, there is a need to select the cell with the maximal value of the probability $P_t(i)$ and scan this cell.
- Scanning the cell with the decreased a posteriori probability of target presence is continued until this probability is less than the threshold value C_2 defined based on a permissible number of cycles required to scan the "empty" cells.
- The cell with the detected target is included in a set of candidates to be scanned after (owing to target moving) the fact when the probability of target presence in this cell becomes less than the threshold C_1.

Thus, the control algorithm ensures a sequential addressing to the cells, including the cells with the low a posteriori probability of target presence. The considered algorithm is optimal based on the viewpoint of minimization of the number of scanning cycles T from the beginning of scan to the instant of detection of all targets. In this case, the a posteriori probabilities of target presence are determined by the following formulas:

- In the υ th cell after regular scanning—the formula (6.22)
- In other no scanning cells

$$P_{t_N}(i) = P^*_{t_N}(i), \quad i \neq \upsilon. \tag{6.25}$$

Determination of the a posteriori probability of target presence at the instant $t_N + \tau$ of next cell scanning taking into consideration the movement and appearance of new targets is carried out based on the following formula:

$$P^*_{t_N+\tau}(i) = \sum_{j=1}^{N} P_\tau(i\,|\,j)P_{t_N}(j) + P_\tau(i). \tag{6.26}$$

These formulas take into consideration a peculiarity of situation model within the limits of radar coverage, in accordance with which the cells are independent and changes in the a posteriori probability of target presence in one cell do not change the a posteriori probability of target presence in other cells. The disadvantage of the considered approach in the synthesis of the optimal scanning control algorithm under detection of targets is the fact that there are no limitations on power resources required to scan the radar coverage. Therefore, it is worthwhile to search another approach to solve the problem of scanning optimization within the limits of the radar coverage.

The criterion maximizing the mean of number of the detected targets during the fixed time T_S is useful in this sense. In the case when the scanning model of radar coverage is in the form of cells, we have

$$E\{T_S\} = \max_{\varphi_i} \sum_{i=1}^{N_{\text{cell}}} P_t(i)P_S(\varphi_i) \tag{6.27}$$

when

$$\sum_{i=1}^{N_{\text{cell}}} \varphi_i = E_{\text{S}}, \quad \varphi_i \geq 0, \tag{6.28}$$

where

E_{S} is the energy during the scanning

N_{cell} is the number of cells within the radar coverage

$P_t(i)$ is the a priori probability of target presence in the ith cell

φ_i is the energy to scan the ith cell

$P_{\text{S}}(\varphi_i)$ is the probability of target detection in the ith cell under the condition that the energy φ_i has been required to do it

The criterion (6.27) allows us to find the optimal distribution of limited power resources by the cells but says nothing about the optimal sequence of scanning all the cells. Naturally, we consider a possibility to combine the criteria (6.21) and (6.27) with the purpose of obtaining the conditions to minimize the time to detect all targets under the limited power resources required to carry out the scanning within the limits of the radar coverage. In this case, the problem of optimal scanning control must be solved in two stages:

- Solving the problem (6.27), there is a need to find the optimal distribution of power resources required to scan the radar coverage expressed, for example, by the number of scanning cycles of each radar coverage cell.
- Using the criterion (6.21) and distribution of power resources by the cells, there is a need to define the optimal sequence of scanning of the radar coverage cells. At the same time, it is permissible to use the queuing theory, in particular, the optimal rule of application service in decreasing order of the ratio $P_t(i)/N_i$, where N_i is the number of scans in the radar coverage along the ith direction.

A sequence of statements for designing the optimal control algorithm follows:

- Let there be the a priori data about the probabilities $P_{t_N}(i)$ $(i = 1, 2, \ldots, N)$ of target presence in each radar coverage cell at the time instant t_N.
- Let the power resources to scan the radar coverage be given by the number N_0 of scanning at the constant cycle T of searching signals.
- If for each cell of radar coverage the maximal radar range $R_{\max_i}$ and the effective scattering area S_{tg_i} of target (the pulse power P_{p} and pulse duration τ_{p} are known) are given, we are able to define the energy of target return signal E_{s_i} under single scanning; the total power spectral density of the interference and noise $\mathcal{N}_{\Sigma_i}$; SNR under scanning by a single pulse (the radar receiver is constructed based on the generalized detector) $\text{SNR}_i = q_i^2 = E_{\text{s}_i}/\mathcal{N}_{\Sigma_i}$; SNR under scanning by N_i signals is given by $\text{SNR}_{N_i} = q_{N_i}^2 = \mathcal{N}_i q_i^2$.
- We assume, for example, that during scanning the ith cell, the target return signal is a noncoherent pulse train with independent fluctuation according to the Rayleigh in-phase and quadrature components. In this case, the conditional probability of false alarm P_{F} and the probability of detection P_{D} of the pulse train consisting of N_i pulses are determined by the following formulas [15]:

$$P_{\text{F}}(N_i) = \frac{1}{2^{N_i}(N_i - 1)!} \int_{K_{\text{g}}}^{\infty} X_i^{N_i - 1} \exp(-0.5 X_i)\, dX_i; \tag{6.29}$$

$$P_D(N_i) = \frac{1}{2^{N_i}(N_i - 1)!} \int\limits_{K_g/(1+q_i^2)}^{\infty} X_i^{N_i-1} \exp(-0.5X_i)dX_i, \tag{6.30}$$

where

X_i is the sum of normalized voltage amplitudes (the signal + the noise) at the GD output

K_g is the normalized threshold used under detection of target return signals by GD determined based on required values of the probability of false alarm P_F

- The problem of optimal distribution of power resources by cells within the limits of the radar coverage for a single scanning is solved by assigning the number of pulses N_i with the purpose to scan each cell, that is,

$$E\{N_0\} = \max_{N_i} \sum_{i=1}^{N_{tg}} P_t(i) P_S(N_i), \quad \sum_{i=1}^{N_{tg}} N_i = N_0. \tag{6.31}$$

  As a result, we obtain a set of pulse trains of the signals $N_1, N_2, \ldots, N_{N_{tg}}$ to scan all cells within the limits of the radar coverage.
- We define the ratios $P_t(i)/N_i$ and arrange them in decreasing order. After that we start a sequential scanning of the cells using the serial numbers of ratios $P_t(i)/N_i$.

Thus, we have a principal possibility to design and realize the optimal control algorithm for scanning in detection of targets within the radar coverage. However, realization of such algorithms in practice is difficult. Difficulties are caused by the following methodological and computational peculiarities:

- The considered cell model of radar coverage is not acceptable in radar. There is a need, at least, to consider a model with a discrete set of scanning directions. At the same time, we should either abandon the cell model of radar coverage or present each direction of scanning by a set of cells that are scanned simultaneously. These approaches are discussed in Ref. [15], but they are very complex to implement in practice.
- The principal problem is to get knowledge about the a posteriori probabilities of target presence in the cells and determination of these probabilities by cell scanning data. The fact is that a computation of the a posteriori probabilities of target presence in the cells after each scanning is very cumbersome and the main thing is that this computation is not matched with the methods to choose the parameters of searching signals and tuning facilities to process these parameters, which are discussed in the previous section. For instance, in this case, it is absolutely unclear how we can control facilities to protect against the passive interferences.
- Preliminary determination of pulse trains to scan each cell of radar coverage by the criterion (6.27) is very cumbersome and not acceptable under signal processing in real time. Moreover, there is a need to know the a priori probabilities of target presence in the cells of radar coverage (directions).

In line with the discussed statements, as a rule, in practice, in designing the control subsystem of signal processing and operation of the complex radar system, we should proceed from the simplest assumptions about spatial searching and distribution of target in this space. First, we assume that the searching space consists of uniform searching zones, within the limits of which the target situation is of the same kind. The intensity of target set in each zone can be estimated a priori based on the purposes of the CRS and conditions to exploit it. The probabilities of target presence in each direction of the selected scanning zone are considered as the same, and the noise regions are assumed to be localized. Because of this, there are no priority directions to search for the targets, and the zone scanning is carried out sequentially and uniformly with the period T_S. In this case, the problem of controlling the scanning is to optimize the zone scanning period, for example, using the criteria to minimize the average time when the target could enter the scanning zone until the target detection under limitations in power resources issued to detect new targets. One of the possible versions of such an algorithm is discussed later.

6.3.4 Example of Scanning Control Algorithm in Complex Radar Systems under Aerial Target Detection and Tracking

As an example, consider a version to organize a scanning control by the corresponding CRS subsystem. In this example, we assume that the power resources must be distributed between the target detection and tracking modes that are realized individually in accordance with the real target and noise environment. At first, we discuss the general statements about designing the scanning control algorithm. After that, we describe the flowchart of the corresponding control algorithm.

First, we note that the power resource of the considered two-mode (in a general case, multimode) CRS required to scan is the random variable depending on the number of tracking targets. For example, at the initial period of operation, when there are no targets to track, all energy of the CRS is utilized to search new targets in the scanning space (the radar coverage). As the targets appear and the number of targets tracking increases, the power required to scan decreases. In doing so, owing to priority of the tracking mode over the detection mode, the power can be very low and close to zero. In the stationary mode of CRS operation, the number of tracking targets also oscillates within wide limits.

The power resources in scanning each direction in the searching mode depend on characteristics and parameters of the targets and the interferences and noise and are also defined by the signal processing algorithm employed in this mode. As noted previously, the minimal expenditure of energy required to scan directions can be reached employing algorithms of sequential analysis under signal processing and two-stage signal detection algorithms. In accordance with the approach considered in this chapter to organize a control of scanning and searching the targets, the number of directional scans under searching is established at each step in passing to a new direction based on analysis of the noise environment and taking into consideration the given probability of detection P_D of targets with the known effective scattering area S_{tg_i} at the far boundary of the radar coverage. Thus, in our case, the number of directional scans is not a controlled parameter in the searching mode.

In practice, it is very difficult to determine the probability of target presence in each direction of searching space. However, in some clearly defined zones of this space, it is possible to compute parameters and characteristics of the set of the targets crossing each zone boundary, and consequently, the probability of crossing by targets these boundaries based on the a priori analysis of target functions in the performance of target tasks. At the same time, the probabilities of target way to the corresponding zone using any arbitrary direction must be considered as the same, naturally. The far boundaries of searching zones correspond to the maximal radar ranges that are characteristics of each zone under detection of targets with approximately the same effective scattering area S_{tg_i}. In accordance with general regulations of optimal control, the scanning process must be considered as a controlled process.

As a rule, the scanned zones possess unequal priorities in service. Priority is related to the importance of the target and with periodicity of searching and computational burdens to carry out a scanning. In particular, the highest priority can be assigned for the ith zone, where the following condition $\bar{t}_{S_i}/T_{S_i} \to \min$ is satisfied. Here $\bar{t}_{S_i}$ is the average time to search the ith zone; T_{S_i} is the period of the ith zone scanning, that is, the relative expenditure of energy for scanning is minimal. Priorities of other zones are decreased in increasing order of the relative expenditure of energy for scanning.

A buffer zone should be provided within the limits of scanning space, in which the energy required for the target detection and tracking is balanced. The strict period of scanning is not set in the buffer zone. Upper boundary T_S^* for the period of scanning is established only, that is, $T_{S_{bz}} < T_S^*$, based on conditions to start new target tracings by the target pips detected within the limits of this zone. In this case, the buffer zone far boundary radar range R_{bz} is the only controlled parameter. In doing so, the far boundary corresponds to the balance between the expenditure of energy, which is issued and used taking into consideration priority zones, to search for the targets. Thus, to control the buffer zone far boundary radar range R_{bz} is the last operation for the scanning control system.

The discussed statements allow us to formulate the problem of optimal scanning control in the following form:

1. *The criterion of optimality*

$$\gamma_{opt} = \max_{T_{S_{bz}}} R_{bz}; \tag{6.32}$$

2. *Limitations*

$$T_{S_{bz}} < T_S^* \quad \text{and} \quad R_{bz}^{min} \le R_{bz} \le R_{bz}^{max}. \tag{6.33}$$

Figure 6.6 shows us the flowchart of the scanning control algorithm. This algorithm operates periodically in accordance with the established control cycle. As a control cycle, that is, the time interval between the control command outputs, we assign the time interval multiple to the scanning period T_{S_i} of the most priority zone. The scanning periods of other priority zones are also multiple to this period T_{S_i}. The present control algorithm starts to operate with calculation of the number directions $N_{bz}^{(k)}$ scanned in the buffer zones during the kth control cycle $T_{c_k} = t_k - t_{k-1}$ (blocks 1 and 2). After that the time required to scan the priority zones is determined by the following formula (block 3)

$$t_\Sigma^{(k)} = \sum_{j=1}^{\upsilon} t_{S_j}^{(k)}, \tag{6.34}$$

where υ is the number of priority zones;

$$t_{S_j}^{(k)} = \sum_{l=1}^{\nu_j} N_{j_l}^{(k)} T_j \tag{6.35}$$

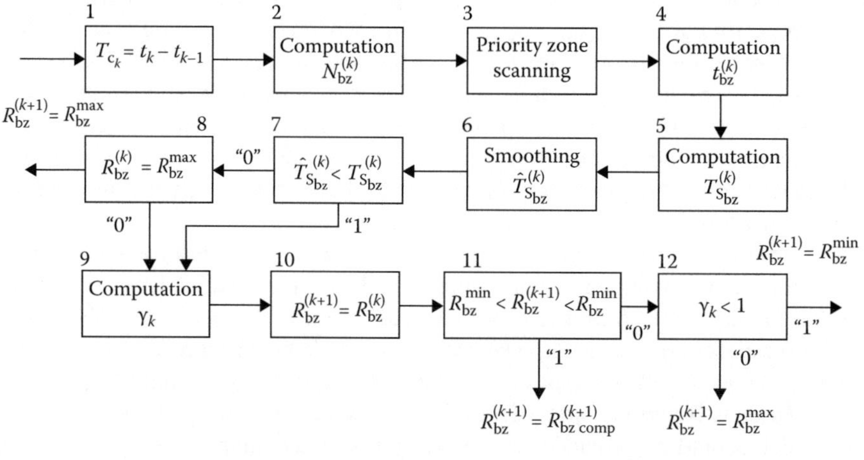

FIGURE 6.6 Flowchart of scanning control algorithm.

is the time required to scan the jth priority zone in the kth cycle with the period of T_{c_k}; ν_j is the number of scans of the jth priority zone within the limits of the period T_{c_k}; $N_{jl}^{(k)}$ is the number of sweeps during the sth scanning of the jth priority zone at the kth control cycle; T_j is the period of searching pulses under scanning of the jth priority zone. The time to scan the buffer zone during the kth control cycle

$$t_{\text{bz}}^{(k)} = T_{c_k} - t_{\Sigma}^{(k)} \tag{6.36}$$

is determined by block 4. The searching period of the buffer zone based on data of the kth control cycle

$$T_{S_{\text{bz}}}^{(k)} = \frac{N_{\text{bz}}}{N_{\text{bz}}^{(k)}} t_{\text{bz}}^{(k)}, \tag{6.37}$$

where N_{bz} is the number of scanned directions within the limits of the buffer zone that is determined by block 5. Henceforth, the computed scan period is smoothed by the totality of some control cycles (block 6). At the same time, as a first approximation, we can assume that the scan period is changed slowly within the limits of a small observation interval, and we can use the formula for exponential smoothing in the simplest form [17]:

$$\hat{T}_{S_{\text{bz}}}^{(k)} = (1-\iota)T_{S_{\text{bz}}}^{(k)} + \iota \hat{T}_{S_{\text{bz}}}^{(k-1)}, \tag{6.38}$$

where ι is the constant with a sense of the smoothing coefficient lying within the limits $0 < \iota < 1$. In future, the scan period $\hat{T}_{S_{\text{bz}}}^{(k)}$ is compared with the permissible value (block 7), that is,

$$\hat{T}_{S_{\text{bz}}}^{(k)} < T_{S_{\text{bz}}}. \tag{6.39}$$

If the condition (6.39) is satisfied, there is a need to check the buffer zone radar range, that is, whether $R_{\text{bz}}^{\max}$ is a maximal value or not (block 8). If "yes," then the control algorithm of CRS functioning receives a command for the next cycle:

$$R_{\text{bz}}^{(k+1)} = R_{\text{bz}}^{\max}. \tag{6.40}$$

If the condition

$$T_{S_{\text{bz}}}^{(k)} > T_{S_{\text{bz}}}^{*} \tag{6.41}$$

is satisfied, block 9 computes the coefficient γ_k of the buffer zone radar range changing $R_{\text{bz}}^{(k+1)}$ to ensure a restriction on the scan period duration. This coefficient depends on the previous buffer zone radar range $R_{\text{bz}}^{(k)}$ and deviation

$$\Delta T_{S_{\text{bz}}}^{(k)} = |\hat{T}_{S_{\text{bz}}}^{(k)} - T_{S_{\text{bz}}}^{*}|, \tag{6.42}$$

that is,

$$\gamma_k = f\left(R_{\text{bz}}^{(k)}, \Delta T_{S_{\text{bz}}}^{(k)}\right). \tag{6.43}$$

In this case, block 10 computes the buffer zone radar range as given by

$$R_{\text{bz}}^{(k+1)} = R_{\text{bz}}^{\max}\gamma_k. \tag{6.44}$$

Computation of the buffer zone radar range $R_{\text{bz}}^{(k+1)}$ is also carried out in the case when (6.39) is satisfied and $R_{\text{bz}}^{(k)}$ is not maximal. In this case, first of all, we compute γ_k; after that we make more exact the buffer zone radar range $R_{\text{bz}}^{(k+1)}$. Evidently, if (6.39) is not satisfied, $\gamma_k < 1$, and if the criterion of optimization (6.44) is not satisfied, $\gamma_k > 1$. Henceforth, we check for all cases a double restriction on the buffer zone radar range. If these restrictions are satisfied, block 11 computes $R_{\text{bz}}^{(k+1)}$. If these restrictions are not satisfied, block 12 checks the inequality $\gamma_k < 1$. If this inequality is satisfied, block 12 yields the equality

$$R_{\text{bz}}^{(k+1)} = R_{\text{bz}}^{\min}. \tag{6.45}$$

Otherwise, the equality (6.44). At this stage the control algorithm cycle is stopped.

There is a need to emphasize that the considered example is one of the possible ways of the scanning control algorithm subjected to analysis and comparison with other versions of scanning control algorithms in the course of designing the CRS.

6.4 POWER RESOURCE CONTROL UNDER TARGET TRACKING

6.4.1 Control Problem Statement

The target tracking mode can be divided conditionally into two stages. The first is the pretracking stage that starts immediately after detection of new target tracing. There is a need to specialize the parameters of new target tracing in such a way that it is possible to evaluate the importance of the target (danger or not danger). In other words, there is a need to define parameters and characteristics of target movement with respect to the defended object, served airdrome of departure/arrival, or other controlled reference points. Additionally, based on obtained target tracking and signal processing data there is a need to define a predictable line of target service, for example, airdrome for air target or air-based, submarine-based, and land-based location. The required accuracy of target information to evaluate the target importance depends on the kind of target and system assigning. To reduce the time required to evaluate target importance, there is a need to select a variable location rate and the value of location rate must be as high as possible.

The second stage is that of stationary target tracking. This stage starts after evaluation of target importance and making the decision about the location of target service line. Since the qualified service and target tracking are possible only after reaching the given accuracy of target tracking estimation at the boundary line, the main task of the second stage is to accumulate information about the target tracking parameters and to extrapolate this information at the required instants, when the target tracking information is in stationary mode. Accuracy of evaluation of the target tracking parameters depends on accuracy of individual measurements of target coordinates and the number of observations (measurements). A choice of signal processing algorithm, in particular, the filtering algorithm, is very essential. In a general case, the problem is to get the required accuracy at minimal expenditure of energy and minimal digital signal processing system loading. Naturally, these requirements look as a contradiction, since increasing accuracy and target tracking subsystem reliability for real targets are possible only under more sophisticated signal processing or filtering algorithm; for example, the implementation of adaptive signal processing algorithms, which leads us to increasing the digital signal processing subsystem loading, or with increasing the frequency of coordinate measurements (increasing the number of observations or measurements), which leads us to an increase in CRS energy expenditure.

In the process of designing the CRS digital signal processing subsystem, sometimes there is a need to implement simple filtering algorithms or other kinds of simple digital signal processing algorithms with the purpose of reducing the digital signal processing subsystem loading. However, this simplification leads us to an increase in CRS energy expenditure for each tracking target. In other words, a throughput of the number of tracking targets by "CRS–digital signal processing subsystem" channel is decreased. Evidently, a simplification of digital signal processing algorithm has a sense if the additional digital signal processing subsystem loading is little and acceptable with CRS specifications.

We return again to the problem of required assurance of accuracy of target information at a target service line and discuss briefly what must be considered as a measure of this accuracy, that is, as a measure of figure of merit of target tracking problem at the considered stage. As we know, an accuracy of target tracing is characterized by the correlation matrix of errors in the definition of target tracking parameters. However, the correlation matrix of errors owing to its multidimensionality cannot be considered as a united measure of target tracking accuracy at a target service line. For this purpose, there is a need to choose any scalar factor that takes into consideration the most complete information contained in the error correlation matrix of estimation. The parameters listed here can serve as such a factor.

Determinant of the diagonal or factored correlation matrix of errors given by

$$\det \mathbf{\Psi}_{n_f} = |\mathbf{\Psi}_{n_f}| = \prod_{l=1}^{s} \sigma_{l_n}^2, \tag{6.46}$$

where $\sigma_{l_n}^2$ is the variance of error of lth parameter estimation at nth filtering step. In the case of the factored correlation matrix, the diagonal elements $\sigma_{l_n}^2$ are equal to squared lengths of the main axles of corresponding unit error ellipsoid. Because of this, (6.46) is a characteristic of the error ellipsoid volume. If the error ellipsoid is the oblong ellipsoid, the determinant (6.46) can be low by value in the case even when the variance value by one of the main axles is impermissibly high. This is a great disadvantage of the considered factor.

Tracing of the correlation matrix of errors can be presented in the following form:

$$\operatorname{tr} \mathbf{\Psi}_{n_f} = \sum_{l=0}^{s} \sigma_{l_n}^2. \tag{6.47}$$

In this case, a high accuracy of estimation by some parameters does not make a great contribution to factor value as a whole, and an action of poor estimations by one or several coordinates is significant.

The highest eigenvalue of the correlation matrix of errors can be presented in the following form:

$$\max \lambda_{l_n} = \max_{l} \sigma_{l_n}^2, \quad l = 1, 2, \ldots, s. \tag{6.48}$$

This factor defines some spherical estimation of error ellipsoid dimensions, since $\max_{l} \sigma_{l_n}^2$ is equal to the square of ball radius, inside of which the unity error ellipsoid is fitted.

6.4.2 Example of Control Algorithm under Target Tracking Mode

In accordance with the adopted initial premises, the target information processing in target tracking mode is started after making a decision about new target track detection. In this mode, first of all, there is a need to accumulate information and refine the parameters to solve the problem of target classification by importance. After that, at the second stage, there is a need to organize an individual target tracking for that targets, for which a numeric measure of target importance exceeds the given threshold.

A simple logical flowchart of a digital signal processing algorithm in the case of the target tracking mode with necessary control elements using the criterion of minimization of energy expenditure and computer cost is presented in Figure 6.7. Based on the mode control algorithm commands, the antenna beam is oriented in the direction β_{tg_j}, $\varepsilon_{\mathrm{tg}_j}$ of the expected location of the jth target at the instant t_n^{scan} given by the pulse train with simultaneous radar range sampling (block 1). In the course of digital signal processing within the limits of the gate, a target pip belonging to target tracing is selected. After that, the coordinates of this target pip are transformed into the rectangular coordinate system (block 2). Further, we check a value of classification factor $\mathcal{P}$ (the block 3). At $\mathcal{P} = 1$, the target is tracked in steady state. At $\mathcal{P} = 0$, there is a transient state. In the last case, the parameters of target tracing are specified by the linear filter F1 (block 4). After that, the variance of target track parameter estimation or another generalized accuracy factor is checked to satisfy the sufficient condition for solution of the target importance evaluation problem (block 5). If the accuracy is not satisfied, the next scanning is assigned at the instant

$$t_{n+1}^{\mathrm{scan}} = t_n^{\mathrm{scan}} + T_{\mathrm{renew}}, \tag{6.49}$$

where T_{renew} is the period of refreshment at the preliminary target tracking stage (block 6). After that, we compute extrapolated parameter values at the instant t_{n+1}^{scan} (block 7) and transform these parameters into the complex radar coordinate system (block 8) and generate the application for a new measurement (block 9).

If an accuracy sufficient for classification is obtained, we can assign $\mathcal{P} = 1$ (the block 10) and evaluate the tactical importance of the target (block 11). The simplest factors for evaluation are the velocity and direction of target moving. The target importance $\mathcal{M}_{\mathrm{tg}}$ is evaluated by

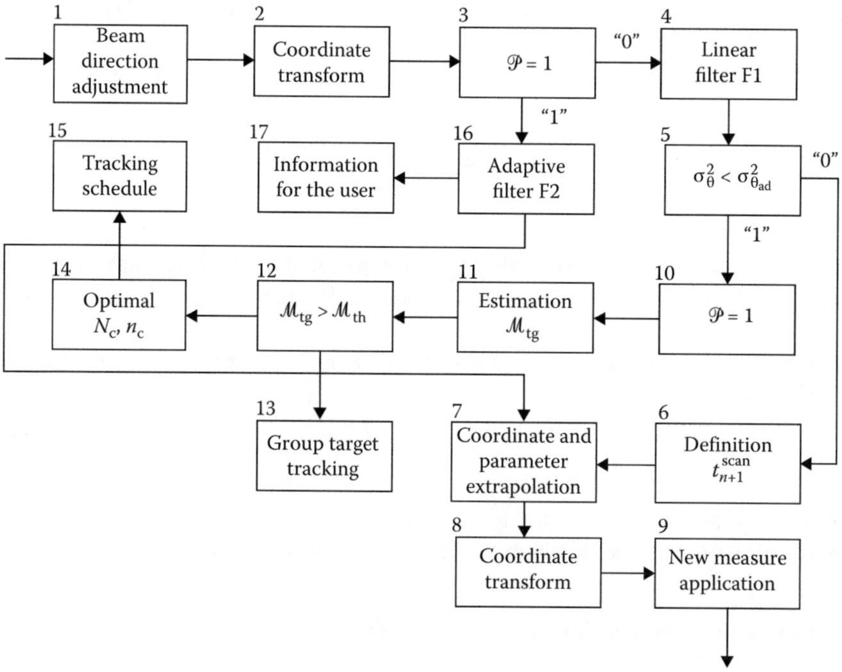

FIGURE 6.7 Flowchart of digital signal processing algorithm in the case of target tracking mode.

any quantitative assessment that is compared with the threshold $\mathcal{M}_{th}$ (block 12). If the condition $\mathcal{M}_{tg} < \mathcal{M}_{th}$ is satisfied, the target is assigned for the system of group target tracking using the periodical scanning data of radar coverage (block 13). If the condition $\mathcal{M}_{tg} > \mathcal{M}_{th}$ is satisfied, the target is considered as the important target and an individual target tracking is organized. For this purpose, first of all, there is a need to solve an optimization problem to define the number N_{sc} of scans in each direction (the number of pulses in pulse train or, in other words, a volume of pulse train) and the number n_{sc} of points of location on the target tracking from the instant of starting the individual target tracking until the instant of reaching the target service line (block 14). In a general case, the choice of the optimal values N_{sc} and n_{sc} is made using the criterion of CRS power resource minimization taking into consideration restrictions assigned under the problem definition. Computation results containing the refreshment period and number of scans at each location point are registered by corresponding blocks of the digital signal processing algorithm and used for CRS operation mode schedule.

At the stage of individual target tracking, it is worthwhile to carry out a parameter filtering using a more complex digital signal processing algorithm in comparison with the previous stage, for example, the adaptive digital signal processing algorithm discussed in Chapter 5. The use of adaptive digital signal processing algorithms makes a realization of the filter F2 (block 16) more complex. However, this action allows us to keep the constant period or information update about the maneuvering target tracing at acceptable dynamical errors of smoothing. Information about the target tracing parameters goes from the adaptive filter F2 output to the display system, user (block 17), and antenna control block (block 7 and further).

Thus, in the discussed example, a direct control of important target tracing is reduced to draw a rational schedule of measurements for each target based on CRS power resource minimization. In doing so, we assume that computer costs allow us to realize the discussed control algorithm. To consider this problem in a general case, there is a need to use the following restrictions on expenditure of energy and computer costs:

- The use of two-target tracking modes—the mode of individual target tracking for very important targets and the mode of target group tracking (this mode does not need additional power resources) for target, the importance of value of which is lower than the threshold
- The use of the linear smoothing filter at the stage of preliminary target tracking that allows us to save computer costs
- Minimization of expenditure of energy that is necessary to obtain the given accuracy at the line of getting information about the important targets

In conclusion, we can make two remarks that are useful in designing the target tracking algorithms in specific conditions.

- For the two-mode CRSs discussed in detail, the target tracking process must be organized in such a way that all targets, independently of their importance, would be processed and the important targets must be processed using an individual mode additionally. In this case, the organization of digital signal processing control becomes simpler. Moreover, the reliability of important target tracking is increased. However, at the same time, the digital signal processing algorithm loading is increased.
- When the given accuracy is reached, the important targets are assigned to a system of direct service. However, these targets cannot be removed from tracking mode and must be tracked and observed, at least, until when information about their full service is available or these targets leave the scanning area. The period of target refreshment for this final stage of target tracking must support precision characteristics of output parameters at the given level or to ensure a stable (without faults) target tracking.

6.4.3 Control of Energy Expenditure under Accuracy Aligning

Control of target tracking mode under the same conditions is optimized by the criterion of energy expenditure minimization up to that moment when the accuracy of parameter estimation of each target coincides with the given value obtained at the line of release of information. At the same time, the duration of scanning signals τ_{scan}, the period of searching signals T, the width of directional diagram of transmit and receive antennas, the number N_{sc} of scans in each target location point, and the number of points of location N_{l} within the limits of measured interval of target tracing can be considered as controlled parameters. In practice, we try to decrease the number of controlled parameters and consider the number N_{sc} of scans or the volume of scanning pulse train in the course of a single determination of target coordinate. The number of measurements (updates) of target coordinates for each target track until reaching the given accuracy is considered as the controlled parameters. In this case, the problem of target tracking mode control is reduced to assign the optimal schedule of CRS operation by each target taking into consideration a set of tracking targets. The control process is an instruction issue consisting of the coordinate measurement instants t_j^{measure} for each jth target, where $j = 1, 2,\ldots, N_{\text{tg}}$ and N_{tg} is the number of targets within the limits of scanning area, and the duration of scanning $N_{c_{ij}}T$ of the jth target under the ith measurement session $(i = 1,2,\ldots,N_{l_j})$.

Solution of the assigned optimal control problem of target tracking mode can be delivered by mathematical programming methods (software). However, these methods are cumbersome and not always appropriate, especially, under the dynamical mode of CRS operation. Consider the simplest and, consequently, easiest to implement version of the assigned problem solution. Let the lines of release of information and the required accuracy of information release at the line be given. Let the maximal variance of estimation of the target coordinates be selected as a measure of accuracy. We assume that the target tracking is organized in such a way that a sequence of measurements is uniformly precise and equally discrete. There is a need to choose N_{sc_j} and N_{l_j} for each target ensuring a given accuracy at minimal expenditure of CRS energy.

We choose the following restrictions directly acting on the assigned optimal control problem:

- The target tracing is linear within the limits of considered interval.
- The minimum number $N_{\text{sc}_j}^{\min}$ is defined by the requirement of measure reliability, that is, the requirement to accumulate the value

$$\text{SNR}_{\Sigma} = N_{\text{sc}_j} q_{1_j}^2 > q_{\text{th}}^2, \tag{6.50}$$

  at which the formulas associating a potential measure accuracy with the SNR by energy are correct [18–24], where $q_{1_j}^2$ is the SNR by energy under receiving a single target return signal from the jth target.
- The period of measurements is bounded up by permissible errors of target tracing extrapolation.

Further, for example, we are limited by consideration of a single angle coordinate under the target moving by circle with the constant radial velocity. In this case, the potential measure accuracy is given by

$$\sigma_{\theta_j}^2 = \frac{\theta_{0.5}^2}{\pi q_{1_j}^2 N_{c_j} \mathcal{L}_{\text{ac}}}, \tag{6.51}$$

where $\mathcal{L}_{ac}$ is the coefficient taking into consideration losses under the accumulation process. If we know CRS characteristics and parameters, targets, noise, and interferences, then

$$\sigma^2_{\theta_j} = \frac{\text{const}}{N_{c_j}}. \tag{6.52}$$

On the other hand, when our measurements are uniformly precise and equally discrete, the variance of filtering error of linear target track coordinates is determined by the following formula (see Chapter 5):

$$\sigma^2_{\hat{\theta}_j} = \frac{2(2N_{1_j} - 1)}{N_{1_j}(N_{1_j} + 1)} \sigma^2_{\theta_j}. \tag{6.53}$$

In the considered case, we must satisfy the condition

$$\sigma^2_{\hat{\theta}_j} \le \sigma^2_{\hat{\theta}_j\text{accept}}, \tag{6.54}$$

where $\sigma^2_{\hat{\theta}_j\text{accept}}$ is the acceptable value of filtering error variance at the line of release of information. Taking into consideration (6.52) and (6.53), we obtain

$$\frac{2(2N_{1_j} - 1)c}{N_{1_j}(N_{1_j} + 1)N_{c_j}} = \sigma^2_{\hat{\theta}_j\text{accept}}. \tag{6.55}$$

Thus, we have a relation between the unknown parameters N_{sc_j} and N_{1_j}. If one parameter is fixed, then based on (6.55) we can obtain another parameter and, in this case, we are able to determine an expenditure of energy. As a measure of expenditure of energy, it is natural to choose the value

$$\mathcal{K}_j = N_{sc_j} N_{1_j}. \tag{6.56}$$

In this case, the optimal control problem is reduced to selection of the pair $\{N^*_{sc_j}, N^*_{1_j}\}$ minimizing $\mathcal{K}_j$:

$$\mathcal{K}_{j\min} = \min_{\{N^*_{sc\,j}, N^*_{1j}\}} (N_{sc_j} N_{1_j}) \quad \text{at} \quad N^*_{sc_j} \ge N^{\min}_{sc_j}, \quad N^*_{1_j} \ge N^{\min}_{1_j}, \quad \sigma^2_{\hat{\theta}_j} \le \sigma^2_{\hat{\theta}_j\text{accept}}. \tag{6.57}$$

Computation is easy and can be done in real time. Figure 6.8 represents the results of computation (6.57). CRS energy parameters are presented conditionally by the SNR by energy under a single measurement. As we can see from Figure 6.8, in the considered case, there is a need to select the permissible minimum number N_{sc} and calculate the corresponding number N_1 to optimize the target tracking process since an increase in accuracy of a single measurement leads us to increase in total losses $\mathcal{K}$. Thus, the optimization of the tracking mode of the jth target based on the considered criterion is reduced, in the simplest case, to performing two operations:

- Based on SNR by energy, there is a need to compute the number of scanning signals N_{sc_j} to satisfy the condition $\sigma^2_{\hat{\theta}_j} \le \sigma^2_{\hat{\theta}_j\text{accept}}$.
- Knowledge of $\sigma^2_{\hat{\theta}_j\text{accept}}$ and $\sigma^2_{\hat{\theta}_j}$ allows us to compute N_{1_j} using (6.53).

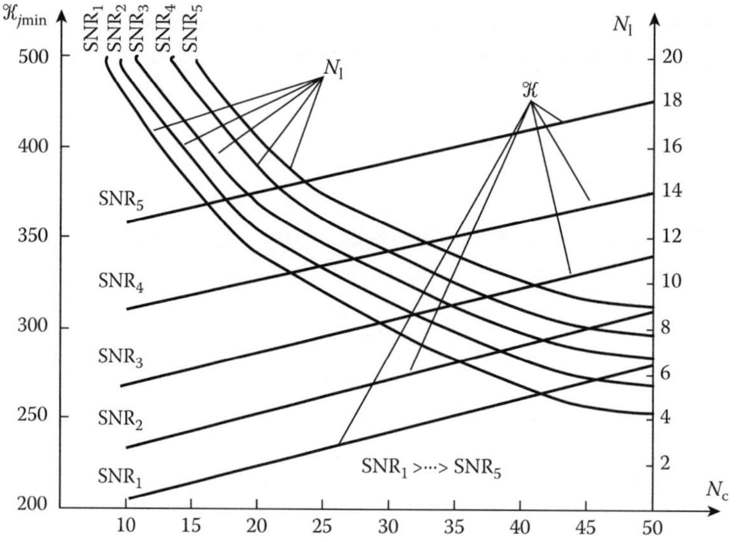

FIGURE 6.8 Results of computation based on (6.57).

6.5 DISTRIBUTION OF POWER RESOURCES OF COMPLEX RADAR SYSTEM UNDER COMBINATION OF TARGET SEARCHING AND TARGET TRACKING MODES

When CRSs work in the mode of individual important target tracking detecting a new target simultaneously, the problem of optimal distribution of limited power resources between these modes arises. In this case, the criterion of optimal distribution must take into account the quality factor of digital signal processing for each mode, for example, a maximization of the tracking targets number with the given precision characteristics at information issue lines with simultaneous detection of new targets with the given probability of detection P_{D} at the given line; an assurance of qualitative (with the user view point) target tracking of the given number of targets and minimization of target detection time in searching mode, simultaneously.

In a general case, a solution of the problem of optimal distribution of power resources between the target scanning modes and target tracking is absent. In this section, we consider one of the possible engineering approaches to the distribution problem of power resources of the two-mode CRS with programming control of the antenna directional diagram beam. In the simplest case, when there are no zones with the high a priori probability of new target appearance in the radar coverage, the target tracking mode is considered as priority mode and, consequently, all requirements of this mode for CRS power resource consumption are satisfied in the first place. In this case, the remaining power resources are used for the new target searching mode. The limitation for maximum permissible period of scanning $T_{\text{scan}}^{\text{per}}$ are fixed for this mode, for example, based on the condition that the probability of true target tracking (without confusion of target tracing) in group target tracking mode with defined scanning period is given.

When the number of tracked target is small, an excess of energy is spent to reduce the period of radar coverage scanning and, consequently, used to improve conditions of new target detection. As the number of tracked target increases, the energy consumption for target tracking increases and under definite task conditions can reach such a value when a remainder of CRS power resources assigned for searching new targets cannot satisfy the condition $T_{\text{scan}} \le T_{\text{scan}}^{\text{per}}$. From this point, the process of limitation and redistribution of power resources issued for two modes is started. At first, for example, the zone to search new targets is limited until some permissible dimension that allows us to save a part of the energy and reduce the scanning period

T_{scan}. Further, if it is impossible to reduce the scanning period T_{scan} in such a way, there is a need to move a part of the less important targets from the individual important target tracking to the group target tracking and so on.

Now, let us define the elementary relations allowing us to reach a balance of CRS power resources at two-mode operation. Let targets be tracked under the ith scanning M_i. Denote

$$t_{ij}^{\text{search}} = N_{ij}^{\text{search}} T \tag{6.58}$$

as the power resource expressed by the time of direction scanning spent on a single coordinate measurement of the jth target under the ith scanning; $j = 1, 2,\ldots, M_i$, T_j^{search} is the required period of refreshment by the jth target. Then, the power resource part of the system required for tracking of all targets can be determined in the following form:

$$\Delta E_i^{\text{search}} = \sum_{j=1}^{M_i} \frac{t_{ij}^{\text{search}}}{T_j^{\text{search}}}. \tag{6.59}$$

Let the number of directions that must be scanned within the limits of radar coverage be N_{d}, and the number of targets inside the radar coverage at the ith scanning is subjected to the condition $M_{\text{d}_i} \geq M_i$. Assume that there is one target along each scanning direction only. Scanning is the uniform process, that is, all the radar coverage cells are scanned uniformly. Then, the average time required to scan all directions during the ith scanning can be determined in the following form:

$$\overline{t_i^{\text{total}}} = (N_{\text{d}} - M_{\text{d}_i})\overline{\tau}_n + M_{\text{d}_i}\overline{\tau}_{\text{scan}} = N_{\text{d}}\overline{\tau}_n + M_{\text{d}_i}(\overline{\tau}_{\text{scan}} - \overline{\tau}_n), \tag{6.60}$$

where

$\overline{\tau}_n$ is the average time of directional analysis, in which there are no targets (the noise direction)
$\overline{\tau}_{\text{scan}}$ is the average time of directional analysis, in which there are targets

Now, based on the condition

$$\Delta E_i^{\text{total}} = 1 - \Delta E_i^{\text{search}} = \frac{t_i^{\text{total}}}{T_i^{\text{total}}}, \tag{6.61}$$

we can define the scanning time at the ith scanning of radar coverage:

$$T_i^{\text{total}} = \overline{t_i^{\text{total}}}\left(1 - \sum_{j=1}^{M_i} \frac{t_{ij}^{\text{search}}}{T_j^{\text{search}}}\right). \tag{6.62}$$

If the obtained scanning time satisfies the following condition $T_{\text{scan}_i} \leq T_{\text{scan}}^{\text{per}}$, then there is no power resource imbalance in the considered CRS and there is no need to redistribute the power resource on the next scanning step. If the condition mentioned earlier is not satisfied, then during the next scanning there is a need to reduce a searching zone or reschedule a service of targets in the individual target tracking mode. The solution of these tasks must be assigned to a control block of higher level.

Now, consider a general case of the distribution problem for limited CRS power resources. Let there be several zones in the radar coverage, namely, $Z_1, Z_2,\ldots, Z_l$, for which the periods

$T_1^{\text{total}}, T_2^{\text{total}}, \ldots, T_l^{\text{total}}$ are assigned, and there is at least the l+1th zone, for which there is the following restriction to scan the zone $T_{\text{scan}_l} \leq T_{\text{scan}}^{\text{per}}$. Target tracking in all scanning zones is carried out in the same way. In the considered case, there is the following procedure of service. Target tracking, as before, has the highest priority. Priorities of scanned zones are decreased with an increase in the sequence number of the zone. Scanning in the l + 1th zone has the lowest priority. The requirement to scan the zone with high priority interrupts a scanning process of the zone with low priority. In this case, the CRS energy balance is obtained, in the first place, owing to the l + 1th zone. However, the zones with high priority can participate in this process.

The part of CRS power resources required for scanning of the kth zone is given by

$$\Delta E_{k_i}^{\text{total}} = \frac{\overline{t_{k_i}^{\text{total}}}}{T_k^{\text{total}}}, \tag{6.63}$$

where

$\overline{t_{k_i}^{\text{total}}}$ is the average time required to scan the priority zone at the ith scanning computed analogously as in (6.30)

T_k^{total} is the assigned period of scanning for the kth zone

The part of CRS power resources required to scan all priority zones at the ith scanning is determined by

$$\Delta E_i^{\text{total}} = \sum_{k=1}^{l} \Delta E_{k_i}^{\text{total}}. \tag{6.64}$$

As a result, the part of CRS power resources remaining to scan the no priority l + 1th zone is determined by

$$\Delta E_{l+1}^{\text{total}} = 1 - \sum_{k=1}^{l} \Delta E_{k_i}^{\text{total}} - \Delta E_i^{\text{search}}. \tag{6.65}$$

In turn,

$$\Delta E_l^{\text{total}} = \frac{t_{(l+1)_i}^{\text{total}}}{T_{(l+1)_i}^{\text{total}}}, \tag{6.66}$$

where

$t_{(l+1)_i}^{\text{total}}$ is the time required to scan the no priority l + 1th zone at the ith scanning

$T_{(l+1)_i}^{\text{total}}$ is the period of the ith scanning for the l + 1th zone

Thus, in the given case, the scanning period $T_{(l+1)_i}^{\text{total}}$ can be determined as

$$\frac{t_{(l+1)_i}^{\text{total}}}{T_{(l+1)_i}^{\text{total}}} = 1 - \sum_{k=1}^{l} \Delta E_{k_i}^{\text{total}} - \Delta E_i^{\text{search}}. \tag{6.67}$$

If the condition $T_{(l+1)_i}^{\text{total}} > T_{\text{scan}}^{\text{per}}$ is satisfied, there is a need to redistribute CRS power resources both owing to a decrease in the tracked targets and owing to the area of priority scanning zones.

6.6 SUMMARY AND DISCUSSION

The radar engineer has a resource time that allows good Doppler processing, bandwidth for good range resolution, space that allows a large antenna, and energy for long-range performance and accurate measurements. External factors affecting radar performance include the target characteristics; external noise that might enter via the antenna; unwanted clutter echoes from land, sea, birds, or rain; interference from other electromagnetic radiators; and propagation effects due to the Earth's surface and atmosphere. These factors are highly important in CRS design and application.

The hierarchy of the control process is characterized by a sequential release of information from a complex radar subsystem of higher level to lower level subsystems. At the same time, the results of particular problem solutions at lower levels are used as the basis to make decisions at the higher level. Thus, information about the energy consumption for scanning in each direction, the number of detected targets, and the energy consumption under target tracking mode comes to the second-level subsystem as a result of the first-level task solutions. The information about the number of tracked targets $n_{\text{tg}}^{\text{track}}$, their importance, and also, about a scanning interval of new target zones is transferred to the subsystem of the third level as a result of task solutions at the second level. Simultaneously, the subsystems of the second and third levels of complex radar control subsystem transfer information about an ability of facilities and quality to solve the tasks to the fourth level subsystem where CRS administration is carried out, as a whole.

Under directional scanning in the mode of searching new targets, the control process is carried out by two stages. At the first stage, the calculation and choice of scanning signal parameters are carried out. The scanning signals must correspond to the functional mode of searching targets and maximal radar range in the given direction. Additionally, the receiver performance and signal processing algorithms for signal preprocessing subsystems should be defined and computed. This stage is carried out before scanning pulsing in the given direction. The control of accumulated received signals up to making decisions about the target detection/nondetection using all resolution elements in the radar range (the case of sequential analysis) or up to accumulation of determined number of the target return signals using detectors with the fixed sample sizes is carried out at the second stage. In the course of accumulation of the target return signals, the initial parameter values of scanning signals are the same except the carrier frequency f_c (see Figure 6.3).

After finishing the accumulation stage and making a decision about the scanned direction, the results of decision making are issued and expenditure of energy and timetable are computed to observe the scanned direction (block 9, Figure 6.3). The radar antenna beam control is assigned for an algorithm of scanning management in the searching mode or mode controller. In conclusion, we note once more that the discussed logical flowchart is the way only to assign a control process under the searching new target mode. Naturally, there are possible other ways, but a general idea to control, namely, parametric matching of CRS subsystems and signal and data processing subsystems with environmental parameters and solved problems and tasks, is the same.

The control algorithm under refreshment of target in target tracing mode operates interactively in a close way with algorithms of the main operations of signal preprocessing and reprocessing subsystems within the limits of the gate coordinating these algorithms into the unique signal processing subsystem in the individual target tracking mode by CRS with the control beam of radar antenna directional diagram.

In radar, a control process is based on to obtain the radar coverage scanning, selection of angle direction and interval on radar range scanning, shape of radar antenna directional diagram, type of searching signal, and so on. In our case, a control process of scanning is reduced to control the radar antenna directional diagram position, the shape and angle dimensions of which are fixed under the condition of realization of the discussed algorithms. Under synthesis of observation control algorithms, it is natural to follow the principle of the average risk minimization, that is, minimization of losses arising under the decision error. However, in this case, a choice of the corresponding function of losses is a very hard problem. The approximate approach is based on a set of assumptions, the important of which are the following: the problem is solved for a single target or set of uncorrelated targets; the observation space consists of cells and we can scan a single cell only within the limits of observation step (the scanning period); in this case, a control process is the integer scalar value υ coinciding with the number of observed cell; and the υth cell random value X observed under scanning is a scalar and subjected to the pdf $p_0(x)$ in the case when the target is absent and the pdf $p_1(x)$ if the target is present; these pdfs are considered as known.

The main result of analysis of optimal control methods to scan a single target within the limits of radar coverage consisting of N cells is the optimal control algorithm of radar coverage scanning that is reduced to get the a posteriori probabilities $P_t(i)$ of target presence in the cells ($i = 1, 2,\ldots, N$) and to select the cell for next scanning ($i = \upsilon$), within the limits of which the a posteriori probabilities $P_t(i)$ of target presence are maximal, that is, $P_t(\upsilon) = \max_i P_t(i)$. In the absence of the a priori data, the initial values of the probability $P_0(i)$ at $t = 0$ are chosen based on the condition that the probability of target presence in any radar coverage cell is the same. After regular scanning (e.g., the υth cell) at the time instant t_N the a posteriori probabilities of target presence in cells can be determined by the corresponding formulas using the a posteriori probability of target presence in the υth cell determined at the instant t_N. Thus, after scanning the υth cell, the probabilities of target presence both in the scanned cell and in the nonscanned cells are changed and, for example, an increase in the probability of target presence in a single cell leads us to a decrease in the probability of target presence in other cells.

As a model of radar coverage situation, we consider an uncorrelated set of targets that appear and move within the limits of the radar coverage (from cell to cell) independently of each other. The probability of a presence in each cell of more than one target is equal to zero, which corresponds to a model of rarefied flow. As earlier, the probability of target appearance in the ith cell during the time τ is equal to $P_\tau(i)$ and replacement of targets from cell to cell is described by the passing probability $P_\tau(i\,|\,j)$. A procedure of optimal scanning is to choose that cell at the next step, for which the a posteriori probability of target presence is maximal. However, the following fact plays a very important role in this procedure: there are several targets within the limits of radar coverage. By this reason, there is a need to ensure a mode of sequential addressing to different cells including the cells with the minimal probability of target presence. Thus, the control algorithm ensures a sequential addressing to the cells including the cells with the low a posteriori probability of target presence. The considered algorithm is optimal based on the viewpoint of minimization of the number of scanning cycles T from the beginning to scan to the instant of detection of all targets. We have a principal possibility to design and realize the optimal control algorithm for scanning under detection of targets within the radar coverage. However, realization of such algorithms in practice is very difficult. Difficulties are caused by the methodological and computational peculiarities.

In practice, in designing the control subsystem of signal processing and CRS operation we should proceed from the simplest assumptions about spatial searching and distribution of target in this space. First, we assume that the searching space consists of uniform searching zones, within the limits of which the target situation is of the same kind. The intensity of target set in each zone can be estimated a priori based on purposes of the CRS and conditions to exploit it. The probabilities of target presence in each direction of the selected scanning zone are considered as the same, and the noise regions are assumed to be localized. Because of this, there are no priority directions to search for the targets and the zone scanning is carried out sequentially and uniformly with the period T_S. In this case, the problem of controlling the scanning is to optimize the zone scanning period, for example, using the criteria to minimize the average time when the target could enter the scanning zone until the target detection under limitations in power resources issued to detect new targets.

When CRSs work in the mode of individual important target tracking detecting a new target simultaneously, the problem of optimal distribution of limited power resources between these modes arises. In this case, the criterion of optimal distribution must take into account a quality factor of digital signal processing for each mode, for example, a maximization of the tracking targets number with the given precision characteristics at information issue lines with simultaneous detection of new targets with the given probability of detection P_D at the given line; an assurance of qualitative (with the user view point) target tracking of the given number of targets and minimization of target detection time in searching mode, simultaneously.

When the number of tracked target is small, an excess of energy is spent to reduce the period of radar coverage scanning and, consequently, used to improve conditions of new target detection. As the number of tracked targets increases, the energy consumption for target tracking increases and under definite task conditions can reach such a value when a remainder of CRS power resources assigned for searching new targets cannot satisfy the condition $T_{\text{scan}} \leq T_{\text{scan}}^{\text{per}}$. From this point, the process of limitation and redistribution of power resources issued for two modes is started. At first, for example, the zone to search new targets is limited until some permissible dimension that allows us to save a part of the energy and reduce the scanning period T_{scan}. Further, if it is impossible to reduce the scanning period T_{scan} in such a way, there is a need to move a part of the less important targets from the individual important target tracking to the group target tracking and so on. If the obtained scanning time satisfies the following condition $T_{\text{scan}_i} \leq T_{\text{scan}}^{\text{per}}$, then there is no power resource imbalance in the considered CRS and there is no need to redistribute the power resource in the next scanning step. If the condition mentioned earlier is not satisfied, then during the next scanning there is a need to reduce the searching zone or reschedule the service of targets in the individual target tracking mode. The solution of these tasks must be assigned to a control block of a higher level.

REFERENCES

1. Scolnik, M. 2008. *Radar Handbook*. 3rd edn. New York: McGraw Hill, Inc.
2. Scolnik, M. 2002. *Introduction to Radar Systems*. 3rd edn. New York: McGraw Hill, Inc.
3. Hansen, R.C. 1998. *Phased Array Antenna*. New York: John Wiley & Sons, Inc.
4. Mailloux, R.J. 2005. *Phased Array Antenna Handbook*. Norwood, MA: Artech House, Inc.
5. Cantrell, B., de Graaf, J., Willwerth, F., Meurer, G., Leibowitz, L., Parris, C., and R. Stableton. 2002. Development of a digital array radar. *IEEE AEES Systems Magazine*, 17(3): 22–27.
6. Scott, M. 2003. Sampson MFR active phased array antenna, in *IEEE International Symposium on Phased Array Antenna Systems and Technology*, October 14–17, Boston, Massachussets, Isle of Wight, U.K., pp. 119–123.
7. Brookner, E. 2006. Phased arrays and radars—Past, present, and future. *Microwave Journal*, 49(1): 24–46.
8. Brookner, E. 1988. *Aspects of Modern Radar*. Norwood, MA: Artech House, Inc.
9. Wehner, D.R. 1995. *High-Resolution Radar*. 2nd edn. Boston, MA: House, Inc.
10. Nathanson, F.E. 1991. *Radar Design Principles*. 2nd edn. New York: McGraw-Hill, Inc.
11. Rastrigin, L.A. 1980. *Modern Control Principles of the Complex Objects*. Moscow, Russia: Soviet Radio.
12. Picinbono, B. 1993. *Random Signals and Noise*. Englewood Cliffs, NJ: Prentice-Hall, Inc.
13. Porat, B. 1994. *Digital Processing of Random Signals*. Englewood Cliffs, NJ: Prentice-Hall, Inc.
14. Therrein, C.W. 1992. *Discrete Random Signals and Statistical Signal Processing*. Englewood Cliffs, NJ: Prentice-Hall, Inc.
15. Bacut, P.A., Julina, Yu.V., and N.A. Ivanchuk. 1980. *Detection of Moving Targets*. Moscow, Russia: Soviet Radio.
16. Kontorov, D.S. and Yu.S. Golubev-Novogilov. 1981. *Introduction to Radar Systems Engineering*. Moscow, Russia: Soviet Radio.
17. Kuzmin, S.Z. 1974. *Foundations of Radar Digital Signal Processing Theory*. Moscow, Russia: Soviet Radio.
18. Helstrom, C.W. 1995. *Elements of Signal Detection and Estimation*. Upper Saddle River, NJ: Prentice-Hall, Inc.
19. Levanon, N. and E. Mozeson. 2004. *Radar Signals*. New York: John Wiley/IEEE Press.
20. Middleon, D. 1996. *An Introduction to Statistical Communication Theory*. New York: McGraw-Hill. Reprinted by IEEE Press, New York.
21. Kay, S.M. 1998. *Fundamentals of Statistical Signal Processing: Detection Theory*. Englewood Cliffs, NJ: Prentice-Hall, Inc.
22. Levanon, N. 1988. *Radar Principles*. New York: John Wiley & Sons, Inc.
23. Di Franco, J.V. and W.L. Rubin. 1980. *Radar Detection*. Norwood, MA: Artech House, Inc.
24. Carrara, W.G., Goodman, R.S., and R.M. Majewski. 1995. *Spotlight Aperture Radar—Signal Processing Algorithms*. Norwood, MA: Artech House, Inc.

Part II

Design Principles of Computer System for Radar Digital Signal Processing and Control Algorithms

7 Design Principles of Complex Algorithm Computational Process in Radar Systems

The exponential growth in digital technology since the 1980s, along with the corresponding decrease in its cost, has had a profound impact on the way radar systems are designed. More and more functions that were historically implemented in analog hardware are now being performed digitally, resulting in increased performance and flexibility and reduced size and cost. Advances in analog-to-digital converter (ADC) and digital-to-analog converter (DAC) technologies are pushing the border between analog and digital processing closer and closer to the antenna. For example, Figure 7.1 shows a simplified flowchart of the receiver front end of a typical radar system that would have been designed around 1990. Note that this system incorporated analog pulse compression (PC). It also included several stages of analog downconversion in order to generate baseband in phase (I) and quadrature (Q) signal components with a small enough bandwidth so that the ADCs of the day could sample them. The digitized signals were then fed into digital Doppler/moving target indicator (MTI) and detection processors.

By contrast, Figure 7.2 depicts a typical digital receiver for a radar front end. The radio frequency (RF) input usually passes through one or two stages of analog downconversion to generate an intermediate frequency (IF) signal that is sampled directly by the ADC. A digital downconverter (DDC) converts the digitized signal samples to complex form at a lower rate for passing through a digital pulse compressor to backend processing. Note that the output of the ADC has a slash through the digital signal line with a letter provided earlier. The letter depicts the number of bits in the digitized input signal and represents the maximum possible dynamic range of the ADC. The use of digital signal processing (DSP) can often improve the dynamic range, stability, and overall performance of the system, while reducing size and cost, compared with the analog approach.

7.1 DESIGN CONSIDERATIONS

In coherent radar systems, all local oscillators (LOs) and clocks that generate system timing are derived from a single reference oscillator. However, this fact alone does not ensure that the transmitted waveform starts at the same RF phase on every pulse, which is a requirement for coherent systems. Consider a system with a 5 MHz reference oscillator, from which is derived a 75 MHz IF center frequency (on transmit and receive) and a complex sample rate of 30 MHz. A rule of thumb is that the clock used to produce the pulse repetition interval (PRI) needs to be a common denominator of the IF center frequencies on transmit and receive and the complex sample frequency in order to assure pulse-to-pulse phase coherency. For this example, with an IF center frequency of 75 MHz and a complex sample rate of 30 MHz, allowable PRI clock frequencies would include 15 and 5 MHz.

In the past, implementing a real-time radar digital signal processor typically required the design of a custom computing machine, using thousands of high-performance integrated circuits (ICs). These machines were very difficult to design, develop, and modify. Digital technology has advanced to the point where several implementation alternatives exist that make the processor more programmable and easier to design and change.

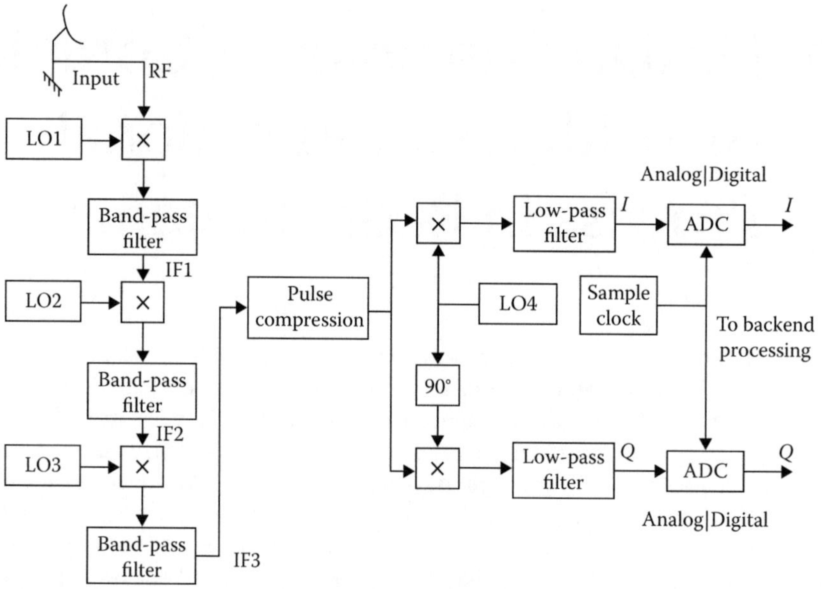

FIGURE 7.1 Typical radar receiver front end design from 1990.

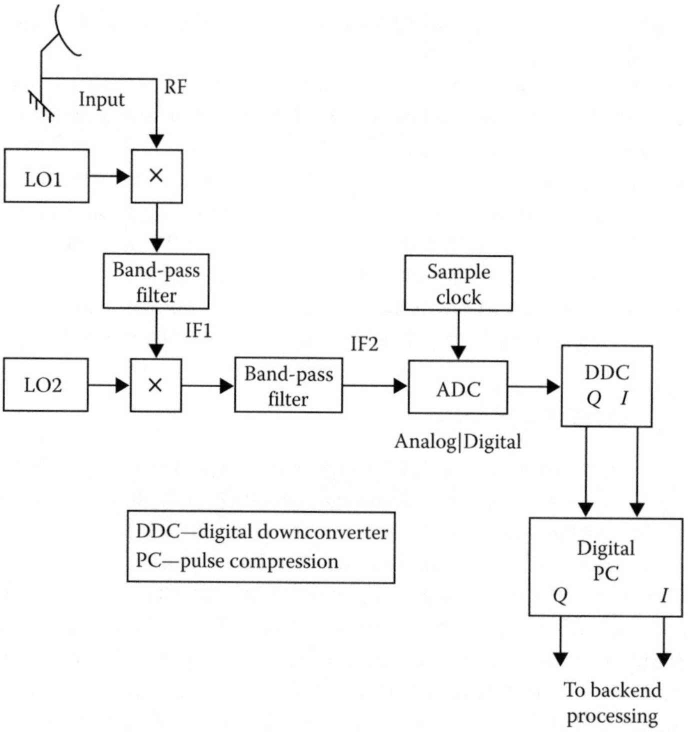

FIGURE 7.2 Typical digital receiver front end.

7.1.1 Parallel General-Purpose Computers

This architecture employs multiple general-purpose processors that are connected via high-speed communication networks. Included in this class are high-end servers and embedded processor architectures. Servers are typically homogeneous processors, where all of the processing nodes are identical and are connected by a very high-performance data bus architecture. Embedded processor architectures are typically composed of single-board computers (blades) that contain multiple general-purpose processors and plug into standard backplane architecture. This configuration offers the flexibility of supporting a heterogeneous architecture, where a variety of different processing blades or interface boards can be plugged into the standard backplane to configure a total system. At this writing, backplanes are migrating from parallel architectures, where data are typically passed as 32 or 64 bit words, to serial data links, which pass single bits at very high clock rates (currently in excess of 3 GB/s [GBps]). These serial data links are typically point-to-point connections. In order to communicate with multiple boards, the serial links from each board go to a high-speed switchboard that connects the appropriate source and destination serial links together to form a *serial fabric*. It is apparent that high-speed serial links will be the primary communication mechanism for multiprocessor systems in the future, with ever-increasing data bandwidths. These parallel processor architectures offer the benefit of being programmable using high-level languages, such as C and C++. A related advantage is that programmers can design the system without knowing the intimate details of the hardware. In addition, the software developed to implement the system can typically be moved relatively easily to new hardware architecture as part of a technology refresh cycle.

On the negative side, these systems can be difficult to program to support real-time signal processing. The required operations need to be split up appropriately among the available processors, and the results need to be properly merged to form the final result. A major challenge in these applications is to support the processing *latency* requirements of the system, which define the maximum length of time allowed to produce a result. The *latency* of the microprocessor system is defined as the amount of time required to observe the effect of a change at a processor's input on its output. Achieving latency goals often requires assigning smaller pieces of the workload to individual processors, leading to more processors and a more expensive system. Another challenge facing these systems in radar application is reset time. In a military application, when a system needs to be reset in order to fix a problem, the system needs to come back to full operation in a very short period of time. These microprocessor systems typically take a long time to reboot from a central program store and, hence, have difficulty meeting reset requirements. Developing techniques to address these deficiencies is an active area of research. Finally, these processors are generally used for non-real time or near-real time data processing, as in target tracking and display processing. Since the 1990s, they have started to be applied to real-time signal processing applications. Although they might be cost-effective for relatively narrowband systems, their use in wideband DSP systems in the early twenty-first century is typically prohibitively expensive due to the large number of processors required. This situation should improve over time as faster and faster processors become available.

7.1.2 Custom-Designed Hardware

Through the 1990s, real-time radar DSP systems were built using discrete logic. These systems were very difficult to develop and modify, but in order to achieve the required system performance, it was the only option available. Many radar systems were built using application-specific integrated circuits (ASICs), which are custom devices designed to perform a particular function. The use of ASICs allowed DSP systems to become very small with high performance. However, they were (and still are) difficult and expensive to develop, often requiring several design interactions before the device was fully operational. If an ASIC-based system needs to be modified, the ASICs need to be redesigned, incurring significant expense. Typically, the use of ASICs makes sense if tens or hundreds of thousands of units are to be sold, so that the development costs can be amortized over the life of the unit. This is rarely the case for radar systems. However, many ASICs have been developed to support the communication industry, such as digital up- and downconverter, which can be utilized in radar systems.

The introduction of the field programmable gate array (FPGA) in the 1980s heralded a revolution in the way real-time DSP systems were designed. FPGAs are integrated circuits that consist of a large array of configurable logic elements that are connected by a programmable interconnect structure. At the time of this writing, FPGAs can also incorporate hundreds of multipliers that can be clocked at rates up to a half billion operations per second, and memory blocks, microprocessors, and serial communication links that can support multigigabit-per-second data transfers. Circuits are typically designed using a hardware description language (HDL), such as VHDL (VHSIC hardware description language) or Verilog. Software tools convert this high-level description of the processor to a file that is sent to the device to tell it how to configure itself. High-performance FPGAs store their configuration in volatile memory, which loses its contents when powered down, making the devices infinitely reprogrammable.

FPGAs allow the designer to fabricate complex signal processing architectures very efficiently. In typical large applications, FPGA-based processors can be a factor of 10 (or more) smaller and less costly than systems based on general-purpose processors. This is due to the fact that most microprocessors only have one or few processing elements, whereas FPGA have an enormous number of programmable logic elements and multipliers. For example, to implement a 16 tap FIR filter in a microprocessor with a single multiplier and accumulator, it would take 16 clock cycles to perform the multiplications. In an FPGA, we could assign 16 multipliers and 16 accumulators to the task, and the filter could be performed in 1 clock cycle. In order to use an FPGA most efficiently, we have to take advantage of all of the resources it offers. These include not only the large numbers of logic elements, multipliers, and memory blocks but also the rate at which the components can be clocked. In the previous example, assume that the data sample rate is 1 MHz and also assume that the multipliers and logic can be clocked at 500 MHz. If we simply assign one multiplier to each coefficient, we would use 16 multipliers clocking at 500 MHz. Since the data rate is only 1 MHz, each multiplier would only perform one significant multiplication every microsecond and then be idle for the other 499 clocks in the microsecond, which is very inefficient. It would be much more efficient, in this case, to use one multiplier to perform as many products as possible. This technique, called *time-domain multiplexing*, requires additional logic to control the system and provide the correct operands to the multiplier at the right time. Since an FPGA can incorporate hundreds of multipliers, one can appreciate the power of this technique.

On the negative side, utilizing an FPGA to its best advantage typically requires the designer to have a thorough understanding of the resources available in the device. This typically makes efficient FPGA-based systems harder to design than radar systems based on general-purpose processors, where a detailed understanding of the microprocessor system architecture is not necessarily required. In addition, FPGA designs tend to be aimed at a particular family of devices and take full advantage of the resources provided by that family. Hardware vendors are constantly introducing new products, invariably incorporating new and improved capabilities. Over time, the older devices become obsolete and need to be replaced during a *technology refresh cycle*. When a technology refresh occurs several years down the road, typically the available resources in the latest FPGAs have changed or a totally different device family is used, which probably requires a redesign. On the other hand, software developed for general-purpose processors may only need to be recompiled in order to move it to a new processor. Tools currently exist that synthesize C or MATLAB® code into an FPGA design, but these tools are typically not very efficient. The evolution of design tools for FPGAs to address these problems is an area of much research and development.

Hybrid processors: Although it would be very desirable to simply write C code to implement a complex radar signal processor, the reality in the early twenty-first century is that, for many radar systems, implementing such a system would be prohibitively expensive or inflict major performance degradation. Although the steady increase in processor throughput may someday come to the rescue, the reality at the time of writing is that high-performance radar signal processors are usually a hybrid of application-specific and programmable processors. Dedicated processors, such as FPGAs or ASICs, are typically used in the high-speed front end of radar signal processors, performing demanding functions such as digital downconversion and pulse compression, followed by programmable processors in the rear, performing the lower-speed tasks such as detection processing. The location of the line that separates the two domains is application-dependent, but over time, it is constantly moving toward the front end of the radar system.

7.2 COMPLEX ALGORITHM ASSIGNMENT

The complex algorithm of a computational process is a set of elementary DSP algorithms for all stages and complex radar system (CRS) modes of operation. Off-line complex algorithms of DSP individual stages that are not associated with each other by information processes and control operations are possible, too. To design the complex algorithm of a computational process, there is a need to have an unambiguous definition of microprocessor system functioning under solution of goal-oriented DSP problems. This definition must include the elementary DSP and control algorithms, a sequence of their application, the conditions of implementation of each elementary DSP and control algorithm, and intercommunication between the DSP and control algorithms using input and output information. A general form of such definition and description can be presented using the logical and graph flowchart of the DSP and control algorithms.

The complex algorithm can be realized by a multiprocessor system. In this case, the elementary operations are distributed among the microprocessors taking into consideration their operation speed, and the complex algorithm must be transformed into the form suitable for realization in parallel microprocessor systems (parallelization of the complex algorithm). The method of algorithmic representation and transformation is the subject of investigation in the theory of algorithms.

7.2.1 Logical and Matrix Algorithm Flowcharts

Universal ways of algorithm assignment are designed by the theory of algorithms. At that time, any general way to assign the complex algorithms is based on the dual character objects, namely, the *counting operators* and the *logical operators* (or recognizers). In our case, the counting operators $A_1, A_2, \ldots, A_i$ are the elementary DSP algorithms. The logical operators (recognizers) $P_1, P_2, \ldots, P_i$ are used to recognize some features of information processed by the complex DSP algorithm and to change a sequence of elementary algorithms depending on results of recognition.

One of the best-known ways to assign the complex DSP algorithm depending on complexity is the logical or formula-logical algorithmic flowchart. The elementary operators and recognizers are presented in the geometrical form in the logical algorithm flowchart (rectangle, jewel boxes, trapeziums, etc.) connected with each other by arrows in accordance with the given sequence of counting operators and recognizers in the complex algorithm. Titles of elementary operations (DSP algorithms) are written inside the geometrical forms. Sometimes the formulas of logical operations carried out and logical conditions under test are written inside the geometrical forms. In this case, the corresponding logical flowchart of the complex DSP algorithm is called the *formula-logical block diagram.*

The complex algorithm can be presented also by operator diagram consisting of elementary counting operators (the elementary DSP algorithms), recognizers, and indicators to carry out operations in sequence. Additionally, we use the specific operators to start A_0 and to stop A_k. For example, we can design the following operator diagram of algorithms based on the counting operators A_1, A_2, A_3, A_4, and A_5 and recognizers P_1, P_2, P_3, and P_4:

$$A_0 \overset{3}{\downarrow} A_1 P_1 \overset{1}{\uparrow} A_2 P_2 \overset{2}{\uparrow}\overset{1}{\downarrow} P_3 \overset{3}{\uparrow} A_3 \overset{2}{\downarrow} A_4 P_4 \overset{4}{\uparrow} A_s \overset{4}{\downarrow} A_k, \tag{7.1}$$

where

↑ means the arrow beginning
↓ means the arrow end

The beginning and the end of the same arrow are designated by the same number. The algorithm starts to work after using the operator A_0 "Start." The order to work for other blocks of the operational flowchart is the following. If the last active block is the counting operator, then the next in step-by-step block works. If the last active block is the recognizer, then two cases are possible. If the test condition is satisfied, then the neighbor at the right side block works. If the test condition is not satisfied, then the block marked by the arrow directed from the recognizer works. The algorithm stops working if the last active operator possesses an assignment to pass to the operator A_k "Stop." The operator flowcharts of algorithms are very compact but not so obvious and require additional explanations and the operator sense decoding.

To write the order of the elementary DSP algorithms functioning as components of the complex algorithm, we use the matrix form:

$$\mathbf{A} = \begin{array}{c} \\ A_0 \\ A_1 \\ A_2 \\ \vdots \\ A_n \end{array} \begin{array}{c} \begin{array}{ccccc} A_1 & A_2 & \cdots & A_n & A_k \end{array} \\ \left\| \begin{array}{ccccc} \alpha_{01} & \alpha_{02} & \cdots & \alpha_{0n} & \alpha_{0k} \\ \alpha_{11} & \alpha_{12} & \cdots & \alpha_{1n} & \alpha_{1k} \\ \alpha_{21} & \alpha_{22} & \cdots & \alpha_{2n} & \alpha_{2k} \\ \ldots & \ldots & \ldots & \ldots & \ldots \\ \alpha_{n1} & \alpha_{n2} & \cdots & \alpha_{nn} & \alpha_{nk} \end{array} \right\| \end{array}, \tag{7.2}$$

where

$$\alpha_{ij} = \alpha_{ij}(P_1, P_2, \ldots, P_l), \quad i = 0, 1, 2, \ldots, n; \quad j = 1, 2, \ldots, n, n+1 \tag{7.3}$$

are the logical functions satisfying the following condition: if after the algorithm A_i performance the function α_{ij} at some set of the logical elements $P_1, P_2, \ldots, P_1$ taking magnitudes $P_i = 1$ or $\bar{P}_i = 0$ is equal to unit, then the algorithm A_j must be the next to operate. If some functions $\alpha_{ij} \equiv 1$, then the algorithm A_j must operate after the algorithm A_i. Conversely, if some functions $\alpha_{ij} \equiv 0$, then under realization of the complex algorithm the operation of the elementary algorithm A_j is impossible after operation of the elementary algorithm A_i. The matrix flowchart of the operator algorithm given by (7.1) takes the following form:

$$\mathbf{A} = \begin{array}{c} \\ A_0 \\ A_1 \\ A_2 \\ A_3 \\ A_4 \\ A_5 \end{array} \begin{array}{c} \begin{array}{cccccc} A_1 & A_2 & A_3 & A_4 & A_s & A_k \end{array} \\ \left\| \begin{array}{cccccc} 1 & 0 & 0 & 0 & 0 & 0 \\ \bar{P}_1\bar{P}_3 & P_1 & \bar{P}_1 P_3 & 0 & 0 & 0 \\ P_2\bar{P}_3 & 0 & P_2 P_3 & \bar{P}_2 & 0 & 0 \\ 0 & 0 & 0 & 1 & 0 & 0 \\ 0 & 0 & 0 & 0 & P_4 & \bar{P}_4 \\ 0 & 0 & 0 & 0 & 0 & 1 \end{array} \right\| \end{array}. \tag{7.4}$$

Let notation $A_i \to A_j$ mean that after the algorithm A_i operation there is a need to carry out the algorithm A_j. Then the notation

$$A_j \to \alpha_{i1}A_1 + \alpha_{i2}A_2 + \cdots + \alpha_{\text{in}}A_n \tag{7.5}$$

means that the operation of one of those algorithms with the functions $\alpha_{ij} \neq 0$ is possible after the algorithm A_i operation. Equation 7.5 is called the transition formula for the algorithm A_i. These formulas can be designed for all elementary DSP algorithms of the complex algorithm given by the matrix flowchart. Thus, in the case of matrix form (7.4), we obtain the following system of transition formulas:

$$\begin{cases} A_0 \to A_1; \\ A_1 \to \bar{P}_1\bar{P}_3A_1 + \bar{P}_1A_2 + \bar{P}_1P_3A_3; \\ A_2 \to P_2\bar{P}_3A_1 + P_2P_3A_3 + \bar{P}_2A_4; \\ A_3 \to A_4; \\ A_4 \to P_4A_s + \bar{P}_4A_k; \\ A_4 \to A_k. \end{cases} \tag{7.6}$$

The matrix flowchart of the complex algorithm allows us to design the table reflecting both information and controlling association between the elementary DSP algorithms. We may change all the matrix elements by magnitudes

$$l_{ij} = \begin{cases} 1, & \text{if } \alpha_{ij} \neq 0 \text{ in (7.2)}; \\ 0, & \text{if } \alpha_{ij} = 0 \text{ in (7.2)}. \end{cases} \tag{7.7}$$

As a result, we obtain the *adjacency matrix* that reflects formally an information association between the elementary DSP algorithms. For instance, the algorithm given by the matrix in (7.4) has the following adjacency matrix:

$$\mathbf{A} = \begin{array}{c} \\ A_0 \\ A_1 \\ A_2 \\ A_3 \\ A_4 \\ A_5 \end{array} \begin{array}{c} \begin{array}{cccccc} A_1 & A_2 & A_3 & A_4 & A_5 & A_k \end{array} \\ \left\| \begin{array}{cccccc} 1 & 0 & 0 & 0 & 0 & 0 \\ 1 & 1 & 1 & 0 & 0 & 0 \\ 1 & 0 & 1 & 1 & 0 & 0 \\ 0 & 0 & 0 & 1 & 0 & 0 \\ 0 & 0 & 0 & 0 & 1 & 1 \\ 0 & 0 & 0 & 0 & 0 & 1 \end{array} \right\| \end{array}. \tag{7.8}$$

We can design the graph flowchart of the complex algorithm of computational process based on the adjacency matrix.

7.2.2 Algorithm Graph Flowcharts

The algorithm graph flowchart is the finite oriented graph satisfying the following conditions:

- There are two marked nodes in the graph: the input node corresponds to the operator "Start" and the only arrow is directed from this operator; and the output node corresponds to the operator "Stop" and no arrow is directed from this node.

- One arrow (the node A) or two arrows (the node P) are directed from each node excepting the input and output nodes; the arrows from the node P are marked by the signs "+" and "–" or by the digits "1" and "0."
- The elementary DSP algorithm A_i is correlated with the A-node and the logical operator P_l is matched with each P-node; in the algorithm graph flowcharts, the A-nodes and the input and output nodes are depicted by circles and the P-nodes are depicted by jewel boxes.

The algorithm graph flowchart i.e., equivalent to the given matrix flowchart, i.e., with the same operation sequence, is designed in the following way:

- The subgraphs equivalent to transition formulas of the given matrix flowchart are designed.
- The equivalent branches of subgraphs are united.
- The same operators are united and the final graph flowchart is formed.

To obtain the graph flowchart with the minimal or near minimum number of P-nodes, there is a need to draw the subgraphs with the minimum number of P-nodes, too, using the given transition formulas. The number of P-nodes can be reduced under union of the equivalent subgraph branches. The procedure to design and transform the algorithm graph flowchart can be illustrated by an example considering the matrix flowchart of algorithm given by (7.4) as the initial form. The transition formulas given by (7.6) are initial to design the graph flowchart, based on which the subgraphs for each elementary DSP algorithms $A_0, A_1, A_2, A_3, A_4, A_5$ must be drawn. The subgraph design process is started from the transition formula transformation to the corresponding form consisting of logical function expansion with respect to each variable. For example, the transition formula for the elementary DSP algorithm A_1 given by (7.6) takes the following form:

$$A_1 \to P_1A_2 + \overline{P}_1(P_3A_3 + \overline{P}_3A_1). \tag{7.9}$$

The formula for the elementary DSP algorithm A_2 given by (7.6) takes the following form:

$$A_2 \to P_2(\overline{P}_3A_1 + P_3A_3) + \overline{P}_2A_4. \tag{7.10}$$

Other transition formulas given by (7.6) do not require representation. Now, it is easy to draw the subgraphs of each elementary DSP algorithm (see Figure 7.3).

The next stage is the search and union of equivalent graph branches. In our case, the equivalent branches are the branches starting from the operator P circled by the dash line. After union of the

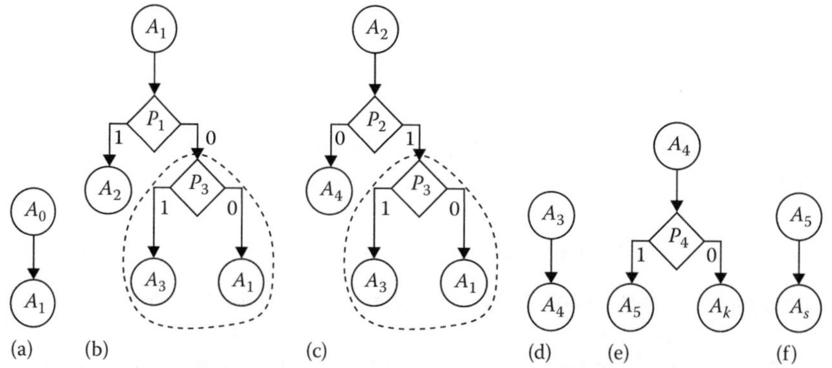

FIGURE 7.3 Subgraphs of each elementary DSP algorithm: (a) the algorithm A_0; (b) the algorithm A_1; (c) the algorithm A_2; (d) the algorithm A_3; (e) the algorithm A_4; and (f) the algorithm A_5.

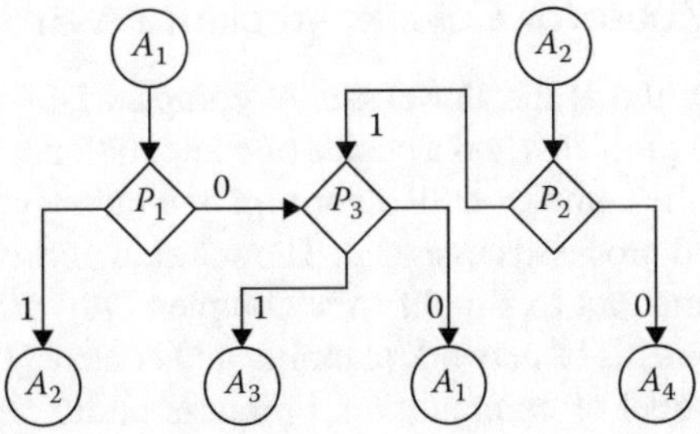

FIGURE 7.4 The subgraph uniting equivalent branches.

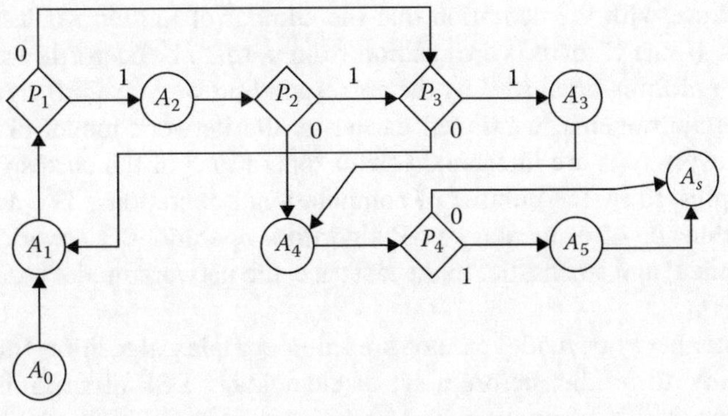

FIGURE 7.5 Final graph flowchart equivalent to the given matrix diagram.

equivalent branches, we obtain the subgraph depicted in Figure 7.4. There are no other equivalent branches for the considered example. Finally, after union of the same counting operators, we obtain the end graph flowchart of algorithm equivalent to the given matrix block diagram (see Figure 7.5). The considered example to design the graph flowchart by the given matrix block diagram allows us to minimize the number of logical operators of complex algorithm, which means to simplify it. The stage of minimization of the operator numbers is the necessary stage for DSP.

Under analysis of features and quality of algorithms presented by the graph flowcharts, it is worthwhile to transform them further with the purpose of uniting the elementary DSP algorithm with the logical operators (in pairs or some elementary DSP algorithm with a single logical operator) into nodes that are used for realization more or less components of the complex algorithm. Two arrows come out from such united node: if the test condition is satisfied as a result of node operation, the arrow is denoted by the sign "+" or "1," otherwise by the sign "–" or "0." For instance, in the case of the algorithm with flowchart presented in Figure 7.5, we can unite the algorithms into blocks (the input and output nodes are not united): $a_1 \sim A_1P_1$, $a_2 \sim A_2P_2$, $a_3 \sim P_3$, $a_4 \sim A_1P_1$, $a_1 \sim A_4P_4$, $a_5 \sim A_3A_4P_4$, $a_6 = A_5$. The obtained flowchart is depicted in Figure 7.6, where the nodes are designated by the light circles.

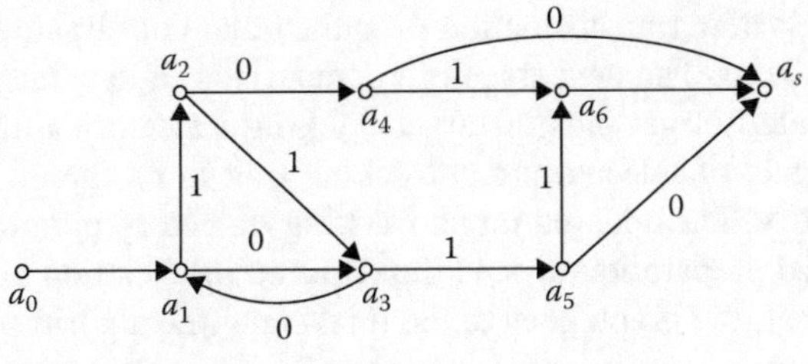

FIGURE 7.6 Union of algorithms into blocks.

7.2.3 Use of Network Model for Complex Algorithm Analysis

The main problem solved by the graph flowcharts of complex DSP algorithms is a definition of rational ways to present these problems and a choice of computational software tools and microprocessor subsystems to realize the complex DSP algorithm of a radar system. In short, the problem of optimization of computational process is assigned. The solution of this problem allows us to reduce significantly a realization time and to simplify the complex DSP algorithm of the radar system. Similar problems are the problems of network planning and control [1,2].

To design the network model of computational process under DSP in CRSs, there is a need to transform the graph flowchart of complex algorithm into the network graph that satisfies the following requirements, namely, the network graph cannot consist of contours, that is, the case when the initial top is matched with the final top, and there must be a strict order of top precedence in accordance with the condition that the number of the top i is less than the number of the top j, that is, $(i < j)$ if there is a transition from a_i to a_j. Henceforth, we call a network of elementary DSP algorithms organized by the corresponding way for CRS signal processing and satisfying all the requirements mentioned earlier as the network model of the complex DSP algorithm. The network tops are interpreted as an "operation" of the corresponding elementary DSP algorithm expressed by the number of computational operations. The network arcs can be interpreted as a sequence of elementary DSP algorithm operations. Network transitions can be deterministic (planned) and stochastic. In the last case, the network model is called the stochastic network model.

The deterministic network model cannot present a complex algorithm functioning in CRSs since it is impossible to predict before a set of elementary DSP algorithms and sequence of realizations for each practical situation. Therefore, the stochastic network model, in which the transitions in the network graph are defined by the corresponding probabilities of transitions given by specific conditions of CRS functioning, is more suitable to image and analyze a realization of the complex DSP algorithm by microprocessor subsystems. When the network model has been constructed, the problem of estimating the time to complete all operations, that is, the time to finish all operations by microprocessor subsystems with the given effective speed of operations, arises. This time cannot be higher than a total duration to finish a complex DSP algorithm operation defined by the most unfavorable way from the initial graph top a_n to the final graph top a_k, that is, along such route that generates a maximal duration of operations. This route is called the extreme route. The extreme route in the stochastic network model cannot be presented in clear form as, for example, in the network model with a given structure. Because of this, the problems of defining the average time or average number of operations required to realize the complex algorithm are assigned under analysis of stochastic network models. To illustrate some principles of designing the stochastic network graph, we consider the following example.

Let the digital signal reprocessing algorithm depicted in Figure 5.13 be considered as the complex DSP algorithm. The network graph of the considered algorithm is shown in Figure 7.7. Special transformations of the algorithm logical flowchart are not needed to design the network graph. Algorithm functioning is started from selection of the immediate target pip of current scanning in the buffer memory device (the elementary DSP algorithm a_n). Later, the algorithm of target pip coordinate system transformation from the polar coordinate system to Cartesian one (the algorithm a_1) is realized. The next stage is a comparison of new target pip coordinates with coordinates of extrapolated target pips of target tracking trajectories (the algorithm a_2). If the target pip is inside the gate of selected target tracking trajectory, then it is considered as a prolongation of this trajectory. The updated target tracking trajectory parameters (the algorithm a_3) are made more exact, and preparation to send the required information to the user (the algorithm a_4) is produced. After that, the graph goes to the final state (the algorithm a_k). If the target pip is outside of the gates of none of target tracking trajectories, then the target pip is checked whether

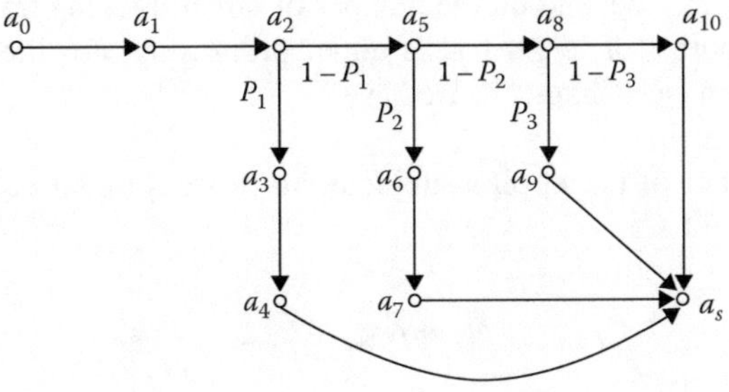

FIGURE 7.7 Flowchart of network graph.

it belongs to the detected target tracks (the algorithm a_5). If a new target pip is able to confirm one of the detected target tracks, we carry out an accurate definition of target track parameters (the algorithm a_6) and test the detection criterion (the algorithm a_7). If the target pip is outside of the gates of none of detected target tracks, we assign it to beginning new target tracks (the algorithm a_8). When the target pip is caught by one of the primary lock-in gates, the beginning of a new target track is carried out and initial values of the new target tracking trajectory are defined (the algorithm a_9). Finally, if the target pip is outside the existing primary lock-in gates it is recorded as the initial point of the new target track (the algorithm a_{10}).

The considered network graph is stochastic without doubt, and there is a need to define the probabilities of transitions between the network graph tops. Let the probability that the selected target pip belongs to the target tracking trajectories be denoted by $P_1 = P_{\text{tr}}$ and processed by the elementary DSP algorithms a_1, a_2, a_3, a_4, a_k. Then,

$$P_1' = 1 - P_1 \tag{7.11}$$

is the probability that the target pip does not belong to the target tracking trajectories and has to be processed further. Let P_{D} be the probability that the target pip belongs to the detected target track. Then, the probability that the target pip processing is finished by realization of the algorithms a_1, a_2, a_5, a_6, a_7, a_k is defined as

$$P_2 = (1 - P_{\text{tr}})P_{\text{D}} \tag{7.12}$$

and the probability to proceed with a target pip processing further is determined in the following form:

$$P_2' = (1 - P_{\text{tr}})(1 - P_{\text{D}}). \tag{7.13}$$

In an analogous way, we can obtain

$$P_3 = (1 - P_{\text{tr}})(1 - P_{\text{D}})P_{\text{beg}} \tag{7.14}$$

and

$$P_3' = (1 - P_{\text{tr}})(1 - P_{\text{D}})(1 - P_{\text{beg}}), \tag{7.15}$$

where

P_{beg} is the probability that the target pip belongs to beginning the new target track
P_3' is the probability that the target pip will be assigned as the beginning point of the new target track

The probabilities P_{tr}, P_0, P_{beg} depend on the number of target tracking trajectories, detected target tracks, and beginning points of target tracks under processing. For the algorithm of target pip beginning by the criterion (see Chapter 4) we have

- The average number of target pips subjected to process under each scanning is determined as

$$\bar{N}_{\Sigma} = \bar{N}_{\text{false}} + \bar{N}_{\text{true}}, \tag{7.16}$$

where
$\bar{N}_{\text{false}}$ is the average number of false target pips
$\bar{N}_{\text{true}}$ is the average number of true target pips appearing within the limits of the scanning period

- The average number of true target pips coming in the target tracking gates under each scanning is given by

$$\bar{n}_{\text{true}}^{\text{scan}} = P_{\text{D}}^{\text{gate}} \bar{N}_{\text{tr}}^{\text{gate}}, \tag{7.17}$$

where $P_{\text{D}}^{\text{gate}}$ is the probability of detection of true target pips within the limits of the target tracking gates, which is the same for all target tracks

- The average number of false target pips inside the target tracking gates under each scanning is determined in the following form:

$$\bar{n}_{\text{false}}^{\text{scan}} = \sum_{j=m+n}^{m+n+k_{\text{th}}-1} P_{\text{F}_j}^{\text{scan}} \bar{N}_j^{\text{scan}}, \tag{7.18}$$

where
$P_{\text{F}_j}^{\text{scan}}$ is the probability of the false target pip hit into the jth target tracking gate
$\bar{N}_j^{\text{scan}}$ is the average number of jth target tracking gates for all true and false target tracking trajectories

Taking into consideration (7.16) through (7.18), the probability of an arbitrary selected target pip belonging to one of the target tracking trajectories can be expressed in the following form:

$$P_{\text{tr}} = \frac{\bar{n}_{\text{true}}^{\text{scan}} + \bar{n}_{\text{false}}^{\text{scan}}}{\bar{N}_{\Sigma}}. \tag{7.19}$$

In an analogous way, we obtain the formula for the probability of target pip belonging to detected target tracks:

$$P_{\text{D}} = \frac{\bar{n}_{\text{D}}^{\text{true}} + \bar{n}_{\text{D}}^{\text{false}}}{\bar{N}_{\Sigma}}, \tag{7.20}$$

where

$$\bar{n}_{\text{D}}^{\text{true}} = P_{\text{DD}} \bar{N}_{\text{D}}^{\text{true}} \tag{7.21}$$

and

$$\bar{n}_{\text{D}}^{\text{false}} = \sum_{j=m}^{m+n-1} P_{\text{D}_j}^{\text{false}} \bar{N}_{\text{D}_j}, \tag{7.22}$$

where

$P_{\text{D}_j}^{\text{false}}$ is the probability of the false target pip hit into the detected target track gates with the number j

P_{DD} is the probability of the true target pip detection into the detected target track gates

$\bar{N}_{\text{D}_j}$ is the average number of gates with the number j formed by all the detected target tracking trajectories

$\bar{N}_{\text{D}}^{\text{true}}$ is the average number of the true target tracks being in the detection process

The probability of target pip belonging to the started target tracks is determined as

$$P_{\text{beg}} = \frac{\sum_{j=1}^{m-n} P_{\text{F}j}^{\text{lock-in}} \bar{N}_j^{\text{lock-in}} + P_{\text{D}}^{\text{lock-in}} \bar{N}_{\text{true}}^{\text{lock-in}}}{\bar{N}_{\Sigma}}, \tag{7.23}$$

where

$P_{\text{F}_j}^{\text{lock-in}}$ is the probability of false target pip hit into the primary lock-in gates with the number j

$\bar{N}_j^{\text{lock-in}}$ is the average number of the primary lock-in gates with the number j

$P_{\text{D}}^{\text{lock-in}}$ is the probability of the true target pip detection into the primary lock-in gates

$\bar{N}_{\text{true}}^{\text{lock-in}}$ is the average number of the primary lock-in gates of true target tracks

Thus, if the statistical characteristics and parameters of noise and target environment inside the radar coverage are known and the target pip beginning algorithm parameters are selected, in the considered case, we are able to determine the probability of transition in the network graph of the complex DSP algorithm. However, we cannot say that this possibility exists forever. In some cases, the probability of transition can be only estimated as a result of computer simulation of the complex DSP algorithm.

7.3 EVALUATION OF WORK CONTENT OF COMPLEX DIGITAL SIGNAL PROCESSING ALGORITHM REALIZATION BY MICROPROCESSOR SUBSYSTEMS

7.3.1 Evaluation of Elementary Digital Signal Processing Algorithm Work Content

We consider the DSP algorithms realizing the main operations and control in a CRS as the elementary DSP algorithms:

- *The signal preprocessing stage*: DSP algorithms of matched filtering in time or frequency domains, cancellation of passive interferences by the digital moving target indicator (MTI), detection and estimation of target return signal parameters, recognition of type and estimation of interference and noise parameters and ranking samples, etc.
- *The reprocessing stage*: DSP algorithms of target track detection, selection of target pips and their beginning to the target tracking trajectories, target track parameters filtering, transformation coordinate system, and so on.
- *The CRS control process*: DSP algorithms of the scanning signal parameter determination under target searching and tracking, determination of power balance between the modes of CRS operations, etc.

Each of the aforementioned DSP algorithms is characterized by the work content expressed by the number of arithmetical operations required to realize it. Preliminary computation of the required number of arithmetical operations is possible only if there is an analytical function between the algorithm input and output. In the case of other algorithms possessing a character of logical operations, mainly, the required number of operations can be obtained only as a result of realization based on microprocessor subsystems.

Results of analytical calculation of the required number of arithmetical operations are obtained individually by the number of additions, subtractions, products, and divisions. Henceforth, there is a need to determine the number of reduced arithmetical operations. As the reduction operation, as a rule, an addition is used (short operation). The number of reduced arithmetical operations is determined for each microprocessor subsystem taking into consideration the known ratio between the time to carry out the ith long and short operations, that is, $\tau_i^{\text{long}}/\tau_i^{\text{short}}$. However, the computation of the work content is not finished at this stage since we must take into consideration other nonarithmetical operations.

The DSP of target return signals has a pronounced information-logical character. Logical operations and transition operations are for about 80% of the total number of elementary DSP operations (cycles) in the process of the complex DSP algorithm realization required for CRS functioning. For example, under realization of the digital signal preprocessing algorithm of two-coordinate radar system by microprocessor subsystems with permanent memory, the following number of operations in percentage is required: forwarding or transferring—45%; reduced arithmetical operations—23%; control transfer—17%; shift—5%; logical operations—3%; information exchange—2%; other operations—5%. Consequently, under computation of the work content of the elementary DSP algorithms there is a need to take into consideration the nonarithmetical operations, too. For this purpose, we can introduce the coefficient K_{na} and write the following relation:

$$\bar{N}_i = \bar{N}_{\text{a}_i} K_{\text{na}}, \quad K_{\text{na}} > 1, \tag{7.24}$$

where $\bar{N}_{\text{a}_i}$ is the average number of arithmetical operations required for the ith algorithm. The number of microprocessor operations depends on a programming mode. Under the use of high-level programming languages, a program length is two to five times greater than that of the optimal program. To take into consideration this fact, we introduce the coefficient $K_{\text{prog}} \approx 2$. Thus, (7.24) can be written in this form:

$$\bar{N}_i = \bar{N}_{\text{a}_i} K_{\text{na}} K_{\text{prog}}. \tag{7.25}$$

This work content definition will be used in our next computations.

7.3.2 Definition of Complex Algorithm Work Content Using Network Model

Under analysis of the work content of complex DSP algorithms, we can use the Markov model of computational process or the stochastic network model [3–6]. The stochastic network model allows us to reduce the number of computations under the work content determination in comparison with the Markov model. Therefore, we prefer to use the stochastic network model. As we noted previously, the network graph model of complex DSP algorithm is the initial condition to evaluate the work content. There is a need that this graph would not contain the cycle paths and the graph tops must be numbered in such a way that the graph top number, to which the transition is carried out, will be higher than any graph top number, from which such transition is possible. Moreover, the graph end top must have the maximal number k. An example of such graph satisfying all the previously mentioned requirements is depicted in Figure 7.8.

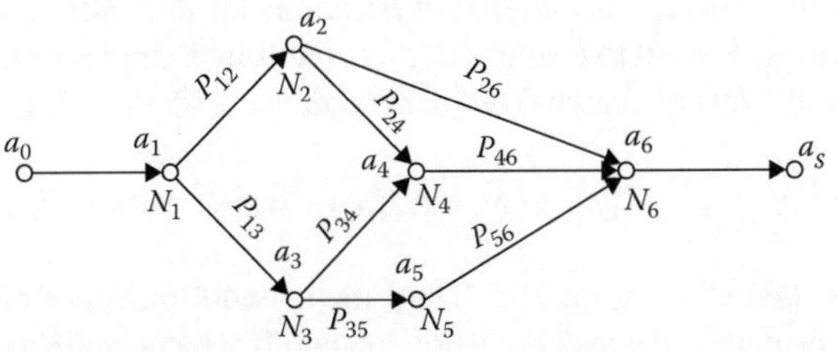

FIGURE 7.8 Example of network graph for complex DSP algorithm.

The average number of addresses to elementary DSP algorithms or the graph tops under a single realization of complex DSP algorithm is denoted by $n_1, n_2, \ldots, n_{k-1}$. Since the graph does not contain the cycles, then at a single realization of the complex DSP algorithm, the average number of hits of the computational process to the graph top with the number i is defined as

$$n_i = \sum_{j=0}^{k-1} n_j P_{ji}, \quad i = 1, 2, \ldots, k. \tag{7.26}$$

Under established order of the top numbering at the instant of computation n_i, the magnitudes of all previous $n_1, n_2, \ldots, n_{i-1}$ are known. The average number of operations under a single realization of the complex DSP algorithm is determined as

$$\bar{M} = \sum_{i=1}^{k} n_i \bar{N}_i, \tag{7.27}$$

where $\bar{N}_i$ is the average number of the elementary DSP algorithm operations corresponding to the ith top of the complex DSP algorithm graph.

Example: There is a need to define the average work content of the complex DSP algorithm presented by the network graph depicted in Figure 7.8 at the following initial conditions: $\bar{N}_1 = 100$; $\bar{N}_2 = 30$; $\bar{N}_3 = 150$; $\bar{N}_4 = 20$; $\bar{N}_5 = 200$; $\bar{N}_6 = 30$; $P_{12} = 0.25$; $P_{13} = 0.75$; $P_{24} = 0.3$; $P_{26} = 0.7$; $P_{34} = 0.2$; $P_{35} = 0.8$; $P_{46} = P_{56} = 1$.

1. Applying (7.26), we obtain

$$\begin{gathered} n_1 = 1,\ n_2 = n_1 P_{12} = 0.25,\ n_3 = n_1 P_{13} = 0.75,\ n_4 = n_2 P_{24} + n_3 P_{34} = 0.225, \\ n_5 = n_3 P_{35} = 0.6,\ n_6 = n_2 P_{26} + n_{46} P_{46} = 1. \end{gathered} \tag{7.28}$$

2. Applying (7.27), we obtain

$$\bar{M} = 100 + 30 \times 0.25 + 150 \times 0.75 + 20 \times 0.225 + 200 \times 0.6 + 30 = 374.5. \tag{7.29}$$

Thus, the average work content of the complex DSP algorithm presented by the graph depicted in Figure 7.8 is 374.5 reduced arithmetical operations.

If the graph of complex DSP algorithm possesses the cycle paths, we cannot apply the considered procedure to determine the work content $\bar{M}$. Owing to this reason, at first, there is a need to exclude the graph cycle paths, that is, to change the graph cycle paths by operator with the equivalent work content. The general procedure to transform the graphs of complex DSP algorithms with the purpose of excluding the cycle paths is discussed in Ref. [7]. As an example, we consider a transformation of the graph depicted in Figure 7.9a. This graph has several cycles different by rank (the order). The cycles that do not contain in the interior none cycle are covered by the rank 1. The number of iterations for this cycle is denoted by $n^{(1)}$. The cycles consisting of one or several cycles with the rank 1 are covered by the rank 2. The number of iterations for this cycle is denoted by $n^{(2)}$, and so on. The graph transform is reduced to changes in cycle paths by a single operator. These changes for the graph shown in Figure 7.9a are carried out in accordance with the following formula:

$$N_2' = \{\bar{N}_2 + [\bar{N}_3 + (\bar{N}_4 + \bar{N}_5)n^{(1)} + \bar{N}_6]n^{(2)} + \bar{N}_7\}r^{(3)}. \tag{7.30}$$

The resulting graph is shown in Figure 7.9b. Thus, the network model of complex DSP algorithm allows us to define, in principle, the average work content. If we know a realization time of a single reduced operation, then we can compute the average realization time of a complex DSP algorithm. Inversely, if a limitation on the average realization time of complex DSP algorithm is given, we are able to determine the required work content of microprocessor subsystems to realize the given complex DSP algorithm. Sometimes, to solve the problems of computational resource analysis we need to know information about the work content variance. The procedure to define the work content variance is very cumbersome and we do not discuss it in this section.

7.3.3 Evaluation of Complex Digital Signal Reprocessing Algorithm Work Content in Radar System

We continue consideration and discussion of the example on analysis of the digital signal reprocessing algorithm in CRSs started in Section 7.1. The initial conditions for analysis are given by the graph flowchart of this algorithm depicted in Figure 7.5. In our analysis, we use the following data:

- The average number of tracking targets $\bar{N}_{\text{tr}}^{\text{gate}} = 80$
- The average number of target tracks under detection $\bar{N}_{\text{D}}^{\text{true}} = 10$
- The average number of started target tracks $\bar{N}_{\text{beg}} = 5$
- The average number of true target pips assigned as the initial points of new target tracks $\bar{N}_{\text{initial}}^{\text{true}} = 5$

Thus, the average number of target pips subjected to processing is equal to $\bar{N}_\Sigma = 100$. False target pips and miss of true target pips are not taken into consideration in this example. In accordance with the initial conditions, we can compute the following probabilities:

- Identification of new target pip among the target tracking trajectories, $P_{\text{tr}} = 0.8$.
- Identification of new target pip among the detected target track, $P_{\text{D}} = 0.1$.
- Identification of new target pip among the begun target tracks, $P_{\text{beg}} = 0.05$.
- The new target pip is assigned as the initial point of new target track, $P_{\text{new}} = 0.05$

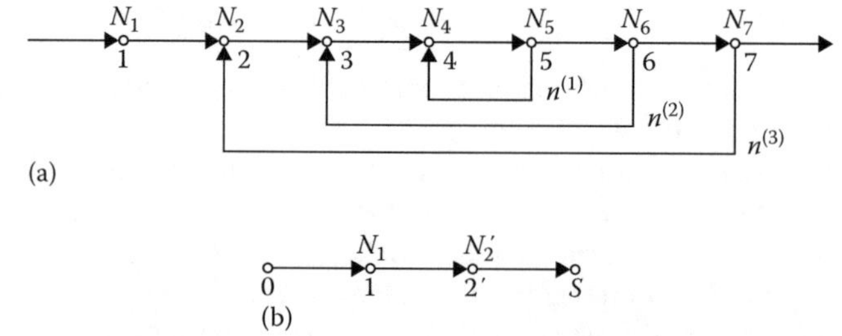

FIGURE 7.9 Example of graph transformation: (a) the graph of cycles different by rank, and (b) the resulting graph.

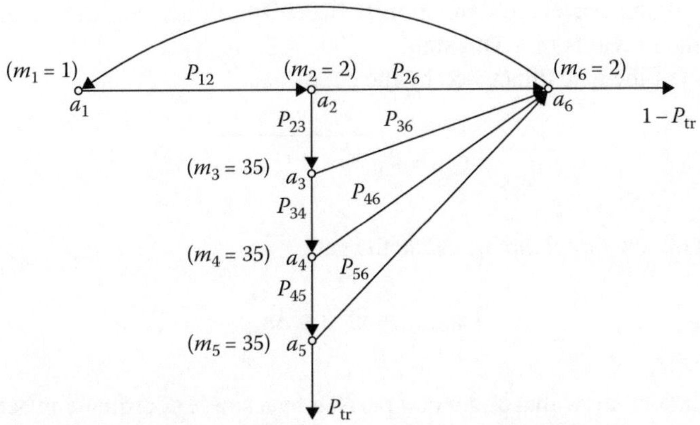

FIGURE 7.10 Typical graph of target and target pip identification algorithm.

Now, we can determine the work content of the elementary DSP algorithms $a_1,\ldots, a_{10}$ including into the complex digital signal reprocessing algorithm presented in Figure 7.7. At first, we consider the typical algorithm of target pip and track identification presented by the graph tops a_2, a_5, a_8. For this purpose, we present this algorithm in the graph form shown in Figure 7.10. The elementary DSP algorithms (operators) are the following:

1. Selection of the next target pip (or target track parameters) from the corresponding scanned area (the algorithm a_1, Figure 7.10).
2. Computation $|t_n - t_{n-1}| \leq \Delta t_n^{\text{acceptable}}$ (the algorithm a_2, Figure 7.10). A sense of this operation is the following: is it a new or old target track? The work content of this operation is equal to $m_2 = 2$ of the reduced arithmetical operations.
3. If the condition 2 is satisfied, a verification of new target pip hit to the gate formed around the extrapolated point of selected target track is carried out. This verification is carried out by several stages. At first, the verification is made by the coordinate x (the algorithm a_3), after that by the coordinate y (the algorithm a_4), and finally, by the coordinate z (the algorithm a_5) under the condition that tests using the previous coordinates give us a positive result. If at the regular step the conditions are not satisfied, we should follow by the algorithm a_6, that is, to verify the fact that all target tracks of scanned area have been tested. If "No," we must carry out a transition to the algorithm a_1; if "Yes," we should make identification with the next set of target tracks (the output algorithm).

Operations carried out under identification by a single coordinate independently of identified target track kind are the following:

1. Extrapolation of the selected target track coordinate by the formula

$$\hat{x}_n^{\text{extr}} = \hat{x}_{n-1} + \hat{\dot{x}}_{n-1}\Delta t_n \quad \text{and} \quad \Delta t_n = t_n - t_{n-1}. \tag{7.31}$$

2. Computation of the extrapolation error variance by the following formula:

$$\sigma^2_{\hat{x}_n^{\text{extr}}} = \sigma^2_{\hat{x}_{n-1}} + 2\Delta t_n R_{\hat{x}\hat{\dot{x}}_{n-1}} + \Delta t_n^2 \sigma^2_{\hat{\dot{x}}_{n-1}}, \tag{7.32}$$

where $R_{\hat{x}\hat{\dot{x}}_{n-1}}$ is the correlation function between the target coordinate estimations and velocity at the previous $(n-1)$th step.
3. Computation of the gate dimension by the coordinate

$$\Delta x_{\text{gate}} = 3\sqrt{\sigma^2_{\hat{x}_n^{\text{extr}}} + \sigma^2_{x_{\text{measure}}}}. \tag{7.33}$$

4. Verification of new target pip hit inside the gate

$$| x_{\text{measure}} - \hat{x}_n^{\text{extr}} | \leq \Delta x_{\text{gate}}. \tag{7.34}$$

Elementary calculations show that under comparison by a single coordinate in accordance with the given formula, there is a need to explore 35 reduced arithmetical operations. Consequently, $m_3 = m_4 = m_5 = 35$. Under definition of the probabilities P_{ij} in the network graph presented in Figure 7.10, we assume the following:

1. The number of cancelled target tracks owing to old information does not exceed 1% from the total number of tracking targets. In accordance with this principle, we have $P_{23} = 0.99$; $P_{26} = 0.01$.
2. For 95% cases, the identification process is finished after comparison by one coordinate. Because of this, $P_{36} = 0.95$; $P_{34} = 0.05$.
3. The probabilities of identification by two and three coordinates are so small that we can neglect these probabilities under determination of the work content of complex DSP algorithm. Under computation of the identification algorithm work content, we are able to obtain, individually by each target track array, the following parameters:
   a. The minimal work content corresponding to the case when the new target pip is within the limits of the first selected target track gate; the number of reduced arithmetical operations independent of the array equals to

$$M_{\min} = m_1 + m_2 + m_3 + m_4 + m_5 \approx 100; \tag{7.35}$$

   b. The maximal work content corresponding to the case when the new target pip is assigned for identification with target tracks of next array after unsuccessful identification with target tracks of the current array; the number of the reduced arithmetical operations for the array of tracking target tracks is given by

$$M_{\text{tr}}^{\max} = (m_3P_{23} + m_4P_{34} + m_5P_{45})N_{\text{tr}}^{\text{gate}} \approx m_3P_{23}N_{\text{tr}}^{\text{gate}}. \tag{7.36}$$

In the case of other arrays, we have

$$M_{\text{D}}^{\max} \approx m_3P_{23}N_{\text{D}}^{\text{true}}, \quad N_{\max}^{\text{lock-in}} = m_3P_{23}N_{\text{tr}}^{\text{lock-in}}. \tag{7.37}$$

The average work content can be determined in the following form:

$$\bar{M} = 0.5(M_{\min} + M_{\max}). \tag{7.38}$$

To simplify further computation, we transform a sequent graph in Figure 7.7 into a parallel graph depicted in Figure 7.11. For this parallel graph, the probabilities of transitions to the graph tops

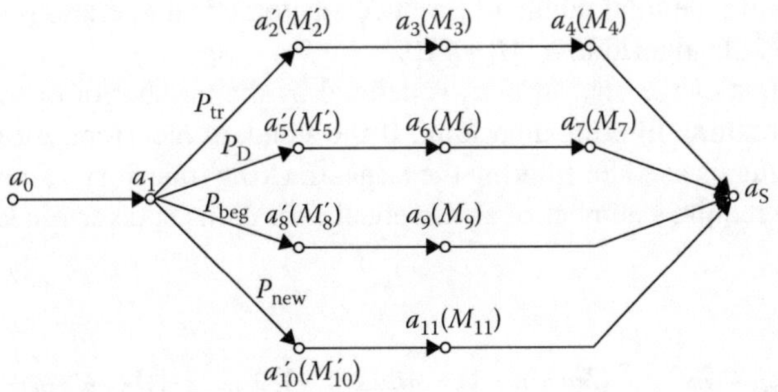

FIGURE 7.11 Transformation to parallel graph.

a_2', a_5', a_8', a_{10}' correspond to the probabilities P_{tr}, P_D, P_{beg}, P_{new}, respectively. The average work content for these graph tops is determined in the following form:

$$\begin{cases} M_2' = 0.5\left(M_2^{max} + M_2^{min}\right) = \bar{M}_2; \\ M_5' = M_2^{max} + 0.5\left(M_5^{max} + M_5^{min}\right) = M_2^{max} + \bar{M}_5; \\ M_8' = M_2^{max} + M_5^{max} + 0.5\left(M_8^{max} + M_8^{min}\right) = M_2^{max} + M_5^{max} + \bar{M}_8; \\ M_{10}' = M_2^{max} + M_5^{max} + M_8^{max}. \end{cases} \tag{7.39}$$

Computational results are presented in Table 7.1.

Now, let us define the work content of other DSP algorithms (the graph tops are depicted in Figure 7.11), taking into consideration the fact that under graph transformation the work content is not changed.

1. The work content of the algorithm a_1 that is responsible to transform the polar coordinate system to the Cartesian one is directly defined by the coordinate transformation formulas in (5.149), and recalculation of the correlation matrices of errors of the coordinate measurements depends mainly on a procedure to calculate the trigonometric functions constituent into the coordinate transformation formulas. For example, under representation of sin x and cos x by the limited series expansion, that is,

$$\sin x = x - \frac{x^3}{3!} + \frac{x^5}{5!} + \cdots \quad \text{and} \quad \cos x = 1 - \frac{x^2}{2!} + \frac{x^4}{4!} + \cdots, \tag{7.40}$$

we need 21 additions and 62 multiplications. Taking into consideration that under reducing multiplications to short operations, we use the coefficient of reduction $K_{red} > 1$, for our case

TABLE 7.1
Work Content after Transformation of Sequent Graph to Parallel Graph

a	M_i^{max}	M_i^{min}	M_i^{av}	M_i'
a_2'	100	3200	1650	1650
a_5'	100	400	250	3450
a_8'	100	200	150	3750
a_{10}'	—	—	—	3800

$K_{red} = 4$, to obtain the total number of reduced arithmetical operations characterizing the work content of the algorithm a_1:$M_1 \approx 270$.

2. The work content of the algorithm a_3 is defined by the number of operations required to realize the smoothing filtering algorithm. If the standard recurrent linear filter, for example, Kalman filter, is used for filtering the target tracking trajectory parameters, then for its realization the required number of arithmetical operations is determined by the following formulas:
   a. Addition

$$n_{add} = 2s^3 + s^2(3m + h - 1) + s[m(2m - 1) + (h^2 - 1)] + m^2(m - 1) \quad (7.41)$$

   b. Multiplication

$$n_{mul} = 2s^3 + s^2(3m + h + 1) + s[2m(m + 1) + h(h + 1)] + m^2(m - 1) \quad (7.42)$$

   c. Division

$$n_{div} = m^2 \quad (7.43)$$

   where s, m, h are the dimensions of smoothing parameter vectors, measured coordinates, and perturbation actions, respectively. Calculations at $s = 6$, $m = 3$, $h = 3$ gives us the following results $n_{add} = 984$, $n_{mul} = 1116$, and $n_{div} = 9$. After reduction of arithmetical operations, the reduction coefficients $K_{mul}^{red} = 4$ and $K_{div}^{red} = 7$, we obtain $M_3 \approx 5500$.

3. The work content of the algorithm a_4 is defined by the number of operations required to prepare the user information. This process may require a regular or specific transformation of coordinates and target track parameters into the user coordinate system and data extrapolation. In the considered example, we think that $M_4 = M_5 \approx 270$.

4. The work content of the algorithm a_5 is defined by the number of operations required to estimate the detected target track parameters based on the fixed sample size. We assume that a target track detection is fixed if the target track beginning defined by two coordinates is confirmed by one more target pip obtained during the next scanning after beginning. Consequently, the sample size of observed data under the target track detection is defined by $n = 3$. In the case of filtering algorithm of polynomial function parameters, the number of operations under the fixed sample size is determined in the following form:
   a. Addition

$$n_{add} = (s + 1)[(n + s)^2 + n - 2] + n^2(n - 1) \quad (7.44)$$

   b. Multiplication

$$n_{mul} = (s + 1)[(n + s)^2 + 3n + s] + n^2(n - 1) \quad (7.45)$$

   c. Division

$$n_{div} = n^2 + (s + 1) \quad (7.46)$$

   where
   s is the polynomial order
   n is the sample size

In the case of linear target track, that is, $s = 1$ and $n = 3$, we have $M_6 \approx 400$ reduced arithmetical operations.

5. The work contents of the algorithm a_7 fixing the fact of target track detection and controlling the rerecording of detected target track array to the array of target tracking trajectories, the algorithm a_9 carrying out the estimation of initial magnitudes of initial magnitudes of target track parameters and transferring this estimation to the array of detected target tracks, and the algorithm a_{10} carrying out a record of target pips as an initial coordinate point of new target tracks can be neglected at this stage owing to simplicity of their realization.

Now, we can determine the average work content of complex DSP algorithm as a whole

$$M = M_1 + P_{\text{tr}}(M_2 + M_3 + M_4) + P_{\text{D}}(M_5 + M_6 + M_7) + P_{\text{beg}}(M_8 + M_9) + P_{\text{new}}(M_{10} + M_{11}). \quad (7.47)$$

Substituting in (7.47) the values obtained before, we obtain $M \approx 6700$. Henceforth, there is a need to take into consideration no arithmetical operations by the corresponding coefficient of reduction $K_{\text{red}}^{\text{na}}$. For example, let $K_{\text{red}}^{\text{na}} = 3$. Then, we obtain that the total number of operations required for microprocessor subsystem in the case of a single realization of the considered complex digital signal reprocessing can be presented as $M \approx 2 \times 10^4$ operations. Thus, in the considered example, there is a need to use 2×10^4 microprocessor operations on average to process a single target pip. Naturally, this number corresponds only to the considered algorithm and can be reduced significantly if we are able to upgrade the algorithm of target pip identification, to simplify the algorithm of target track parameters smoothing, etc. The main purpose of consideration of this example is to present a possibility to calculate the work content of the complex DSP algorithm and indicate simultaneously some problems arising in the course of these calculations.

7.4 PARALLELING OF COMPUTATIONAL PROCESS

The results of work content evaluation of the complex DSP algorithm give us the initial information to select the structure and elements of microprocessor subsystems assigned to realize this complex DSP algorithm in a CRS. To ensure the required work content and operational reliability, the designed computational subsystem must include several microprocessor subsystems, as a rule. The main peculiarity of these subsystems is instrument or programmable parallelism of the computational process. To organize the parallel computational process, there is a need to carry out a paralleling of the complex DSP algorithm. In a general case, the paralleling of the complex DSP algorithms can be considered only for a specific problem taking into consideration the supposed structure of the computational system. Consequently, in the course of designing, the problems of selecting a structure of computational system based on the microprocessor subsystems and algorithmic transformation in accordance with the proposed structure of computational system are closely related. There is a set of general statements and methods of algorithmic solution paralleling, and we consider some of these methods in this section.

7.4.1 Multilevel Graph of Complex Digital Signal Processing Algorithm

The source for paralleling is the graph flowchart of algorithm presented in the multilevel form [8]. The multilevel form is introduced as a generalization of graph flowcharts to illustrate possibilities of serial–parallel operation of algorithms and is a characteristic of the fact that the tops of each level are not related by information features since the final results of operations carried out by one top cannot be considered as initial data for another top. Operations of independent tops (algorithms) can be performed simultaneously. Consequently, there is a possibility to realize the elementary DSP algorithms at the definite level using different microprocessor subsystems.

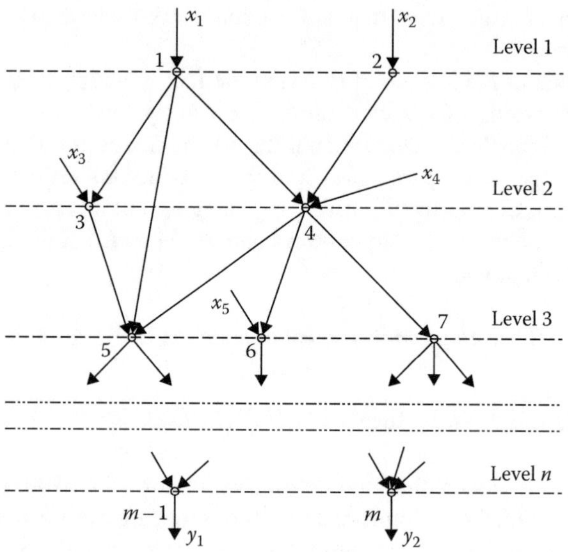

FIGURE 7.12 Multilevel graph of complex DSP algorithm.

The multilevel graph form is obtained in the following way (see Figure 7.12). The first-level top is called the top that has no input arc. The second-level top is the top whose input arc is the output arc of the first-level top. By analogous way, the $(n-1)$th level top is the top whose input arc is the output arc of the nth-level top and the output arcs of some previous level tops. In the realization of elementary DSP algorithms, a paralleling is also possible based on the network graph representation of the complex DSP algorithm for the solved problem. In this case, we should transform the initial graph form to the multilevel one. In the case of multilevel graph form, each level represents independent elementary DSP algorithms or their set, one of which is carried out necessarily in the course of a single realization of the complex DSP algorithm. If an individual microprocessor subsystem is used to realize the elementary DSP algorithms of each level, we obtain the so-called pipeline subsystem, in which several operations are made simultaneously on data passing in sequence. In this case, the DSP is divided into several stages (by the number of graph levels) and each stage is carried out in parallel with other stages.

To evaluate possibilities of parallel computation organization based on the multilevel algorithmic graph, we introduce a set of quantitative characteristics, namely, b_i is the ith level width, that is, the number of independent branches at the ith level; B is the width of the multilevel graph or $\max_i\{b_i\}$; L is the graph length, that is, maximal critical way leading from zero to the final state, and so on. As follows from the multilevel graph, the realization time of a set of DSP algorithms is limited by some threshold value T_{th}. By knowing T_{th}, we can evaluate the required number of the same type of microprocessors N_{mp} to realize the given set of elementary DSP algorithms. In doing so, we obtain

$$N_{\text{mp}} \le \frac{T_{\text{mp}}^{\text{single}}}{T_{\text{th}}}, \tag{7.48}$$

where

$T_{\text{mp}}^{\text{single}} = \sum_{i=1}^{m} t_i$ is the realization time of all algorithms for a single microprocessor subsystem
t_i is the realization time of the ith elementary DSP algorithm
m is the number of elementary DSP algorithms in the complex one

We consider a design procedure of the multilevel graph using an example of algorithm paralleling for the linear recurrent filtering of target track parameters given by the state equation (5.1). In the given case, the linear recurrent filtering algorithm takes the following form:

$$\begin{cases} \hat{\boldsymbol{\theta}}_{\text{ex}_n} = \boldsymbol{\Phi}_n \boldsymbol{\theta}_{n-1} + \boldsymbol{\Gamma}_n \boldsymbol{\eta}_{n-1}; \\ \boldsymbol{\Psi}_{\text{ex}_n} = \boldsymbol{\Phi}_n \boldsymbol{\Psi}_{n-1} \boldsymbol{\Phi}_n^T + \boldsymbol{\Gamma}_n \boldsymbol{\Psi}_\eta \boldsymbol{\Gamma}_n^T; \\ \mathbf{G}_n = \boldsymbol{\Psi}_{\text{ex}_n} \mathbf{H}_n^T \left(\mathbf{H}_n \boldsymbol{\Psi}_{\text{ex}_n} \mathbf{H}_n^T + \mathbf{R}_n \right)^{-1}; \\ \hat{\mathbf{Y}}_{\text{ex}_n} = \mathbf{H}_n \hat{\boldsymbol{\theta}}_{\text{ex}_n}; \\ \hat{\boldsymbol{\theta}}_n = \hat{\boldsymbol{\theta}}_{\text{ex}_n} + \mathbf{G}_n (\mathbf{Y}_n - \hat{\mathbf{Y}}_{\text{ex}_n}); \\ \boldsymbol{\Psi}_n = \boldsymbol{\Psi}_{\text{ex}_n} - \mathbf{G}_n \mathbf{H}_n \boldsymbol{\Psi}_{\text{ex}_n}, \end{cases} \tag{7.49}$$

where

$\hat{\boldsymbol{\theta}}_n$ is the $(s \times 1)$ vector of estimated target track parameters
$\hat{\boldsymbol{\theta}}_{\text{ex}_n}$ is the $(s \times 1)$ vector of extrapolated target track parameters
$\mathbf{Y}_n$ is the $(m \times 1)$ vector of measured coordinate magnitudes
$\hat{\mathbf{Y}}_{\text{ex}_n}$ is the $(m \times 1)$ vector of extrapolated coordinate magnitudes
$\boldsymbol{\eta}_{n-1}$ is the $(h \times 1)$ vector of disturbance of target track parameters
$\boldsymbol{\Phi}_n$ is the transfer $(s \times s)$ matrix of target track model
$\boldsymbol{\Psi}_n$ is the correlation $(s \times s)$ matrix of errors of target track parameter estimation
$\boldsymbol{\Psi}_{\text{ex}_n}$ is the correlation $(s \times s)$ matrix of extrapolation of target track parameters
$\boldsymbol{\Psi}_\eta$ is the correlation $(h \times h)$ matrix of target track random disturbances
$\boldsymbol{\Gamma}_n$ is the $(s \times h)$ matrix (see (5.1))
$\mathbf{H}_n$ is the $(m \times s)$ matrix (see (5.11))
$\mathbf{R}_n$ is the correlation $(m \times m)$ error matrix of target track coordinate measurements

Dimensions of matrices and vectors are required in determination of the work content for the branches of the multilevel complex DSP algorithm graph.

The ordinary complex DSP algorithm graph is the basis for the multilevel graph. In the case of ordinary complex DSP algorithm graph, the two-input functional operators on vector and matrices are considered as the tops and results of operations, and transitions in the graph are considered as the arcs. In doing so, a set of arcs without initial tops is an ensemble of initial arguments, and a totality of arcs without the end tops is a set of output results. The graph, as a rule, is designed manually because to formalize this process is a very difficult problem. The initial graph of linear filtering algorithm is depicted in Figure 7.13. To design the multilevel graph, at first, we should present the initial graph in the form of adjacency matrix with the number of columns and rows equal to the number of initial graph tops. The elements of the adjacency matrix l_{ij} take the value 0 if there is no arc between the top i and the top j and the value 1 if there is an arc between the top i and the top j. The adjacency matrix elements of the multilevel graph shown in Figure 7.13 are represented in Table 7.2.

The transformation process of the initial complex DSP algorithm graph into the multilevel complex DSP algorithm graph is to sort the adjacency matrix rows and columns and is based on an adjacency matrix feature to have zero rows if there are end tops and zero columns if there are initial tops, that is, all arcs for these tops are initial. In our case, the tops "1," "2," "8," and "10" are the initial tops. These tops form the first level of the sorted multilevel graph. The next step is nulling all the nonzero elements of rows with the numbers selected at the first step. At the same time, the first level tops are considered as the end tops and the output arcs are considered as the input arcs for the

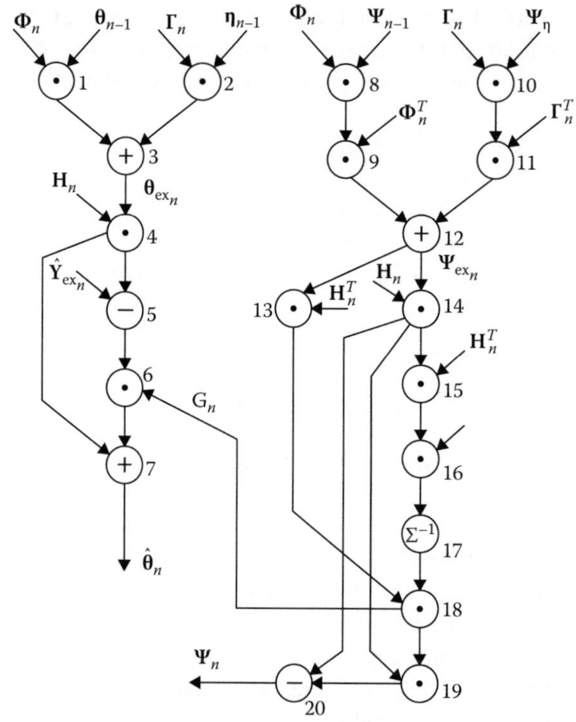

FIGURE 7.13 The multilevel graph of linear filtering algorithm.

TABLE 7.2
Adjacency Matrix Elements of the Multilevel Graph Shown in Figure 7.13

$i \backslash j$	1	2	3	4	5	6	7	8	9	10	11	12	13	14	15	16	17	18	19	20
1	0	0	1	0	0	0	0	0	0	0	0	0	0	0	0	0	0	0	0	0
2	0	0	1	0	0	0	0	0	0	0	0	0	0	0	0	0	0	0	0	0
3	0	0	0	1	0	0	0	0	0	0	0	0	0	0	0	0	0	0	0	0
4	0	0	0	0	1	0	1	0	0	0	0	0	0	0	0	0	0	0	0	0
5	0	0	0	0	0	1	0	0	0	0	0	0	0	0	0	0	0	0	0	0
6	0	0	0	0	0	0	1	0	0	0	0	0	0	0	0	0	0	0	0	0
7	0	0	0	0	0	0	0	0	0	0	0	0	0	0	0	0	0	0	0	0
8	0	0	0	0	0	0	0	0	1	0	0	0	0	0	0	0	0	0	0	0
9	0	0	0	0	0	0	0	0	0	0	0	1	0	0	0	0	0	0	0	0
10	0	0	0	0	0	0	0	0	0	0	1	0	0	0	0	0	0	0	0	0
11	0	0	0	0	0	0	0	0	0	0	0	1	0	0	0	0	0	0	0	0
12	0	0	0	0	0	0	0	0	0	0	0	0	1	1	0	0	0	0	0	0
13	0	0	0	0	0	0	0	0	0	0	0	0	0	0	0	0	0	1	0	0
14	0	0	0	0	0	0	0	0	0	0	0	0	0	0	1	0	0	0	1	0
15	0	0	0	0	0	0	0	0	0	0	0	0	0	0	0	1	0	0	0	1
16	0	0	0	0	0	0	0	0	0	0	0	0	0	0	0	0	1	0	0	0
17	0	0	0	0	0	0	0	0	0	0	0	0	0	0	0	0	0	1	0	
18	0	0	0	0	0	1	0	0	0	0	0	0	0	0	0	0	0	0	1	
19	0	0	0	0	0	0	0	0	0	0	0	0	0	0	0	0	0	0	0	
20	0	0	0	0	0	0	0	0	0	0	0	0	0	0	0	0	0	0	0	

TABLE 7.3
Distribution of Graph Tops along the Levels

Graph Level Number	Graph Tops
1	[1], [2], [8], [10]
2	[3], [9], [11]
3	[4], [12]
4	[5], [13], [14]
5	[15]
6	[16]
7	[17]
8	[18]
9	[6], [19]
10	[7], [20]

higher level (next level). Henceforth, we search for columns not marked previously of the adjacency matrix with zero elements. The numbers of these columns form the second-level tops and so on until all columns will be numbered. As a result, we obtain the table with the numbers of levels and graph tops corresponding to each level. For the considered case, a distribution of tops along the levels is presented in Table 7.3. Using Table 7.3 and taking into consideration relations between the initial graph tops presented in Figure 7.14, the multilevel complex DSP algorithm of linear filtering has been designed. There is a need to note that the multilevel graph form depends on a sequence based on how this graph is designed, namely, starting from a set of tops as in designing the graph depicted in Figure 7.14 or starting from a set of output tops. In the last case, a distribution of tops by levels is presented in Table 7.4 and the corresponding multilevel graph is depicted in Figure 7.15. As we can see, in this case, the graph is different from the previous one.

A simple consideration of multilevel graph indicates that there is a possibility to parallel a computational process under realization of the corresponding complex DSP algorithms. As follows from Figures 7.14 and 7.15, more than one macro-operations (from 1 to 3) can be carried out

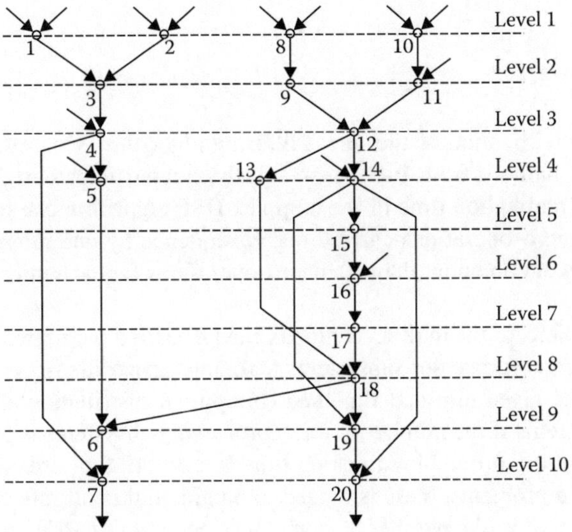

FIGURE 7.14 Tops of initial graph.

TABLE 7.4
Distribution of Graph Tops along the Levels

Graph Level Number	Graph Tops
1	[20], [7]
2	[6], [10]
3	[5], [18]
4	[4], [13], [17]
5	[3], [16]
6	[1], [2], [15]
7	[14]
8	[12]
9	[9], [11]
10	[8], [10]

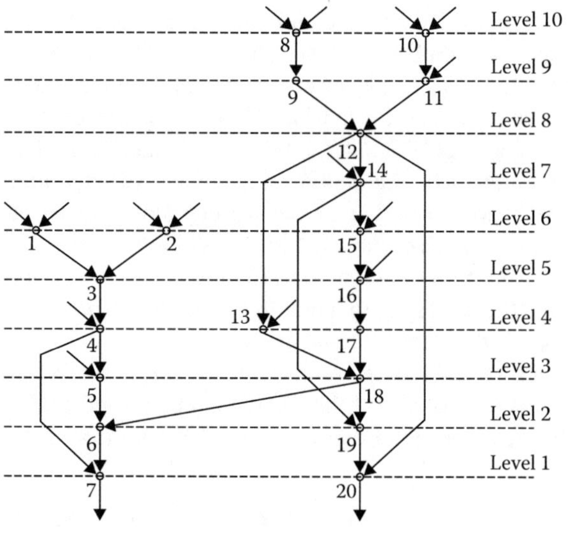

FIGURE 7.15 Multilevel graph.

simultaneously under realization of the linear filtering algorithm. Consequently, several microprocessor subsystems, namely, from 1 to 3, respectively, can participate in the computational process. In doing so, the realization time of the complex DSP algorithm can be reduced essentially, since instead of 20 macro-operations carried out in sequence by one microprocessor subsystem there is a need to carry out not more than 10 macro-operations for each microprocessor subsystem in parallel scheme.

Further transformations of the multilevel graphs can be carried out in two avenues: (a) definition of rational number of microprocessor subsystems realizing the paralleling complex DSP algorithm within the limits of the given minimal time and (b) optimal distribution of macro-operations by microprocessor subsystems if the number of microprocessor subsystems is given and their characteristics are known and the minimal realization time is a criterion of effectiveness. To solve both the first and the second problems, there is a need to obtain additional information concerning the weights of graph tops, that is, the number of elementary operations carried out during a realization of all macro-operations, by which the graph tops are marked.

7.4.2 Paralleling of Linear Recurrent Filtering Algorithm Macro-Operations

Consider the computational process paralleling problem for the given number, in our case two, of the same type of microprocessor systems based on the example of the linear recurrent filtering algorithm. To solve this problem, first, there is a need to compute the work content of the multilevel graph tops of the considered linear recurrent filtering algorithm. The graph depicted in Figure 7.15 is considered as the basis. Number the tops of this graph from right to left and top-down and define the macro-operations carried out under transition to each top. As noted earlier, these macro-operations are the two-input operations on the vectors and matrices and one operation of the matrix inversion. Table 7.5 represents the operations corresponding to the graph tops and expressions to compute the number of arithmetical operations under realization of corresponding operators by the microprocessor systems. The number of reduced arithmetical operations for each graph operator is determined at $s = 6$, $m = 3$, $h = 3$. The arithmetical operation reduction, as earlier, is carried out taking into consideration the fact that the multiplication operation can be accomplished by the microprocessor system in four cycles and the division operation—for seven cycles. Next columns of Table 7.5 present the results of the work content computation for the graph tops taking into consideration nonarithmetical operations. The multilevel graph tops shown in Figure 7.16 are marked by these computational results.

TABLE 7.5
Operations Corresponding to the Graph Tops and Expressions to Compute the Number of Arithmetical Operations

Graph Top Number	Operation	Number of Arithmetical Operations: Addition + Subtraction	Multiplication	Division	M	N	Top Rank
[1]	$\mathbf{\Phi}_n\mathbf{\Psi}_{n-1}$	$(s-1)s^2$	s^3	—	1044	3000	11,230
[2]	$\mathbf{\Gamma}_n\mathbf{\Psi}_\eta$	$(h-1)hs$	sh^2	—	252	750	7,480
[3]	$(\mathbf{\Phi}_n\mathbf{\Psi}_{n-1})\mathbf{\Phi}_n^T$	$(s-1)s^2$	s^3	—	1044	3000	8,230
[4]	$(\mathbf{\Gamma}_n\mathbf{\Psi}_\eta)\mathbf{\Gamma}_n^T$	$(h-1)s^2$	hs^2	—	504	1500	6,730
[5]	[3] + [4]	s^2	—	—	36	100	5,230
[6]	$\mathbf{H}_n\mathbf{\Psi}_{\text{ex}_n}$	$sm(s-1)$	ms^2	—	522	1500	5,130
[7]	$\mathbf{\Phi}_n\hat{\boldsymbol{\theta}}_{n-1}$	$s(s-1)$	s^2	—	174	500	1,060
[8]	$\mathbf{\Gamma}_n\boldsymbol{\eta}_{n-1}$	$(h-1)s$	hs	—	84	250	810
[9]	$(\mathbf{H}_n\mathbf{\Psi}_{\text{ex}_n})\mathbf{H}_n^T$	$m^2(s-1)$	sm^2	—	261	750	3,630
[10]	$[7]+[8]=\hat{\boldsymbol{\theta}}_{\text{ex}_n}$	s	—	—	6	25	560
[11]	$\mathbf{R}_n + [9] = \mathbf{\Sigma}_n$	m^2	—	—	9	30	2,880
[12]	$\mathbf{H}_n\hat{\boldsymbol{\theta}}_{\text{ex}_n}$	$(s-1)m$	sm	—	87	250	535
[13]	$\mathbf{\Psi}_{\text{ex}_n}\mathbf{H}_n^T$	$sm(s-1)$	ms^2	—	522	1500	3,850
[14]	$[\mathbf{\Sigma}_n]^{-1}$	$m^2(m-1)$	$m^2(m-1)$	m^2	153	500	2,850
[15]	$\mathbf{Y}_n-[12]=\Delta\hat{\mathbf{Y}}_n$	m	—	—	3	15	285
[16]	$[13][14] = \mathbf{G}_n$	$sm(m-1)$	sm^2	—	250	750	2,350
[17]	$\mathbf{G}_n[15]$	$(m-1)s$	ms	—	86	250	270
[18]	$\mathbf{G}_n(\mathbf{H}_n\mathbf{\Psi}_{\text{ex}_n})$	$(m-1)s^2$	ms^2	—	504	1500	1,600
[19]	$[10] + [17] = \hat{\boldsymbol{\theta}}_n$	s	—	—	6	20	20
[20]	$[5]–[18] = \mathbf{\Psi}_n$	s^2	—	—	36	100	100

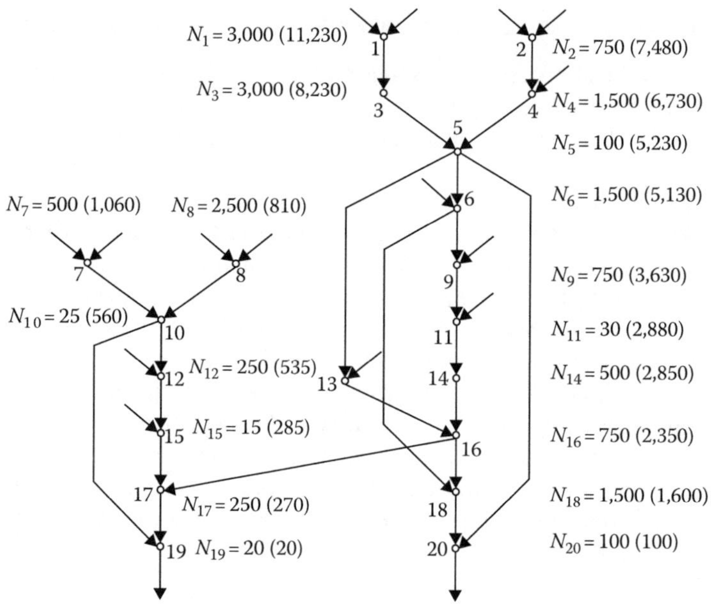

FIGURE 7.16 Multilevel graph.

Now, let us start a direct distribution of the graph tops by microprocessor subsystems. The distribution is carried out in the following sequence:

1. The graph top ranking is carried out—the weight equal to maximal way (by the operation number) leading from the given top to the end top (the top ranks are given in the last column of Table 7.5 and designated by the numbers in parentheses in Figure 7.16) is assigned to each graph top.
2. The graph top with the maximal rank (the top 1) is assigned for the first microprocessor subsystem. The graph top with the maximal rank after loading the first microprocessor subsystem and not requiring results of operations made by the first microprocessor subsystem is assigned for the second microprocessor subsystem (the top 2).
3. The remaining graph tops with maximal rank are assigned to realize the given algorithm under the condition that there are required data. If the required data are not available or absent, a microprocessor subsystem is in the standby mode until obtaining the required data from other microprocessor subsystem.

The loading schedule of microprocessor subsystems is presented in Figure 7.17. As follows from Figure 7.17, the threshold number of microprocessor subsystem operations under paralleling on two microprocessor subsystems is defined as $N_{\text{th}} = 11{,}230$ operations under the condition that the total work content of the considered algorithm is given by $M_{\text{total}} = 16{,}290$ operations. The second microprocessor subsystem is loaded only on 45%. The coefficient of microprocessor subsystem loading, as a whole, is determined in the following form:

$$K_{\text{load}} = \frac{M_{\text{total}}}{2N_{\text{th}}} 0.725. \tag{7.50}$$

Thus, we cannot say that a paralleling of the considered algorithm by this way is an ideal process with the viewpoint of loading the two microprocessor subsystems. To increase the coefficient of loading K_{load} of microprocessor subsystems, there is a need to decrease a length of macro-operations and to carry out paralleling computations inside each macro-operation.

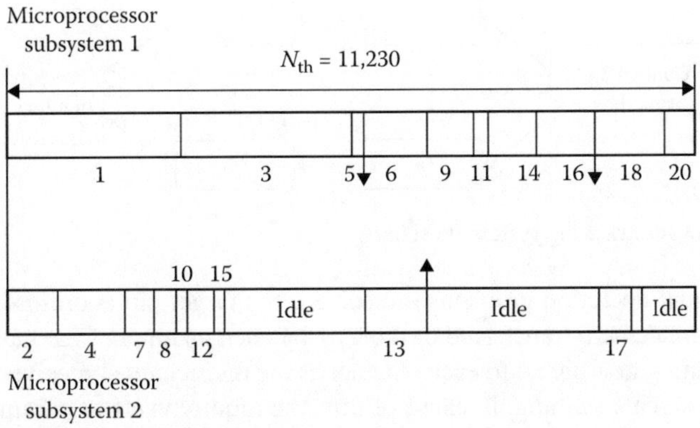

FIGURE 7.17 Loading schedules of microprocessor subsystems.

In conclusion, we note that the considered example is purely illustrative. We do not take into consideration any possibility of reducing the number of operations under realization of some graph tops, for example, owing to matrix set sparseness including in the algorithmic formula flowchart, special procedures in carrying out the specific vector–matrix operations, and so on. The reduction coefficients of multiplication and division operations to short operations are conditional, etc.

7.4.3 Paralleling Principles of Complex Digital Signal Processing Algorithm by Object Set

Under complex DSP, we consider a set of objects, namely, the target return signals, target pips, target tracks, and so on. Information about these objects must be processed using the same complex DSP algorithms. If the considered objects are independent, then information about each object can be processed independently. In this case, we use the independent objects paralleling. Consider some examples when we use the paralleling principles for independent objects at different levels of complex DSP algorithm for the surveillance radar.

Example 1: Consider the multichannel receiver with signal preprocessing assigned to detect the targets and estimate the target pip parameters under the radar coverage scanning (see Figure 7.18). The receiver consists of the multichannel microprocessor subsystem representing a set of specific signal processing processors and the general controlling processor carrying out the functions of new information distribution by channels. We assume that the number of parallel signal processing channels is equal to or less than the number of independent target return signal sources. As the independent target return signal sources, we consider, in this case, the intervals of incoming stochastic process within the limits of each discrete time interval. In doing so, we suppose that a time discretization frequency of the input stochastic process is selected in such a way that the signals from the neighbor discrete time intervals are statistically independent. The signals are accumulated, the target pips

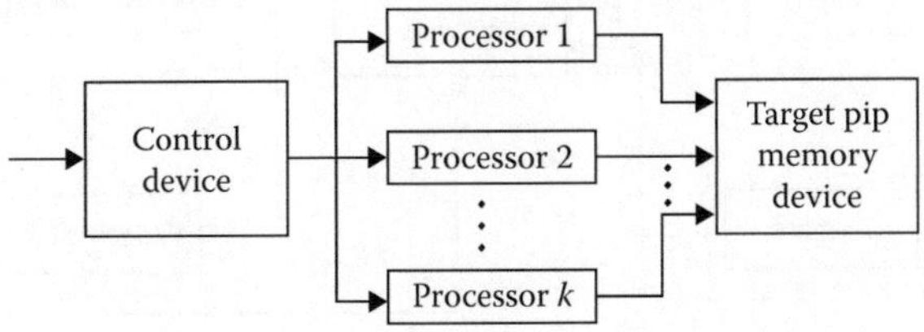

FIGURE 7.18 Multichannel receiver with digital signal reprocessing algorithm.

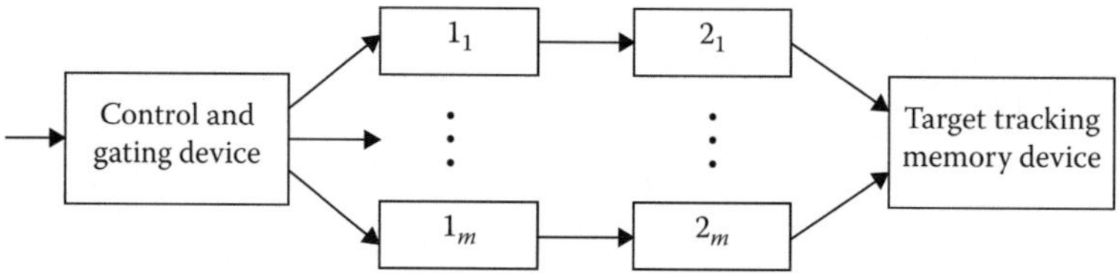

FIGURE 7.19 Target autotracking system flowchart.

are formed, the signal detection problems are solved, the target pip coordinates are defined, and the target pip coordinates are transferred to the user by each channel of the multichannel receiver. The period of incoming new signal to each channel of the multichannel receiver corresponds to the period of scanning signals sending. Because of this, the requirements to parameters of processors possess medium character. Another advantage of the considered multichannel receiver is low power consumption to distribute the problems by channels.

Example 2: Consider an organization of target trajectory autotracking by the autonomous microprocessor subsystems carrying out a whole cycle of signal processing by a single target. This target tracking system (see Figure 7.19) consists of m channels (microprocessor subsystems). The signal detection algorithms, the definition and measurement of signal parameters, and the signal selection in the physical gates of individual target tracking (block 1), the algorithm of estimation of target track parameters, and sending the final information to the system forming a general location situation (block 2) are realized by each channel of the target autotracking system. In a general case, each of the complex DSP algorithms (signal preprocessing and signal reprocessing) can be realized by individual specific microprocessor subsystems. In this case, the number of signal preprocessing and signal reprocessing channels can be considered different. Additionally, the target autotracking system contains the microprocessor subsystems realizing the control and commutation of signal processing channels and gating the receivers. This target autotracking system allows us to organize the complex signal processing algorithm for target return signal in real time by microprocessor subsystems with limited productivity.

Example 3: Consider an organization of paralleling under the DSP of target return signals from M independent CRSs using a computer system based on N microprocessor subsystems. The DSP algorithm is to define the target tracks by the target pips sent by each radar system or signal preprocessing subsystem. All radar systems operate in asynchronous mode, and their radar coverage of each radar system is overlapped with other ones. The computer system under the parallel DSP depicted in Figure 7.20 contains an associative addressing device in addition to microprocessor subsystems, which allows us to carry out simultaneously a comparison of each target pip coordinate coming from radar systems jointly with the gates of extrapolated points of all target trajectories tracked by

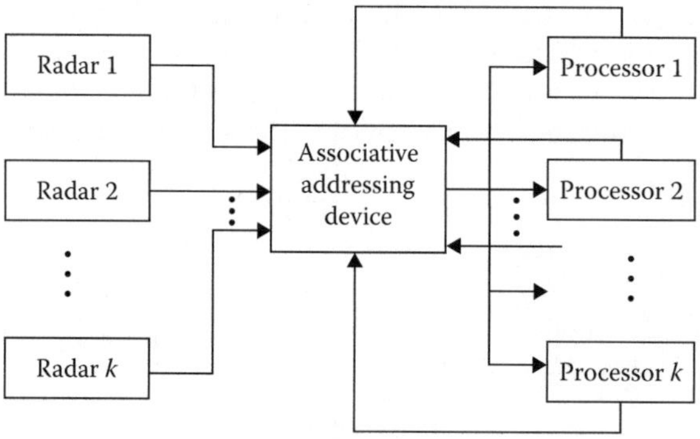

FIGURE 7.20 Computer system with DSP algorithm paralleling.

radar systems simultaneously. On-stream of the computer system, each microprocessor subsystem sends to the associative addressing device the gates of such targets that are processed by the microprocessor subsystem at the current time. All the gates are extrapolated with a definite cycle in such a way that the incoming new target pips and extrapolated points of target tracks are obtained at the same time. Information about the target pips comes in at the associative addressing device input from the CRS output, and the associative addressing device defines which microprocessor subsystem must process the incoming target pip based on an associative comparison of boundaries of all the gates. The target pip is sent to the corresponding microprocessor subsystem to renew information about the specific target. If the target pip belongs to a new target, that is, there is no gate to which this target pip can belong, then the information goes to one from the microprocessor subsystem that is not loaded completely. We can control the loading process of the microprocessor subsystems. For this purpose, there is a need to determine the coefficients of current loading of all the microprocessor subsystems for each scanning cycle T_{scan}:

$$K_{ij}^{\text{load}} = \frac{T_{ij}}{T_{\text{scan}}}, \tag{7.51}$$

where

T_{ij} is the time required for the ith microprocessor subsystem under the jth scanning to process all targets

T_{scan} is the scanning cycle

To increase the control stability, we define the averaged by n scans coefficient of loading in the following form:

$$K_{i}^{\text{load}} = \frac{1}{n}\sum_{j=1}^{n} K_{ij}^{\text{load}}. \tag{7.52}$$

The loading coefficient is defined more exactly after each scanning. If all $K_i^{\text{load}} = 1, i = 1,\ldots,N$, then the microprocessor subsystem and, consequently, the computer system have been loaded completely and under upcoming new target pip belonging to the new target tracks, the computer system is overloaded and, consequently, the microprocessor subsystems are overloaded, too, and the corresponding control signal informs us about this situation. The computer system consisting of the microprocessor subsystems can be designed in such a way that, at first, the first microprocessor system is overloaded, after that the second microprocessor subsystem is overloaded, and so on. Under this type of computer system organization, it is simpler to take into consideration an incomplete microprocessor subsystem capacity and organize the complete capacity.

The discussed paralleling computer system consisting of the microprocessor subsystems and carrying out the digital signal reprocessing algorithms has a high reliability and operability and, additionally, ensures high-quality user information about the tracking targets. The disadvantage of this complex digital signal reprocessing algorithm paralleling is a necessity to use a very complex associative addressing device, especially, if the number of tracking targets is high.

7.5 SUMMARY AND DISCUSSION

The exponential growth in digital technology since the 1980s, along with the corresponding decrease in its cost, has had a profound impact on the way radar systems are designed. More and more functions that were historically implemented in analog hardware are now being performed digitally, resulting in increased performance and flexibility and reduced size and cost. Advances in ADC and DAC technologies are pushing the border between analog and digital processing closer and closer to the antenna. In the past, implementing a real-time radar digital signal processor typically required the design of a custom computing machine, using thousands of high-performance ICs. These machines were very difficult to design, develop, and modify. Digital technology has advanced to the point where several implementation alternatives exist that make the processor more programmable, and easier to design and change.

The parallel microprocessor subsystem architecture employs multiple general-purpose processors that are connected via high-speed communication networks. Included in this class are high-end servers and embedded processor architectures. Servers are typically homogeneous processors, where all of the processing nodes are identical and are connected by a very high-performance data bus architecture. Embedded processor architectures are typically composed of single-board computers (blades) that contain multiple general-purpose processors and plug into standard backplane architecture. This configuration offers the flexibility of supporting a heterogeneous architecture, where a variety of different processing blades or interface boards can be plugged into the standard backplane to configure a total system. It is apparent that high-speed serial links will be the primary communication mechanism for multiprocessor subsystems into the future, with ever-increasing data bandwidths. These parallel microprocessor subsystem architectures offer the benefit of being programmable using high-level languages, such as C and C++. A related advantage is that programmers can design the system without knowing the intimate details of the hardware. In addition, the software developed to implement the system can typically be moved relatively easily to new hardware architecture as part of a technology refresh cycle. On the negative side, these systems can be difficult to program to support real-time signal processing. The required operations need to be split up appropriately among the available processors, and the results need to be properly merged to form the final result. A major challenge in these applications is to support the processing *latency* requirements of the system, which defines the maximum length of time allowed to produce a result. The latency of the microprocessor subsystem is defined as the amount of time required to observe the effect of a change at a processor's input on its output. Achieving latency goals often requires assigning smaller pieces of the workload to individual processors, leading to more processors and a more expensive system. Another challenge facing these systems in radar application is reset time. In a military application, when a system needs to be reset in order to fix a problem, the system needs to come back to full operation in a very short period of time. These microprocessor subsystems typically take a long time to reboot from a central program store and, hence, have difficulty meeting reset requirements. Developing techniques to address these deficiencies is an active area of research. Finally, these processors are generally used for non-real time or near-real-time data processing, as in target tracking and display processing. Since the 1990s, they have started to be applied to real-time signal processing applications. Although they might be cost-effective for relatively narrowband systems, their use in wideband DSP systems in the early twenty-first century is typically prohibitively expensive due to the large number of processors required. This situation should improve over time as faster and faster processors become available.

The introduction of the FPGA in the 1980s heralded a revolution in the way real-time DSP systems were designed. FPGAs are integrated circuits that consist of a large array of configurable logic elements that are connected by a programmable interconnect structure. At the time of this writing, FPGAs can also incorporate hundreds of multipliers that can be clocked at rates up to a half billion operations per second, and memory blocks, microprocessors, and serial communication links that can support multigigabit-per-second data transfers. FPGAs allow the designer to fabricate complex signal processing architectures very efficiently. In typical large applications, FPGA-based processors can be a factor of 10 (or more) smaller and less costly than systems based on general-purpose processors. This is due to the fact that most microprocessors only have one or very few processing elements, whereas FPGA have an enormous number of programmable logic elements and multipliers. On the negative side, utilizing an FPGA to its best advantage typically requires the designer to have a thorough understanding of the resources available in the device. This typically makes efficient FPGA-based systems harder to design than radar systems based on general-purpose processors, where a detailed understanding of the microprocessor subsystem architecture is not necessarily required. In addition, FPGA designs tend to be aimed at a particular family of devices and take full advantage of the resources provided by that family. Hardware vendors are constantly introducing new products, invariably incorporating new and improved capabilities. Over time, the older devices become obsolete and need to be replaced during a *technology refresh* cycle. When a technology refresh occurs several years down the road, typically the available resources in the latest FPGAs have changed or a totally different device family is used, which probably requires a redesign. On the other hand, software developed for general-purpose processors may only need to be recompiled in order to move it to a new processor. Tools currently exist that synthesize C or MATLAB code into a FPGA design, but these tools are typically not very efficient. The evolution of design tools for FPGAs to address these problems is an area of much research and development.

The complex algorithm of the computational process is a set of elementary DSP algorithms for all stages and CRS modes of operation. Off-line complex algorithms of individual stages of DSP, which are not associated with one another by information processes and control operations, are possible, too. To design the complex algorithm of the computational process, there is a need to have an unambiguous definition of microprocessor subsystem functioning in the solution of goal-oriented DSP problems. This definition must include the elementary DSP and control algorithms, a sequence of their application, the conditions of implementation of each elementary DSP and control algorithm, and intercommunication between the DSP and control algorithms using input and output information. A general form of such definition and description can be presented using the logical and graph flowchart of the DSP and control algorithms.

One of the best-known ways to assign the complex algorithm in depending on complexity is the logical or formula-logical algorithmic flowchart. The elementary operators and recognizers are presented in a geometrical form in the logical algorithm flowchart (rectangle, jewel boxes, trapeziums, etc.) connected with each other by arrows in accordance with the given sequence order of counting operators and recognizers in the complex algorithm. Titles of elementary operations (DSP algorithms) are written inside the geometrical forms. Sometimes the formulas of logical operations carried out and logical conditions under test are written inside the geometrical forms. In this case, the corresponding logical flowchart of the algorithm is called the formula-logical block diagram.

The main problems solved by the graph flowcharts of complex algorithms are a definition of rational ways to present these problems and a choice of computational software tools and microprocessor subsystems to realize the complex DSP algorithm of a radar system. In short, the problem of optimization of computational process is assigned. The solution of this problem allows us to reduce significantly the realization time and to simplify the complex DSP algorithm of the radar system. Similar problems are the problems of network planning and control.

The deterministic network model cannot present a complex DSP algorithm functioning in a CRS since it is impossible to predict a set of elementary DSP algorithms and sequence of realizations for each practical situation. Therefore, the stochastic network model, in which the transitions in the network graph are defined by the corresponding probabilities of transitions given by specific conditions of CRS functioning, is more suitable to image and analyze a realization of complex algorithm by microprocessor systems. When the network model has been constructed, the problem to estimate the time to complete all operations, that is, the time to finish all operations by microprocessor subsystems with the given effective speed of operations, arises. This time cannot be higher than a total duration to finish a complex algorithm operation defined in the most unfavorable way from the initial graph top a_n to the final graph top a_k, that is, along such a route that generates a maximal duration of operations. This route is called extreme. The extreme route in the stochastic network model cannot be presented in clear form as, for example, in the network model with a given structure. Because of this, the problems of defining the average time or average number of operations required to realize the complex algorithm are assigned under analysis of stochastic network models. If the statistical characteristics and parameters of noise and target environment inside the radar coverage are known and the target pip beginning algorithm parameters are selected, we are able to determine the probability of transition in the network graph of the complex DSP algorithm. However, we cannot say that this possibility exists forever. In some case, the probability of transition can only be estimated as a result of computer simulation of the complex DSP algorithm.

Results of analytical calculation of the required number of arithmetical operations are obtained separately by the number of additions and subtractions, products, and divisions. There is a need to determine the number of reduced arithmetical operations. As the reduction operation, as a rule, an addition is used (short operation). The number of reduced arithmetical operations is determined for each microprocessor subsystem taking into consideration the known ratio between the time to carry out the *i*th long and short operations. The DSP of target return signals has a pronounced information-logical character. Logical operations and transition operations are for about 80% of the total number of elementary DSP operations (cycles) in the process of realization of the complex DSP algorithm required for CRS functioning. Consequently, under the work content computation of the elementary DSP algorithms there is a need to take into consideration the nonarithmetical operations, too. Under the work content definition, we must take into consideration additionally that the number of microprocessor operations depends on a mode of programming.

The network model of complex DSP algorithm allows us to define, in principle, the average work content. If we know the realization time of a single reduced operation, then we can compute the average realization time of a complex DSP algorithm. Inversely, if a limitation on the average realization time of a complex DSP algorithm is given, we are able to determine the required work content of microprocessor subsystems to realize the given complex DSP algorithm. Sometimes, to solve the problems of computational resource analysis we need to know information about the work content variance.

Henceforth, there is a need to take into consideration the nonarithmetical operations by the corresponding coefficient of reduction $K_{\text{red}}^{\text{na}}$. For example, let $K_{\text{red}}^{\text{na}} = 3$; then we obtain that the total number of operations required for a microprocessor subsystem in the case of a single realization of the considered complex digital signal reprocessing can be presented as $M \approx 2 \times 10^4$ operations. Thus, in the considered example, there is a need to use 2×10^4 microprocessor operations on average to process a single target pip. Naturally, this number corresponds only to the considered algorithm and can be reduced significantly if we are able to upgrade the algorithm of target pip identification, to simplify the algorithm of target track parameters smoothing, etc. The main purpose of consideration of this example is to present a possibility of calculating the work content of the complex DSP algorithm and indicate simultaneously some problems arising in the course of these calculations.

The results of work content evaluation of the complex DSP algorithm give us the initial information to select the structure and elements of microprocessor subsystems assigned to realize this complex DSP algorithm in a CRS. To ensure the required work content and operational reliability, the designed computational system must include several microprocessor subsystems, as a rule. The main peculiarity of these microprocessor subsystems is instrument or programmable parallelism of computational process. To organize the parallel computational process, there is a need to carry out a paralleling of the complex DSP algorithm. In a general case, the paralleling of complex DSP algorithms can be considered only for a specific problem taking into consideration the supposed structure of a computational subsystem. Consequently, in the course of designing, the problems of selecting the structure of a computational subsystem based on the microprocessor subsystems and algorithmic transformation in accordance with the proposed structure of the computational subsystem are closely related. There is a set of general statements and methods of algorithmic solution paralleling.

A simple consideration of the multilevel graph indicates that there is a possibility to parallel a computational process under realization of the corresponding complex DSP algorithms. As follows from Figures 7.14 and 7.15, more than one macro-operations (from 1 to 3) can be carried out simultaneously in the realization of the linear filtering algorithm. Consequently, several microprocessor subsystems, namely, from 1 to 3, respectively, can participate in the computational process. In doing so, the realization time of the complex DSP algorithm can be reduced essentially, since instead of 20 macro-operations carried out in sequence by one microprocessor system there is a need to carry out not more 10 macro-operations for each microprocessor subsystem in parallel scheme.

Further transformations of the multilevel graphs can be carried out in two avenues:

1. Definition of a rational number of microprocessor subsystems realizing the paralleling complex DSP algorithm within the limits of the given minimal time
2. Optimal distribution of macro-operations by microprocessor subsystems if the number of microprocessor subsystems is given and their characteristics are known and the minimal realization time is a criterion of effectiveness

To solve both the first and the second problems, there is a need to obtain additional information concerning the weights of graph tops, that is, the number of elementary operations carried out during the realization of all macro-operations, by which the graph tops are marked.

The loading schedule of microprocessor subsystems is presented in Figure 7.17. As follows from Figure 7.17, the threshold number of microprocessor subsystem operations under paralleling on two microprocessor subsystems is defined as $N_{\text{th}} = 11{,}230$ operations under the condition that the total work content of the considered algorithm is given by $M_{\text{total}} = 16{,}290$ operations. The second microprocessor subsystem is loaded only on 45%. The coefficient of microprocessor subsystem loading K_{load} can be defined by (7.50). Thus, we cannot say that a paralleling of the considered algorithm by this way is an ideal process with the viewpoint of loading the two microprocessor subsystems. To increase the coefficient of loading K_{load} of microprocessor subsystems, there is a need to decrease a length of macro-operations and to carry out paralleling computations inside each macro-operation.

Under complex DSP, we consider a set of objects, namely, the target return signals, target pips, target tracks, and so on. Information about these objects must be processed using the same complex DSP algorithms. If the considered objects are independent, then information about each object can be processed independently. In this case, we use the independent objects paralleling.

The discussed paralleling computer system consisting of the microprocessor subsystems and carrying out the digital signal reprocessing algorithms has a high reliability and operability and, additionally, ensures high-quality user information about the tracking targets. The disadvantage of this complex digital signal reprocessing algorithm paralleling is the necessity to use a very complex associative addressing device, especially if the number of tracking targets is high.

REFERENCES

1. Rahnema, M. 2007. *UMTS Network Planning, Optimization, and Interoperation with GSM*. New York: John Willey & Sons, Inc.
2. Laiho, J., Wacker, A., and T. Novosad. 2006. *Radio Network Planning and Optimization for UMTS*. 2nd edn. New York: John Wiley & Sons, Inc.
3. Woolery, J. and K. Crandall. 1983. Stochastic network model for planning scheduling. *Journal of Construction Engineering and Management*, 109(3): 342–354.

4. Butler, R. and A. Huzurbazar. 1997. Stochastic network models for survival analysis. *Journal of the American Statistical Association*, 92(437): 246–257.
5. Tsitsiashvili, G. and M. Osipova. 2008. *Distributions in Stochastic Network Models*. New York: Nova Publishers.
6. Neely, M. 2010. *Stochastic Network Optimization with Application to Communication and Queuing Systems.* Synthesis Lectures on Communication Networks. Los Angeles, CA: Morgan & Claypool Publishers.
7. Creebery, D. and D. Golenko-Ginzburg. 2010. Upon scheduling and controlling large-scale stochastic network project. *Journal of Applied Quantitative Methods*, 5(3): 382–388.
8. Pospelov, D. 1982. *Introduction to Theory of Computational Systems*. Moscow, Russia: Soviet Radio (in Russian).

8 Design Principles of Digital Signal Processing Subsystems Employed by a Complex Radar System

8.1 STRUCTURE AND MAIN ENGINEERING DATA OF DIGITAL SIGNAL PROCESSING SUBSYSTEMS

At the present time, the digital signal processing subsystems employed by complex radar systems (CRSs) are used to solve various problems. We aim to consider the digital signal processing subsystems implementing the following:

- Solution of the problems to accumulate and process information files in real time
- Information exchange between sensors and users in the course of functional problem solution
- Long-time continuous processing
- Simultaneous realizations of wide-range signal processing and control problems with relative consistency in the problems solved during the exploiting period

To satisfy the listed requirements, the microprocessor subsystems are designed for each case. In this section, we discuss the main design principles, parameters, and performance of microprocessor subsystems employed by digital signal processing subsystems in a CRS.

8.1.1 Single-Computer Subsystem

The control microprocessor (see Figure 8.1) is the central component of single-computer subsystem with the following main constituents:

- Central processor consisting of the arithmetic and logic unit (ALU) and the central control device (CCD)
- Random access memory (RAM) assigned to store information (programs, intermediate and end computations) directly used at the time of digital signal processing operations
- External memory (EM) assigned to store large information arrays for a long time and exchange with RAM by these arrays
- Input–output (I/O) devices assigned to organize the exchange of information between RAM, control microprocessor, and other facilities
- Adapters, including control console, display, and so on

In management information systems, including CRSs, a single-computer subsystem possesses some devices providing an interface between sensors and users and specific subsystems of digital signal preprocessing. In Figure 8.1, these devices are united in the same block. The control microprocessor is characterized by the total engineering data. We consider the

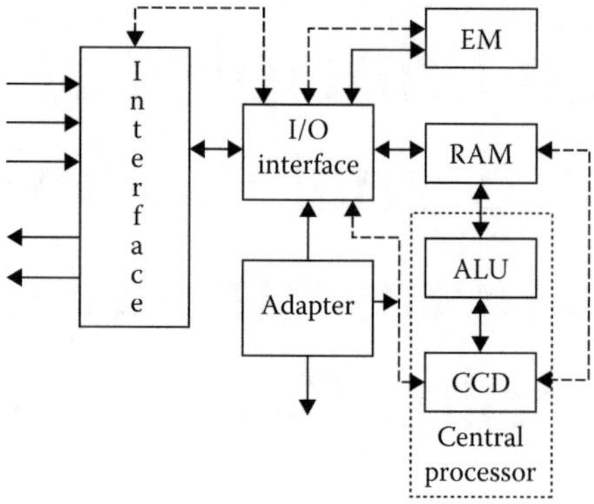

FIGURE 8.1 Control microprocessor.

following features of engineering data that are used only to compare the microprocessor systems between each other.

Addressness: The number of address codes used by the instruction code. There are one-, two-, three-address instructions and zero-address instructions. The control microprocessor uses the one-address instructions.

Number representation capacity: The control microprocessor subsystems are characterized by 16, 24, 32, and 64 digits per word.

Number representation form: There are two number representation forms: the *fixed* and the *floating point*. The control microprocessor with the fixed point represents the numbers in the form of proper fraction, and the point is fixed before the first more significant digit (MSD). The control microprocessor with floating point represents the numbers by the sign, mantissa, and order (the normal number representation form).

Microprocessor speed of operation: To characterize the speed of operation of the microprocessor subsystem, the statement of "rated speed" is introduced independently of the solved problem class in accordance with the formula

$$V_{\text{rs}} = \tau_{\text{short}}, \tag{8.1}$$

where τ_{short} is the duration of short operation (e.g., addition).

Effective speed of operation: The average number of operations per second for obtaining the specific digital signal processing algorithm is as follows:

$$V_{\text{eff}} = \mu \frac{N_{\text{total}}}{T_{\text{sol}}}, \tag{8.2}$$

where

- N_{total} is the total average number of operations carried out for a single realization of digital signal processing algorithm
- μ is the coefficient depending on microprocessor addressness (in the case of a single-address microprocessor $\mu = 1$)
- T_{sol} is the time to solve the problem given by

$$T_{\text{sol}} = \sum_{i=1}^{n} N_i \tau_i \tag{8.3}$$

where

n is the number of operation type, namely, addition, multiplication, division, addressing to RAM, and so on, carried out in the course of the digital signal processing algorithm

N_i is the number of the ith type operations

τ_i is the execution time of the ith type operation

Taking into consideration (8.3), we can write

$$V_{\text{eff}} = \frac{1}{\sum_{i=1}^{n} \omega_i \tau_i} \tag{8.4}$$

where ω_i is the average execution frequency (probability) of the ith type operation.

Memory device size may be expressed in bit (memory cell), byte (8 bits), kbit (1024 bits), and kbyte. The memory device size may be expressed also using the computer word, that is, the number of words, a length of which in bits corresponds to the number representation capacity in microprocessor subsystem memory. Speed of operation of memory device is characterized by a memory device cycle time τ_{cycle}. Thus, the cycle time differs between recording and reading and is defined in the following form:

$$\tau_{\text{cycle}}^{\text{rec}} = \tau_{\text{search}} + \tau_{\text{rec}} + \tau_{\text{clean}}; \tag{8.5}$$

$$\tau_{\text{cycle}}^{\text{read}} = \tau_{\text{search}} + \tau_{\text{read}} + \tau_{\text{rebuild}}, \tag{8.6}$$

where

τ_{search} is the time required to search information

τ_{rec} is the time required to record information

τ_{clean} is the time required to clean the memory cells for preparing corresponding cells to record new information

τ_{read} is the time required to read information

τ_{rebuild} is the time required to restore information destroyed while reading

In general, the microprocessor subsystem reliability is defined by the probability of instruction issued for a correct problem solution with a single-time use of digital signal processing algorithm:

$$P(t_{\text{real}}) = [1 - P_{\text{fail}}(t_{\text{real}})][1 - P_{\text{circuit}}(t_{\text{real}})], \tag{8.7}$$

where

t_{real} is the time of a single realization of digital signal processing algorithm for the solved problem

$P_{\text{fail}}(t_{\text{real}})$ is the probability of microprocessor subsystem failure within the limits of t_{real}, which can be rebuilt without shutting down the operation of microprocessor subsystem

$P_{\text{circuit}}(t_{\text{real}})$ is the probability of microprocessor subsystem circuit drop-in within the limits of t_{real}

In the course of comparative analysis, the microprocessor subsystem's reliability is evaluated by the average time between the microprocessor subsystem failures within the limits of t_{real}. Here, we understand failures associated with both the hardware and the software as the failure of microprocessor subsystem.

8.1.2 Multicomputer Subsystem

Until recently, the effective speed of operation has mainly improved due to an increase in the speed of element base operation and by designing the most efficient microprocessor subsystem functioning algorithms. At the present time, the speed of the element base operation is very close to being achieved. The use of parallel microprocessor subsystems plays a main role in increasing the effective speed of microprocessor subsystem functioning. The idea of parallel computations is very simple: several microprocessor subsystems try to solve the same problem. Technical realization of this idea depends largely on the nature of the problems to be solved (a possibility to parallel effectively a computational process) and on the level of efficacy of the modern parallel microprocessor subsystems.

One of the possible ways to increase the microprocessor subsystem performance and reliability during the digital signal processing in real time is to design and construct the multicomputer subsystems; these are used in CRSs to use effectively the digital signal processing algorithms. Computational process in the multicomputer subsystems is organized using new principles, namely, a parallel digital signal processing by several microprocessor subsystems. The main factors defining such a multicomputer subsystem structure are the end use, required performance, and memory size for the solution of the given problem in totality and functional reliability taking into consideration external environment and economical factors—the maximum permissible cost and energy consumption and so on.

To provide a solution for a single target problem, we need to organize the exchange of information between the microprocessor subsystems. This operation requires adequate memory size and speed of microprocessor subsystem operation and that microprocessor subsystems be combined in computer subsystem to be used in a CRS. Another aspect is the dead time, which is caused by expectation of final solutions of some problems at the previous levels of computer subsystem. These circumstances lead to a decrease in the multicomputer subsystems' performance compared to a single-computer subsystem's performance. However, other parameters such as the reliability and system survival are increased essentially. Thus, we have an opportunity to exploit the digital signal processing algorithms requiring larger memory size compared to a single-microprocessor subsystem, given the structural modifications of microprocessor subsystems with limited effective speed of operation and memory size.

In general, the multicomputer subsystem performance defined by the effective speed of multicomputer subsystem operation can be given by

$$\mathcal{Q}_{\text{eff}} = K(M)\sum_{i=1}^{M} V_{\text{eff}\,i}, \tag{8.8}$$

where

$V_{\text{eff}\,i}$ is the effective operation speed of the ith microprocessor
M is the number of microprocessors in the subsystem
$K(M) < 1$ is the coefficient taking into account the multicomputer subsystem operating costs depending on the number of microprocessors combined into a multicomputer subsystem

Microprocessors in the multicomputer subsystem may process the digital signal processing algorithms in *off-line mode* or while interacting with each other. In accordance with this statement, there are two types of multicomputer subsystems. The first type of multicomputer subsystems using only an information exchange between the autonomous microprocessors consist of, as a rule, the microprocessors of the same type. Each microprocessor in this multicomputer subsystem has the processor and RAM and interacts with other microprocessors through specific interface and information channels. An example of such multicomputer subsystem is the multiplexed multicomputer subsystem, in which $M - m$ microprocessors are working and m microprocessors are redundant and M is the total number of microprocessors in the multicomputer subsystem. An important task of this multicomputer subsystem is to facilitate information exchange between the microprocessors. The information exchange can occur

- Between RAM of microprocessors using the common jump-address fields
- Between RAM of microprocessors using the standard information channels based on the "channel–channel" type adapters
- Between EM devices using the standard information channels based on the total EM device control

The second type of multicomputer subsystems are assigned to increase the system performance by a simultaneous solution of independent paths of parallel digital signal processing algorithm. These subsystems possess programmable structure. As a rule, the second type of multicomputer subsystems consist of the same type microprocessors or they are homogeneous. Regularized programmable communication channels that are organized using the standard information channels and switchboards carry out functional interaction between microprocessors. The switchboard and the microprocessor with a block of subsystem operation realizations (BSOR) represent the elementary cell of the homogeneous multicomputer subsystem [1–3]. Combining elementary cells in the multicomputer subsystem can be carried out by ring coupling or by routing switching. One example of the ring coupling homogeneous multicomputer subsystem is shown in Figure 8.2. Switchboards S_i consist of gates opening or closing the communication channel running to a neighbor cell at the right side. BSOR consists of the adjustment register and block realizing the subsystem operations. The adjustment register content defines the switchboard function type and the degree of participation of the corresponding elementary cell while carrying out some instructions in accordance with the digital signal processing algorithm at each step. The homogeneous multicomputer subsystem operations are as follows:

- *Adjustment operation* assigned to program a structure of connections between the elementary cells
- *Operation of information exchange* assigned to organize the information exchange between the elementary cells of multicomputer subsystem
- *Operation of generalized conditional jump* to control a computation process during joint functioning of microprocessors in the multicomputer subsystem
- *Operation of generalized unconditional jump* to provide a hyphenation of computational data from one cell to other cells of the homogeneous multicomputer subsystem

The homogeneous multicomputer subsystem constructed based on these principles allows us to realize any digital signal processing algorithms. In other words, this is a universal subsystem in algorithmic sense. These homogeneous multicomputer subsystems do not have any limitations in the performance of computing complex digital signal processing algorithms and allow us to ensure the required reliability and subsystem survival in case there are unlimited number of elementary cells (microprocessors). A main task with designing homogeneous multicomputer subsystems is

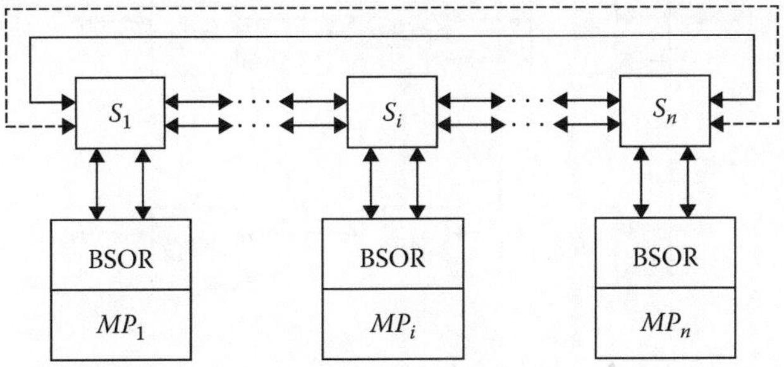

FIGURE 8.2 Ring coupling homogeneous multimicroprocessor subsystem.

to define the number of elementary cells, which ensures the realization of complex digital signal processing algorithms in real time and required operational reliability taking into consideration the multicomputer subsystem losses in the effective speed of operations.

8.1.3 Multimicroprocessor Subsystems for Digital Signal Processing

Computer subsystem consisting of several microprocessors coupled with each other by the common RAM and interface is called the *multimicroprocessor subsystem*. As a rule, the multimicroprocessor subsystem is designed based on the homogeneous single-microprocessor subsystems. Homogeneity ensures their interchangeability that allows us to increase efficiently the reliability and survival of the multimicroprocessor subsystem. The presence of common RAM that is equally available for all microprocessors reduces costs of the multimicroprocessor subsystem, which are required for data exchange in the course of parallel computation. The multimicroprocessor subsystem performance increases compared to the single-microprocessor subsystem due to simultaneous digital signal processing of several problems or parallel digital signal processing of some parts of the same problem. To use effectively the multimicroprocessor subsystem we need to separate a whole set of problems that must be solved on a set of subprograms that can be used in parallel and, in this case, the information delivered from the same RAM. In this case, a computational ganging is carried out by the unified control instruction stream for all microprocessors compared to the multicomputer subsystems. The multimicroprocessor subsystem programming differs from the multicomputer subsystem one and is a specific problem associated with systems programming theory and technique.

At the present time, the homogeneous multimicroprocessor subsystems for digital signal processing are designed based on two structural types: in the case of parallel digital signal processing—the matrix and associative structure, and in the case of sequential digital signal processing—the backbone structure. The matrix of homogeneous multimicroprocessor subsystems possess a single control block (CB) and a set of microprocessors combined into a matrix form (see Figure 8.3). Each microprocessor has its own RAM and operates with internal data, and the multimicroprocessor subsystem operates with larger arrayed data. There are the central control block (CCB) and memory device that store the data and programs in the homogeneous multimicroprocessor subsystem structure. In addition, the common memory may be included in the homogeneous multimicroprocessor subsystem structure

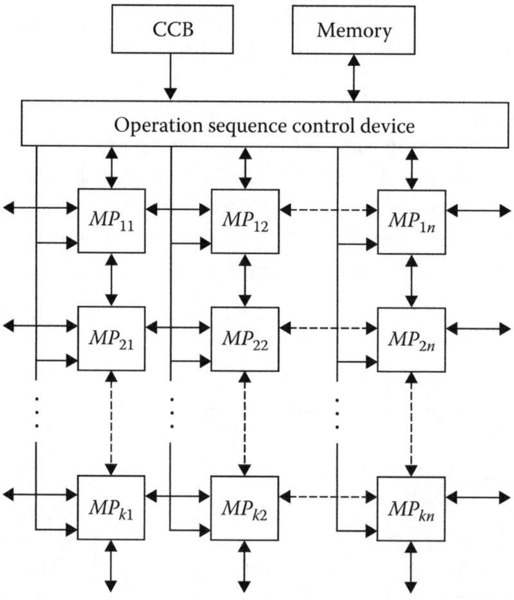

FIGURE 8.3 Matrix homogeneous multimicroprocessor subsystem.

and all microprocessors are able to address the common memory following the assigned regulations. Description of the matrix homogeneous multimicroprocessor systems is widely presented in literature [4–8]. Associative parallel homogeneous multimicroprocessor subsystems differ from the matrix ones by the presence of the so-called associative memory, that is, the memory where the data are selected based on the data content, not data address. The homogeneous multimicroprocessor subsystems with parallel structure are assigned to solve the problems possessing a natural parallelism [9] since all microprocessors of this system carry out the same operation simultaneously (each microprocessor uses its own data). These subsystems can be used in CRSs used for digital signal processing operations, including spatial-time signal processing, filtering of the target trajectory parameters by recurrent filters realized in the matrix form, and others where the linear algebra operations must be done, for example, the multiplication of vectors and matrices, the matrix inversion, and so on.

The backbone multimicroprocessor subsystem combines several independent microprocessors coupled with each other in such a way that information at the output of one microprocessor comes in at the input of another microprocessor; that is, the microprocessors process information sequentially or in conveyer style. The conveyer style of digital signal processing is based on partition of the digital signal processing algorithm on a set of steps and matching in time these steps when the digital signal processing algorithm is accomplished. Advantages with implementing the backbone multimicroprocessor subsystems compared to the matrix homogeneous multimicroprocessor ones are the moderate requirements with respect to inner coupling and the simplicity with which the backbone multimicroprocessor subsystem's computational power can be increased. Further disadvantages are removed by organizing priority exchange of information between the system cells.

Two types of backbone multimicroprocessor subsystems are widely used for digital signal processing in real time [10–12]: The first type is a one-dimensional inner long-distance channel operating in the time-sharing mode (see Figure 8.4), and the second type is a multidimensional inner long-distance channel coupled with two-input RAM, in particular (see Figure 8.5). In the backbone multimicroprocessor subsystem of the first kind, all RAM and read-only memory (ROM), central microprocessor (CMP), and specific microprocessors (SMPs) are coupled by the same channel. There is a block to control the channel. This block is used in the case of conflicts when several microprocessors address RAM simultaneously and for controlling data upload and extraction using the channel units to exchange data.

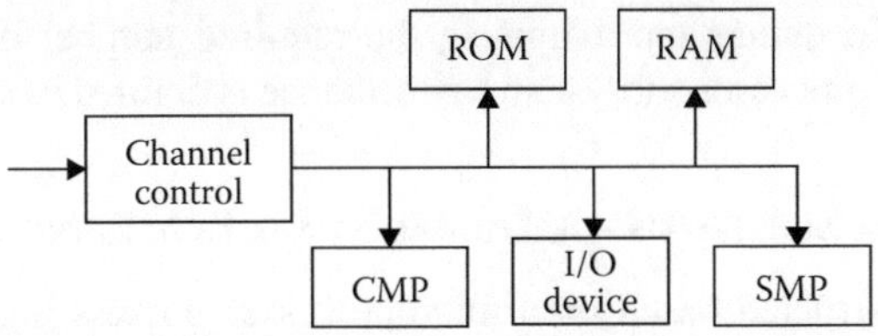

FIGURE 8.4 Backbone multimicroprocessor subsystem with a single inner long-distance channel.

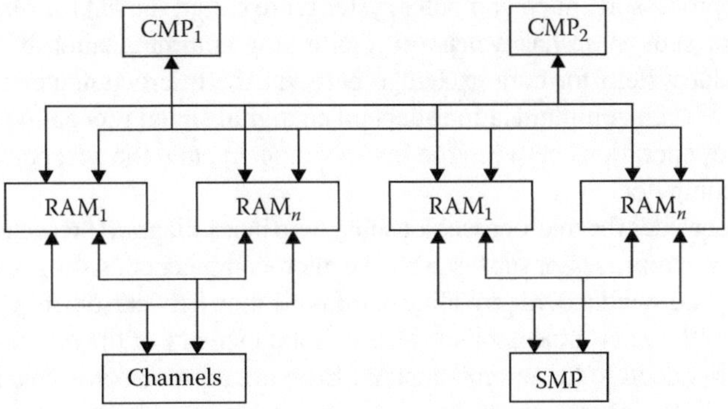

FIGURE 8.5 Backbone multimicroprocessor subsystem with multidimensional inner long-distance channel.

Operation of a single backbone multimicroprocessor subsystem is carried out in the following way. Each microprocessor submits, as needed, an application to the channel to address to RAM or ROM. If the channel is free then the microprocessor can access RAM or ROM immediately. Otherwise, the microprocessor is in the idle mode. The microprocessor can access RAM or ROM when the previous queries from other microprocessors are ended and there is no query from the microprocessors with high priority. Consequently, there are high requirements of the information channel with respect to the data throughput. Determination of the required throughput is carried out for each specific case, using the queuing theory methods. Note that the mentioned structure of the backbone multimicroprocessor subsystem has a limited speed of operation defined by the data channel throughput and, in addition, with the data channel failure the backbone multimicroprocessor subsystem stops any operation. A great advantage of the backbone multimicroprocessor subsystem is its simplicity.

In the backbone multimicroprocessor subsystem with the multidimensional inner long-distance channel (see Figure 8.5) all microprocessors operate using the independent asynchronous mode and all conflict situations are practically excluded. This system has high reliability and allows us to increase the performance without any limitations. Disadvantage of the backbone multimicroprocessor subsystem with the multidimensional inner long-distance channel is the high complexity caused by the multiinput RAM and ROM use. While designing the multimicroprocessor subsystem, the main problem is to provide the required performance by selecting the number of microprocessors to be included in this subsystem. With an increase in the number of microprocessors in the multimicroprocessor subsystem structure, a portion of overhead caused by a waiting time while the microprocessor addresses RAM or ROM and simultaneously addressing common tables and operational systems also causes an increase in the timetables related to supervisory routing. The total performance of multimicroprocessor subsystem is determined in the following form:

$$\mathcal{Q} = MV'_{\text{eff}}(1-\eta), \tag{8.9}$$

where

V'_{eff} is the effective speed of operation of a single microprocessor
η is the coefficient of relative performance loss

Equation 8.9 allows us to define approximately the required number of microprocessors if the required total multimicroprocessor subsystem performance is defined in advance.

8.1.4 Microprocessor Subsystems for Digital Signal Processing in Radar

The microcomputer constructed based on a microprocessor possess high reliability and flexible universality; however, speed of operation is low. In this case, the required performance in the course of the digital signal processing in a CRS can be reached by multiple microcomputer systems using several microprocessors. Such computer systems are called the *microprocessor systems.* Any such microprocessor subsystem has a network facilitating communication between the elements of microprocessor subsystem for data exchange between the microcomputers. Configuration and level of complexity of such communications depend on digital signal processing algorithms used by CRSs, distribution of operations between the microcomputers, and the acceptable number of RAM used by one microcomputer.

For an example, consider the microprocessor subsystem block diagram presented in Figure 8.6. The main element of this microprocessor subsystem is the microcomputer consisting of the microprocessor, RAM, I/O interface, and switchboard providing communication between the microcomputer elements and other elements of the microprocessor subsystem. Total memory of the microprocessor subsystem is based on the RAM added to the microcomputer. Each microprocessor is able to address the local RAM or RAM of other microcomputers by means of the address translation controller (ATC). The process of any microprocessor addressing the microprocessor subsystem's total memory is facilitated

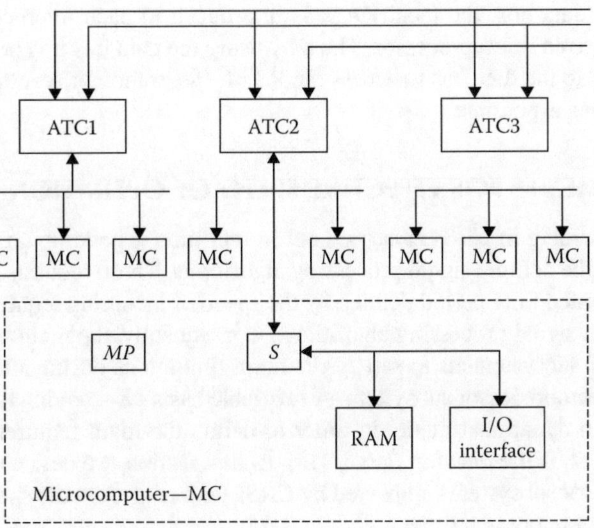

FIGURE 8.6 Example of microprocessor subsystem.

in such ways. Addressing mechanism is absolutely independent of RAM topology in the microprocessor subsystem. Access time is a function of distance to the addressed RAM cell. Communication of microcomputers with ATC is carried out by the one-channel data bus in the time-sharing mode. In doing so, we assume that the microprocessor will address mainly the local RAM; that is, the microprocessor will not use the data bus between microcomputers. However, to ensure the required reliability of the microprocessor subsystem and the possibility to increase it, the microcomputers are coupled into blocks consisting of 1–14 microcomputers.

The considered microprocessor subsystem structure possesses a high level of complexity of communication between the microprocessor and RAM, which leads to losses in the effective speed of operation. The high performance can be reached only when there is high efficiency of interaction between the microcomputers. In particular, the data bus overloading should be prevented. For this purpose, we need to ensure that, on priority, each microprocessor addresses local RAM in the course of the parallel digital signal processing. Another example of the microprocessor subsystem is depicted in Figure 8.7. In this microprocessor subsystem, an asynchronous communication between the microcomputers and RAM is used. If several microprocessors are coupled to the data bus they work using the time-sharing mode. Operational efficacy depends on channel throughput.

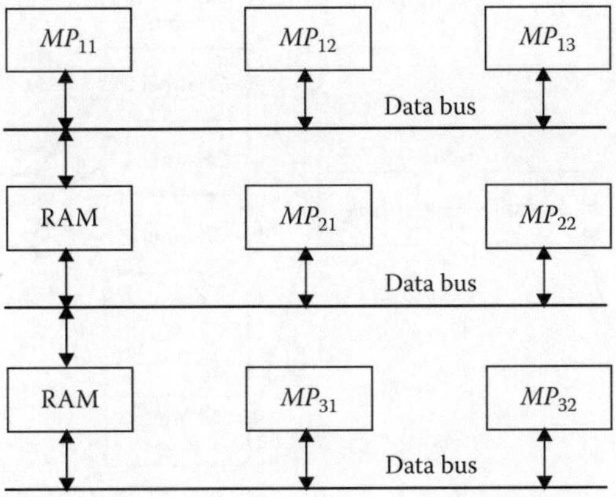

FIGURE 8.7 Example of microprocessor subsystem.

For unloading the data bus, the local RAM is introduced to each microcomputer accessible for addressing by the given microprocessor. Thus, by using the data bus it is possible for other microprocessors coupled to the data bus to access the RAM. Integration of several one-channel data bus into a single data bus is possible.

8.2 REQUIREMENTS FOR EFFECTIVE SPEED OF OPERATION

Digital signal processing in CRSs is carried out in real time depending on the speed of incoming requests to realize the definite signal processing algorithms. These requests would be satisfied by a CRS within the limited time period defined by the speed of incoming requests from corresponding stages of the digital signal processing algorithms. Consequently, the main problem with designing the microprocessor subsystems is to satisfy the main limitations during attended time of requests coming into the microprocessor subsystem. A favorable basis for studying the microprocessor subsystem operation in dynamical mode, in order to define the main requirements to structure and technical parameters, is the *queuing theory* [13]. In this section, we define the main specifications of the microprocessor subsystems employed by CRSs with respect to the speed of operation during digital signal processing in real time.

8.2.1 Microprocessor Subsystem as a Queuing System

The queuing system (QS) interacts with sources of requests for queuing. In the discussed case, the targets and other interferences in the radar coverage are considered as the sources of queuing requests (see Figure 8.8). These sources interact with the QS through the request sensor—the CRS plays this role. The radar system transforms the queuing request into signals that are subjected to process. Time sequence of these signals, which is ordered during radar sensing and scanning of the controlled space, forms an input request flux for the QS.

In general, the input request flux is considered as a stochastic process given by the probability density function of interval duration between the instants of two neighboring requests. The initial input flux of requests for a CRS is the target set in the radar coverage. At the initial designing stage, we can think that pdf of time intervals τ_{tg} between the neighboring targets is given by the exponential function:

$$p(\tau_{tg}) = S_{tg}\exp\{-S_{tg}\tau_{tg}\}, \tag{8.10}$$

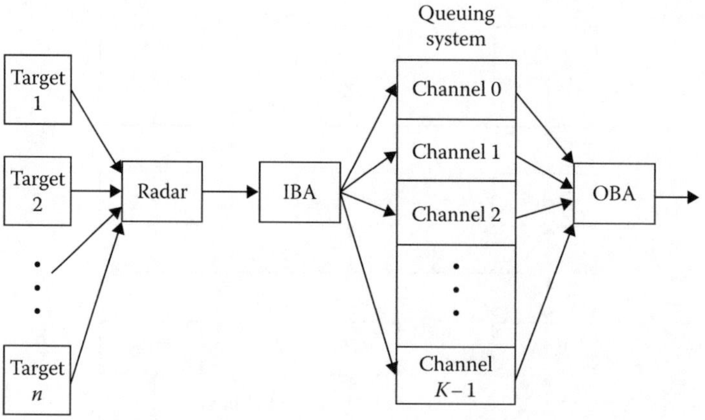

FIGURE 8.8 Radar coverage as a queuing system.

where

$$S_{tg} = \frac{1}{\overline{\tau}_{tg}} \tag{8.11}$$

is the information density equal to the average number of target movements within the radar coverage along the external border per unit time; $\overline{\tau}_{tg}$ is the average time interval between two considered events (the target traverses). The flux with pdf of time intervals τ_{tg} given by (8.10) is called the Poisson flow. If additional requirements of stationarity, ordinariness, and absence of sequence are satisfied, then the Poisson flow is called the *simplest* Poisson flow [14–16]. The average number of targets in the radar coverage is given by

$$\overline{N}_{tg} = S_{tg}\overline{t}_{rc}, \tag{8.12}$$

where $\overline{t}_{rc}$ is the average time of target being within the limits of radar coverage. Furthermore, we assume that the targets are distributed uniformly within the limits of radar coverage with the same pdf per unit volume. At the input of a digital signal processing subsystem (the radar receiver) of a CRS, the spatial pattern of targets distribution within the limits of the radar coverage is transformed into the time sequence of signals adhering to process. Henceforth, this signal flux is considered as the simplest with the intensity

$$\gamma_{tg} = \overline{N}_{tg}T_{scan}, \tag{8.13}$$

where T_{scan} is the period of radar coverage scanning.

In many cases, the hypothesis about the simplest flux of requests at the digital signal processing subsystem does not correspond to parameters and characteristics of real incoming signal flux. Nevertheless, an implementation of this hypothesis is considered acceptable for designing radar systems by the following reasons:

- First, analysis of QS is very simple.
- Second, because the simplest incoming signal flux is a very hard problem for such QSs, the CRS designed based on the simplest flux representation ensures successful service of other possible signal fluxes with the same intensity.

Along with the target return signals, the interferences (the false signals) caused by the internal and external noise sources come in at the input digital signal processing subsystem as requests for service. This interference (the false signals) flux can also be considered as the simplest flux with definite assumptions.

The incoming flux request service is organized by the QS. The QS can be considered as both the device carrying out a direct request service and the device assigned to store a request queue adhering to service. The QS is characterized by (see Figure 8.8) the following:

- The number of devices (cells) assigned to store a request queue of the incoming signal flux (the input buffer accumulator (IBA) memory size)
- The number of devices (channels) K that can serve multiple requests simultaneously
- The number of devices (cells) that accumulate and store the digital signal processing results (the IBA memory size)

The request queue is stored by the microprocessor subsystem IBA. The waiting time and the number of queued requests (the request queue length) are the random variables depending on statistical parameters of the incoming signal flux and service rate. Because the IBA size (memory capacity) is limited, the request queue length and, consequently, the waiting time of request queue are limited too. This QS is called the *queuing system with the limited queue* (the limited waiting time). Thus, the digital signal processing subsystem can be considered as the subsystem belonging to the class of QSs with limited queue.

The request queue in the digital signal processing subsystem of a CRS is realized by the microprocessors that can belong to a single or a set of parallel channels and is reduced to a single program realization of the corresponding digital signal processing algorithms. The time requested for each realization of the digital signal processing algorithm is the random variable owing to many reasons and, consequently, can be given statistically only. If we denote the time of a single request queue as τ_{queue}, then the sufficient characteristic of τ_{queue}, as a random value, is the pdf $p(\tau_{\text{queue}})$. QSs with the exponential pdf of τ_{queue}

$$p(\tau_{\text{queue}}) = \mu \exp\{-\mu\tau_{\text{queue}}\} \tag{8.14}$$

play a specific role in the queuing theory, where

$$\mu = \frac{1}{\bar{\tau}_{\text{queue}}} \tag{8.15}$$

is the queue intensity. The request queue approximation by the exponential pdf in the digital signal processing subsystem of a CRS allows us to carry out an analytical estimation of statistical characteristics of the request queue process during the simplest incoming signal flux using a sufficiently easy way.

If the request queue in the digital signal processing subsystem is carried out by K identical channels or microprocessors coupled with these channels and each channel or microprocessor has the exponential pdf of τ_{queue}, then the pdf of τ_{queue} for whole digital signal processing (microprocessor) subsystem is determined in the following form:

$$p(\tau_{\text{queue}}) = \frac{K}{\bar{\tau}_{\text{queue}}} \times \frac{(\tau_{\text{queue}})^{K-1}}{(K-1)!} \exp\left\{-\mu^K \tau_{\text{queue}}\right\}, \tag{8.16}$$

where

$$\bar{\tau}_{\text{queue}} = \frac{1}{\mu K} \tag{8.17}$$

is the average time queue. Distribution law given by (8.16) is called the *Erlang distribution* (pdf). At $K = 1$, the Erlang distribution is transformed into exponential pdf. Other distribution laws that can be represented by some exponential functions with different decay rates are used too. Presentation of distribution laws based on combination principle of exponential components allows us to approximate the request queue by Markov process and to obtain an analytical solution of this problem. Unfortunately, in CRS digital signal processing subsystems, the pdf of τ_{queue} differs from the exponential pdf and cannot be reduced to Erlang distribution law. For this reason, the pdfs of τ_{queue} must be analyzed for each specific case of designing the CRS digital signal processing subsystem.

Finally, let us discuss some observations about the flux forming at the digital signal processing subsystem output, that is, the output flux. At first, the output flux is stored by the output buffer accumulator (OBA). Thereafter, the output flux with the given rate is transferred to the user. As a rule, the digital signal processing of target return signals is realized by a set of serial microprocessor subsystems. In the course of the digital signal processing, we need to make transformations of fluxes; for instance, the signal flux must be transformed into the coordinate flux or the detected target pip flux that must be transformed into the flux of target tracking trajectory parameters and so on. In doing so, in the course of the digital signal processing, the output flux of the previous stage is the input flux of the next stage.

In the considered case, the output fluxes at each stage of the digital signal processing, as well as the input fluxes, can be considered as the simplest fluxes and their intensities can be determined using the estimated parameters of target and noise situations within the limits of the radar coverage. For example, the flux density of true target pips at the digital signal preprocessing subsystem output is defined by the number of targets within the limits of the radar coverage, the probability of target detection, and the scanning rate. Analogously, the flux density of false target pips is defined by the noise environment within the limits of the radar coverage, the scanning rate, and so on. Knowledge about the output fluxes is required to establish the necessary IBA size (capacity) for designing the information buses to exchange information between the microprocessors in the microprocessor subsystem and the communication channels to transfer information to the user.

8.2.2 Functioning Analysis of Single-Microprocessor Control Subsystem as Queuing System

Assume that the digital signal processing subsystem is realized based on a single-microprocessor subsystem (see Figure 8.9). We think that the request flux comes in at the input of the digital signal processing subsystem. The request flux is the multidimensional process consisting of M components corresponding to the given number of digital signal processing algorithms assigned to be realized

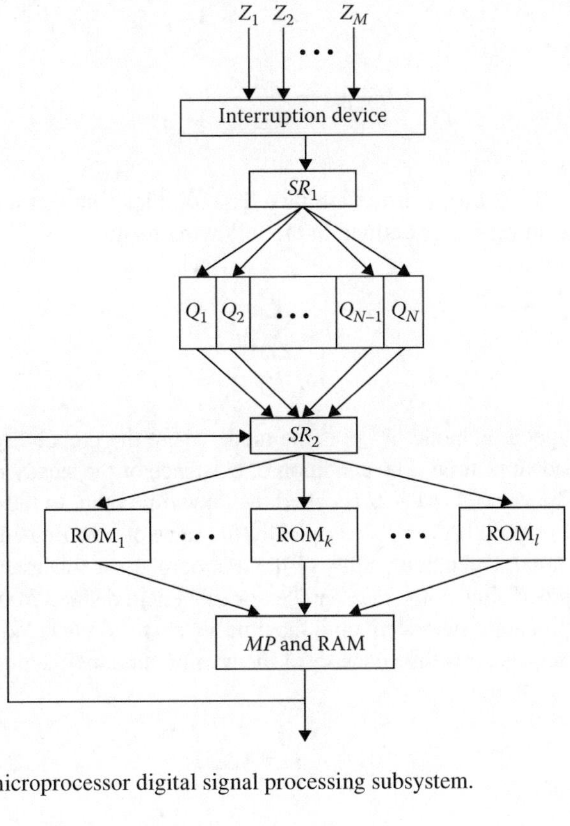

FIGURE 8.9 Single-microprocessor digital signal processing subsystem.

by a single-microprocessor subsystem. The requests $z_1, z_2, \ldots, z_M$ come in at the input of interruption device included into the microprocessor structure. When the request z_i comes in at the input of interruption device, this device interrupts the computation process and transfers control to the supervisory routine SR_1 that organizes the receiving process and the request z_i queue in accordance with assigned priority if this request z_i does not belong to immediate processing. The queuing process is stored by the buffer accumulator (BA). In doing so, the groups of cells $Q_1, Q_2, \ldots, Q_N$ form regions where the requests will be stored with corresponding priority. Within the limits of each priority zone, the requests are written on the first come first ordering basis. After receiving and queuing, the control is transferred to the supervisory routine SR_2 that organizes the digital signal processing of requests from queuing using the microprocessor.

The process of selecting a new request from a set of requests waiting in queue is the following. After the digital signal processing of immediate request, the supervisory routine SR_2 investigates sequentially queuing $Q_1, Q_2, \ldots, Q_N$ and selects for service the request z_k possessing the greatest priority. After initiation of the corresponding routine R_k, it is realized by the microprocessor. The queued request z_k leaves the system and the control is transferred to the supervisory routine SR_2 again. If there are no further requests, then the supervisory routine SR_2 switches on the microprocessor in the idle mode. At each instance, the microprocessor can run a single program only. For this reason, the considered QS is called a single-channel or a single-microprocessor QS.

One of the important quality parameters of the microprocessor subsystem as a QS is the capacity factor that characterizes the time interval within the limits of which the QS (microprocessor) processes a request and simultaneously the probability that the QS works (no idle mode) at the present time. The input flux with the density λ_k can be determined in the following form:

$$\rho_k = \frac{\lambda_k}{\mu_k}, \tag{8.18}$$

where

$$\mu_k = \frac{1}{\tau_{\text{queue}\,k}} \tag{8.19}$$

is the intensity of the request queue of the kth flux. The total loading factor of the microprocessor subsystem by all M incoming flux is defined in the following form:

$$U = \sum_{k=1}^{M} \rho_k. \tag{8.20}$$

The steady stationary operation mode of QS is the mode where the probability characteristics of QS functioning are independent of time. The condition of existence of the steady mode is defined by the loading factor $U < 1$. The value $\mathfrak{Q} = 1 - U$ is called the downtime ratio. In the steady QS functioning mode the down ratio is positive, that is, $\mathfrak{Q} > 0$. Logically, the value of $\mathfrak{Q}$ facilitated by the microprocessor subsystem must be minimal. Operation quality of the microprocessor subsystem as the QS is defined by the time of request processing by the microprocessor subsystem defined from the instant of request coming in at the microprocessor subsystem until the time a service is concluded by the microprocessor subsystem. For the kth request, this time consists of the waiting time for queuing $t_{\text{wait}\,k}$ and the queuing time $\tau_{\text{queue}\,k}$:

$$\overline{\tau}_{\Sigma k} = \overline{t}_{\text{wait}\,k} + \overline{\tau}_{\text{queue}\,k}, \tag{8.21}$$

where $\overline{t}_{\text{wait } k}$ and $\overline{\tau}_{\text{queue } k}$ are the average values. The waiting time $\overline{t}_{\text{wait } k}$ depends on the regulation applied to the request queuing—the order of multidimensional flux request selection for queuing from the total number of requests. There are the following regulations for request queuing:

- Nonpriority request queuing: The queuing in the order of request coming in at the QS, that is, in other words, "who could come early, he is the first."
- Relative priority request queuing: Priority for queuing is taken into consideration only at the instant to be served.
- Absolute priority: The incoming request with higher priority interrupts the queued request with lower priority.

During nonpriority request queuing, the average waiting time for all requests is the same and given by

$$\overline{t}_{\text{wait}} = \sum_{k=1}^{M} \frac{1+\upsilon_k^2}{2(1-U)} \rho_k \overline{\tau}_{\text{queue } k}, \tag{8.22}$$

where

$$\upsilon_k = \frac{\sigma_{\text{queue } k}}{\overline{\tau}_{\text{queue } k}} \tag{8.23}$$

is the variation factor defined by the ratio between the root mean square deviations of $\tau_{\text{queue } k}$ to its average value $\overline{\tau}_{\text{queue } k}$. In accordance with (8.22), the value of $\overline{t}_{\text{wait}}$ will be minimum at the constant value of $\tau_{\text{queue } k}$, that is, at $\upsilon_k = 0$. In the case of exponential pdf of $\tau_{\text{queue } k}$, $\upsilon_k = 1$ and, consequently, there is a twofold increase in $\overline{t}_{\text{wait}}$ when $\upsilon_k = 0$.

The widely used regulation in the information subsystem with the request queuing in CRSs is the regulation with the fixed relative priorities for each component of request flux. In this regulation, an appearance of the request with high priority does not interrupt the request queuing process with lower priority if it has been started. In this case, the average waiting time for the ith request independently on distribution law of $\tau_{\text{queue } i}$ is given by

$$\overline{t}_{\text{wait } i} = \sum_{k=1}^{M} \frac{1+\upsilon_k^2}{2(1-U_{i-1})(1-U_i)} \rho_k \overline{\tau}_{\text{queue } k}, \tag{8.24}$$

where

$$U_{i-1} = \rho_1 + \rho_2 + \cdots + \rho_{i-1} \tag{8.25}$$

is the loading factor of the microprocessor subsystem created by signal fluxes with priorities higher than i; the higher priority, the lesser i;

$$U_i = \rho_1 + \rho_2 + \cdots + \rho_i \tag{8.26}$$

is the loading factor created by signal fluxes with priorities not lower than i. Analysis of (8.24) shows that with a decrease in priority the waiting time to start the request queuing increases monotonically. Comparing the waiting time of requests with the relative priorities with the waiting time for the nonpriority request queuing, we can see that an introduction of relative priorities leads to a decrease in the waiting time for the request queuing with high priority owing to an increase in the waiting time for the request queuing with low priorities.

Level of total loading of the microprocessor subsystems with the QSs stimulates a waiting time for the request queuing with relative priorities. If the loading level of the microprocessor subsystems with the QSs is low, that is, $U = 0.2–0.3$, the presence of relative priorities for the request queuing has a very small effect on the waiting time between the requests of various signal fluxes. If the loading factor is close to unit, this effect becomes more essential and leads essentially to a reduction in the waiting time for the request queuing with high priorities owing to an increase in the waiting time for request queuing with low priority. In addition to the discussed regulations of requests queuing, in some cases, the regulations of requests queuing with absolute priorities are used for some components of the signal flux. Employing the request QSs with absolute priorities, the average waiting time is defined in the following form:

$$\overline{t}_{\text{wait } i} = \frac{U_{i-1}\overline{\tau}_{\text{queue } k}}{1-U_{i-1}} + \frac{1}{2(1-U_{i-1})(1-U_i)}\sum_{k=1}^{M}\left(1+\upsilon_k^2\right)\rho_k\overline{\tau}_{\text{queue } k}. \tag{8.27}$$

Evidently, an introduction of absolute priorities in the request QS leads to a reduction in the waiting time for the request queuing with high priorities, however with a simultaneous increase in the waiting time for the request queuing with low priorities.

Dependences illustrating the average waiting time of the request queuing as a function of priorities of the request queuing are shown in Figure 8.10. During the nonpriority request queuing (the curve 1, Figure 8.10), the average waiting time is constant. In the case of relative (the curve 2) and absolute (the curve 3) priorities, there takes place an increase in the waiting time for the high priority request queuing due to an increase in the waiting time for the low priority request queuing.

As a rule, we need to follow strong limitations with respect to the waiting time for request queuing of individual signal fluxes that require assigning them by absolute priorities, for example, the requests to process the target return signals. Other requests have excess waiting time and may be assigned with relative priorities. Several requests can be served by the simple queuing procedure. Thus, we see that we need to apply mixed regulations for request queuing, the investigation of which is carried out for each specific case by simulation procedures.

Referring to (8.21), we need to pay attention to the following fact. The average waiting time $\overline{t}_{\text{wait } k}$ is the constituent of the total request queuing time as well as the average request queuing time $\overline{\tau}_{\text{queue } k}$. However, the request queuing time does not vary for different regulations of the request QS. For this reason, the main time characteristic of the microprocessor subsystem operation is the average

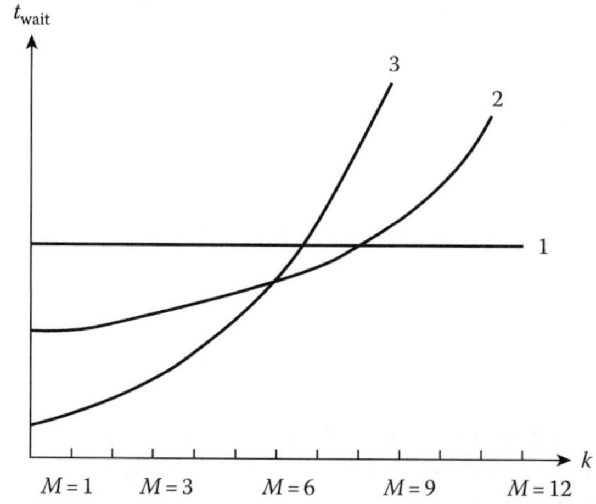

FIGURE 8.10 The average request queuing waiting time versus the request queuing priorities.

waiting time $\bar{t}_{\text{wait }k}$. Other parameters and characteristics of the request QS will be introduced as needed.

8.2.3 Specifications for Effective Speed of Microprocessor Subsystem Operation

In the considered case, the initial data that define the specifications for effective speed of the microprocessor subsystem operation include the following:

- The digital signal processing algorithm represented in the form of flow graph for each of the signal or request flux constituents
- The work content of constituents of the complex computational process algorithm expressed by the average number and variance of number of operations carried out for a single realization of these algorithms
- The type of the request QS of the designed microprocessor subsystem

Representation of the complex computational process algorithms and determination of the work content of these algorithms are carried out by methods discussed in Chapter 7.

The request queuing type or the type of microprocessor subsystem is defined by time requirements to process the requests by request QS or, independent of the request queuing type or the type of microprocessor subsystem, by the acceptable waiting time of the request QS. Accordingly, we call the microprocessor subsystems of the first type such microprocessor subsystems in which there are no limitations on the request queuing waiting time for all constituents of incoming signal flux. These microprocessor subsystems require the effective microprocessor operation speed that satisfies the request queuing within the finite time interval limits, the limiting value of which is unlimited in the considered case.

The request queuing waiting time is finite if the microprocessor subsystem works in the stationary mode, that is, if the total loading factor U of the microprocessor subsystem is less than the unit. The condition of stationarity takes the following form:

$$\sum_{k=1}^{M} \rho_k = \sum_{k=1}^{M} \lambda_k \bar{\tau}_{\text{queue }k} < 1. \tag{8.28}$$

The average queuing duration $\bar{\tau}_{\text{queue }k}$ is related to the work content $\bar{N}_{0k}$ and effective speed V_{eff} of the microprocessor subsystem operation in the following form:

$$\bar{\tau}_{\text{queue }k} = \frac{\bar{N}_{0k}}{V_{\text{eff}}}, \quad k = 1, 2, \ldots, M. \tag{8.29}$$

Substituting (8.29) into (8.28), we obtain

$$\frac{1}{V_{\text{eff}}} \sum_{k=1}^{M} \lambda_k \bar{N}_{0k} < 1 \quad \text{or} \quad V_{\text{eff}} > \sum_{k=1}^{M} \lambda_k \bar{N}_{0k}. \tag{8.30}$$

Equation 8.30 defines the minimum required speed of the microprocessor subsystem operation to establish the stationary operation mode of the microprocessor subsystem to realize the given set of digital signal processing algorithms in a CRS. If existing or assumed to be employed in the digital signal processing subsystems of CRS types of control single-microprocessor subsystems do not satisfy the main requirement given by (8.30), then the computation can be provided by the multimicroprocessor subsystems only. In this case, the parallel problem of computational process arises and we need to design the corresponding tools and facilities of parallel digital signal processing.

If the control single-microprocessor subsystem can satisfy the specifications and requirements for the effective speed of operation of the request QSs of the first type, we should study the possibilities of organizing the request QSs of the second and third types employing this control single-microprocessor systems. We call the control single-microprocessor subsystem of the second class subsystem if there are limitations on the average request queuing waiting time for all or several constituents of signal flux. The limitations are given in the following form:

$$\bar{t}_{\text{wait } k} \le t_k^*, \quad k = 1,2,\ldots,L, \quad L \le M, \tag{8.31}$$

where t_k^* is the limiting value of the average request-QS time for requests of the kth incoming signal flux.

Now, let us define the problem of the effective operation speed determination in the control single-microprocessor subsystem providing the request queuing with the given limitations and taking into consideration the fact that, similar to the request QSs of the first type, all requests are queued, with the so-called request QSs without loss and the average request queuing waiting time remaining limited. In the considered case, the minimal required speed of operation depends strongly on the request queuing regulation. If, for example, there is a nonpriority request QS, in which the average waiting time for all requests is the same and given by (8.22), the required effective speed of operation of the control single-microprocessor subsystem of the second type is determined by the following inequality:

$$\frac{\sum_{k=1}^{M} \lambda_k \bar{\tau}_{\text{queue } k} < 1\left(1 + \upsilon_k^2\right)}{2\left\{1 - \sum_{k=1}^{M} \lambda_k \bar{\tau}_{\text{queue } k}\right\}} \le t^* . \tag{8.32}$$

Solution of (8.32) can be presented in the following form:

$$V_{\text{eff}} \ge \frac{1}{2}\sum_{k=1}^{M} \lambda_k \bar{N}_{0k} + \sqrt{0.25\left\{\sum_{k=1}^{M} \lambda_k \bar{N}_{0k}\right\}^2 + (t^*)^{-1}\sum_{k=1}^{M} \lambda_k \bar{N}_{0k}\left(1 + \upsilon_k^2\right)}. \tag{8.33}$$

Based on (8.33), we are able to define the required speed of operation of the control single-microprocessor subsystem by the given (the same for the request queuing of all signal flux constituents) limitation on the average request QS waiting time during the nonpriority request queuing at the time of signal flux constituents coming in at the single-microprocessor subsystem input. In the limiting case, when there are no limitations in waiting time ($t^* \to \infty$), (8.33) transforms into the following form:

$$V_{\text{eff}} \ge \sum_{k=1}^{M} \lambda_k \bar{N}_{0k}, \tag{8.34}$$

that is, it coincides with (8.30). At $t^* < \infty$, the required effective speed of operation of the second-type control single-microprocessor subsystem is greater than that of the control single-microprocessor subsystem of the first type.

To organize the priority request queuing, for example, the relative priority, we need to take into consideration a difference in limitations applied to the average request QS waiting time of each priority in the second type of request QS. In principle, this problem can be solved by the following way. All requests are divided into groups with the same value of average waiting time. The loading factor and required effective speed of operation V_{eff} of the control single-microprocessor subsystem is defined for each request group. Making summation of partial values of the effective speed of operation, we obtain

$$V_{\text{eff}} = \sum_{p=1}^{P} V_{\text{eff}\,p}, \tag{8.35}$$

where P is the number of request groups with approximately the same acceptable average waiting time of request queuing. In the considered case of the second-type request QS, as well as in the case of the first-type request QS, a decision to realize the control single-microprocessor subsystem is made as a result of comparison of requirements and specifications to effective speed of operations of the request QS with effective speed of the CMP operation. The third type of request QS is called the microprocessor subsystem that generates limitations on acceptable (absolute) time to process the request by the request QS. In these cases, the limitations are given in the following form:

$$P\left(t_{\text{wait}\,k} > T_k^*\right) \le P^*, \tag{8.36}$$

where

T_k^* is the allowed waiting time for requests of the kth signal flux

P^* is the allowable probability to satisfy the inequality $t_{\text{wait}\,k} > T_k^*$

The request queuing waiting time probability distribution function of the total signal flux can be approximated with satisfactory accuracy by the following formula [17–19]:

$$P(t_{\text{wait}} \le \tau) = 1 - R\ \exp\left\{-\frac{R\tau}{\bar{t}_{\text{wait}}}\right\}. \tag{8.37}$$

In this case, the probability of exceeding the allowed waiting time T_k can be determined in the following form:

$$P(t_{\text{wait}} > T^*) = 1 - R\ \exp\left\{-\frac{RT^*}{\bar{t}_{\text{wait}}}\right\}. \tag{8.38}$$

For example, in the case of the nonpriority request queuing regulation, for which $\bar{t}_{\text{wait}}$ is given by (8.22), based on solution of the transcendent equation we can write

$$1 - R\ \exp\left\{-\frac{RT^*}{\bar{t}_{\text{wait}}}\right\} = P^* \tag{8.39}$$

We can obtain the required effective speed of operation of the control third-type single-microprocessor subsystem. Mathematical notation of (8.39) with respect to V_{eff} is tedious, and solution can be obtained only by numerical procedures.

The required effective speed down boundary definition of the third-type control single-microprocessor subsystem operation can be considered as the initial stage of designing the microprocessor subsystem for CRS digital signal processing. The next stage is a choice of the optimal request queuing regulation. At the system design stage, we will not have sufficient information to choose the specific request queuing regulation, but we should take into consideration the following circumstances:

- The request queuing regulation based on request queuing in turn to come in at the control single-microprocessor subsystem possesses the best performance among all the nonpriority request queuing regulations; in particular, the variance of the request queuing waiting time is minimal compared to other nonpriority request queuing regulations.
- Introduction of the priority request queuing regulation allows us to decrease essentially the request queuing waiting time for the most important requests of signal flux owing to increase in the delay during the queuing of signal minor fluxes.
- In general, an introduction of the priority request queuing regulations provides an equivalent performance of the control single-microprocessor subsystem compared to the request queuing one in turn. Consequently, the determined values of the required minimum speed of the control single-microprocessor subsystem operation makes it possible to choose the priority request queuing regulation without any additional increase in the effective speed of operation.

Now, let us discuss some ways of choosing the optimal speed of operation of the control single-microprocessor subsystem. The minimum value of the effective speed of operation ensures the boundary values of the request queuing waiting time of the earlier-considered types of the request queuing regulations for the control single-microprocessor subsystem and related to a queue length that can be defined for the kth signal flux as $l_k = \lambda_k t_{\text{wait }k}$. Queue existence and its storage are related to definite losses. To decrease these losses, the effective speed of control single-microprocessor subsystem operation must be greater than the minimal one. However, in this case, the loading factor U decreases and, correspondingly, the idle time factor $Q = 1 - U$ increases, which is not good from the point of view of hardware costs.

Evidently, we can select the optimal speed of operation taking into consideration these two contradictory factors. As the criterion of effectiveness we use the functional in the following form:

$$C_V = \beta_Q Q(V_{\text{eff}}) + \sum_{k=1}^{M} \beta_k \lambda_k \overline{t}_{\text{wait }k}(V_{\text{eff}}), \tag{8.40}$$

where

β_Q is the loss caused by idle mode

β_k, $k = 1, 2, \ldots, M$ are the losses caused by the queue length for each signal flux request queuing

The optimal value of the effective speed of control single-microprocessor subsystem operation can be obtained from solution of the following equation:

$$\frac{dC_V}{dV_{\text{eff}}} = 0 \quad \text{at} \quad \frac{d^2 C_V}{dV_{\text{eff}}^2} > 0. \tag{8.41}$$

Solution of (8.41) is not difficult, but a choice of losses β_Q and β_k is difficult. A single acceptable procedure for this choice is the procedure of expert judgments.

8.3 REQUIREMENTS FOR RAM SIZE AND STRUCTURE

Requirements for RAM size and structure are defined based on the end use of system, character and digital signal processing algorithms, and an analysis of input and output signal fluxes. As a first approximation, the total RAM size (memory capacity) is defined from the formula

$$Q_{\Sigma} = Q_{\text{routine}} + Q_{\text{digit}}, \tag{8.42}$$

where

Q_{routine} is the RAM cell array assigned to store the routines of digital signal processing algorithms, control routine of computational process and operation of whole system, interruption routines, and routines controlling calculations

Q_{digit} is the RAM cell array assigned to store a numerical information

In turn, Q_{digit} can be presented in the following form:

$$Q_{\text{digit}} = Q_{\text{in}} + Q_{\text{working}} + Q_{\text{out}}, \tag{8.43}$$

where

Q_{in} is the RAM cell array assigned to receive external information

Q_{working} is the RAM cell array participating at the calculation process

Q_{out} is the RAM cell array assigned to store the digital signal processing results

In many cases, the relative independence of routine information on numerical one leads us to expediently use individual permanent memory or ROM to store the routine information. ROM uses only a reading mode. Implementation of ROM gives us a great advantage because information stored by ROM is not lost even if the power is off. In addition, ROM possesses high reliability. As for ROM size (capacity), at the initial stage of designing the digital signal processing subsystem it is possible to estimate this parameter approximately. The exact estimation is possible only after program debugging.

Now consider some ways of choosing the size (capacity) of memory to be assigned to store the numerical information. The cell array Q_{out} forms a buffer memory. The buffer memory size depends on the number of external objects, information content incoming from each object within the limits of digital signal processing cycle, microprocessor operation speed, request queuing regulation, and so on. Consider the example of the control single-microprocessor subsystem receiving information from a 3D CRS with the space periodical scanning. Assume that the number of scanned targets is $\bar{N}_{\text{scan}}^{\text{tg}}$. Information about each target at the control single-microprocessor subsystem input (the target pip) includes the spherical coordinates of the current target position and the variance of error for the measurement of these coordinates. We assume that the coordinate code width and coordinate measurement errors are established at the previous stages of digital signal processing and is characterized by the number of binary digits n_{p} for each target pip. Digital signal processing procedure must be organized in such a way that information obtained during the previous cycle would be completely processed during the next cycle. Thus, during the cycle equal to the scanning period, $\bar{N}_{\text{scan}}^{\text{tg}}$ of n_{p}-digit words will be recorded in the buffer memory. If we assume that a target distribution in the radar coverage is uniform, the target pip flux at the buffer memory input conforms to Poisson law, and there is the nonpriority request queuing regulation at the control single-microprocessor subsystem, then the queue length of request queuing can be determined in the following form:

$$\bar{n}_{\text{wait}} = \bar{N}_{\text{scan}}^{\text{tg}} \bar{t}_{\text{wait}}, \tag{8.44}$$

where $\bar{t}_{\text{wait}}$ is the average waiting time of request queuing given by (8.22). Since the buffer memory can be considered as the multichannel request QS with losses, then, for the considered case of the Poisson input signal flux, we can use the Erlang formula to compute the number of buffer memory cells [13]:

$$P_{Q_{\text{buf}}} = \frac{\left\{\bar{N}^{\text{tg}}_{\text{scan}}\,\bar{t}_{\text{wait}}\right\}^{Q_{\text{buf}}}}{Q_{\text{buf}}!} = \frac{\dfrac{\left\{\bar{N}^{\text{tg}}_{\text{scan}}\,\bar{t}_{\text{wait}}\right\}^{Q_{\text{buf}}}}{Q_{\text{buf}}!}}{\sum_{k=0}^{Q_{\text{buf}}}\dfrac{\left\{\bar{N}^{\text{tg}}_{\text{scan}}\,\bar{t}_{\text{wait}}\right\}^{k}}{k!}}, \tag{8.45}$$

which relates the probability of request losses (target pips) $P_{Q\text{buf}}$ with the required number of buffer memory cells Q_{buf}. For the given $P_{Q\text{buf}}$ at the buffer memory input, we can determine Q_{buf} in n_{p}-digit words using (8.45). The buffer memory in the control single-microprocessor subsystem can be realized as individual module or as RAM array.

The cell array Q_{working} is the main constituent of RAM and is assigned to store the constants, initial data, and results of previous computations used in the next operations. In addition, there are working cells in the cell array Q_{working}, which are assigned to store the intermediate computational results during a sequential realization of individual constituents of the digital signal processing algorithm. Preliminary calculation of the required RAM size (capacity) for the realization of specific digital signal processing algorithm can be carried out in the course of the system design stage.

As an example, consider a determination of the RAM size (capacity) for the realization of the digital signal processing algorithm of recurrent linear filtering, that is, the Kalman filter. The formulation given by the equation system (7.49) can be considered as the initial data for the Kalman filter. All initial data are represented in Table 8.1. Based on the data from Table 8.1 we are able to determine the required number of RAM cells for the filtering of a single target

$$Q'_{\text{RAM}} = 2s^2 + s(2m + h + 1) + m(m + 1) + h. \tag{8.46}$$

If the digital signal reprocessing subsystem is assigned to track $\bar{N}^{\text{tg}}_{\text{scan}}$ targets, then the total number of required RAM cells is determined as $Q_{\text{RAM}} = Q'_{\text{RAM}}\bar{N}^{\text{tg}}_{\text{scan}}$. A decrease in this number can be reached owing to the constant parameters of targets; sparseness of the matrices $\mathbf{\Phi}_n$, $\mathbf{\Gamma}_n$, $\mathbf{\Psi}_\eta$, $\mathbf{H}_n$; and

TABLE 8.1
Initial Data and Required Number of RAM Cells to Store the Computational Results

Initial Data	Dimension	Number of RAM Cells
$\mathbf{\Phi}_n$	$s \times s$	s^2
$\mathbf{\Gamma}_n$	$s \times h$	sh
$\mathbf{\Psi}_{\eta-1}$	$h \times h$	h^2
$\mathbf{R}_n$	$m \times m$	m^2
$\boldsymbol{\eta}_n$	$h \times 1$	h
$\mathbf{H}_n$	$m \times s$	ms
$\mathbf{\Psi}_{n-1}$	$s \times s$	s^2
$\mathbf{G}_{n-1}$	$s \times m$	ms
$\boldsymbol{\theta}_{n-1}$	$s \times s$	s
$\mathbf{Y}_n$	$s \times 1$	m

specific procedures to form arrays. In addition to the required number of RAM cells determined by (8.46), an array of working cells to store the intermediate results of computations needs to be envisaged. In the case of the considered example, during the sequential realization of the linear filtering algorithm in accordance with the graph depicted in Figure 7.13, the number of cells in this array is not more than s^2. RAM is the main memory of any microprocessor subsystem and is constructed in the module form.

The cell array Q_{out} serves to store the numerical information assigned to transfer results of the digital signal processing of the target return signals to external objects. This cell array, as well as the cell array Q_{in}, is the BA to communicate with users. For this reason, a determination of the number of array cells, that is, the size (capacity) of the array Q_{out}, as well as the number of array cells of the array Q_{in} is carried out analogously. Finally, note that the specification of requirements to the RAM size is carried out simultaneously (in parallel) with the specification of requirements to the microprocessor subsystem operation speed.

8.4 SELECTION OF MICROPROCESSOR FOR DESIGNING THE MICROPROCESSOR SUBSYSTEMS

The task of choosing the special-purpose microprocessors is accomplished based on the requirements for making demands to the designed microprocessor subsystem in terms of speed of operation, RAM and ROM sizes, technical characteristics, reliability, overall dimensions, use, cost, and other requirements. Choosing the appropriate microprocessor depends on the degree of correspondence between a set (vector) of quality of service (QoS) of the selected microprocessor and a set (vector) of requirements to these quality indices. A numerical measure or criterion of effectiveness must be established to compare the selected microprocessors. However, at the initial stages of designing it is very difficult, as a rule, to establish a function between the parameters of microprocessor and generalized criterion of effectiveness as a whole. Moreover, selection of the generalized criterion, demonstrable and convenient in computational sense, is not easy. In what follows, we consider a simple procedure for selecting the appropriate microprocessor.

Let a nomenclature of microprocessors produced by industry form a set

$$\mathbf{M} = \{M_1, M_2, \ldots, M_i, \ldots, M_m\}, \tag{8.47}$$

where m is the number of microprocessor. Each element of the set (8.47) is defined by the totality

$$\mathbf{K}_{ij}^{(M)} = \{K_{i1}^{(M)}, K_{i2}^{(M)}, \ldots, K_{ij}^{(M)}, \ldots, K_{in}^{(M)}\}, \tag{8.48}$$

where n is the number of parameters taken into consideration. Let

$$\mathbf{K} = \{K_1, K_2, \ldots, K_j, \ldots, K_n\} \tag{8.49}$$

be the set of requirements to the microprocessor subsystem.

Selection of serviceable microprocessors is to find in the set $\mathbf{M} = \{M_i\}_1^m$ one or several microprocessors satisfying the given specifications and requirements $\{K_j\}_1^m$. At the same time, the following results of comparison between the microprocessor parameters and set of requirements $\mathbf{K}$ are possible:

- The only type of microprocessors from existing nomenclature satisfies all requirements.
- None of the microprocessors from the existing nomenclature satisfies all requirements.
- There are several types of microprocessors from existing nomenclature satisfying all requirements.

In the first case, the only microprocessor satisfying all requirements is selected to design the microprocessor subsystem. In the second case, we need either to correct the requirements based on simplification of digital signal processing algorithms for solved problems and change an environment or to make a decision to construct the multimicroprocessor subsystem based on the implementation of microprocessor of the same or different types produced by industry. In the third case, the problem of selecting the best microprocessor satisfying all requirements is solved. In this case, there are several ways by which one can make this choice. We consider the simplest way—the ranking way that is described as follows.

Let the microprocessor requirements be ranked in decreasing order of their importance, for example, in the following sequence: K_1 is the reliability, K_2 is the weight, K_3 is the power, and so on. The parameters of compared microprocessors are ranked in the corresponding order. Then if the first by rank (importance) parameter of some microprocessor is essentially better than others, then independent of the remaining parameters this microprocessor is considered the best. If several microprocessors have the same first parameter, then the second parameter is analyzed and the microprocessor with the highest second-rank parameter is considered as the best microprocessor and so on. This selection is continued until there is only one microprocessor left to be selected. Optimization procedure is multistage and involves sequential reduction of the number of considered microprocessors. In the theory of optimal system designing, this method is called the sequential increasing of resolution level of the applied criterion of effectiveness.

8.5 STRUCTURE AND ELEMENTS OF DIGITAL SIGNAL PROCESSING AND COMPLEX RADAR SYSTEM CONTROL MICROPROCESSOR SUBSYSTEMS

For designing the microprocessor subsystem structure, first of all, we should take into consideration the content and characteristics of problems that must be solved. Recall that the digital signal processing subsystem in an automated CRS must carry out the following problems:

- Intraperiod digital signal processing of the target return signals with the speed defined by the period of discretization t_d
- Interperiod digital signal processing of the target return signals with the speed defined by the period T of scanning the signals pulsing
- Intersurveillance digital signal processing of the target return signals with the speed defined by the period T_{sc} of scanning the radar coverage during target detection or the period of refreshment T_{new} during target tracking

Thus, for designing the microprocessor subsystem structure for CRS digital signal processing of target return signals we need to consider a sequence of digital signal processing step realization and difference in the scale of real time for each step in digital signal processing. An important initial condition for the synthesis of microprocessor subsystems is the selection of hardware and software that as a unit helps to solve the problems assigned for CRSs. A total set of microprocessor subsystem hardware can be divided on the following groups:

- Microprocessor subsystem facilities providing a realization of digital signal processing algorithms
- Communication facilities providing transmission of information from sources to users
- Facilities of transmission of information
- Interface and commutation facilities assigned to unify microprocessor subsystem facilities into multimicroprocessor subsystems for the purpose of increasing speed of operation and reliability of computations and numerical calculations

Naturally, the main microprocessor subsystem element defining its structure is a system of computing facilities. Two types of computing facilities are employed by digital signal processing subsystems, namely,

- The control microprocessor subsystem and multimicroprocessor subsystems for specific applications providing a realization of the main digital signal processing and control algorithms
- The special-purpose high-performance microprocessors assigned mainly for target return signal filtering at the intra- and inter-period digital signal processing stages [20–22]

Other foregoing hardware facilities are special purpose and assigned for SMP subsystems implemented in special-purpose CRSs. Questions of rational selection of hardware facilities for designing the microprocessor subsystem are essential because, in the case of special-purpose applications, dimensions and cost of such SMP subsystems outweigh the corresponding characteristics and performance of usual microprocessor systems.

Software is the programmable facility system assigned to increase the effectiveness with which the microprocessor subsystem is used and decrease the work content of preliminary operation for the solution of problems by the microprocessor subsystem. Software can be divided into internal and external software. In the control microprocessor, the internal software consists of the automized programming routine, operating system routine, that is, computational process control routine, and functioning control routine. The external software consists mainly of application program library and specific programs of CRS digital signal processing. Since the cost of designing software exceeds the cost of designing hardware, one of the main areas in microprocessor subsystem development is the realization of some typical software functions by hardware.

In general, the microprocessor subsystem structure is defined by the hardware facilities, for example, microprocessors, controllers, interface, and so on, and ways to combine the hardware in a system, to organize a computational process, to exchange information between some elements of the microprocessor subsystem, to expand the microprocessor subsystem for obtaining high performance, to organize logically combined operation of various elements of the microprocessor subsystem. On the basis of these general thoughts and taking into consideration the specific character of the solved problems, we can classify the microprocessor subsystems of CRS digital signal processing in the following form:

1. The microprocessor subsystems with complete host subsystem structure, for which there are the CMPs controlling digital signal processing and a set of interface processing modules (the microprocessors, the memories, etc.) operating under control of the CMP. In this case, the CMP must possess an extremely effective speed of operation (we consider an example of such microprocessor subsystem in the next section).
2. The microprocessor subsystems with the so-called *federal structure*, for which some special-purpose microprocessors assigned for digital signal processing operate in the free-running mode and can be considered as maintainable CRS equipment. Combination of information with the outputs of-line microprocessors and digital signal processing ending taking into consideration all interests of users are carried out by the CMP. Federated control of the microprocessor subsystems is carried out by the CMP.

   As an example, a microprocessor subsystem version with the federal structure for digital signal processing and control in CRSs for detection and target tracking is depicted in Figure 8.11. Independent devices of the microprocessor subsystem are the signal microprocessors carrying out a digital signal processing of target return signals received using an interface device. There are several such microprocessors and each of them operates in its own timescale

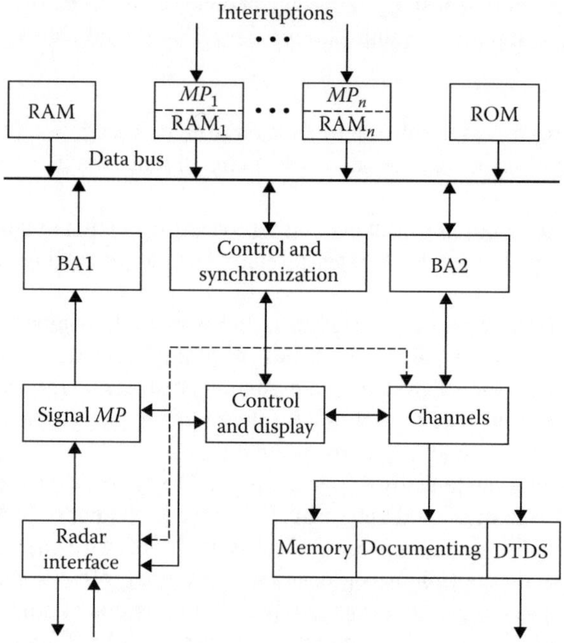

FIGURE 8.11 Microprocessor subsystem with federal structure.

in accordance with the realized step of digital signal processing. The processed information, that is, target pips, comes in at the input of central block over the buffer accumulator (BA1), in our case, the microprocessor special-purpose subsystem. The CMP solves a digital signal reprocessing and other problems interested for users are solved. The processed information comes over the BA2 and data bus at the inputs of the control and display subsystems, documenting and transferring data subsystems (DTDS). Additional digital signal processing is also realized by these subsystems using the off-line units. Control process in the considered microprocessor subsystem is carried out only for the purpose of operation plans and adaptation to the changing environment and character of solved problems. Control operation can be carried out by the individual control processors and synchronization of CRS hardware.

3. The microprocessor subsystems with a completely decentralized structure, both in space and functionality, in which the local digital signal processing blocks solve the problems in the off-line mode or under the control of one block. This microprocessor subsystem allows us to increase principally the digital signal processing performance owing to the specialization of blocks carrying out the numerical calculation and best matching their structure with specific character of the realized digital signal processing algorithms. An obvious disadvantage of this microprocessor subsystem is a complexity involved in controlling this subsystem, but there are no doubts that this disadvantage can be overcome and the distributed microprocessor subsystem of digital signal processing of target return signal would be widely used in practice.

The main problems with improving the structure and elements of the considered microprocessor subsystems are as follows:

- To design high-performance signal microprocessors for digital signal processing of target return signals taking into consideration solutions for both the considered and discussed problems and new problems. Obviously, one way to solve this problem is further specialization of the special-purpose microprocessors and introduction of parallel algorithms to solve the digital signal processing of target return signal problems.

- To design and construct high-performance parallel (matrix, conveyer, and other types) microprocessors based on the modern and perspective element base providing the required performance for the automation of all main problems of the digital signal processing of target return signal and control process.
- To design and construct the microprocessor subsystems with both uniformly and nonuniformly distributed structure satisfying the requirements of digital signal processing hardware unification oriented to the special-purpose applications.

Considered in this section perspective avenues to develop the hardware and structure of CRS microprocessor subsystems assume the maximal parallelism in computational processes that, in the final analysis, is reduced to constructive solution of operational problems of parallel programming. Planning methods of parallel numerical calculation processes in the multimicroprocessor subsystems are related to the modern problems.

8.6 HIGH-PERFORMANCE CENTRALIZED MICROPROCESSOR SUBSYSTEM FOR DIGITAL SIGNAL PROCESSING OF TARGET RETURN SIGNALS IN COMPLEX RADAR SYSTEMS

As an example of the high-performance centralized microprocessor subsystem for digital signal processing of target return signals, we consider the microprocessor subsystem with an ensemble of parallel microprocessors [23,24]. Parallelism of independent objects during the target tracking is used by this microprocessor subsystem. Computational subsystem consists of three main constituents (see Figure 8.12): the principal microprocessor, the N independent identical microprocessors called the microprocessor elements, and the central controller of the microprocessor subsystem. The principal microprocessor is the central element of the whole microprocessor subsystem and is assigned to solve all problems that are not related to the digital signal reprocessing of target

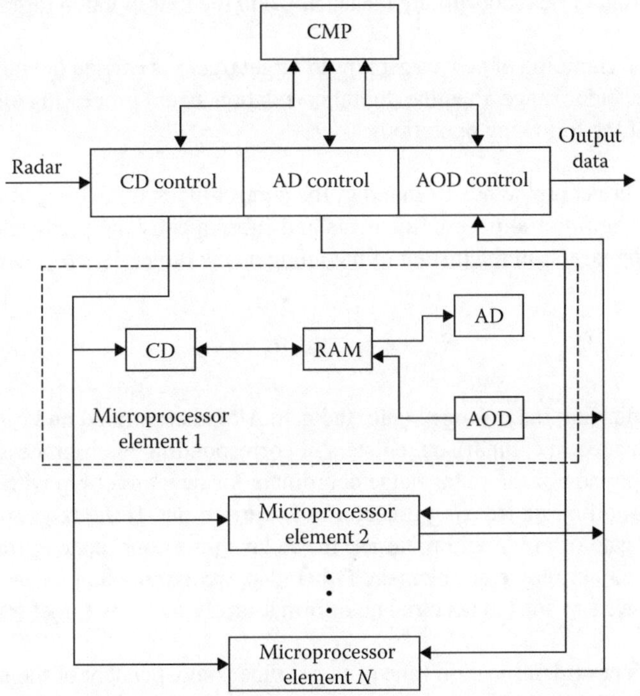

FIGURE 8.12 Microprocessor subsystem with ensemble of parallel microprocessors.

return signals. In addition, the principal microprocessor ensures all control functions of computational subsystem programs, including a translation of programs for all microprocessor elements. Independent and identical microprocessor elements, the number of which defines the microprocessor system performance, operate in parallel under the common control of the principal microprocessor. Each microprocessor element includes the following:

- The arithmetical device (AD) that is used for computations made by the microprocessor element.
- The associative output device (AOD) containing necessary circuits to select (to activate) the microprocessor element.
- The correlation device (CD) is the high-speed data input device with addressing by content, which is designed specially for target return signal data input.
- The RAM with arbitrary addressing.

The central control microprocessor coupling an ensemble of microprocessor elements with the principal microprocessor consists of three CBs that operate simultaneously. This way, a parallel digital signal processing of target return signal in AD can be matched in time with the associative data input over CD and data output over AOD. Thus, the microprocessor subsystem can realize $3N$ operations in parallel. The microprocessor subsystem shown in Figure 8.12 is assigned for the individual multiple targets tracking. For this purpose, the individual microprocessor element is assigned for each tracked target. During digital signal processing of the target return signals the problems of data input, linear filtering, CRS control, and data output are solved. New information about each tracked target in the form of the target pip coordinates is stored by the memory of corresponding microprocessor element using CD. Distribution of RAMs for each microprocessor element is the same. Each RAM is divided into three sections: to store new unprocessed data waiting for filtering, that is, the buffer of unprocessed data; to store the tracked target trajectory parameters obtained at the previous step of filtering; to store the requests queuing, that is, the instant of next coordinate measuring and the extrapolated target track parameters at this instant.

If we assume that clamping of new target pips to target tracks is carried out only by a single coordinate, for example, radar range, then the digital correlation signal processing algorithm is reduced to the fulfillment of the following operations:

- When a new target pip comes from radar, the control block of CD interrupts arithmetical operations for all microprocessor elements and subsequently the predictable radar ranges for all tracking targets at the instant of incoming a new target pip are determined:

$$\hat{R}_n^{\text{el}} = \hat{R}_{n-1} + R_{n-1}(t_n - t_{n-1}). \tag{8.50}$$

- For each predictable radar range value, the gate ΔR is determined and dimensions of the gate ΔR are loaded in comparison registers of corresponding microprocessor elements.
- Parallel comparison of the radar range coordinate for new target pip with the predictable coordinates of all target tracking trajectories is carried out. If the target pip is within the limits of the gate of any tracking target, the radar range coordinate is transferred to the corresponding microprocessor element. Otherwise, we assume that a new target has been detected and we can start to accumulate information about a new target track.

Thus, in this case, a new identification time of target pips is independent of the number of tracking targets. The target track filtering is carried out by each microprocessor element using iteration of the digital signal processing algorithms. For the realization of operation of selecting the next target for queuing, the AOD is used. Checking of all tracking target by any feature is carried out by the AOD in parallel. For this reason, the time required to search the most priority target for target queuing is the same as that is required for checking a single target.

8.7 PROGRAMMABLE MICROPROCESSOR FOR DIGITAL SIGNAL PREPROCESSING OF TARGET RETURN SIGNALS IN COMPLEX RADAR SYSTEMS

As an example of the high-performance CMP subsystem assigned to solve the main problems of digital signal preprocessing of target return signals we consider one possible realization version of the programmable signal microprocessor, the structure and software of which possess a definite degree of universality that allows us to interface this CMP subsystem with CRSs of different types, including radar systems of old standards during their modernization. The programmable signal microprocessor subsystem is represented in Figure 8.13 and consists of the following main elements:

1. The microprocessor of intraperiod digital signal processing, that is, the input microprocessor, consisting of analog–digital converter (ADC), two-channel filter to compress the linear-frequency-modulated signals in frequency domain (fast Fourier transform [FFT]), and CB. The ADC allows us to carry out a discretization in time of input signals (target return signals). The FFT microprocessor carries out FFT, convolution in frequency domain, inverse FFT, and combination of in-phase and quadrature components. The microprocessor has a structure allowing us to increase its performance.
2. The arithmetical microprocessor consists of two parallel general-purpose microprocessors with RAM and CB. The arithmetical microprocessor allows us to realize parallel, including conveyer, computations by two identical channels and is assigned to solve all problems of interperiod digital signal processing of target return signals, namely, moving target indication, constant false alarm, target detection and resolution, target coordinate evaluation,

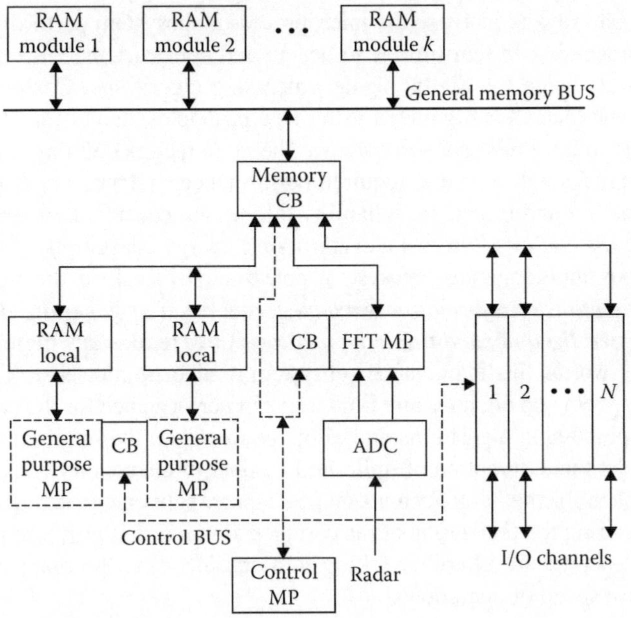

FIGURE 8.13 Programmable signal microprocessor subsystem.

and so on. The speed of operation of the microprocessor subsystem allows us to solve the problems of Doppler filtering. For this purpose, the FFT microprocessor is used.
3. The RAM as a set of RAM modules. These RAM modules can be coupled with other microprocessor subsystem elements using CB.
4. The interface with external objects includes 8 bidirectional I/O channels ensuring transmission of 32 digits per word. In addition, there is a wideband channel of direct memory access.
5. The control microprocessor assigned to solve the control problems of the microprocessor subsystem is realized as the 32-digits-per-word programmable microprocessor including RAM of larger size (capacity) and microinstruction memory.

Thus, the considered programmable microprocessor subsystem provides for the individual realization of digital signal processing using two real timescales: the intraperiod digital signal processing using the input programmable microprocessor of real time and the interperiod digital signal processing using the arithmetical microprocessor. The programmable microprocessor subsystem can be implemented in combination with a wide range of CRSs for target detection and tracking with mechanically rotating antenna. If requirements for speed of operation and memory size (capacity) cannot be satisfied by using a single-programmable-microprocessor subsystem, we can apply the programmable multimicroprocessor subsystem with controller.

8.8 SUMMARY AND DISCUSSION

To compare the microprocessor systems between each other we can consider the following engineering data: *addressness*—the number of address codes used by the instruction code; *number representation capacity*—the control microprocessor systems are characterized by 16, 24, 32, and 64 digits per word; *number representation form*—there are two number representation forms: the fixed and the floating point; *microprocessor speed of operation*—to characterize the speed of operation of the microprocessor subsystem the statement of "rated speed" is introduced independent of the solved problem class; *effective speed of operation*—the average number of operations per second for the realization of the specific digital signal processing algorithm.

One of the possible ways to increase the microprocessor subsystem performance and reliability for digital signal processing in real time is to design and construct the CRS multicomputer subsystems for effectively using the digital signal processing algorithms. Computational process in the multicomputer subsystems is organized using new principles, namely, a parallel digital signal processing by several microprocessor subsystems. The main factors defining such a multicomputer subsystem structure include the end use, required performance, and memory size for the solution of given problem in totality and functional reliability taking into consideration external environment and economical factors—admissible cost and admissible energy consumption.

The homogeneous multicomputer subsystem constructed based on the principles of *adjustment operation*, *operation of information exchange*, *operation of generalized conditional jump*, and *operation of generalized unconditional jump* allows us to realize any digital signal processing algorithms. In other words, this is universal subsystem in algorithmic sense. These homogeneous multicomputer subsystems do not have any limitations in performance for the computation of complex digital signal processing algorithms and allow us to ensure the required reliability and subsystem survival under the assumption of unlimited number of elementary cells (microprocessors). The designing problem for the homogeneous multicomputer subsystems is to define the number of elementary cells ensuring the realization of the complex digital signal processing algorithms in real time and required operational reliability taking into consideration the multicomputer subsystem losses in the effective speed of operations.

The backbone multimicroprocessor subsystem combines several independent microprocessors coupled with each other in such a way that information at the output of one microprocessor comes in at the input of another microprocessor; that is, the microprocessors process information sequentially or in conveyer style. The conveyer style of digital signal processing is based on partition of the digital signal processing algorithm on a set of steps and matching in time these steps when the digital signal processing algorithm is accomplished. Advantages with the implementation of the backbone multimicroprocessor subsystems compared to the matrix homogeneous multimicroprocessor ones are the moderate requirements with respect to inner coupling and simplicity with which the backbone multimicroprocessor subsystem's computational power can be increased. Disadvantages are removed by organizing a priority exchange of information between the system cells.

Two types of the backbone multimicroprocessor subsystems are widely used for digital signal processing in real time: the first type, a one-dimensional inner long-distance channel operating in the time-sharing mode, and the second type, a multidimensional inner long-distance channel coupled with two-input RAM. In the backbone multimicroprocessor subsystem of the first kind all RAM and ROM, CMP, and SMP are combined by the same channel. There is a block to control the channel. This block is used in the case of conflict situations when several microprocessors address RAM simultaneously and for controlling data upload and extraction using the channel units to exchange the data. Operation of a single backbone multimicroprocessor subsystem is carried out in the following way. Each microprocessor submits, as needed, an application to the channel to address to RAM or ROM. If the channel is free then the microprocessor can access RAM or ROM immediately. Otherwise, the microprocessor is in the idle mode. The microprocessor can access RAM or ROM when the previous queries from other microprocessors are ended and there are no further high-priority queries from microprocessors. Consequently, there are high requirements to the information channel with respect to the data channel throughput. Determination of the required throughput is carried out for each specific case using methods of the queuing theory. Finally note that the said structure of the backbone multimicroprocessor subsystem has limited speed of operation defined by the data channel throughput and, in addition, with the data channel failure, the backbone multimicroprocessor subsystem stops any operation. A great advantage of the backbone multimicroprocessor subsystem is its simplicity. In the backbone multimicroprocessor subsystem with the multidimensional inner long-distance channel, all microprocessors operate using independent asynchronous mode and all conflict situations are practically excluded. This system has very high reliability and allows us to increase the performance without any limitations. Disadvantage of the backbone multimicroprocessor subsystem with the multidimensional inner long-distance channel is the high complexity caused by the use of multiinput RAM and ROM. While designing the multimicroprocessor subsystem, the main problem is to provide the required performance by selecting the number of microprocessors to be included in the structure of this subsystem. With an increase in the number of microprocessors in the multimicroprocessor subsystem structure, a portion of overhead caused by a waiting time while microprocessor addresses RAM or ROM, simultaneously addressing common tables and operational systems, and the timetables caused by operation of the supervisory routing increase too.

The microcomputer constructed based on a microprocessor possesses high reliability and flexible universality; however, speed of operation is low. In this case, the required performance for CRS digital signal processing can be realized by organization of multiple microcomputer systems using several microprocessors. Such computer systems are called the microprocessor systems. In any microprocessor subsystem there must be an organized network providing communication between the microprocessor subsystem elements for data exchange between microcomputers. Configuration and the degree of complexity of such communications depend on the CRS digital signal processing algorithms, the distribution of operations between the microcomputers, and the acceptable number of RAM used by one microcomputer.

The QS interacts with sources of requests for queuing. In the discussed case, the targets and other interferences being in the radar coverage are considered as the sources of queuing requests. These sources interact with the QS by a request sensor—the CRS plays this role. Radar system transforms the queuing request into signals that are subjected to process. Time sequence of these signals, which is ordered during radar sensing and scanning of the controlled space, forms an input request flux for the QS. The incoming flux request service is organized by the QS. The QS can be considered as both the device carrying out a direct request service and the device assigned to store a request queue meant for service. The QS is characterized by the following: (a) the number of devices (cells) assigned to store a request queue of the incoming signal flux (the IBA memory size); (b) the number of devices (channels) K that can serve the requests simultaneously; (c) the number of devices (cells) to accumulate and store the results of digital signal processing (the IBA memory size). The request queue is stored by IBA of microprocessor subsystem. The waiting time and the number of queued requests (the request queue length) are the random variables depending on statistical parameters of the incoming signal flux and service rate. Because the IBA size (memory capacity) is limited, the request queue length and, consequently, the waiting time of request queue are limited too. This QS is called the QS with the limited queue (the limited waiting time). Thus, the digital signal processing subsystem can be considered as the subsystem belonging to the class of QSs with limited queue.

The process of selecting a new request from a set of requests waiting queuing involves the following. After digital signal processing of immediate request, the supervisory program investigates sequentially queuing and selects for service the request possessing the greatest priority. After initiation of the corresponding routine, it is realized by the microprocessor. The queued request leaves the system, and a control is transferred to the supervisory routine again. If there are no further requests, then the supervisory routine switches on the microprocessor in the idle mode. At each instant, the microprocessor can run a single program only. For this reason, the considered QS is called a single-channel or a single-microprocessor QS.

As a rule, we need to exercise strong limitations with respect to the waiting time for request queuing of individual signal fluxes that require assigning them by absolute priorities, for example, the requests to process the target return signals. Other requests have excess waiting time and may be assigned by relative priorities. Several requests can be served by the simple queuing procedure. Thus, we see that we need to apply mixed regulations for request queuing, investigation of which is carried out for each specific case by simulation procedures.

The required effective speed down boundary definition of the third type of control single-microprocessor subsystem operation can be considered as the initial stage of designing the microprocessor subsystem for CRS digital signal processing. The next stage involves choosing the optimal request queuing regulation. At the system design stage, we will not have sufficient information to choose the specific request queuing regulation, but we should take into consideration the following circumstances: (a) the request queuing regulation based on request queuing in turn to come in at the control single-microprocessor subsystem possesses the best performance among all the nonpriority request queuing regulations; in particular, in this case, the variance of the request queuing waiting time is minimum compared to other nonpriority request queuing regulations; (b) introduction of the priority request queuing regulation allows us to decrease essentially the request queuing waiting time for the most important requests of signal flux owing to increased delay during the queuing of signal minor fluxes; (c) in general, introduction of the priority request queuing regulations provides an equivalent performance of the control single-microprocessor subsystem compared to the request queuing one in turn. Consequently, the aforementioned values of the required minimum speed of control single-microprocessor subsystem operation provides a possibility to choose the priority request queuing regulation without any additional increase in the effective speed of operation. The minimum value of the effective speed of operation ensures the boundary values of the request queuing waiting time of the earlier-considered types of the request queuing regulations for the control single-microprocessor subsystem and related to a queue length that can be defined for the kth signal flux as $l_k = \lambda_k t_{\text{wait}\,k}$. Queue existence and its store are related to definite losses. To decrease these losses, the effective speed of control single-microprocessor subsystem operation must be greater than minimal one. However, in this case, the loading factor U decreases and, correspondingly, the idle time factor $Q = 1 - U$ increases, which is not good from the point of view of hardware costs.

Requirements for RAM size and structure are defined based on the end use of the system, character and digital signal processing algorithms, and analysis of input and output signal fluxes. As a first approximation, the total RAM size (memory capacity) is defined from the formulas $Q_\Sigma = Q_{\text{routine}} + Q_{\text{digit}}$, where Q_{routine} is the RAM cell array assigned to store the routines of digital signal processing algorithms, control routine of computational process and operation of whole system, interruption routines, and routines controlling calculations; Q_{digit} is the RAM cell array assigned to store a numerical information. In turn, Q_{digit} can be presented in the following form: $Q_{\text{digit}} = Q_{\text{in}} + Q_{\text{working}} + Q_{\text{out}}$, where Q_{in} is the RAM cell array assigned to receive external information; Q_{working} is the RAM cell array participating at the calculation process; Q_{out} is the RAM cell array assigned to store the results of digital signal processing. In many cases, the relative independence of routine information on numerical one leads us to expediently use individual permanent memory or ROM to store the routine information. ROM uses only a reading mode. Implementation of ROM gives us a great advantage because information stored by ROM is not lost even if the power is off. In addition, ROM possesses high reliability. As for ROM size (capacity), at the initial stage of designing the digital signal processing subsystem it is possible to estimate this parameter approximately. The exact estimation is possible only after program debugging.

The problem with choosing the special-purpose microprocessors is solved based on requirements for making demands to the designed microprocessor subsystem on speed of operation, RAM and ROM sizes, and technical characteristics, reliability, overall dimensions, use, cost, and other requirements. Selection of the appropriate microprocessor depends on the degree of correspondence between a set (vector) of QoS of the selected microprocessor and a set (vector) of requirements to these quality indices. A numerical measure or criterion of effectiveness must be established to compare the selected microprocessors. However, at the initial stages of designing it is very difficult, as a rule, to establish a function between the parameters of microprocessor and generalized criterion of effectiveness as a whole. Moreover, selection of the generalized criterion, demonstrable and convenient in computational sense, is not easy. We can obtain the following results of comparison between the microprocessor parameters: (a) only one type of microprocessors from existing nomenclature satisfies all requirements; (b) none of the microprocessors from existing nomenclature satisfies all requirements; and (c) there are several types of microprocessors from existing nomenclature satisfying all requirements. In the first case, only the microprocessor that satisfies all requirements is selected to design the microprocessor subsystem. In the second case, we need either to correct the requirements based on simplification of digital signal processing algorithms for solved problems and change an environment or to make a decision to construct a multimicroprocessor subsystem based on implementation of microprocessor of the same or different types produced by industry. In the third case, the problem with selecting the best microprocessor satisfying all requirements is solved. In this case, there are several ways to make this choice. We consider the simplest way—the ranking way.

While designing the structure of microprocessor subsystem for CRS digital signal processing of target return signals we need to consider a sequence of digital signal processing step realization and the difference in the scale of real time for each step involved in digital signal processing. An important initial condition for the synthesis of microprocessor subsystems is the selection of appropriate hardware and software that as a unit helps to solve the problems associated with CRSs. A total set of the microprocessor subsystem hardware can be divided into the following groups: (a) the microprocessor subsystem facilities providing a realization of CRS digital signal processing algorithms; (b) the communication facilities providing transmission of information from sources to users; (c) the facilities of transmission of information; and (d) the interface and commutation facilities assigned to unify the microprocessor subsystem facilities into the multimicroprocessor subsystems for the purpose of increasing the speed of operation and reliability of computations and numerical calculations. Naturally, the main microprocessor subsystem element defining its structure is a system of computing facilities. Two types of computing facilities are employed by the digital signal processing subsystems, namely, the control microprocessor subsystem and the multimicroprocessor subsystems for specific applications providing a realization of the main digital signal processing and control algorithms, and the special-purpose high-performance microprocessors assigned mainly for target return signal filtering at the stages of the intra- and inter-period digital signal processing. The other aforementioned hardware facilities are special purpose and are assigned for SMP subsystems implemented in CRSs of special-purpose application. Questions of rational selection of hardware facilities for designing the microprocessor subsystem are essential because, in the case of special-purpose applications, dimensions and cost of such SMP subsystems outweigh the corresponding characteristics and performance of usual microprocessor systems.

Software is the programmable facility system assigned to increase the effectiveness of employing a microprocessor subsystem and decrease the work content of preliminary operation for the solution of problems by microprocessor subsystem. Software can be divided as internal and external software. In the control microprocessor, the internal software consists of the automized programming routine, operating system routine, that is, computational process control routine, and functioning control routine. The external software consists mainly of the application program library and the specific programs of CRS digital signal processing. Because the cost of designing software exceeds the cost of designing hardware, one of the main avenues of the microprocessor subsystem development is the realization of some typical software functions by hardware.

The main problems with improving the structure and elements of the considered microprocessor subsystems are the following: (a) to design the high-performance signal microprocessors for digital signal processing of target return signals, taking into consideration solutions for both considered and discussed problems and new problems (obviously, one way to solve this problem is to further specialize the special-purpose microprocessors and to introduce parallel algorithms for solving the digital signal processing of target return signal problems); (b) to design and construct high-performance parallel (matrix, conveyer, and other types) microprocessors based on modern and perspective element base providing the required performance during the automation of all main problems of CRS digital signal processing of target return signal and control process; and (c) to design and construct the microprocessor subsystems with uniformly and nonuniformly distributed structure satisfying requirements of unification of digital signal processing hardware oriented to the special-purpose applications.

REFERENCES

1. Evreinov, A.V. and V.G. Choroshevskiy. 1978. *Homogeneous Computer Systems*. Novosibirsk, Russia: Nauka.
2. Corree, E., de Castro Dutra, I., Fiallos, M., and L.F.G. da Silva. 2010. *Models for Parallel and Distributed Computation: Theory, Algorithmic Techniques and Applications*. New York: Springer, Inc.
3. Dandamudi, S. 2003. *Hierarchical Scheduling in Parallel and Cluster Systems*. New York: Springer, Inc.
4. Milutinovic, V. 2000. *Surviving the Design of Microprocessor and Multimicroprocessor Systems: Lessons Learned*. New York: Wiley Interscience, Inc.
5. Shen, J.P. and M.H. Lipasti. 2004. *Modern Processor Design: Fundamentals of Superscalar Processors*. New York: McGraw Hill, Inc.
6. Conte, G. and D. de Corso. 1985. *Multi-Microprocessor Systems for Real-Time Applications*. New York: Springer, Inc.
7. Gupta, A. 1987. *Multi-Microprocessors*. New York: IEEE Press, Inc.
8. Parker, Y. 1984. *Multi-Microprocessor Systems*. San Diego, CA: Academic Press, Inc.
9. Cartsev, M.A. and V.A. Brick. 1981. *Computer Systems and Synchronous Arithmetics*. Moscow, Russia: Radio and Svyaz.
10. Yamanaka, N., Shiomoto, K., and E. Ok. 2005. *GMPLS Technologies: Broadband Backbone Networks and Systems*. Boca Raton, FL: CRC Press, Inc.
11. Kartalopoulos, S. 2010. *Next Generation Intelligent Optical Networks from Access to Backbone*. New York: Springer, Inc.

12. Williams, M. 2010. *Broadband for Africa: Developing Backbone Communications Networks*. New York: World Bank Publications, Inc.
13. Gnedenko, V.V. and I.N. Kovalenko. 1966. *Queuing Theory*. Moscow, Russia: Nauka.
14. Kobayqashi, H., Mark, B.L., and W. Turin. 2011. *Probability, Random Processes, and Statistical Analysis: Applications to Communications, Signal Processing, Queuing Theory, and Mathematical Finance*. Cambridge, U.K.: The Cambridge University Press, Inc.
15. Furmans, K. 2012. *Material Handling and Production Systems Modeling—Based on Queuing Models*. New York: Springer, Inc.
16. Alfa, A.S. 2010. *Queuing Theory for Telecommunications*. New York: Springer, Inc.
17. Tolk, A. and L.C. Jain. 2009. *Complex Systems in Knowledge-Based Environments: Theory, Models, and Applications*. New York: Springer, Inc.
18. Cornelius, T.L. Ed. 1996. *Digital Control Systems Implementation and Computational Techniques*. San Diego, CA: Academic Press, Inc.
19. Nedjah, N. 2010. *Multi-Objective Swarm Intelligent Systems: Theory & Experiences*. New York: Springer, Inc.
20. Yu, H.H. Ed. 2001. *Programmable Digital Signal Processors: Architecture, Programming, and Applications*. Boca Raton, FL: CRC Press, Inc.
21. Baese, M. 2007. *Digital Signal Processing with Field Programmable Gate Arrays*, 3rd edn. New York: Springer, Inc.
22. Parhi, K.K. 1999. *VLSI Digital Signal Processing Systems: Design and Implementation*. New York: Wiley–Interscience Publication.
23. Kirk, D.B. and W.H. Wen-Mei. 2010. *Programming Massively Parallel Processors*. Burlington, MA: Morgan Kaufman Publishers.
24. McCormick, J.W., Singhoff, F., and J. Huques. 2011. *Building Parallel, Embedded, and Real-Time Applications with Ada*. Cambridge, U.K.: The Cambridge University Press, Inc.

9 Digital Signal Processing Subsystem Design (Example)

9.1 GENERAL STATEMENTS

In this chapter we consider and discuss the main stages of system design of the digital target return signal processing subsystem employed by a complex surveillance radar system using a phased array. However, this is not considered a practical case. Preliminarily, we will discuss some general considerations associated with the designing and construction of a digital target return signal processing subsystem, particularly in the context of an automated complex "radar–digital signal processing and control subsystem" (complex radar system or CRS).

The basic idea of any CRS digital signal processing and control subsystem is defined, first of all, by the type and purpose of the CRS. Before starting the designing process, at the initial stage of analysis of the problems to be solved by a higher-order system, the specifications of the corresponding CRS support must be first defined. The first step will be to consider and select which type of CRS to construct, including type of radar to be used; thus it is critical to define the problems typically associated with digital signal processing and control subsystems and embark on a design process that is prepared to solve the many problems that might arise during the design and construction of the subsystem. At the same time, there is a need to adhere to the following principal positions. The CRS, as a source of information, is very complicated and expensive. It is designed and constructed in a way to ensure compliance of the highly technical specifications that form part of the design procedure. Evidently, the CRS digital signal processing and control subsystem must be able to stabilize the radar performance under hard conditions that involve high-precision operations and applications.

From the viewpoint of automated CRS digital signal processing and control processes, we can distinguish between the automatic and automated radar systems. The automatic radar systems are required, for example, to monitor the air condition in regions that are difficult to access. However, the complete automatization of digital signal processing and control subsystems is worthwhile in the case of CRSs operating in global systems on control, earth-based guidance and tracking, landing, and so on. In doing so, all available methods and tools to cancel the interference and noise must be used. The digital signal processing and control subsystem must be able to keep stable the probability of false alarm to prevent overloading on the central microprocessor system. If it is found that complete automation of a CRS is not worthwhile or impossible on account of specific technical or tactical considerations, then keeping such requirements in mind, different types of systems are designed customizing the system to meet specific requirements, especially where some of the processes are manually carried out by an operator, for example, even signal processing and other specific controls. For example, the operator is responsible for blanking the intensive interference and noise zones, primary target lock-in with the purpose of target tracking, switching on the security equipment and tools of interference and noise protection, semiautomatic target tracking, and so on.

The level of development and production status of the digital computing system element base are critical for effective designing and construction of the digital signal processing and control systems. By this, we only mean computer-driven applications deployed to solve specific problems encountered in the signal processing and control, not computer techniques in general. Thus, there is increased interest in a new element base, namely, graphene.

While designing a specific digital signal processing and control system, theoretical and technical investigations on signal processing methods and algorithms, the technical team, including engineers and subject experts, communications skills, and so on play a very important role.

9.2 DESIGN OF DIGITAL SIGNAL PROCESSING AND CONTROL SUBSYSTEM STRUCTURE

9.2.1 Initial Statements

In accordance with the basic idea, let the designed CRS be assigned for air target detection and tracking with a highly effective reflective surface $S_{tg} \geq 1\,\text{m}^2$. Scanning area is omnidirectional. The maximum radar range is $R_{max} \leq 150\,\text{km}$ ($T = 1\,\text{ms}$). Information is presented to users via the smoothed polar coordinates $\hat{\rho}_{tg}$ and $\hat{\beta}_{tg}$, the target course $\hat{Q}_{tg}$, and the velocity scalar of vector $\hat{V}_{tg}$. The accuracy of smoothed coordinates and parameters are $\sigma_\rho = 500\,\text{m}$, $\sigma_\beta = 0.5°$, $\sigma_Q = 2°$, and $\sigma_V = 50\,\text{m/s}$.

The CRS must distinguish the targets from the background of passive interferences with the coefficient of distinguishability that is sufficient for target detection and target tracking while the target and passive interference are resolved. Moreover, the system must be able to detect a stationary target, or a moving target with the so-called blind velocities, for which $f_D = kT^{-1}$ ($k = 1, 2, \ldots$). The problem with detecting the target at the border of scanning range with the probability of detection $P_D = 0.95$ is set within the time limits of 15 s. The probability of target-tracking failure within the limits of scanning range is $P_{failure} \leq 0.05$. The maximum number of targets tracking simultaneously is $N_{tg} = 20$. All CRS operations, namely the target detection, the lock-in for target tracking, and the target tracking by trajectory, must be automated completely. The reliability control measures allowing the CRS to operate without a regular labor force must be provided.

The first stage of any radar system design involves the selection of radar structure and energy parameters constituent of a CRS. The radar antenna type, shape and width of radar antenna directional diagram, method and period of scanning coverage, transmitter power, duration and scanning signal modulation technique, period of scanning pulsing, resources and methods needed for protection of the system from active interferences, and other radar parameters must be defined and justified at this stage. Thus, as a result of the first-stage activities, we have the following:

- The cylindrical antenna is selected as the transmit–receive antenna, which makes discrete scanning possible as facilitated by the radar antenna's directional diagram beam in omnidirectional scanning mode. The radar antenna directional diagram beam is fan-shaped in the vertical plane and covers all scanning range by tilt angle. The radar antenna directional diagram beam width in the horizontal plane is $\theta_\beta = 3°$. Scanning resolution is equal to 2.5°. The number of fixed positions of the radar antenna directional diagram under the omnidirectional scanning is equal to 144. The omnidirectional scanning period is $T_{scan} = 4.5$ s.
- The linear-frequency-modulated pulse with duration $\tau_{scan} = 64\,\mu\text{s}$ and spectrum bandwidth $\Delta f_{scan} = 0.5\,\text{MHz}$ is considered as the scanning signal. The scanning signal base is $\tau_{scan}\Delta f_{scan} = 32$. Duration of the compressed signal at the GD output is $\tau_{scan}^{comp} = 2\,\mu\text{s}$.
- The power of scanning signal is selected in such a way that each direction is scanned by the pulse bursts consisting of 30 pulses divided on groups, each of 10 pulses at three convertible in series frequencies $f_{0_1}, f_{0_2}, f_{0_3}$ diverging on the constant interval Δf_0 to satisfy the required signal-to-noise ratio (SNR). The coherent accumulation of reflected signals for each group of 10 pulses and the noncoherent accumulation of corresponding total signals in each resolution element by radar range, azimuth, and Doppler frequency between groups must be provided.

Note that specific magnitudes of some radar parameters are presented here only because they are directly used in the designing and construction of a CRS digital signal processing and control subsystem.

At the second stage, the design process involves the following key steps: specifics related to the CRS structure are firmed up, the parameters of digital signal processing subsystem are justified, and the ways of specific realizations are defined. Foremost, the main problems and tasks of digital signal processing and control subsystem should be discussed and the procedures to solve these problems should be defined.

9.2.2 Main Problems of Digital Signal Processing and Control Subsystem

The task of running a CRS operation in automatic mode is carried out based on the data gathered at the initial stage of designing. Successful running of this operation is foremost achieved by high-quality cancellation of the passive interference formed due to reflections from the underlying surface, local objects, and hydrometeors. It is universally accepted that in the automatic mode the coefficient of cancellation η_{can} of the passive interference caused by reflections from the underlying surface and local object should not be less than 50 dB and, by reflections from the hydrometeors, it should be approximately 30 dB.

To cancel the passive interference, the moving target indicator systems based on the rejector filters with interperiod subtraction of the order ν are widely used. However, in the case of high-density nonstationary passive interferences, the moving target indicator systems based on the rejector filters with the interperiod subtraction of the low order, $\nu = 2/3$, cannot ensure the required SNR at the output. Using the moving target indicator systems based on the rejector filters with the interperiod subtraction of the high order ν increases blind velocity zone that adversely affects the detection performance, especially for targets moving in directions tangential to the radar antenna directional diagram main lobe. The coherent signal processing technique of target return pulse bursts is effective for improving the detection performance of moving targets and decreasing the rate of stimulation of blind velocities. This signal processing method can be carried out using the fast Fourier transform (FFT) processors or filters. Thus, it is worthwhile to use the rejector filters with the interperiod subtraction of the order ν and the filters of coherent accumulation in the form of FFT processors or filters in series to ensure the required quality of passive interference cancellation and, thus, better performance in detecting moving targets.

The number of pulses in the scanning pulse burst at each of the three carrier frequencies $f_{0_1}, f_{0_2}, f_{0_3}$ is defined based on the condition to use the rejector filters with the interperiod subtraction of the order $\nu = 2$ and 8-point FFT for coherent accumulation of target return signals within the limits of the pulse burst in the digital signal processing subsystem. In this case, the noncoherent accumulation of target return signals of three pulse bursts for each resolution element by radar range and for each Doppler channel must be provided. Variations in the carrier frequency of scanning pulses from burst to burst lead to a corresponding shift in the Doppler frequency of moving target return signals. The main effect of this shift is that the task of identifying the return signals from the same target becomes difficult, particularly because these return signals can appear in different Doppler channels. Thus, to eliminate this shift in the Doppler frequency of return signals from the same target at different channels of frequency, we need to select the period of scanning signals at each pulse burst in such a way that the following condition

$$f_{0i}T_i = \text{const}, \quad i = 1, 2, 3 \tag{9.1}$$

could be satisfied. Actually, at $V_{\text{tg}} = \text{const}$ for each frequency f_{0i}, the Doppler frequency is defined as

$$f_{\text{D}} = \frac{2V_{\text{tg}} f_{0i}}{c}. \tag{9.2}$$

On the other hand, the maxima position of the impulse response from the N-channel FFT processor or filter depends on the scanning pulse frequency

$$f_l^{(i)} = \frac{kl}{NT_i}, \quad k = 0,1,2,\ldots \quad \text{and} \quad l = 0,1,\ldots,N-1. \tag{9.3}$$

To ensure the coincidence of the Doppler frequency with the tuning frequency of the lth channel, the following inequality needs to be satisfied:

$$\frac{kl}{NT_i} = \frac{2V_{\text{tg}} f_{0i}}{c} \quad \text{or} \quad f_{0i}T_i = \frac{klc}{2V_{\text{tg}}N}. \tag{9.4}$$

At l = constant, the first part in (9.4) is the constant value and the condition $f_{0i}T$ = const follows.

As noted in Chapter 3, to decrease the side-lobe level of amplitude–frequency characteristic of the Doppler channels, the signal weighting at the output of the FFT processor or filter is applied using the windows with the symmetric negative-going to end characteristics. In practice, the Hemming window is often used for signal weighting [1], wherein the computation of weighted signal readings at the in-phase and quadrature channels at the FFT processor or filter output is carried out as shown in the following algorithm:

$$f_l^{\text{weight}} = -0.25 f_{l-1} + 0.5 f_l - 0.25 f_{l+1}, \tag{9.5}$$

where

f_l^{weight} is the weighted signal at the output of lth Doppler channel

f_{l-1}, f_l, f_{l+1} are the nonweighted signals at the outputs of $l-1$th, lth, and $l+1$th Doppler channels, respectively

In this case, the signal weighting based on the algorithm given by (9.5) is carried out at the Doppler channels with $l = 2$–6. Further, there is a need to note that the FFT processor or filter accomplishes the coherent accumulation of target return signals if the targets move with zero radial velocity, which makes the target detection an easy task.

Maintaining stability of the probability of false alarm is also an important task. The use of adaptive threshold control under signal detection is a practice commonly followed in solving this problem. The functioning principle of adaptive detector at the FFT processor or filter Doppler channel output is as follows. Estimation of the interference and noise variance inside the moving window of the width −0.5 to 0.5 m is defined at the outputs of all Doppler channels, excluding the zero Doppler channel, by the following formula:

$$\sigma_{n_j}^2 = \frac{1}{m-3} \sum_{i=-0.5m}^{0.5m} \alpha_i Z_{ij}^2, \tag{9.6}$$

where i is the number of radar range resolution element with respect to signal element, for which the average power level of interference and noise is estimated (the signal element is the central one);

$$\alpha_i = \begin{cases} 1, & i = j-0.5m, j-0.5m+1,\ldots, j-2, j+2, j+3,\ldots, j+0.5m; \\ 0, & i = j-1, j, j+1; \end{cases} \tag{9.7}$$

Z_{ij}^2 is the squared amplitude envelope of the signal at the cell ij, and $j = 0.5m,\ldots, M_R - 0.5m$, where M_R is the number of discrete resolution elements by radar range. As we can see from (9.7), under averaging we do not take into consideration the signal cell and the neighboring cells at right and left. This action is directed to suppress the effect of signal peak and the first signal side lobes on the estimation of interference and noise variance.

The interference and noise map is formed for zero Doppler channel [2–4]. The interference and noise map is a characteristic of the average by the set of power observations of the signals reflected by the underlying surface and local objects for each resolution element by radar range and azimuth. This map is stored by the specific memory. The current magnitude of the interference and noise power at zero Doppler channel is defined by signals forming at the output of the Doppler channel for zero velocity. Periodical, with interval equal to the scanning period, update of the interference and noise map is carried out by rearranging the previously averaged and current power of interference and noise for each resolution element by radar range and azimuth, for example, using the formula for exponential smoothing.

Detection of targets with zero or very low radial velocity is made employing the interference and noise map. Target return signals for these targets will appear and accumulate at zero Doppler channel. The detection threshold for each resolution element by radar range R and azimuth β is formed taking into consideration the average power of signals reflected by the underlying surface and local objects. If the signal reflected by the slowly moving target or target moving with the blind velocity exceeds this threshold, this signal will be detected.

Thus, adaptation of the interference and noise power can maintain the stability of false alarm probability while detecting target return signals, but, in this case, the average number of false detections is not controlled and can be estimated only during the simulation and digital signal processing and control subsystem sample debugging.

The next topic to be discussed involves the following: automatic target lock-in, target trajectory tracking, and target trajectory reset. Given the technical specifications, the number of target tracking trajectories, $N_{tg} = 20$, and the probability of target trajectory tracking without reset, $P_{tt} \geq 0.95$, can be considered as a moderate mode. However, if the radar system has to function on the automatic mode under conditions of high-level interference, it entails prevention of overloading at the digital signal processing subsystem specifically in the context of nondetection of moving targets or detection of false targets. To thwart such effects, the following measures are recommended:

- Effective algorithm to detect the target trajectories providing the methods and procedures to decrease the probability of false target track beginning
- Selection of target pips in target tracking gates, taking into consideration the high density of false target pips
- Filtering the target trajectory parameters providing tracking of both nonmaneuvering and maneuvering targets
- Implementation of specific algorithms of target classification providing a selection of the most important targets for target tracking
- Ensuring speed of operation and better utilization of resources such as memory capacity of central computer system

When the radar system operates under automatic mode, it is important to ensure effective control of the radar system as a whole and functional synchronization of all elements, with the central microprocessor system. Algorithmic designing and realization of digital signal processing and control subsystem are specific tasks under construction of central computer system.

We have just considered only the main types of subsystems of the CRS central computer system. Naturally, other subsystems or constituents of the central computer system also play an important role in functioning, for example, the digital signal processing and control subsystem, the subsystem to communicate with users, and so on. However, given the limited scope and space in this chapter, it is not possible to discuss in detail the features and operations of all subsystems inherent to the radar system.

9.2.3 Central Computer System Structure for Signal Processing and Control

The principle of signal processing is divided into several stages: coherent digital signal preprocessing, noncoherent digital signal preprocessing, and digital signal reprocessing used as a basis for designing the structure of digital signal processing and control system of the radar complex. In addition, individual elements of the central computer system can be considered as subsystems to be used for the signal processing in cases, for example, of the "native–foreigner" mode subsystem and CRS control. Taking into account these considerations, an example of the structure of global digital signal processing and control system is provided in Figure 9.1. We see that the following subsystems solving the direct problems of digital signal processing and control (the rectangles with bold lines) and carrying out the transmission, receiving, and preliminary signal processing are the constituents of the global digital signal processing and control system. Data flow is represented by continuous lines and the interaction between subsystems is shown by dashed lines. Each subsystem included into the structure of the global digital signal processing and control system solves the following problems:

1. The digital coherent target return signal preprocessing subsystem:
   a. Matched filtering and compression of the linear-frequency-modulated pulses
   b. Double interperiod subtraction of the target return signals by rejector filters
   c. Eight-point FFT of target return pulse bursts
   d. Weighting of target return signal amplitudes at the FFT processor or filter output
   e. Definition of target return signal amplitudes at each resolution element by radar range and Doppler frequency

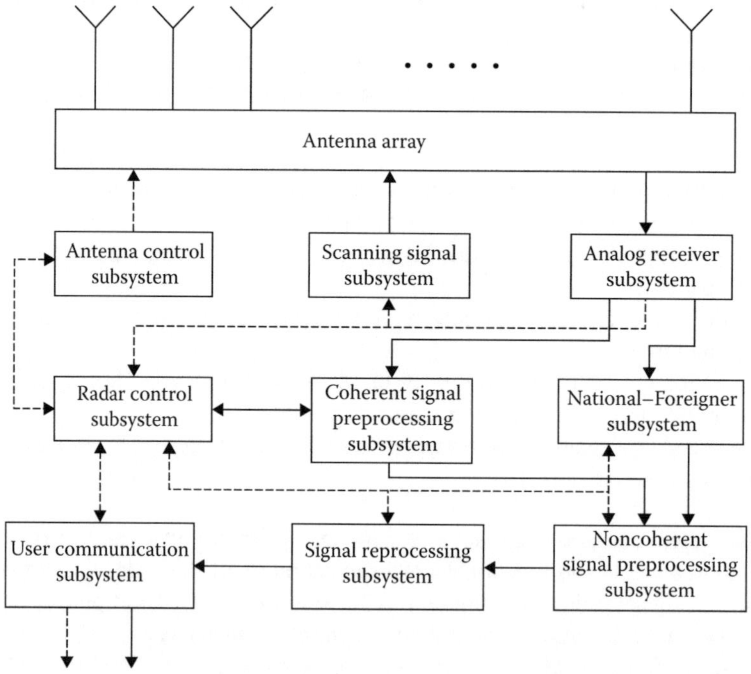

FIGURE 9.1 Global structure of the digital signal processing and control subsystem employed by the radar complex.

2. The digital noncoherent target return signal preprocessing subsystem:
   a. Noncoherent accumulation of target return signals of the three processed pulse bursts in the corresponding cells
   b. Corrections of the map of passive interferences formed by reflections from the underlying surface and local objects
   c. Adaptive detection of target return signals
   d. Estimation of target range
   e. Estimation of target azimuth
3. The digital signal reprocessing subsystem:
   a. Detection of target tracks
   b. Tracking the detected target trajectories
   c. Estimation of coordinates and parameters of moving target trajectories
   d. Estimation of target importance
   e. Providing information to the user
4. Digital subsystem of CRS control provides detailed and prompt information:
   a. Control of the CRS switching on
   b. Variation in scanning pulse frequency
   c. Control of the radar antenna directional diagram beam
   d. Control of the detection thresholds and so on

The control signals are generated both in real-time and in asynchronous mode to control the system operation, the request "native–foreigner" mode subsystem, and so on. Furthermore, the foregoing list of subsystems will be discussed and considered in more detail.

9.3 STRUCTURE OF COHERENT SIGNAL PREPROCESSING MICROPROCESSOR SUBSYSTEM

The digital coherent signal preprocessing subsystem ensures an interface between the analog receiver linear tract and the central computer system of radar complex. The following important operations are carried out by the digital coherent signal preprocessing subsystem:

- Analog-to-digital conversion
- Signal detection based on the generalized approach to signal processing in noise of linear-frequency-modulated target return signals
- Suppression of passive interferences
- Coherent accumulation of target return signals
- Data provision to assist with solving the problems of stabilization of the probability of false alarm

The digital coherent signal preprocessing subsystem is realized by an individual microprocessor sub-subsystem or a distributed set of specific microprocessor sub-subsystems.

A key part of the design process involves the design and construction of the digital coherent signal preprocessing subsystem and justification of structure of the corresponding microprocessor sub-subsystem or set of microprocessor sub-subsystems. Further, we assume that the following alternatives are key to designing and constructing the digital coherent signal preprocessing microprocessor subsystem:

- Design and construction of specific microprocessor sub-subsystem based on very-large-scale integration (VLSI) circuits
- Microcomputer based on a set of microprocessor sub-subsystems
- Specific processor based on analog charge-coupled device components

The use of microcomputer based on a set of microprocessor sub-subsystems is not worthwhile because these cannot satisfy the functional requirements when the speed of operation takes priority, for example, during conditions requiring matched filtering and compression of linear-frequency-modulated target return signals. Microprocessor sub-subsystems with analog-to-digital conversion based on charge-coupled devices, in principle, may satisfy the main requirements on speed of operation and low power consumption, but, at the current level of research, the main elements of the digital coherent signal preprocessing subsystem, for example, the GD linear tract filters, have not been studied thoroughly. Thus, an effective alternative is the digital coherent signal preprocessing subsystem designed based on VLSI microprocessor sub-subsystem.

It is worthwhile to employ hardware in the specific digital coherent signal preprocessing subsystem designed based on VLSI microprocessor sub-subsystem to solve the following problems:

- Signal detection based on the generalized approach to signal processing in noise
- Suppression of passive interferences
- Computation of target return signal amplitudes
- Estimation of interference and noise power

To solve the problems of Doppler filtering it is worthwhile to implement the FFT processor or filter using the effective and simple realization methods and procedures of complex multiplication. All elements of the specific digital coherent signal preprocessing subsystem designed based on VLSI microprocessor sub-subsystem must possess high reliability and low power consumption. High reliability can be ensured by equipment reservation or information redundancy. Low power consumption can be achieved by very careful circuit designing and implementation of VLSI with small power dissipation.

Now, consider an interaction between the elements of the specific digital coherent signal preprocessing subsystem designed based on VLSI microprocessor sub-subsystem (see Figure 9.2) and discuss the principles of realization of the main objectives of digital coherent target return signal preprocessing subsystems and the main technical specifications of hardware.

The first element of the global structure of digital coherent target return signal preprocessing subsystem is the analog-to-digital converter (ADC) of target return signals at the phase detector output. Because the duration of the compressed linear-frequency scanning signal is $\tau_{scan}^{comp} = 2\,\mu s$,

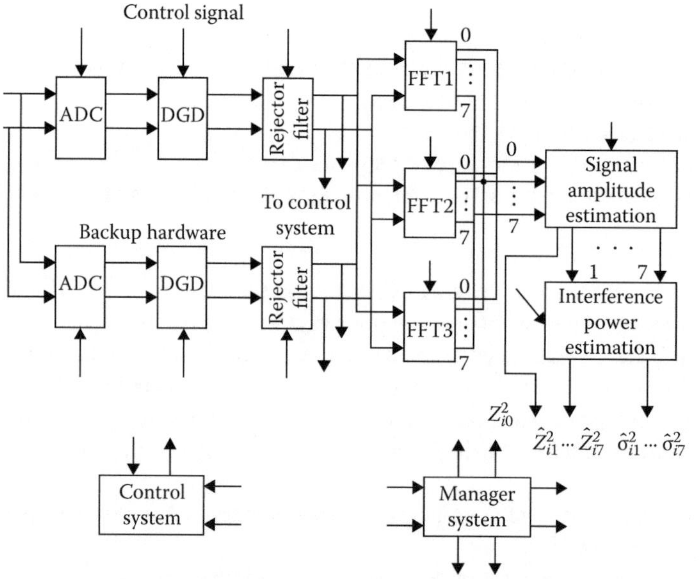

FIGURE 9.2 Structure of the specific microprocessor subsystem for coherent signal processing.

the limiting frequency of sampling in time must be $f_s = 500\,\text{kHz}$ both for in-phase and quadrature channels. However, to improve the digital GD (DGD) characteristics it is worthwhile to increase this frequency, at least twice, that is, $f_s = 1\,\text{MHz}$ (see Chapter 2). The number of amplitude quantization bits is selected based on linear transformation of the target return signals exceeding the receiver noise on 50–60 dB. For this purpose, we need to use 10-bit ADC and/or ADC with capacity of more than 10 bits. Thus, ADC must issue the signal codes at the outputs of in-phase and quadrature channels with resolution $\tau_s = 1\,\mu\text{s}$ and capacity $N_{\text{capacity}} = 10$ bits.

Next element of the total structure of digital coherent target return signal preprocessing subsystem is the DGD for linear-frequency-modulated target return signals that can be realized in time domain employing the nonrecursive filter or in frequency domain using the FFT processor or filter. In doing so, while using the sequential scheme of DGD in time domain to compute a single magnitude of the output signal, we need to use $4f_s\tau_{\text{scan}} = 4 \times 1 \times 64 = 256$ multiplications and 252 additions. Consequently, the required speed of operation of DGD is equal to 256×10^6 multiplications per second, which is very difficult to realize. Under realization of DGD in frequency domain we obtain some benefits in speed of operations (see Chapter 2), but these benefits are not so essential that we could use them without any doubts.

To improve the DGD speed of operation we can use properties of parallelism of matched filtering algorithms. For this reason, to realize DGD with high speed of operation we can use the parallel mixers based on ROM, parallel adders with simultaneous addition of several numbers, and parallel registers. For example, for parallel nonrecursive digital filter, coincidence-type adders, and parallel registers the minimal time of convolution both in the in-phase channel and in the quadrature channel is 75 ns. For a four-channel DGD (see Figure 2.7) the minimal time to obtain a single magnitude of the output signal is 100 ns.

The nonrecursive smoothing filter with short impulse response is placed after DGD. The main tasks of this filter are, first, to suppress the signal side lobes at the DGD output and, second, to decrease the sampling rate to the frequency corresponding to the sampling theorem requirements. Decrease in the sampling rate in m times (m is the integer) is carried out by the element providing a sample of each mth element from the sequence of input sampled target return signals $\{x(kT_s)\}$. As a result, we obtain the output signal $\{x(kmT_s)\}$ with the sampling period $T_s' = mT_s$. Naturally, the sampling period T_s' must satisfy the following condition:

$$T_s' \leq \frac{1}{\Delta f_{\text{scan}}^{\max}} \tag{9.8}$$

both at the in-phase channel and at the quadrature channel.

The cancellation of correlated interference is realized after the signal detection based on the generalized approach to signal processing in noise. For this purpose, the samples of in-phase and quadrature channels in each resolution element by radar range (the number of resolution elements by radar range is $M_R = 500$) are stored during 10 periods by the buffer memory with capacity of $Q_{BM} = 2 \times 500 \times 10 = 10^4$ of 10-bit words. After that, 10-pulse bursts corresponding to each resolution element by radar range are processed by the rejector filters with the interperiod subtraction of the order $\nu = 2$ individually both in the in-phase channel and in the quadrature channel. Owing to the simplicity of the rejector filters with the interperiod subtraction of the order $\nu = 2$ algorithm, its realization in real time is not difficult. Thus far, we have considered and discussed operations during the first stage of digital coherent signal processing. To ensure the required reliability, all hardware elements of the first stage must be doubled. The output signals of the main and reserved hardware set are compared by a specific module. Comparison results are used both by the searching system and by the default system.

Signals at the outputs of the rejector filters with the interperiod subtraction of the order $\nu = 2$ in the form of 8-pulse bursts come in individually by the in-phase and quadrature channels at the input of filter carrying out a coherent accumulation of the target return signals within the limits of 8-pulse

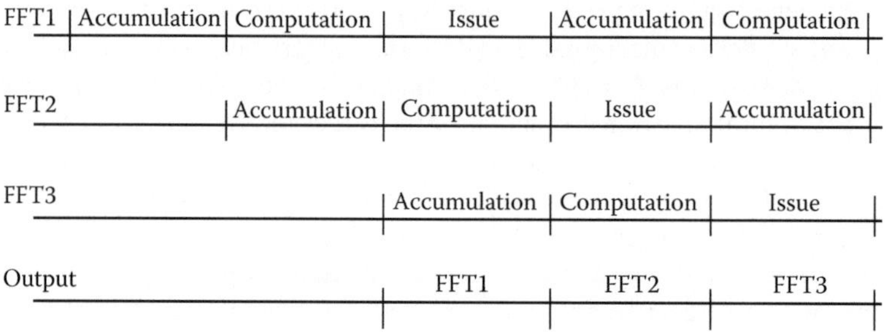

FIGURE 9.3 Timing diagram for three FFT processors.

bursts. In the considered case, the target return signal coherent accumulation is realized in the frequency domain using the FFT processors or filters. Using the FFT processor or filter is the most appropriate choice considering the simplicity of usage and tuning convenience while processing pulse bursts with variable width. Information at the FFT processor or filter output is accumulated during eight scanning periods, that is, 8×10^{-3} s, $T_{scan} = 10^{-3}$ s. The number of resolution elements by radar range is $M_R = 500$. Consequently, 8-point FFT must be finished within the limits of 16 μs. With current level of technology, it is possible to use different schemes of FFT realization that guarantee faster operations without any problems. Moreover, to increase the efficiency of coherent signal processing and the reliability of its corresponding hardware module, three FFT processors or filter in series with the time are required, as shown in Figure 9.3. This functional capability of FFT processors or filters reduces requirements for faster operation of FFT processors or filters and for the high reliability of the corresponding hardware module.

Each FFT processor or filter possesses a buffer memory to store the input data. The buffer memory capacity must store the 8-pulse burst signals at each of the 500 resolution elements by radar range and take into consideration the in-phase and quadrature channels, that is, $Q_{BM} = 2 \times 8 \times 500 = 8 \times 10^3$ 10-bit words. At the FFT processor or filter output, the signal weighting at the Doppler channels 2–6 is carried out to reduce the effect of signal side lobes and expand zero part width of amplitude–frequency characteristics of Doppler filters. Weighting is carried out based on the algorithm given by (9.5). By this operation, the coherent signal processing is concluded, and the envelope amplitude of target return signals needs to be selected, for example, to unify signals at the outputs of in-phase and quadrature channels.

Estimation of envelope amplitudes at each resolution element by radar range for all Doppler channels is to be carried out in accordance with signal processing algorithms discussed in Chapter 3. Estimations of squared envelope amplitudes obtained in the zero Doppler channel are used to make corrections in the interference and noise power map. Estimations of squared envelope amplitudes obtained in other Doppler channels come in at the input module of definition and estimation of interference and noise power. For this purpose, the signal processing algorithm used to estimate the sampled interference and noise variance within the limits of the window, including eight resolution elements by radar range (approximately 2.5 km) from both sides of the central (signal) element given by (9.6), is realized. Estimations of sampled interference and noise variance are defined at the outputs of each Doppler channel, except for zero Doppler channel. The corresponding magnitudes of squared amplitudes $\hat{x}_1^2,\ldots,\hat{x}_7^2$ and variances $\hat{\sigma}_1^2,\ldots,\hat{\sigma}_7^2$ are transmitted to the noncoherent signal preprocessing subsystem.

9.4 STRUCTURE OF NONCOHERENT SIGNAL PREPROCESSING MICROPROCESSOR SUBSYSTEM

9.4.1 Noncoherent Signal Preprocessing Problems

Based on the global structure of the digital signal processing and control system employed by radar complex (see Figure 9.1) the noncoherent signal preprocessing subsystem solves the following problems:

1. *Noncoherent accumulation of signals after coherent signal processing of each from three 8-pulse bursts by the FFT processor or filter.* Accumulation can be presented by adding the squared envelope amplitudes at each ijth cell ($i = 1, 2,\ldots, M_R$—are the numbers of discrete elements by radar range; $j = 0, 1,\ldots, 7$—are the numbers of Doppler channels). Total number of such cells is $M_R = 500 \times 8 = 4000$. Addition of two numbers by specific central computer system requires 3–5 reduced arithmetical operations. To exclude the information losses, the noncoherent accumulation of signal containing the next pulse burst must be done within the limits of coherent signal preprocessing of subsequent pulse bursts, that is, in our case, within the limits of 8th scanning period (8 ms).
2. *Noncoherent accumulation of interference and noise power estimations for all Doppler channels, except for zero Doppler channel, and for all resolution elements by radar range.* Noncoherent accumulation of the interference and noise estimations is carried out analogously as the noncoherent accumulation of signals and requires approximately the same level of central computer system performance.
3. *Corrections of the interference and noise map.* This map is stored by the specific noncoherent signal preprocessing microprocessor system memory and holds the average estimations of the squared signal amplitudes received by zero Doppler channel for each resolution element by radar range and for each azimuth direction. Periodical update of these estimations, including the period of air surveillance, is carried out using the signals obtained at the Doppler channel for zero velocity output by the following formula:

$$\hat{Z}^2_{nil} = (1-\zeta)\hat{Z}^2_{(n-1)il} + \zeta\bar{Z}^2_{nil}, \tag{9.9}$$

where
$\hat{Z}^2_{(n-1)il}$ is the previous estimation of squared signal amplitude at ith resolution element by radar range at lth azimuth direction
$\bar{Z}^2_{nil}$ is the squared signal amplitude on the next (nth) update step derived based on the data of three Doppler zero velocity filters at ith resolution element by radar range at lth azimuth direction
ζ is the smoothing coefficient, as a rule $\zeta = 0.2$–0.3

In the case of a single realization of the signal processing algorithm given by (9.9), two multiplications on the constant and one addition are required. Taking into consideration all nonarithmetical operations, the total number of reduced arithmetical operations will be for about 10. If the radar antenna directional diagram beam is delayed at each azimuth direction on 24 ms (3 × 8 ms) and the only string of the interference and noise power map is updated (500 cells), we can assume that the operation on updating the interference and noise power map is not critical for designing the specifications of the noncoherent signal preprocessing microprocessor system.
4. *Forming the adaptive detection thresholds.* To form the detection thresholds we can use the interference and noise power estimations at Doppler channels 1–7 and the average signal power estimations at each Doppler zero channel (the interference and noise power map) estimations. Formation of thresholds to detect the signals received by all Doppler channels except for Doppler zero channel is carried out based on the current magnitudes of variance estimations σ^2_{ij} for each resolution element by radar range and Doppler frequency. Threshold is formed by multiplication of σ^2_{ij} on the coefficient α_1 defined based on the requirements of the probability of false alarm. Forming the thresholds for signals received by the Doppler zero channel is carried out by multiplication of corresponding magnitude of the interference and noise power map on the coefficient α_2 defined based on

the requirements of the probability of false alarm for the target with zero velocity—this probability of false alarm can differ from the admissible probability of false alarm for moving target.

5. *Signal detection is provided by a comparison between the signals and their corresponding thresholds computed for each cell "radar range–Doppler frequency."* To reduce the number of references to the interference and noise power map, the received signal of Doppler zero frequency channel is first compared with the constant threshold defined based on the permissible magnitude of the probability of exceeding the receiver noise power by this signal. If exceeding takes place, then the threshold is computed using the interference and noise power map and the signal is compared with the threshold. As a result of this comparison with the thresholds, we can observe one or several signals exceeding the detection threshold at some resolution elements by radar range. For a single realization of threshold formation and generalized signal detection algorithms one multiplication on the constant value and one comparison of two magnitudes are required. In doing so, the number of computer operations does not exceed 7–8. Within the limits of a single period of duration of 24 ms, 4000 realizations of the generalized signal processing algorithms must be done.
6. *Estimation of the target azimuth by a set of signals exceeding the detection threshold at three neighboring positions of the radar antenna directional diagram by azimuth.* For this purpose, at first, we need to select signal groups exceeding the detection threshold and related to the same radar range. If there is a single signal that exceeds the detection threshold into the group, then the target azimuth is defined by azimuth direction of this signal. If there are two or three signals that exceed the detection threshold into the group, then, at first, we need to choose the greatest signal, that is, the signal with maximum amplitude. The azimuth direction of this signal is $\beta[i]$. Azimuth adjustment can be provided taking into consideration a single additional side signal using the following formula:

$$\hat{\beta}_{\text{tg}} = \beta[i] + \gamma\frac{Z_i - Z_j}{Z_i + Z_j}\varphi(\vartheta), \tag{9.10}$$

where Z_i is the amplitude of maximum signal;

$$Z_j = \begin{cases} Z_{i-1}, & \text{if} \quad Z_{i-1} > Z_{i+1}, \\ Z_{i+1}, & \text{if} \quad Z_{i+1} > Z_{i-1}; \end{cases} \tag{9.11}$$

$$\gamma = \begin{cases} -1, & \text{if} \quad j = i-1, \\ +1, & \text{if} \quad j = i+1; \end{cases} \tag{9.12}$$

$\varphi(\vartheta)$ is the function characterizing the shape of radar antenna directional diagram.

For a single realization of the generalized signal processing algorithm given by (9.10) through (9.12) about 30 reduced microprocessor operations are required. For this purpose, the duration required is 24 ms. The number of realizations per cycle is the random variable characterizing the possible number of targets at three azimuth directions close to each other. The problems considered and discussed so far related to the noncoherent signal preprocessing and are solved using the corresponding partial signal processing algorithms. The total partial generalized signal processing algorithm is the global noncoherent target return signal preprocessing algorithm.

9.4.2 Noncoherent Signal Preprocessing Microprocessor Subsystem Requirements

In contrast to the coherent signal preprocessing algorithm, all main operations of the noncoherent signal preprocessing are programmable and cyclical by periodicity. Each problem of the noncoherent signal preprocessing possesses its own cycle, and, consequently, specification of speed of the microprocessor subsystem operation differs from other microprocessor subsystems used by the CRS. The microprocessor of highest throughput is required to fulfill the noncoherent target return signal accumulation. The period of this operation is equal to the period of the coherent target return signal preprocessing, that is, $T_{\text{cycle}} = 8\,\text{ms}$. The addition of previous sums with new magnitudes of target return signal amplitudes must be carried out at $500 \times 8 = 4000$ cells of "radar range–Doppler frequency" complex within the limits of 8 ms. The number of elementary operations of single summation is equal to 3, for example. Then, within the limits of 8 ms there must be $4000 \times 3 = 12 \times 10^3$ operations per cycle produced, which means the effective speed of operation of the noncoherent target return signal preprocessing microprocessor subsystems must be

$$\eta_{\text{eff}} = \frac{12 \times 10^3}{8 \times 10^{-3}} = 1.5 \times 10^6 \text{ operations per s.} \tag{9.13}$$

This requirement can be made less stringent if we consider that the real number of resolution elements by radar range will be less than 500, since the maximum radar range is less than the predetermined radar range, which is based on the scanning pulse frequency. Nevertheless, the required speed of operations exceeds 10^6 operations per second. Computations of the required effective speed of operation for all tasks accomplished by the noncoherent target return signal preprocessing microprocessor subsystem are presented in Table 9.1.

As follows from Table 9.1, the required effective speed of operation of the noncoherent target return signal preprocessing microprocessor subsystem under realization of the partial generalized signal processing algorithms is not more than 2.5×10^6 operations per second. If we assume that all

TABLE 9.1
Effective Operation of Speed to Carry Out All Operations by the Noncoherent Target Return Signal Preprocessing Subsystem

Operation	Number of Commands	Number of Elements	Period ms	Required Speed of Operation, 10^6 Operations per s
Noncoherent signal accumulation	3–5	4000	8	1.5–2.5
Noncoherent interference power accumulation	3–5	3500	8	1.3–2.2
Interference power map correction	≈10	500	24	≈0.35
Threshold forming and signal detection	7–8	3500	24	1.0–1.2
Target azimuth estimation	≈30	500	245	0.6

operations of the considered signal processing algorithm are executed by a single microprocessor, then the required speed of operations must be defined as

$$\eta_{\Sigma}^{\text{eff}} = \frac{N_{\Sigma}^{\text{op}}}{T_{\text{cycle}}^{\max}}, \tag{9.14}$$

where N_{Σ}^{op} is the total number of commands performed per $T_{\text{cycle}}^{\max} = 24 \times 10^{-3}$ s as shown in Table 9.1, for the minimum number of commands required for a single realization of the considered generalized signal processing algorithm:

$$N_{\Sigma}^{\text{op}} = 2 \times 12 \times 10^3 + 2 \times 10{,}500 + 5{,}000 + 24{,}500 + 500 = 75 \times 10^3 \text{ operations.} \tag{9.15}$$

For calculating N_{Σ}^{op} we considered that the noncoherent accumulation of the target return signals and interference and noise are realized twice, only within the limits of the cycle $T_{\text{cycle}}^{\max} = 24 \times 10^{-3}$ s. Using (9.13) we obtain

$$\eta_{\Sigma}^{\text{eff}} = \frac{75 \times 10^3}{24 \times 10^{-3}} \approx 3 \times 10^6 \text{ operations per s.} \tag{9.16}$$

Thus, as we can see, the requirement for speed of operation of the noncoherent target return signal preprocessing microprocessor subsystem is certainly high. However, the noncoherent target return signal preprocessing operations can be paralleled on two microprocessor subsystems. For example, the first microprocessor subsystem must perform the noncoherent accumulation of target return signals, corrections in the interference and noise power map, and the target azimuth estimation. The second microprocessor subsystem must carry out the noncoherent accumulation of interference and noise power, threshold forming, and signal detection. In this case, the required speed of operation for each microprocessor subsystem is not more than 2×10^3 operations per s.

Now, let us estimate the requirements to the microprocessor subsystem memory capacity. Elementary considerations give us the following values of required memory capacity or the number of cells:

- To store signal amplitudes—4000
- To store intermediate results of noncoherent accumulation—4000
- To store the estimations of interference and noise power—4000
- To store the intermediate results of noncoherent accumulation of interference and noise power estimations—4000
- To store the output signals of three azimuth directions—1500
- To store the interference and noise power map—$500 \times 144 = 72{,}000$

The total memory capacity is $Q_{\Sigma} = 89{,}500$.

9.5 SIGNAL REPROCESSING MICROPROCESSOR SUBSYSTEM SPECIFICATIONS

The information about the detected target pip coordinates comes in at the input of digital signal reprocessing subsystem (see Figure 9.4). The main tasks of digital signal reprocessing subsystem are to achieve the target trajectory detection, target tracking and target trajectory tracking, filtering of target trajectory parameters, and other digital signal processing algorithms derived during the target processing stages and to provide definite information to the user. In the said example, we assume

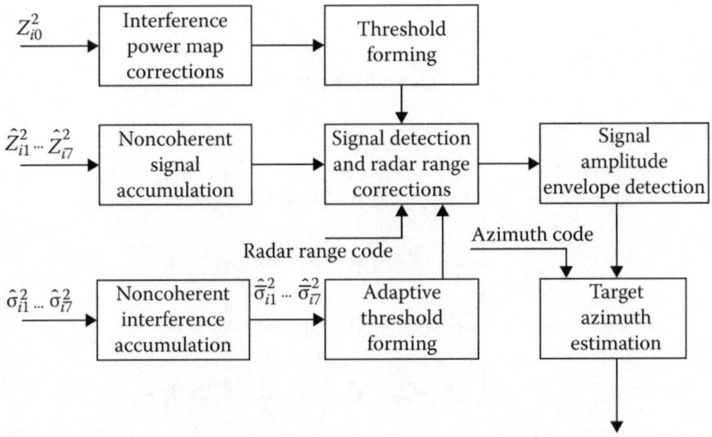

FIGURE 9.4 Complex algorithm of the noncoherent target return signal reprocessing.

that all tasks of the digital signal reprocessing subsystem are carried out by a single microprocessor subsystem. We also assume that the typical digital signal processing algorithms, which we discussed previously, modified for the case of two-coordinate surveillance radar are realized in the course of digital signal reprocessing. For this reason, we do not present in this section a detailed description of operations and global digital signal reprocessing algorithm. Some specific features of these digital signal processing algorithms are defined by specific application of the considered two-coordinate surveillance radar and are reduced to the following ones.

1. To reduce the required number of operations for solving the problem of new target pip binding to target tracked trajectories, there is a need to perform a target trajectory picking within the limits of ±15° from a direction corresponding to the azimuth of new target pip. This rough narrow-beam pulse sampling by azimuth allows us to select only such target trajectories for comparison that can be continued using a new target pip taking into consideration their displacements in radial direction with maximum velocity and the acceptable number (three) of missing target pips on the target trajectory. Coordinate extrapolation at the instant of getting a new target pip, gating, and a verification of new target pip present within the limits of gate are carried out sequentially for the target trajectories thus selected. Gate dimension depends on the number of confirmation omissions that the target trajectory was tested in the previous scanning periods.
2. The target trajectory binding is carried out using the criterion "2 from 3." The target trajectory binding can be also considered as a decision about the target trajectory detection. Thus, in the considered case, the operations of confirmation about the target trajectory bindings using one or several target pips are excluded.
3. Each detected target is estimated by the principle "important–not important," and a decision "to apply–not apply" a signal processing to the detected target is made.
4. The information generated must be precise for it allows us to implement an individual smoothing of the Cartesian coordinates X and Y without taking into consideration the correlation between them. In this case, laborious operations on the vector and matrices are excluded while computing, which essentially reduces the number of operations under realization of the digital signal reprocessing algorithms.
5. Determination of the target velocity scalar of vector $\hat{V}_n^{\text{tg}}$ and the target course Q_n^{tg} at nth step of filtering the Cartesian coordinates X and Y is made by the following formulas:

$$\hat{V}_n^{\text{tg}} = \sqrt{\left(\hat{V}_{X_n}^{\text{tg}}\right)^2 + \left(\hat{V}_{Y_n}^{\text{tg}}\right)^2}; \tag{9.17}$$

$$\hat{Q}'_n = \text{arctg}\frac{\left|\hat{\dot{Y}}_n\right|}{\left|\hat{\dot{X}}_n\right|}; \tag{9.18}$$

$$\hat{Q}_n^{\text{tg}} = \begin{cases} \hat{Q}'_n & \text{if } \hat{\dot{Y}}_n > 0,\ \hat{\dot{X}}_n > 0; \\ \pi - \hat{Q}'_n & \text{if } \hat{\dot{Y}}_n > 0,\ \hat{\dot{X}}_n > 0; \\ \pi + \hat{Q}'_n & \text{if } \hat{\dot{Y}}_n > 0,\ \hat{\dot{X}}_n > 0; \\ 2\pi - \hat{Q}'_n & \text{if } \hat{\dot{Y}}_n > 0,\ \hat{\dot{X}}_n > 0; \end{cases} \tag{9.19}$$

where $\hat{V}_{X_n}^{\text{tg}}$ and $\hat{V}_{Y_n}^{\text{tg}}$ are the estimations of target velocity by the Cartesian coordinates X and Y. Logical flowchart of the digital signal reprocessing algorithm is shown in Figure 9.5. Based on this flowchart, the sequence and interaction of operations under realization of the digital signal reprocessing algorithm can be defined.

Now, consider the main requirements for the speed of operation and the memory capacity of digital signal reprocessing microprocessor subsystem. In the case of a single realization of the digital signal reprocessing algorithm partly responsible for processing a single new target pip, a reduced number of approximately 2×10^3 operations are required. Considering that for each scanning period about 20 new true and 5 false target pips come in at the digital signal reprocessing microprocessor subsystem input, we obtain $N_{\Sigma}^{\text{op}} = 5 \times 10^4$ operations per scanning period. The scanning period is $T_{\text{scan}} = 4.5$ s. The required effective speed of operation is $\eta_{\Sigma}^{\text{eff}} \approx 10^4$ operations per second. Thus, we can conclude that the speed expected of the digital signal reprocessing microprocessor subsystem is not high.

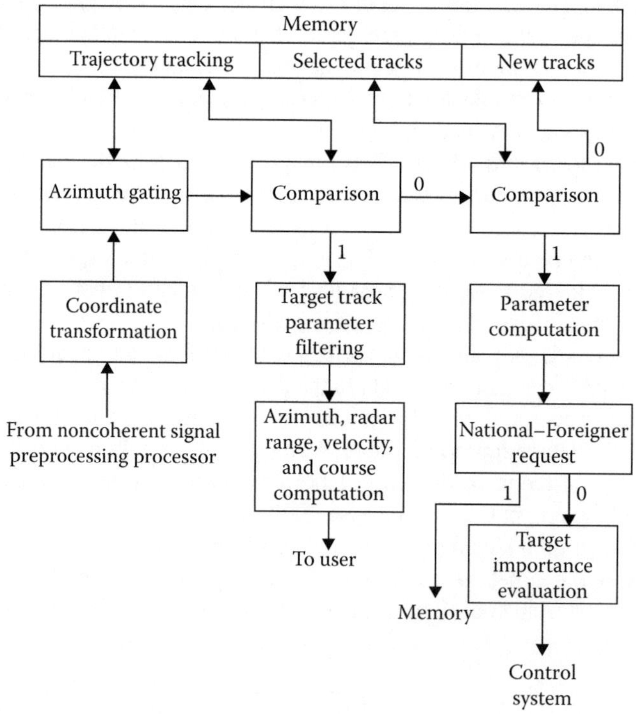

FIGURE 9.5 Digital signal reprocessing algorithm flowchart.

To determine the required memory capacity of digital signal reprocessing microprocessor subsystem, we assume that 10 false target pips are stored by the memory device in addition to the 20 true target pips. Our estimation shows that data about a single target trajectory requires approximately fifteen 32-bit words. Consequently, to store data about 30 tracked trajectories approximately four hundred and fifty 32-bit cells are required. Moreover, to store data on the detected target trajectories and the new target trajectories taking into consideration a reserved memory capacity, it is reasonable to have approximately 100 more 32-bit cells. The memory capacity requires about 400–500 cells. Thus, to solve the digital signal reprocessing problems we need about 10^3 memory cells.

The foregoing analysis shows the digital signal reprocessing problems do not require high computational resources compared to the previous stages of target return signal processing. This allows to assign for the digital signal reprocessing microprocessor subsystem the problems to control over all elements and parameters of the CRS and the problems of the current hardware and computational process control, additionally. Considering that the control optimization problems are not solved in dynamic mode in the automatic "radar complex–global digital signal processing complex" system and the control system is constructed as the system with a predetermined fixed sequence of control commands, we do not consider such system in the present chapter. It is assumed that the scheduled problems are considered and analyzed at the next stage of designing and construction of the CRS.

9.6 STRUCTURE OF DIGITAL SIGNAL PROCESSING SUBSYSTEM

Functional requirements, cost, reliability, conveniences in fault lookup, maintainability—all these factors are taken into consideration in the course of designing the global digital signal processing structure of the CRS. The functional requirements to individual elements of this structure, which have been discussed in the previous sections, are evidences of differences in the intensity of information flows processed by the central computer system of the radar complex. The digital coherent target return preprocessing subsystem working in real time has the greatest service load. For this reason, given the latest technological breakthroughs in signal detection and signal processing theory and in computer applications, there is a need to design and construct a specific nonprogrammable microprocessor subsystem for the digital coherent target return preprocessing subsystem working in real time, which consists of specific adders, mixers, convolution networks, memory devices based on VLSI, and charge injection devices. Without any doubts, we can state that the possibility of designing and constructing such specific microprocessor subsystems is the foundation stone to developing solutions for the problems encountered in automatic digital target return signal processing under real conditions of CRS functioning.

At the output of digital coherent target return preprocessing microprocessor subsystem the information density is essentially reduced, and at the stage of noncoherent digital signal preprocessing, we can use the programmable microprocessor subsystems. Analysis of requirements to specific subsystems of global computer system of radar complex and existing general avenues applicable to the microprocessor subsystems in digital target return signal processing systems favor using a module structure of target return signal processing systems, the main body of which consists of several identical microprocessor networks. Such microprocessor networks enhance ease of use, reliability, and effectiveness of the digital signal processing system as a whole. However, such microprocessor networks require that the problems of some stages of target return signal processing be parallel. In particular, we have considered such requirement while discussing the noncoherent target return signal preprocessing subsystem. Moreover, there is a need to focus on organizing the control process in such multimicroprocessor network system.

In our case, the control process must be hard; that is, it must be carried out in accordance with a program assigned before and must be subject to the given time sequence. This can be organized by way of instructions to carry out the signal processing and control problems using the specific stack memory with subsequent task choice in accordance with the time diagram of global digital signal processing system. After fulfillment of the immediate task, each microprocessor subsystem can select from the programmable control device memory the next task, indicating the following:

- The operation that must be done
- Initial data that must be used
- Memory cells that should store the obtained results
- Way for further computational process

Thus, carrying out the foregoing steps will enable the microprocessor networks to operate independently of each other. Increasing the number of microprocessor networks will enhance the reliability and speed of data processing without any changes in administration principles.

In accordance with general principles discussed thus far and needing consideration during the design and construction of the central computer system for global signal processing and control employed by radar complex, one variant of a structural flowchart is shown in Figure 9.6. These are the following subsystem boxes:

1. Coherent signal preprocessing microprocessor subsystem with the buffer memory to store the information data issued for the noncoherent signal preprocessing subsystem
2. Programmable control subsystem to control the CRS issuing the time sequence of control command and operations, such as
   a. Radar antenna scanning control
   b. Changing the transmitter carrier frequency
   c. Computation of detection thresholds
   d. Definition of operation mode—scanning mode or request "native–foreigner" mode
3. Radar synchronizer forming the signals controlling the global digital signal processing system and radar complex
4. Memory device to store the power map of interference formed by the underlying surface and local objects; this memory device is represented by individual box owing to large capacity and specific problems solved

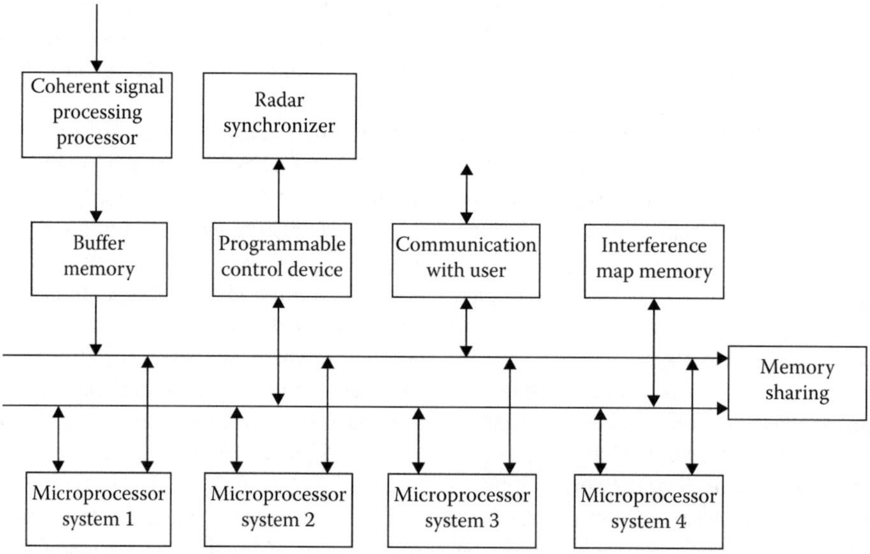

FIGURE 9.6 Structure of the central computer system for global signal processing and control.

5. Module of communication with the user, with interface functions "input–output," preparation of information for display, and sending information by the communication channel
6. Four central microprocessor networks (three microprocessor networks are working and one microprocessor network is reserve) assigned to solve all problems and tasks of global signal processing and control system, except for the problems and tasks assigned for the coherent target return signal preprocessing subsystem
7. Memory device for cooperative use assigned for the execution of central computer system functions during data processing and control distributed in space and time
8. The structure of the central computer system for global signal processing and control employed by the radar complex gives us an original approximation under solution of the assigned problem. Furthermore, we need to consider all partial signal processing and control algorithms in detail, incorporate these algorithms into the complex signal processing and control algorithm, and design the functional time diagram of the system. This stage of designing must be started after selecting specific types of microprocessor networks and the decision whether it is practicable to continue with a specific process of CRS design and construction

9.7 SUMMARY AND DISCUSSION

The first stage of system design of any radar system is the selection of structure and energy parameters of radar being a constituent of the CRS. The radar antenna type, shape, and width of radar antenna directional diagram, method and period of scanning coverage, transmitter power, duration and scanning signal modulation technique, period of scanning pulsing, resources and methods of protection from active interferences, and other radar parameters must be defined and justified at this stage. At the second stage of CRS design the structure is designed, the parameters of digital signal processing subsystem are justified, and the ways of specific realizations are defined. At the initial stages, the main problems and tasks of digital signal processing and control subsystem should be discussed and the procedures to solve these problems should be defined.

It is worthwhile to employ the rejector filters with the interperiod subtraction of the order ν and the filters of coherent accumulation in the form of FFT processors or filters in series to ensure the required quality of passive interference cancellation and to improve the detection performance of moving targets. Stabilization of the probability of false alarm is also a very important problem. The methods of adaptive threshold control under signal detection are widely used to solve this problem. The implementation of adaptation to the interference and noise power can stabilize the probability of false alarm under detection of target return signals, but, in this case, the average number of false detections is not controlled and can be estimated only during simulation and digital signal processing and control subsystem sample debugging.

Functions such as automatic target lock-in, target trajectory tracking, and target trajectory reset play a vital role and require thorough understanding. A demand to operate in automatic mode under conditions of the high-level interference requires sufficiently effective measures that prevent overloading of the digital signal reprocessing subsystem while processing both the targets that are not captured by the CRS and the false targets. These effective measures are as follows: effective algorithm to detect the target trajectories providing the methods and procedures to decrease the probability of false target track beginning; selection of target pips in target tracking gates taking into consideration the high density of false target pips; filtering the target trajectory parameters providing tracking of both the nonmaneuvering targets and maneuvering targets; implementation of specific algorithms of target classification providing a selection of the most important targets for target tracking; ensuring appropriate speed of operation and memory capacity of central computer system. Under the automatic mode of operation, the most important problem associated with the central microprocessor system is effective control of the CRS as a whole and functional synchronization of operations of all elements. Algorithmic designing and realization of digital signal processing and control subsystem are other problems to be considered while designing the central computer system.

The principle of division of signal processing procedure into various stages, that is, coherent digital signal preprocessing, noncoherent digital signal preprocessing, and digital signal reprocessing, forms the basis of designing the structure of global digital signal processing and control system of radar complex, which consists of the digital coherent target return signal preprocessing subsystem, digital noncoherent target return signal preprocessing subsystem, digital signal reprocessing subsystem, and digital control subsystem.

The digital coherent signal preprocessing subsystem ensures an interface between the analog receiver linear tract and the central computer system of radar complex. The following important operations are carried out by the digital coherent signal preprocessing subsystem: analog-to-digital conversion; generalized signal processing algorithm of linear-frequency-modulated target return signals; suppression of passive interferences; coherent accumulation of target return signals; and generation of data to solve the problems related to stabilization of the probability of false alarm. The digital coherent signal preprocessing subsystem is realized by individual microprocessor sub-subsystem or distributed set of specific microprocessor subsystems. Choosing the right approach to design and construct the digital coherent signal preprocessing subsystem and justification of structure of the corresponding microprocessor sub-subsystem or set of microprocessor sub-subsystems is the first and foremost requirement of the system design process.

It is worthwhile to employ hardware in the specific digital coherent signal preprocessing subsystem designed based on VLSI microprocessor sub-subsystem to solve the following problems: generalized signal processing algorithm, suppression of passive interferences, computation of target return signal amplitudes, and estimation of interference and noise power. To solve the problems of Doppler filtering, it is worthwhile to implement the FFT processor or filter using effective and simple methods and procedures such as complex multiplication. All elements of the specific digital coherent signal preprocessing subsystem designed based on VLSI microprocessor sub-subsystem must possess high reliability and require low power consumption. High reliability can be achieved by equipment reservation or information redundancy. Low power consumption can be achieved by very careful circuit designing and implementation of VLSI with small power dissipation.

The first element of the global structure of digital coherent target return signal preprocessing subsystem is the ADC of the target return signals at the phase detector output. Next element of the total structure of the digital coherent target return signal preprocessing subsystem is the DGD for linear-frequency-modulated target return signals that can be realized in the time domain employing the nonrecursive filter or in the frequency domain using the FFT processor or filter. To improve the DGD speed of operation, we can use properties of parallelism applied to the matched filtering algorithms. For this reason, to realize the DGD with high speed of operation we can use the parallel mixers based on ROM, parallel adders with simultaneous addition of several numbers, and parallel registers. The nonrecursive smoothing filter with short impulse response is placed after DGD. The main tasks of this filter include the following: first, suppressing the signal side lobes at the DGD output and, second, decreasing the sampling rate to frequency corresponding to requirements of the sampling theorem. A decrease in the sampling rate in m times (m is the integer) is carried out by the element providing a sample of each mth element from sequence of input sampled target return signals $\{x(kT_s)\}$. The cancellation of correlated interference is realized after the matched filtering. The operations discussed are part of the first stage of the digital coherent signal processing. To ensure the required reliability all hardware elements of the first stage must be doubled. The output signals of the main and reserved hardware set are compared by specific modules. Comparison results are used by both the searching system and the default system.

In accordance with the global structure of digital signal processing and control system used by the radar complex (see Figure 9.1), the noncoherent signal preprocessing subsystem solves the following problems: (a) noncoherent accumulation of signals after the coherent signal processing of each from three 8-pulse bursts by the FFT processor or filter; (b) noncoherent accumulation of interference and noise power estimations for all Doppler channels, except for the zero Doppler channel, and for all resolution elements by radar range; (c) corrections of the interference and noise power map, forming the adaptive detection thresholds; (d) signal detection to be provided by comparison between the signals and the corresponding thresholds computed for each cell's "radar range–Doppler frequency"; (e) estimation of the target azimuth by a set of signals exceeding the detection threshold at three neighboring positions of the radar antenna directional diagram by azimuth.

The requirements to speed of operation of the noncoherent target return signal preprocessing microprocessor subsystem are sufficiently high. However, the noncoherent target return signal preprocessing operations can be conducted in parallel using two microprocessor subsystems simultaneously. For example, the first microprocessor subsystem must perform the noncoherent accumulation of target return signals, corrections in the interference and noise power map, and the target azimuth estimation. The second microprocessor subsystem must carry out the noncoherent accumulation of interference and noise power, threshold forming, and signal detection. In this case, the required speed of operation of each microprocessor subsystem is not more than 2×10^3 operations per second.

The main tasks of the digital signal reprocessing subsystem include the following: to realize the target trajectory detection, the target tracking and target trajectory tracking, the filtering of target trajectory parameters, and other digital signal processing algorithms accomplished during the target processing stages, and to provide precise information to the user. The analysis performed thus shows that the digital signal reprocessing problems do not require the high computational resources in comparison with the previous stages of target return signal processing. This fact allows us to assign for the digital signal reprocessing microprocessor subsystem the problems of control over all elements and parameters of the CRS and the current hardware and computational process control problems. Taking into consideration that the control optimization problems are not solved in dynamic mode in the automatic "radar complex–global digital signal processing complex" system and the control system is constructed as the system with a fixed, predetermined sequence of control commands.

Functional requirements, cost, reliability, conveniences in fault lookup, maintainability are the important factors that are taken into consideration in the course of designing the global digital signal processing structure of the CRS. The functional requirements to individual elements of this structure discussed and considered previously are evidences of differences in the density of information flow processed by the central computer system of radar complex. The digital coherent target return preprocessing subsystem working in real time has the greatest service load. For this reason, and given the technological breakthroughs in the signal detection and signal processing theory and computer applications, there is a need to design and construct a specific nonprogrammable microprocessor subsystem for the digital coherent target return preprocessing subsystem working in real time, which consists of the specific adders, mixers, convolution networks, memory devices based on VLSI and charge-injection devices. Without any doubts, we can state that the possibility of designing and constructing such specific microprocessor subsystems is the foundation stone for developing processes that solve the problems encountered in automatic digital target return signal processing in the functioning of the CRS under real-time conditions.

REFERENCES

1. Moon, T.K. and W.C. Stirling. 2000. *Mathematical Methods and Algorithms for Signal Processing*. Upper Saddle River, NJ: Prentice Hall, Inc.
2. Billingsley, L.B. 2002. *Low-Angle Radar Land Clutter—Measurements and Empirical Models*. Norwich, NY: William Andrew Publishing, Inc.
3. Richards, M.A. 2005. *Fundamentals of Radar Signal Processing*. New York: McGraw Hill, Inc.
4. Levy, B.C. 2008. *Principles of Signal Detection and Parameter Estimation*. New York: Springer Science + Business Media, LLC.

10 Global Digital Signal Processing System Analysis

10.1 DIGITAL SIGNAL PROCESSING SYSTEM DESIGN

10.1.1 Structure of Digital Signal Processing System

We consider the following problem—to design and construct a complex radar system (CRS) based on the all-round surveillance radar with the uniform antenna rotation. The main task of this CRS is to search for all the detected targets, to integrate information, and to make a generalization of the air situation. An additional task is target tracking with the high accuracy of important targets from a user viewpoint. In this case, the following version of a global digital signal processing system structure may be used to design and construct the CRS based on the all-round surveillance radar with the uniform antenna rotation (see Figure 10.1) [1,2].

Detection and tracking of all targets in an all-round surveillance radar coverage with accuracy that is sufficient to reproduce and estimate data of the air situation are carried out by the so-called rough channel of a global digital signal processing system covering the whole scanned area. The rough channel consists of the following subsystems: the binary signal quantization, the specific microprocessor network for target return signal preprocessing, and the microprocessor network for digital signal reprocessing and control. Employment of binary quantization and simplified versions of digital signal processing algorithms allows us to implement the microprocessor networks and sets in this channel of global digital signal processing system of the all-round surveillance radar with the uniform antenna rotation.

Accurate tracking of the targets that are important from a user viewpoint is carried out by radar measurers. The radar measurer is the digital device assigned to solve the signal processing problems for one or several targets. The process of accurate target tracking can be organized at least by two ways:

1. An individual radar measurer is assigned for each target and it is capable of target tracking within the period when the target is within the limits of radar coverage. These radar measurers are constructed by the principle of automatic tracking systems and are called moving target indicators (MTIs). The MTI number corresponds to the number of targets subjected to tracking, i.e., $N_{\text{MTI}} = N_{\text{tg}}$. In this case, we have the digital signal processing system with computational parallelism by a set of objects subjected to processing (see Chapter 7).
2. The MTI system operates based on the main principles of queuing theory. The time sequence of signals received by each target tracking gates is considered as a request queue. In doing so, the targets subjected to tracking can be served by the lesser number of nontracking at the present time MTI, i.e., $N_{\text{MTI}} < N_{\text{tg}}$. Interaction between the rough and "accurate" channels of global digital signal processing system can be organized in the following form. All targets are checked based on their importance by the queuing system using the digital signal reprocessing microprocessor network. The coordinates of center and dimensions of the preliminary target lock-in gate are sent to the device of physical gating by targets, an important criterion of which exceeds the given threshold and trajectory parameters of the target subjected to accurate tracking are sent to the dispatch device of MTI system simultaneously. Under scanning of the corresponding direction, the device of physical gating organizes a selection of signals

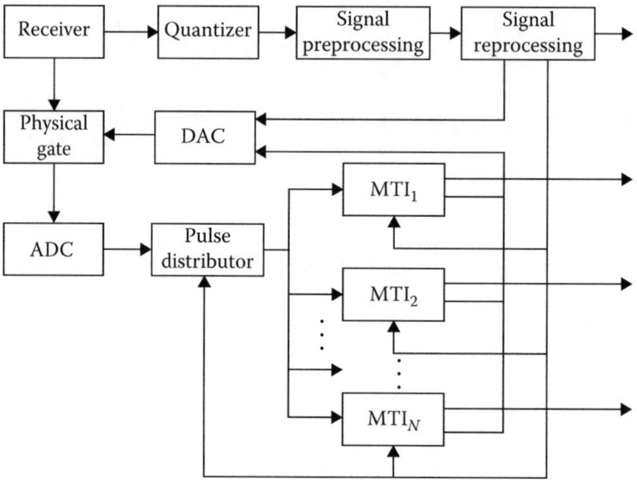

FIGURE 10.1 Global digital signal processing system structure.

of the gated area of radar coverage. The target return signals received by gates after an analog-to-digital converter (ADC) come in at the input of MTI selected by the dispatch device and are processed by MTI. Initial values of target trajectory parameters determined by the digital signal reprocessing microprocessor network are used by MTI at the first stage of accurate definition of the target trajectory parameters. Furthermore, the extrapolated coordinates of gate centers and dimensions are computed and are sent to the device of physical gating by corresponding MTI networks. Under unspecified reset action of target trajectory tracked by MTI, the target relock-in is carried out using the global digital signal processing system rough channel. The considered structure of the global digital signal processing system possesses a high reliability and requires moderate designing charges and operating costs.

10.1.2 Structure and Operation of Nontracking MTI

Let us consider the nontracking MTI system. The nontracking MTI is the microprocessor network for digital signal processing. The signals come in at the nontracking MTI input from the radar coverage zone limited by the physical gate. The nontracking MTI structure is shown in Figure 10.2, which shows that the nontracking MTI possesses the same functions as those under the target tracking. The essential difference is matching the signal preprocessing operations with a selection of target pips in the target tracking gate. In doing so, owing to limiting gate dimensions it is possible to realize the signal quantization and signal processing algorithms. As to the target trajectory parameter smoothing operations, the nontracking MTI ensures the maneuvering of target tracking.

Consider in detail the signal preprocessing and selection in the target tracking gate algorithm. The problems of target detection, definition of target position, and selection of target return signals within the limits of gate are reduced to checking several hypotheses. Moreover, a signal absence in the gate is considered as the hypothesis $\mathcal{H}_0$, and the alternative hypotheses are the hypotheses about the signal presence in a single (or several) gate cell. We consider the gate cell as the gate volume (or square) element limited by sampling intervals on the radar range, azimuth, and tilt angle coordinates. An optimal procedure under processing the observation results with the purpose of checking the statistical hypotheses is a generation of the likelihood ratio and its comparison with the threshold that is selected from acceptable losses attributed to correct or incorrect decisions. Furthermore, we consider the signal processing algorithm of binary signals in two-dimensional

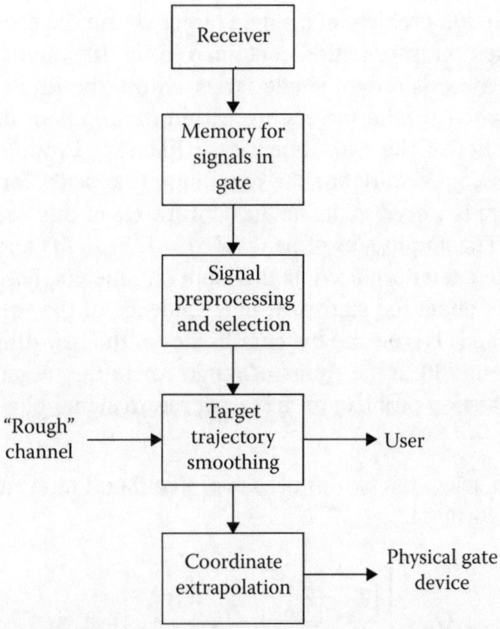

FIGURE 10.2 Nontracking MTI structure.

gate, as an example. Under optimal binary signal processing within the limits of two-dimensional (by radar range and azimuth) gate, the logarithm of likelihood ratio takes the form [3]

$$\ln l = \sum_{i,j} d_{ij}\varpi(ij \mid i_0 j_0), \tag{10.1}$$

where

d_{ij} are the values of binary signals, "1" or "0," in ijth gate cell; $i = 1,\ldots, n$ and $j = 1,\ldots, m$ are the number of discrete gate elements by the radar range and azimuth, respectively

$\varpi(ij \mid i_0 j_0)$ is a set of the weight coefficients, with which the signals 1 and 0 are added in the gate cells under formation of the likelihood ratio

The concrete form of the function $\varpi(ij \mid i_0 j_0)$ depends on the shape of amplitude envelope of two-dimensional signal that is processed. If we consider the signal model in the form of two-dimensional pdf surface in the polar coordinate system, the coordinates r and β, the weighting function takes the following form:

$$\varpi(ij \mid i_0 j_0) = C_1 \exp\left[-\frac{C_2\Delta_r^2(i-i_0)^2}{2\delta_r^2}\right]\exp\left[-\frac{C_2\Delta_\beta^2(i-i_0)^2}{2\delta_\beta^2}\right], \tag{10.2}$$

where

C_1 and C_2 are the constant values

Δ_r and Δ_β are the gate sampling intervals by the coordinates r and β

δ_r and δ_β are the half signal bandwidth by the coordinates r and β at the level exp (−0.5)

i_0 and j_0 are the coordinates of the maximal signal amplitude envelope

The two-dimensional likelihood ratio surface, which is the initial likelihood ratio to detect and select the target return signal pips within the limits of gate, is obtained as a weighting the binary target return signals by the weight function (10.2). Moreover, in this case, the two-dimensional likelihood ratio surface peaks contain all information about the signal presence and signal coordinates.

The detection and selection problem of a single target within the limits of the gate is assigned in the following way based on information contained in the likelihood ratio maxima. First, we take the hypothesis that there is only a single target within the limits of the gate. In the considered case, the event, when several targets are within the limits of the gate, is possible but is improbable. Using the relief of the two-dimensional likelihood ratio surface with M peaks of different heights, there is a need to define the maximum (the peak) formed by the target return signal and, if a "yes," there is a need to define the coordinates of this two-dimensional likelihood ratio surface maximum. The amplitudes of peaks Z_l ($l = 1, 2, \ldots, M$) and their coordinates ξ_l and η_l with respect to the gate center are used as the input parameters, based on which the decision is made. If the hypothesis about the statistical independence of the two-dimensional likelihood ratio surface peak amplitudes is true and the coordinates of the two-dimensional likelihood ratio surface maxima are known within the limits of range where the target return signal is present, the optimal detection–selection problem of the target return signal pips within the limits of gate is solved in two steps [4].

Step 1: There is a need to select the two-dimensional likelihood ratio surface maximum with the number l^* with quadratic form

$$Q_{l^*} = \left[\frac{\left(Z_{l^*} - \bar{Z}\right)^2}{2\sigma_Z^2} + \frac{\xi_{l^*}^2}{2\sigma_\xi^2} + \frac{\eta_{l^*}^2}{2\sigma_\eta^2}\right] = \min_{\{l\}}, \tag{10.3}$$

where

$\bar{Z}$ is the average amplitude of the two-dimensional likelihood ratio surface in the signal domain
σ_Z^2 is the amplitude variance of the two-dimensional likelihood ratio surface in the signal domain

Step 2: Compare the obtained quadratic form Q_{l^*} with the threshold defined based on the acceptable probability of error decisions and, in the case of exceeding the threshold, issue a decision about the target pip detection. The coordinates of the two-dimensional likelihood ratio surface maximum are considered as the coordinates of the detected target pip within the limits of gate.

The described optimal algorithm is difficult to realize in practice owing to the high work content to estimate the average value $\bar{Z}$ of amplitude and amplitude variance σ_Z^2 of the two-dimensional likelihood ratio surface in the signal domain. Therefore, the following simplified algorithms of target pip detection and selection are used in practice:

- Algorithm of detection and selection using the two-dimensional likelihood ratio surface maximum—at first, there is a need to define the amplitude of the two-dimensional likelihood ratio surface maximum within the limits of gate and compare with the threshold. Target detection pip is fixed if the amplitude of the two-dimensional likelihood ratio surface maximum exceeds the threshold and the coordinates are defined by a position of the two-dimensional likelihood ratio surface maximum within the limits of gate.
- Algorithm of detection and selection using the two-dimensional likelihood ratio surface maximum amplitude exceeding the threshold and possessing a minimal deviation with respect to the gate center.

The first algorithm possesses the best selecting features evaluated by the probability of detection P_D and selection at the fixed probability of false alarm P_F. For each case, there is a need to carry out a detailed analysis and find the trade-off, taking into consideration the requirements of the problem solution quality and available computational resources with the purpose of selecting the acceptable signal detection and selection algorithms within the limits of the target tracking gate of the nontracking MTI.

10.1.3 MTI as Queuing System

Each MTI consists of the following blocks solving their own functions (see Figures 10.3 through 10.5):

- Static memory of the digital target return signals within the limits of physical target tracking gate
- Detector–selector assigned to detect and select the target return signals within the limits of physical target tracking gate
- Measurer assigned to estimate the target trajectory parameters, to extrapolate the target trajectory coordinates, and to compute dimensions of physical target tracking gate

Static memory can be realized in the matrix form (the two-dimensional case) or as a set of matrices (under the target return signal processing within the limits of three-dimensional physical target tracking gate) of memory cells. Each memory cell stores information obtained as a result of the target return signal sampling within the limits of the corresponding volume or area element of the physical target tracking gate. Processing of information about the target return signals stored by the memory is carried out after filling all physical target tracking gate cells. After processing of the stored information about the target return signals, the corresponding memory matrix is ready to receive new information. One or several specific microprocessor networks can be used as the *detector–selector*. Taking into consideration the large computation content under realization of signal processing algorithms to estimate the target trajectory parameters and coordinates in the course of the target return signal reprocessing and, additionally, the necessity to store previous information about each target tracked trajectory, it is worth constructing the MTI based on a set of microprocessor networks.

Under the target tracking by several MTIs, we are able to reduce a set of blocks owing to an efficient structural organization. Now, consider the following versions:

- *System* "$n-1-1$" (see Figure 10.3). The memory is the n-channel queuing system with losses and the detector–selector and MTI are the one-channel queuing systems with request queue waiting; the system $n-1-1$ is the three-phase queuing system with input failures.
- *System* "$n-n-1$" (see Figure 10.4). This system is different from the previous one in that each detector–selector has own memory and the "memory–detector–selector" devices connected in series can be considered as a single queuing system channel. The totality of the memory–detector–selector devices is the n-channel queuing system with losses. The MTI, as earlier, is considered as the one-channel queuing system with request queue waiting.
- *System* "$n-m-1$" (see Figure 10.5). For this system, the queuing request forming at the n-channel memory output comes in at the m-channel detector–selector passing a splitter. Generally, the splitter operates as the associate device or probabilistic automation unit, a mode of which is defined by characteristics and parameters of the output request queue incoming from the memory and by a mode of the second device. The second device is the m-channel queuing system with request queue waiting. The data forming at the second device output come in at the input of the third device, which is the one-channel queuing system with request queue waiting.

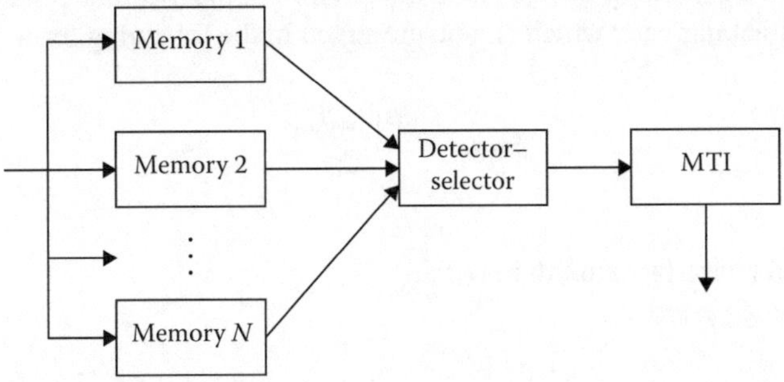

FIGURE 10.3 Nontracking MTI blocks organizing the system "$n-1-1$."

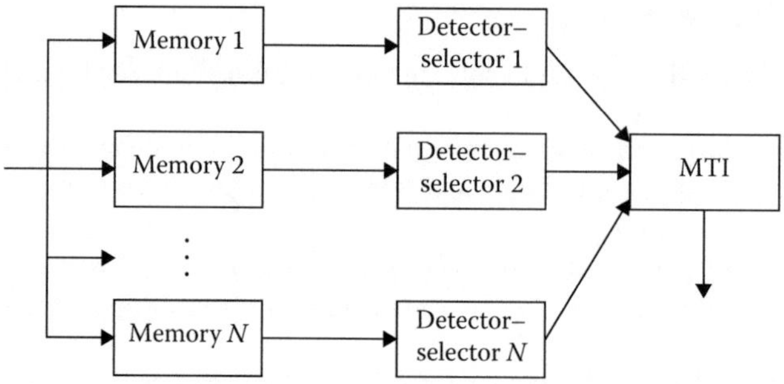

FIGURE 10.4 Nontracking MTI blocks organizing the system "$n - n - 1$."

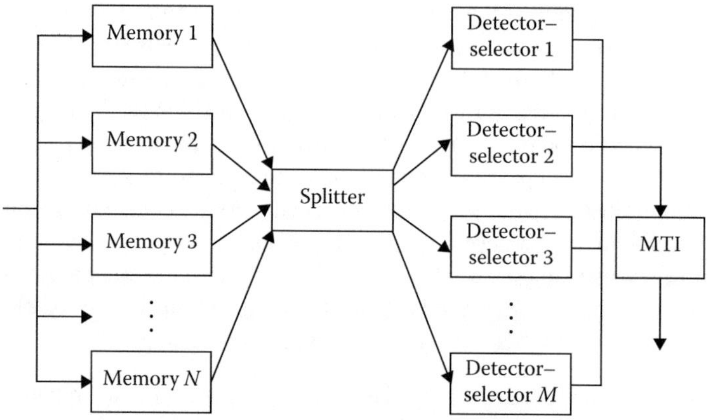

FIGURE 10.5 Nontracking MTI blocks organizing the system "$n - m - 1$."

All considered and discussed versions of target tracking by several MTI systems are the *three-phase queuing systems*. Each phase is represented by one or several devices of the queuing system connected in series. In a general case, the request queuing time for each device is the random variable with pdf available to investigate and define. It is assumed that there is a simple request queue at the first-phase input of the queuing system. The request queue coming in at the first device input is immediately served if, at least, one channel of the queuing system is free from service; otherwise, the request is rejected and flushed. The requests carried out by the first device are processed sequentially at the next phases, in other words, the request loss at each next phase or stage is inadmissible. In line with this, the memory device with the purpose of storing requests in line should be provided before the second and third phases of the queuing system.

Now, let us discuss the pdf of request queuing time at different phases of the queuing system. The request queuing time for memory device is the time to fill out the memory cells within the limits of the physical target tracking gate, which can be presented in the following form:

$$\tau_{\text{memory}} = \frac{\Delta\beta_{\text{gate}} T_{\text{scan}}}{2\pi}, \tag{10.4}$$

where

$\Delta\beta_{\text{gate}}$ is the angular size by azimuth in radian
T_{scan} is the scanning period

Dimensions of the physical target tracking gate are determined based on the required probability of target pip hitting within the limits of gate taking into consideration the random errors in the course of target trajectory coordinate measurement, random error of coordinate extrapolation, and dynamical errors caused by maneuvering targets. Errors of measurements and extrapolation by each individual coordinate are subjected to the normal Gaussian pdf with zero mean and known variance.

As noted in Chapter 4, under stable target tracking, the physical target tracking gate dimensions by each coordinate are minimal, i.e., $\Delta Z_{\min}$ $\{Z = \{r, \beta\}\}$. The target maneuver, regular and random deviations of target return signal power, and misses of target pips on target trajectory track lead to an increase in the physical target tracking gate dimensions in comparison with minimal ones. At the same time, the probability of event that the physical target tracking gate dimensions are minimal or close to minimal is the highest. Distribution corresponding to described process of changes in dimensions of the physical target tracking gate can be presented in the following form:

$$p(\Delta Z) = \begin{cases} \dfrac{2}{\sqrt{2\pi}\sigma_{\Delta Z}} \exp\left[-\dfrac{(\Delta Z - \Delta Z_{\min})^2}{2\sigma_{\Delta Z}^2}\right], & \Delta Z \geq \Delta Z_{\min}, \\ 0, & \Delta Z < \Delta Z_{\min}, \end{cases} \tag{10.5}$$

where σ_Z^2 is the variance of changes in the physical target tracking gate dimensions by the coordinate Z. In this case, the pdf of request queuing time takes the following form:

$$p(\tau_{\text{memory}}) = \begin{cases} \dfrac{2}{\sqrt{2\pi}\sigma_{\tau_{\text{memory}}}} \exp\left[-\dfrac{(\tau_{\text{memory}} - \tau_{\text{memory}_{\min}})^2}{2\sigma_{\tau_{\text{memory}}}^2}\right], & \tau_{\text{memory}} \geq \tau_{\text{memory}_{\min}}, \\ 0, & \tau_{\text{memory}} < \tau_{\text{memory}_{\min}}. \end{cases} \tag{10.6}$$

Sometimes the physical target tracking gate dimensions are chosen as constant values determined at the maximum total error of extrapolation. For this case, the request queuing time by memory will be constant.

When the simplest signal processing algorithm for definition of the maximal weighted sum of the target return signals is used in the detector–selector, the signal processing operations are the following:

- Definition of the weighted sum amplitude of the target return signals for each physical target tracking gate cell
- Sequential comparison of amplitudes of the target return signals for each physical target tracking gate cell with the purpose of choosing the maximum one
- Comparison of selected amplitudes of the target return signals for each physical target tracking gate cell with the threshold and making a decision about detection of the target pip within the limits of the physical target tracking gate

In this case, the analysis time is defined by dimensions of the physical target tracking gate. Furthermore, we consider the case of two-dimensional physical target tracking gate. According to (10.5), the distribution of normalized dimensions of the two-dimensional physical target tracking gate by each coordinate is defined as

$$p(x) = \frac{2}{\sqrt{2\pi}} \exp\left[-\frac{(x - x_0)^2}{2}\right], \quad x \geq x_0, \tag{10.7}$$

$$p(y) = \frac{2}{\sqrt{2\pi}} \exp\left[-\frac{(y - y_0)^2}{2}\right], \quad y \geq y_0, \tag{10.8}$$

where

$$x = \frac{\Delta\beta_{\text{gate}}}{\sigma_{\beta_{\text{gate}}}}; \quad y = \frac{\Delta r_{\text{gate}}}{\sigma_{r_{\text{gate}}}}; \quad x_0 = \frac{\Delta\beta_{\text{gate}}^{\min}}{\sigma_{\beta_{\text{gate}}}}; \quad y_0 = \frac{\Delta r_{\text{gate}}^{\min}}{\sigma_{r_{\text{gate}}}}. \tag{10.9}$$

In this case, the pdf of two-dimensional physical target tracking gate area S_{gate} is defined as the pdf of product between the random variables x and y with the pdf given by (10.7) and (10.8), respectively. The cumulative probability distribution function of the two-dimensional physical target tracking gate area S_{gate} is given by

$$F(S_{\text{gate}}) = \int_{x_0}^{\infty}\int_{\frac{S_{\text{gate}}}{x}}^{\infty} p(x)p(y)dxdy = \frac{2}{\pi}\int_{x_0}^{\infty}\int_{\frac{S_{\text{gate}}}{x}}^{\infty} \exp\left[-\frac{(x - x_0)^2}{2}\right]\exp\left[-\frac{(y - y_0)^2}{2}\right]\frac{dx}{x}. \tag{10.10}$$

Differentiating (10.10) by S_{gate}, we obtain the pdf in the following form:

$$p(S_{\text{gate}}) = \frac{2}{\pi}\int_{x_0}^{\infty} \exp\left\{-0.5(x - x_0)^2\right\}\exp\left[-0.5\frac{(S_{\text{gate}} - S_0)^2}{x^2}\right]\frac{dx}{x}. \tag{10.11}$$

Equation 10.11 is not integrated in an explicit form. Results of numerical integration show that at small values of S_0 the pdf of S_{gate} can be approximated by exponential pdf with the shift given by the following form:

$$p(S_{\text{gate}}) = \gamma \exp\{-\gamma(S_{\text{gate}} - S_0)\}, \quad S_{\text{gate}} > S_0. \tag{10.12}$$

Under increasing S_0, the pdf of the two-dimensional physical target tracking gate area S_{gate} is approximated by the truncated normal Gaussian pdf:

$$f(S_{\text{gate}}) = \frac{2}{\sqrt{2\pi}} \exp\{-0.5(S_{\text{gate}} - S_0)\}, \quad S_{\text{gate}} > S_0. \tag{10.13}$$

In accordance with (10.12) and (10.13), the pdf of request queuing time in the detector–selector can also be approximated either by the truncated exponential pdf with the shift in the following form:

$$p(\tau_{\text{DS}}) = \mu \exp\{-\mu(\tau_{\text{DS}} - \tau_0)\}, \quad \tau_{\text{DS}} \geq \tau_0, \tag{10.14}$$

where μ is the intensity of request queuing by the detector–selector or by the truncated and shifted normal Gaussian pdf given by

$$p(\tau_{\text{DS}}) = \frac{2}{\sqrt{2\pi\sigma_{\tau_{\text{DS}}}^2}} \exp\left[-\frac{(\tau_{\text{DS}} - \tau_0)^2}{2\sigma_{\tau_{\text{DS}}}^2}\right], \quad \tau_{\text{DS}} > \tau_0. \tag{10.15}$$

Since generally, the exact definition of the request queuing time pdf is impossible, we use two types of approximation given by (10.14) and (10.15) in further analysis of the detector–selectors. Finally, in the considered case we think that the time required to complete all signal processing operations is a constant value.

Now, consider and discuss the quality of service (QoS) of the multiphase queuing system. We can think that QoS is based on the probability of failure under a service of the next request as a function of input memory capacity and the average time of request processing by the queuing system:

$$\overline{\tau}_{\Sigma}^{QS} = \sum_{i=1}^{3} \overline{\tau}_i + \sum_{i=2}^{3} \overline{t}_i^{\text{wait}}, \tag{10.16}$$

where

$\overline{\tau}_i$ is the average request queuing time during the ith phase
$\overline{t}_i^{\text{wait}}$ is the average waiting time for request queue before the ith phase or stage

Taking into consideration these QoS indicators under the known number of operations that are required for a single request queuing, we can define and estimate the required speed of microprocessor network operation realizing each phase of queuing systems.

Difficulties under analysis of multiphase queuing systems are the following. At all cases of practical importance, the output stream of phase takes more complex form in comparison with the incoming request queue. In some cases, the output request queue can be approximated by the simplest incoming stream with the same parameters. Then we can use analytical procedures and methods of the queuing theory to analyze the next phase or stage. If this approximation is impossible, then the only method to investigate the stream is the simulation. Rational combination of analytical and simulation methods and procedures allows us to solve the problem of the three-phase signal processing system analysis using the MTI system for any design and construction version.

10.2 ANALYSIS OF "n – 1 – 1" MTI SYSTEM

10.2.1 Required Number of Memory Channels

Since according to the operational conditions of MTI system the digital signal processing within the limits of the physical target tracking gate can be started only after filling out all memory cells of this gate, the request queuing time in memory is equal to the time of scanned angle by radar antenna corresponding to the azimuth dimension of the physical target tracking gate. This time is distributed according to (10.6). In this case, the average memory request queuing time is given by

$$\overline{\tau}_{\text{memory}} = \tau_{\text{memory min}} + \frac{2\sigma_{\tau_{\text{memory}}}}{\sqrt{2\pi}}. \tag{10.17}$$

The acceptable probability of request losses in memory is given and equal to, as a rule, $P_{\text{loss}} = 10^{-3} - 10^{-4}$. It is assumed that the request queue at the memory input is the simplest with the density γ_{in} that is given. The density γ_{in} is assigned based on the possible number of targets liable to tracking.

Using the Erlang formula [4]

$$P_{\text{loss}} = \frac{(\gamma_{\text{in}}\overline{\tau}_{\text{memory}})^N / N!}{\sum_{k=0}^{N} (1/k!)(\gamma_{\text{in}}\overline{\tau}_{\text{memory}})^k}, \tag{10.18}$$

we can determine the required number of channels N (gates). The output request queue can be considered as an iteration of the input request queue, i.e., the simplest at the low probability of failure.

10.2.2 Performance Analysis of Detector–Selector

QoS factors of the detector–selector as the one-channel queuing system are the average request queuing time $\overline{\tau}_{DS}$ and the average waiting time for service $\overline{t}_{DS}^{wait}$. At first, consider the case of the exponential with shift pdf for $\overline{\tau}_{DS}$. In this case,

$$\overline{\tau}_{DS} = \int_{\tau_0}^{\infty} \tau_{DS} p(\tau_{DS}) d\tau_{DS} = \tau_0 + \sigma_{\tau_{DS}}, \tag{10.19}$$

where $\sigma_{\tau_{DS}} = \mu^{-1}$. The variance of request queuing time is determined in the following form:

$$\text{Var}(\tau_{DS}) = \overline{\tau}_{DS}^2 + \sigma_{\tau_{DS}}^2 = \tau_0^2 + 2\tau_0\sigma_{\tau_{DS}} + 2\sigma_{\tau_{DS}}^2 = \tau_0^2\left[1 + \frac{2\sigma_{\tau_{DS}}}{\tau_0} + 2\frac{\sigma_{\tau_{DS}}^2}{\tau_0^2}\right]. \tag{10.20}$$

Denote $\nu = \tau_0\sigma_{\tau_{DS}}^{-1}$. Then, we obtain

$$\overline{\tau}_{DS} = \tau_0(1 + \nu^{-1}); \tag{10.21}$$

$$\text{Var}(\tau_{DS}) = \tau_0^2(1 + 2\nu^{-1} + 2\nu^{-2}). \tag{10.22}$$

Under the unpriority queuing, the average waiting time is given by

$$\overline{t}_{DS}^{wait} = \frac{\gamma_{in}}{2(1 - \gamma_{in}\overline{\tau}_{DS})}\text{Var}(\tau_{DS}). \tag{10.23}$$

Substituting (10.22) into (10.23), we obtain

$$\overline{t}_{DS}^{wait} = \frac{\gamma_{in}\tau_0^2}{2(1 - \gamma_{in}\overline{\tau}_{DS})}(1 + 2\nu^{-1} + 2\nu^{-2}). \tag{10.24}$$

Expressing τ_0 over $\overline{\tau}_{DS}$ based on (10.21), after elementary algebra we obtain

$$\overline{t}_{DS}^{wait} = \frac{\gamma_{in}\tau_0^2}{2(1 - \gamma_{in}\overline{\tau}_{DS})}\left[1 + \frac{1}{(1 + \nu)^2}\right]. \tag{10.25}$$

Denoting $\chi_{DS} = \gamma_{in}\overline{\tau}_{DS}$, we obtain finally

$$\overline{t}_{DS}^{wait}\upsilon_{DS} = \frac{\chi_{DS}}{2(1-\chi_{DS})}\left[1+\frac{1}{(1+\nu)^2}\right], \tag{10.26}$$

where
$\upsilon_{DS} = \overline{\tau}_{DS}^{-1}$
χ_{DS} is the loading factor of the detector–selector

Formula (10.26) allows us to determine the detector–selector QoS factors as a function of its loading and relative shift ν of the pdf.

If the request queuing time τ_{DS} is distributed according to (10.15), then

$$\overline{\tau}_{DS} = \tau_0 + \frac{2\sigma_{\tau_{DS}}}{\sqrt{2\pi}}; \tag{10.27}$$

$$\text{Var}(\tau_{DS}) = \tau_0^2 + \sigma_{\tau_{DS}}^2 + \frac{4\tau_0\sigma_{\tau_{DS}}}{\sqrt{2\pi}}. \tag{10.28}$$

Taking into consideration that $\nu = \tau_0\sigma_{\tau_{DS}}^{-1}$ and $2\times\left(\sqrt{2\pi}\right)^{-1}$, we obtain

$$\overline{\tau}_{DS} = \tau_0\left[\frac{\nu+0.8}{\nu}\right]; \tag{10.29}$$

$$\text{Var}(\tau_{DS}) = \overline{\tau}_{DS}^2\frac{1+1.6\nu+\nu^2}{(\nu+0.8)^2}. \tag{10.30}$$

The average waiting time is determined as

$$\overline{t}_{DS}^{wait} = \frac{\gamma_{in}\text{Var}(\tau_{DS})}{2(1-\chi_{DS})} = \frac{\gamma_{in}\overline{\tau}_{DS}^2}{2(1-\chi_{DS})}\times\frac{1+1.6\nu+\nu^2}{(\nu+0.8)^2} \tag{10.31}$$

or

$$\overline{t}_{DS}^{wait}\upsilon_{DS} = \frac{\chi_{DS}}{2(1-\chi_{DS})}\left[1+\frac{0.36}{(\nu+0.8)^2}\right]. \tag{10.32}$$

Formula (10.32) allows us to determine also the detector–selector performance at the corresponding pdf of request queuing time. In particular, (10.26) and (10.32) show that an increase in relative shift of the pdf of request queuing time at the same loading factor χ_{DS} leads to a decrease in the average relative waiting time. As $\nu \to \infty$, the limiting value of this time tends to approach the value corresponding to the constant request queuing time. In this case, the request queue is stored in the input memory device. Computing the memory capacity by (10.30), there is a need to take into consideration the average waiting time for request queue in the detector–selector adding the average request queuing time for queuing requests in the memory device:

$$\overline{\tau}'_{memory} = \overline{\tau}_{memory} + \overline{t}_{DS}^{wait}. \tag{10.33}$$

Further, knowing the average request queuing time $\bar{\tau}_{DS}$ and the number of reduced operations required for a single realization of the target detection and selection algorithm, we are able to determine the effective speed of operation of the detector–selector:

$$V_{DS}^{effective} = \frac{\bar{N}_{DS}}{\bar{\tau}_{DS}} = \frac{\gamma_{in}\bar{N}_{DS}}{\chi_{DS}}. \quad (10.34)$$

Now, consider the request queue at the detector–selector output taking into consideration that the loading factor χ_{DS} is close to unit, i.e., $\chi_{DS} = 0.9 \div 0.95$. The detector–selector output request queue is defined by instants of the request incoming for service, the request queuing time, and the waiting time to start a service. Let $t_1, t_2,\ldots, t_{i-1}, t_i, t_{i+1}$ be the instants of the request queue incoming at the detector–selector input, t_i^{wait} be the waiting time for request queue coming in at the instant t_i, and $\tau_i = \tau_0 + \xi_i$ be the request queue time of the ith request, where τ_0 is the constant and ξ_i is the random component of the request queue time. In the course of service of the request queue, two cases are possible.

First case: The request queue comes in at the instant t_i when the detector–selector serves the previous request and stands in a queue (see Figure 10.6a and b). The waiting time to start the given request queue is t_i^{wait}. Denote the time intervals between two requests at the detector–selector input and output $\Delta t_i = t_i - t_{i-1}$ and $\Delta t_i' = t_i' - t_{i-1}$, respectively. Then, in the considered case (Figure 10.6b), the time interval at the detector–selector output between queuing requests is equal to the request queue time:

$$\Delta t_i < t_{i-1}^{wait} + \tau_{i-1}, \quad (10.35)$$

$$\Delta t_i' = \tau_0 + \xi_i = \tau_i. \quad (10.36)$$

Second case: The request queue comes in for service at the instant t_{i+1} when the detector–selector is free and accepts the request queue immediately (Figure 10.6c). In this case, the time interval between two neighboring queuing requests at the detector–selector output is greater than the request queue time interval on the value of downtime Δt:

$$\Delta t_{i-1}' = t_{i+1}' - t_i' = \tau_{i+1} + \Delta t. \quad (10.37)$$

Since, by the initial condition, the loading factor χ_{DS} is high, the probability of the second case is proportional to $1 - \chi_{DS}$ and low by magnitude. Therefore, we are able to think that there is a request queue at the detector–selector input, i.e., *Case 1* is realized with the high probability.

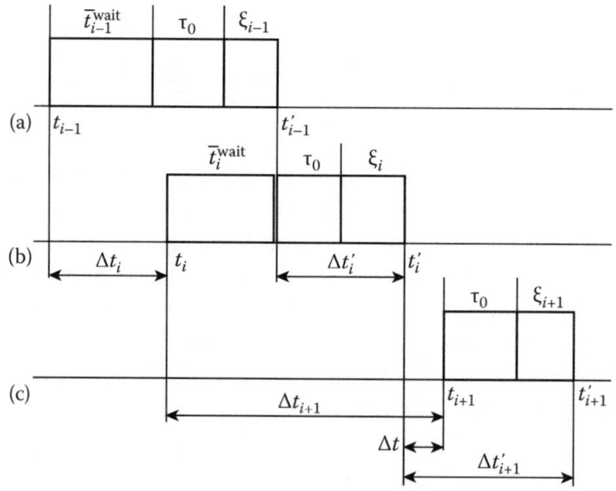

FIGURE 10.6 Time diagram of request queue processed by the detector–selector: (a) service of previous request; (b) the request stands in queue; and (c) the request queue is accepted and processed.

In accordance with the earlier discussion, a distribution of time intervals between the queuing requests served by the detector–selector (the detector–selector input) can coincide with distribution of the request queue time in the detector–selector. The parameter of request queue at the detector–selector output coincides with the parameter of the incoming request queue.

10.2.3 Analysis of MTI Characteristics

The MTI is considered as the one-channel queuing system with waiting time. The request queue served by the detector–selector with the parameter γ_{in} and the pdf of time intervals between the requests coinciding with the pdf of request queue time in the detector–selector given by (10.14) and (10.15) comes in at the MTI input. The request queue time in MTI is constant and equal to $\tau_{\text{MTI}} = a$. Depending on the relation between the constant constituent τ_0 in (10.36) and the duration of the request queuing time interval in MTI $\tau_{\text{MTI}} = a$, the following cases are possible.

Case 1: $a \le \tau_0$. MTI can forever serve the request queue before the next request queue comes in, and the downtime interval is even possible. The average duration of downtime depends on the variance of the random component ξ_i of the time interval between the requests at the MTI input. If the pdf of time intervals between the requests coming in at the MTI input is subjected to the exponential pdf with the shift given by (10.14), then the average downtime is defined as

$$\bar{t}_{\text{down}} = \bar{\tau}_{\text{DS}} - a \ge \sigma_{\tau_{\text{DS}}}. \quad (10.38)$$

If the pdf of time intervals between the requests coming in at the MTI input is subjected to the pdf given by (10.15), we have

$$\bar{t}_{\text{down}} = \bar{\tau}_{\text{DS}} - a \ge \frac{2\sigma_{\tau_{\text{DS}}}}{\sqrt{2\pi}}. \quad (10.39)$$

Case 2: $\tau_0 < a \le \gamma_{\text{in}}^{-1}$. There is a request queue at the MTI input. It is impossible to compute the request queue length or the waiting time using analytical methods because the input request queue is not simple. To define the mentioned characteristics there is a need to apply a simulation. The greatest difficulty is to simulate the request queue with the pdf given by (10.14) or (10.15). The curves of relative waiting time for request queue by MTI as a function of the loading factor and the shift coefficient $\alpha = \tau_0\sigma_{\tau_{\text{DS}}}^{-1}$ as the parameter are shown in Figure 10.7. As we can see from Figure 10.7,

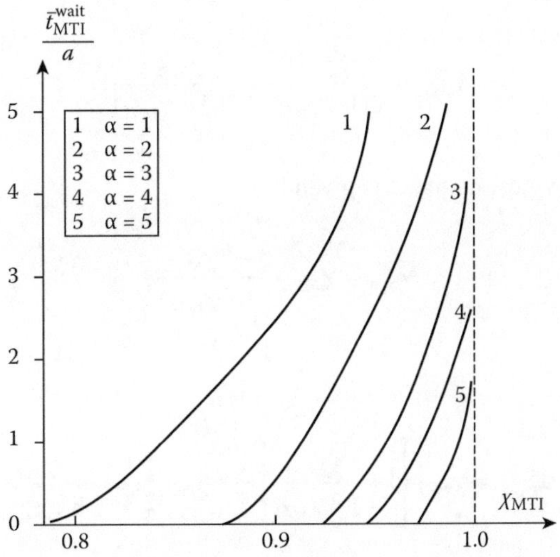

FIGURE 10.7 Relative waiting time versus the loading factor at different values of shift coefficient α.

the average waiting time at the MTI input under the fixed loading factor is a function of the shift coefficient α. If $\alpha \geq 3$, the input request queue degenerates into a regular request queue and the MTI request queue time is approximately equal to τ_0. In this case, the request queue at the MTI input is absent and the input register is used as the buffer memory.

10.3 ANALYSIS OF "$n - n - 1$" MTI SYSTEM

The "$n - n - 1$" MTI system is presented in Figure 10.4. For this system the request queuing by the detector–selector is started immediately after filling out the memory matrix. If, as before, we assume that the pdf of time to fill out the memory is given by (10.6), then the pdf of request queuing time by each channel of the system memory–detector–selector is a combination of the pdfs given by (10.6) and (10.14) or (10.6) and (10.15) and the average request queuing time is defined as

$$\overline{\tau}_{\text{memory-DS}} = \overline{\tau}_{\text{memory}} + \overline{\tau}_{\text{DS}}, \tag{10.40}$$

where

$\overline{\tau}_{\text{memory}}$ is the average time to fill out the memory
$\overline{\tau}_{\text{DS}}$ is the average time for request queuing by the detector–selector given by (10.17) or (10.29)

Now, if the request queuing parameter is given (the input request queuing is considered as the simplest, as before), the number of channels of the system memory–detector–selector can be determined using the Erlang formula (10.18) at the given probability of failure P_{failure}.

Let us discuss the request queuing pdf at the output of n-channel system memory–detector–selector. Action of the considered system memory–detector–selector on the incoming request queue can be presented as an expansion of the simplest request queue on the elementary request queues, the number of which is equal to the number of channels of the system memory–detector–selector. In a general case, these elementary request queues may not be the simplest. The output request queue of the system memory–detector–selector is a superposition of elementary request queues and can be considered as the simplest request queue with the parameter equal to the parameter of the incoming or input request queue at low values of P_{failure} based on Sevastyanov's theorem [5]. The request queue time in MTI is constant and equal to a, as earlier. Therefore, the average waiting time for request queue by MTI is defined as

$$\overline{t}_{\text{MTI}}^{\text{wait}} = \frac{\gamma_{\text{in}} a^2}{2(1-\gamma_{\text{in}} a)} = \frac{\chi_{\text{MTI}} a}{2(1-\chi_{\text{MTI}})}. \tag{10.41}$$

The average number of request in queue is given by

$$\overline{N}_{\text{wait}} = \chi_{\text{MTI}} \frac{3-2\chi_{\text{MTI}}}{2(1-\chi_{\text{MTI}})}, \tag{10.42}$$

and the variance

$$D_{N_{\text{wait}}} = \sigma^2_{N_{\text{wait}}} = \chi_{\text{MTI}} \left\{ 1 + \frac{\chi_{\text{MTI}}}{1-\chi_{\text{MTI}}} \left[\frac{1}{2} + \chi_{\text{MTI}} \left(\frac{1}{3} + \frac{\chi_{\text{MTI}}}{4(1-\chi_{\text{MTI}})} \right) \right] \right\}, \tag{10.43}$$

where χ_{MTI} is the MTI loading factor.

Knowing $\overline{N}_{\text{wait}}$ and $\sigma^2_{N_{\text{wait}}}$, we can define the buffer memory capacity at the MTI input under the given acceptable probability to loss the request and assuming, for example, that the pdf of the number of requests in queue is the normal Gaussian. This approximation makes sense only at high values of $\overline{N}_{\text{wait}}$. In this case, the probability of losing the request at the MTI input is defied as [6]

$$P_{\text{MTI}}^{\text{loss}} = \int_{Q_{\text{BM}}}^{\infty} \frac{1}{\sqrt{2\pi\sigma^2_{N_{\text{wait}}}}} \exp\left\{-\frac{(N_{\text{wait}} - \overline{N}_{\text{wait}})^2}{2\sigma^2_{N_{\text{wait}}}}\right\} dN_{\text{wait}}, \tag{10.44}$$

where Q_{BM} is the required buffer memory. There is a need to note that the acceptable probability of losing the request at the MTI input $P_{\text{MTI}}^{\text{loss}}$ must be, at least, on the order less than the acceptable probability of losing the request at the system input. Only in this case the main requirement to the system operation will be satisfied, namely, all queue requests that pass the first phase of queue must be served by MTI.

10.4 ANALYSIS OF "$n - m - 1$" MTI SYSTEM

The considered system is the three-phase queuing system with losses at the input. The first phase is the n-channel queuing system with losses. Definition of the required number of channels of this system is carried out according to the procedure discussed earlier. The second phase is the m-channel queuing system with waiting. Analysis of QoS factors of this system is the subject of the present section. The input request queue is the simplest one with the parameter γ_{in}. We assume that the splitter operates as a counter with respect to the base m sending to the ith channel, $i = 1, 2,\ldots, m$, the requests with numbers $i, i + m, i + 2m$. This allocation method of requests is called the *cyclic way*. The simplest request queue thinned in $m - 1$ times, i.e., the Erlang request queue of the $(m - 1)$th order, with the following pdf

$$p_{m-1}(t) = \frac{\gamma_{\text{in}}(\gamma_{\text{in}}t)^{m-1}\exp(-\gamma_{\text{in}}t)}{(m-1)!} \tag{10.45}$$

comes in at the input of each channel of the one-channel queuing system. The condition of the m-channel queuing system stationary mode is the following:

$$\sum_{i=1}^{m} \chi_i = \chi_m = \gamma_{\text{in}}\overline{\tau}_{\text{DS}} < m. \tag{10.46}$$

At this time we assume that the average request queue time $\overline{\tau}_{\text{DS}}$ is the same for all channels. Thus, in the considered case, the one-channel queuing system with the Erlang incoming request queue possessing the pdf equal to $\gamma_{\text{in}}m^{-1}$ and the request queue time subjected to the pdf given by (10.14) and (10.15) with the parameters $\overline{\tau}_{\text{DS}}$ and $\sigma^2_{\tau_{\text{DS}}}$ is investigated. The result of investigation must be the statistical characteristics of the request queue waiting time, namely, the average time $\overline{t}_{\text{wait}}$ and the variance of this time $\sigma^2_{t_{\text{wait}}}$.

As noted earlier, the analytical investigation of queuing systems with the incoming request queue different from the simplest one is a complicated problem. A general approach to solving this problem is discussed in Ref. [5]. However, an implementation of this general approach to specific conditions of our problem is difficult, not obvious, and does not lead to final formulae for $\overline{t}_{\text{wait}}$ and $\sigma^2_{t_{\text{wait}}}$. The explicit solutions for $\overline{t}_{\text{wait}}/\overline{\tau}_{\text{DS}}$ obtained by numerical simulation are shown in Figure 10.8, where the solid lines represent the exponential with shift pdf of request queue time and the dashed lines represent the truncated and shifted normal Gaussian pdf. As we can see from Figure 10.8, with an increase in m the

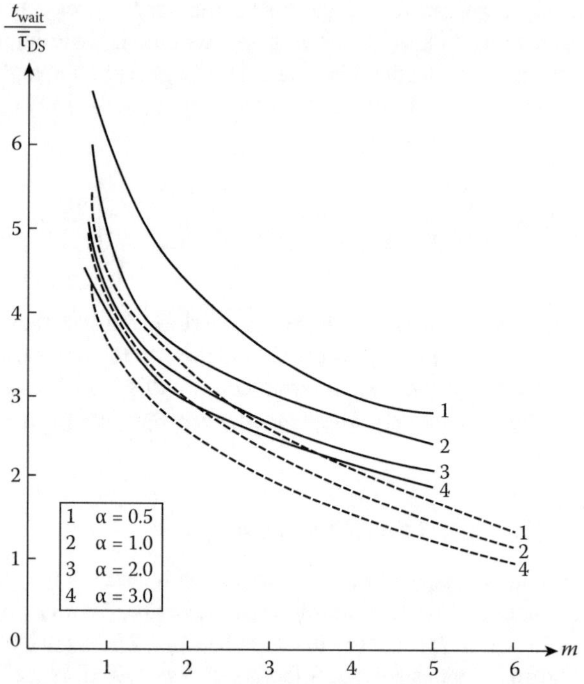

FIGURE 10.8 Average waiting time versus the number of queuing system channels.

relative average waiting time decreases exponentially. The shift coefficient α affects the average waiting time by an analogous way as for previously discussed systems. At the given parameters γ_{in}, m, α, we can compute $\bar{\tau}_{DS}$ based on the curves presented in Figure 10.8. We are able to define the average waiting time of request queue and, taking into consideration the average waiting time of request queue, we are able to define the required number of channels n for the first-phase queuing system.

Dependences of the required speed of operation for each mth channel of the second-phase queuing system for several values of the input request queue parameter γ_{in} as a function of the average number of reduced operations $\bar{N}$ required to process a single request under the loading factor of each mth channel $\chi_i = 0.9m$ are shown in Figure 10.9 for the truncated and shifted normal Gaussian pdf given by (10.15). The required speed of operation is computed by the following formula:

$$V_{ef} = \frac{\bar{N}\gamma_{in}}{0.9m}. \tag{10.47}$$

Analysis of (10.47) and dependences in Figure 10.9 show that at the fixed parameters of the input request queue and the already-given number of operations for a single realization of digital signal processing algorithm, the required speed of operation of the second-phase queuing system decreases proportionally to the number of channels in this system.

A set of requests processed by the second-phase queuing system form the incoming request queue for the third-phase queuing system that is a one-channel queuing system with waiting. According to the limiting theorem for summarized flows, the incoming request queue at the third-phase queuing system input is very close to the simplest one with the parameter γ_{in}. The request queue time in the third-phase queuing system is constant, as earlier. Therefore, a computation of required characteristics of the three-phase queuing system and buffer memory capacity is carried out by the procedure discussed in Section 10.3.

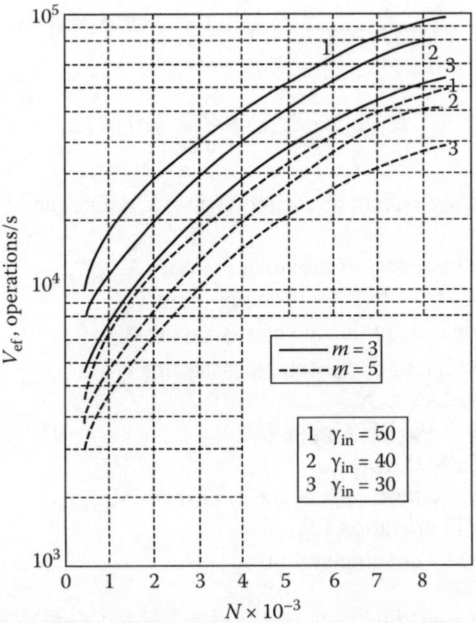

FIGURE 10.9 Required speed of operation for each *m*th channel of the second-phase queuing system for several values γ_{in} versus the average number of operations under the loading factor $\chi_i = 0.9m$; truncated and shifted normal Gaussian pdf.

10.5 COMPARATIVE ANALYSIS OF TARGET TRACKING SYSTEMS

The total QoS factors of discussed versions of the target tracking systems are the following:

- The required number of input memory channels under the given probability of failure $P_{failure}$
- The required effective speed of operation
- The average time required to process the request by the queuing system

Based on the discussions provided in the previous sections, we are able to compare the considered target tracking systems under the following initial data:

- The request queue at the target tracking system input is the simplest with $\gamma_{in} = 40\ s^{-1}$.
- The acceptable probability of failure at the target tracking system input $P_{failure} = 10^{-3}$.
- The loading factor of the detector–selector and MTI is $\chi_{DS\ (MTI)} = 0.9$.
- The pdf of request queue time in the detector–selector is given by (10.15) with $\alpha = 2$.
- The average number of reduced short operations required to process a single request by the detector–selector system is $\bar{N}_{DS} = 1500$.
- The number of reduced short operations required to process a single request by the MTI is $\bar{N}_{MTI} = 2500$.
- Scanning by azimuth with the uniform velocity $V_0 = 250°\ s^{-1}$.
- Minimal size of the target tracking gate by azimuth is $\Delta\beta_{gate}^{min} = 2°$.
- Minimal time to fill out the memory matrix (the constant constituent of the request queue time in memory) $\tau_{memory}^{min} = \Delta\beta_{gate}^{min}/V_0 = 0.008$ s.

If we assume in (10.6) $\alpha = \tau_{\text{memory}}^{\min}/\sigma_{\tau_{\text{memory}}} = 2$, then the average request queue time in the memory is defined as

$$\bar{\tau}_{\text{memory}} = \tau_{\text{memory}}^{\min} + \frac{2\sigma_\tau}{\sqrt{2\pi}} = 0.011 \text{ s}. \tag{10.48}$$

To compare the target tracking system of various types we define the following:

- Number of channels (matrices) of the input memory N_{memory}
- Average waiting time at the detector–selector input $\bar{t}_{\text{DS}}^{\text{wait}}$
- Average request queue time in the detector–selector $\bar{\tau}_{\text{DS}}$
- Required speed of operation of the detector–selector $V_{\text{DS}}^{\text{ef}}$
- Number of detector–selectors N_{DS}
- Average waiting time at the MTI input $\bar{t}_{\text{MTI}}^{\text{wait}}$
- Request queue time in MTI τ_{MTI}
- Number of the buffer memory cells at the MTI input $N_{\text{BM (MTI)}}$
- Required speed of MTI operation $V_{\text{MTI}}^{\text{ef}}$
- Average request delay by queuing system τ_Σ

Computational results are presented in Table 10.1, which shows that the system "$n - 1 - 1$" ensures the minimal average time of signal processing by one target equal to $\tau_\Sigma = 0.16$s at the speed of operation of MTI equal to $V_{\text{MTI}}^{\text{ef}} = 150 \times 10^3$ operations per second and the speed of operation of detector–selector equal to $V_{\text{DS}}^{\text{ef}} = 66.7 \times 10^3$ operations per second, i.e., providing a refreshment on nine targets in the course of each scanning. The same average time of signal processing is provided by the system "$n - n - 1$" when there are six detector–selectors with the speed of operation equal to $V_{\text{DS}}^{\text{ef}} = 66.7 \times 10^3$ operations per second but at the lesser speed of MTI operation. Obviously, a realization of this version is expensive in comparison with the system "$n - 1 - 1$." Other versions of the system can be realized using the detector–selectors with the low speed of operation that can be a clincher in their favor [7]. The system "$n - m - 1$" can also be realized using the detector–selectors with a low speed of operation. In doing so, the number of detector–selectors is not high (two or three), which is an advantage of this system in comparison with the second type of system. The disadvantage is an increase in the request queue time.

Selection of definite version of the system is defined by computational resources and acceptable time of signal processing. Evidently, the most effective system from considered ones is the system "$n - m - 1$" with two or three detector–selectors.

TABLE 10.1
Comparative Analysis of Target Tracking Systems

		Detector–Selector				MTI				
System	N_{memory}	$\bar{\tau}_{\text{DS}}$ s	$\bar{t}_{\text{DS}}^{\text{wait}}$ s	$V_{\text{DS}}^{\text{ef}}$, 1000 Operations/s	N_{DS}	τ_{MTI} s	$\bar{t}_{\text{MTI}}^{\text{wait}}$ s	$N_{\text{BM (MTI)}}$	$V_{\text{MTI}}^{\text{ef}}$, 1000 Operations/s	τ_Σ s
$n - 1 - 1$	14	0.0225	0.107	66.7	1	0.0167	0	1	150	0.16
$n - n - 1$	6	0.0225	0	66.7	6	0.0225	0.1	15	110	0.16
	8	0.045	0	33.3	8	0.0225	0.1	15	110	0.18
	15	0.15	0	10	15	0.0225	0.1	15	110	0.28
$n - m - 1$	14	0.045	0.12	33.3	2	0.0225	0.1	15	110	0.3
	14	0.0685	0.123	22.2	3	0.0225	0.1	15	110	0.325
	13	0.09	0.112	16.7	4	0.0225	0.1	15	110	0.335
	13	0.113	0.107	13	5	0.0225	0.1	15	110	0.355

10.6 SUMMARY AND DISCUSSION

Detection and tracking of all targets in the all-round surveillance radar coverage with accuracy sufficient to reproduce and estimate data of the air situation are carried out by the so-called rough channel of global digital signal processing system covering the whole scanned area. The rough channel consists of the following subsystems: the binary signal quantization, the specific microprocessor network for target return signal preprocessing, and the microprocessor network for digital signal reprocessing and control. Employment of binary quantization and simplified versions of digital signal processing algorithms allows us to implement the microprocessor networks and sets in this channel of global digital signal processing system of the all-round surveillance radar with the uniform antenna rotation.

The detection and selection problem of a single target within the limits of the gate is assigned in the following way based on information contained in the likelihood ratio maxima. First, we take the hypothesis that there is only a single target within the limits of the gate. In the considered case, the event, when several targets are within the limits of the gate, is possible but it is improbable. Using the relief of the two-dimensional likelihood ratio surface with M peaks of different heights, there is a need to define the maximum (the peak) formed by the target return signal and, if a "yes," there is a need to define the coordinates of this two-dimensional likelihood ratio surface maximum. The amplitudes of peaks $Z_l(l = 1, 2,\ldots, M)$ and their coordinates ξ_l and η_l with respect to the gate center are used as the input parameters, based on which the decision is made. If the hypothesis about the statistical independence of the two-dimensional likelihood ratio surface peak amplitudes is true and the coordinates of the two-dimensional likelihood ratio surface maxima are known within the limits of range where the target return signal is present, the optimal detection–selection problem of the target return signal pips within the limits of gate is solved in two steps.

All considered and discussed versions of target tracking by several MTI systems are three-phase queuing systems. Each phase is represented by one or several devices of queuing system connected in series. Generally, the request queuing time for each device is the random variable with pdf available to investigate and define. It is assumed that there is a simple request queue at first-phase input of the queuing system. The request queue coming in at the first device input is immediately served if, at least, one channel of the queuing system is free from service; otherwise, the request is rejected and flushed. The requests carried out by the first device are processed sequentially at the next phases; in other words, the request loss at each next phase or stage is inadmissible. In line with this, the memory device with the purpose storing requests in line should be provided before the second and third phases of the queuing system.

Difficulties under analysis of multiphase queuing systems are the following. At all cases of practical importance, the output stream of phase takes a more complex form in comparison with the incoming request queue. In some cases, the output request queue can be approximated by the simplest incoming stream with the same parameters. Then we can use analytical procedures and methods of the queuing theory to analyze the next phase or stage. If this approximation is impossible, then the only method to investigate the stream is the simulation. A rational combination of analytical and simulation methods and procedures allows us to solve the problem of the three-phase signal processing system analysis using the MTI system for any design and construction version.

REFERENCES

1. Lyons, R.G. 2004. *Understanding Digital Signal Processing*. 2nd edn. Upper Saddle River, NJ: Prentice Hall, Inc.
2. Harris, F.J. 2004. *Multirate Signal Processing for Communications Systems*. Upper Saddle River, NJ: Prentice Hall, Inc.
3. Barton, D.K. 2005. *Modern Radar System Analysis*. Norwood, MA: Artech. House, Inc.

4. Skolnik, M.I. 2008. *Radar Handbook*. 3rd edn. New York: McGraw-Hill, Inc.
5. Gnedenko, V.V. and I.N. Kovalenko. 1966. *Introduction to Queueing Theory*. Moscow, Russia: Nauka.
6. Skolnik, M.I. 2001. *Introduction to Radar Systems*. 3rd edn. New York: McGraw-Hill, Inc.
7. Hall, T.M. and W.W. Shrader. 2007. Statistics of clutter residue in MTI radars with IF limiting, in *IEEE Radar Conference*, April 2007, Boston, MA, pp. 01–06.

Part III

Stochastic Processes Measuring in Radar Systems

11 Main Statements of Statistical Estimation Theory

11.1 MAIN DEFINITIONS AND PROBLEM STATEMENT

In a general form, the problem of estimates of stochastic process parameters can be formulated in the following manner. Let the incoming realization $x(t)$ or a set of realizations $x_i(t)$, $i = 1,\ldots, \nu$ of the random process $\xi(t)$ be observed within the limits of the fixed time interval $[0, T]$. In general, the multidimensional (one-dimensional or n-dimensional) probability density function (pdf) of the random process $\xi(t)$ contains μ unknown parameters $\mathbf{l} = \{l_1, l_2,\ldots, l_\mu\}$ to be estimated. We assume that the estimated multidimensional parameter vector $\mathbf{l} = \{l_1, l_2,\ldots, l_\mu\}$ is the continuous function within the limits of some range of possible values $\mathcal{L}$. Based on observation and analysis of the incoming realization $x(t)$ or realizations $x_i(t)$, $i = 1,\ldots, \nu$, we need to decide what magnitudes from the given domain of possible values possessing the parameters $\mathbf{l} = \{l_1, l_2,\ldots, l_\mu\}$ are of interest for the user. Furthermore, only the realization $x(t)$ of the random process $\xi(t)$ is considered if the special conditions are not discussed. In other words, we need to define the estimate of the required multidimensional parameter $\mathbf{l} = \{l_1, l_2,\ldots, l_\mu\}$ on processing the observed realization $x(t)$ within the limits of the fixed time interval $[0, T]$.

Estimate of the parameter $\mathbf{l} = \{l_1, l_2,\ldots, l_\mu\}$ of the random process $\xi(t)$ is a set of functions or a single function of the observed data $x(t)$. The values of these functions for the fixed realization $x(t)$ estimate, that is, defined by a given way, are the unknown parameters of stochastic process. Depending on the requirements of estimation process and estimates of parameters, various estimation procedures are possible. In doing so, each estimate is characterized by its own quality performance that, in majority of cases, indicates a measure of the estimate's closeness to the true value of estimated random process parameter. The quality performance, in turn, is defined by a choice of estimation criterion. Because of this, before estimate definition, we need to define a criterion of estimation. Selection of estimate criterion depends on the end problem, for which this random process parameter estimate is used. Because these problems can differ substantially, it is impossible to define and use the integrated criterion and one common estimate for the given random process parameter. This circumstance makes it difficult to compare various estimates.

In many practical applications, the estimate criteria are selected based on assumption about the purpose of estimate. At the same time, a design of process to choose definite estimate criteria is of great interest because such approach allows us to understand more clearly the essence and characteristic property of the problem to estimate the random process parameter and, that is very important, gives us a possibility to define the problem of valid selection of the estimate criterion more fairly. Owing to the finite time of observation and presence of noise and interference concurrent to observation, specific errors in the course of estimate definition arise. These errors are defined both by the quality performance and by conditions under which an estimation process is carried out. Because of this, the problem of optimal estimation of the parameter $\mathbf{l} = \{l_1, l_2,\ldots, l_\mu\}$ is to define a procedure that allows minimizing the errors in estimation of the parameter $\mathbf{l} = \{l_1, l_2,\ldots, l_\mu\}$. In general, the requirement of minimizing estimation error is not based on one particular aspect. However, if the criterion of estimation is given, the quality performance is measured using that criterion. Then the problem of obtaining the optimal estimation is reduced to a definition of solution procedure that minimizes or maximizes the quality performance. In doing so, the parameter estimate must be close, in a certain sense, to the true value of estimated parameter, and the optimal estimate must minimize this measure of closeness in accordance with the chosen criterion.

To simplify the written form and discussion in future, we assume, if it is not particularly fixed, that the unknown parameter of the random process $\xi(t)$ is the only parameter $\mathbf{l} = \{l_1, l_2,\ldots, l_\mu\}$. Nevertheless, all conclusions made based on our analysis of estimation process of the parameter $\mathbf{l} = \{l_1, l_2,\ldots, l_\mu\}$ of the random process $\xi(t)$ are correct with respect to joint estimation process of several parameters of the same random process $\xi(t)$. Thus, it is natural to obtain one function from the observed realization $x(t)$ to estimate a single parameter $\mathbf{l} = \{l_1, l_2,\ldots, l_\mu\}$ of the random process $\xi(t)$. Evidently, the more knowledge we have about the characteristics of the analyzed random process $\xi(t)$ and noise and interference in the received realization $x(t)$, the more accurate will be our estimation of possible values of the parameters of the random process $\xi(t)$, and thus more accurate will be the solution based on synthesis of the devices designed using the chosen criterion with the minimal errors in the estimation of random process parameters of interest.

More specifically, the estimated parameter is a random variable. Under this condition, the most complete data about the possible values of the parameter $\mathbf{l} = \{l_1, l_2,\ldots, l_\mu\}$ of the random process $\xi(t)$ are given by the a posteriori pdf $p_{\text{post}}(l) = p\{l|x(t)\}$, which is the conditional pdf if the given realization $x(t)$ is received. The formula of a posteriori pdf can be obtained based on the theorem about conditional probabilities of two random variables l and X, where $X\{x_1, x_2,\ldots, x_n\}$ is the multidimensional (n-dimensional) sample of the realization $x(t)$ within the limits of the interval $[0, T]$. According to the theorem about the conditional probabilities [1]

$$p(l,X) = p(l)p(l|X) = p(X)p(l|X), \tag{11.1}$$

we can write

$$p_{\text{post}}(l) = p(l|X) = \frac{p(l)p(X|l)}{p(X)}. \tag{11.2}$$

In (11.1) and (11.2) $p(l) \equiv p_{\text{prior}}(l)$ is the a priori pdf of the estimated parameter l; $p(X)$ is the pdf of multidimensional sample X of the realization $x(t)$. The pdf $p(X)$ does not depend on the current value of the estimated parameter l and can be determined based on the condition of normalization of $p_{\text{post}}(l)$:

$$p(X) = \int_{\mathcal{L}} p(X|l)p_{\text{prior}}(l)dl. \tag{11.3}$$

Integration is carried out by the a priori region (interval) $\mathcal{L}$ of all possible values of the estimated parameter l. Taking into consideration (11.3), we can rewrite (11.2) in the following form:

$$p_{\text{post}}(l) = \frac{p_{\text{prior}}(l)p(X|l)}{\int_{\mathcal{L}} p(X|l)p_{\text{prior}}(l)dl}. \tag{11.4}$$

The conditional pdf of the observed data sample X, under the condition that the estimated parameter takes a value l, has the following form:

$$p(X|l) = p(x_1, x_2,\ldots, x_n|l), \tag{11.5}$$

and can be considered as the function of l and is called the *likelihood function*. For the fixed sample X, this function shows that one possible value of parameter l is more likely in comparison with other value.

The likelihood function plays a very important role in the course of solution of signal detection problems, especially in radar systems. However, in a set of applications, it is worthwhile to consider the *likelihood ratio* instead of the likelihood function:

$$\Lambda(l) = \frac{p(x_1, x_2, \ldots, x_n | l)}{p(x_1, x_2, \ldots, x_n | l_{\text{fix}})}, \tag{11.6}$$

where $p(x_1, x_2, \ldots, x_n | l_{\text{fix}})$ is the pdf of observed data sample at some fixed value of the estimated random process parameter l_{fix}. As applied to analysis of continuous realization $x(t)$ within the limits of the interval $[0, T]$, we introduce the likelihood functional in the following form:

$$\hat{\Lambda}(l) = \lim_{n\to\infty} \frac{p(x_1, x_2, \ldots, x_n | l)}{p(x_1, x_2, \ldots, x_n | l_{\text{fix}})}, \tag{11.7}$$

where the interval between samples is defined as

$$\Delta = \frac{T}{n}. \tag{11.8}$$

Using the introduced notation, the a posteriori pdf takes the following form:

$$p_{\text{post}}(l) = \kappa p_{\text{prior}}(l)\Lambda(l), \tag{11.9}$$

where κ is the normalized coefficient independent of the current value of the estimated parameter l;

$$\kappa = \frac{1}{\int_{\mathcal{L}} p_{\text{prior}}(l)\Lambda(l)dl}. \tag{11.10}$$

We need to note that the a posteriori pdf $p_{\text{post}}(l)$ of the estimated parameter l and the likelihood ratio $\Lambda(l)$ are the random functions depending on the received realization $x(t)$.

In theory, for statistical parameter estimation, two types of estimates are used:

- Interval estimations based on the definition of confidence interval
- Point estimations, that is, the estimate defined at the point

Employing the interval estimations, we need to indicate the interval, within the limits of which there exists the true value of unknown random process parameter with the probability that is not less than the value given before. This earlier-given probability is called the confidence factor, and the indicated interval of possible values of the estimated random process parameter is called the *confidence interval*. The upper and lower bounds of the confidence interval, which are called the *confidence limits*, and the confidence interval are the functions to be considered for both digital signal processing (a discretization) and analog signal processing (continuous function) of the received realization $x(t)$. In the point estimation case, we assign one parameter value to the unknown parameter from the interval of possible parameter values; that is, some value is obtained based on the analysis of the received realization $x(t)$ and we use this value as the true value of the evaluated parameters.

In addition to the procedure of analysis of the random process parameter based on the value of received realization $x(t)$, there is a sequential estimation method. This method essentially involves the sequential statistical analysis that estimates the random process parameter [2,3]. The basic idea of sequential estimation is to define the time of analysis of the received realization $x(t)$, within the limits of which we are able to obtain the estimate of parameter with the earlier-given reliability. In the case of point estimate, the root-mean-square deviation of estimate or other convenience function characterizing a deviation of estimate from the true value of estimated random process parameter can be considered as the measure of reliability. From the viewpoint of interval sequential estimation, the estimate reliability can be defined using the length of the confidence interval with a given confidence coefficient.

11.2 POINT ESTIMATE AND ITS PROPERTIES

To make the point estimation means that some number $\gamma = \gamma[x(t)]$ from the interval $\mathcal{L}$ of possible values of the estimated random process parameter l must correspond to each possible received realization $x(t)$. This number $\gamma = \gamma[x(t)]$ is called the *point estimate*. Owing to the random nature of the point estimate of random process parameter, it is characterized by the conditional pdf $p(\gamma|l)$. This is a general and total characteristic of the point estimate. The shape of this pdf defines the quality of point estimate definition and, consequently, all properties of the point estimate. At the given estimation rule $\gamma = \gamma[x(t)]$, the conditional pdf $p(\gamma|l)$ can be obtained from the pdf of received realization $x(t)$ based on the well-known transformations of pdf [4]. We need to note that a direct determination of the pdf $p(\gamma|l)$ is very difficult for many application problems. Because of this, if there are reasons to suppose that this pdf is a unimodal function and is very close to symmetrical function, then the bias, dispersion, and variance of estimate that can be determined without direct definition of the $p(\gamma|l)$ are widely used as characteristics of the estimate γ.

In accordance with definitions, the bias, dispersion, and variance of estimate are defined as follows:

$$b(\gamma|l) = \langle(\gamma - l)\rangle = \int_X [\gamma(X) - l]p(X|l)dX; \tag{11.11}$$

$$D(\gamma|l) = \langle(\gamma - l)^2\rangle = \int_X [\gamma(X) - l]^2 p(X|l)dX; \tag{11.12}$$

$$\mathrm{Var}(\gamma|l) = \langle[\gamma - \langle\gamma\rangle]^2\rangle = \int_X [\gamma(X) - \langle\gamma\rangle]^2 p(X|l)dX. \tag{11.13}$$

Here and further $\langle\cdots\rangle$ means averaging by realizations. The estimate obtained taking into consideration of a priori pdf is called the *unconditional estimate*. The unconditional estimates are obtained as a result of averaging (11.11) through (11.13) on possible values of the variable l with a priori pdf $p_{\mathrm{prior}}(l)$; that is, the unconditional bias, dispersion, and variance of estimate are determined in the following form:

$$b(\gamma) = \int_X b(\gamma|l)p_{\mathrm{prior}}(l)dl; \tag{11.14}$$

$$D(\gamma) = \int_X D(\gamma|l)p_{\mathrm{prior}}(l)dl; \tag{11.15}$$

$$\mathrm{Var}(\gamma|l) = \int_X \mathrm{Var}(\gamma|l)p_{\mathrm{prior}}(l)dl. \tag{11.16}$$

Since the conditional and unconditional estimate characteristics have different notations, we will drop the term "conditional" when discussing a single type of characteristics.

The estimate of random process parameters, for which the conditional bias is equal to zero, is called the *conditionally unbiased estimate*; that is, in this case, the mathematical expectation of the estimate coincides with the true value of estimated parameter: $\langle\gamma\rangle = l$. If the unconditional bias is equal to zero, then the estimate is unconditionally unbiased estimate; that is, $\langle\gamma\rangle = l_{\text{prior}}$, where l_{prior} is the a priori mathematical expectation of the estimated parameter. Evidently, if the estimate is conditionally unbiased, then we can be sure that the estimate is unconditionally unbiased. Inverse proposition, generally speaking, is not correct. In practice, the conditional unbiasedness often plays a very important role. During simultaneous estimation of several random process parameters, for example, estimation of the vector parameter $\mathbf{l} = \{l_1, l_2, \ldots, l_\mu\}$, we need to know the statistical relationship between estimates in addition to introduced conditional and unconditional bias, dispersion, and variance of estimate. For this purpose, we can use the mutual correlation function of estimates.

If estimations of the random process parameters $l_1, l_2, \ldots, l_\mu$ are denoted by $\gamma_1, \gamma_2, \ldots, \gamma_\mu$, then the conditional mutual correlation function of estimations of the parameters l_i and l_j is defined in the following form:

$$R_{ij}(\mathrm{v}|l) = \left\langle \left[\left(\mathrm{v}_i - \langle \mathrm{v}_i \rangle\right)\left(\mathrm{v}_j - \langle \mathrm{v}_j \rangle\right) \right] \right\rangle. \tag{11.17}$$

The correlation matrix is formed based on these elements $R_{ij}(\mathrm{v}|l)$; moreover, the matrix diagonal elements are the *conditional variances* of estimations. By averaging the conditional mutual correlation function using possible a priori values of the estimated random process parameters, we obtain the *unconditional mutual correlation function* of estimations.

There are several approaches to define the properties of the point estimations. We consider the following requirements to properties of the point estimations in terminology of conditional characteristics:

- It is natural to try to define such point estimate γ so that the conditional pdf $p(\gamma|l)$ stays very close to the value l.
- It is desirable that while increasing the observation interval, that is, $T \to \infty$, the estimation would coincide with or approach stochastically the true value of estimated random process parameter. In this case, we can say that the estimate is the *consistent estimate*.
- The estimate must be unbiased $\langle\gamma\rangle = l$ or, in extreme cases, asymptotically unbiased, that is, $\lim_{T\to\infty}\langle\gamma\rangle = l$.
- The estimate must be the best by some criterion; for example, it must be characterized by minimal values of dispersion or variance at zero or constant bias.
- The estimate must be statistically sufficient.

The statistics, that is, in the considered case, the function or functions of the observed data, is sufficient if all statements about the estimated random process parameter can be defined based on the considered statistical data without any additional observation of received realization data. Evidently, the a posteriori pdf is always a sufficient statistic. Conditions of estimation sufficiency can be formulated in terms of the likelihood function: The necessary and sufficient condition of such estimation means the possibility to present the likelihood function in the form of product between two functions [5,6]:

$$p(X|l) = h[x(t)]\, g(\gamma|l). \tag{11.18}$$

Here, $h[x(t)]$ is the arbitrary function of the received realization $x(t)$ independent of the current value of the estimated random process parameter l. Since the parameter l does not enter into the function

$h[x(t)]$, we cannot use this function to obtain any information about the parameter l. The factor $g(\gamma|l)$ depends on the received realization $x(t)$ over the estimation $\gamma = \gamma[x(t)]$ only. For this reason, all information about the estimated random process parameter l must be contained into $\gamma[x(t)]$.

11.3 EFFECTIVE ESTIMATIONS

One of the main requirements is to obtain an estimate with minimal variance or minimal dispersion. Accordingly, a statement of effective estimations was introduced in the mathematical statistics. As applied to the bias estimations of the random process parameter, the estimation l_{ef} is considered effective if the mathematical expectation of its squared deviation from the true value of the estimated random process parameter l does not exceed the mathematical expectation of quadratic deviation of any other estimation γ; in other words, the following condition

$$D_{ef}(l) = \langle (l_{ef} - l)^2 \rangle \le \langle (\gamma - l)^2 \rangle. \tag{11.19}$$

must be satisfied. Dispersion of the unbiased estimate coincides with its variance and, consequently, the effective unbiased estimate is defined as the estimation with the minimal variance.

Cramer–Rao lower bound [5] was defined for the conditional variance and dispersion of estimations that are the variance and dispersion of effective estimations under the condition that they exist for the given random process parameters. Thus, in particular, the biased estimate variance is defined as

$$\text{Var}\left(\gamma|l\right) > \frac{\left[\dfrac{1 + db\left(\gamma|l\right)}{dl}\right]^2}{\left\langle\left[\dfrac{d}{dl}\ln\Lambda(l)\right]^2\right\rangle}. \tag{11.20}$$

The variance of unbiased estimations and estimations with the constant bias is simplified and takes the following form:

$$\text{Var}\left(\gamma|l\right) > \frac{1}{\left\langle\left[\dfrac{d}{dl}\ln\Lambda(l)\right]^2\right\rangle}. \tag{11.21}$$

We need to note that in the case of analog signal processing of all possible realizations $x(t)$ the averaging is carried out using a multidimensional sample of the observed data X and the derivatives are taken at the point where the estimated random process parameter has a true value. Equality in (11.20) and (11.21) takes place only in the case of effective estimations and if the two conditions are satisfied. The first condition is the condition that estimation remains sufficient (11.18). The second condition is the following: The likelihood function or likelihood ratio logarithm derivative should satisfy the equality [5]

$$\frac{d}{dl}\ln\Lambda(l) = q(l)(\gamma - \langle\gamma\rangle), \tag{11.22}$$

where the function $q(l)$ does not depend on the estimate δ and sample of observed data but depends on the current value of the estimated random process parameter l. At the same time, the condition (11.22) exists if and only if the estimate is sufficient; that is, the condition (11.18) is satisfied and the condition of sufficiency can exist when (11.22) is not satisfied. Analogous limitations are applied to the effective unbiased estimations, at which point the sign of inequality in (11.21) becomes the sign of equality.

11.4 LOSS FUNCTION AND AVERAGE RISK

There are two ways to make a decision in the theory of statistical estimations: nonrandom and random. In the case of nonrandom decision making (estimation), the definite decision is made by each specific realization of the received data $x(t)$; that is, there is a deterministic dependence between the received realization and the decision made. However, owing to the random nature of the observed data, the values of estimation are viewed as random variables. The probability of definite decision making is assigned in the case of random decision making using each specific realization $x(t)$; that is, the relationship between the received realization and the decision made has a probabilistic character. Furthermore, we consider only the nonrandom decision-making rules.

Based on the random character of the observed realization, there are errors for any decision-making rules; that is, the decision made γ does not coincide with the true value of parameter l. Evidently, by applying various decision-making rules different errors will appear with the various probabilities. Since the nonzero probability of error exists forever, we need to characterize the quality of different estimations in one way or another. For this purpose, the *loss function* is introduced in the theory of decisions. This function defines a definite loss $\mathcal{L}(\gamma, l)$ for each combination from the decision γ and parameter l. As a rule, the losses are selected as positive values and the correct decisions are assigned zero or negative losses. The physical sense of the loss function is described as follows. A definite nonnegative weight is assigned to each incorrect decision. In doing so, depending on targets, for which the estimate is defined, the most undesirable decisions are assigned the greatest weights. A choice of definite loss function is made depending on a specific problem of estimation of the random process parameter l. Unfortunately, there is no general decision-making rule to select the loss function. Each decision-making rule is selected based on a subjective principle. A definite arbitrariness in selecting losses leads to definite difficulties with applying the theory of statistical decisions. The following types of loss functions are widely used (see Figures 11.1 through 11.4):

- Simple loss function (see Figure 11.1)

$$\mathcal{L}(\gamma,l) = 1-\delta(\gamma-l); \tag{11.23}$$

  where $\delta(z)$ is the Dirac delta function;
- Linear modulo loss function (see Figure 11.2)

$$\mathcal{L}(\gamma,l) = |\gamma-l|; \tag{11.24}$$

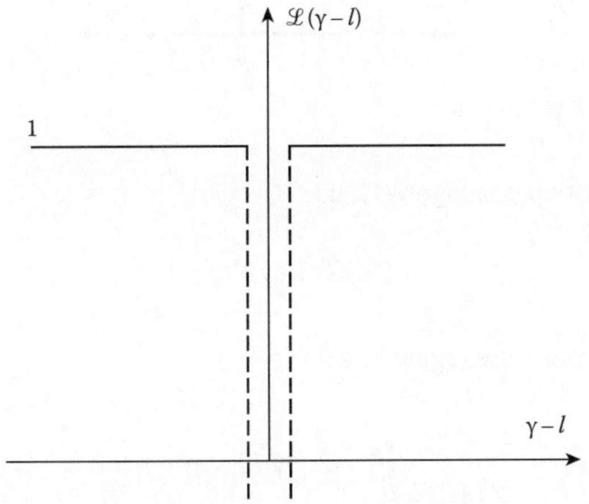

FIGURE 11.1 Simple loss function.

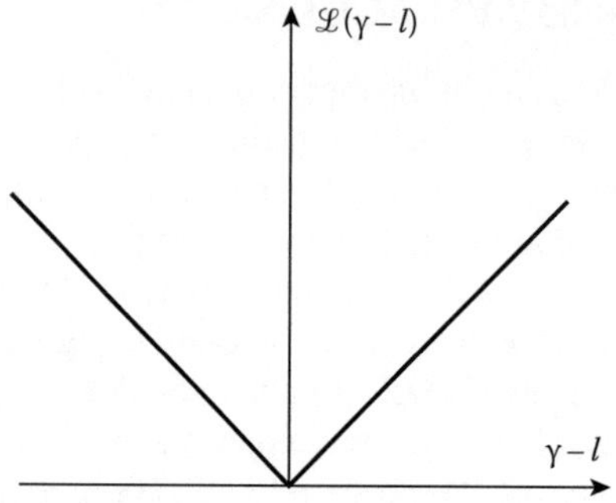

FIGURE 11.2 Linear modulo loss function.

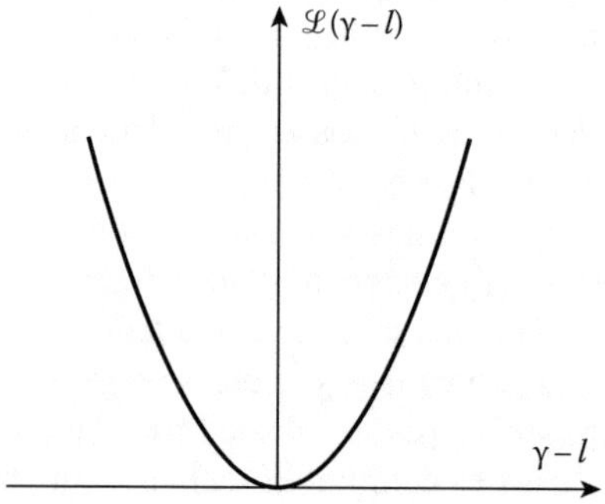

FIGURE 11.3 Quadratic loss function.

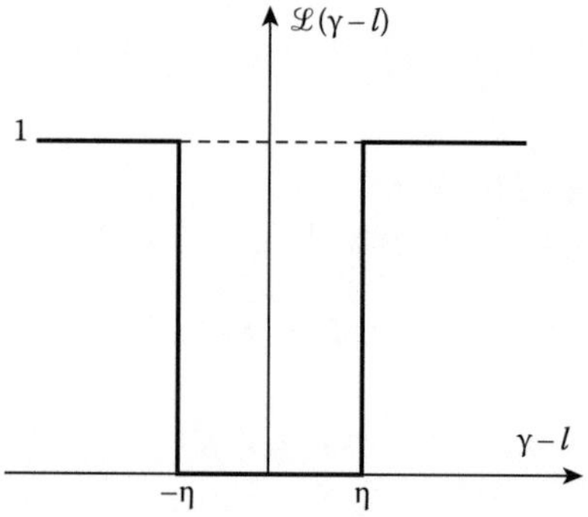

FIGURE 11.4 Rectangle loss function.

- Quadratic loss function (see Figure 11.3)

$$\mathcal{L}(\gamma, l) = (\gamma - l)^2; \tag{11.25}$$

- Rectangle loss function (see Figure 11.4)

$$\mathcal{L}(\gamma, l) = \begin{cases} 0 & \text{if} \quad |\gamma - l| < \eta, \quad \eta > 0, \\ 1 & \text{if} \quad |\gamma - l| > \eta, \quad \eta > 0. \end{cases} \tag{11.26}$$

In general, there may be a factor on the right side of loss functions given by (11.23) through (11.24). These functions are the symmetric functions of difference $|\gamma - l|$. In doing so, deviations of the parameter estimate with respect to the true value of estimated random process parameter are undesirable. In addition, there are some application problems when the observer ignores the sign of estimate. In this case, the loss function is not symmetric.

Based on the random nature of estimations γ and random process parameter l, the losses are random for any decision-making rules and cannot be used to characterize an estimate quality. To characterize an estimate quality, we can apply the mathematical expectation of the loss function that takes into consideration all incorrect solutions and relative frequency of their appearance. Choice of the mathematical expectation to characterize a quality of estimate, not another statistical characteristic, is rational but a little bit arbitrary. The mathematical expectation (conditional or unconditional) of the loss function is called the *risk* (conditional or unconditional). Conditional risk is obtained by averaging the loss function over all possible values of multidimensional sample of the observed data that are characterized by the conditional pdf $p(X|l)$

$$\mathcal{R}(\gamma \mid l) = \int_X \mathcal{L}(\gamma, l) p(X|l) dX. \tag{11.27}$$

As we can see from (11.27), the preferable estimations are the estimations with minimal conditional risk. However, at various values of the estimated random process parameter l, the conditional risk will have different values. For this reason, the preferable decision-making rules can be various, too. Thus, if we know the a priori pdf of the estimated parameter, then it is worthwhile to define the best decision-making rule for the definition of estimation based on the condition of the unconditional average minimum risk, which can be written in the following form:

$$\mathcal{R}(\gamma) = \int_X p(X) \left[\int_{\mathcal{L}} \mathcal{L}(\gamma, l) p_{\text{post}}(l) dl \right] dX, \tag{11.28}$$

where $p(X)$ is the pdf of the observed data sample.

Estimations obtained by the criterion of minimum conditional and unconditional average risk are called the conditional and unconditional *Bayes estimates*. The unconditional Bayes estimate is often called the Bayes estimate. Furthermore, we will understand that for the Bayes estimate γ_m of the parameter l the estimate ensuring the minimum average risk at the given loss function is $\mathcal{L}(\gamma, l)$. The minimal value of average risk corresponding to the Bayes estimate is called the *Bayes risk*

$$\mathcal{R}_m = \left\langle \int_{\mathcal{L}} \mathcal{L}(\gamma, l) f_{\text{post}}(l) dl \right\rangle. \tag{11.29}$$

Here, the averaging is carried out by samples (digital signal processing) of the observed data X or by realizations $x(t)$ (analog signal processing). The average risk can be determined for any given decision-making rule and because of the definition of Bayes estimate the following condition is satisfied forever:

$$\mathcal{R}_m \le \mathcal{R}(\gamma). \tag{11.30}$$

Computing the average risk by (11.28) for different estimations and comparing each of these risks with Bayes risk, we can evaluate and conclude how one estimate can be better compared to another estimate and that the best estimate is the one close to the optimal Bayes estimate. Since the conditional or unconditional risk has a varying physical sense depending on the shape and physical interpretation of the loss function $\mathcal{L}(\gamma, l)$, the sense of criterion of optimality depends on the shape of the loss function, too.

The pdf $p(X)$ is the nonnegative function. For this reason, a minimization of (11.28) by γ at the fixed sample of observed data is reduced to minimization of the function

$$\mathcal{R}_{\text{post}}(\gamma) = \int_{\mathcal{L}} \mathcal{L}(\gamma, l) p_{\text{post}}(l) dl, \tag{11.31}$$

called the a posteriori risk. If the a posteriori risk $\mathcal{R}_{\text{post}}(\gamma)$ is differentiated by γ, then the Bayes estimate γ_m can be defined as a solution of the following equation:

$$\left[\frac{d\mathcal{R}_{\text{post}}(\gamma)}{d\gamma}\right]_{\gamma_m} = 0. \tag{11.32}$$

To ensure a global minimum (minimum minimorum) of the a posteriori risk $\mathcal{R}_{\text{post}}(\gamma)$, we need to use the root square.

The criterion of minimum average risk is based on the knowledge of the total a priori information about the estimated parameter, which gives the answer to the question about how we can use all a priori information to obtain the best estimate. However, the absence of the total a priori information about the estimated parameter that takes place in the majority of applications leads us to definite problems (a priori problems) with applying the methods of the theory of statistical estimations. Several approaches are found to solve the problems of definition of the optimal estimations at the unknown a priori distribution of estimated parameters. One of them is based on definition of the Bayes estimations that are invariant with respect to the sufficiently wide class of a priori distributions. In other problems, we are limited by a choice of estimation based on the minimization of the conditional risk or we make some assumptions with respect to a priori distribution of estimated parameters. The least favorable distribution, with the Bayes risk at the maximum, is considered as the a priori distribution of estimated parameters; the obtained estimation of the random process parameter is called the *minimax estimate.* The minimax estimate defines the upper bound of the Bayes risk that is called the *minimax risk.* In spite of the fact that minimax estimate can generate heavy losses compared to other possible estimate, it can be useful if losses under the most unfavorable a priori conditions are avoided.

In accordance with the definition, the minimax estimate can be defined in the following way. In the case of a priori distribution of estimated parameter in accordance with the given loss function, we need to define the Bayes estimate $\gamma_m = \gamma_m[x(t)]$. After that we select such a priori distribution of estimated parameter, at which the minimal value of the average risk (Bayes risk) reaches the maximum. The Bayes estimate will be minimax at such a priori distribution. We need to note that a rigorous definition of the least favorable a priori distribution of estimated parameter is related to great mathematical problems. However, in the majority of applications, including the problems of random parameter estimations, the least favorable pdf is the *uniform distribution* within the limits of the given interval.

11.5 BAYESIAN ESTIMATES FOR VARIOUS LOSS FUNCTIONS

Now, we discuss the properties of Bayes estimations for some loss functions mentioned previously.

11.5.1 Simple Loss Function

Substituting the simple loss function given by (11.23) into (11.31) and using the filtering property of the delta function

$$\int_{-\infty}^{\infty} \varphi(z)\delta(z-z_0)dz = \varphi(z_0). \tag{11.33}$$

we obtain

$$\mathcal{R}_{\text{post}}(\gamma) = 1 - p_{\text{post}}(\gamma). \tag{11.34}$$

The a posteriori risk $\mathcal{R}_{\text{post}}(\gamma)$ and, consequently, the average risk $\mathcal{R}(\gamma)$ are minimum if the a posteriori pdf $p_{\text{post}}(\gamma)$ is the maximum for the given estimate. In other words, the a posteriori pdf $p_{\text{post}}(\gamma)$ is maximum, if there is a single maximum, or maximum maximorum, if there are several local maxima. This means that the probable value γ_m, under which the following condition

$$p_{\text{post}}(\gamma_m) \geq f_{\text{post}}(l), \quad \gamma_m, \quad l \in \mathcal{L} \tag{11.35}$$

is satisfied, can be considered as the Bayes estimate. If the a posteriori pdf $p_{\text{post}}(\gamma)$ is differentiable with respect to the parameter l, then the estimate γ_m can be defined based on the following equation:

$$\left[\frac{dp_{\text{post}}(l)}{dl}\right]_{\gamma_m} = 0 \quad \text{at} \quad \left[\frac{d^2 p_{\text{post}}(l)}{dl^2}\right]_{\gamma_m} < 0, \tag{11.36}$$

and we take into consideration only the solution of this equation satisfying (11.35). Substituting the simple loss function in (11.29), we obtain the Bayes risk

$$\mathcal{R}_m = 1 - \langle \max p_{\text{post}}(l) \rangle = 1 - \langle p_{\text{post}}(\gamma_m) \rangle. \tag{11.37}$$

The second term on the right-hand side of (11.37) is the average probability of correct decision making that is accurate within the constant factor. Consequently, in the case of the simple loss function, the probability of correct decision-making is maximum at the Bayes estimate. It is evident that, in the case of the simple loss function, the probability of incorrect decision-making is minimal when applying the Bayes estimate to random process parameter. At the same time, all errors have the same weight that means that all errors are undesirable independent of their values. In the case of simple loss function, the Bayes estimate is well known in literature as the estimate by maximum maximorum of the a posteriori pdf.

If the a priori pdf $p_{\text{pr}}(l)$ is constant within the limits of interval of possible values of the estimated random process parameter, then, according to (11.9), the a posteriori pdf is matched accurately within the constant factor with the likelihood ratio $\Lambda(l)$. The estimate by maximum of the a posteriori pdf γ_m transforms into the estimate of the maximum likelihood l_m. The maximum likelihood estimate l_m is defined as a maximum maximorum position of the likelihood ratio. As a rule, the maximum likelihood ratio estimations are applied in the following practical cases:

- The a priori pdf of estimated random process parameter is unknown.
- It is complicated to obtain the a posteriori pdf in comparison with the likelihood ratio or function.

The method of maximum likelihood has the following advantages compared to other methods.

1. In practice, the maximum likelihood estimation is very close to the estimation by maximum a posteriori pdf for a wide class of a priori pdfs of the estimated random process parameter. In other words, the maximum likelihood estimation is the Bayes estimate of the simple loss function. Indeed, for highly accurate measurement of random process parameters, the a priori pdf about the estimate can be considered very often as the constant value and the a posteriori pdf about the estimate is matched by shape with the likelihood function or likelihood ratio. This property is very important when the a priori pdf of estimated random process parameter is unknown and a definition of the least favorable a priori pdf is associated with mathematical difficulties.
2. Estimation of random process parameter by the maximum likelihood is independent of the mutual one-valued no inertial transformation by estimated parameter of the signal at the receiver output in the form $F[\Lambda(t)]$, since the maximum likelihood point remains invariant during these transformations. This property is very important for practical realizations of optimal measurers.
3. Analytical definition of random process parameter estimate quality using the maximum likelihood ratio requires less mathematical problems and difficulties compared to other methods of estimation. As a rule, the other methods of estimation require a mathematical or physical simulation to define the characteristics of estimate quality that, in turn, complicates the design and construction of the corresponding random process parameter measurers.
4. It is defined in the mathematical statistics [5] that if there is an effective estimate, then the maximum likelihood estimate is effective too.
5. If the random process observation interval tends to approach infinity, the maximum likelihood estimate is asymptotically effective and unbiased and, in this case, it is the limiting form of Bayes estimations for a wide class of a priori pdfs and loss functions.

These and other particular advantages of the maximum likelihood estimate method are the reasons for its wide application. We need to note that the estimation of random process parameters by the maximum likelihood method has a major disadvantage if there is noise, because the noise will be analyzed jointly with the considered random signal parameters. In this case, the maximum maximorum caused by the presence of noise can appear.

11.5.2 Linear Module Loss Function

According to (11.31), the a posteriori risk for the loss function given by (11.24) can be determined based on the following equation:

$$\mathcal{R}_{\text{post}}(\gamma) = \int_{-\infty}^{\infty} |\gamma - l|\, p_{\text{post}}(l)dl. \tag{11.38}$$

We can divide the integral on two integrals with the following limits of integration $-\infty < l \le \gamma$ and $\gamma < l \le \infty$. Based on the definition of the extremum of $\mathcal{R}_{\text{post}}(\gamma)$, we obtain

$$\left[\frac{d\mathcal{R}_{\text{post}}(\gamma)}{d\gamma}\right]_{\gamma_m} = \int_{-\infty}^{\gamma_m} p_{\text{post}}(l)dl - \int_{\gamma_m}^{\infty} p_{\text{post}}(l)dl = 0. \tag{11.39}$$

It follows that the Bayes estimate is the median of the a posteriori pdf.

11.5.3 Quadratic Loss Function

According to (11.31), in the case of the quadratic loss function, we can write

$$\mathcal{R}_{\text{post}}(\gamma) = \int_{-\infty}^{\infty} (\gamma - l)^2 p_{\text{post}}(l) dl. \tag{11.40}$$

From the condition of extremum of the function $\mathcal{R}_{\text{post}}(\gamma)$, we obtain the following estimate:

$$\gamma_m = \int_{-\infty}^{\infty} l p_{\text{post}}(l) dl = l_{\text{post}}. \tag{11.41}$$

Thus, the average of the a posteriori pdf of l_{post}, that is, the main point of the a posteriori pdf, is considered as the estimate of the random process parameter l at the quadratic loss function. The value of $\mathcal{R}_{\text{post}}(\gamma)$ characterizes the minimum dispersion of random process parameter estimate. Since $\mathcal{R}_{\text{post}}(\gamma)$ depends on a specific form of the received realization $x(t)$, the conditional dispersion is a random variable. In the case of the quadratic loss function, the Bayes risk coincides with the unconditional dispersion of estimate given by (11.15):

$$\mathcal{R}_m = \text{Var}(\gamma_m) = \text{Var}(l_{\text{post}}) = \int_{\mathcal{L}} \int_{X} (l - l_{\text{post}})^2 p_{\text{post}}(l) p(X) dl dX. \tag{11.42}$$

The estimate (11.41) is defined based on average risk minimum condition. For this reason, we can state that the Bayes estimate for the quadratic loss function makes minimum the unconditional dispersion of estimate. In other words, in the case of quadratic loss function, the Bayes estimate ensures the minimum estimate variance value with respect to the true value of estimated random process parameter among all possible estimations. We need to note one more property of the Bayes estimate when the loss function is quadratic. Substituting the a posteriori pdf (11.2) into (11.41) and averaging by the sample of observed data X, we can write

$$\langle l_{\text{post}} \rangle = \int_{-\infty}^{\infty} \int_{X} l p_{\text{prior}}(l) p(X|l) dl dX. \tag{11.43}$$

Changing the order of integration and taking into consideration the condition of normalization, we obtain

$$\langle l_{\text{post}} \rangle = \int_{-\infty}^{\infty} l p_{\text{prior}}(l) dl = l_{\text{prior}}, \tag{11.44}$$

that is, in the case of the quadratic loss function, the Bayes estimate is unconditional and unbiased, forever.

The quadratic loss function is proportional to the deviation square of the estimate from the true value of estimated random process parameter; that is, the weight is assigned to deviations and this weight increases as the square function of deviation values. These losses take place very often in various applications of mathematical statistics and theory of estimations. However, although the quadratic loss function has a set of advantages, that is, it is very convenient from the mathematical viewpoint and takes into consideration efficiently the highest value of big deviations compared to small ones, it is seldom used for the solution of optimal estimation problems compared to the simplest loss function. This is associated with the fact that the device that generates the estimate given by (11.41) has very complex construction and functioning. In the quadratic loss function case, definite difficulties can be met during computation of variance of the Bayes estimations.

11.5.4 Rectangle Loss Function

In the case of rectangle loss function (11.26), all estimate deviations from the true value of estimated random process parameter, which are less than the given modulo value η, are not equally dangerous and do not affect the quality of random process parameter estimation. Deviations, the instantaneous values of which exceed the modulo value η, are not desirable and are assigned the same weight. Substituting (11.26) into (11.31) and dividing the limits of integration on three intervals, namely, $-\infty < l < \gamma - \eta$, $\gamma - \eta < l < \gamma + \eta$, and $\gamma + \eta < l < \infty$, we obtain

$$\mathcal{R}_{\text{post}}(\gamma) = 1 - \int_{\gamma-\eta}^{\gamma+\eta} p_{\text{post}}(l)dl. \tag{11.45}$$

As we can see from (11.45), in the case of rectangular loss function, the value $\gamma = \gamma_m$ is the Bayes estimate, for which the probability to satisfy the condition $|\gamma - l| \leq \eta$ is maximum. Based on the condition of the a posteriori risk $\mathcal{R}_{\text{post}}(\gamma)$ extremum, we obtain the equation for estimation

$$p_{\text{post}}(\gamma_m - \eta) = p_{\text{post}}(\gamma_m + \eta). \tag{11.46}$$

This equation shows that we need to use that value of the parameter γ_m as the Bayes estimate, under which the values of a posteriori pdf corresponding to random process parameters deviating from the estimate on the value η both at right and left sides are the same.

If the a posteriori pdf $p_{\text{post}}(l)$ is differentiable, then, at small values η, there can be a limitation by the first three terms of Taylor series expansion with respect to the point γ_m of the a posteriori pdfs given by (11.46). As a result, we obtain the equation coinciding with (11.36). Consequently, at small insensibility areas η, the Bayes estimations for the simplest and rectangular loss functions coincide. The Bayes risk given by (11.29) for the rectangular loss function takes the following form:

$$\mathcal{R}_m = 1 - \int_X \left[\int_{\gamma_m-\eta}^{\gamma_m+\eta} p_{\text{post}}(l)p(X)dl \right] dX. \tag{11.47}$$

The random variable given by (11.45) is the a posteriori probability of the event that the true value of estimated random process parameter of the given realization is not included within the limits of the interval $-\eta < \gamma_m < \eta$. In doing so, in the case of the Bayes estimate γ_m, this probability is less compared to that of other types of estimations. The Bayes risk given by (11.47) defines the minimum average probability of incorrect decision making by the given criterion.

11.6 SUMMARY AND DISCUSSION

In many practical applications, the estimate criteria are selected based on assumptions about the purpose of estimate. At the same time, during the process design, choosing a definite estimate criteria is of great interest because such approach allows us to understand more clearly the essence and characteristic property of the problem to estimate the random process parameter, which is very important, giving us a possibility to define the problem of valid selection of the estimate criterion more fairly. Owing to the finite time of observation and presence of noise and interference concurrent to observation, specific errors in the course of estimate definition arise. These errors are defined both by the quality performance and by conditions under which an estimation process is carried out. In general, the requirement of minimization of estimation error by value does not depend on just one assumption. However, if the criterion of estimation is given, the quality performance is measured based on this criterion. Then the problem of obtaining the optimal estimation is reduced to a definition of solution procedure that minimizes or maximizes the quality performance. In doing so, the parameter estimate must be close, in a certain sense, to the true value of estimated parameter and the optimal estimate must minimize this measure of closeness in accordance with the chosen criterion.

In theory, for statistical parameter estimation, two types of estimates are used: the interval estimations based on the definition of confidence interval, and the point estimation, that is, the estimate defined at the point. Employing the interval estimations, we need to indicate the interval, within the limits of which there is the true value of unknown random process parameter with the probability that is not less than the predetermined value. This predetermined probability is called the confidence factor and the indicated interval of possible values of estimated random process parameter is called the confidence interval. The upper and lower bounds of the confidence interval, which are called the confidence limits, and the confidence interval are the functions to be considered during digital signal processing (a discretization) or during analog signal processing (continuous function) of the received realization $x(t)$. In the case of point estimation, we assign one parameter value to the unknown parameter from the interval of possible parameter values; that is, some value is obtained based on the analysis of the received realization $x(t)$ and we use this value as the true value of the evaluated parameters.

In addition to the procedure of analysis of the estimation random process parameter based on an analysis of the received realization $x(t)$, a sequential estimation method is used. Essentially, this method is used for the sequential statistical analysis to estimate the random process parameter. The basic idea of sequential estimation is to define the time of analysis of the received realization $x(t)$, within the limits of which we are able to obtain the estimate of parameter with the preset criteria for reliability. In the case of the point estimate, the root-mean-square deviation of estimate or other convenience function characterizing a deviation of estimate from the true value of estimated random process parameter can be considered as the measure of reliability. From a viewpoint of interval sequential estimation, the estimate reliability can be defined using a length of the confidence interval with the given confidence coefficient.

To make the point estimation means that some number from the interval of possible values of the estimated random process parameter must correspond to each possible received realization. This number is called the point estimate. Owing to the random character of the point estimate of random process parameter, it is characterized by the conditional pdf. This is a general and total characteristic of the point estimate. A shape of this pdf defines a quality of point estimate definition and, consequently, all properties of the point estimate.

There are several approaches to define the properties of the point estimations: (a) It is natural to try to define such point estimate that the conditional pdf would be grouped very close to the estimate value; (b) it is desirable that while increasing the observation interval, the estimation would coincide with or approach stochastically the true value of the estimated random process parameter (in this case, we can say that the estimate is the consistent estimate); (c) the estimate must be unbiased or, in an extreme case, asymptotically unbiased; (d) the estimate must be the best by some criterion; for example, it must be characterized by minimal values of dispersion or variance at zero or constant bias; and (e) the estimate must be a statistically sufficient measure.

Based on the random character of the observed realization, we can expect errors in any decision-making rules; that is, such a decision does not coincide with the true value of a parameter.

By applying various decision-making rules, different errors will appear with different levels of probability. Since the nonzero probability of error exists forever, we need to characterize the quality of different estimations in one way or another. For this purpose, the loss function is introduced in the theory of decisions. This function defines a definite loss for each combination from the decision and parameter. The physical sense of the loss function is the following. A definite nonnegative weight is assigned to each incorrect decision. In doing so, depending on targets, for which the estimate is defined, the most undesirable decisions are assigned the greatest weights. A choice of definite loss function is made depending on a specific problem of estimation of the random process parameter. There is no general decision-making rule to select the loss function. Each decision-making rule is selected based on a subjective principle. Any arbitrariness with selecting losses leads to definite difficulties with applying the theory of statistical decisions. The following types of loss functions are widely used in practice: the simple loss function, the linear modulo loss function, the quadratic loss function, and the rectangle loss function.

REFERENCES

1. Kay, S.M. 2006. *Intuitive Probability and Random Processes Using MATLAB*. New York: Springer + Business Media, Inc.
2. Govindarajulu, Z. 1987. *The Sequential Statistical Analysis of Hypothesis Testing Point and Interval Estimation, and Decision Theory (American Series in Mathematical and Management Science)*. New York: American Sciences Press, Inc.
3. Sieqmund, D. 2010. *Sequential Analysis: Test and Confidence Intervals* (Springer Series in Statistics). New York: Springer + Business Media, Inc.
4. Kay, S.M. 1993. *Fundamentals of Statistical Signal Processing: Estimation Theory*. Upper Saddle River, NJ: Prentice Hall, Inc.
5. Cramer, H. 1946. *Mathematical Methods of Statistics*. Princeton, NJ: Princeton University Press.
6. Cramer, H. and M.R. Leadbetter. 2004. *Stationary and Related Stochastic Processes: Sample Function Properties and Their Applications*. Mineola, NY: Dover Publications.

12 Estimation of Mathematical Expectation

12.1 CONDITIONAL FUNCTIONAL

Let the analyzed Gaussian stochastic process $\xi(t)$ possess the mathematical expectation defined as

$$E(t) = E_0 s(t), \tag{12.1}$$

the correlation function $R(t_1, t_2)$, and be observed within the limits of the finite time interval $[0, T]$. We assume that the law of variation of the mathematical expectation $s(t)$ and correlation function $R(t_1, t_2)$ are known. Thus, the received realization takes the following form:

$$x(t) = E(t) + x_0(t) = E_0 s(t) + x_0(t), \quad 0 \le t \le T, \tag{12.2}$$

where

$$x_0(t) = x(t) - E(t) \tag{12.3}$$

is the centered Gaussian stochastic process. The problem with estimating the mathematical expectation is correlated to the problem with estimating the amplitude $E(t)$ of the deterministic signal in the additive Gaussian noise.

The pdf functional of the Gaussian process given by (12.2) takes the following form:

$$F[x(t) | E_0] = B_0 \exp\left\{-0.5 \int_0^T \int_0^T \vartheta(t_1, t_2)[x(t_1) - E_0 s(t_1)] \times [x(t_2) - E_0 s(t_2)] dt_1 dt_2\right\}, \tag{12.4}$$

where B_0 is the factor independent of the estimated parameter E_0; the function

$$\vartheta(t_1, t_2) = \vartheta(t_2, t_1) \tag{12.5}$$

is defined from the integral equation

$$\int_0^T R(t_1, t)\vartheta(t, t_2) dt = \delta(t_2 - t_1). \tag{12.6}$$

Introduce the following function:

$$\upsilon(t) = \int_0^T s(t_1)\vartheta(t_1, t) dt_1, \tag{12.7}$$

which is a solution of the following integral equation:

$$\int_0^T R(t,\tau)\upsilon(\tau)d\tau = s(t). \tag{12.8}$$

Since the received realization of the Gaussian stochastic process does not depend on the current value of estimation E_0, the pdf functional of Gaussian stochastic process can be written in the following form:

$$F[x(t)|E_0] = B_1 \exp\left\{E_0\int_0^T x(t)\upsilon(t)dt - 0.5E_0^2\int_0^T s(t)\upsilon(t)dt\right\}, \tag{12.9}$$

where B_1 is the factor independent of the estimated parameter E_0.

As applied to analysis of the stationary stochastic process, we can write

$$\begin{cases} s(t) = 1, \\ R(t_1,t_2) = R(t_2 - t_1) = R(t_1 - t_2). \end{cases} \tag{12.10}$$

In doing so, the function $\upsilon(t)$ and the pdf functional are determined as

$$\int_0^T R(t-\tau)\upsilon(\tau)d\tau = 1; \tag{12.11}$$

$$F[x(t)|E_0] = B_1 \exp\left\{E_0\int_0^T x(t)\upsilon(t)dt - 0.5E_0^2\int_0^T \upsilon(t)dt\right\}. \tag{12.12}$$

In practice, the stationary stochastic processes occur very often and their correlation functions can be written in the following form:

$$R(\tau) = \sigma^2 \exp\{-\alpha|\tau|\}, \tag{12.13}$$

$$R(\tau) = \sigma^2 \exp\{-\alpha|\tau|\}\left[\cos\omega_1\tau + \frac{\alpha}{\omega_1}\sin\omega_1|\tau|\right], \tag{12.14}$$

where σ^2 is the variance of stationary stochastic process. These correlation functions correspond to the stationary stochastic processes obtained as a result of excitation of the RC-circuit, $\alpha = (RC)^{-1}$, and the RLC-circuit, $\omega_1^2 = \omega_0^2 - \alpha^2$, $\omega_0^2 = (LC)^{-1} > \alpha = R(2L)^{-1}$ inputs by the "white" noise.

Solution of (12.8) for the correlation functions given by (12.13) and (12.14) takes the following form, respectively:

$$\upsilon(t) = \frac{\alpha}{2\sigma^2}[s(t) - \alpha^{-2}s''(t)] + \frac{1}{\sigma^2}\{[s(0) - \alpha^{-1}s'(0)]\delta(t) + [s(T) + \alpha^{-1}s'(T)]\delta(T-t)\}, \tag{12.15}$$

$$\upsilon(t) = \frac{1}{4\sigma^2\alpha\omega_0^2}\left[s''''(t) + 2\left(\omega_0^2 - 2\alpha^2\right)s''(t) + \omega_0^4 s(t)\right]$$

$$+\frac{1}{2\sigma^2\alpha\omega_0^2}\left\{\left[s'''(0) + \left(\omega_0^2 - 4\alpha^2\right)s'(0) + 2\alpha\omega_0^2 s(0)\right]\delta(t)\right.$$

$$-\left[s'''(T) + \left(\omega_0^2 - 4\alpha^2\right)s'(T) - 2\alpha\omega_0^2 s(T)\right]\delta(t-T) + \left[s''(0) - 2\alpha s'(0) + \omega_0^2 s(0)\right]\delta'(t)$$

$$\left.-\left[s''(T) + 2\alpha s'(T) + \omega_0^2 s(T)\right]\delta'(t-T)\right\}. \tag{12.16}$$

The notations $s'(t)$, $s''(t)$, $s'''(t)$, $s''''(t)$ mean the derivatives of the first, second, third, and fourth order with respect to t, respectively. If the function $s(t)$ and its derivatives at $t = 0$ and $t = T$ become zero, then (12.15) and (12.16) have a simple form. As applied to stochastic process at $s(t) = 1 = \text{const}$, (12.15) and (12.16) have the following form:

$$\upsilon(t) = \frac{\alpha}{2\sigma^2} + \frac{1}{\sigma^2}[\delta(t) + \delta(T-t)], \tag{12.17}$$

$$\upsilon(t) = \frac{\omega_0^2}{4\alpha\sigma^2} + \frac{1}{\sigma^2}[\delta(t) + \delta(t-T)] + \frac{1}{2\alpha\sigma^2}[\delta'(t) - \delta'(t-T)]. \tag{12.18}$$

The following spectral densities

$$S(\omega) = \frac{2\alpha\sigma^2}{\alpha^2 + \omega^2} \tag{12.19}$$

and

$$S(\omega) = \frac{4\alpha\sigma^2\left(\omega_1^2 + \alpha^2\right)}{\omega^4 - 2\omega^2\left(\omega_1^2 - \alpha^2\right) + \left(\omega_1^2 + \alpha^2\right)^2} \tag{12.20}$$

correspond to the correlation functions given by (12.13) and (12.14), respectively. It is necessary to note that there is no general procedure to solve (12.8). However, if the correlation function of stochastic process depends on the absolute value of difference of arguments $|t_2 - t_1|$ and the observation time T is much more than the correlation interval defined as

$$\tau_{\text{cor}} = \frac{1}{\sigma^2}\int_0^\infty |R(\tau)|d\tau = \int_0^\infty |\mathcal{R}(\tau)|\,d\tau, \tag{12.21}$$

where

$$\mathcal{R}(\tau) = \frac{R(\tau)}{\sigma^2} \tag{12.22}$$

is the normalized correlation function, and the function $s(t)$ and its derivatives at $t = 0$ and $t = T$ become zero, it is possible to define the approximate solution of the integral equation (12.8) using the Fourier transform. Applying the Fourier transform to the left and right side of the following equation

$$\int_{-\infty}^{\infty} R(t-\tau)\tilde{\upsilon}(\tau)d\tau = s(t), \tag{12.23}$$

it is not difficult to use the inverse Fourier transform in order to obtain

$$\tilde{\upsilon}(t)=\frac{1}{2\pi}\int_{-\infty}^{\infty}\frac{\mathbf{S}(\omega)}{S(\omega)}\exp\{j\omega t\}d\omega, \tag{12.24}$$

where

$S(\omega)$ is the Fourier transform of the correlation function $R(\tau)$

$\mathbf{S}(\omega)$ is the Fourier transform of mathematical expectation of the function $s(t)$, which can be defined as

$$S(\omega)=\int_{-\infty}^{\infty}R(\tau)\exp\{-j\omega\tau\}d\tau, \tag{12.25}$$

$$\mathbf{S}(\omega)=\int_{-\infty}^{\infty}s(t)\exp\{-j\omega t\}dt. \tag{12.26}$$

The inverse Fourier transform gives us the following formulae:

$$R(\tau)=\frac{1}{2\pi}\int_{-\infty}^{\infty}S(\omega)\exp\{j\omega\tau\}d\omega, \tag{12.27}$$

$$s(t)=\frac{1}{2\pi}\int_{-\infty}^{\infty}\mathbf{S}(\omega)\exp\{j\omega\tau\}d\omega. \tag{12.28}$$

If the function $s(t)$ and its derivatives do not become zero at $t=0$ and $t=T$ and the function $\mathbf{S}(\omega)$ is a ratio of two polynomials of pth and dth orders, respectively, with respect to ω^2 and $d>p$, then there is a need to add the delta function $\delta(t)$ and its derivative $\delta'(t)$ taken at $t=0$ and $t=T$. Thus, there is a need to define the solution for Equation 12.8 in the following form:

$$\upsilon(t)=\tilde{\upsilon}(t)+\sum_{\mu=0}^{d-1}\left[b_\mu\delta^\mu(t)+c_\mu\delta^\mu(t-T)\right]. \tag{12.29}$$

Here, the coefficients b_μ and c_μ are defined from the solutions of equations obtained under the substitution of (12.29) in (12.8); $\delta^\mu(t)$ is the delta function derivative of μth order with respect to the time.

In the case of stationary stochastic process, we have $s(t)=1$. In this case, the spectral density takes the following form:

$$\mathbf{S}(\omega)=2\pi\delta(\omega). \tag{12.30}$$

From (12.24) and (12.29), we have

$$\upsilon(t)=S^{-1}(\omega=0)+\sum_{\mu=0}^{d-1}\left[b_\mu\delta^\mu(t)+c_\mu\delta^\mu(t-T)\right]. \tag{12.31}$$

As applied to the stationary stochastic process with the spectral density given by (12.19), we have that $d = 1$. For this reason, we can write

$$\upsilon(t) = \frac{\alpha}{2\sigma^2} + b_0\delta(t) + c_0\delta(t-T). \tag{12.32}$$

Substituting (12.13) and (12.32) into (12.5), we obtain

$$0.5\left[\alpha\int_0^T \exp\{-\alpha|t-\tau|\}d\tau + b_0\sigma^2\exp\{-\alpha t\} + c_0\sigma^2\exp\{-\alpha(T-t)\}\right] = 1. \tag{12.33}$$

Dividing the integration intervals on two intervals, namely, $0 \le \tau < t$ and $t \le \tau \le T$, after integration we obtain

$$(b_0\sigma^2 - 1)\exp\{-\alpha t\} + (c_0\sigma^2 - 1)\exp\{-\alpha(t-T)\} = 0. \tag{12.34}$$

This equality is correct if the coefficient of the terms $\exp\{-\alpha t\}$ and $\exp\{-\alpha(t-T)\}$ is equal to zero, that is, $b_0 = c_0 = \sigma^{-2}$. Substituting b_0 and c_0 into (12.32), we obtain the formula given by (12.17).

Now, consider the exponential function in (12.9). The formula

$$\rho_1^2 = \int_0^T s(t)\upsilon(t)dt \tag{12.35}$$

is the deterministic component or, in other words, the signal when the estimated parameter $E_0 = 1$. The random component

$$\int_0^T x_0(t)\upsilon(t)dt \tag{12.36}$$

is the noise component. The variance of the noise component taking into consideration (12.8) is defined as

$$\left\langle\left[\int_0^T x_0(t)\upsilon(t)dt\right]^2\right\rangle = \int_0^T\int_0^T \langle x_0(t_1)x_0(t_2)\rangle\upsilon(t_1)\upsilon(t_2)dt_1dt_2 = \int_0^T s(t)\upsilon(t)dt = \rho_1^2. \tag{12.37}$$

As we can see from (12.37), ρ_1^2 is the ratio between the power of the signal and the power of the noise. Because of this, we can say that (12.37) is the signal-to-noise ratio (SNR) when the estimated parameter value $E_0 = 1$.

12.2 MAXIMUM LIKELIHOOD ESTIMATE OF MATHEMATICAL EXPECTATION

Consider the conditional functional given by (12.9) of the observed stochastic process. Solving the likelihood equation with respect to the parameter E, we obtain the *mathematical expectation estimate*

$$E_E = \frac{\int_0^T x(t)\upsilon(t)dt}{\int_0^T s(t)\upsilon(t)dt}. \tag{12.38}$$

As applied to analysis of stationary stochastic process, (12.38) becomes simple, namely,

$$E_{\mathrm{E}} = \frac{\int_0^T x(t)\upsilon(t)dt}{\int_0^T \upsilon(t)dt}. \tag{12.39}$$

In doing so, at the condition $T\tau_{\mathrm{cor}}^{-1} \to \infty$ we can neglect the values of stochastic process and its derivatives at $t = 0$ and $t = T$ under estimation of the mathematical expectation; that is, in other words, we can think that the following approximation is correct:

$$\upsilon(t) = S^{-1}(\omega = 0). \tag{12.40}$$

In this case, we obtain the asymptotical formula for the mathematical expectation estimate of stationary stochastic process, namely,

$$E_{\mathrm{E}} = \lim_{T\to\infty} \frac{1}{T}\int_0^T x(t)dt, \tag{12.41}$$

which is widely used in the theory of stochastic processes to define the mathematical expectation of ergodic stochastic processes with arbitrary pdf. At the large and finite values $T\tau_{\mathrm{cor}}^{-1}$, we can neglect an effect of values of the stochastic process and its derivatives at $t = 0$ and $t = T$ on the mathematical expectation estimate. As a result, we can write

$$E_{\mathrm{E}} \approx \frac{1}{T}\int_0^T x(t)dt. \tag{12.42}$$

Although the obtained formulae for the mathematical expectation estimate are optimal in the case of Gaussian stochastic process, these formulae will be optimal for the stochastic process differed from the Gaussian pdf in the class of linear estimations. Equations 12.38 and 12.39 are true if the a priori interval of changes of the mathematical expectation is not limited. Equation 12.38 allows us to define the optimal device structure to estimate the mathematical expectation of the stochastic process (Figure 12.1). The main function is the linear integration of the received realization $x(t)$ with the weight $\upsilon(t)$ that is defined based on the solution of the integral equation (12.8). The decision device issues the output process at the instant $t = T$. To obtain the current value of the mathematical

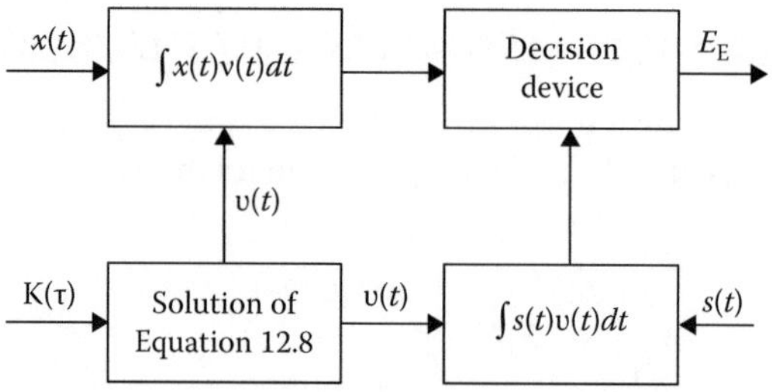

FIGURE 12.1 Optimal structure to define the mathematical expectation estimate.

expectation estimate, the limits of integration in (12.38) must be $t - T$ and t, respectively. Then the parameter estimation is defined as

$$E_{\mathrm{E}}(t) = \frac{\int_{t-T}^{t} x(\tau)\upsilon(\tau)d\tau}{\int_{t-T}^{t} s(\tau)\upsilon(\tau)d\tau}. \tag{12.43}$$

The weight integration can be done by the linear filter with corresponding impulse response. For this purpose, we introduce the function

$$\upsilon(\tau) = h(t-\tau) \quad \text{or} \quad h(\tau) = \upsilon(t-\tau) \tag{12.44}$$

and substitute this function into (12.43) instead of $\upsilon(t)$ introducing a new variable $t - \tau = z$. Then (12.43) can be transformed to the following form:

$$E_{\mathrm{E}}(t) = \frac{\int_{0}^{T} x(t-z)h(z)dz}{\int_{0}^{T} s(t-z)h(z)dz}. \tag{12.45}$$

The integrals in (12.45) are the output responses of the linear filter with the impulse response $h(t)$ given by (12.44) when the filter input is excited by $x(t)$ and $s(t)$, respectively.

The mathematical expectation of estimate

$$\langle E_{\mathrm{E}} \rangle = \frac{1}{\rho_1^2}\int_{0}^{T} \langle x(t) \rangle \upsilon(t)dt = E_0, \tag{12.46}$$

that is, the estimate of the maximum likelihood of the mathematical expectation of stochastic process is both the conditionally and unconditionally unbiased estimate. The conditional variance of the mathematical expectation estimate can be presented in the following form:

$$\mathrm{Var}\{E_{\mathrm{E}} \mid E_0\} = \left\langle E_{\mathrm{E}}^2 \right\rangle - \langle E_{\mathrm{E}} \rangle^2 = \frac{1}{\rho_1^4}\int_{0}^{T}\int_{0}^{T} \langle x_0(t_1)x_0(t_2)\rangle \upsilon(t_1)\upsilon(t_2)dt_1dt_2 = \rho_1^{-2}, \tag{12.47}$$

that is, the variance of estimate is unconditional. Since, according to (12.38), the integration of Gaussian stochastic process is a linear operation, the estimate E_{E} is subjected to the Gaussian distribution.

Let the analyzed stochastic process be a stationary process and possess the correlation function given by (12.13). Substituting instead of the function $\upsilon(t)$ its value from (12.17) into (12.47) and integrating with delta functions, we obtain

$$\mathrm{Var}\{E_{\mathrm{E}}\} = \frac{2\sigma^2}{2 + (T/\tau_{\mathrm{cor}})} = \frac{2\sigma^2}{2 + \alpha T} = \frac{2\sigma^2}{2 + p}, \tag{12.48}$$

where p is a ratio between the time required to analyze the stochastic process and the correlation interval of the same stochastic process. In doing so, according to (12.38), the formula for the optimal estimate takes the following form:

$$E_{\text{E}} = \frac{x(0) + x(T) + \alpha \int_0^T x(t)dt}{2 + p}. \tag{12.49}$$

If $p \gg 1$, we have

$$\text{Var}\{E_{\text{E}}\} \approx \frac{2\sigma^2}{p}. \tag{12.50}$$

Formulae (12.48) and (12.49) can be obtained without determination of the pdf functional. For this purpose, the value defined by the following equation:

$$E^* = \int_0^T h(t)x(t)dt \tag{12.51}$$

can be considered as the estimate. Here $h(t)$ is the weight function defined based on the condition of unbiasedness of the estimate that is equivalent to

$$\int_0^T h(t)dt = 1, \tag{12.52}$$

and minimization of the variance of estimate,

$$\text{Var}\{E^*\} = \int_0^T \int_0^T h(t_1)h(t_2)R(t_1, t_2)dt_1 dt_2. \tag{12.53}$$

Transform the formula for the variance of estimate into a convenient form. For this purpose, introduce new variables in the double integral, namely,

$$\tau = t_2 - t_1 \quad \text{and} \quad t_1 = z, \tag{12.54}$$

and change the order of integration. Taking into consideration that $R(\tau) = R(-\tau)$, we obtain

$$\text{Var}\{E^*\} = 2\int_0^T R(\tau) \int_0^{T-\tau} h(z)h(z+\tau)dz d\tau. \tag{12.55}$$

As was shown in Ref. [1], a definition of optimal form of the weight function $h(t)$ is reduced to a solution of the integral Wiener–Hopf equation

$$\int_0^T h(\tau)R(\tau - s)d\tau - \text{Var}_{\min}\{E\} = 0, \quad 0 \le s \le T, \tag{12.56}$$

where $\text{Var}_{\min}\{E\}$ is the minimal estimate variance, jointly with the condition given by (12.52). However, the solution of (12.56) is complicated.

Define the formula for an optimal estimate of mathematical expectation of the stationary stochastic process possessing the correlation function given by (12.14) and weight function given by (12.18). Substituting (12.18) into the formula for mathematical expectation estimate of the stochastic process defined as (12.38) and calculating the corresponding integrals, the following is obtained:

$$E_{\mathrm{E}} = \frac{\dfrac{1}{T}\displaystyle\int_0^T x(t)dt + \dfrac{2\alpha}{\omega_0^2 T}[x(0)+x(T)] + \dfrac{1}{\omega_0^2 T}[x'(0)-x'(t)]}{1+\dfrac{4\alpha}{\omega_0^2 T}}. \tag{12.57}$$

In doing so, the variance of the mathematical expectation estimate is defined as

$$\text{Var}\{E_{\mathrm{E}}\} = \frac{4\alpha\sigma^2}{\omega_0^2 T + 4\alpha}. \tag{12.58}$$

If $\alpha \ll \omega_0$ and $\omega_0 T \gg 1$, the formula for the mathematical expectation estimate of the stationary Gaussian stochastic process transforms to the well-known formula of the mathematical expectation definition of the ergodic stochastic process given by (12.42), and the variance of the mathematical expectation estimate is defined as

$$\text{Var}\{E_{\mathrm{E}}\} \approx \frac{4\alpha\sigma^2}{\omega_0^2 T}. \tag{12.59}$$

At $\omega_1 = 0$ ($\omega_0 = \alpha$), the correlation function given by (12.14), can be transformed into the following form

$$R(\tau) = \sigma^2 \exp\{-\alpha|\tau|\}(1+\alpha|\tau|), \quad \tau_{\mathrm{cor}} = \frac{2}{\alpha} \tag{12.60}$$

by limiting process. In particular, the given correlation function corresponds to the stationary stochastic process at the output of two *RC* circuits connected in series when the "white" noise excites the input. In this case, the formulae for the mathematical expectation estimate and variance take the following form:

$$E_{\mathrm{E}} = \frac{\dfrac{1}{T}\displaystyle\int_0^T x(t)dt + \dfrac{2}{\alpha T}[x(0)+x(T)] + \dfrac{1}{\alpha^2 T}[x'(0)-x'(T)]}{1+\dfrac{4}{\alpha T}}, \tag{12.61}$$

$$\text{Var}\{E_{\mathrm{E}}\} = \frac{4\sigma^2}{\alpha T + 4}. \tag{12.62}$$

Relationships between the definition of the estimate and the estimate variance of the mathematical expectation of the stochastic processes with other types of correlation functions can be defined analogously.

As we assumed before, the a priori domain of definition of the mathematical expectation is not limited. Thus, we consider a domain of possible values of the mathematical expectation as a function of the mathematical expectation estimate. Let the a priori domain of definition of the mathematical expectation be limited both by the upper bound and by the lower bound, that is,

$$E_L \le E \le E_U. \tag{12.63}$$

In the considered case, the mathematical expectation estimate $\hat{E}$ cannot be outside the considered interval given by (12.63), even though it is defined as a position of the absolute maximum of the likelihood functional logarithm (12.9). The likelihood functional logarithm reaches its maximum at $E = E_E$. As a result, when $E_E \le E_L$ the likelihood functional logarithm becomes a monotonically decreasing function within the limits of the interval $[E_L, E_U]$ and reaches its maximum value at $E = E_L$. If $E_E \ge E_U$, the likelihood functional logarithm becomes a monotonically increasing function within the limits of the interval $[E_L, E_U]$ and, consequently, reaches its maximum value at $E = E_U$. Thus, in the case of the limited a priori domain of definition of the mathematical expectation, the estimate of mathematical expectation of stochastic process can be presented in the following form:

$$\hat{E} = \begin{cases} E_U & \text{if} \quad E_E > E_U, \\ E_E & \text{if} \quad E_L \le E_E \le E_U, \\ E_L & \text{if} \quad E_E < E_L. \end{cases} \tag{12.64}$$

Taking into consideration the last relationship, the structure of optimal device for the mathematical expectation estimate determination in the case of the limited a priori domain of mathematical expectation definition can be obtained by the addition of a linear limiter with the following characteristic:

$$g(z) = \begin{cases} E_U & \text{if} \quad z > E_U, \\ z & \text{if} \quad E_L \le z \le E_U, \\ E_L & \text{if} \quad z < E_L \end{cases} \tag{12.65}$$

to the circuit shown in Figure 12.1. Using the well-known relationships [2] to transform the Gaussian random variable pdf of by a nonlinear inertialess system with the chain characteristic $g(z)$, we can define the conditional pdf of the mathematical expectation estimate as follows:

$$p(\hat{E}|E_0) = \begin{cases} P_L\delta(\hat{E} - E_L) + P_U\delta(\hat{E} - E_U) + \dfrac{1}{\sqrt{2\pi \text{Var}(E_E|E_0)}} \exp\left\{-\dfrac{(\hat{E} - E_0)^2}{2\text{Var}(E_E|E_0)}\right\} & \text{at } E_L \le E_E \le E_U, \\ 0, \quad \text{at } \hat{E} < E_L, \quad \hat{E} > E_U. \end{cases} \tag{12.66}$$

Here

$$\begin{cases} P_L = 1 - Q\left(\dfrac{E_L - E_0}{\sqrt{\text{Var}(E_E|E_0)}}\right), \\ P_U = Q\left(\dfrac{E_L - E_0}{\sqrt{\text{Var}(E_E|E_0)}}\right); \end{cases} \tag{12.67}$$

where

$$Q(z) = \frac{1}{\sqrt{2\pi}} \int_{z}^{\infty} \exp\{-0.5y^2\} dy \tag{12.68}$$

is the Gaussian Q function [3,4]; $\text{Var}(E_\text{E} | E_0)$ is the variance given by (12.47). The conditional bias is defined as

$$\begin{aligned} b(\hat{E} | E_0) = \langle \hat{E} \rangle - E_0 &= \int_{-\infty}^{\infty} (\hat{E} - E_0) p(\hat{E} | E_0) d\hat{E} \\ &= P_\text{L}(E_\text{L} - E_0) + P_\text{U}(E_\text{U} - E_0) \\ &+ \sqrt{\frac{\text{Var}(E_\text{E} | E_0)}{2\pi}} \left\{ \exp\left[-\frac{(E_\text{L} - E_0)^2}{2\text{Var}(E_\text{E} | E_0)}\right] - \exp\left[-\frac{(E_\text{U} - E_0)^2}{2\text{Var}(E_\text{E} | E_0)}\right] \right\}. \end{aligned} \tag{12.69}$$

Thus, in the case of the limited a priori domain of possible values of the mathematical expectation of stochastic process, the maximum likelihood estimate of the stochastic process mathematical expectation is conditionally biased. However, at small variance values of the maximum likelihood estimate of stochastic process mathematical expectation, that is, $\text{Var}(E_\text{E} | E_0) \to 0$, as it follows from (12.67) and (12.69), we obtain the asymptotical expression

$$\lim_{\text{Var}(E_\text{E} | E_0) \to 0} b(E_\text{E} | E_0) = 0; \tag{12.70}$$

that is, at $\text{Var}(E_\text{E} | E_0) \to 0$, the maximum likelihood estimate of mathematical expectation of stochastic process is asymptotically unbiased. At the high variance values of the maximum likelihood estimate of stochastic process mathematical expectation, that is, $\text{Var}(E_\text{E} | E_0) \to \infty$, the bias of the maximum likelihood estimate of stochastic process mathematical expectation tends to approach

$$b(E_\text{E} | E_0) = 0.5(E_\text{L} + E_\text{U} - 2E_0). \tag{12.71}$$

The conditional dispersion of the maximum likelihood estimate of stochastic process mathematical expectation is defined as

$$\begin{aligned} D(E_\text{E} | E_0) = \int_{-\infty}^{\infty} (\hat{E} - E_0)^2 f(E_\text{E} | E_0) d\hat{E} &= P_\text{L}(E_\text{L} - E_0)^2 + P_\text{U}(E_\text{L} - E_0)^2 + \text{Var}(1 - P_\text{U} - P_\text{L}) \\ &+ \sqrt{\frac{\text{Var}(E_\text{E} | E_0)}{2\pi}} \left\{ (E_\text{L} - E_0)\exp\left[-\frac{(E_\text{L} - E_0)^2}{2\text{Var}(E_\text{E} | E_0)}\right] - (E_\text{U} - E_0)\exp\left[-\frac{(E_\text{U} - E_0)^2}{2\text{Var}(E_\text{E} | E_0)}\right] \right\}. \end{aligned} \tag{12.72}$$

At small variance values of the maximum likelihood estimate of stochastic process mathematical expectation

$$\frac{\text{Var}(E_\text{E} | E_0)}{E_\text{U} - E_\text{L}} \ll 1 \quad \text{and} \quad E_\text{L} < E < E_\text{U} \tag{12.73}$$

if the limiting process is carried out at $E_L \to -\infty$ and $E_U \to \infty$, the conditional dispersion of the maximum likelihood estimate of stochastic process mathematical expectation coincides with the variance of estimate given by (12.47). If the true value of the mathematical expectation coincides with one of two bounds of the a priori domain of possible values of the mathematical expectation, then the following approximation is true:

$$D(E_E|E_0) \approx 0.5\text{Var}(E_E|E_0); \tag{12.74}$$

that is, the dispersion of estimate is twice as less compared to the unlimited a priori domain case. With increasing variance of the maximum likelihood estimate of stochastic process mathematical expectation $\text{Var}(E_E|E_0) \to \infty$, the conditional dispersion of the maximum likelihood estimate of the stochastic process mathematical expectation tends to approach the finite value since $P_L = P_U = 0.5$

$$D(E_E|E_0) \to 0.5[(E_L - E_0)^2 + (E_U - E_0)^2], \tag{12.75}$$

whereas the dispersion of the maximum likelihood estimate of stochastic process mathematical expectation within the unlimited a priori domain of possible values of the maximum likelihood estimate of stochastic process mathematical expectation is increased without limit as $\text{Var}(E_E|E_0) \to \infty$. It is important to note that although the bias and dispersion of the maximum likelihood estimate of stochastic process mathematical expectation are defined as the conditional values, they are nevertheless independent of the true value of the mathematical expectation E_0 and are the unconditional estimates simultaneously.

Determine the unconditional bias and dispersion of maximum likelihood estimate of stochastic process mathematical expectation in the case of the limited a priori domain of possible estimate values. For this purpose, it is necessary to average the conditional characteristics given by (12.69) and (12.72) with respect to possible values of estimated parameter, assuming that the a priori pdf of estimated parameter is uniform within the limits of the interval $[E_L, E_U]$. In this case, we observe that the unconditional estimate is unbiased, and the unconditional dispersion is determined in the following form:

$$D(\hat{E}) = \text{Var}\left\{1 - 2Q\left[\frac{E_U - E_L}{\sqrt{\text{Var}(E_E|E_0)}}\right]\right\} + \frac{2}{3}(E_U - E_L)^2 Q\left[\frac{E_U - E_L}{\sqrt{\text{Var}(E_E|E_0)}}\right]$$

$$- \frac{2\text{Var}(E_E|E_0)\sqrt{\text{Var}(E_E|E_0)}}{3\sqrt{2\pi}(E_U - E_L)}\left\{1 - \exp\left\{-\frac{(E_U - E_L)^2}{2\text{Var}(E_E|E_0)}\right\}\right\}$$

$$- \frac{2\sqrt{\text{Var}(E_E|E_0)}(E_U - E_L)}{3\sqrt{2\pi}}\exp\left\{-\frac{(E_U - E_L)^2}{2\text{Var}(E_E|E_0)}\right\}. \tag{12.76}$$

At the same time, it is not difficult to see that at small values of the variance, that is, $\text{Var}(E_E|E_0) \to 0$, the unconditional dispersion transforms into a dispersion of the estimate obtained under the unlimited a priori domain of possible values, $D(\hat{E}) \to \text{Var}(E_E|E_0)$. Otherwise, at high values of variance, that is, $\text{Var}(E_E|E_0) \to \infty$, the dispersion of the estimate given by (12.47) increases without limit and the unconditional dispersion given by (12.76) has a limit equal to the average square of the a priori domain of possible values of the estimate, that is, $(E_U - E_L)^2/3$.

12.3 BAYESIAN ESTIMATE OF MATHEMATICAL EXPECTATION: QUADRATIC LOSS FUNCTION

As before, we analyze the realization $x(t)$ of stochastic process given by (12.2). The a posteriori pdf of estimated stochastic process parameter E can be presented in the following form:

$$p_{\text{post}}(E) = \frac{p_{\text{prior}}(E)\exp\left\{E\int_0^T x(t)\upsilon(t)dt - \frac{E^2}{2}\int_0^T s(t)\upsilon(t)dt\right\}}{\int_{-\infty}^{\infty} p_{\text{prior}}(E)\exp\left\{E\int_0^T x(t)\upsilon(t)dt - \frac{E^2}{2}\int_0^T s(t)\upsilon(t)dt\right\}dE}, \tag{12.77}$$

where

$p_{\text{prior}}(E)$ is the a priori pdf of estimated stochastic process parameter
$\upsilon(t)$ is the solution of the integral equation given by (12.8)

In accordance with the definition given in Section 11.4, the Bayesian estimate γ_E is the estimate minimizing the unconditional average risk given by (11.29) at the given loss function. As applied to the quadratic loss function defined as

$$\mathcal{L}(\gamma, E) = (\gamma - E)^2, \tag{12.78}$$

the average risk coincides with the dispersion of estimate. In doing so, the Bayesian estimate γ_E is obtained based on minimization of the a posteriori risk at each fixed realization of observed data

$$\gamma_E = \int_{-\infty}^{\infty} E p_{\text{post}}(E)dE. \tag{12.79}$$

To define the estimate characteristics, that is, the bias and dispersion, it is necessary to determine two first moments of the random variable γ_E. However, in the case of the arbitrary a priori pdf of estimated stochastic process parameter E, it is impossible to determine these moments in a general form. In accordance with this statement, we consider the discussed problem for the case of a priori Gaussian pdf of estimated parameter; that is, we assume [5]

$$p_{\text{prior}}(E) = \frac{1}{\sqrt{2\pi \text{Var}_{\text{prior}}(E)}}\exp\left\{-\frac{(E - E_{\text{prior}})^2}{2\text{Var}_{\text{prior}}(E)}\right\}, \tag{12.80}$$

where E_{prior} and $\text{Var}_{\text{prior}}(E)$ are the a priori values of the mathematical expectation and variance of the mathematical expectation estimate. Substituting (12.80) into the formula defining the Bayesian estimate and carrying out the integration, we obtain

$$\gamma_E = \frac{\text{Var}_{\text{prior}}(E)\int_0^T x(t)\upsilon(t)dt + E_{\text{prior}}}{\text{Var}_{\text{prior}}(E)\int_0^T s(t)\upsilon(t)dt + 1}. \tag{12.81a}$$

It is not difficult to note that if $\text{Var}_{\text{prior}}(E) \to \infty$, the a priori pdf of estimate is approximated by the uniform pdf of the estimate and the estimate becomes the maximum likelihood estimate (12.38).

In the opposite case, that is, $\mathrm{Var}_{\mathrm{prior}}(E) \to 0$, the a priori pdf of estimate degenerates into the Dirac delta function $\delta(E - E_{\mathrm{prior}})$ and, naturally, the estimate γ_{E} will match with E_{prior}.
The mathematical expectation of estimate can be presented in the following form:

$$\langle\gamma_{\mathrm{E}}\rangle = \frac{\mathrm{Var}_{\mathrm{prior}}(E)\rho_1^2 E_0 + E_{\mathrm{prior}}}{\mathrm{Var}_{\mathrm{prior}}(E)\rho_1^2 + 1}, \tag{12.81b}$$

where ρ_1^2 is given by (12.35). In doing so, the conditional bias of the considered estimate is defined as

$$b(\gamma_{\mathrm{E}} \mid E_0) = \langle\gamma_{\mathrm{E}}\rangle - E_0 = \frac{E_{\mathrm{prior}} - E_0}{\mathrm{Var}_{\mathrm{prior}}(E)\rho_1^2 + 1}. \tag{12.82}$$

Averaging the conditional bias by all possible a priori values E_0, we obtain that in the case of the quadratic loss function the Bayesian estimate for the Gaussian a priori pdf is the unconditionally unbiased estimate.

The conditional dispersion of the obtained estimate can be presented in the following form:

$$D(\gamma_{\mathrm{E}} \mid E_0) = \langle(\gamma_{\mathrm{E}} - E_0)^2\rangle = \frac{(E_{\mathrm{prior}} - E_0)^2 + \mathrm{Var}_{\mathrm{prior}}^2(E)\rho_1^2}{\left\{\mathrm{Var}_{\mathrm{prior}}^2(E)\rho_1^2 + 1\right\}^2}. \tag{12.83}$$

We see that the unconditional dispersion coincides with the unconditional variance and is defined as

$$\mathrm{Var}(\gamma_{\mathrm{E}}) = D(\gamma_{\mathrm{E}}) = \frac{\mathrm{Var}_{\mathrm{prior}}(E)}{\mathrm{Var}_{\mathrm{prior}}(E)\rho_1^2 + 1}. \tag{12.84}$$

If $\mathrm{Var}_{\mathrm{prior}}(E)\rho_1^2 \gg 1$, then the variance of the considered Bayesian estimate coincides with the variance of the maximum likelihood estimate given by (12.47). In the opposite case, if $\mathrm{Var}_{\mathrm{prior}}(E)\rho_1^2 \ll 1$, the variance of estimate tends to approach

$$\mathrm{Var}(\gamma_{\mathrm{E}}) \approx \mathrm{Var}_{\mathrm{prior}}(E)\left\{1 - \mathrm{Var}_{\mathrm{prior}}(E)\rho_1^2\right\}. \tag{12.85}$$

As applied to arbitrary pdfs of estimate, we can obtain the approximated formulae for the bias and dispersion of estimate. For this purpose, we can transform (12.77) by substituting the realization $x(t)$ given by (12.2). Then, we can write

$$E\int_0^T x(t)\upsilon(t)dt - \frac{E^2}{2}\int_0^T s(t)\upsilon(t)dt = \rho^2 S(E) + \rho N(E), \tag{12.86}$$

where

$$\rho^2 = E_0^2\rho_1^2; \tag{12.87}$$

$$S(E) = \frac{E(2E_0 - E)}{2E_0^2}; \tag{12.88}$$

$$N(E) = \frac{E}{E_0\rho_1}\int_0^T x_0(t)\upsilon(t)dt. \tag{12.89}$$

The introduced function $S(E)$ and $N(E)$ can be called *normalized signal* and *noise* components, respectively. In doing so, they are normalized in such a way that the function $S(E)$ can reach the maximum equal to 0.5 at $E = E_0$:

$$S(E)_{\max} = S(E = E_0) = 0.5. \tag{12.90}$$

The noise component $N(E)$ has zero mathematical expectation, and its correlation function is defined as

$$\langle N(E_1)N(E_2)\rangle = \frac{E_1 E_2}{E_0^2}. \tag{12.91}$$

At $E = E_0$, the variance of noise component can be presented in the following form:

$$\langle N^2(E_0)\rangle = 1. \tag{12.92}$$

As a result, the Bayesian estimate of the mathematical expectation of stochastic process can be written in the following form:

$$\gamma_E = \frac{\int_{-\infty}^{\infty} Ep_{\text{prior}}(E)\exp\{\rho^2 S(E) + \rho N(E)\}dE}{\int_{-\infty}^{\infty} p_{\text{prior}}(E)\exp\{\rho^2 S(E) + \rho N(E)\}dE}. \tag{12.93}$$

Consider two limiting cases: the weak and powerful signals or, in other words, the low and high SNR ρ^2.

12.3.1 Low Signal-to-Noise Ratio ($\rho^2 \ll 1$)

As we can see from (12.93), at low values of the SNR ($\rho \to 0$), the exponential function tends to approach the unit and, as a result, the Bayesian estimate γ_E coincides with the a priori mathematical expectation

$$\gamma_E(\rho \to 0) = \gamma_0 = \int_{-\infty}^{\infty} Ep_{\text{prior}}(E)dE = E_{\text{prior}}. \tag{12.94}$$

At finite values of the SNR, the difference $\gamma_E - \gamma_0$ is not equal to zero. Closeness γ_E to γ_0 at $\rho \ll 1$ allows us to find the estimate characteristics if we are able to define a deviation of γ_E from γ_0 in the form of corresponding approximations and, consequently, the deviation of γ_E from the true value of the estimated parameter E_0, since in general $E_{\text{prior}} \neq E_0$. At $\rho \ll 1$, the estimate γ_E can be defined in the following approximated form [6]:

$$\gamma_E = \gamma_0 + \rho\gamma_1 + \rho^2\gamma_2 + \rho^3\gamma_3 + \cdots. \tag{12.95}$$

Considering the exponential function $\exp\{\rho^2 S(E) + \rho N(E)\}$ in (12.93) as a function of ρ, we can expand it in Maclaurin series by ρ. Then, neglecting the terms with an order of more than 4, we can write

$$\int_{-\infty}^{\infty}(\gamma_E - E)p_{\text{prior}}(E)\exp\{\rho^2 S(E)+\rho N(E)\}dE = \int_{-\infty}^{\infty}(\gamma_0 - E+\rho\gamma_1+\rho^2\gamma_2+\rho^3\gamma_3+\cdots)p_{\text{prior}}(E)$$

$$\times\left\{1+\rho N(E)+\frac{1}{2}\rho^2[N^2(E)+2S(E)]+\frac{1}{6}\rho^3[N^3(E)+6N(E)S(E)]+\cdots\right\}dE = 0. \tag{12.96}$$

Equating with zero the coefficients of terms with the same order ρ, we obtain the formulae for corresponding approximations:

$$\gamma_0 = \int_{-\infty}^{\infty} Ep_{\text{prior}}(E)dE = E_{\text{prior}}; \tag{12.97}$$

$$\gamma_1 = \int_{-\infty}^{\infty}(E - E_{\text{prior}})p_{\text{prior}}(E)N(E)dE; \tag{12.98}$$

$$\gamma_2 = \int_{-\infty}^{\infty}(E - E_{\text{prior}})p_{\text{prior}}(E)\left[0.5N^2(E)+S(E)\right]dE - \gamma_1\int_{-\infty}^{\infty}p_{\text{prior}}(E)N(E)dE; \tag{12.99}$$

$$\gamma_3 = \frac{1}{6}\int_{-\infty}^{\infty}(E - E_{\text{prior}})p_{\text{prior}}(E)\left[N^3(E)+6N(E)S(E)\right]dE - \gamma_2\int_{-\infty}^{\infty}p_{\text{prior}}(E)N(E)dE$$

$$-\gamma_1\int_{-\infty}^{\infty}p_{\text{prior}}(E)\left[0.5N^2(E)+S(E)\right]dE. \tag{12.100}$$

To define the approximate values of bias and dispersion of the estimate, it is necessary to determine the corresponding moments of approximations γ_1, γ_2, and γ_3. Taking into consideration that all odd moments of the stochastic process $x_0(t)$ are equal to zero, we obtain

$$\langle\gamma_1\rangle = \langle\gamma_3\rangle = 0, \tag{12.101}$$

$$\langle\gamma_2\rangle = \frac{\text{Var}_{\text{prior}}(E_0 - E_{\text{prior}})}{E_0^2}, \tag{12.102}$$

$$\left\langle\gamma_1^2\right\rangle = \frac{\text{Var}_{\text{prior}}^2}{E_0^2}, \tag{12.103}$$

where

$$\mathrm{Var}_{\mathrm{prior}} = \int_{-\infty}^{\infty} E^2 p_{\mathrm{prior}}(E)dE - E_{\mathrm{prior}}^2 \tag{12.104}$$

is the variance of a priori distribution.

Based on (12.101) through (12.104), we obtain the conditional estimate bias in the following form:

$$b(\gamma_{\mathrm{E}} | E_0) = E_{\mathrm{prior}} + \rho^2 \mathrm{Var}_{\mathrm{prior}} \frac{E_0 - E_{\mathrm{prior}}}{E_0^2} - E_0 = (E_{\mathrm{prior}} - E_0)\left(1 - \rho_1^2 \mathrm{Var}_{\mathrm{prior}}\right). \tag{12.105}$$

Formula for the conditional bias coincides with the approximation given by (12.82) at low values of SNR ρ_1^2, that is, $\mathrm{Var}_{\mathrm{prior}}\rho_1^2 \ll 1$. We can see that the unconditional estimate of mathematical expectation that is averaged with respect to all possible values E_0 is unbiased. The conditional dispersion of estimate with accuracy of the order ρ^4 and higher is defined in the following form:

$$D(\gamma_{\mathrm{E}} | E_0) = \langle(\gamma_{\mathrm{E}} - E_0)^2\rangle \approx (E_{\mathrm{prior}} - E_0)^2 + \rho^2\left[\left\langle\gamma_1^2\right\rangle + 2(E_{\mathrm{prior}} - E_0)\langle\gamma_2\rangle\right]. \tag{12.106}$$

Substituting the determined moments, we obtain

$$D(\gamma_{\mathrm{E}} | E_0) \approx (E_{\mathrm{prior}} - E_0)^2\left(1 - 2\mathrm{Var}_{\mathrm{prior}}\rho_1^2\right) + \rho_1^2 \mathrm{Var}_{\mathrm{prior}}^2. \tag{12.107}$$

Averaging (12.107) by all possible values of estimated parameter E_0 with the a priori pdf $p_{\mathrm{prior}}(E_0)$ matched with the pdf $p_{\mathrm{prior}}(E)$, we can define the unconditional dispersion of the mathematical expectation estimate defined by approximation in (12.85)

12.3.2 High Signal-to-Noise Ratio ($\rho^2 \gg 1$)

Bayesian estimate of the stochastic process mathematical expectation given by (12.93) can be written in the following form:

$$\gamma_{\mathrm{E}} = \frac{\int_{-\infty}^{\infty} E p_{\mathrm{prior}}(E)\exp\{-\rho^2 Z(E)\}dE}{\int_{-\infty}^{\infty} p_{\mathrm{prior}}(E)\exp\{-\rho^2 Z(E)\}dE}, \tag{12.108}$$

where

$$Z(E) = \left[S(E_{\mathrm{E}}) + \rho^{-1}N(E_{\mathrm{E}})\right] - \left[S(E) + \rho^{-1}N(E)\right]; \tag{12.109}$$

E_{E} is the maximum likelihood estimate given by (12.38). We can see that at the maximum likelihood point $E = E_{\mathrm{E}}$, the function $Z(E)$ reaches its minimum and is equal to zero, that is, $Z(E) = 0$.

At high values of the SNR ρ^2, we can use the asymptotic Laplace formula [7] to determine the integrals in (12.108)

$$\lim_{\lambda\to\infty} \int_a^b \varphi(x)\exp\{\lambda h(x)\}dx \approx \sqrt{\frac{2\pi}{\lambda h''(x_0)}} \exp\{\lambda h(x_0)\}\varphi(x_0), \tag{12.110}$$

where $a < x_0 < b$ and the function $h(x)$ has a maximum at $x = x_0$. Substituting (12.110) into an initial equation for the Bayesian estimate (12.108), we obtain $\gamma_E \approx E_E$. Thus, at high values of the SNR, the Bayesian estimate of the stochastic process mathematical expectation coincides with the maximum likelihood estimate of the same parameter.

12.4 APPLIED APPROACHES TO ESTIMATE THE MATHEMATICAL EXPECTATION

Optimal methods to estimate the stochastic process mathematical expectation envisage the need for having accurate and complete knowledge of other statistical characteristics of the considered stochastic process. Therefore, as a rule, various nonoptimal procedures based on (12.51) are used in practice. In doing so, the weight function is selected in such a way that the variance of estimate tends to approach asymptotically the variance of the optimal estimate.

Thus, let the estimate be defined in the following form:

$$E^* = \int_0^T h(t)x(t)\,dt. \tag{12.111}$$

The function of the following form

$$h(t) = \begin{cases} T^{-1} & \text{if} \quad 0 \le t \le T, \\ 0 & \text{if} \quad t < 0, \; t > T, \end{cases} \tag{12.112}$$

is widely used as the weighted function $h(t)$. In doing so, the mathematical expectation estimate of stochastic process is defined as

$$E^* = \frac{1}{T}\int_0^T x(t)dt. \tag{12.113}$$

Procedure of the estimate definition given by (12.113) coincides with approximation in (12.42) that was delivered based on the optimal rule of estimation in the case of large interval in comparison with the interval of correlation of the considered stochastic process. A device operating according to the rule given by (12.113) is called the *ideal integrator.*

The variance of mathematical expectation estimate is defined as

$$\text{Var}(E^*) = \frac{1}{T^2}\int_0^T\int_0^T R(t_2 - t_1)dt_1 dt_2. \tag{12.114}$$

We can transform the double integral introducing the new variables, namely, $\tau = t_2 - t_1$ and $t_2 = t$. Then,

$$\text{Var}(E^*) = \frac{1}{T^2}\int_0^T \left\{ \int_{-t}^{0} R(\tau)d\tau + \int_0^{T-t} R(\tau)d\tau \right\} dt. \tag{12.115}$$

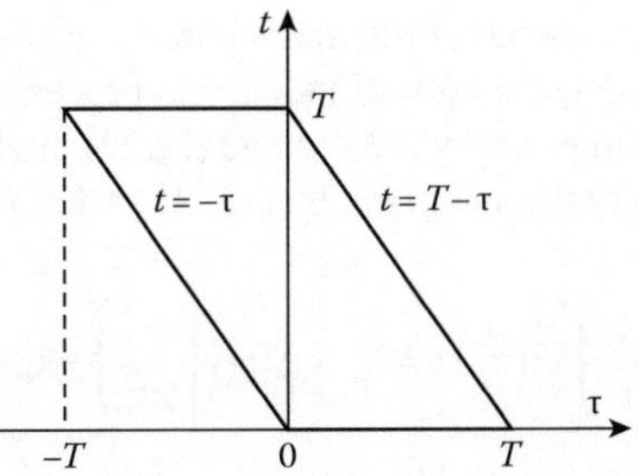

FIGURE 12.2 Integration domains.

The integration domain is shown in Figure 12.2. Changing the integration order, we obtain

$$\mathrm{Var}(E^*) = \frac{2}{T}\int_0^T \left(1 - \frac{\tau}{T}\right) R(\tau) d\tau. \tag{12.116}$$

If the interval of observation [0, T] is much more than the correlation interval τ_{cor}, we can change the upper integration limit in (12.116) by infinity and neglect the integrand term τ/T in comparison with unit. Then

$$\mathrm{Var}(E^*) = \frac{2\mathrm{Var}(E_E | E_0)}{T}\int_0^\infty R(\tau) d\tau. \tag{12.117}$$

If the normalized correlation function $R(\tau)$ is not a sign-changing function of the argument τ, the formula (12.117) takes a simple and obvious form:

$$\mathrm{Var}(E^*) = \frac{2\mathrm{Var}(E_E | E_0)}{T}\tau_{cor}. \tag{12.118}$$

Consequently, if the ideal integrator integration time is sufficiently large in comparison with the correlation interval of stochastic process, then to determine the variance of mathematical expectation estimate of stochastic process there is a need to know only the values of variance and the ratio between the observation interval and correlation interval.

In the case of the sign-changing normalized correlation function with the argument τ, we can write

$$\int_0^T | R(\tau) | d\tau > \int_0^\infty R(\tau) d\tau. \tag{12.119}$$

Thus, at $T \gg \tau_{cor}$ the formula for the variance of the mathematical expectation estimate in the case of arbitrary correlation function of stationary stochastic process can be written in the following form:

$$\mathrm{Var}(E^*) \le \frac{2\mathrm{Var}(E_E | E_0)}{T}\tau_{cor}. \tag{12.120}$$

If $T \gg \tau_{cor}$, the formula for the variance of the mathematical expectation estimate can be presented in the form using the spectral density $S(\omega)$ of stochastic process that is related to the correlation function by the Fourier transform given by (12.25) and (12.27). In doing so, the formula for variance of the mathematical expectation estimate given by (12.117) takes the following form:

$$\mathrm{Var}(E^*) \approx \frac{1}{T}\int_{-\infty}^{\infty} R(\tau)d\tau = \frac{1}{T}\int_{-\infty}^{\infty} S(\omega)\left\{\frac{1}{2\pi}\int_{-\infty}^{\infty}\exp\{j\omega\tau\}d\tau\right\}d\omega. \tag{12.121}$$

Taking into consideration that

$$\delta(\omega) = \frac{1}{2\pi}\int_{-\infty}^{\infty}\exp\{j\omega\tau\}d\tau, \tag{12.122}$$

we obtain

$$\mathrm{Var}(E^*) \approx \frac{1}{T}S(\omega)\Big|_{\omega=0}. \tag{12.123}$$

Thus, the variance of the mathematical expectation estimate of stochastic process is proportional to the spectral density value of fluctuation component of the considered stochastic process at $\omega = 0$ when the ideal integrator is used as a smoothing circuit. In other words, in the considered case, the variance of the mathematical expectation estimate of stochastic process is defined by spectral components about zero frequency. To obtain the current value of the mathematical expectation estimate and to investigate the realization of stochastic process within the limits of large interval of observation, we use the following estimate:

$$E^*(t) = \int_{0}^{T} h(\tau)x(\tau)d\tau. \tag{12.124}$$

Evidently, this estimate has the same statistical characteristics as the estimate defined by (12.111).

In practice, the linear low-pass filters with constant parameters defined by the impulse response

$$h(t) = \begin{cases} h(t) & \text{at} \quad t \geq 0, \\ 0 & \text{at} \quad t < 0, \end{cases} \tag{12.125}$$

are used as averaging devices. In this case, the formula describing the process at the low-pass filter output, taking into consideration the unbiasedness of estimate, takes the following form:

$$E^*(t) = c\int_{0}^{T} h(\tau)x(T-\tau)d\tau, \tag{12.126}$$

where the constant factor c can be determined from the following condition:

$$c = \frac{1}{\int_0^T h(\tau)d\tau}. \tag{12.127}$$

If a difference between the measurement instant and instant of appearance of stochastic process at the low-pass filter input is much more than the correlation interval τ_{cor} of the considered stochastic process and the low-pass filter time constant, then we can write

$$E^*(t) = \frac{\int_0^\infty h(t-\tau)x(\tau)d\tau}{\int_0^\infty h(\tau)d\tau}. \tag{12.128}$$

The variance of the mathematical expectation estimate of stochastic process is defined as

$$\text{Var}(E^*) = c^2 \int_0^T \int_0^T R(\tau_1 - \tau_2)h(\tau_1)h(\tau_2)d\tau_1 d\tau_2. \tag{12.129}$$

Introducing new variables $\tau_1 - \tau_2 = \tau$ and $\tau_2 = t$ and changing the order of integration, the formula for the variance of the mathematical expectation estimate can be presented in the following form:

$$\text{Var}(E^*) = 2c^2 \int_0^T R(\tau)r_h(\tau)d\tau, \tag{12.130}$$

where, if $T \gg \tau_{cor}$ we can change the upper integration limit on infinity, and the introduced function

$$r_h(\tau) = \int_0^{T-\tau} h(t)h(t+\tau)dt, \quad \tau > 0, \tag{12.131}$$

corresponds to the correlation function of stochastic process forming at the output of filter with the impulse response $h(t)$ when the "white" noise with the correlation function $R(\tau) = \delta(\tau)$ [8] excites the filter input.

When the process at the low-pass filter is stationary, that is, duration of exciting input stochastic process is much more in comparison with the low-pass filter constant time, the formula for the variance of the mathematical expectation estimate of stochastic process can be written in the following form:

$$\text{Var}(E^*) = \frac{1}{2\pi} \int_{-\infty}^{\infty} S(\omega) | \mathsf{S}(j\omega) |^2 d\omega, \tag{12.132}$$

using the spectral density

$$\mathsf{S}(j\omega) = \int_0^\infty h(t)\exp\{-j\omega t\}dt, \tag{12.133}$$

that is, the Fourier transform of the impulse response or frequency characteristic of the low-pass filter.

Consider an example where we compute the normalized variance of the mathematical expectation estimate $\mathrm{Var}(E^*)/\sigma^2$ as a function of the ratio T/τ_{cor} by averaging the investigated stochastic process by the ideal integrator with the pulse response given by (12.112) and *RC*-circuit, the impulse response of which takes the following form:

$$h(t)=\begin{cases}\beta\exp\{-\beta\tau\} & \text{at} \quad 0\le t\le T,\\ 0 & \text{at} \quad \tau<0\,,\ \tau>T.\end{cases} \tag{12.134}$$

The corresponding frequency characteristics take the following form:

$$\mathrm{S}(j\omega)=\frac{1-\exp\{-j\omega T\}}{j\omega T}; \tag{12.135}$$

$$\mathrm{S}(j\omega)=\frac{\beta\{1-\exp\{-(\beta+j\omega)T\}\}}{\beta+j\omega}. \tag{12.136}$$

As an example, consider the stationary stochastic process with the exponential correlation function given by (12.13) where the parameter α is inversely proportional to the correlation interval, that is,

$$\alpha=\frac{1}{\tau_{\mathrm{cor}}}. \tag{12.137}$$

Substituting (12.13) into (12.116) and (12.130), we obtain the normalized variance of the mathematical expectation estimate for the ideal integrator and *RC* filter:

$$\frac{\mathrm{Var}_1(E^*)}{\sigma^2}=\frac{2}{p^2}[p-1+\exp\{-p\}]; \tag{12.138}$$

$$\frac{\mathrm{Var}_2(E^*)}{\sigma^2}=\lambda\frac{1-\lambda+2\lambda\exp\{-p(1+\lambda)\}-(1+\lambda)\exp\{-2p\lambda\}}{(1-\lambda^2)[1-\exp\{-\lambda p\}]^2}, \tag{12.139}$$

where

$$p=\alpha T=\frac{T}{\tau_{\mathrm{cor}}} \quad \text{and} \quad \lambda=\frac{\beta}{\alpha}=\frac{\tau_{\mathrm{cor}}}{\tau_{\mathrm{RC}}} \tag{12.140}$$

are the ratio between the observation interval duration and the correlation interval and the ratio between the correlation interval and the *RC* filter time constant, respectively. The *RC* filter time constant τ_{RC} is defined as the correlation interval (see (12.21)).

If the observation time interval is large, that is, the condition $\lambda p \gg 1$ is satisfied, for example, for the RC filter, then the normalized variance of the mathematical expectation estimate will be limited by the *RC* filter, that is,

$$\frac{\mathrm{Var}_2(E^*)}{\sigma^2}\approx\frac{\lambda}{1+\lambda}. \tag{12.141}$$

At $\lambda \ll 1$, that is, when the spectral bandwidth of considered stochastic process is much more than the averaging *RC* filter bandwidth, we can think that the *RC* filter plays the role of the ideal integrator and the estimate variance formula (12.139) under limiting transition, that is, $\lambda \to 0$, takes a form given by (12.138).

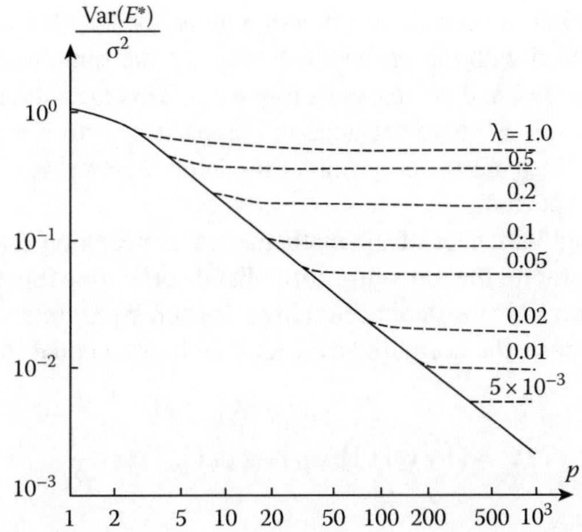

FIGURE 12.3 Normalized variance of the mathematical expectation estimate versus p at various values of λ.

Dependences of the normalized mathematical expectation estimate variances $\mathrm{Var}(E^*)/\sigma^2$ versus the ratio p between the observation time interval and the correlation interval for the ideal integrator (the continuous line) and the *RC* filter (the dashed lines) are presented in Figure 12.3, where λ serves as the parameter. As we can see from Figure 12.3, in the case of the ideal integrator, the variance of estimate decreases proportionally to the increase in the observation time interval, but in the case of the *RC* filter, the variance of estimate is limited by the value given by (12.141). Taking into consideration $\lambda \ll 1$, the normalized variance of the mathematical expectation estimate is limited by the value equal to the ratio between the correlation interval and the *RC* filter time constant in the limiting case.

It is worthwhile to compare the mathematical expectation estimate using the ideal integrator with the optimal estimate, the variance of which $\mathrm{Var}(E_{\mathrm{E}})$ is given by (12.48). Relative increase in the variance using the ideal integrator in comparison with the optimal estimate is defined as

$$\kappa = \frac{\mathrm{Var}_1(E^*) - \mathrm{Var}(E_{\mathrm{E}})}{\mathrm{Var}(E_{\mathrm{E}})} = \frac{(2+p)[p-1+\exp\{-p\}]}{p^2} - 1. \tag{12.142}$$

Relative increase in the variance as a function of T/τ_{cor} is shown in Figure 12.4. As we can see from Figure 12.4, the relative increase in the variance of the mathematical expectation estimate of

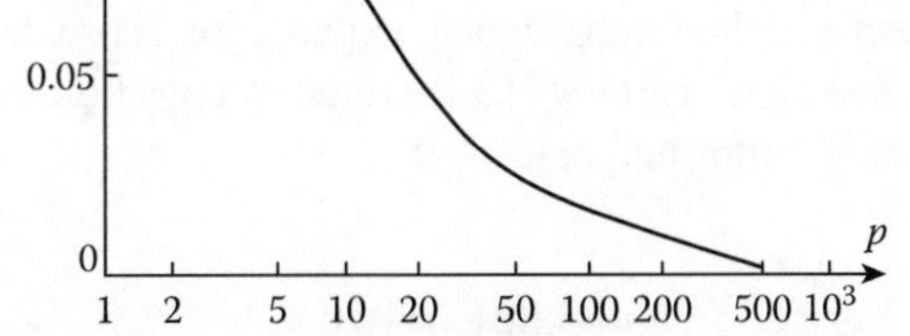

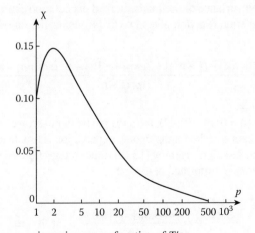

FIGURE 12.4 Relative increases in variance as a function of T/τ_{cor}.

stochastic process possessing the correlation function given by (12.13) using the ideal integrator is less than 0.01, in comparison with the optimal estimate. At the same time, the maximum relative increase in the variance is 0.14 and corresponds to $p \approx 2.7$. This maximum increase is caused by a rapid decrease in the optimal estimate variance in comparison with the estimate obtained by the ideal integrator at small values of the observation time interval. However, as $p \to \infty$, both estimates are equivalent, as it was expected.

Consider the normalized variances of the mathematical expectation estimates of stochastic process using the ideal integrator for the following normalized correlation functions that are widely used in practice. We analyze two *RC* filters connected in series and the "white" noise excites the input of this linear system. In this case, the normalized correlation function takes the following form:

$$R(\tau) = \left(1 + \alpha|\tau|\right)\exp\left\{-\alpha|\tau|\right\}, \quad \alpha = \frac{1}{RC}. \tag{12.143}$$

In doing so, the normalized variance of the mathematical expectation estimate of stochastic process is defined as

$$\frac{\mathrm{Var}_3(E^*)}{\sigma^2} = \frac{2[2p_1 - 3 + (3 + p_1)\exp\{-p_1\}]}{p_1^2}, \tag{12.144}$$

where

$$p_1 = \alpha T = \frac{2T}{\tau_{\mathrm{cor}}} \quad \text{and} \quad \tau_{\mathrm{cor}} = \frac{2}{\alpha}. \tag{12.145}$$

The set of considered stochastic processes has the normalized correlation functions that are approximated in the following form:

$$R(\tau) = \exp\left\{-\alpha|\tau|\right\}\cos\varpi\tau. \tag{12.146}$$

Depending on the relationships between the parameters α and ϖ, the normalized correlation function (12.146) describes both the low-frequency ($\alpha \gg \varpi$) and high-frequency ($\alpha \ll \varpi$) stochastic processes. The normalized variance of the mathematical expectation estimate of stochastic process with the normalized correlation function given by (12.146) takes the following form:

$$\frac{\mathrm{Var}_4(E^*)}{\sigma^2} = \frac{2[p_1(1+\eta^2) - (1-\eta^2)] + 2\exp\{-p_1\}[(1-\eta^2)\cos p_1\eta - 2\eta\sin p_1\eta]}{p_1^2(1+\eta^2)^2}, \tag{12.147}$$

where $\eta = \varpi\alpha^{-1}$. At $\varpi = 0$ ($\eta = 0$) in (12.147), as in the particular case, we obtain (12.138); that is, we obtain the normalized variance of the mathematical expectation estimate of stochastic process with the exponential correlation function given by (12.13) under integration using the ideal integrator. In this case, (12.147) is essentially simplified at $\varpi = \alpha$:

$$\frac{\mathrm{Var}_4(E^*)}{\sigma^2} = \frac{p_1 - \exp\{-p_1\}\sin p_1}{p_1^2}, \quad \text{at } \varpi = \alpha. \tag{12.148}$$

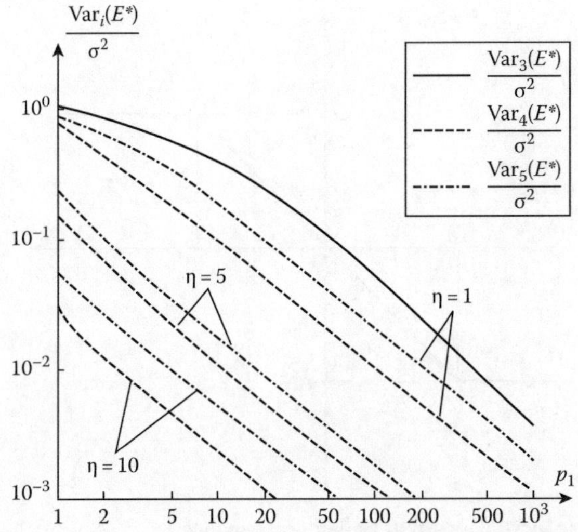

FIGURE 12.5 Normalized variances given by (12.144), (12.147), and (12.149) as functions of p_1 with the parameter η.

As applied to the correlation function given by (12.14), the normalized variance of the mathematical expectation estimate of stochastic process is defined by

$$\frac{\text{Var}_5(E^*)}{\sigma^2} = \frac{2\left[2p_1\left(1+\eta_1^2\right)-\left(3-\eta_1^2\right)\right]+2\exp\{-p_1\}\left[\left(3-\eta_1^2\right)\cos p_1\eta_1 - 3\eta_1 + \eta_1^{-1}\sin(p_1\eta_1)\right]}{p_1^2\left(1+\eta_1^2\right)^2}, \tag{12.149}$$

where $\eta_1 = \varpi_1\alpha^{-1}$. As $\varpi_1 \to 0$ ($\eta_1 \to 0$), the correlation function given by (12.14) can be written as the correlation function given by (12.143), and the formula (12.149) is changed to (12.144).

The normalized variances of the mathematical expectation estimate of stochastic process given by (12.144), (12.147), and (12.149) as a function of the parameter p_1 with the parameter η, are shown in Figure 12.5. As expected, at the same value of the parameter p_1, the normalized variance of the mathematical expectation estimate of stochastic process decreases corresponding to an increase η characterizing the presence of quasiharmonical components in the considered stochastic process.

Discussed procedures to measure the mathematical expectation assume that there are no limitations of instantaneous values of the considered stochastic process in the course of measurement. Presence of limitations leads to additional errors while measuring the mathematical expectation of stochastic process.

Determine the bias and variance of estimate applied both to the symmetrical inertialess signal limiter (see Figure 12.6) and to the asymmetrical inertialess signal limiter (see Figure 12.7) when the input of the signal limiter is excited by the Rayleigh stochastic process. In doing so, we assume that the mathematical expectation is defined according to (12.113) where we use $y(t) = g[x(t)]$ instead of $x(t)$ and $g(x)$ as the characteristic functions of transformation. The variance of the mathematical expectation estimate of stochastic process is defined by (12.116) where under the correlation function $R(\tau)$ we should understand the correlation function $R_y(\tau)$ defined as

$$R_y(\tau) = \int_{-\infty}^{\infty}\int_{-\infty}^{\infty} g(x_1)g(x_2)p_2(x_1,x_2;\tau)dx_1dx_2 - E_y^2. \tag{12.150}$$

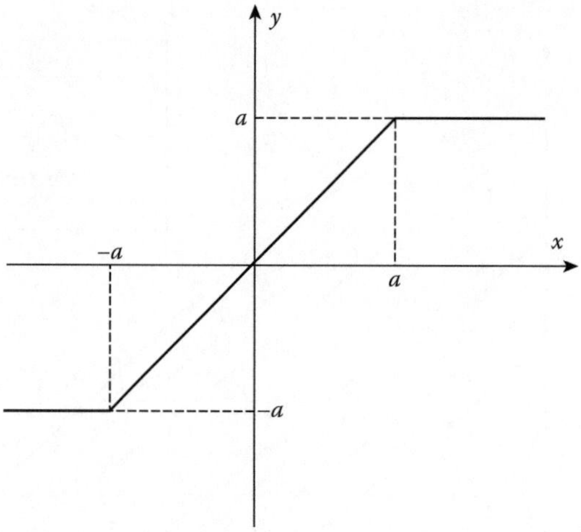

FIGURE 12.6 Symmetric inertialess signal limiter performance.

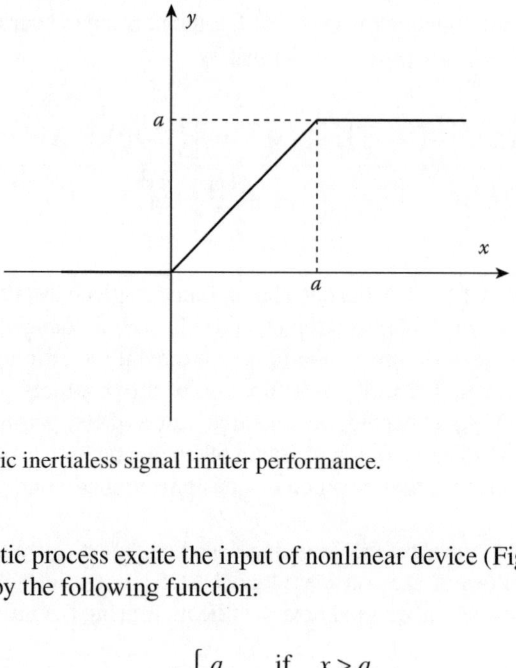

FIGURE 12.7 Asymmetric inertialess signal limiter performance.

Let the Gaussian stochastic process excite the input of nonlinear device (Figure 12.6) and the transformation be described by the following function:

$$y = g(x) = \begin{cases} a & \text{if} \quad x > a, \\ x & \text{if} \quad -a \le x \le a, \\ -a & \text{if} \quad x < -a. \end{cases} \tag{12.151}$$

The bias of estimate is defined as

$$b(E^*) = \int_{-\infty}^{\infty} g(x)p(x)dx - E_0 = -a\int_{-\infty}^{-a} p(x)dx + \int_{-a}^{a} xp(x)dx + a\int_{a}^{\infty} p(x)dx - E_0. \tag{12.152}$$

Substituting the one-dimensional pdf of Gaussian stochastic process

$$p(x) = \frac{1}{\sqrt{2\pi\,\mathrm{Var}(x)}}\exp\left\{-\frac{(x-E_0)^2}{2\,\mathrm{Var}(x)}\right\} \tag{12.153}$$

into (12.152), we obtain the bias of the mathematical expectation estimate of stochastic process in the following form:

$$b(E^*) = \sqrt{\mathrm{Var}(x)}\left\{\chi[Q(\chi-q)-Q(\chi+q)] - q[Q(\chi+q)+Q(\chi-q)] - \right.$$

$$\left. + \frac{1}{\sqrt{2\pi}}\left\{\exp[-0.5(\chi+q)^2] - \exp[-0.5(\chi-q)^2]\right\}\right\}, \tag{12.154}$$

where

$$\chi = \frac{\alpha}{\sqrt{\mathrm{Var}(x_0)}} \quad \text{and} \quad q = \frac{E_0}{\sqrt{\mathrm{Var}(x_0)}} \tag{12.155}$$

are the ratio of the limitation threshold and the mathematical expectation for the square root of the variance of the observed realization of stochastic process; $Q(x)$ is the Gaussian Q function given by (12.68).

To determine the variance of mathematical expectation estimate of stochastic process according to (12.116) and (12.150), there is a need to expand the two-dimensional pdf of Gaussian stochastic process in series [9]

$$p_2(x_1,x_2;\tau) = \frac{1}{\mathrm{Var}(x)}\sum_{\nu=0}^{\infty} Q^{(\nu+1)}\left(\frac{x_1-E_0}{\sqrt{\mathrm{Var}(x)}}\right) Q^{(\nu+1)}\left(\frac{x_2-E_0}{\sqrt{\mathrm{Var}(x)}}\right)\frac{\mathcal{R}^{\nu}(\tau)}{\nu!}, \tag{12.156}$$

where $\mathcal{R}(\tau)$ is the normalized correlation function of the initial stochastic process $\xi(t)$;

$$Q^{(\nu+1)}(z) = \frac{d^{\nu}}{dz^{\nu}}\left[\frac{\exp\{-0.5z^2\}}{\sqrt{2\pi}}\right], \qquad \nu = 0,1,2,\ldots. \tag{12.157}$$

are the derivatives of $(\nu + 1)$th order of the Gaussian Q function.

Substituting (12.156) into (12.150) and (12.1167) and taking into consideration that

$$E_y = \frac{1}{\sigma^2}\int_{-\infty}^{\infty} g(x)Q'\left(\frac{x-E_0}{\sigma}\right)dx, \tag{12.158}$$

we obtain

$$\mathrm{Var}(E^*) = \frac{1}{\sigma^2}\sum_{\nu=1}^{\infty}\frac{1}{\nu!}\left\{\int_{-\infty}^{\infty} g(x)Q^{(\nu+1)}\left(\frac{x-E_0}{\sigma}\right)dx\right\}^2 \frac{2}{T}\int_0^T\left(1-\frac{\tau}{T}\right)\mathcal{R}^{\nu}(\tau)d\tau. \tag{12.159}$$

Computing the integral in the braces, we obtain

$$\text{Var}(E^*) = \sigma^2 \sum_{\nu=1}^{\infty} \frac{1}{\nu!}[Q^{(\nu-1)}(\chi - q) - Q^{(\nu-1)}(-\chi - q)]^2 \frac{2}{T}\int_0^T \left(1 - \frac{\tau}{T}\right)\mathcal{R}^{\nu}(\tau)d\tau. \quad (12.160)$$

As $\chi \to \infty$, the derivatives of the Gaussian Q function tend to approach zero. As a result, only the term at $\nu = 1$ remains and we obtain the initial formula (12.116).

In practical applications, the stochastic process measurements are carried out, as a rule, under the conditions of "weak" limitations of instantaneous values, that is, under the condition $(\chi - |q|) \geq 1.5 \div 2$. In this case, the first term at $\nu = 1$ in (12.160) plays a very important role:

$$\text{Var}(E^*) \approx [1 - Q(\chi - q) - Q(\chi + q)]^2 \frac{2\sigma^2}{T}\int_0^T \left(1 - \frac{\tau}{T}\right)\mathcal{R}(\tau)d\tau, \quad (12.161)$$

where, at sufficiently high values, that is, $(\chi - |q|) \geq 3$, the term in square brackets is very close to unit and we may use (12.116) to determine the variance of mathematical expectation estimate of stochastic process.

In practice, the Rayleigh stochastic processes are widely employed because this type of stochastic processes have a wide range of applications. In particular, the envelope of narrow-band Gaussian stochastic process described by the Rayleigh pdf can be presented in the following form:

$$z(t) = x(t)\cos[2\pi f_0 t + \varphi(t)], \quad (12.162)$$

where

$x(t)$ is the envelope

$\varphi(t)$ is the phase of stochastic process

Representation in (12.162) assumes that the spectral density of narrow-band stochastic process is concentrated within the limits of narrow bandwidth Δf with the central frequency f_0 and the condition $f_0 \gg \Delta f$. As applied to the symmetrical spectral density, the correlation function of stationary narrow-band stochastic process takes the following form:

$$R_z(\tau) = \sigma^2 \mathcal{R}(\tau)\cos(2\pi f_0 \tau). \quad (12.163)$$

In doing so, the one-dimensional Rayleigh pdf can be written as

$$f(x) = \frac{x}{\sigma^2}\exp\left\{-\frac{x^2}{2\sigma^2}\right\}, \quad x \geq 0. \quad (12.164)$$

The first and second initial moments and the normalized correlation function of the Rayleigh stochastic process can be presented in the following form:

$$\begin{cases} \langle \xi(t) \rangle = \sqrt{\dfrac{\pi}{2}\sigma^2}, \\ \langle \xi^2(t) \rangle = 2\sigma^2, \\ \rho(\tau) \approx \mathcal{R}^2(\tau). \end{cases} \quad (12.165)$$

As applied to the Rayleigh stochastic process and nonlinear transformation (see Figure 12.7) given as

$$y = g(x) = \begin{cases} a & \text{if} \quad x > a, \\ x & \text{if} \quad 0 \le x \le a, \end{cases} \tag{12.166}$$

the bias of the mathematical expectation estimate takes the following form:

$$b(E^*) = E_y - E_0 = \int_0^\infty g(x)f(x)dx - E_0 = \sqrt{2\pi\sigma^2}Q(\chi), \tag{12.167}$$

where χ is given by (12.155). As $\chi \to \infty$, the mathematical expectation estimate, as it would be expected, is unbiased.

Determining the variance of mathematical expectation estimate of stochastic process in accordance with (12.116) and (12.150) is very difficult in the case of Rayleigh stochastic process. It is evident that to determine the variance of mathematical expectation estimate of stochastic process in the first approximation, the formula (12.116) has to be true, provided the condition $\chi \ge 2$–3 is satisfied, which is analogous to the Gaussian stochastic process at weak limitation.

12.5 ESTIMATE OF MATHEMATICAL EXPECTATION AT STOCHASTIC PROCESS SAMPLING

In practice, we use digital measuring devices to measure the parameters of stochastic processes after sampling. Naturally, we do not use a part of information that is outside a sample of stochastic process.

Let the Gaussian stochastic process $\xi(t)$ be observed at some discrete instants t_i. Then, there are a set of samples $x_i = x(t_i)$, $i = 1, 2,\ldots, N$ at the input of digital measuring device. As a rule, a sample clamping of the observed stochastic process is carried out over equal time intervals $\Delta = t_{i+1} - t_i$. Each sample value can be presented in the following form:

$$x_i = E_i + x_{0_i} = Es_i + x_{0_i} \tag{12.168}$$

as in (12.2), where $E_i = Es_i = Es(t_i)$ is the mathematical expectation and $x_{0_i} = x_0(t_i)$ is the realization of the centralized Gaussian stochastic process at the instant $t = t_i$. A set of samples x_i are characterized by the conditional N-dimensional pdf

$$f_N(x_1,\ldots,x_N|E) = \frac{1}{(2\pi)^{-0.5N}\sqrt{\det\|R_{ij}\|}}\exp\left\{-0.5\sum_{i=1}^{N}\sum_{j=1}^{N}(x_i - E_i)(x_j - E_j)C_{ij}\right\}, \tag{12.169}$$

where

det $\|R_{ij}\|$ is the determinant of the correlation matrix $\|R_{ij}\| = R$ of the $N \times N$ order

C_{ij} is the elements of the matrix $\|C_{ij}\| = C$, which is the reciprocal matrix with respect to the correlation matrix and the elements C_{ij} are defined from the following equation:

$$\sum_{l=1}^{N} C_{il}R_{lj} = \delta_{ij} = \begin{cases} 1 & \text{if} \quad i = j, \\ 0 & \text{if} \quad i \ne j. \end{cases} \tag{12.170}$$

The conditional multidimensional pdf in (12.169) is the multidimensional likelihood function of the parameter E of stochastic process. Solving the likelihood equation with respect to the parameter E, we obtain the formula for the mathematical expectation estimate of stochastic process:

$$E_{\text{E}} = \frac{\sum_{i=1,j=1}^{N} x_i s_j C_{ij}}{\sum_{i=1,j=1}^{N} s_i s_j C_{ij}}. \tag{12.171}$$

This formula can be written in a simple form if we introduce the weight coefficients

$$\upsilon_i = \sum_{j=1}^{N} s_j C_{ij}, \tag{12.172}$$

which satisfy, as the function $\upsilon(t)$ given by (12.7), the system of equations

$$\sum_{l=1}^{N} R_{il}\upsilon_l = s_i, \quad i = 1,2,\ldots,N. \tag{12.173}$$

In doing so, the mathematical expectation estimate can be presented in the following form:

$$E_{\text{E}} = \frac{\sum_{i=1}^{N} x_i \upsilon_i}{\sum_{i=1}^{N} s_i \upsilon_i}. \tag{12.174}$$

The mathematical expectation of estimate takes the following form:

$$\langle E_{\text{E}} \rangle = \frac{\sum_{i=1}^{N} \langle x_i \rangle \upsilon_i}{\sum_{i=1}^{N} s_i \upsilon_i} = E_0. \tag{12.175}$$

The variance of estimate in accordance with (12.172) can be presented in the following form:

$$\text{Var}(E_{\text{E}}) = \frac{\sum_{i=1,j=1}^{N} R_{ij}\upsilon_i \upsilon_j}{\left[\sum_{i=1}^{N} s_i \upsilon_i\right]^2} = \frac{1}{\sum_{i=1}^{N} s_i \upsilon_i}. \tag{12.176}$$

The weight coefficients are determined using a set of linear equations:

$$\begin{cases} \sigma^2\upsilon_1 + R_1\upsilon_2 + R_2\upsilon_3 + \cdots + R_{N-1}\upsilon_N = s_1, \\ R_1\upsilon_1 + \sigma^2\upsilon_2 + R_1\upsilon_3 + \cdots + R_{N-2}\upsilon_N = s_2, \\ \vdots \qquad\qquad\qquad\qquad\qquad \vdots \\ R_{N-1}\upsilon_1 + R_{N-2}\upsilon_2 + R_{N-3}\upsilon_3 + \cdots + \sigma^2\upsilon_N = s_N, \end{cases} \tag{12.177}$$

where $R_l = R(l\Delta)$ are the elements of the correlation matrix of difference in time $|i-j|\Delta = l\Delta$, $l = 0, 1,\ldots, N-1$. The solution derived from this system of linear equations can be presented in the following form:

$$\upsilon_j = \frac{\det\|G_{ij}\|}{\det\|R_{ij}\|}, \quad j = 1,2,\ldots,N, \tag{12.178}$$

where det $\|G_{ij}\|$ is the determinant of matrix obtained from the matrix $\|R_{ij}\| = R$ by substituting the column containing elements $s_1, s_2,\ldots, s_N$ instead of the jth column. In the case of independent samples of stochastic process, that is, $R_{ij} = 0$ at $i \neq j$ and $R_{ii} = \sigma^2$, the correlation matrix and its reciprocal matrix will be diagonal. In doing so, for all $i = 1, 2,\ldots, N$ the weight coefficients are defined as

$$\upsilon_i = \frac{s_i}{\sigma^2}. \tag{12.179}$$

Substituting (12.179) into (12.174) and (12.176) we obtain

$$E_{\mathrm{E}} = \frac{\sum_{i=1}^{N} x_i s_i}{\sum_{i=1}^{N} s_i^2}, \tag{12.180}$$

$$\mathrm{Var}(E_{\mathrm{E}}) = \frac{\sigma^2}{\sum_{i=1}^{N} s_i^2}. \tag{12.181}$$

If the observed stochastic process is stationary, that is, $s_i = 1\ \forall i$, $i = 1, 2,\ldots, N$, even in the presence of independent samples, we obtain the mean and variance of the mean

$$E^* = \frac{1}{N}\sum_{i=1}^{N} x_i; \tag{12.182}$$

$$\mathrm{Var}(E^*) = \frac{\sigma^2}{N}, \tag{12.183}$$

respectively.

Now, consider the estimate of stationary stochastic process mathematical expectation with the correlation function given by (12.13). Denote $\psi = \exp\{-\alpha\Delta\}$. In this case, the correlation matrix takes the following form:

$$\|R_{ij}\| = \sigma^{2N}\begin{vmatrix} 1 & \psi & \psi^2 & \cdots & \psi^{N-1} \\ \psi & 1 & \psi & \cdots & \psi^{N-2} \\ \psi^2 & \psi & 1 & \cdots & \psi^{N-3} \\ \vdots & & & & \vdots \\ \psi^{N-1} & \psi^{N-2} & \psi^{N-3} & \cdots & 1 \end{vmatrix}. \tag{12.184}$$

The determinant of this matrix and its reciprocal matrix are defined in the following form [10]:

$$\det\|R_{ij}\| = \sigma^{2N}(1-\psi^2)^{N-1}, \tag{12.185}$$

$$\|C_{ij}\| = \frac{1}{(1-\psi^2)\sigma^2}\begin{vmatrix} 1 & -\psi & 0 & \ldots & 0 \\ -\psi & 1+\psi^2 & -\psi & \ldots & 0 \\ 0 & -\psi & 1+\psi^2 & \ldots & 0 \\ \vdots & & & & \vdots \\ 0 & 0 & 0 & \ldots & 1 \end{vmatrix}. \tag{12.186}$$

It is important to note that all elements of the reciprocal matrix are equal to zero, except for the elements of the main diagonal and the elements flanking the main diagonal from right and left. As we can see from (12.172) and (12.186), the optimal values of weight coefficients are defined as

$$\begin{cases} \upsilon_1 = \upsilon_N = \dfrac{1}{(1+\psi)\sigma^2}; \\ \upsilon_2 = \upsilon_3 = \cdots = \upsilon_{N-1} = \dfrac{1-\psi}{(1+\psi)\sigma^2}. \end{cases} \tag{12.187}$$

Substituting the obtained weight coefficients into (12.174) and (12.176), we have

$$E_{\mathrm{E}} = \frac{(x_1 + x_N) + (1-\psi)\sum_{i=2}^{N-1} x_i}{N-(N-2)\psi}; \tag{12.188}$$

$$\operatorname{Var}(E_{\mathrm{E}}) = \sigma^2 \frac{1+\psi}{N-(N-2)\psi}. \tag{12.189}$$

Dependence of the normalized variance of the optimal mathematical expectation estimate versus the values ψ of the normalized correlation function between the samples and the various numbers of samples N is shown in Figure 12.8. As we can see from Figure 12.8, starting from $\psi \geq 0.5$, the variance of estimation increases rapidly corresponding to an increase in the value of the normalized correlation function, which tends to approach the variance of the observed stochastic process as $\psi \to 1$.

We can obtain the formulae (12.188) and (12.189) by another way without using the maximum likelihood method. For this purpose, we suppose that

$$E^* = \sum_{i=1}^{N} x_i h_i \tag{12.190}$$

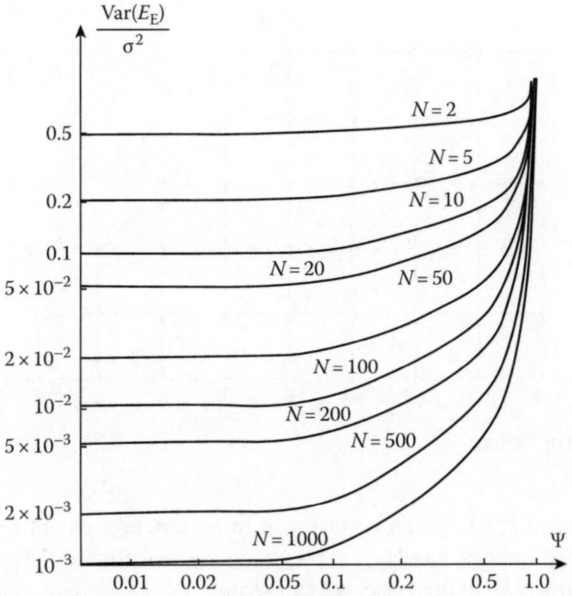

FIGURE 12.8 Normalized variance of the optimal mathematical expectation estimate versus ψ and the number of samples N.

can be used as the estimate, where h_i are the weight coefficients satisfying the following condition

$$\sum_{i=1}^{N} h_i = 1 \tag{12.191}$$

for the unbiased estimations. The weight coefficients are chosen from the condition of minimization of the variance of mathematical expectation estimate. As applied to observation of stationary stochastic process possessing the correlation function given by (12.13), the weight coefficients h_i are defined in Ref. [11] and related with the obtained weight coefficients (12.187) by the following relationship:

$$h_i = \frac{\upsilon_i}{\sum_{i=1}^{N} \upsilon_i}. \tag{12.192}$$

In the limiting case, as $\Delta \to 0$ the formulae in (12.188) and (12.189) are changed into (12.48) and (12.49), respectively. Actually, as $\Delta \to 0$ and if $(n-1)\Delta = T = \text{const}$ and $\exp\{-\alpha\Delta\} \approx 1 - \alpha\Delta$, the summation in (12.188) is changed by integration and x_1 and x_N are changed in $x(0)$ and $x(T)$, respectively. In practice, the equidistributed estimate (the mean) given by (12.182) is widely used as the mathematical expectation estimate of stationary stochastic process that corresponds to the constant weight coefficients $h_i \approx N^{-1}$, $i = 1, 2, \ldots, N$ given by (12.190).

Determine the variance of the mathematical expectation estimate assuming that the samples are equidistant from each other on the value Δ. The variance of the mathematical expectation estimate of stochastic process is defined as

$$\text{Var}(E^*) = \frac{1}{N^2} \sum_{i=1,j=1}^{N} R(t_1 - t_j) = \frac{1}{N^2} \sum_{i=1,j=1}^{N} R[(i-j)\Delta]. \tag{12.193}$$

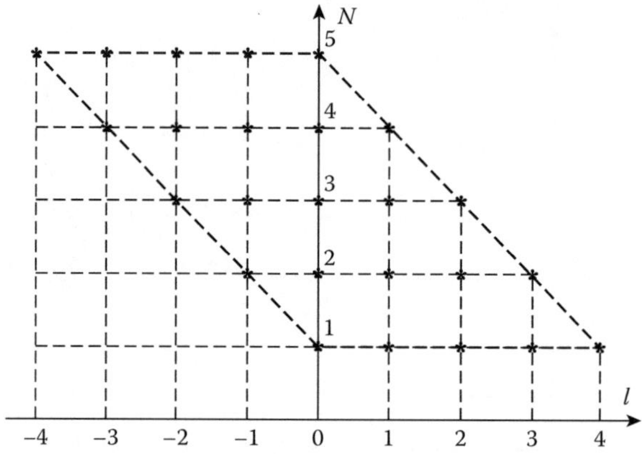

FIGURE 12.9 Domain of indices.

The double summation in (12.193) can be changed in a more convenient form. For this purpose, there is a need to change indices, namely, $l = i - j$ and $j = j$, and change the summation order in the domain shown in Figure 12.9. In this case, we can write

$$\mathrm{Var}(E^*) = \frac{1}{N^2}\sum_{j=1}^{N}\sum_{l=-j}^{N-j}\mathcal{R}(l\Delta) = \frac{1}{N^2}\left\{N\sigma^2 + 2\sum_{i=1}^{N-1}(N-i)\mathcal{R}(i\Delta)\right\} = \frac{\sigma^2}{N}\left\{1 + 2\sum_{i=1}^{N-1}\left(1-\frac{i}{N}\right)\mathcal{R}(i\Delta)\right\}, \tag{12.194}$$

where $\mathcal{R}(i\Delta)$ is the normalized correlation function of observed stochastic process. As we can see from (12.194), if the samples are not correlated the formula (12.183) can be considered as a particular case.

If the correlation function of observed stochastic process is described by (12.13), the variance of the equidistributed estimate of mathematical expectation is defined as

$$\mathrm{Var}(E^*) = \sigma^2\frac{N(1-\psi^2) + 2\psi(\psi^N - 1)}{N^2(1-\psi)^2}, \tag{12.195}$$

where, as before, $\psi = \exp\{-\alpha\Delta\}$. We have just obtained (12.195) by taking into consideration the formula for summation [12]

$$\sum_{i=0}^{N-1}(a+ir)q^i = \frac{a - [a+(N-1)r]q^N}{1-q} + \frac{rq(1-q^{N-1})}{1-q^2}. \tag{12.196}$$

Computations made by the formula (12.195) show that the variance of the equidistributed estimate of mathematical expectation differs from the variance of the optimal estimate (12.189). Figure 12.10 represents a relative increase in the variance of the equidistributed estimate of mathematical expectation in comparison with the variance of the optimal estimate

$$\varepsilon = \frac{\mathrm{Var}(E^*) - \mathrm{Var}(E_E)}{\mathrm{Var}(E_E)} \tag{12.197}$$

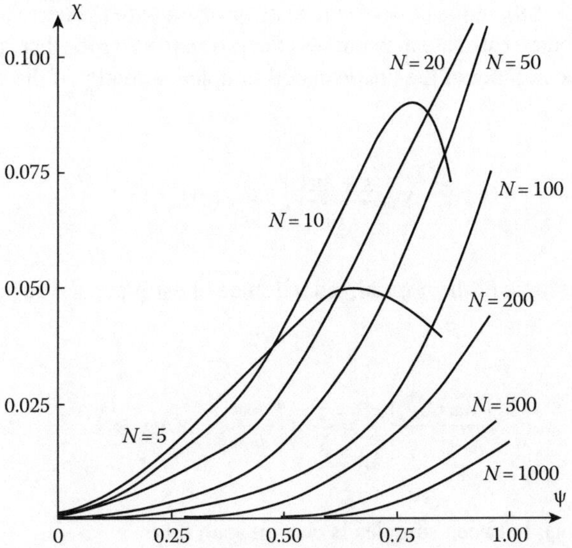

FIGURE 12.10 Relative increase in the variance of equidistributed estimate of mathematical expectation as a function of the normalized correlation function between samples.

as a function of values of the normalized correlation function between the samples ψ for various numbers of samples. Naturally, if the relative increase in the variance is low, then the magnitude of the normalized correlation function between samples will be low as well. Similar to the mathematical expectation estimate by the continuous realization of stochastic process, the presence of maxima is explained by the fact that in the case of small numbers of samples N and sufficiently large magnitude ψ, the variance of the optimal estimate decreases rapidly in comparison with the variance of the equidistributed estimate of mathematical expectation. As we can see from Figure 12.10, the magnitude of the normalized correlation function between samples is less than $\psi = 0.5$, then the optimal and equidistributed estimates coincide practically.

As applied to the normalized correlation function (12.146), the normalized variance of estimate is defined as

$$\frac{\mathrm{Var}(E^*)}{\sigma^2} = \frac{N(1-\psi^2)[1+\psi^2-2\psi\cos(\Delta\varpi)]+2\psi^{(2N+1)}\{\cos[(N+1)\Delta\varpi]-2\psi\cos(N\Delta\varpi)+\psi^2\cos[(N-1)\Delta\varpi]\}}{N^2[1+\psi^2-2\psi\cos(\Delta\varpi)]^2} - \frac{2\psi^2[(1+\psi^2)\cos(\Delta\varpi)-2\psi]}{N^2[1+\psi^2-2\psi\cos(\Delta\varpi)]^2}, \tag{12.198}$$

where, as before, $\psi = \exp\{-\alpha\Delta\}$. At $\varpi = 0$, we obtain the formula (12.195). In the case of large numbers of samples, the formula (12.198) is simplified

$$\frac{\mathrm{Var}(E^*)}{\sigma^2} \approx \frac{(1-\psi^2)}{N[1+\psi^2-2\psi\cos(\Delta\varpi)]}, \quad N\Delta\alpha \gg 1. \tag{12.199}$$

As we can see from (12.199), in the case of stochastic process with the correlation function given by (12.146) the equidistributed estimate may possess the estimate variance that is less than the variance of estimate by the same number of the uncorrelated samples. Actually, if the samples are taken over the interval

$$\Delta = \frac{\pi + 2\pi k}{\varpi}, \quad k = 0, 1, \ldots, \tag{12.200}$$

then the minimal magnitude of the normalized variance of estimate can be presented in the following form:

$$\left.\frac{\operatorname{Var}(E^*)}{\sigma^2}\right|_{\min} \approx \frac{1}{N} \times \frac{1-\psi}{1+\psi}, \quad N\Delta\alpha \gg 1. \tag{12.201}$$

Otherwise, if the interval between samples is chosen such that

$$\Delta = \frac{2\pi k}{\varpi}, \quad k = 0, 1, \ldots. \tag{12.202}$$

then the maximum value of the normalized variance of estimate takes the following form:

$$\left.\frac{\operatorname{Var}(E^*)}{\sigma^2}\right|_{\max} \approx \frac{1}{N} \times \frac{1+\psi}{1-\psi}. \tag{12.203}$$

Thus, for some types of correlation functions, the variance of the equidistributed estimate of mathematical expectation by the correlated samples can be lesser than the variance of estimate by the same numbers of uncorrelated samples.

If the interval between samples Δ is taken without paying attention to the conditions discussed previously, then the value $\Delta\varpi = \varphi$ can be considered as the random variable with the uniform distribution within the limits of the interval $[0, 2\pi]$. Averaging (12.199) with respect to the random variable φ uniformly distributed within the limits of the interval $[0, 2\pi]$, we obtain the variance of the mathematical expectation estimate of stochastic process by N uncorrelated samples

$$\left\langle \frac{\operatorname{Var}(E^*)}{\sigma^2} \right\rangle_{\varphi} = \frac{1-\psi^2}{N} \int_0^{2\pi} \frac{d\varphi}{1+\psi^2 - 2\psi\cos\varphi} = \frac{1}{N}. \tag{12.204}$$

Of definite interest for the definition of the mathematical expectation estimate is the method to measure the stochastic process parameters by additional signals [13,14]. In this case, the realization $x(t_i) = x_i$ of the observed stochastic process $\xi(t_i) = \xi_i$ is compared with the realization $v(t_i) = v_i$ of the additional stochastic process $\zeta(t_i) = \zeta_i$. A distinct feature of this measurement method is that the values x_i of the observed stochastic process realization must be with the high probability within the limits of the interval of possible values of the additional stochastic process. Usually, it is assumed that the values of the additional stochastic process are independent from each other and from the values of the observed stochastic process.

To further simplify an analysis of the stochastic process parameters and the definition of the mathematical expectation, we believe that the values x_i are independent of each other and the random variables ζ_i are uniformly distributed within the limits of the interval $[-A, A]$, that is,

$$f(v) = \frac{1}{2A}, \quad -A \le v \le A. \tag{12.205}$$

As applied to the pdf given by (12.203), the following condition must be satisfied

$$P[-A \le \xi \le A] \approx 1 \tag{12.206}$$

in the case of the considered method to measure the stochastic process parameters. As a result of comparison, a new independent random variable sequence ς_i is formed:

$$\varsigma_i = x_i - v_i. \tag{12.207}$$

The random variable sequence ς_i can be transformed to the new independent random variable sequence η_i by the nonlinear inertialess transformation $g(\varepsilon)$

$$\eta_i = g(\varepsilon_i) = \operatorname{sgn}[\varsigma_i = \xi_i - \zeta_i] = \begin{cases} 1, & \xi_i \ge \zeta_i, \\ -1, & \xi < \zeta_i. \end{cases} \tag{12.208}$$

Determine the mathematical expectation of random variable η_i under the condition that the random variable ξ_i takes the fixed value x and the following condition $|x| \le A$ is satisfied:

$$\langle(\eta_i | x)\rangle = 1 \times P(v < x) - 1 \times P(v > x) = 2 \times P(v < x) - 1 = \frac{x}{A}. \tag{12.209}$$

The unconditional mathematical expectation of the random variable η_i can be presented in the following form:

$$\langle\eta_i\rangle = \int_{-A}^{A} \langle(\eta_i | x)\rangle p(x)dx \approx \frac{1}{A}\int_{-\infty}^{\infty} xp(x)dx = \frac{E_0}{A}. \tag{12.210}$$

Based on the obtained representation, we can consider the following value

$$\tilde{E} = \frac{A}{N}\sum_{i=1}^{N} y_i, \tag{12.211}$$

as the mathematical expectation estimate of random variable, where y_i is the sample of random sequence η_i. At that point, it is not difficult to see that the considered mathematical expectation estimate is unbiased for the accepted conditions. The structural diagram of device measuring the mathematical expectation using the additional signals is shown in Figure 12.11. The counter defines a difference between the positive and negative pulses forming at the transformer output $g(\varepsilon)$. The functional purpose of other elements is clear from the diagram.

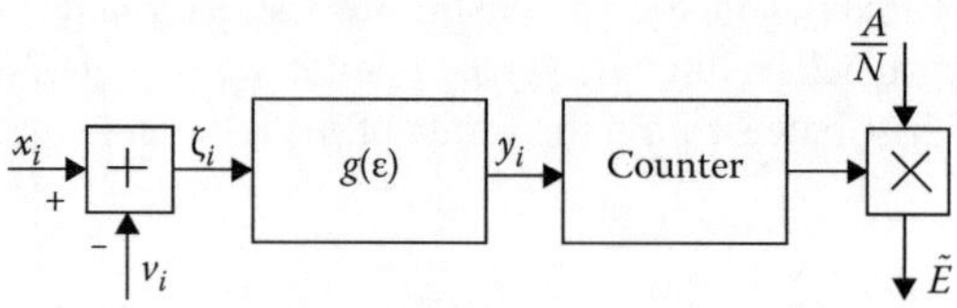

FIGURE 12.11 Measurer of mathematical expectation.

If we take into consideration the condition that $P(x < -A) \neq 0$ and $P(x > A) \neq 0$, then the mathematical expectation estimate given by (12.211) has a bias defined as

$$b(\tilde{E}) = \tilde{E} - E_0 = -\left\{\int_{-\infty}^{-A} xp(x)dx + \int_{A}^{\infty} xp(x)dx\right\}. \tag{12.212}$$

The variance of the mathematical expectation given by (12.211) can be presented in the following form:

$$\operatorname{Var}(\tilde{E}) = \frac{A^2}{N^2}\sum_{i=1,j=1}^{N}\langle y_i y_j\rangle - E_0^2. \tag{12.213}$$

Taking into consideration the statistical independence of the samples y_i, we have

$$\langle y_i y_j\rangle = \begin{cases}\langle y_i^2\rangle, & i = j,\\ \langle y_i\rangle\langle y_j\rangle, & i \neq j.\end{cases} \tag{12.214}$$

According to (12.208),

$$\langle \eta_i^2\rangle = \langle(\eta_i\,|\,x)^2\rangle = \eta_i^2 = y_i^2 = 1. \tag{12.215}$$

Consequently, the variance of the mathematical expectation estimate is simplified and takes the following form:

$$\operatorname{Var}(\tilde{E}) = \frac{A^2}{N}\left(1 - \frac{E_0^2}{A^2}\right). \tag{12.216}$$

As we can see from (12.216), since $E_0^2 < A^2$, the variance of the mathematical expectation of stochastic process is defined completely by half-interval of possible values of the additional random sequence.

Comparing the variance of the mathematical expectation estimate given by (12.216) with the variance of the mathematical expectation estimate by N independent samples given by (12.183)

$$\frac{\operatorname{Var}(\tilde{E})}{\operatorname{Var}(E^*)} = \frac{A^2}{\sigma^2}\left(1 - \frac{E_0^2}{A^2}\right), \tag{12.217}$$

we see that the considered procedure to estimate the mathematical expectation possesses the high variance since $A^2 > \sigma^2$ and $E_0^2 < A^2$.

If we know a priori that the observed stochastic sequence $v(t_i) = v_i$ is a positive value, then we can use the following pdf:

$$p(v) = \frac{1}{A}, \quad 0 \le v \le A, \tag{12.218}$$

and the following function:

$$\eta_i = g(\varepsilon_i) = \begin{cases} 1, & \xi_i \ge \zeta_i, \\ 0, & \xi < \zeta_i \end{cases} \tag{12.219}$$

as the nonlinear transformation $\eta = g(\varepsilon)$. In doing so, the following condition must be satisfied:

$$P[0 \le \xi \le A] \approx 1. \tag{12.220}$$

As we can see from (12.220), this condition is analogous to the condition given by (12.206).

The conditional mathematical expectation of random variable η_i at $\xi_i = x$ takes the following form:

$$\langle(\eta_i \| x|)\rangle = 1 \times P(v < x) + 0 \times P(v > x) = \int_0^x p(v)dv = \frac{x}{A}. \tag{12.221}$$

In doing so, the unconditional mathematical expectation of the random variable η_i is defined in the following form:

$$\langle\eta_i\rangle \approx \frac{1}{A}\int_0^\infty xp(x)dx = \frac{E_0}{A}. \tag{12.222}$$

For this reason, if the mathematical expectation estimate of the random sequence ξ_i is defined by (12.211) it will be unbiased at the first approximation.

The variance of the mathematical expectation estimate, as we discussed previously, is given by (12.213). In doing so, the conditional second moment of the random variable η_i is determined analogously as shown in (12.221):

$$\langle(\eta_i | x)^2\rangle = \int_0^x p(v)dv = \frac{x}{A}. \tag{12.223}$$

The unconditional moment is given by

$$\langle\eta_i^2\rangle = \langle y_i^2\rangle = \int_0^\infty \langle(\eta_i | x)^2\rangle p(x)dx = \frac{E_0}{A}. \tag{12.224}$$

Taking into consideration (12.223) and (12.224) under definition of the variance of the mathematical expectation estimate, we have

$$\text{Var}(\tilde{E}) = \frac{AE_0}{N}\left(1 - \frac{E_0}{A}\right). \tag{12.225}$$

Thus, in the considered case, the variance of the mathematical expectation estimate is defined by the interval of possible values of the additional stochastic sequence and is independent of the variance of the observed stochastic process and is forever more than the variance of the equidistributed estimate of the mathematical expectation by independent samples. For example, if the observed stochastic sequence subjected to the uniform pdf coinciding in the limiting case with (12.218), then the variance of the mathematical expectation for the considered procedure is defined as

$$\mathrm{Var}(\tilde{E}) = \frac{A^2}{4N} \tag{12.226}$$

and the variance of the mathematical expectation in the case of equidistributed estimate of the mathematical expectation is given by

$$\mathrm{Var}(E^*) = \frac{A^2}{12N}; \tag{12.227}$$

that is, the variance of the mathematical expectation estimate is three times more than the variance of the mathematical expectation in the case of equidistributed estimate of the mathematical expectation under the use of additional stochastic signals in the considered limiting case when the observed and additional random sequences are subjected to the uniform pdf. At other conditions, a difference in variances of the mathematical expectation estimate is higher.

As applied to (12.219), the flowchart of the mathematical expectation measurer using additional stochastic signals shown in Figure 12.11 is the same, but the counter defines the positive pulses only in accordance with (12.219).

12.6 MATHEMATICAL EXPECTATION ESTIMATE UNDER STOCHASTIC PROCESS AMPLITUDE QUANTIZATION

Define an effect of stochastic process quantization by amplitude on the estimate of its mathematical expectation. With all this going on, we assume that a quantization can be considered as the inertialess nonlinear transformation with the constant quantization step and the number of quantization levels is so high that the quantized stochastic process cannot be outside the limits of staircase characteristic of the transform $g(x)$, the approximate form of which is shown in Figure 12.12. The pdf $p(x)$ of observed stochastic process possessing the mathematical expectation that does not match with the middle between the quantization thresholds x_i and x_{i+1} is presented in Figure 12.12.

Transformation or quantization characteristic $y = g(x)$ can be presented in the form of summation of the rectangular functions shown in Figure 12.13, the width and height of which are equal to the quantization step:

$$g(x) = \sum_{k=-\infty}^{\infty} k\Delta a(x - k\Delta), \tag{12.228}$$

where $a(z)$ is the rectangular function with unit height and the width equal to Δ. Hence, we can use the following mathematical expectation estimate

$$E = \sum_{k=-\infty}^{\infty} \frac{k\Delta T_k}{T}, \tag{12.229}$$

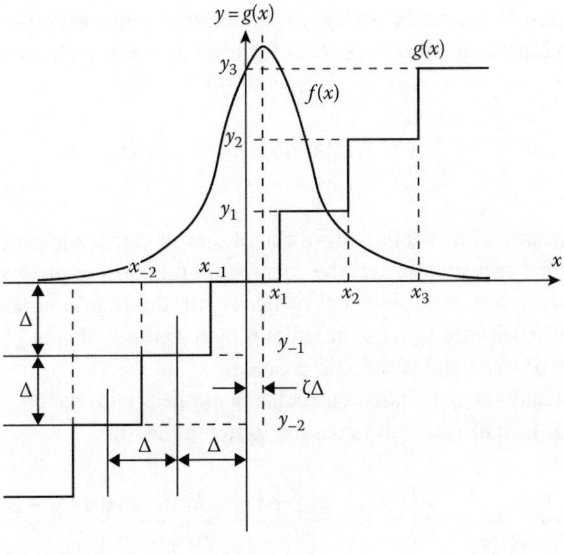

FIGURE 12.12 Staircase characteristic of quantization.

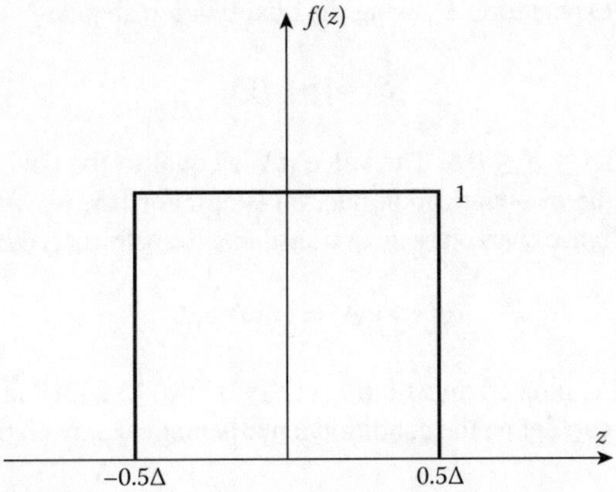

FIGURE 12.13 Rectangular function.

where $T_k = \Sigma_i \tau_i$ is the total time when the observed realization is within the limits of the interval $(k \pm 0.5)\Delta$ during its observation within the limits of the interval $[0, T]$. In doing so, $\lim_{T\to\infty} \frac{T_k}{T}$ is the probability that the stochastic process is within the limits of the interval $(k \pm 0.5)\Delta$.

The mathematical expectation of the realization $y(t)$ forming at the output of inertialess element (transducer) with the transform characteristic given by (12.228) when the realization $x(t)$ of the stochastic process $\xi(t)$ excites the input of inertialess element is equal to the mathematical expectation of the mathematical expectation estimate given by (12.229) and is defined as

$$\langle E \rangle = \int_{-\infty}^{\infty} g(x)p(x)dx = \Delta \sum_{k=-\infty}^{\infty} k \int_{(k-0.5)\Delta}^{(k+0.5)\Delta} p(x)dx, \tag{12.230}$$

where $p(x)$ is the one-dimensional pdf of the observed stochastic process. In general, the mathematical expectation of estimate $\langle E \rangle$ differs from the true value E_0; that is, as a result of quantization we obtain the bias of the mathematical expectation of the mathematical expectation estimate defined as

$$b(E) = \langle E \rangle - E_0. \tag{12.231}$$

To determine the variance of the mathematical expectation estimate according to (12.116), there is a need to define the correlation function of process forming at the transducer output, that is,

$$R_y(\tau) = \int_{-\infty}^{\infty}\int_{-\infty}^{\infty} g(x_1)g(x_2)p_2(x_1, x_2; \tau)dx_1dx_2 - \langle E\rangle^2, \quad (12.232)$$

where $p_2(x_1, x_2; \tau)$ is the two-dimensional pdf of the observed stochastic process.

Since the mathematical expectation and the correlation function of process forming at the transducer output depend on the observed stochastic process pdf, the characteristics of the mathematical expectation estimate of stochastic process quantized by amplitude depend both on the correlation function and on the pdf of observed stochastic process.

Apply the obtained results to the Gaussian stochastic process with the one-dimensional pdf given by (12.153) and the two-dimensional pdf takes the following form:

$$p_2(x_1, x_2; \tau) = \frac{1}{2\pi\sigma^2\sqrt{1 - R^2(\tau)}} \exp\left\{-\frac{(x_1 - E_0)^2 + (x_2 - E_0)^2 - 2R(\tau)(x_1 - E_0)(x_2 - E_0)}{2\sigma^2[1 - R^2(\tau)]}\right\}. \quad (12.233)$$

Define the mathematical expectation E_0 using the quantization step Δ:

$$E_0 = (c + d)\Delta, \quad (12.234)$$

where c is the integer; $-0.5 \leq d \leq 0.5$. The value $d\Delta$ is equal to the deviation of the mathematical expectation from the middle of quantization interval (step). Further, we will take into consideration that for the considered staircase characteristic of transducer (transformer) the following relation is true:

$$g(x + w\Delta) = g(x) + w\Delta, \quad (12.235)$$

where $w = 0, \pm1, \pm2, \ldots$ is the integer. Substituting (12.153) into (12.230) and taking into consideration (12.234) and (12.235), we can define the conditional mathematical expectation in the following form:

$$\langle E|d\rangle = \frac{1}{\sqrt{2\pi\sigma^2}} \int_{-\infty}^{\infty} g(x)\exp\left\{-\frac{[(x - c\Delta) - d\Delta]^2}{2\sigma^2}\right\}dx$$

$$= \sum_{k=-\infty}^{\infty} k\Delta\{Q[(k - 0.5 - d)\lambda] - Q[(k + 0.5 - d)\lambda]\} + c\Delta, \quad (12.236)$$

where

$$\lambda = \frac{\Delta}{\sigma} \quad (12.237)$$

is the ratio between the quantization step and root-mean-square deviation of stochastic process; $Q(x)$ is the Gaussian Q function given by (12.68).

The conditional bias of the mathematical expectation estimate can be presented in the following form:

$$b(E|d) = \Delta\sum_{k=-\infty}^{\infty} k\{Q[(k - 0.5 - d)\lambda] - Q[(k + 0.5 - d)\lambda]\} - d\Delta. \quad (12.238)$$

It is easy to see that the conditional bias is the odd function d, that is,

$$b(E|d) = -b(E|-d), \tag{12.239}$$

and at that, if $d = 0$ and $d = \pm 0.5$, the mathematical expectation estimate is unbiased. If $\lambda \gg 1$, in practice at $\lambda \geq 5$, (12.238) is simplified and takes the following form:

$$b(E|d) \approx \left\{Q[0.5\lambda(1-2d)] - Q[0.5\lambda(1+2d)] - d\right\}\Delta. \tag{12.240}$$

At $\lambda < 1$, the conditional bias can be simplified. For this purpose, we can expand the function in braces in (12.238) into the Taylor series about the point $(k - d)\lambda$, but limit to the first three terms of expansion. As a result, we obtain

$$b(E|d) = \frac{\lambda\Delta}{\sqrt{2\pi}} \sum_{k=-\infty}^{\infty} k \exp\left\{-0.5(k-d)^2\lambda^2\right\} - d\Delta. \tag{12.241}$$

If $\lambda \ll 1$, the sum in (12.241) can be changed by the integral. Denoting $x = \lambda k$ and $dx = \lambda$, we obtain

$$\sum_{k=-\infty}^{\infty} k\lambda \exp\left\{-0.5(k-d)^2\lambda^2\right\} \approx \frac{1}{\lambda}\int_{-\infty}^{\infty} x \exp\left\{-0.5(x-d\lambda)^2\right\}dx = \sqrt{2\pi}d. \tag{12.242}$$

As we can see from (12.241) and (12.242), at $\lambda \ll 1$, that is, the quantization step is much lower than the root-mean-square deviation of stochastic process, the mathematical expectation estimate is unbiased for all practical training.

To obtain the unconditional bias we assume that d is the random variable uniformly distributed within the limits of the interval [–0.5; 0.5]. Let us take into consideration that

$$\int Q(x)dx = xQ(x) - \int xQ'(x)dx. \tag{12.243}$$

Averaging (12.238) by all possible values of d we obtain

$$b(E) = \lambda\Delta \sum_{k=-\infty}^{\infty} k\left\{2kQ(k\lambda) - (k+1)Q[(k+1)\lambda] - (k-1)Q[(k-1)\lambda]\right.$$
$$\left. + \frac{1}{\sqrt{2\pi}\lambda}\left[-2\exp\{-0.5\lambda^2k^2\} + \exp\{-0.5\lambda^2(k+1)^2\} + \exp\{-0.5\lambda^2(k-1)^2\}\right]\right\}. \tag{12.244}$$

The terms of expansion in series at $k = p$ and $k = -p$ are equal by module and are inverse by sign. Because of this, $b(E) = 0$; that is, the mathematical expectation estimate of stochastic process quantized by amplitude is unconditionally unbiased. Substituting (12.233) into (12.232), introducing new variables

$$\begin{cases} z_1 = x_1 - c\Delta, \\ z_2 = x_2 - c\Delta, \end{cases} \tag{12.245}$$

and taking into consideration (12.235), we obtain

$$R_y(\tau) = \int_{-\infty}^{\infty}\int_{-\infty}^{\infty} g(z_1)g(z_2)p_2(z_1, z_2; \tau)dz_1dz_2, \tag{12.246}$$

where $p_2(z_1, z_2; \tau)$ is the two-dimensional pdf of Gaussian stochastic process with zero mathematical expectation. To determine (12.246) we can present the two-dimensional pdf as expansion in series by derivatives of the Gaussian Q function (12.156) assuming that $x = z$ and $E_0 = 0$ for the last formula. Substituting (12.156) and (12.228) into (12.246) and taking into consideration a parity of the function $Q^{(1)}(z/\sigma)$ and oddness of the function $g(z)$, we obtain

$$R_y(\tau) = \frac{1}{\sigma^2}\sum_{\nu=1}^{\infty}\frac{R^{\nu}(\tau)}{\nu!}\left\{\int_{-\infty}^{\infty} g(z)Q^{(\nu+1)}\left(\frac{z}{\sigma}\right)dz\right\}^2. \tag{12.247}$$

Taking the integral in the braces by parts and taking into consideration that $Q^{(\nu)}(\pm\infty) = 0$ at $\nu \geq 1$, we obtain

$$\int_{-\infty}^{\infty} g(z)Q^{(\nu+1)}\left(\frac{z}{\sigma}\right)dz = -\sigma\int_{-\infty}^{\infty} g'(z)Q^{(\nu)}\left(\frac{z}{\sigma}\right)dz. \tag{12.248}$$

According to (12.228),

$$g'(z) = \frac{dg(z)}{dz} = \sum_{k=1}^{\infty}\Delta\delta[z-(k-0.5)\Delta] + \sum_{k=-\infty}^{-1}\Delta\delta[z-(k+0.5)\Delta], \tag{12.249}$$

where $\delta(z)$ is the Dirac delta function. Then the correlation function given by (12.247) takes the following form:

$$R_y(\tau) = \Delta^2\sum_{\nu=1}^{\infty}\frac{a_\nu^2 R^{\nu}(\tau)}{\nu!}, \tag{12.250}$$

where

$$a_\nu = \sum_{k=1}^{\infty} Q^{(\nu)}[(k-0.5)\lambda] + \sum_{k=-\infty}^{-1} Q^{(\nu)}[(k+0.5)\lambda], \tag{12.251}$$

and at that the coefficients a_ν are equal to zero at even ν; that is, we can write

$$a_\nu = \begin{cases} 2\sum_{k=1}^{\infty} Q^{(\nu)}[(k-0.5)\lambda] & \text{at odd } \nu, \\ 0 & \text{at even } \nu. \end{cases} \tag{12.252}$$

Substituting (12.250) into (12.116) and taking into consideration (12.252), we obtain the normalized variance of the mathematical expectation estimate of stochastic process:

$$\frac{\mathrm{Var}(E)}{\sigma^2} = 4\lambda^2 \sum_{\nu=1}^{\infty} \frac{C_{2\nu-1}^2}{(2\nu-1)!} \times \frac{2}{T} \int_0^T \left(1 - \frac{\tau}{T}\right) R^{2\nu-1}(\tau) d\tau, \tag{12.253}$$

where

$$C_{2\nu-1} = \sum_{k=1}^{\infty} Q^{(2\nu-1)}[(k-0.65)\lambda]. \tag{12.254}$$

In the limiting case as $\Delta \to 0$, we have

$$\begin{aligned} \lim_{\Delta\to 0} \frac{\Delta}{\sigma} C_{2\nu-1} &= \lim_{\Delta\to 0} \frac{\Delta}{\sigma} \sum_{k=1}^{\infty} Q^{(2\nu-1)}\left[(k-0.5)\frac{\Delta}{\sigma}\right] \\ &= \int_0^{\infty} Q^{(2\nu-1)}(z) dz = Q^{(2\nu-2)}(x)\Big|_0^{\infty} = \begin{cases} 0.5 & \text{if } \nu = 1, \\ 0 & \text{if } \nu \ge 2. \end{cases} \end{aligned} \tag{12.255}$$

As we can see from (12.253), based on (12.255) we obtain (12.116). Computations carried out, $\lambda C_{2\nu-1}$, for the first five magnitudes ν show that at $\lambda \le 1.0$ the formula (12.255) is approximately true with the relative error less than 0.02. Taking into consideration this statement, we observe a limitation caused by the first term in (12.253) for the indicated magnitudes λ, especially for the reason that the contribution of terms with higher order in the total result in (12.253) decreases proportionally to the factors $\lambda^2 C_{2\nu-1}^2/(2\nu-1)!$ Thus, if the quantization step is not more than the root-mean-square deviation of the observed Gaussian stochastic process, (12.116) is approximately true for the definition of the variance of mathematical expectation estimate.

12.7 OPTIMAL ESTIMATE OF VARYING MATHEMATICAL EXPECTATION OF GAUSSIAN STOCHASTIC PROCESS

Consider the estimate of varying mathematical expectation $E(t)$ of Gaussian stochastic process $\xi(t)$ based on the observed realization $x(t)$ within the limits of the interval $[0, T]$. At that time, we assume that the centralized stochastic process $\xi_0(t) = \xi(t) - E(t)$ is the stationary stochastic process and the time-varying mathematical expectation of stochastic process can be approximated by a linear summation in the following form:

$$E(t) \approx \sum_{i=1}^{N} \alpha_i \varphi_i(t), \tag{12.256}$$

where

α_i indicates unknown factors

$\varphi_i(t)$ is the given function of time

If the number of terms in (12.256) is finite, that is, if N is finite, there will be a difference between $E(t)$ and expansion in series. However, with increasing N, the error of approximation tends to approach zero:

$$\varepsilon^2 = \int_0^T \left[E(t) - \sum_{i=1}^{N} \alpha_i \varphi_i(t) \right]^2 dt. \tag{12.257}$$

In this case, we say that the series given by (12.256) approaches the average.

Based on the condition of approximation error square minimum, we conclude that the factors α_i are defined from the system of linear equations:

$$\sum_{i=1}^{N} \alpha_i \int_0^T \varphi_i(t)\varphi_j(t)dt = \int_0^T E(t)\varphi_j(t)dt, \quad j = 1,2,\ldots,N. \tag{12.258}$$

In the case of representation $E(t)$ in the series form (12.256), the functions $\varphi_i(t)$ are selected in such a way to ensure fast series convergence. However, in some cases, the main factor in selection of the functions $\varphi_i(t)$ is the simplicity of physical realization (generation) of these functions. Thus, the problem of definition of the mathematical expectation $E(t)$ of stochastic process $\xi(t)$ by a single realization $x(t)$ within the limits of the interval $[0, T]$ is reduced to estimation of the coefficients α_i in the series given by (12.256). In doing so, the bias and dispersion of the mathematical expectation estimate $E^*(t)$ of the observed stochastic process caused by measurement errors of the coefficients α_i are given in the following form:

$$b_E(t) = E(t) - E^*(t) = \sum_{i=1}^{N} \varphi_i(t)\left(\alpha_i - \left\langle \alpha_i^* \right\rangle\right); \tag{12.259}$$

$$D_E(t) = \sum_{i=1,j=1}^{N} \varphi_i(t)\varphi_j(t)\left\langle \left(\alpha_i - \alpha_i^*\right)\left(\alpha_j - \alpha_j^*\right) \right\rangle, \tag{12.260}$$

where α_i^* is the estimate of the coefficients α_i. Statistical characteristics (the bias and dispersion) of the mathematical expectation estimate of stochastic process averaged within the limits of the observation interval $[0, T]$ take the following form:

$$b_E(t) = \frac{1}{T}\int_0^T b_E(t)dt = \frac{1}{T}\sum_{i=1}^{N}\left(\alpha_i - \left\langle \alpha_i^* \right\rangle\right)\int_0^T \varphi_i(t)dt; \tag{12.261}$$

$$D_E(t) = \frac{1}{T}\int_0^T D_E(t)dt = \sum_{i=1,j=1}^{N} \left\langle \left(\alpha_i - \alpha_i^*\right)\left(\alpha_j - \alpha_j^*\right) \right\rangle \frac{1}{T}\int_0^T \varphi_i(t)\varphi_j(t)dt. \tag{12.262}$$

The functional of the observed stochastic process pdf with accuracy, until the terms remain independent of the unknown mathematical expectation $E(t)$, can be presented by analogy with (12.4) in the following form:

$$F\left[x(t) | E(t)\right] = B_1 \exp\left\{ \sum_{i=1}^{N} \alpha_i y_i - \frac{1}{2}\sum_{i=1,j=1}^{N} \alpha_i \alpha_j c_{ij} \right\}, \tag{12.263}$$

where

$$y_i = \int_0^T x(t)\upsilon_i(t)dt = \int_0^T\int_0^T x(t_1)\varphi_i(t_2)\vartheta(t_1,t_2)dt_1dt_2; \tag{12.264}$$

$$c_{ij} = c_{ji} = \int_0^T \varphi_i(t)\upsilon_i(t)dt = \int_0^T\int_0^T \varphi_i(t_1)\varphi_j(t_2)\vartheta(t_1,t_2)dt_1dt_2, \tag{12.265}$$

and the function $\upsilon_i(t)$ is defined by the following integral equation:

$$\int_0^T R(t,\tau)\upsilon_i(\tau)d\tau = \varphi_i(t). \tag{12.266}$$

Solving the likelihood equation with respect to unknown coefficients α_i

$$\frac{\partial F[x(t) \mid E(t;\alpha_1,\ldots,\alpha_N)]}{\partial \alpha_j} = 0, \tag{12.267}$$

we can find the system of N linear equations

$$\sum_{i=1}^{N} \alpha_i^* c_{ij} = y_j, \quad j = 1,2,\ldots,N \tag{12.268}$$

with respect to the estimates α_i^*. Based on (12.268), we can obtain the estimates of coefficients

$$\alpha_m^* = \frac{A_m}{A} = \frac{1}{A}\sum_{i=1}^{N} A_{m_i} y_i, \quad m = 1,2,\ldots,N, \tag{12.269}$$

where

$$A = \| c_{ij} \| = \begin{vmatrix} c_{11} & c_{12} & \cdots & c_{1m} & \cdots & c_{1N} \\ c_{21} & c_{22} & \cdots & c_{2m} & \cdots & c_{2N} \\ \vdots & \vdots & & \vdots & & \vdots \\ c_{N1} & c_{N2} & \cdots & c_{Nm} & \cdots & c_{NN} \end{vmatrix} \tag{12.270}$$

is the determinant of the system of linear equations given by (12.268),

$$A_m = \begin{vmatrix} c_{11} & c_{12} & \cdots & y_1 & \cdots & c_{1N} \\ c_{21} & c_{22} & \cdots & y_2 & \cdots & c_{2N} \\ \vdots & \vdots & & \vdots & & \vdots \\ c_{N1} & c_{N2} & \cdots & y_N & \cdots & y_{NN} \end{vmatrix} \tag{12.271}$$

is the determinant obtained from the determinant A given by (12.270) by changing the column c_{im} by the column y_i; A_{im} is the algebraic supplement of the mth column elements (the column y_i). In doing so, the following relationship

$$\sum_{j=1}^{N} c_{ij}A_{kj} = \sum_{j=1}^{N} c_{ij}A_{jk} = A\delta_{ik} \tag{12.272}$$

is true for the quadratic matrix $\|c_{ij}\|$.

A flowchart of the optimal measurer of varying mathematical expectation of Gaussian stochastic process is shown in Figure 12.14. The measurer operates in the following way. Based on the previously mentioned system of functions $\varphi_i(t)$ generated by the "Gen_φ" and a priori information about the correlation function $R(\tau)$ of observed stochastic process, the generator "Gen_v" forms the system of linear functions $v_i(t)$ in accordance with the integral equation (12.266). According to (12.265), the generator "Gen_C" generates the system of coefficients c_{ij} sent to "Microprocessor." The magnitudes y_i come in at the input of "Microprocessor" and also from the outputs of the "Integrator." Based on a solution of N linear equations given by (12.268) with respect to unknown coefficients α_i, the "Microprocessor" generates their estimates. Using the estimates α_i^* and system of functions $\varphi_i(t)$, the estimate $E^*(t)$ of time-varying mathematical expectation $E(t)$ is formed according to (12.256). The generated estimate $E^*(t)$ will have a delay with respect to the true value on $T + \Delta T$ that is required to compute the random variables y_i and to solve the m linear equations by "Microprocessor." The delay blocks T and ΔT are used by the flowchart with this purpose.

Substituting $x(t)$ given by (12.2) into (12.264), we can write

$$y_i = \int_0^T x_0(t)\upsilon_i(t)dt + \int_0^T E(t)\upsilon_i(t)dt = n_i + \alpha_m c_{im} + \sum_{q=1,q\neq m}^{N} \alpha_q c_{iq}, \tag{12.273}$$

where

$$n_i = \int_0^T x_0(t)\upsilon_i(t)dt. \tag{12.274}$$

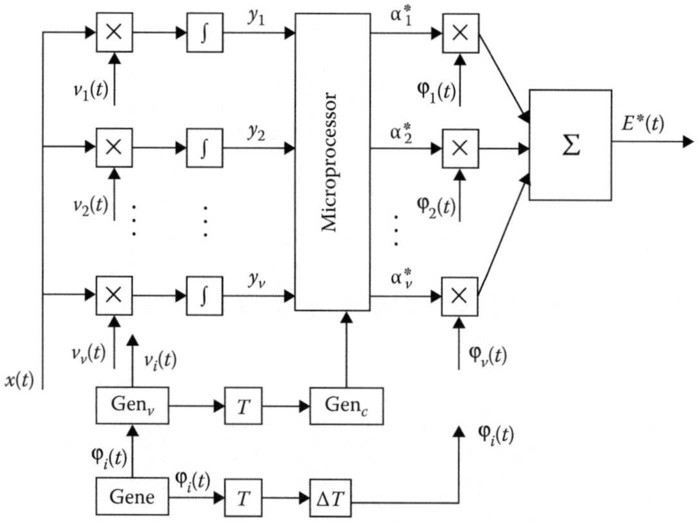

FIGURE 12.14 Optimal measurer of time-varying mathematical expectation estimate.

Taking into consideration (12.272) and (12.273), the determinant given by (12.271) can be presented in the form of sum of two determinants

$$A_m = B_m + C_m, \tag{12.275}$$

where

$$B_m = \begin{vmatrix} c_{11} & c_{12} & \cdots & (n_1 + \alpha_m c_{1m}) & \cdots & c_{1N} \\ c_{21} & c_{22} & \cdots & (n_2 + \alpha_m c_{2m}) & \cdots & c_{2N} \\ \vdots & \vdots & & \vdots & & \vdots \\ c_{N1} & c_{N2} & \cdots & (n_N + \alpha_m c_{Nm}) & \cdots & c_{NN} \end{vmatrix}, \tag{12.276}$$

The determinant C_m is obtained from the determinant B_m by changing the mth column on the column consisting of the terms $\sum_{q=1,q\neq m}^{N} \alpha_q c_{iq}$ at $q \neq m$. Since the mth column of the determinant C_m consists of two elements that are the linear combination of elements of other columns, then $C_m = 0$.

Let us present the determinant $A_m = B_m$ in the form of sum of products of all elements of the mth column on their algebraic supplement A_m, that is,

$$A_m = \sum_{i=1}^{N} (n_i + \alpha_m c_{im}) A_{im}. \tag{12.277}$$

Taking into consideration that

$$\sum_{i=1}^{N} A_{im} c_{im} = A, \tag{12.278}$$

the estimate of the coefficients α_m takes the following form:

$$\alpha_m^* = \frac{1}{A} \sum_{i=1}^{N} A_{im} n_i + \alpha_m, \quad m = 1, 2, \ldots, N. \tag{12.279}$$

Since

$$\langle n_i \rangle = \int_0^T \langle x_0(t) \rangle \upsilon_i(t) dt = 0, \tag{12.280}$$

the estimates of the coefficients α_i of series given by (12.256) are unbiased. The correlation function of estimates of the coefficients α_m and α_q is defined by the following form:

$$R(\alpha_m, \alpha_q) = \frac{1}{A^2} \sum_{i=1, j=1}^{N} \langle n_i n_j \rangle A_{im} A_{jq} = \frac{A_{mq}}{A}. \tag{12.281}$$

While delivering (12.281), we have taken into consideration the integral equation (12.266) and the formula given by (12.272). Now, we are able to define the variance of estimate of the coefficients α_m, namely,

$$\mathrm{Var}(\alpha_m) = \frac{A_{mm}}{A}. \tag{12.282}$$

In practice, we can assume that the frequency band Δf_E of the varying mathematical expectation is much less the effective frequency band Δf_{ef} of the spectrum $G(f)$ of the observed centralized stochastic process $\xi_0(t)$, and the spectrum $G(f)$ is not changed for all practical training within the limits of the frequency band Δf_E. In this case, the centralized stochastic process $\xi_0(t)$ can be considered as the "white" noise with the effective spectral density

$$\mathcal{N}_{ef} = \int_0^{\Delta f_E} \frac{G(f)}{\Delta f_E} df \tag{12.283}$$

for further analysis. In doing so, the effective spectrum bandwidth can be defined as

$$\Delta f_{ef} = \int_0^{\infty} \frac{G(f)}{G_{max}(f)} df. \tag{12.284}$$

In the case of accepted approximation, the correlation function of the centralized stochastic process $\xi_0(t)$ is approximated by

$$R(\tau) = \frac{\mathcal{N}_{ef}}{2}\delta(\tau). \tag{12.285}$$

Substituting (12.285) into the integral equation (12.266), we obtain

$$\upsilon_i(t) = \frac{2}{\mathcal{N}_{ef}}\varphi_i(t). \tag{12.286}$$

In doing so, the matrix of the coefficients

$$c_{ij} = \frac{2}{\mathcal{N}_{ef}}\delta_{ij} \tag{12.287}$$

is the diagonal matrix and the determinant and algebraic supplement of this matrix are defined, correspondingly

$$A = \left\{\frac{2}{\mathcal{N}_{ef}}\right\}^N; \tag{12.288}$$

$$A_{im} = \begin{cases} A_{mm} = \left\{\dfrac{2}{\mathcal{N}_{ef}}\right\}^{N-1}, & \text{at} \quad i = m, \\ 0, & \text{at} \quad i \neq m. \end{cases} \tag{12.289}$$

As we can see from (12.281), the correlation function of the estimates α_m^* and α_q^* takes the following form:

$$R\left(\alpha_m^*, \alpha_q^*\right) = \frac{\mathcal{N}_{\text{ef}}}{2}\delta_{mq}. \tag{12.290}$$

Based on (12.260), (12.262), and (12.290), we can note that the current and averaged variances of the varying mathematical expectation estimate $E^*(t)$ can be presented in the following form:

$$\text{Var}_{E^*}(t) = \frac{\mathcal{N}_{\text{ef}}}{2}\sum_{i=1}^{N}\varphi_i^2(t), \tag{12.291}$$

$$\text{Var}(E^*) = \frac{1}{T}\sum_{i=1}^{N}\text{Var}(\alpha_i) = \frac{\mathcal{N}_{\text{ef}}^N}{2T}. \tag{12.292}$$

As we can see from (12.292), the higher the number of terms under expansion in series in (12.256) used for approximation of the mathematical expectation, the higher the variance of time-varying mathematical expectation estimate at the same conditions averaged within the limits of the observation interval. In doing so, there is a need to note that in general, the number N of series expansion terms essentially increases corresponding to the increase in the observation interval $[0, T]$, within the limits of which the approximation is carried out.

As applied to the correlation function given by (12.13), the effective spectral density of the centralized stochastic process $\xi_0(t)$ takes the following form:

$$\mathcal{N}_{\text{ef}} = \frac{2\sigma^2}{\pi\Delta f_{\text{E}}}\text{arctg}\left(\frac{2\pi\Delta f_{\text{E}}}{\alpha}\right). \tag{12.293}$$

If the observed stochastic process is stationary, then

$$\mathcal{N}_{\text{ef}} = \frac{4\sigma^2}{\alpha} \quad \text{and} \quad \nu = 1, \tag{12.294}$$

and the variance of the mathematical expectation estimate takes a form given by (12.50). The formulae for the variance of estimates of the coefficients α_m given by (12.282) and the variance of the time-varying mathematical expectation $E(t)$ given by (12.260) and (12.262) are simplified essentially if the functions $\varphi_i(t)$ satisfy the integral Fredholm equation of the second kind:

$$\varphi_i(t) = \lambda_i\int_0^T R(t,\tau)\varphi_i(\tau)d\tau. \tag{12.295}$$

In the considered case, the coefficients λ_i and the functions $\varphi_i(t)$ are called the eigenvalues and eigenfunctions of the integral equation, respectively. Comparing (12.295) and (12.266), we can see

$$\upsilon_i(t) = \lambda_i\varphi_i(t). \tag{12.296}$$

In theory of stochastic processes [15–17], it is proved that if the functions $\varphi_i(t)$ were to satisfy the Equation 12.295, then these functions are the orthogonal normalized (orthonormalized) functions and the eigenvalues $\lambda_i > 0$. In this case, the following equation

$$\int_0^T \varphi_i(t)\varphi_j(t)dt = \delta_{ij} = \begin{cases} 1 & \text{if} \quad i = j, \\ 0 & \text{if} \quad i \neq j \end{cases} \quad (12.297)$$

is true for the eigenfunctions $\varphi_i(t)$, and the correlation function $R(t_1, t_2)$ can be presented by the following expansion in series

$$R(t_1, t_2) = \sum_{i=1}^{\infty} \frac{\varphi_i(t_1)\varphi_i(t_2)}{\lambda_i}, \quad (12.298)$$

and the following equality is satisfied:

$$\sum_{i=1}^{\infty} \frac{1}{\lambda_i} = \sigma^2 T. \quad (12.299)$$

Substituting (12.295) into (12.265) and taking into consideration (12.297), we obtain

$$c_{ij} = \begin{cases} \lambda_i & \text{if} \quad i = j, \\ 0 & \text{if} \quad i \neq j. \end{cases} \quad (12.300)$$

At that, the matrix of coefficients $c_{ij} = \lambda_i$ is the diagonal matrix. The determinant A and the algebraic supplements A_m of this matrix can be presented in the following form, respectively,

$$A = \prod_{i=1}^{N} \lambda_i, \quad (12.301)$$

$$A_{im} = \begin{cases} \dfrac{A}{\lambda_m} & \text{if} \quad i = m, \\ 0 & \text{if} \quad i \neq m. \end{cases} \quad (12.302)$$

As we can see from (12.301), (12.302), (12.269), and (12.256), the estimates of the coefficients α_i^* and the time-varying mathematical expectation estimate $E^*(t)$ can be presented in the following form:

$$\alpha_i^* = \int_0^T x(\tau)\varphi_i(\tau)d\tau; \quad (12.303)$$

$$E^*(t) = \sum_{i=1}^{N} \alpha_i^* \varphi_i(t). \quad (12.304)$$

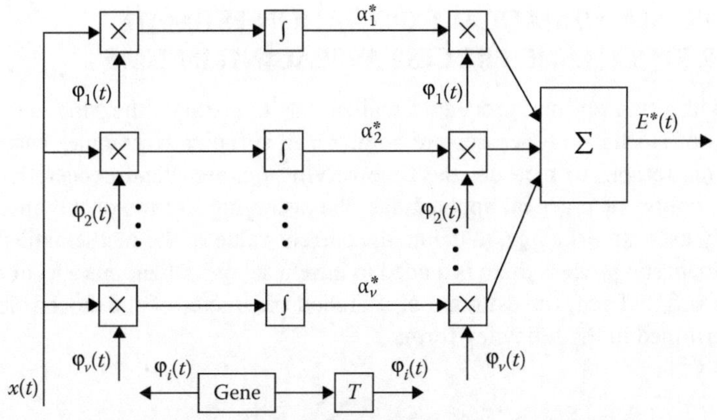

FIGURE 12.15 Optimal measurer of time-varying mathematical expectation estimate in accordance with (12.304); particular case of the flowchart in Figure 12.14.

The optimal measurer of the time-varying mathematical expectation of Gaussian stochastic process operating in accordance with (12.304) is shown in Figure 12.15. The flowchart presented in Figure 12.15 is a particular case of the measurer depicted in Figure 12.14. Based on the well-known correlation function of the observed stochastic process, in accordance with (12.295), the generator "Gene" forms the functions $\varphi_i(t)$ that are employed to form the coefficients α_i^*. The estimates of coefficients are multiplied by the functions $\varphi_i(t)$ again and obtained products come in at the summator input. The expected estimate $E^*(t)$ of the time-varying mathematical expectation is formed at the summator output. Delay is required to compute the coefficients α_i^*.

The variance of estimate of the coefficients α_i^* and the current and averaged variances of estimates $E^*(t)$ of the time-varying mathematical expectation are transformed in accordance with (12.282), (12.260), and (12.262) in the following form:

$$\operatorname{Var}\left(\alpha_m^*\right) = \frac{1}{\lambda_m}; \tag{12.305}$$

$$\operatorname{Var}\{E^*(t)\} = \sum_{i=1}^{N} \frac{\varphi_i^2(t)}{\lambda_i}; \tag{12.306}$$

$$\operatorname{Var}(E^*) = \sum_{i=1}^{N} \frac{1}{\lambda_i T}. \tag{12.307}$$

As we can see from (12.307), with increase in the number of terms under approximation $E(t)$ by the series given by (12.256), the averaged variance of the time-varying mathematical expectation estimate E^* also increases. As $N \to \infty$, taking into consideration (12.299), we obtain

$$\operatorname{Var}(E^*) = \sigma^2. \tag{12.308}$$

Thus, at a sufficiently large number of the eigenfunctions in the sum given by (12.256), the averaged variance of the time-varying mathematical expectation estimate $E(t)$ is equal to the variance of the initial stochastic process. In doing so, the estimate bias caused by the finite number of terms in series given by (12.256) tends to approach zero. However, in practice, there is a need to choose the number of terms in series given by (12.256) in such a way that a dispersion of the time-varying mathematical expectation estimate caused both by the bias and by the estimate variance would be minimal.

12.8 VARYING MATHEMATICAL EXPECTATION ESTIMATE UNDER STOCHASTIC PROCESS AVERAGING IN TIME

The problems with using optimal procedure realizations to estimate the time-varying mathematical expectations of stochastic processes are common in sufficiently complex mathematical computations. For this reason, to measure the time-varying mathematical expectations of stochastic processes in the course of practical applications, the averaging in time of the observed stochastic process is widely used. In principle, to define the current value of the mathematical expectation of nonstationary stochastic process there is a need to have a set of realizations $x_i(t)$ of the investigated stochastic process $\xi(t)$. Then, the estimate of searched parameter of the stochastic process at the instant t_0 is determined in the following form:

$$E^*(t_0) = \frac{1}{N}\sum_{i=1}^{N} x_i(t_0), \tag{12.309}$$

where N is the number of investigated stochastic process realizations. As we can see from (12.309), the mathematical expectation estimate is unbiased and the variance of the mathematical expectation estimate can be presented in the following form:

$$\mathrm{Var}[E^*(t)] = \frac{\sigma^2(t_0)}{N}, \tag{12.310}$$

in the case of independent realizations $x_i(t)$, where $\sigma^2(t_0)$ is the variance of investigated stochastic process at the instant $t = t_0$. Thus, the estimate given by (12.309) is the consistent since $N \to \infty$ the variance of estimate tends to approach zero. However, as a rule, a researcher does not use a sufficient number of realizations of stochastic process; thus, there is a need to carry out an estimation of the mathematical expectation based on an analysis of the limited number of realizations and, sometimes, based on a single realization.

Under definition of estimation of the time-varying mathematical expectation of stochastic process by a single realization, we meet with difficulties caused by the definition of optimal time of averaging (integration) or the time constant of smoothing filter at the earlier-given filter impulse response. In doing so, two conflicting requirements arise. On one hand, there is a need to decrease the variance of estimate caused by finite time interval of measuring; this time interval must be large. On the other hand, for better distinguishing the mathematical expectation variations in time, there is a need to choose the integration time as short as possible. Evidently, there is an optimal averaging time or the bandwidth of the smoothing filter under the given impulse response, which corresponds to minimal dispersion of the mathematical expectation estimate of stochastic process caused by the aforementioned factors.

The simplest way to define the mathematical expectation of stochastic process at the instant $t = t_0$ is an averaging of ordinates of stochastic process realization within the limits of time interval about the given magnitude of argument $t = t_0$. In doing so, the mathematical expectation estimate is defined as

$$E^*(t_0,T) = \frac{1}{T}\int_{t_0-0.5T}^{t_0+0.5T} x(t)dt = \frac{1}{T}\int_{-0.5T}^{0.5T} x(t_0+t)dt. \tag{12.311}$$

Averaging (12.311) by realizations, we obtain the mathematical expectation of estimate at the instant $t = t_0$:

$$\langle E^*(t_0,T)\rangle = \frac{1}{T}\int_{-0.5T}^{0.5T} E(t_0+t)dt, \tag{12.312}$$

where $E(t_0 + t)$ is the true mathematical expectation value of the investigated stochastic process at the instant $t = t_0$. Thus, the mathematical expectation of estimate of the time-varying mathematical expectation of stochastic process in contrast to the stationary case is obtained by smoothing the estimate within the limits of time interval $[t_0 - 0.5T; t_0 + 0.5T]$.

In general, as a result of considered averaging, there is the mathematical expectation bias that can be presented in the following form:

$$b[E^*(t_0,T)] = \frac{1}{T}\int_{-0.5T}^{0.5T}[E(t_0+t) - E(t_0)]dt. \quad (12.313)$$

If the magnitude $E(t)$ is described about the point $t = t_0$ within the limits of time interval $[t_0 - 0.5T;$ $t_0 + 0.5T]$ by the series with odd powers in the following form

$$E(t+t_0) \approx E(t_0) + \sum_{k=1}^{N}\frac{t^{2k-1}}{(2k-1)!}\left[\frac{d^{(2k-1)}E(t)}{dt^{(2k-1)}}\right]_{t_0}, \quad (12.314)$$

the mathematical expectation estimate bias would be minimal. Then

$$b[E^*(t_0,T)] \approx 0. \quad (12.315)$$

The variance of the mathematical expectation estimate of the investigated stochastic process is defined in the following form:

$$\mathrm{Var}[E^*(t_0,T)] = \frac{1}{T^2}\int_{-0.5T}^{0.5T}\int_{-0.5T}^{0.5T}R(t_0+t_1, t_0+t_2)dt_1dt_2, \quad (12.316)$$

where $R(t_1, t_2)$ is the correlation function of the investigated stochastic process $\xi(t)$.

In practice, the nonstationary stochastic processes with time-varying mathematical expectation or variance or both of them simultaneously are widely used. In doing so, the mathematical expectation and variance vary slowly in comparison with variations of the investigated stochastic process. In other words, the mathematical expectation and variance of stochastic process are constant within the limits of the correlation interval. In this case, to define the variance of the time-varying mathematical expectation estimate we can assume that the centralized stochastic process $\xi_0(t) = \xi(t) - E(t)$ is the stationary stochastic process within the limits of the interval $t_0 \pm 0.5T$ with the correlation function that can be presented in the following form:

$$R(\tau) = \langle\xi_0(t)\xi_0(t+\tau)\rangle \approx \sigma^2(t_0)\mathcal{R}(\tau). \quad (12.317)$$

Taking into consideration the given approximation, the variance of the mathematical expectation estimate given by (12.316) after transformation of the double integral by introducing new variables $\tau = t_2 - t_1$, $t_2 = t$ and changing the order of integration takes the following form:

$$\mathrm{Var}[E^*(t_0,T)] \approx \sigma^2(t_0) \times \frac{2}{T}\int_0^T\left(1 - \frac{\tau}{T}\right)\mathcal{R}(\tau)d\tau. \quad (12.318)$$

As we can see from (12.313) and (12.318), the dispersion of the mathematical expectation estimate is defined in the following form:

$$D[E^{*}(t_0,T)] = b^2[E^{*}(t_0,T)] + \mathrm{Var}[E^{*}(t_0,T)]. \tag{12.319}$$

In principle, we can define the optimal integration time T, under which the dispersion will be minimum at the instant t_0 minimizing the dispersion of estimate by the parameter T. However, we can present a solution to this problem in an acceptable analytical form by giving a specific function of $E(t)$.

Evaluate how the mathematical expectation estimate varies when the mathematical expectation $E(t)$ deviates from the linear function. At the same time, we assume that the estimated mathematical expectation $E(t)$ possesses the first and second continuous derivatives with respect to the time t. Then, according to the Taylor formula, we can write

$$E(t) = E(t_0) + (t - t_0)E'(t_0) + 0.5(t - t_0)^2 E''[t_0 + \vartheta(t - t_0)], \tag{12.320}$$

where $0 < \vartheta < 1$. Substituting (12.320) into (12.313), we obtain

$$b[E^{*}(t_0,T)] = \frac{1}{2T}\int_{-0.5T}^{0.5T} t^2 E''[t_0 + t\vartheta]dt. \tag{12.321}$$

Denoting M as the maximum value of the second derivative of the mathematical expectation $E(t)$ with respect to the time t, we obtain the top bound of the mathematical expectation estimate bias by module

$$|b[E^{*}(t_0,T]| \le \frac{T^2 M}{24}. \tag{12.322}$$

As a rule, the maximum magnitude of the second derivative of the mathematical expectation $E(t)$ with respect to the time t can be evaluated based on an analysis of specific physical problems.

To estimate the optimal time of integration T minimizing the dispersion of estimate given by (12.319), we assume that the correlation interval of the investigated stochastic process is much less than the integration time, that is, $\tau_{\mathrm{cor}} \ll T$. Then, the following written form is true:

$$\int_0^T \left(1 - \frac{\tau}{T}\right)\mathcal{R}(\tau)d\tau \approx \int_0^\infty \mathcal{R}(\tau)d\tau \le \int_0^\infty |\mathcal{R}(\tau)|\,d\tau = \tau_{\mathrm{cor}}. \tag{12.323}$$

Taking into consideration (12.322) and (12.323) and based on the condition of minimization of the estimate dispersion given by (12.319), we obtain the optimal estimation of the integration time:

$$T \approx 2\left[\frac{9\sigma^2(t_0)\tau_{\mathrm{cor}}}{M^2}\right]^{\frac{1}{5}}. \tag{12.324}$$

As we can see from (12.324), the larger the integration time, the larger the correlation interval and the variance of the investigated stochastic process. The lesser the integration time, the larger the maximum absolute value of the second derivative of the mathematical expectation measured. This statement agrees well with the physical interpretation of measuring the time-varying mathematical expectation.

In some applications, the time-varying mathematical expectation can be approximated by the series given by (12.256). The values minimizing the function

$$\varepsilon^2(\alpha_1,\alpha_2,\ldots,\alpha_N)=\frac{1}{T}\int_0^T\left[x(t)-\sum_{i=1}^{N}\alpha_i\varphi_i(t)\right]^2 dt \tag{12.325}$$

can be considered as the estimates of the coefficients α_i^*. This representation of the coefficients α_i^* is possible only if the mathematical expectation $E(t)$ and the functions $\varphi_i(t)$ vary in time slowly in comparison with the variation of the first derivative of the function $x_0(t)$ with respect to the time; that is, the following condition must be satisfied:

$$\left.\begin{array}{l}\left|\dfrac{E'(t)}{E(t)}\right|_{\max}\\[2ex]\left|\dfrac{\varphi_i'(t)}{\varphi_i(t)}\right|_{\max}\end{array}\right\}\ll\sqrt{\frac{\left\langle[x_0'(t)]^2\right\rangle}{\sigma^2(t)}}. \tag{12.326}$$

In other words, the condition (12.325) to define the coefficients α_i^* is true if the frequency band Δf_{E} of the mathematical expectation $E(t)$ is much less than the effective bandwidth of energy spectrum of the stochastic component $x_0(t)$. Based on the condition of minimization the function ε^2, that is,

$$\frac{d\varepsilon^2}{\partial\alpha_m}=0, \tag{12.327}$$

we obtain the system of equations to estimate the coefficients α_m^*

$$\sum_{i=1}^{N}\alpha_i^*\int_0^T\varphi_i(t)\varphi_m(t)dt=\int_0^T x(t)\varphi_m(t)dt,\quad m=1,2,\ldots,N. \tag{12.328}$$

Denote

$$\int_0^T\varphi_i(t)\varphi_m(t)dt=c_{im}; \tag{12.329}$$

$$\int_0^T x(t)\varphi_m(t)dt=y_m. \tag{12.330}$$

Then the estimations of the coefficients α_m^* can be presented in the following form:

$$\alpha_m^*=\frac{A_m}{A},\quad m=1,2,\ldots,N, \tag{12.331}$$

where the determinant A of the system of linear equations given by (12.328) and the determinant A_m obtained by changing the mth column c_{im} of the determinant A by the column y_i are determined based on (12.270) and (12.271).

The flowchart of measurer of the time-varying mathematical expectation estimate is similar to the block diagram shown in Figure 12.14, but the difference is that a set of functions $\varphi_i(t)$ are assigned for the sake of convenience in generating them and the coefficients c_{ij} and values y_i are formed according to (12.329) and (12.330), correspondingly. Definition of the coefficients $\overset{*}{\alpha}_m$ is simplified essentially if the functions $\varphi_i(t)$ are orthonormal functions; that is, the formula (12.297) is true. In this case,

$$\overset{*}{\alpha}_m = \int_0^T x(t)\varphi_m(t)dt. \tag{12.332}$$

The flowchart of measurer of the time-varying mathematical expectation estimate differs from the block diagram shown in Figure 12.15, and the difference is that a set of functions $\varphi_i(t)$ are assigned for the sake of convenience in generating them; thus, there is no need to solve the integral equation (12.295).

Compute the estimate bias and mutual correlation functions between the estimates of the coefficients $\overset{*}{\alpha}_m$ and $\overset{*}{\alpha}_q$. Based on investigation carried out in Section 12.7, we can conclude that the estimations of the coefficients $\overset{*}{\alpha}_m$ of expansion in series given by (12.256) are unbiased estimates and the correlation functions and variances of estimates of the coefficients $\overset{*}{\alpha}_m$ are defined in the following form:

$$R\left(\overset{*}{\alpha}_m, \overset{*}{\alpha}_q\right) = \frac{1}{A^2}\sum_{i=1,j=1}^{N} A_{im}A_{jq}\mathcal{C}_{ij}, \tag{12.333}$$

$$\text{Var}\left(\overset{*}{\alpha}_m\right) = \frac{1}{A^2}\sum_{i=1,j=1}^{N} A_{im}A_{jm}\mathcal{C}_{ij}, \tag{12.334}$$

where A_{im} is the algebraic supplement of the determinant given by (12.270),

$$\mathcal{C}_{ij} = \int_0^T\int_0^T R(t_1,t_2)\varphi_i(t_1)\varphi_j(t_2)dt_1dt_2, \tag{12.335}$$

and $R(t_1, t_2)$ is the correlation function of the investigated stochastic process.

With $\varphi_i(t)$ used as the orthonormal function, the coefficients c_{im} given by (12.329) take the following form:

$$c_{im} = \delta_{im}. \tag{12.336}$$

In doing so, the matrix $\|c_{ij}\|$ is transformed to the diagonal matrix and the determinant A of this matrix and algebraic supplements A_{ij} are defined as follows:

$$A = 1, \quad A_{ij} = \delta_{ij}. \tag{12.337}$$

Based on (12.336) and (12.333), the correlation function of estimation of the coefficients $\overset{*}{\alpha}_m$ and $\overset{*}{\alpha}_q$ can be presented in the following form:

$$R\left(\overset{*}{\alpha}_m, \overset{*}{\alpha}_q\right) = \mathcal{C}_{mq}, \tag{12.338}$$

where $\mathcal{C}_{ij}$ is given by (12.335) at $i = m, j = q$. In doing so, the current variance of estimate given by (12.260) and the averaged variance of estimate (12.262) of the time-varying mathematical expectation take the following form:

$$\operatorname{Var}\{E^*(t)\} = \sum_{i=1,j=1}^{N} \mathcal{C}_{ij}\varphi_i(t)\varphi_j(t); \tag{12.339}$$

$$\operatorname{Var}\{E^*\} = \frac{1}{T}\sum_{i=1}^{N} \mathcal{C}_{ii}. \tag{12.340}$$

correspondingly.

If it is possible to approximate the centralized stochastic process $\xi_0(t)$ by the "white" noise with the effective spectral density given by (12.283) in addition to the orthonormal functions $\varphi_i(t)$, then we can write

$$\mathcal{C}_{ij} = \frac{\mathcal{N}_{\text{ef}}}{2}\delta_{ij}. \tag{12.341}$$

Based on (12.341), we are able to define the current and averaged variances of the time-varying mathematical expectation estimates coinciding with the optimal estimations given by (12.336) and (12.338), which are applied to the observation of the Gaussian stochastic process with the time-varying mathematical expectation.

12.9 ESTIMATE OF MATHEMATICAL EXPECTATION BY ITERATIVE METHODS

Currently, the iterative methods or procedures of step-by-step approximation are widely used to estimate the parameters of stochastic processes. These procedures and methods are also called the recurrent procedures or methods of stochastic approximation. Essence of the iterative method applied to estimation of scalar parameter l by discrete sample with the size N is to form the recurrent relationship in the following form [18]:

$$l^*[N] = l^*[N-1] + \gamma[N]\{f(x[N]) - l^*[N-1]\}, \tag{12.342}$$

where

$l^*[N-1]$ and $l^*[N]$ are the estimates of stochastic process parameter based on the observation of $N-1$ and N samples, respectively

$f(x[N])$ is the function of received sample related with the transformation required to obtain the searched stochastic process parameter

$\gamma[N]$ is the factor defining a value of next step to make accurate the estimate of parameter l, depending on the number of step N and satisfying the following conditions:

$$\begin{cases} \gamma[N] > 0, \\ \sum_{N=1}^{\infty} \gamma[N] \to \infty, \\ \sum_{N=1}^{\infty} \gamma^2[N] < \infty. \end{cases} \tag{12.343}$$

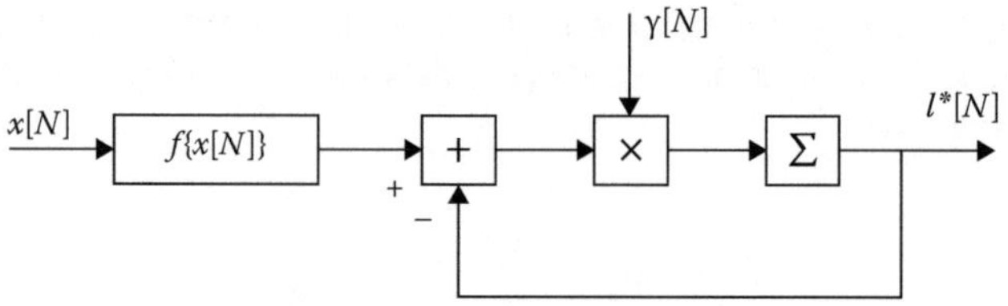

FIGURE 12.16 Iterative measurer of mathematical expectation.

The relationship (12.342) allows us to image the flowchart of the iterative measurer of stochastic process parameter.

The discrete algorithm given by (12.342) can be transformed to the continuous algorithm using the limiting process for the difference equation

$$l[N]-l[N-1]=\Delta l[N]=\gamma[N]\{f(x[N])-l[N-1]\} \tag{12.344}$$

to differential equation

$$\frac{dl(t)}{dt}=\gamma(t)\{f[x(t)]-l(t)\}. \tag{12.345}$$

The flowchart of measurer corresponding to (12.345) is similar to the block diagram shown in Figure 12.16 where there is a need to change $\gamma[N]$ on $\gamma(t)$ and the summator on the integrator.

As applied to the mathematical expectation estimate of stationary stochastic process, the recurrent algorithm of measurement takes the following form:

$$E^*[N]=E^*[N-1]+\gamma[N]\{x[N]-E^*[N-1]\}. \tag{12.346}$$

The optimal magnitude of the factor $\gamma[N]$ to estimate the mathematical expectation of stochastic process by uncorrelated samples can be obtained from (12.182). This optimal value must ensure the minimal variance of the mathematical expectation estimate over the class of linear estimations given by (12.183). Actually, (12.182) can be presented in the following form:

$$E^*[N]=\frac{1}{N}\sum_{i=1}^{N}x_i=E^*[N-1]+\frac{1}{N}\{x[N]-E^*[N-1]\}. \tag{12.347}$$

Comparing (12.346) and (12.347), we obtain the optimal magnitude of the factor $\gamma[N]$:

$$\gamma_{\text{opt}}[N]=\frac{1}{N}. \tag{12.348}$$

The flowchart of iterative measurer of the mathematical expectation is similar to the block diagram shown in Figure 12.16, in which there is a need to exclude block $f(x[N])$ responsible for transformation of stochastic process.

Since the iterative algorithm (12.347) is equivalent to the algorithm (12.182), we can state that, in the considered case, the mathematical expectation estimate is unbiased and the variance of the mathematical expectation estimate is given by (12.183). In practice, we sometimes use the constant values of the factor, $\gamma[N]$ that is,

$$\gamma[N] = \gamma = \text{const}, \quad 0 < \gamma < 1. \tag{12.349}$$

In this case, the estimation given by (12.346) can be presented in the following form [19,20]:

$$E^*[N] = (1-\gamma)^N x[1] + \gamma \sum_{i=2}^{N} (1-\gamma)^{N-i} x_i = (1-\gamma)^N \left\{ x[1] + \gamma \sum_{i=2}^{N} \frac{x_i}{(1-\gamma)^i} \right\}. \tag{12.350}$$

As we can see from (12.350), the mathematical expectation estimate is unbiased. To define the variance of estimate, we assume that the samples x_i are uncorrelated. Then, the variance of mathematical expectation estimate takes the following form:

$$\text{Var}\{E^*\} = \frac{2(1-\gamma)^{2N-1} + \gamma}{2-\gamma} \sigma^2. \tag{12.351}$$

As $1 - \gamma < 1$ and $N \to \infty$ or $N \gg 1$, (12.351) can be simplified and takes the limiting form:

$$\text{Var}\{E^*\} \approx \frac{\gamma}{2-\gamma} \sigma^2, \tag{12.352}$$

that is, the considered estimate is not consistent.

The ratio between the variance of the mathematical expectation estimate defined by (12.351) and the variance of the optimal mathematical expectation estimate given by (12.183) as a function of the number of uncorrelated samples N and various magnitudes of the factor γ is shown in Figure 12.17. As we can see from Figure 12.17, for each value N there is a definite magnitude γ, at which the ratio of variances reaches the maximum.

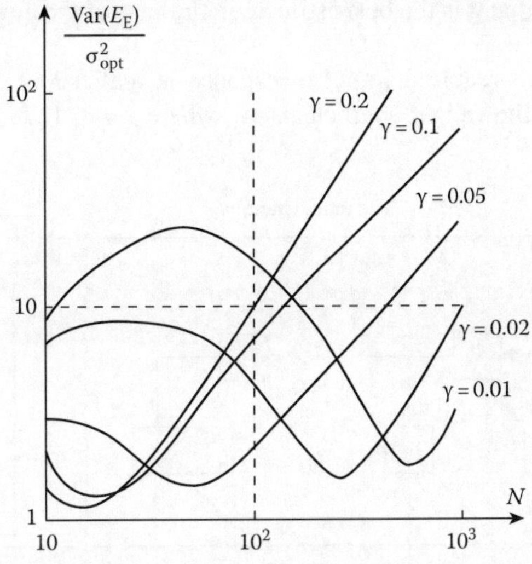

FIGURE 12.17 Ratio between the variance of mathematical expectation estimate given by (12.351) and the variance of optimal mathematical expectation estimate given by (12.183) as a function of the number of uncorrelated samples N and various values of the factor γv.

12.10 ESTIMATE OF MATHEMATICAL EXPECTATION WITH UNKNOWN PERIOD

In some applications, we can suppose that the time-varying mathematical expectation $E(t)$ of the stochastic process $\xi(t)$ is the periodic function

$$E(t) = E(t + kT_0), \quad k = 0, 1, \ldots, \tag{12.353}$$

and the value of the period T_0 is unknown. At the same time, the practical case, when the period T_0 is much more than the correlation interval τ_{cor} of the observed stochastic process, is of interest for us. We employ the adaptive filtering methods widely used in practice under interference cancellation to measure the time-varying mathematical expectation estimate [21]. We consider the discrete sample $x(t_i) = x_i = x(iT_s)$ where the sampling period is T_s. The sample can be presented in the form discussed in (12.168).

The observed samples x_i come in at the inputs of the main and reference (the adaptive filter) channels (see Figure 12.18). The delay $\tau = kT_s$, where k is integer, is chosen in such a way that the samples at the main and reference channels would be uncorrelated. There is a filter with varying parameters in the reference channel. The incoming samples x_i and the process y_i forming at the adaptive filter output are sent at the subtractor input. At the subtractor output, the following process

$$\varepsilon_i = x_i - y_i = x_{0_i} + (E_i - y_i) \tag{12.354}$$

takes place. Taking into consideration that the samples are uncorrelated, the mathematical expectation of quadratic signal in the main and reference channels is defined as

$$\langle \varepsilon^2 \rangle = \sigma_0^2 + \langle (E_i - y_i)^2 \rangle. \tag{12.355}$$

The minimum of $\langle \varepsilon^2 \rangle$ corresponds to the minimum of the second term in the right side (12.355). Thus, if the parameters of adaptive filter are changed before definition of the minimum $\langle \varepsilon^2 \rangle$, there is a need to use the signal y_i at the adaptive filter output as the estimate E_i^* of time-varying mathematical expectation. As was shown in Ref. [21], under the given structure of interference and noise canceller, the value y_i is the best estimate in the case of the quadratic loss function given by (11.25).

The adaptive filter with required impulse response is realized in the form of linear vector summing of signals with the weight coefficients W_j, where $j = 0, 1, \ldots, P - 1$ are the numbers of

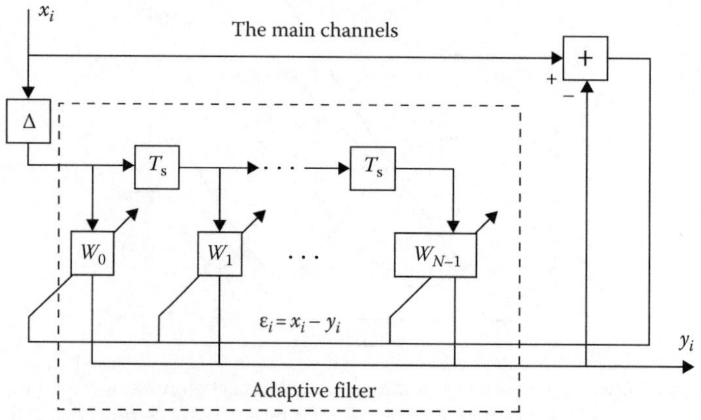

FIGURE 12.18 Adaptive filter flowcharts.

parallel channels, at that, the delay between neighboring channels is equal to the sampling period T_s. Tuning of the weight coefficients W_j^* is carried out in accordance with the recurrent Widrow–Hopf algorithm [22]

$$W_j^*[N+1] = W_j^*[N] + 2\mu\left\{x[N]x[N-d-j] - x[N-d-j]\sum_{l=0}^{P-1} W_l^*[N]x[N-d-l]\right\}, \tag{12.356}$$

where
$l = 0, 1, \ldots, P-1$
μ is the adaptation parameter characterizing the algorithmic convergence rate and the tuning accuracy of the weight coefficients

As was shown in Ref. [22], if the parameter μ satisfies the condition $0 < \mu < \lambda_{\max}^{-1}$, where $\lambda_{\max}^{-1}$ is the largest eigenvalue [3] of the covariance matrix consisting of elements $C_{ij} = \langle c_i c_j \rangle$, then the algorithm (12.356) is converged. In the physical sense the eigenvalues of covariance matrix characterize the power of input stochastic process and, under the other equal conditions, the larger λ, the more power of input stochastic process. It was proved that

$$\lim_{N\to\infty} \left\langle W_j^*[N] \right\rangle = W_j, \tag{12.357}$$

where W_j are the elements of optimal vector of the weight coefficients satisfying the Wiener–Hopf equation that takes the following form:

$$\sum_{j=0}^{P-1} C_{lj} W_j = C_{l+d}, \quad l = 0,1,\ldots,P-1. \tag{12.358}$$

in the stationary mode. The block diagram of computation algorithm for the weight coefficients of adaptive filter is shown in Figure 12.19. To stop the adaptation process we can use the procedures discussed in Ref. [22]. The most widely used procedure applied to the considered problem is based on the following inequality:

$$\varepsilon[N] = \left| \frac{W_j^*[N] - W_j^*[N-1]}{W_j^*[N]} \right| \le \nu, \tag{12.359}$$

where ν is the number given before.

The estimated mathematical expectation E_i is the periodical function and can be approximated by the Fourier series with finite number of terms

$$E_i \approx a_0 + \sum_{\mu=1}^{M} [a_\mu \cos(\omega\mu\kappa) + b_\mu \sin(\omega\mu\kappa)], \tag{12.360}$$

where

$$\omega = \frac{2\pi}{T_0} \tag{12.361}$$

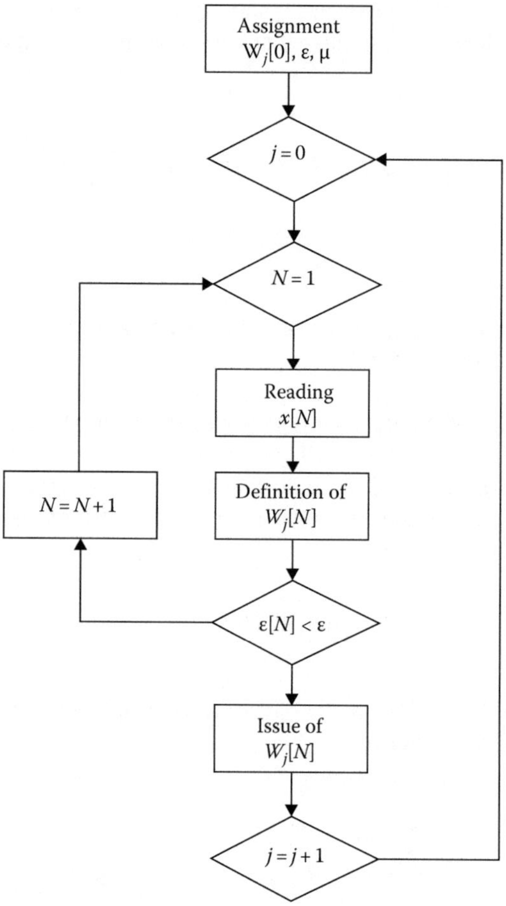

FIGURE 12.19 Block diagram of algorithm for determination of the weight coefficients of adaptive filters.

is the radial frequency of the first harmonic. In addition, we assume that the sampling interval T_s is chosen in such a way that the sample readings x_i are uncorrelated between each other. Taking into consideration the orthogonality between components of the mathematical expectation and a definition of the covariance function (the ambiguity function) of the deterministic signal with constant component as

$$C_E(k) = \lim_{k\to\infty} \frac{1}{2k} \sum_{i=-k}^{k} E(\kappa)E(\kappa+k), \tag{12.362}$$

the covariance matrix elements given by (12.358) can be written in the following form:

$$C(k) = \sigma^2\delta(k) + \sum_{\mu=0}^{M} A_\mu^2 \cos(k\omega\mu), \tag{12.363}$$

where $\delta(k)$ is the discrete analog of the Dirac delta function given by (12.170);

$$A_0^2 = a_0^2; \tag{12.364}$$

$$A_\mu^2 = \frac{1}{2}\left(a_\mu^2 + b_\mu^2\right), \quad \mu = 1,2,\ldots,M. \tag{12.365}$$

Substituting (12.363) into (12.358), we obtain

$$\sum_{j=0}^{P-1}\left\{\sigma^2\delta(l-j) + \sum_{\mu=1}^{M} A_\mu^2 \cos[\omega\mu(l-j)]\right\} W_j = \sum_{\mu=0}^{M} A_\mu^2 \cos[\omega\mu(l-d)]. \tag{12.366}$$

Denoting

$$\begin{cases} \varphi_\mu = \sum_{j=0}^{P-1} W_j \cos(j\omega\mu); \\ \psi_\mu = \sum_{j=0}^{P-1} W_j \sin(j\omega\mu), \end{cases} \tag{12.367}$$

we obtain a solution for the weight coefficient W_j in the case of stationary mode in the following form:

$$W_j = \frac{1}{\sigma^2}\sum_{\mu=0}^{M} A_\mu^2\{[\cos(d\omega\mu) - \varphi_\mu]\cos(j\omega\mu) - [\sin(d\omega\mu) - \psi_\mu]\sin(j\omega\mu)\}. \tag{12.368}$$

Substituting (12.368) into (12.366), we obtain the system of equations with respect to the unknown variables φ_μ and ψ_μ in the following form:

$$\varphi_\chi = \frac{1}{\sigma^2}\sum_{\mu=0}^{M} A_\mu^2\{[\cos(d\omega\mu) - \varphi_\mu]\,\alpha_{\mu\chi} - [\sin(d\omega\mu) - \psi_\mu]\,\gamma_{\mu\chi}\}; \tag{12.369}$$

$$\psi_\chi = \frac{1}{\sigma^2}\sum_{\mu=0}^{M} A_\mu^2\{[\cos(d\omega\mu) - \varphi_\mu]\,\beta_{\mu\chi} - [\sin(d\omega\mu) - \psi_\mu]\vartheta_{\mu\chi}\}; \tag{12.370}$$

$$\chi = 0,1,\ldots,M. \tag{12.371}$$

where

$$\alpha_{\mu\chi} = \sum_{j=0}^{P-1} \cos(j\omega\mu)\cos(j\omega\chi); \tag{12.372}$$

$$\beta_{\mu\chi} = \sum_{j=0}^{P-1} \cos(j\omega\mu)\sin(j\omega\chi); \tag{12.373}$$

$$\gamma_{\mu\chi} = \sum_{j=0}^{P-1} \sin(j\omega\mu)\cos(j\omega\chi); \tag{12.374}$$

$$\vartheta_{\mu\chi} = \sum_{j=0}^{P-1} \sin(j\omega\mu)\sin(j\omega\chi). \tag{12.375}$$

With the purpose of subsequent simplifying, we are limited by the case when the number of channels N is sufficiently high and in (12.372) through (12.375) we can use the integrals. Moreover, we assume that the constant component of the estimated mathematical expectation is equal to zero, that is, $a_0 = 0$. As a result of limiting process, we have

$$\begin{cases} \alpha_{\mu\chi} = \vartheta_{\mu\chi} = 0.5P\delta(\mu - \chi); \\ \gamma_{\mu\chi} = \beta_{\mu\chi} = 0. \end{cases} \tag{12.376}$$

Using the approximation given by (12.376), the solutions of the equation system given by (12.369) and (12.370) take the following form:

$$\begin{cases} \varphi_\mu = \dfrac{\cos(d\omega\mu)}{1 + \dfrac{2\sigma^2}{PA_\mu^2}}; \\ \psi_\mu = \dfrac{\sin(d\omega\mu)}{1 + \dfrac{2\sigma^2}{PA_\mu^2}}. \end{cases} \tag{12.377}$$

In the case of stationary mode and high number of channels, the weight coefficients given by (12.368) can be presented in the following form:

$$W_j = \sum_{\mu=1}^{M} \frac{2A_\mu^2}{2\sigma^2 + A_\mu^2 P} \cos[\omega\mu(d + j)]. \tag{12.378}$$

If the following condition is satisfied

$$P \gg \frac{2\sigma^2}{A_\mu^2}, \tag{12.379}$$

we obtain

$$W_j = \frac{2}{P}\sum_{\mu=1}^{M} \cos\omega\mu(d + j) = \frac{2}{P} \times \frac{\sin\dfrac{\pi(M+1)(d+j)}{2M}\cos\dfrac{\pi(d+j)}{2}}{\sin\dfrac{\pi(d+j)}{2M}}. \tag{12.380}$$

As we can see from (12.380), the adaptive filter transfer characteristic in the stationary mode will be a sequence of maxima with magnitudes $2M/P$ and period $2M$.

Consider the statistical characteristics of discrete magnitudes of the mathematical expectation estimate E_i^* at the adaptive filter output in the stationary mode. The mathematical expectation of estimate E_i^* is defined as

$$\langle E_i^* \rangle = \sum_{j=0}^{N-1} W_j E_{i-j-d}. \tag{12.381}$$

Changing E_i and W_i on their magnitudes from (12.360) and neglecting the sum of fast oscillating terms, we obtain

$$\langle E_i^* \rangle = 0.5P \sum_{\mu=1}^{M} \frac{1}{q_\mu^{-1} + 0.5P} [a_\mu \cos(i\omega\mu) + b_\mu \sin(i\omega\mu)], \tag{12.382}$$

where

$$q_\mu = \frac{a_\mu^2 + b_\mu^2}{2\sigma^2} \tag{12.383}$$

is the SNR for the μth component (or harmonic) of the mathematical expectation. As we can see from (12.383), as the number of channels P tends to approach infinity, that is, $P \to \infty$, the mathematical expectation estimate is unbiased.

By analogy, determining the second central moment that is defined completely by the centralized component of stochastic process, it is easy to define the variance of estimate

$$\text{Var}\{E_i^*\} = \sigma^2 \sum_{j=0}^{P-1} W_j^2. \tag{12.384}$$

We can see from (12.384) that the variance of the investigated mathematical expectation estimate E_i^* decreases concomitant to a decrease in the number of harmonics M in the observed stochastic process. In the limiting case, given the high number of channels ($P \to \infty$) the variance of the mathematical expectation estimate tends to approach zero in stationary mode; that is, the considered estimate of the periodically changing mathematical expectation E_i^* is consistent.

To illustrate the obtained results, a simulation of the described adaptation algorithm based on the example of the mathematical expectation estimate $E(t) = a\cos(\omega t)$ given in the form of readings E_i and when the period T_0 corresponds to four sampling periods, that is, $T_0 = 4T_s$, is carried out. As the centralized component x_{0_i} of stochastic process realization, the uncorrelated samples of Gaussian stochastic process are used. To obtain the uncorrelated samples in the main and reference channels, the delay corresponding to one sampling period ($d = 1$) is introduced. Initial magnitudes of the weight coefficients, except for the channel with $j = 0$, are chosen equal to zero. The initial magnitude of the weight coefficient at $j = 0$ is taken equal to unit, that is, $W_0[0] = 1$.

The memory of microprocessor system continuously updates the discrete sample $x_i = x_{0_i} + E_i$ and in accordance with the algorithm given by (12.357) a tuning of the weight coefficients is carried out. The tuning is considered to be complete when the components of the vector differ from each other by no more than 10% on two neighboring steps of adaptation. Thus, the realization formed at the output of adaptive interference and noise canceller is investigated.

The normalized variance of estimate of the harmonic component amplitude, that is, $\text{Var}\{a^*\}/a^2$, at the SNR equal to unit, that is, $q = 1$, as a function of the number of adaptation cycles N and two values of the number of parallel channels ($P = 4$; $P = 16$) is depicted in Figure 12.20.

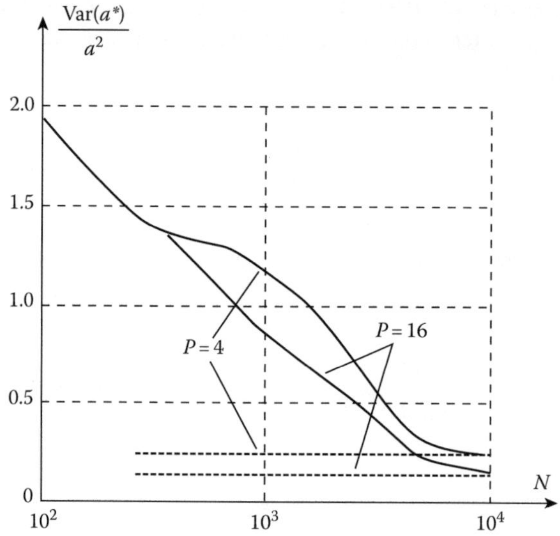

FIGURE 12.20 The normalized variance of amplitude estimation as a function of the adaptation cycles number.

The dashed lines correspond to the theoretical asymptotical values of variance of the mathematical expectation estimate computed according to (12.378) and (12.384). As we can see from Figure 12.20, the adaptation process is evident and its beginning depends on the number of parallel channels of the adaptive filter.

12.11 SUMMARY AND DISCUSSION

Let us summarize briefly the main results discussed in this chapter.

In spite of the fact that the formulaes for the mathematical expectation estimate (12.41) and (12.42) are optimal in the case of Gaussian stochastic process, these formulae are also optimal for the stochastic process that is different from the Gaussian pdf in the class of linear estimations. Equations 12.38 and 12.39 are true if the a priori interval of changes of the mathematical expectation is not limited. Equation 12.38 allows us to define the optimal device structure to estimate the mathematical expectation of stochastic process (Figure 12.1). The main function is the linear integration of the received realization $x(t)$ with the weight $\upsilon(t)$ that is defined based on the integral equation (12.8). The decision device issues the output process at the instant $t = T$. To obtain the current value of the mathematical expectation estimate, the limits of integration in (12.38) must be $t - T$ and t, respectively, then the parameter estimation is given by (12.43).

The mathematical expectation estimate of the maximum likelihood of stochastic process is both the conditionally and unconditionally unbiased estimate. Conditional variance of the mathematical expectation estimate can be presented by (12.47), from which we can see that the variance of estimate is unconditional. Since, according to (12.38), the integration of Gaussian stochastic process is a linear operation, the estimate E_E is subjected to Gaussian distribution.

The procedure to define the optimal estimate of mathematical expectation of stationary stochastic process is presented in Figure 12.1 in the case of the limited a priori domain of definition of the mathematical expectation. In this case, the maximum likelihood estimate of the mathematical expectation of stochastic process is conditionally biased. The unconditional estimate is unbiased and the unconditional dispersion is given by (12.76).

The Bayesian estimate of the mathematical expectation of stochastic process is a function of the SNR. At low SNR, the conditional estimate bias coincides with the approximation given by (12.82). We can see that the unconditional estimate of mathematical expectation averaged with respect to all possible values E_0 is unbiased. At high SNR, the Bayesian estimate of the mathematical expectation of stochastic process coincides with the maximum likelihood estimate of the same parameter.

Optimal methods to estimate the mathematical expectation of stochastic process envisage the need for accurate and complete knowledge of other statistical characteristics of the considered stochastic process. For this reason, as a rule, various nonoptimal procedures are used in practice. In doing so, the weight function is selected in such a way that the variance of estimate tends to approach asymptotically the variance of the optimal estimate. If the integration time of ideal integrator is sufficiently large in comparison with the correlation interval of stochastic process, then to determine the variance of mathematical expectation estimate of stochastic process there is a need to know only the values of variance and the ratio between the observation interval and correlation interval. The variance of the mathematical expectation estimate of stochastic process is proportional to the spectral density value of fluctuation component of the considered stochastic process at $\omega = 0$ when the ideal integrator is used as a smoothing circuit. In other words, in the considered case, the variance of the mathematical expectation estimate of stochastic process is defined by the spectral components in the case of zero frequency. To obtain the current value of the mathematical expectation estimate and to investigate the realization of stochastic process within the limits of large interval of observation, we use the estimate given by (12.124). Evidently, this estimate has the same statistical characteristics as the estimate defined by (12.111). The previously discussed procedures that measure the mathematical expectation suppose that there are no limitations due to instantaneous values of the considered stochastic process in the course of measurement. Presence of limitations leads to additional errors while measuring the mathematical expectation of stochastic process.

The variance of the mathematical expectation estimate is defined by the interval of possible values of the additional stochastic sequence and is independent of the variance of the observed stochastic process and is forever more than the variance of the equidistributed estimate of the mathematical expectation by independent samples. For example, if the observed stochastic sequence subjected to the uniform pdf coinciding in the limiting case with (12.218), then the variance of the mathematical expectation for the considered procedure is defined by (12.226) and the variance of the mathematical expectation in the case of equidistributed estimate of the mathematical expectation is given by (12.227), that is, the variance of the mathematical expectation estimate in three times more than the variance of the mathematical expectation in the case of equidistributed estimate of the mathematical expectation under the use of additional stochastic signals in the considered limiting case when the observed and additional random sequences are subjected to the uniform pdf. At other conditions, a difference in variances of the mathematical expectation estimate is higher.

Under the definition of stochastic process quantization effect by amplitude on the estimate of its mathematical expectation, we assume that quantization can be considered as the inertialess nonlinear transformation with the constant quantization step and the number of quantization levels is so high that the quantized stochastic process cannot be outside the limits of staircase characteristic of the transform $g(x)$, the approximate form of which is shown in Figure 12.12. The pdf $p(x)$ of observed stochastic process possessing the mathematical expectation that does not match with the middle between the quantization thresholds x_i and x_{i+1} is presented in Figure 12.12 for obviousness. The mathematical expectation of the realization $y(t)$ forming at the output of inertialess element (transducer) with the transform characteristic given by (12.228) when the realization $x(t)$ of the stochastic process $\xi(t)$ excites the input of inertialess element is equal to the mathematical expectation of the mathematical expectation estimate given by (12.229) and is defined by (12.230). In general, the mathematical expectation of estimate $\langle E \rangle$ differs from the true value E_0, that is, as a result of quantization we obtain the bias of the mathematical expectation of the mathematical expectation estimate given by (12.231). Since the mathematical expectation and the correlation function of process forming at the transducer output depend on the observed stochastic process pdf, the characteristics of the mathematical expectation estimate of stochastic process quantized by amplitude depend both on the correlation function and on the pdf of observed stochastic process.

In the case of the time-varying mathematical expectation of stochastic process, the problem of definition of the mathematical expectation $E(t)$ of stochastic process $\xi(t)$ by a single realization $x(t)$ within the limits of the interval $[0, T]$ is reduced to an estimation of the coefficients α_i of the series given by (12.256). In doing so, the bias and dispersion of the mathematical expectation estimate $E^*(t)$ of the observed stochastic process caused by measurement errors of the coefficients α_i are given by (12.259) and (12.260). Statistical characteristics (the bias and dispersion) of the mathematical expectation estimate of stochastic process averaged within the limits of the observation interval are defined by (12.261) and (12.262), respectively. The higher the number of terms under expansion in series in (12.256) used for approximation of the mathematical expectation, the higher at the same conditions the variance of time-varying mathematical expectation estimate averaged within the limits of the observation interval. In doing so, there is a need to note that in general the number N of series expansion terms essentially increases parallel to an increase in the observation interval $[0, T]$, within the limits of which the approximation is carried out.

Thus, at sufficiently large number of the eigenfunctions in the sum given by (12.256) the average time-varying mathematical expectation estimate $E(t)$ is equal to the variance of the initial stochastic process. In doing so, the estimate bias caused by the finite number of terms in series given by (12.256) tends to approach zero. However, in practice, there is a need to choose the number of terms in series given by (12.256) in such a way that a dispersion of the time-varying mathematical expectation estimate caused both by the bias and by the estimate variance would be minimal.

Under definition of estimation of the time-varying mathematical expectation of stochastic process by a single realization, we meet difficulties caused by the definition of optimal time of averaging (integration) or the time constant of smoothing filter at the filter impulse response given before. In doing so, two conflicting requirements arise. On one hand, there is a need to decrease the variance of estimate caused by finite time interval of measuring; this time interval must be large. On the other hand, for better distinguishing the mathematical expectation variations in time, there is a need to choose the integration time to be as short as possible. Evidently, there is an optimal averaging time or the bandwidth of the smoothing filter under the given impulse response, which corresponds to minimal dispersion of the mathematical expectation estimate of stochastic process caused by the factors listed previously.

In practice, the nonstationary stochastic processes with time-varying mathematical expectation or variance or both of them simultaneously are widely used. In doing so, the mathematical expectation and variance vary slowly in comparison with variations of investigated stochastic process. In other words, the mathematical expectation and variance of stochastic process are constant within the limits of the correlation interval. In this case, to define the variance of the time-varying mathematical expectation estimate we can assume that the centralized stochastic process $\xi_0(t) = \xi(t) - E(t)$ is the stationary stochastic process within the limits of the interval $t_0 \pm 0.5T$ with the correlation function given by (12.317).

In some applications, we can suppose that the time-varying mathematical expectation of stochastic process is the periodic function and the value of the period is unknown. At the same time, in a practical case, when the period is much longer than the correlation interval of the observed stochastic process is of interest to us. We employ the adaptive filtering methods widely used in practice under interference cancellation to measure. We consider the discrete sample that can be presented in the form discussed in (12.168). The estimated mathematical expectation is the periodical function and can be approximated by the Fourier series with finite number of terms. We can assume that the sampling interval is chosen in such a way that the sample readings remain uncorrelated with each other. Taking into consideration the orthogonality between the components of mathematical expectation and a definition of the covariance function (the ambiguity function) of the deterministic signal with constant component given by (12.362), the covariance matrix elements given by (12.358) can be presented in the form (12.363).

REFERENCES

1. Lindsey, J.K. 2004. *Statistical Analysis of Stochastic Processes in Time*. Cambridge, U.K.: Cambridge University Press.
2. Ruggeri, F. 2011. *Bayesian Analysis of Stochastic Process Models*. New York: Wiley & Sons, Inc.
3. Van Trees, H. 2001. *Detection, Modulation, and Estimation Theory. Part 1*. New York: Wiley & Sons, Inc.
4. Taniguchi, M. 2000. *Asymptotic Theory of Statistical Inference for Time Series*. New York: Springer + Business Media, Inc.
5. Franceschetti, M. 2008. *Random Networks for Communication: From Physics to Information Systems*. Cambridge, U.K.: Cambridge University Press.
6. Le Cam, L. 1986. *Asymptotic Methods in Statistical Decision Theory*. New York: Springer + Business Media, Inc.
7. Anirban DasGupta. 2008. *Asymptotic Theory of Statistics and Probability*. New York: Springer + Business Media, Inc.
8. Berger, J. 1985. *Statistical Decision Theory and Bayesian Analysis*. New York: Springer + Business Media, Inc.
9. Le Cam, L. and G.L. Yang. 2000. *Asymptotics in Statistics: Some Basic Concepts*. New York: Springer + Business Media, Inc.
10. Liese, F. and K.J. Miescke. 2008. *Statistical Decision Theory: Estimation, Testing, and Selection*. New York: Springer + Business Media, Inc.
11. Schervish, M. 1996. *Theory of Statistics*. New York: Springer + Business Media, Inc.
12. Lehmann, E.L. 2005. *Testing Statistical Hypothesis*, 3rd edn. New York: Springer + Business Media, Inc.
13. Jesbers, P., Chu, P.T., and A.A. Fettwers. 1962. A new method to compute correlations. *IRE Transactions on Information Theory*, 8(8): 106–107.
14. Mirskiy, G. Ya. 1972. *Hardware Definition of Stochastic Process Characteristics*, 2nd edn. Moscow, Russia: Energy.
15. Gusak, D., Kukush, A., Kulik, A., Mishura, Y., and A. Pilipenko. 2010. *Theory of Stochastic Processes*. New York: Springer + Business Media, Inc.
16. Gikhman, I., Skorokhod, A., and S. Kotz. 2004. *The Theory of Stochastic Processes I*. New York: Springer + Business Media, Inc.
17. Gikhman, I., Skorokhod, A., and S. Kotz. 2004. *The Theory of Stochastic Processes II*. New York: Springer + Business Media, Inc.
18. Tzypkin, Ya. 1968. *Adaptation and Training in Automatic Systems*. Moscow, Russia: Nauka.
19. Cox, D.R. and H.D. Miller. 1977. *Theory of Stochastic Processes*. Boca Raton, FL: CRC Press.
20. Brzezniak, Z. and T. Zastawniak. 2004. *Basic Stochastic Processes*. New York: Springer + Business Media, Inc.
21. Ganesan, S. 2009. *Model Based Design of Adaptive Noise Cancellation*. New York: VDM Verlag, Inc.
22. Zeidler, J.R., Satorius, E.H., Charies, D.M., and H.T. Wexler. 1978. Adaptive enhancement of multiple sinusoids in uncorrelated noise. *IEEE Transactions on Acoustics, Speech, and Signal Processing*, 26(3): 240–254.

13 Estimation of Stochastic Process Variance

13.1 OPTIMAL VARIANCE ESTIMATE OF GAUSSIAN STOCHASTIC PROCESS

Let the stationary Gaussian stochastic process $\xi(t)$ with the correlation function

$$R(t_1,t_2) = \sigma^2\mathcal{R}(t_1,t_2) \tag{13.1}$$

be observed at N equidistant discrete time instants t_i, $i = 1, 2,\ldots, N$ in such a way that

$$t_{i+1} - t_i = \Delta = \text{const.} \tag{13.2}$$

Then, at the measurer input we have a set of samples $x_i = x(t_i)$. Furthermore, we assume that the mathematical expectation of observed stochastic process is zero. Then, the conditional N-dimensional pdf of Gaussian stochastic process can be presented in the following form:

$$p(x_1, x_2,\ldots, x_N|\sigma^2) = \frac{1}{(2\pi\sigma^2)^{N/2}\sqrt{\det\|\mathcal{R}_{ij}\|}}\exp\left\{-\frac{1}{2\sigma^2}\sum_{i=1,j=1}^{N} x_i x_j C_{ij}\right\} \tag{13.3}$$

where

det $\|\mathcal{R}_{ij}\|$ is the determinant of matrix consisting of elements of the normalized correlation function $\mathcal{R}(t_i, t_j) = \mathcal{R}_{ij}$

C_{ij} are the elements of the matrix that is reciprocal to the matrix $\|\mathcal{R}_{ij}\|$

In doing so, the elements C_{ij} are defined from the equation that is analogous to (12.170)

$$\sum_{l=1}^{N} C_{il}\mathcal{R}_{lj} = \delta_{ij}. \tag{13.4}$$

The conditional multidimensional pdf (13.3) is the *multidimensional likelihood function* of the parameter σ^2. Solving the likelihood equation, we obtain the variance estimate

$$\text{Var}_E = \frac{1}{N}\sum_{i=1,j=1}^{N} x_i x_j C_{ij} = \frac{1}{N}\sum_{i=1}^{N} x_i v_i, \tag{13.5}$$

where

$$v_i = \sum_{j=1}^{N} x_j C_{ij}. \tag{13.6}$$

The random variables v_i possess zero mean and are subjected to the Gaussian pdf. Prove that the random variables v_i are dependent on each other and are mutually independent of the samples x_p at $p \neq i$. Actually,

$$\langle v_i v_q \rangle = \sum_{j=1,p=1}^{N} C_{ij} C_{qp} \langle x_j x_p \rangle = \sigma^2 \sum_{j=1}^{N} C_{ij} \sum_{p=1}^{N} C_{qp} \mathcal{R}_{jp} = \sigma^2 \sum_{j=1}^{N} C_{ij} \delta_{jq} = \sigma^2 C_{iq}; \tag{13.7}$$

$$\langle x_p v_i \rangle = \sum_{j=1}^{N} C_{ij} \langle x_j x_p \rangle = \sigma^2 \sum_{j=1}^{N} C_{ij} \mathcal{R}_{jp} = \sigma^2 \delta_{ip}, \tag{13.8}$$

where σ^2 is the true variance of the stochastic process $\xi(t)$.

Determine the statistical characteristics of variance estimate, namely, the mathematical expectation and variance. The mathematical expectation of variance estimate is given by

$$E\{\mathrm{Var}_E\} = \frac{1}{N} \sum_{i=1,j=1}^{N} C_{ij} \langle x_i x_j \rangle = \sigma^2, \tag{13.9}$$

that is, the variance estimate is unbiased.

The variance of variance estimate can be presented in the following form:

$$\mathrm{Var}\{\mathrm{Var}_E\} = \frac{1}{N^2} \sum_{i=1,j=1}^{N} \sum_{p=1,q=1}^{N} \langle x_i x_j x_p x_q \rangle \, C_{ij} C_{pq} - \sigma^4. \tag{13.10}$$

Determining the mixed fourth moment of the Gaussian random variable x

$$\langle x_i x_j x_p x_q \rangle = \sigma^4 [\mathcal{R}_{ij} \mathcal{R}_{pq} + \mathcal{R}_{ip} \mathcal{R}_{jq} + \mathcal{R}_{iq} \mathcal{R}_{jp}] \tag{13.11}$$

and substituting (13.11) into (13.10) and taking into consideration (13.4), we obtain

$$\mathrm{Var}\{\mathrm{Var}_E\} = \frac{2\sigma^4}{N}. \tag{13.12}$$

As we can see from (13.12), the variance of the optimal variance estimate is independent of the values of the normalized correlation function between the samples of observed stochastic process. This fact may lead to some results that are difficult to explain with the physical viewpoint or cannot be explained absolutely. Actually, increasing the number of samples within the limits of the finite small time interval, we can obtain the estimate of variance with the infinitely small estimate variance according to (13.12). Approaching the variance of optimal variance estimate of Gaussian stochastic process to zero is especially evident while passing from discrete to continuous observation of stochastic process. Considering

$$N = \sum_{i=1}^{N} \sum_{j=1}^{N} C_{ij} \mathcal{R}_{ij} \tag{13.13}$$

and applying the limiting process to (13.5) from discrete observation to continuous one at $T = \text{const}$ ($\Delta \to 0, N \to \infty$), we obtain

$$\text{Var}_E = \frac{\int_0^T \int_0^T \vartheta(t_1, t_2) x(t_1) x(t_2) dt_1 dt_2}{\int_0^T \int_0^T \vartheta(t_1, t_2) \mathcal{R}(t_1, t_2) dt_1 dt_2}, \tag{13.14}$$

where the function $\vartheta(t_1, t_2)$, similar to (12.6), can be defined from the following equation:

$$\int_0^T \vartheta(t_1, t) \mathcal{R}(t_1, t_2) dt = \delta(t_2 - t_1). \tag{13.15}$$

Determining the mathematical expectation of variance estimate, we can see that it is unbiased. The variance of the variance estimate can be presented in the following form:

$$\text{Var}\{\text{Var}_E\} = \frac{2\sigma^4}{\int_0^T \int_0^T \vartheta(t_1, t_2) \mathcal{R}(t_1, t_2) dt_1 dt_2}. \tag{13.16}$$

Since

$$\lim_{\tau \to t_2} \int_0^T \vartheta(t_1, t_2) \mathcal{R}(\tau, t_1) dt_1 = \delta(0) \to \infty, \tag{13.17}$$

we can see from (13.16) that the variance of the optimal variance estimate of Gaussian stochastic process approaches symmetrical to zero for any value of the observation interval $[0, T]$.

Analogous statement appears for a set of problems in statistical theory concerning optimal signal processing in high noise conditions. In particular, given the accurate measurement of the stochastic process variance, it is possible to detect weak signals in powerful noise within the limits of a short observation interval $[0, T]$. In line with this fact, in [1] it was assumed that for the purpose of resolving detection problems, including problems related to zero errors, we should reject the accurate knowledge of the correlation function of the observed stochastic processes or we need to reject the accurate measurement of realizations of the input stochastic process. Evidently, in practice these two factors work. However, depending on which errors are predominant in the analysis of errors, limitations arise due to insufficient knowledge about the correlation function or due to inaccurate measurement of realizations of the investigated stochastic process.

It is possible that errors caused by inaccurate measurement of the investigated stochastic process have a characteristic of the additional "white" noise with Gaussian pdf. In other words, we think that the additive stochastic process

$$y(t) = x(t) + n(t) \tag{13.18}$$

comes in at the input of measurer instead of the realization $x(t)$ of stochastic process $\xi(t)$, where $n(t)$ is the realization of additional "white" noise with the correlation function defined as $R_n(\tau) = 0.5\mathcal{N}_0\delta(\tau)$ and $\mathcal{N}_0$ is the one-sided power spectral density of the "white" noise.

To define the characteristics of the optimal variance estimate in signal processing in noise we use the results discussed in Section 15.3 for the case of the optimal estimate of arbitrary parameter of the correlation function $R(\tau, l)$ of the Gaussian stochastic process combined with other Gaussian stochastic processes with the known correlation function. As applied to a large observation interval compared to the correlation interval of the stochastic process and to the variance estimate ($l \equiv \sigma^2$), the variance estimate, in the first approximation, is unbiased. The variance of optimal variance estimate of the correlation function parameter given by (15.152) can be simplified and presented in the following form:

$$\mathrm{Var}\left\{\frac{\mathrm{Var}_E}{\sigma^2}\right\} = \frac{1}{\left\{\dfrac{T}{4\pi}\displaystyle\int_{-\infty}^{\infty}\frac{S^2(\omega)d\omega}{[\sigma^2 S(\omega) + 0.5\mathcal{N}_0]^2}\right\}^2}, \tag{13.19}$$

where $S(\omega)$ is the spectral power density of the stochastic process $\xi(t)$ with unit variance.

To investigate the stochastic process possessing the power spectral density given by (12.19), the variance of the variance estimate in the first approximation is determined in the following form:

$$\mathrm{Var}\left\{\frac{\mathrm{Var}_E}{\sigma^2}\right\} = \frac{4\sigma^2\sqrt{\alpha\mathcal{N}_0\sigma^2}}{\alpha T}. \tag{13.20}$$

Denote

$$P_n = \mathcal{N}_0\Delta f_{\mathrm{eff}} \tag{13.21}$$

as the noise power within the limits of effective spectrum band of investigated stochastic process, and according to (12.284) $\Delta f_{\mathrm{eff}} = 0.25\alpha$, and

$$q^2 = \frac{P_n}{\sigma^2} \tag{13.22}$$

as the ratio of the noise power to the variance of investigated stochastic process, in other words, the noise-to-signal ratio. As a result, we obtain the relative variation in the variance estimate of the stochastic process with an exponential normalized correlation function:

$$r = \frac{\mathrm{Var}\{\mathrm{Var}_E\}}{\sigma^4} = 8\frac{q}{p}, \tag{13.23}$$

where $p = \alpha T = T\tau_{\mathrm{cor}}^{-1}$, as mentioned earlier, is the ratio of the observation time to correlation interval of stochastic process. In practice, the measurement errors of instantaneous values $x_i = x(t_i)$ can be infinitely small. At the same time, the measurement error of the normalized correlation function depends, in principle, on measurement conditions and, first of all, the observation time of stochastic process. Note that in practice, as a rule, the normalized correlation function is defined in the course of the joint estimate of the variance and normalized correlation function.

It is interesting to compare the earlier-obtained optimal variance estimate given by (13.3) based on the sample with the variance estimate obtained according to the widely used in mathematics statistical rule:

$$\mathrm{Var}^* = \frac{1}{N}\sum_{i=1}^{N} x_i^2. \tag{13.24}$$

We can easily see that the variance estimate of stationary Gaussian stochastic process with zero mean carried out according to (13.24) is unbiased, and the variance of the variance estimate is defined as

$$\mathrm{Var}\{\mathrm{Var}^*\} = \frac{2\sigma^4}{N^2}\sum_{i=1,j=1}^{N} \mathcal{R}[(i-j)\Delta] = \frac{2\sigma^4}{N^2}\left\{1 + 2\sum_{i=1}^{N-1}\left(1-\frac{i}{N}\right)\mathcal{R}^2(i\Delta)\right\}. \tag{13.25}$$

By analogy with (12.193) and (12.194), the new indices were introduced and the order of summation was changed; $\Delta = t_{i+1} - t_i$ is the time interval between samples. If the samples are uncorrelated (in a general case, independent) then (13.25) is matched with (13.12).

Thus, the optimal variance estimate of Gaussian stochastic process based on discrete sample is equivalent to the error given by (13.24) for the same sample size (the number of samples). This finding can be explained by the fact that under the optimal signal processing, the initial sample multiplies by the newly formed uncorrelated sample. However, if the normalized correlation function is unknown or is found to be inaccurate, then the optimal variance estimate of the stochastic process has the finite variance depending on the true value of the normalized correlation function. To simplify the investigation of this problem, we need to compute the variation in the variance estimate of Gaussian stochastic process with zero mean by two samples applied to the estimate by using the maximum likelihood function for the following cases: the normalized correlation function or, as it is often called, the correlation coefficient [2], in completely known, unknown, and erroneous conditions.

Applying the known correlation coefficient ρ to the optimal variance estimate of the stochastic process with the variance estimate yields the definition

$$\mathrm{Var}_E = \frac{x_1^2 + x_2^2 - 2\rho x_1 x_2}{2(1-\rho^2)}, \tag{13.26}$$

where

$$\rho = \frac{\langle x_1 x_2\rangle}{\sigma^2} \tag{13.27}$$

is the correlation coefficient between the samples. The variance of the optimal variance estimate can be defined based on (13.12)

$$\mathrm{Var}\{\mathrm{Var}_E\} = \sigma^4. \tag{13.28}$$

If the correlation coefficient ρ is unknown, then it must be considered as the variance or the unknown parameter of pdf. In this case, we need to solve the problem to obtain the joint estimate of the variance Var and the correlation coefficient ρ. As applied to the considered two-dimensional Gaussian sample, the conditional pdf (the likelihood function) takes the following form:

$$p_2(x_1, x_2 | \mathrm{Var}, \rho) = \frac{1}{2\pi \mathrm{Var}\sqrt{1-\rho^2}} \exp\left\{-\frac{x_1^2 + x_2^2 - 2\rho x_1 x_2}{2\,\mathrm{Var}(1-\rho^2)}\right\}. \tag{13.29}$$

Solving the likelihood equation

$$\begin{cases} \dfrac{\partial p(x_1, x_2 | \mathrm{Var}, \rho)}{\partial \mathrm{Var}} = 0; \\ \dfrac{\partial p(x_1, x_2 | \mathrm{Var}, \rho)}{\partial \rho} = 0 \end{cases} \tag{13.30}$$

with respect to the estimations Var_E and ρ_E, we obtain the following estimations of the variance and the correlation coefficient:

$$\mathrm{Var}_E = \frac{x_1^2 + x_2^2}{2}, \tag{13.31}$$

$$\rho_E = \frac{2x_1 x_2}{x_1^2 + x_2^2}. \tag{13.32}$$

As we can see from (13.31), the variance estimate is unbiased and the variance of the variance estimate can be presented in the following form:

$$\mathrm{Var}\{\mathrm{Var}_E\} = \sigma^4(1 + \rho^2), \tag{13.33}$$

that is, in the case when the correlation coefficient is unknown, the variance of the variance estimate of the Gaussian stochastic process depends on the absolute value of the correlation coefficient.

Let the correlation coefficient ρ be known with the random error ε, that is,

$$\rho = \rho_0 + \varepsilon, \tag{13.34}$$

where ρ_0 is the true value of the correlation coefficient between the samples of the investigated stochastic process. At the same time, it can be assumed that the random error ε is statistically independent of specific sample values x_1 and x_2. Furthermore, we assume that the mathematical expectation of the random error ε is zero, that is, $\langle \varepsilon \rangle = 0$ and $\mathrm{Var}_\varepsilon = \langle \varepsilon^2 \rangle$ is the variance to be used for defining the true value ρ_0.

Using the assumptions made based on (13.26), it is possible to conclude that the variance estimate is unbiased and the variance of the variance estimate can be presented in the following form:

$$\mathrm{Var}\{\mathrm{Var}_E^*\} = \sigma^4\left[1 + \mathrm{Var}_\varepsilon \frac{1 + 2\rho_0^2}{1 - \rho_0^2}\right]. \tag{13.35}$$

If the random error ε is absent under the definition of the correlation coefficient and we can use only the true value ρ_0, that is, $\mathrm{Var}_\varepsilon = 0$, the variance of variance estimate by two samples is defined

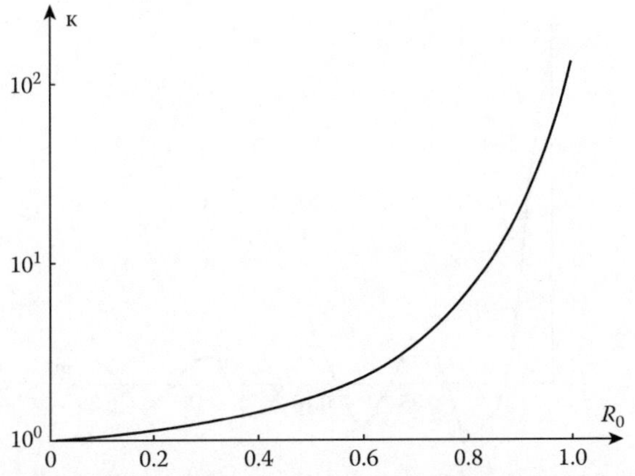

FIGURE 13.1 Variance of the variance estimate versus the true value of the correlation coefficient.

as σ^2 and is independent of the true value ρ_0 of the correlation coefficient. The random error under definition of the variance of variance estimate

$$\kappa = \frac{1+2\rho_0^2}{1-\rho_0^2} \tag{13.36}$$

as a function of the true value ρ_0 of the correlation coefficient is shown in Figure 13.1. The value κ shows how much the variance of variance estimate increases with the increase in the error under definition of the correlation coefficient ρ. As we can see from (13.35) and Figure 13.1, with increase in the absolute value of the correlation coefficient, the errors in variance estimate also rise rapidly. In doing so, if the module of the correlation coefficient between samples tends to approach unit, then independent of the small variance Var_ε of the definition of the correlation coefficient the variance of the variance estimate of stochastic process increases infinitely. Compared to $\rho_0 = 0$, at $|\rho_0| = 0.95$, the value κ increases 30 times and at $|\rho_0| = 0.99$ and $|\rho_0| = 0.999$, 150 and 1500 times, respectively. For this reason, at sufficiently high values of the module of correlation coefficient between samples we need to take into consideration the variance of definition of the correlation coefficient ρ. Qualitatively, this result is correct in the case of the optimal variance estimate of the Gaussian stochastic process under multidimensional sampling.

Although in applications we define the correlation function with errors, in particular cases, when we carry out a simulation, we are able to satisfy the simulation conditions, in which the normalized correlation function is known with high accuracy. This circumstance allows us to verify experimentally the correctness of the definition of optimal variance estimate of Gaussian stochastic process by sampling with high values given by the module of correlation coefficient between samples.

Experimental investigations of the optimal (13.5) and nonoptimal (13.24) variance estimations by two samples with various values of the correlation coefficient between them were carried out. To simulate various values of the correlation coefficient between samples, the following pair of values was formed:

$$x_i = \frac{1}{L}\sum_{p=1}^{L} y_{i-p}, \tag{13.37}$$

$$x_{i-k} = \frac{1}{L}\sum_{p=1}^{L} y_{i-k-p}, \quad k = 0,1,\ldots,L. \tag{13.38}$$

As we can see from (13.37) and (13.38), the samples x_i and x_{i-k} are the sums of independent samples y_p obtained from the stationary Gaussian stochastic process with zero mathematical expectation.

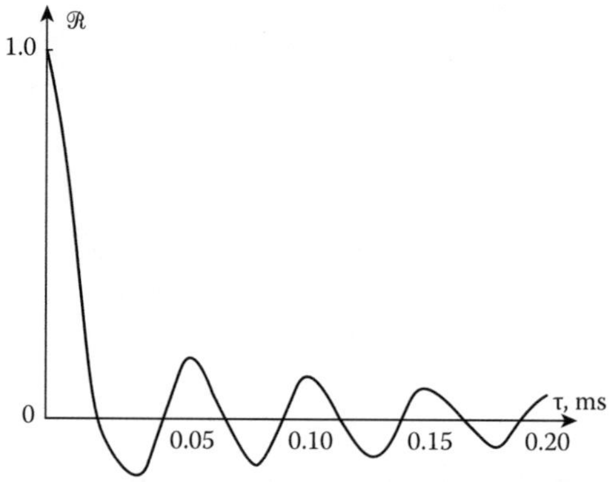

FIGURE 13.2 Normalized correlation function of stationary Gaussian stochastic process.

The samples x_i and x_{i-k} are distributed according to the Gaussian pdf and possess zero mathematical expectations. The correlation function between the newly formed samples can be presented in the following form:

$$R_k = \langle x_i x_{i-k} \rangle = \frac{1}{L^2} \sum_{p=1}^{L-|k|} \langle y_p^2 \rangle = \frac{\sigma^2}{L} \left(1 - \frac{|k|}{L} \right). \tag{13.39}$$

To obtain the samples x_i and x_{i-k} subjected to the Gaussian pdf we use the stationary Gaussian stochastic process with the normalized correlation function presented in Figure 13.2. As we can see from Figure 13.2, the samples of stochastic process y_p with the sampling period exceeding 0.2 ms are uncorrelated practically. In the course of test, to ensure a statistical independence of the samples y_p we use the sampling rate equal to 512 Hz, which is equivalent to 2 ms between samples. Experimental test with the sample size equal to 10^6 shows that the correlation coefficient between neighboring readings of the sample y_p is not higher than 2.5×10^{-3}.

Statistical characteristics of estimations given by (13.5) and (13.24) at $N = 2$ are computed by the microprocessor system. The samples x_i and x_{i-k} are obtained as a result of summing 100 independent readings of y_{i-p} and y_{i-k-p}, respectively, which guarantees a variation of the correlation coefficient from 0 to 1 at the step 0.01 when the value k changes from 100 to 0. The statistical characteristics of estimates of the mathematical expectation $\langle \text{Var}_j \rangle$ and the variance $\text{Var}\{\text{Var}_j\}$ at $j = 1$ corresponding to the optimal estimate given by (13.5) and at $j = 2$ corresponding to the estimate given by (13.24) are determined by 3×10^5 pair samples x_i and x_{i-k}. In the course of this test, the mathematical expectation and variance have been determined based on 10^6 samples of x_i. In doing so, the mathematical expectation has been considered as zero, not high 1.5×10^{-3}, and the variance has been determined as $\sigma^2 = 2.785$.

Experimental statistical characteristics of variance estimate obtained with sufficiently high accuracy are matched with theoretical values. The maximal relative errors of definition of the mathematical expectation estimate and the variance estimate do not exceed 1% and 2.5%, correspondingly. Table 13.1 presents the experimental data of the mathematical expectations and variations in the variance estimate at various values of the correlation coefficient ρ_0 between samples. Also, the theoretical data of the variance of the variance estimate based on the algorithm given by (13.24) and determined in accordance with the formula

$$\text{Var}\{\text{Var}^*\} = \sigma^2 \left(1 + \rho_0^2 \right) \tag{13.40}$$

are presented for comparison in Table 13.1. Theoretical value of the variance of the optimal variance estimate is equal to $\text{Var}\{\text{Var}_E\} = 7.756$ according to (13.5). Thus, the experimental data prove the previously mentioned theoretical definition of estimates and their characteristics, at least at $N = 2$.

TABLE 13.1
Experimental Data of Definition of the Mathematical Expectation and Variance of the Variance Estimate

	Algorithm (13.5)		Algorithm (13.4)		
				Variance of Estimate	
R_0	Mean	Variance	Mean	Experimental	Theoretical
0.00	2.770	7.865	2.770	7.865	7.756
0.01	2.773	7.871	2.773	7.867	7.757
0.03	2.774	7.826	2.773	7.822	7.763
0.05	2.773	7.824	2.770	7.826	7.776
0.07	2.769	7.797	2.766	7.818	7.794
0.10	2.763	7.765	2.760	7.825	7.834
0.15	2.764	7.743	2.760	7.883	7.931
0.20	2.765	7.777	2.759	8.019	8.067
0.25	2.762	7.688	2.758	8.085	8.241
0.30	2.766	7.707	2.759	8.303	8.454
0.35	2.775	7.735	2.765	8.583	8.706
0.40	2.778	7.735	2.765	8.820	8.997
0.45	2.782	7.683	2.760	9.125	9.327
0.50	2.780	7.742	2.769	9.584	9.695
0.55	2.777	7.718	2.766	9.984	10.103
0.60	2.782	7.804	2.767	10.451	10.548
0.65	2.780	7.637	2.765	10.818	11.003
0.70	2.785	7.710	2.772	11.379	11.557
0.75	2.778	7.636	2.777	12.040	12.119
0.80	2.770	7.574	2.777	12.706	12.720
0.85	2.768	7.624	2.777	13.301	13.360
0.90	2.787	7.767	2.776	14.033	14.039
0.91	2.793	7.782	2.779	14.208	14.179
0.93	2.794	7.751	2.776	14.507	14.465
0.95	2.789	7.806	2.770	14.794	14.756
0.97	2.776	7.724	2.775	15.192	15.054
0.99	2.792	7.826	2.777	15.483	15.358

13.2 STOCHASTIC PROCESS VARIANCE ESTIMATE UNDER AVERAGING IN TIME

In practice, under investigation of stationary stochastic processes, the value

$$\mathrm{Var}^* = \int_0^T h(t)[x(t) - E]^2 dt \tag{13.41}$$

is considered as the variance estimate, where $x(t)$ is the realization of observed stochastic process; E is the mathematical expectation; $h(t)$ is the weight function with the optimal form defined based on the condition of unbiasedness of the variance estimate

$$\int_0^T h(t)dt = 1 \tag{13.42}$$

and minimum of the variance of the variance estimate. As applied to the stationary Gaussian stochastic process, the variance of the variance estimate can be presented in the following form:

$$\mathrm{Var}\{\mathrm{Var}^*\} = 4\sigma^2 \int_0^T \mathcal{R}^2(\tau) r_h(\tau) d\tau, \tag{13.43}$$

when the mathematical expectation is known by analogy with (12.130), where σ^2 is the true value of variance, $\mathcal{R}(\tau)$ is the normalized correlation function of the observed stochastic process, and the function $r_h(\tau)$ is given by (12.131).

The optimal weight function applied to stochastic process with the exponential normalized correlation function given by (12.13) is discussed in Ref. [3]. The optimal variance estimate in the sense of the rule given by (13.41) and the minimal value of the variance of the variance estimate take the following form:

$$\mathrm{Var}^*_{\mathrm{opt}} = \frac{x_0^2(0) + x_0^2(T) + 2\alpha \int_0^T x^2(t) dt}{2(1+\alpha T)}; \tag{13.44}$$

$$\mathrm{Var}\{\mathrm{Var}^*_{\mathrm{opt}}\} = \frac{2\sigma^4}{1+\alpha T}. \tag{13.45}$$

To obtain the considered optimal estimate we need to know the correlation function with high accuracy that is not possible forever. For this reason, as a rule, the weight function is selected based on the simplicity of realization, but the estimate would be unbiased and with increasing observation time interval, the variance of the variance estimate would tend to approach zero monotonically decreased. The function given by (12.112) is widely used as the weight function.

We assume that the investigated stochastic process is a stationary one and the current estimate of the stochastic process variance within the limits of the observation time interval [0, T] has the following value:

$$\mathrm{Var}^*(t) = \frac{1}{T}\int_0^T \left\{ x(\tau) - \frac{1}{T}\int_0^T x(z)dz \right\}^2 = \frac{1}{T}\int_0^T [x(\tau) - E^*(t)]^2 d\tau, \tag{13.46}$$

where $E^*(t)$ is the mathematical expectation estimate at the instant t. The flowchart of measurer operating in accordance with (13.46) is shown in Figure 13.3.

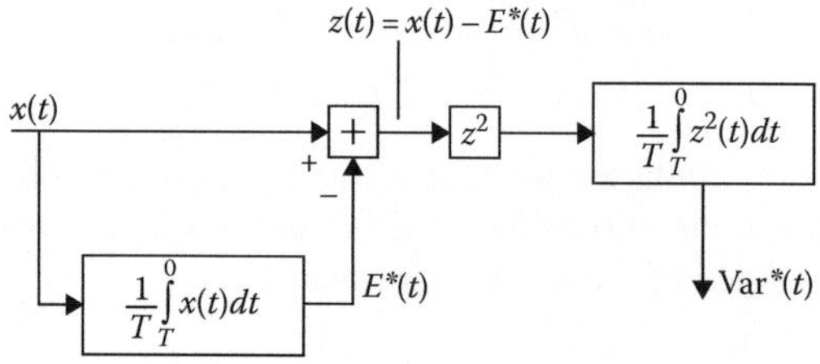

FIGURE 13.3 Flowchart of measurer operating in accordance with (13.46).

Determine the mathematical expectation and the variance of the variance estimate assuming that the investigated stochastic process is a stationary Gaussian process with unknown mathematical expectation. The mathematical expectation of variance estimate is given by

$$\langle \mathrm{Var}^* \rangle = \frac{1}{T}\int_0^T \langle x^2(\tau)\rangle \, d\tau - \frac{1}{T^2}\int_0^T\int_0^T \langle x(\tau_1)x(\tau_2)\rangle \, d\tau_1 d\tau_2. \tag{13.47}$$

Taking into consideration that

$$x_0(t) = x(t) - E_0 \tag{13.48}$$

and transforming the double integral into a single one introducing new variables, we obtain

$$\langle \mathrm{Var}^* \rangle = \sigma^2 \left\{ 1 - \frac{2}{T}\int_0^T \left(1 - \frac{\tau}{T}\right) \mathcal{R}(\tau) d\tau \right\}. \tag{13.49}$$

As we can see from (13.49), while estimating the variance of stochastic process with the unknown mathematical expectation, there is a bias of estimate defined as

$$b\{\mathrm{Var}^*\} = \frac{2\sigma^2}{T}\int_0^T \left(1 - \frac{\tau}{T}\right) \mathcal{R}(\tau) d\tau. \tag{13.50}$$

In other words, the bias of variance estimate coincides with the variance of mathematical expectation estimate of stochastic process (12.116). The bias of variance estimate depends on the observation time interval and the correlation function of investigated stochastic process. If the observation time interval is much more than the correlation interval of the investigated stochastic process and the normalized correlation function is not an alternative function, the bias of variance estimate is defined by (12.118).

Analogous effect is observed in the case of sampling the stochastic process. We consider the following value as the estimate of stochastic process variance:

$$\mathrm{Var}^* = \frac{1}{N-1}\sum_{i=1}^{N} \left\{ x_i - \frac{1}{N}\sum_{j=1}^{N} x_j \right\}^2, \tag{13.51}$$

where N is the total number of samples of stochastic process at the instants $t = i\Delta$, $i = 1, 2, \ldots, N$. The mathematical expectation of variance estimate is determined in the following form:

$$\langle \mathrm{Var}^* \rangle = \frac{\sigma^2}{N-1}\left\{ N - \frac{1}{N}\sum_{i=1,k=1}^{N} \mathcal{R}[(i-k)\Delta] \right\}. \tag{13.52}$$

Carrying out the transformation as it was done in Section 12.5, we obtain

$$\langle \text{Var}^* \rangle = \sigma^2 \left\{ 1 - \frac{2}{N-1} \sum_{i=1}^{N} \left(1 - \frac{i}{N} \right) \mathcal{R}(i\Delta) \right\}. \tag{13.53}$$

If the samples are uncorrelated, we have

$$\langle \text{Var}^* \rangle = \sigma^2. \tag{13.54}$$

In other words, we see from (13.54) that under the uncorrelated samples the variance estimate (13.51) is unbiased.

In a general case of the correlated samples, the bias of estimate of stochastic process variance is given by

$$b\{\text{Var}^*\} = \frac{2\sigma^2}{N-1} \sum_{i=1}^{N} \left(1 - \frac{i}{N} \right) \mathcal{R}(i\Delta). \tag{13.55}$$

In doing so, if the number of investigated samples is much more than, the ratio of the correlation interval of stochastic process to the sampling period, (13.55) is simplified and takes the following form:

$$b\{\text{Var}^*\} = \frac{2\sigma^2}{N-1} \sum_{i=1}^{N} \mathcal{R}(i\Delta). \tag{13.56}$$

For measuring the variance of stochastic processes, we can often assume that the mathematical expectation of stochastic process is zero and the variance estimate is carried out based on smoothing the squared stochastic process by the linear filter with the impulse response $h(t)$, namely,

$$\text{Var}^*(t) = \frac{\int_0^T h(\tau) x^2(t-\tau) d\tau}{\int_0^T h(\tau) d\tau}. \tag{13.57}$$

If the observation time interval is large, that is, a difference between the instants of estimation and beginning to excite the filter input by stochastic process much more than the correlation interval of the investigated stochastic process and filter time constant, then

$$\text{Var}^*(t) = \frac{\int_0^\infty h(\tau) x^2(t-\tau) d\tau}{\int_0^\infty h(\tau) d\tau}. \tag{13.58}$$

Under sampling the stochastic process with known mathematical expectation, the value

$$\text{Var}^* = \frac{1}{N} \sum_{i=0}^{N} (x_i - E_0)^2 \tag{13.59}$$

is taken as the variance estimate that ensures the estimate unbiasedness.

Let us define the variance of the variance estimate caused by the finite observation time interval of stochastic process. Let the investigated stochastic process be the stationary one and we use the ideal integrator as the smoothing filter. In this case, the variance estimate is carried out in accordance with (13.46) and the variance of the variance estimate takes the following form:

$$\mathrm{Var}\{\mathrm{Var}^*\} = \frac{1}{T^2}\int_0^T\int_0^T \langle x^2(t_1)x^2(t_2)\rangle dt_1 dt_2 - \frac{2}{T^3}\int_0^T\int_0^T\int_0^T \langle x^2(t_1)x(t_2)x(t_3)\rangle dt_1 dt_2 dt_3$$
$$+\frac{1}{T^4}\int_0^T\int_0^T\int_0^T\int_0^T \langle x(t_1)x(t_2)x(t_3)x(t_4)\rangle dt_1 dt_2 dt_3 dt_4 - \sigma^4\left\{1-\frac{2}{T}\int_0^T\left(1-\frac{\tau}{T}\right)\mathcal{R}(\tau)d\tau\right\}^2. \tag{13.60}$$

Determination of moments in the integrands (13.60) is impossible for the arbitrary pdf of the investigated stochastic process. For this reason, to define the main regularities we assume that the investigated stochastic process is Gaussian with the unknown mathematical expectation, the true value of which is E_0. Then, after the corresponding transformations, we have

$$\mathrm{Var}\{\mathrm{Var}^*\} = \frac{4\sigma^4}{T}\left\{\int_0^T\left(1-\frac{\tau}{T}\right)\mathcal{R}^2(\tau)d\tau + \frac{2}{T}\left[\int_0^T\left(1-\frac{\tau}{T}\right)\mathcal{R}(\tau)d\tau\right]^2\right.$$
$$\left.-\frac{1}{T^2}\int_0^T\int_0^T\int_0^T \mathcal{R}(t_1-t_2)\mathcal{R}(t_1-t_3)dt_1dt_2dt_3\right\}. \tag{13.61}$$

If the mathematical expectation of the investigated stochastic process is known accurately, then according to (13.46) the variance estimate is unbiased and the variance of the variance estimate takes the following form:

$$\mathrm{Var}\{\mathrm{Var}^*\} = \frac{4\sigma^4}{T}\int_0^T\left(1-\frac{\tau}{T}\right)\mathcal{R}^2(\tau)d\tau. \tag{13.62}$$

When the observation time interval is sufficiently large, that is, the condition $T \gg \tau_{\mathrm{cor}}$ is satisfied, (13.62) is simplified and takes the following form:

$$\mathrm{Var}\{\mathrm{Var}^*\} \approx \frac{4\sigma^4}{T}\int_0^\infty \mathcal{R}^2(\tau)d\tau \le \frac{4\sigma^4\tau_{\mathrm{cor}}}{T}. \tag{13.63}$$

Using (12.27) and (12.122), at the condition $T \gg \tau_{\mathrm{cor}}$, variance of the variance estimate can be presented in the following form:

$$\mathrm{Var}\{\mathrm{Var}^*\} \approx \frac{2}{T}\int_{-\infty}^{\infty} R^2(\tau)d\tau = \frac{2}{\pi T}\int_0^\infty S^2(\omega)d\omega. \tag{13.64}$$

As applied to the exponential correlation function given by (12.13), the normalized variance of the variance estimate can be presented in the following form:

$$\frac{\mathrm{Var}\{\mathrm{Var}^*\}}{\sigma^4} = \frac{2p - 1 + \exp\{-2p\}}{p^2}, \tag{13.65}$$

where p is given by (12.48). Analogous formulae for the variance of the variance estimate can be obtained when the stochastic process is sampling. Naturally, at that time, the simplest expressions are obtained in the case of independent readings of the investigated stochastic process.

Determine the variance of the variance estimate of stochastic process according to (13.51) for the case of independent samples. For this case, we transform (13.51) in the following form:

$$\mathrm{Var}^* = \frac{1}{N-1}\left\{\sum_{i=1}^{N} y_i^2 - \frac{1}{N}\left[\sum_{p=1}^{N} y_p\right]^2\right\}, \tag{13.66}$$

where $y_i = x_i - E$. As we can see, the variance estimate is unbiased. The variance of the variance estimate of stochastic process can be determined in the following form:

$$\mathrm{Var}\{\mathrm{Var}^*\} = \frac{1}{(N-1)^2}\left\{\sum_{i=1,j=1}^{N}\left\langle y_i^2 y_j^2\right\rangle - \frac{2}{N}\sum_{i=1,p=1,q=1}^{N}\left\langle y_i^2 y_p y_q\right\rangle + \frac{1}{N^2}\sum_{i=1,j=1,p=1,q=1}^{N}\langle y_i y_j y_p y_q\rangle\right\} - \sigma^4. \tag{13.67}$$

To compute the sums in the brackets we select the terms with the same indices. Taking into consideration the independence of the samples, we have

$$\sum_{i=1,j=1}^{N}\left\langle y_i^2 y_j^2\right\rangle = \sum_{i=1}^{N}\left\langle y_i^4\right\rangle + \sum_{\substack{i=1,j=1\\ i\neq j}}^{N}\left\langle y_i^2\right\rangle\left\langle y_j^2\right\rangle = N\mu_4 + N(N-1)\mu_2^2, \tag{13.68}$$

where

$$\mu_{\nu_i} = \left\langle y_i^{\nu}\right\rangle = \left\langle (x_i - E)^{\nu}\right\rangle \tag{13.69}$$

is the central moment of the νth order and, naturally, $\mu_2 = \sigma^2$. Analogously, we obtain

$$\sum_{i=1,p=1,q=1}^{N}\left\langle y_i^2 y_p y_q\right\rangle = N\mu_4 + N(N-1)\mu_2^2; \tag{13.70}$$

$$\sum_{i=1,j=1,p=1,q=1}^{N}\langle y_i y_j y_p y_q\rangle = N\mu_4 + 3N(N-1)\mu_2^2. \tag{13.71}$$

Substituting (13.70) and (13.71) into (13.67), we obtain

$$\mathrm{Var}\{\mathrm{Var}^*\} = \frac{\mu_4 - \{1-(2/(N-1))\}\sigma^2}{N}. \tag{13.72}$$

In the case of Gaussian stochastic process we have $\mu_4 = 3\sigma^4$. As a result we obtain

$$\mathrm{Var}\{\mathrm{Var}^*\} = \frac{2\sigma^4}{N-1}. \tag{13.73}$$

If the mathematical expectation is known, then the unbiased variance estimate is defined as

$$\mathrm{Var}^* = \frac{1}{N}\sum_{i=1}^{N}(x_i - E_0)^2, \tag{13.74}$$

and the variance of the variance estimate takes the following form:

$$\mathrm{Var}\{\mathrm{Var}^*\} = \frac{\mu_4 - \sigma^4}{N}. \tag{13.75}$$

In the case of Gaussian stochastic process, variation in the variance estimate has the following form:

$$\mathrm{Var}\{\mathrm{Var}^*\} = \frac{2\sigma^4}{N}. \tag{13.76}$$

Comparing (13.76) with (13.73), we can see that at $N \gg 1$ these formulas are coincided.

As applied to the iterative variance estimate of stochastic process with zero mathematical expectation, based on (12.342) and (12.345) we obtain

- In the case of discrete stochastic process

$$\mathrm{Var}^*[N] = \mathrm{Var}^*[N-1] + \gamma[N]\{x^2[N] - \mathrm{Var}^*[N-1]\}; \tag{13.77}$$

- In the case of continuous stochastic process

$$\frac{d\,\mathrm{Var}^*(t)}{dt} = \gamma(t)\{x^2(t) - \mathrm{Var}^*(t)\}. \tag{13.78}$$

As shown in Section 12.9, we can show that the optimal value of the factor $\gamma[N]$ is equal to N^{-1}.

13.3 ERRORS UNDER STOCHASTIC PROCESS VARIANCE ESTIMATE

As we can see from formulas for the variance estimate of stochastic process, a squaring of the stochastic process realization (or its samples) is a very essential operation. The device carrying out this operation is called the *quadratic transformer* or *quadrator*. The difference in quadrator performance from square-law function leads to additional errors arising under measurement of the stochastic process variance. To define a character of these errors we assume that the stochastic process possesses zero mathematical expectations and the variance estimate is carried out based on investigation of independent samples. The characteristic of transformation $y = g(x)$ can be presented by the polynomial function of the μth order

$$y = g(x) = \sum_{k=0}^{\mu} a_k x^k. \tag{13.79}$$

Substituting (13.79) into (13.59) instead of x_i and carrying out averaging, we obtain the mathematical expectation of variance estimate:

$$\langle \text{Var}^* \rangle = \sum_{k=0}^{\mu} a_k \langle x^k \rangle = a_0 + a_2\sigma^2 + \sum_{k=3}^{\mu} a_k \langle x^k \rangle. \tag{13.80}$$

The estimate bias caused by the difference between the transformation characteristic and the square-law function can be presented in the following form:

$$b\{\text{Var}^*\} = \sigma^4(a_2 - 1) + a_0 + \sum_{k=3}^{\mu} a_k \langle x^k \rangle. \tag{13.81}$$

For the given transformation characteristic the coefficients a_k can be defined before. For this reason, the bias of variance estimate caused by the coefficients a_0 and a_2 can be taken into consideration. The problem is to take into consideration the sum in (13.81) because this sum depends on the shape of pdf of the investigated stochastic process:

$$\langle x^k \rangle = \int_{-\infty}^{\infty} x^k f(x) dx. \tag{13.82}$$

In the case of Gaussian stochastic process, the moment of the kth order can be presented in the following form:

$$\langle x^k \rangle = \begin{cases} 1\cdot 3\cdot 5\cdots(k-1)\sigma^{0.5k}, & \text{if} \quad k \text{ is even;} \\ 0, & \text{if} \quad k \text{ is odd.} \end{cases} \tag{13.83}$$

Difference between the transformation characteristic and the square-law function can lead to high errors under definition of the stochastic process variance. Because of this, while defining the stochastic process variance we need to pay serious attention to transformation performance. Verification of the transformation performance is carried out, as a rule, by sending the harmonic signal of known amplitude at the measurer input.

While measuring the stochastic process variance, we can avoid the squaring operation. For this purpose, we need to consider two sign functions (12.208)

$$\begin{cases} \eta_1(t) = \text{sgn}[\xi(t) - \mu_1(t)] \\ \eta_2(t) = \text{sgn}[\xi(t) - \mu_2(t)] \end{cases}, \tag{13.84}$$

instead of the initial stochastic process $\xi(t)$ with zero mathematical expectation, where $\mu_1(t)$ and $\mu_2(t)$ are additional independent stationary stochastic processes with zero mathematical expectations and the same pdf given by (12.205). At the same time, the condition given by (12.206) is satisfied.

The functions $\eta_1(t)$ and $\eta_2(t)$ are the stationary stochastic processes with zero mathematical expectations. In doing so, if the fixed value $\xi(t) = x$ the conditional stochastic processes $\eta_1(t|x)$ and $\eta_2(t|x)$ are statistically independent, that is,

$$\langle \eta_1(t|x)\eta_2(t|x)\rangle = \langle \eta_1(t|x)\rangle\langle \eta_2(t|x)\rangle. \tag{13.85}$$

Taking into consideration (12.209), we obtain

$$\langle \eta_1(t|x)\eta_2(t|x)\rangle = \frac{x^2}{A^2}. \tag{13.86}$$

The unconditional mathematical expectation of product can be presented in the following form:

$$\langle \eta_1(t|x)\eta_2(t|x)\rangle = \frac{1}{A^2}\int_{-\infty}^{\infty} x^2 p(x)dx = \frac{\sigma^2}{A^2}. \tag{13.87}$$

Denote the realizations of stochastic processes $\eta_1(t)$ and $\eta_2(t)$ by the functions $y_1(t)$ and $y_2(t)$, respectively. Consequently, if we consider the following value as the variance estimate,

$$\widetilde{\text{Var}} = \frac{A^2}{T}\int_0^T y_1(t)y_2(t)dt, \tag{13.88}$$

then the variance estimate is unbiased.

The realizations $y_1(t)$ and $y_2(t)$ take the values equal to ±1. For this reason, the multiplication and integration in (13.88) can be replaced by individual integration of new unit functions obtained as a result of coincidence and noncoincidence of polarities of the realizations $y_1(t)$ and $y_2(t)$:

$$\int_0^T y_1(t)y_2(t)dt = \int_0^T z_1(t)dt - \int_0^T z_2(t)dt, \tag{13.89}$$

where

$$z_1(t) = \begin{cases} 1 \quad \text{if} \begin{cases} y_1(t) > 0, \; y_2(t) > 0; \\ y_1(t) < 0, \; y_2(t) < 0; \end{cases} \\ 0 \quad \text{otherwise}; \end{cases} \tag{13.90}$$

$$z_2(t) = \begin{cases} 1 \quad \text{if} \begin{cases} y_1(t) > 0, \; y_2(t) < 0; \\ y_1(t) < 0, \; y_2(t) > 0; \end{cases} \\ 0 \quad \text{otherwise}. \end{cases} \tag{13.91}$$

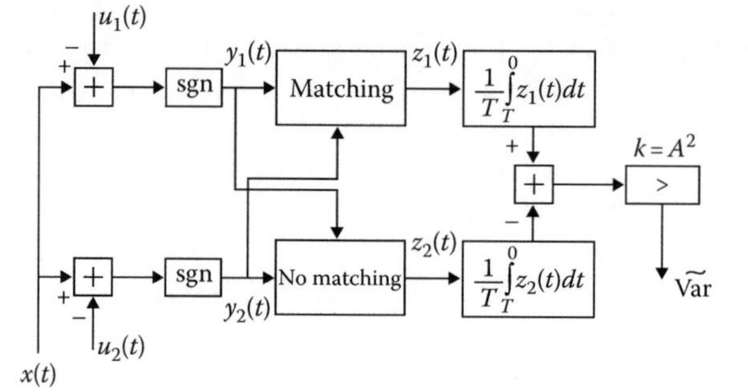

FIGURE 13.4 Measurer based on additional signals.

The flowchart of measuring device based on the implementation of additional signals is shown in Figure 13.4.

In the case of discrete process, the variance estimate can be presented in the following form:

$$\widetilde{\text{Var}} = \frac{A^2}{N}\sum_{i=1}^{N} y_{1_i} y_{2_i}, \tag{13.92}$$

where $y_{1_i} = y_1(t_i)$ and $y_{2_i} = y_2(t_i)$. In this case, the integrator can be replaced by the summator in Figure 13.4.

Determine the variance of the variance estimate $\widetilde{\text{Var}}$ applied to investigation of stochastic process at discrete instants for the purpose of simplifying further analysis that the samples y_{1i} and y_{2i} are independent. Otherwise, we should know the two-dimensional pdfs of additional uniformly distributed stochastic processes $\mu_1(t)$ and $\mu_2(t)$. The variance of the variance estimate can be presented in the following form:

$$\text{Var}\{\widetilde{\text{Var}}\} = \frac{A^4}{N^2}\sum_{i=1,j=1}^{N} \langle y_{1_i} y_{1_j} y_{2_i} y_{2_j}\rangle - \langle \widetilde{\text{Var}}\rangle^2. \tag{13.93}$$

In double sum, we can select the terms with $i = j$. Then

$$\sum_{i=1,j=1}^{N} \langle y_{1_i} y_{1_j} y_{2_i} y_{2_j}\rangle = \sum_{i=1}^{N} \langle (y_{1_i} y_{2_i})^2\rangle + \sum_{\substack{i=1,j=1\\ i\neq j}}^{N} \langle y_{1_i} y_{1_j} y_{2_i} y_{2_j}\rangle, \tag{13.94}$$

where

$$\langle y_{1_i} y_{1_j} y_{2_i} y_{2_j}\rangle = \langle \eta_{1_i}\eta_{1_j}\eta_{2_i}\eta_{2_j}\rangle. \tag{13.95}$$

Based on a definition of sign functions, we can obtain from (13.84) the following condition:

$$[y_1(t) y_2(t)]^2 = 1. \tag{13.96}$$

Define the cumulative moment $\langle \eta_{1_i}\eta_{1_j}\eta_{2_i}\eta_{2_j} | x_i x_j\rangle$ at the condition that $\xi(t_i) = x_i$ and $\xi(t_j) = x_j$. Taking into consideration the statistical independence of the samples η_1 and η_2, the statistical mutual independence of the stochastic values $\eta_1(t|x)$ and $\eta_2(t|x)$, and (12.209), the conditional cumulative moment can be presented in the following form:

$$\langle \eta_{1_i}\eta_{1_j}\eta_{2_i}\eta_{2_j}|x_i x_j\rangle = \frac{x_i^2 x_j^2}{A^4}. \tag{13.97}$$

Averaging (13.97) by possible values of independent random variables x_i and x_j, we obtain

$$\langle \eta_{1_i}\eta_{1_j}\eta_{2_i}\eta_{2_j}\rangle = \frac{\langle x_i^2 x_j^2\rangle}{A^4} = \frac{\langle x_i^2\rangle\langle x_j^2\rangle}{A^4} = \frac{\sigma^4}{A^4}. \tag{13.98}$$

Substituting (13.96) and (13.98) into (13.93), we obtain the variance of the variance estimate (13.92)

$$\mathrm{Var}\{\mathrm{Var}^{\bullet}\} = \frac{A^4}{N}\left(1 - \frac{\sigma^4}{A^4}\right). \tag{13.99}$$

The variance estimate according to (13.92) envisages that the condition (12.206) is satisfied. Consequently, the inequality $\sigma^2 \ll A^2$ must be satisfied. Because of this, as in the case of estimation of the mathematical expectation with employment of additional signals, the variance of the variance estimate $\widetilde{\mathrm{Var}}$ is completely defined by the half intervals of possible values of additional random sequences. Comparing the obtained variance (13.99) with the variance of the variance estimate in (13.76) by N independent samples, we can find that

$$\frac{\mathrm{Var}\{\widetilde{\mathrm{Var}}\}}{\mathrm{Var}\{\mathrm{Var}^*\}} = \frac{A^4}{2\sigma^4}\left(1 - \frac{\sigma^4}{A^4}\right). \tag{13.100}$$

Thus, we can conclude that the method of measurement of the stochastic process variance using the additional signals is characterized by the higher variance compared to the algorithm (13.74). This is based on the example of definition of the variance estimate of stochastic sample with the uniform pdf coinciding with (12.205), $\sigma^2 = A^2/3$. As a result, we have

$$\frac{\mathrm{Var}\{\widetilde{\mathrm{Var}}\}}{\mathrm{Var}\{\mathrm{Var}^*\}} = 4. \tag{13.101}$$

The methods that are used to measure the variance assume the absence of limitations of instantaneous values of the investigated stochastic process. The presence of these limitations leads to additional errors while measuring the variance. Determine the bias of variance estimate of the Gaussian stochastic process when there is a limiter of the type (12.151) presented in Figure 12.6 applied to zero mathematical expectation and the true value of the variance σ^2. The variance estimate is defined by (13.41). The ideal integrator $h(t) = T^{-1}$ plays a role in averaging or smoothing filter. Substituting the realization $y(t) = g[x(t)]$ into (13.41) instead of $x(t)$ and carrying out averaging of the obtained formula with respect to $x(t)$, the bias of variance estimate of Gaussian stochastic process can be written in the following form:

$$b\{\mathrm{Var}^*\} = \int_{-\infty}^{\infty} g^2(x)p(x)dx - \sigma^2 = -2\sigma^2\left\{(1-\gamma^2)Q(\gamma) - \frac{\gamma}{\sqrt{2\pi}}\exp\{-0.5\gamma^2\}\right\}, \tag{13.102}$$

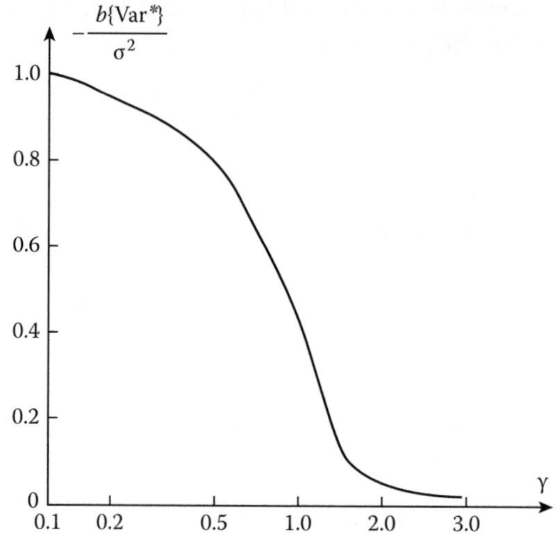

FIGURE 13.5 Bias of variance estimate of Gaussian stochastic process as a function of the normalized limitation level.

where

$$\gamma = \frac{a}{\sigma} \tag{13.103}$$

is the normalized limitation level. At $\gamma \gg 1$, we obtain that the bias of variance estimate tends to approach zero, which is expected. The bias of variance estimate of Gaussian stochastic process as a function of the normalized limitation level is shown in Figure 13.5.

13.4 ESTIMATE OF TIME-VARYING STOCHASTIC PROCESS VARIANCE

To define the current value of nonstationary stochastic process variance there is a need to have a set of realizations $x_i(t)$ of this process. Then the variance estimate of stochastic process at the instant t_0 can be presented in the following form:

$$\text{Var}^*(t_0) = \frac{1}{N}\sum_{i=1}^{N}[x_i(t_0) - E(t_0)]^2, \tag{13.104}$$

where
- N is the number of realizations $x_i(t)$ of stochastic process
- $E(t_0)$ is the mathematical expectation of stochastic process at the instant t_0

As we can see from (13.104), the variance estimate is unbiased and the variance of the variance estimate under observation of N independent realizations can be presented in the following form:

$$\text{Var}\{\text{Var}^*(t_0)\} = \frac{\text{Var}^2(t_0)}{N}; \tag{13.105}$$

that is, the variance estimate according to (13.104) is the consistent estimate.

In most cases, the researcher defines the variance estimate on the basis of investigation of a single realization of stochastic process. While estimating time-varying variance based on a single realization of time-varying stochastic process, there are similar problems as seen with the estimation of the time-varying mathematical expectation. On the one hand, to decrease the variance estimate caused by the finite observation time interval, the last must be as soon as large. On the other hand, we need to choose the integration time as short as possible for the best definition of variations in variance. Evidently, there must be a compromise.

The simplest way to define the time-varying stochastic process variance at the instant t_0 is averaging the transformed input data of stochastic process within the limits of finite time interval. Thus, let $x(t)$ be the realization of stochastic process $\xi(t)$ with zero mathematical expectation. Measurement of variance of this stochastic process at the instant t_0 is carried out by averaging the quadrature ordinates $x(t)$ within the limits of the interval about the given value of argument $[t_0 - 0.5T, t_0 + 0.5T]$. In this case, the variance estimate takes the following form:

$$\text{Var}^*(t_0,T) = \frac{1}{T}\int_{t_0-0.5T}^{t_0+0.5T} x^2(t)dt = \frac{1}{T}\int_{-0.5T}^{0.5T} x^2(t+t_0)dt. \tag{13.106}$$

Averaging the variance estimate by realizations, we obtain

$$\langle \text{Var}^*(t_0,T)\rangle = \frac{1}{T}\int_{-0.5T}^{0.5T} \text{Var}(t+t_0)dt. \tag{13.107}$$

Thus, as in the case of time-varying mathematical expectation, the estimate of the variance of the time-varying mathematical expectation does not coincide with its true value in a general case, but it is obtained by smoothing the variance within the limits of finite time interval $[t_0 - 0.5T, t_0 + 0.5T]$. As a result of such averaging, the bias of variance of stochastic process can be presented in the following form:

$$b\{\text{Var}^*(t_0,T)\} = \frac{1}{T}\int_{-0.5T}^{0.5T} [\text{Var}(t+t_0) - \text{Var}(t_0)]dt. \tag{13.108}$$

If we would like to have the unbiased estimate, a variance of the variance $\text{Var}(t)$ within the limits of time interval $[t_0 - 0.5T, t_0 + 0.5T]$ must be the odd function and, in the simplest case, the linear function of time

$$\text{Var}(t_0+t) \approx \text{Var}(t_0) + \text{Var}'(t_0)t, \quad -0.5T \le t \le 0.5T. \tag{13.109}$$

Evaluate the influence of deviation of the current variance $\text{Var}(t)$ from linear function on its estimate. In doing so, we assume that the estimated current variance has the continuous first and second derivatives with respect to time. Then, according to the Taylor expansion we can write

$$\text{Var}(t) = \text{Var}(t_0) + (t-t_0)\text{Var}'(t) + 0.5(t-t_0)^2\text{Var}''[t_0 + \theta(t-t_0)], \tag{13.110}$$

where $0 < \theta < 1$. Substituting (13.110) into (13.108), we obtain

$$b\{\text{Var}^*(t_0,T)\} = \frac{1}{2T}\int_{-0.5T}^{0.5T} t^2\text{Var}''(t_0 + t\theta)dt. \tag{13.111}$$

Denoting M_{Var} the maximal absolute value of the second derivative of the current variance Var(t) with respect to time, we obtain the high bound value of the variance estimate bias by absolute value

$$\left|b\{\text{Var}^*(t_0,T)\}\right| \le \frac{T^2 M_{\text{Var}}}{24}. \tag{13.112}$$

The maximum value of the second derivative of the time-varying variance Var(t) can be evaluated, as a rule, based on the analysis of specific physical problems.

To estimate the difference in the time-varying variance a square of the investigated realization of the stochastic process can be presented in the following form of two sums:

$$\xi^2(t) = \text{Var}(t) + \zeta(t), \tag{13.113}$$

where

- Var(t) is the true value of the stochastic process variance at the instant t
- $\zeta(t)$ are the fluctuations of square of the stochastic process realization with respect to its mathematical expectation at the same instant t

Then the variation in the variance estimate of the investigated nonstationary stochastic process can be presented in the following form:

$$\text{Var}\{\text{Var}^*(t_0,T)\} = \frac{1}{T^2}\int_{-0.5T}^{0.5T}\int_{-0.5T}^{0.5T} R_\zeta(t_1 + t_0, t_2 + t_0)dt_1 dt_2, \tag{13.114}$$

where

$$R_\zeta(t_1,t_2) = \left\langle\left[x^2(t_1) - \text{Var}(t_1)\right]\left[x^2(t_2) - \text{Var}(t_2)\right]\right\rangle \tag{13.115}$$

is the correlation function of the random component of the investigated squared stochastic process with respect to its variance at the instant t.

To define the approximate value of the variance of slowly varying in time variance estimate we can assume that the centralized stochastic process $\zeta(t)$ within the limits of the finite time interval $[t_0 - 0.5T, t_0 + 0.5T]$ is the stationary stochastic process with the correlation function defined as

$$R_\zeta(t, t+\tau) \approx \text{Var}_\zeta(t_0)\mathcal{R}_\zeta(\tau). \tag{13.116}$$

Taking into consideration the approximation given by (13.116), the variance of the variance estimate presented in (13.114) can be presented in the following form:

$$\mathrm{Var}\{\mathrm{Var}^*(t_0,T)\} = \mathrm{Var}_\zeta(t_0)\frac{2}{T}\int_0^T\left(1-\frac{\tau}{T}\right)\mathcal{R}_\zeta(\tau)d\tau. \tag{13.117}$$

As applied to the Gaussian stochastic process at the given assumptions, we can write

$$R_\zeta(t,t+\tau) \approx 2\sigma^4(t_0)\mathcal{R}^2(\tau), \tag{13.118}$$

and the variance of the variance estimate at the instant t_0 takes the following form:

$$\mathrm{Var}\{\mathrm{Var}^*(t_0,T)\} = \frac{4\sigma^4(t_0)}{T}\int_0^T\left(1-\frac{\tau}{T}\right)\mathcal{R}^2(\tau)d\tau. \tag{13.119}$$

Consider the characteristics of the estimate of the time-varying stochastic process variance for the case, when the investigated stochastic process variance can be presented in the series expansion form

$$\mathrm{Var}(t) \approx \sum_{i=1}^N \beta_i\psi_i(t), \tag{13.120}$$

where

- β_i are some unknown numbers
- $\psi_i(t)$ are the given functions of time

With increase in the number of terms in series given by (13.120), the approximation errors can decrease to an infinitesimal value. Similar to (12.258), we can conclude that the coefficients β_i are defined by the system of linear equations

$$\sum_{i=1}^N \beta_i\int_0^T \psi_i(t)\psi_j(t)dt = \int_0^T \mathrm{Var}(t)\psi_j(t)dt, \quad j=1,2,\ldots,N \tag{13.121}$$

based on the condition of minimum of quadratic error of approximation (13.120). Thus, the problem with definition of the stochastic process variance estimate by a single realization observed within the limits of the interval [0, T] can be reduced to estimation of the coefficients β_i in the series given by (13.120). In doing so, the bias and the variance of the variance estimate of the investigated stochastic process caused by errors occurring while measuring the coefficients β_i take the following form:

$$b\{\mathrm{Var}^*(t)\} = \mathrm{Var}(t) - \mathrm{Var}^*(t) = \sum_{i=1}^N \psi_i(t)[\beta_i - \langle\beta_i^*\rangle], \tag{13.122}$$

$$\mathrm{Var}\{\mathrm{Var}^*(t)\} = \sum_{i=1,j=1}^N \psi_i(t)\psi_j(t)\langle(\beta_i-\beta_i^*)(\beta_j-\beta_j^*)\rangle. \tag{13.123}$$

The bias and the variance of the variance estimate averaged within the limits of the time interval of observation of stochastic process can be presented in the following form, respectively:

$$b\{\mathrm{Var}^*(t)\} = \frac{1}{T}\sum_{i=1}^{N}\left[\beta_i - \langle\beta_i^*\rangle\right]\int_0^T \psi_i(t)dt, \tag{13.124}$$

$$\mathrm{Var}\{\mathrm{Var}^*(t)\} = \frac{1}{T}\sum_{i=1,j=1}^{N}\left\langle\left(\beta_i - \beta_i^*\right)\left(\beta_j - \beta_j^*\right)\right\rangle\int_0^T \psi_i(t)\psi_j(t)dt. \tag{13.125}$$

The values minimizing the function

$$\varepsilon^2(\beta_1,\beta_2,\ldots,\beta_N) = \frac{1}{T}\int_0^T\left\{x^2(t) - \sum_{i=1}^{N}\beta_i\psi_i(t)\right\}^2 dt \tag{13.126}$$

can be considered as estimations of the coefficients β_i. As we can see from (13.126), the estimation of the coefficients β_i is possible if we have a priori information that Var(t) and $\psi_i(t)$ are slowly varying in time functions compared to the averaged velocity of component

$$z(t) = x^2(t) - \mathrm{Var}(t). \tag{13.127}$$

The function $z(t)$ is the realization of stochastic process $\zeta(t)$ and possesses zero mathematical expectation. In other words, (13.126) is true for the definition of coefficients β_i if the frequency band Δf_{Var} of the time-varying variance is lower than the effective spectrum bandwidth of the investigated stochastic process $\zeta(t)$. This fact corresponds to the case when the correlation function of the investigated stochastic process can be written in the following form:

$$R(t,t+\tau) \approx \mathrm{Var}(t)\mathcal{R}(t). \tag{13.128}$$

The following stochastic process corresponds to this correlation function:

$$\xi(t) = a(t)\eta(t), \tag{13.129}$$

where

$\eta(t)$ is the stationary stochastic process
$a(t)$ is the slow time-varying deterministic function time in comparison with the function $\eta(t)$

Based on the condition of minimum of the function ε^2, that is,

$$\frac{\partial\varepsilon^2}{\partial\beta_m} = 0, \tag{13.130}$$

we obtain the system of equations to estimate the coefficients β_m

$$\sum_{i=1}^{N}\beta_i^*\int_0^T \psi_i(t)\psi_m(t)dt = \int_0^T x^2(t)\psi_m(t)dt, \quad m = 1,2,\ldots,N. \tag{13.131}$$

Denote

$$\int_0^T \psi_i(t)\psi_m(t)dt = c_{im}; \tag{13.132}$$

$$\int_0^T x^2(t)\psi_m(t)dt = y_m. \tag{13.133}$$

Then

$$\beta_m^* = \frac{1}{A}\sum_{i=1}^{N} A_{im}y_i = \frac{A_m}{A}, \quad m = 1,2,\ldots,N, \tag{13.134}$$

where the determinant A of linear equation system (13.131) and the determinant A_p are defined in accordance with (12.270) and (12.271).

Flowchart of measurer of the time-varying variance Var(t) of the investigated stochastic process is shown in Figure 13.6. The measurer operates in the following way. The coefficients c_{ij} are generated by the generator "Gen$_c$" in accordance with (13.132) based on the functions $\psi_i(t)$ that are issued by the generator "Gen$_\psi$." Microprocessor system solves the system of linear equations with respect to the estimations β_i^* of coefficients β_i based on the coefficients c_{ij} and values y_i obtained according to (13.133). The estimate of time-varying variance Var*(t) is formed based on previously obtained data. The variance estimate has a delay $T + \Delta T$ with respect to the true value, and this delay is used to compute the values y_i and to solve the system of N linear equations, respectively. The delay blocks denoted as T and ΔT are used for this purpose. Definition of coefficient estimations β_i^* is essentially simplified if the functions $\psi_i(t)$ are orthonormalized:

$$\int_0^T \psi_i(t)\psi_j(t)dt = \begin{cases} 1 & \text{if} \quad i = j, \\ 0 & \text{if} \quad i \neq j. \end{cases} \tag{13.135}$$

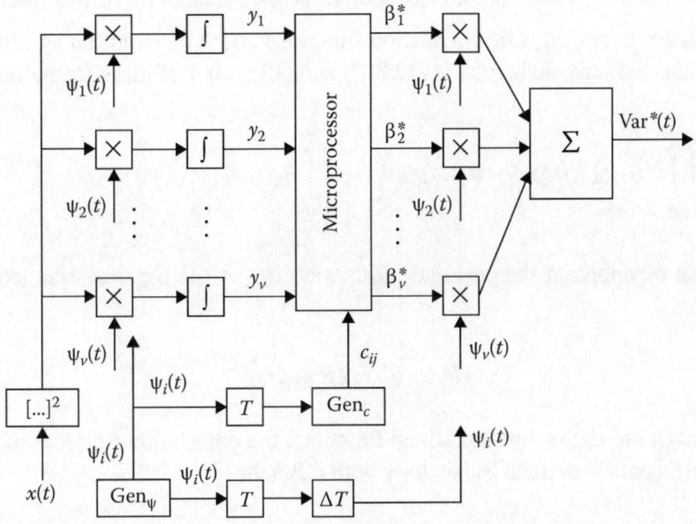

FIGURE 13.6 Measurer of variance in time.

In this case, the series coefficient estimations are defined by a simple integration of the quadratic realization of stochastic process with the corresponding weight function $\psi_m(t)$:

$$\beta_m^* = \int_0^T x^2(t)\psi_m(t)dt. \tag{13.136}$$

In doing so, the flowchart presented in Figure 13.6 is essentially simplified because there is no need to generate the coefficients c_{ij} and to solve the system of linear equations. The number of delay block is decreased too.

Determine the bias and mutual correlation functions between the estimates β_m^* and β_q^*. Taking into consideration (13.127) we can rewrite (13.133) in the following form:

$$y_i = \int_0^T z(t)\psi_i(t)dt + \int_0^T \text{Var}(t)\psi_i(t)dt = g_i + \beta_m c_{im} + \sum_{q=1,q\neq m} \beta_q c_{iq}, \tag{13.137}$$

where

$$g_i = \int_0^T z(t)\psi_i(t)dt. \tag{13.138}$$

As in Section 12.7, we can write the estimate β_m^* of the coefficients in the following form:

$$\beta_m^* = \frac{1}{A}\sum_{i=1}^{N} g_i A_{im} + \beta_m, \quad m = 1,2,\ldots,N, \tag{13.139}$$

where A_{im} are the algebraic complement of the determinant $A_m \equiv B_m$ (12.276) in which c_{ip} and g_i are given by (13.132) and (13.138).

Since $\langle\zeta(t)\rangle = 0$, then $\langle g_i\rangle = 0$ and, consequently, the estimations β_m^* of the coefficients in series given by (13.120) are unbiased. The correlation functions $R\left(\beta_m^*, \beta_q^*\right)$ and the variance $\text{Var}\left(\beta_m^*\right)$ are defined by formulas that are analogous to (12.333) and (12.334). For these formulas, we have

$$B_{ij} = \int_0^T\int_0^T \langle z(t_1)z(t_2)\rangle\, \psi_i(t_1)\psi_j(t_2)dt_1dt_2 = \int_0^T\int_0^T R_\zeta(t_1,t_2)\psi_i(t_1)\psi_j(t_2)dt_1dt_2. \tag{13.140}$$

We can show that by applying the correlation function $R(t_1, t_2)$ to the Gaussian stochastic process we have

$$R_\zeta(t_1,t_2) = 2R^2(t_1,t_2). \tag{13.141}$$

If the functions $\psi_i(t)$ are the orthonormalized functions the correlation function of the estimations β_m^* and β_q^* of coefficients is defined by analogy with (12.338):

$$R\left(\beta_m^*, \beta_q^*\right) = B_{mq}. \tag{13.142}$$

At that time, the formulas for the current and averaged variance of the time-varying variance estimate can be presented as in (12.339) and (12.240)

$$\mathrm{Var}\{\mathrm{Var}^*(t)\} = \sum_{i=1,j=1}^{N} B_{ij}\psi_i(t)\psi_j(t), \tag{13.143}$$

$$\mathrm{Var}\{\mathrm{Var}^*\} = \frac{1}{T}\sum_{i=1}^{N} B_{ij}. \tag{13.144}$$

If in addition to the condition of orthonormalization of the functions $\psi_i(t)$, we assume that the frequency band of time-varying variance is much lower than the effective spectrum bandwidth of the investigated stochastic process $\xi(t)$, the correlation function of the centralized stochastic process $\zeta(t)$ can be presented in the following form:

$$R_\zeta(t_1,t_2) \approx \mathrm{Var}(t_1)\delta(t_1 - t_2). \tag{13.145}$$

Then, in the case of arbitrary functions $\psi_i(t)$, based on (13.140), we obtain

$$B_{ij} = \int_0^T \mathrm{Var}(t)\psi_i(t)\psi_j(t)dt. \tag{13.146}$$

If the functions $\psi_i(t)$ are the orthonormalized functions satisfying the Fredholm equation of the second type

$$\psi_i(t) = \lambda_i \int_0^T R_\zeta(t_1,t_2)\psi_i(\tau)d\tau, \tag{13.147}$$

we obtain from (13.140) that

$$B_{ij} = \begin{cases} \dfrac{1}{\lambda_i} & \text{if } i = j, \\ 0 & \text{if } i \neq j. \end{cases} \tag{13.148}$$

In doing so, the variance of estimations β_m^* of the coefficients and the current and averaged variations in the variance estimate can be presented in the following form:

$$\mathrm{Var}\{\beta_m^*\} = \frac{1}{\lambda_m}, \tag{13.149}$$

$$\mathrm{Var}\{\mathrm{Var}^*(t)\} = \sum_{i=1}^{N} \frac{\psi_i^2(t)}{\lambda_i}, \tag{13.150}$$

$$\mathrm{Var}\{\mathrm{Var}^*\} = \frac{1}{T}\sum_{i=1}^{N} \frac{1}{\lambda_i}. \tag{13.151}$$

As we can see from (13.151), at the fixed time interval T, with increasing number of terms under approximation of the variance Var(t) by the series given by (13.120) the averaged variance of time-varying variance estimate increases, too.

13.5 MEASUREMENT OF STOCHASTIC PROCESS VARIANCE IN NOISE

The specific receivers called the radiometer-type receiver are widely used to measure the variance or power of weak noise signals [4]. Minimal increment in the variance of the stochastic signal is defined by the threshold of sensitivity. In doing so, the threshold of sensitivity of methods that measure the stochastic process variance is called the value of the investigated stochastic process variance equal to the root-mean-square deviation of measurement results. The threshold of sensitivity of radiometer depends on many factors, including the intrinsic noise and random parameters of receiver and the finite time interval of observation of input stochastic process.

Radiometers are classified into four groups by a procedure to measure the stochastic process variance: compensation method, method of comparison with variance of reference source, correlation method, and modulation method. Discuss briefly a procedure to measure the variance of stochastic process by each method.

13.5.1 Compensation Method of Variance Measurement

Flowchart of compensation method to measure the stochastic process variance is shown in Figure 13.7. While carrying out the compensation method, the additive mixture of the realization $s(t)$ of the investigated stochastic process $\zeta(t)$ and the realization $n(t)$ of the noise $\varsigma(t)$ is amplified and squared. The component $z(t)$ that is proportional to the variance of total signal is selected by the smoothing low-pass filter (or integrator). In doing so, the constant component z_{const} formed by the intrinsic noise of amplifier is essentially compensated by constant bias of voltage or current. There is a need to note that under amplifier we understand the amplifiers of radio and intermediate frequencies.

Under practical realization of compensation method to measure the stochastic process variance, the squarer, low-pass filter, and compensating device are often realized as a single device based on the lattice network and the squarer transformer is included in one branch of this lattice network. Branches of lattice network are chosen in such a way that a diagonal indicator could present zero voltage, for example, when the investigated stochastic process is absent. Presence of slow variations in the variance of intrinsic receiver noise and random variations of amplifier coefficient leads to imbalance in the compensation condition that, naturally, decreases the sensitivity of compensation method to measure the stochastic process variance. We consider the low-pass filter as an ideal integrator with the integration time equal to T for obtaining accurate results.

Define the dependence of the variation in the stochastic process variance estimate on the main characteristics of investigated stochastic process at the measurer input by the compensation method. We assume that the measurement errors of compensating device are absent with the compensation of the amplifier noise constant component and the spectral densities of the

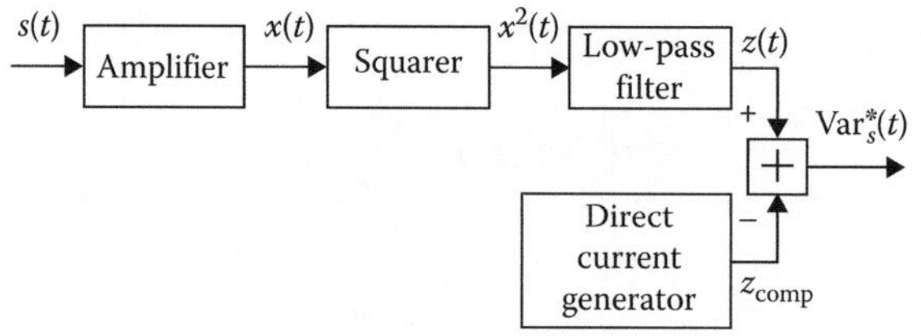

FIGURE 13.7 Compensation method.

investigated and measured stochastic process $\zeta(t)$ and receiver noise $\varsigma(t)$ are distributed uniformly within the limits of amplifier bandwidth.

The realization $x(t)$ of stochastic process $\xi(t)$ at the squarer input can be presented in the following form:

$$x(t) = [1 + \upsilon(t)]\,[s(t) + n(t)] \tag{13.152}$$

accurate within the constant coefficient characterizing the average value of amplifier coefficient, where $\upsilon(t)$ is the realization of random variations of the receiver amplifier coefficient $\beta(t)$. Since the amplifier coefficient is a positive characteristic, the pdf of the process $1 + \beta(t)$ must be approximated by the positive function, too. In practice, measurement of weak signals is carried out, as a rule, under the condition of small value of the variance Var_β of variations of the amplifier coefficient compared to its mathematical expectation that is equal to unit in our case. In other words, the condition $\mathrm{Var}_\beta \ll 1$ must be satisfied.

Taking into consideration the foregoing statements and to simplify analysis, we assume that all stochastic processes $\zeta(t)$, $\varsigma(t)$, and $\beta(t)$ are the stationary Gaussian stochastic processes with zero mathematical expectation and correlation function defined as

$$\langle \zeta(t_1)\zeta(t_2)\rangle = R_s(t_2 - t_1) = \mathrm{Var}_s \mathcal{R}_s(t_2 - t_1) = \mathrm{Var}_s r_s(t_2 - t_1)\cos[\omega_0(t_2 - t_1)]; \tag{13.153}$$

$$\langle \varsigma(t_1)\varsigma(t_2)\rangle = R_n(t_2 - t_1) = \mathrm{Var}_n \mathcal{R}_n(t_2 - t_1) = \mathrm{Var}_n r_n(t_2 - t_1)\cos[\omega_0(t_2 - t_1)]; \tag{13.154}$$

$$\langle \beta(t_1)\beta(t_2)\rangle = R_\beta(t_2 - t_1) = \mathrm{Var}_\beta \mathcal{R}_\beta(t_2 - t_1) = \mathrm{Var}_\beta r_\beta(t_2 - t_1). \tag{13.155}$$

We assume that the measured stochastic process and receiver noise are the narrow-band stochastic processes. The stochastic process $\beta(t)$ characterizing the random variations of the amplifier coefficient is the low-frequency stochastic process. In addition, we consider situations when the stochastic processes $\zeta(t)$, $\varsigma(t)$, and $\beta(t)$ are mutually independent.

Realization of stochastic process at the ideal integrator output can be presented in the following form:

$$z(t) = \frac{1}{T}\int_{T-t}^{t} x^2(t)dt. \tag{13.156}$$

The variance estimate of the investigated stochastic process after cancellation of amplifier noise takes the following form:

$$\mathrm{Var}_s^*(t) = z(t) - z_{\mathrm{const}}. \tag{13.157}$$

To define the sensitivity of compensation procedure we need to determine the mathematical expectation and the variance of estimate $z(t)$. We can obtain that

$$E_z = (1 + \mathrm{Var}_\beta)(\mathrm{Var}_s + \mathrm{Var}_n). \tag{13.158}$$

After cancellation of variance of the amplifier noise $(1 + \mathrm{Var}_\beta)\mathrm{Var}_n$, the mathematical expectation of the output signal with accuracy within the coefficient $1 + \mathrm{Var}_\beta$ corresponds to the true value of variance of the observed stochastic process. Thus, in the case of random variations of the amplification coefficient, the variance possesses the following bias:

$$b\{\mathrm{Var}_s^*\} = \mathrm{Var}_\beta \mathrm{Var}_s. \tag{13.159}$$

Determine the variance of the variance estimate that limits the sensitivity of compensation procedure to measure a variance of the investigated stochastic process. Given that the considered stochastic processes are stationary, we can change the integration limits in (13.156) from 0 to T. In this case, the variance of the variance estimate takes the following form:

$$\mathrm{Var}\{\mathrm{Var}_s^*\} = \frac{2}{T^2}(1+q)^2 \mathrm{Var}_s^2 \int_0^T\int_0^T \{[1+2\mathcal{R}^2(t_2,t_1)]\left[R_\beta^2(t_2,t_1)+2R_\beta(t_2,t_1)\right]$$
$$+(1+\mathrm{Var}_\beta)^2\mathcal{R}^2(t_2,t_1)\}dt_1dt_2, \tag{13.160}$$

where

$$q = \frac{\mathrm{Var}_n}{\mathrm{Var}_s} \tag{13.161}$$

is the ratio between the amplifier noise and the noise of investigated stochastic process within the limits of amplifier bandwidth. Double integral in (13.160) can be transformed into a single integral by introducing new variables and changing the order of integration. Taking into consideration the condition $\mathrm{Var}_\beta \ll 1$ and neglecting the integrals with double frequency $2\omega_0$, we obtain

$$\mathrm{Var}\{\mathrm{Var}_s^*\} = \frac{2(1+q)^2\mathrm{Var}_s^2}{T}\int_0^T\left(1-\frac{\tau}{T}\right)[r^2(\tau)+4r_\beta(\tau)\mathrm{Var}_\beta]d\tau. \tag{13.162}$$

The time interval of observation corresponding to the sensitivity threshold is determined as

$$D\{\mathrm{Var}_s^*\} = b^2\{\mathrm{Var}_s^*\} + \mathrm{Var}\{\mathrm{Var}_s^*\} = \mathrm{Var}_s^2. \tag{13.163}$$

As applied to the compensation procedure of measurement, we obtain

$$\mathrm{Var}_\beta^2 + \frac{2(1+q)^2}{T}\int_0^T\left(1-\frac{\tau}{T}\right)[r^2(\tau)+4r_\beta(\tau)\mathrm{Var}_\beta]d\tau = 1. \tag{13.164}$$

As an example, the stochastic processes with exponential normalized correlation functions can be considered:

$$\begin{cases} r(\tau) = \exp\{-\alpha|\tau|\}; \\ r_\beta(\tau) = \exp\{-\gamma|\tau|\}, \end{cases} \tag{13.165}$$

where α and γ are the characteristics of effective spectrum bandwidth of their corresponding stochastic processes. As a result, we have

$$\mathrm{Var}\left\{\mathrm{Var}_s^*\right\} = 2\mathrm{Var}_s^2(1+q)^2\left\{\frac{2\alpha T - 1 + \exp\{-2\alpha T\}}{4\alpha^2 T^2} + 4\mathrm{Var}_\beta \frac{\gamma T - 1 + \exp\{-\gamma T\}}{\gamma^2 T^2}\right\}. \tag{13.166}$$

The obtained general and particular formulas for the variation in the variance estimate of stochastic process are essentially simplified in practice since the time interval of observation is much more than the correlation interval of stochastic processes $\zeta(t)$ and $\varsigma(t)$. In other words, the inequalities $\alpha T \gg 1$ and $\alpha \gg \gamma$ are satisfied. In this case, (13.162) and (13.166) take the following form, correspondingly

$$\mathrm{Var}\left\{\mathrm{Var}_s^*\right\} = \frac{2(1+q)^2\mathrm{Var}_s^2}{T}\left\{\int_0^\infty r^2(\tau)d\tau + 4\mathrm{Var}_\beta \int_0^T \left(1 - \frac{\tau}{T}\right) r_\beta(\tau)d\tau\right\}, \tag{13.167}$$

$$\mathrm{Var}\left\{\mathrm{Var}_s^*\right\} = \frac{(1+q)^2\mathrm{Var}_s^2}{\alpha T}\left\{1 + 8\frac{\alpha}{\gamma}\mathrm{Var}_\beta \frac{\gamma T - 1 + \exp\{-\gamma T\}}{\gamma T}\right\}. \tag{13.168}$$

When the random variations of the amplifier coefficient are absent ($\mathrm{Var}_\beta = 0$) the variance of the variance estimate can be presented in the following form:

$$\mathrm{Var}\left\{\mathrm{Var}_s^*\right\} = \frac{(1+q)^2}{\alpha T}\mathrm{Var}_s^2. \tag{13.169}$$

Consequently, when there are random variations of the amplification coefficient, there is an increase in the variance of the variance estimate of the investigated stochastic process on the value

$$\Delta\mathrm{Var}\left\{\mathrm{Var}_s^*\right\} = \frac{8(1+q)^2\mathrm{Var}_s^2\mathrm{Var}_\beta}{(\gamma T)^2}[\gamma T - 1 + \exp\{-\gamma T\}]. \tag{13.170}$$

As we can see from (13.170), with an increase in average time (the parameter γT) the additional random errors decrease correspondingly, and in the limiting case at $\gamma T \gg 1$ we have

$$\Delta\mathrm{Var}\left\{\mathrm{Var}_s^*\right\} = \frac{8(1+q)^2\mathrm{Var}_s^2\mathrm{Var}_\beta}{\gamma T}. \tag{13.171}$$

Let T_0 be the time required to measure the variance of investigated stochastic process with the given root-mean-square deviation if random variations of the amplification coefficient are absent. Then, T_β is the time required to measure the variance of investigated stochastic process with the given root-mean-square deviation if random variations of the amplification coefficient are present and is given by

$$\frac{1}{T_0} = \left[1 + 8\frac{\alpha}{\gamma}\mathrm{Var}_\beta \frac{\gamma T_\beta - 1 + \exp\{-\gamma T_\beta\}}{\gamma T_\beta}\right]\frac{1}{T_\beta}. \tag{13.172}$$

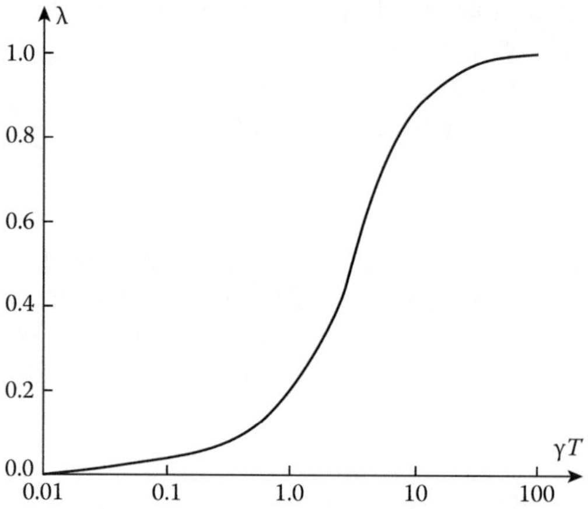

FIGURE 13.8 Relative increase of the variance of the variance estimate as a function of the parameter $\gamma T = 1$ at $8(\alpha/\gamma)\text{Var}_\beta = 1$.

The relative increase in the variance of variance estimate

$$\lambda = \frac{\Delta\text{Var}\{\text{Var}_s^*\}}{\text{Var}\{\text{Var}_s^*\}} = 8\frac{\alpha}{\gamma}\text{Var}_\beta \frac{\gamma T - 1 + \exp\{-\gamma T\}}{\gamma T} \tag{13.173}$$

as a function of the parameter γT when $8(\alpha/\gamma)\text{Var}_\beta = 1$ is shown in Figure 13.8. Since $\text{Var}_\beta \ll 1$, this case corresponds to the condition $(\alpha/\gamma) \gg 1$; that is, the spectrum bandwidth of the investigated stochastic processes is much more than the spectrum bandwidth of variations in amplification coefficient. Formula (13.173) is simplified for two limiting cases $\gamma T \ll 1$ and $\gamma T \gg 1$. At $\gamma T \ll 1$; in this case, the correlation interval of the amplification coefficient is greater than the time interval of observation of stochastic process, that is,

$$\lambda \approx 4\alpha T\, \text{Var}_\beta. \tag{13.174}$$

In the opposite case, that is, $\gamma T \gg 1$, we have

$$\lambda = 8\frac{\alpha}{\gamma}\text{Var}_\beta. \tag{13.175}$$

As we can see from Figure 13.8, at definite conditions the random variations in the amplification coefficient increase essentially the variance of the variance estimate of stochastic process; that is, the radiometer sensitivity is decreased.

Formula (13.169) allows us to obtain a value of the time interval of observation corresponding to the sensitivity threshold at $\text{Var}_\beta = 0$. This time can be defined as

$$T = \frac{(1+q)^2}{\alpha}. \tag{13.176}$$

As we can see from (13.176), the time interval of observation essentially increases with an increase in the variance of amplifier noise. The amplifier intrinsic noise can be presented in the form of product between the independent stationary Gaussian stochastic processes, namely, the narrow-band stochastic process $\chi(t)$ and the low-frequency stochastic process $\theta(t)$ with zero mathematical expectations. In doing so, we assume that the stochastic process $\theta(t)$ possesses the unit variance, that is, $\langle\theta^2(t)\rangle = 1$, and the correlation functions can be presented in the following form:

$$\langle\chi(t_1)\chi(t_2)\rangle = \text{Var}_n \exp\{-\alpha|\tau|\}\cos\omega_0\tau; \tag{13.177}$$

$$\langle\theta(t_1)\theta(t_2)\rangle = \exp\{-\eta|\tau|\}, \quad \tau = t_2 - t_1. \tag{13.178}$$

Moreover, we assume that $\alpha T \gg 1$, $\alpha \gg \eta$, $\alpha \gg \gamma$. In this case, the variation in the variance estimate or the variance of the signal at the ideal integrator output can be presented in the following form:

$$\text{Var}\{\text{Var}_s^*\} = \frac{(1+q)^2\text{Var}_s^2}{\alpha T}\left\{1 + 8\frac{\alpha}{\gamma}\text{Var}_\beta \frac{\gamma T - 1 + \exp\{-\gamma T\}}{\gamma T}\right\}$$
$$+ \text{Var}_n^2\left\{\frac{2\eta T - 1 + \exp\{-2\eta T\}}{\eta^2 T^2} + 16\text{Var}_\beta \frac{(2\eta+\gamma)T - 1 + \exp\{-(2\eta+\gamma)T\}}{(2\eta+\gamma)^2 T^2}\right\}. \tag{13.179}$$

Comparing (13.179) with (13.168), we see that owing to the low-frequency fluctuations of amplifier intrinsic noise the variance of the variance estimate increases as a result of the value yielded by the second term of (13.179).

13.5.2 Method of Comparison

This method is based on a comparison of constant signal components formed at the output of two-channel amplifier by the investigated stochastic process and the generator signal coming in at the amplifier input. Moreover, the generator signal is calibrated by power or variance. The wideband Gaussian stochastic process with known power spectral density or the deterministic harmonic signal can be employed as the calibrated generator signal. Flowchart of stochastic process variance measurer employing a comparison of signals at the low-pass filter output is presented in Figure 13.9.

Assuming that the receiver channels are identical and independent, the signal at the squarer inputs can be presented in the following form:

$$\begin{cases} x_1(t) = [1 + \upsilon_1(t)]\,[s(t) + n_1(t)]; \\ x_2(t) = [1 + \upsilon_2(t)]\,[s_0(t) + n_2(t)], \end{cases} \tag{13.180}$$

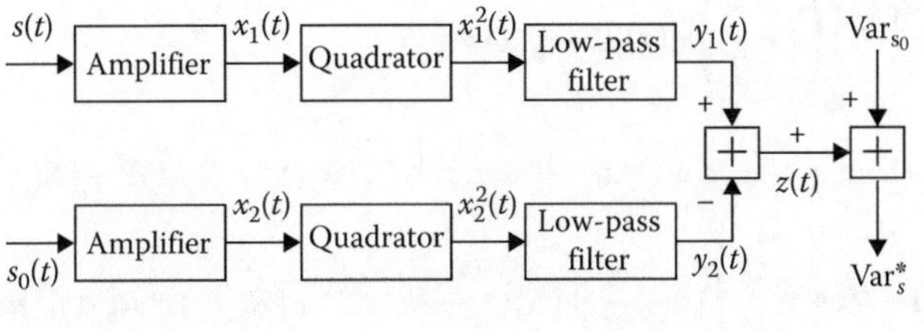

FIGURE 13.9 Flowchart of the stochastic process variance measurer.

where

$\upsilon_1(t)$ and $\upsilon_2(t)$ are the realizations of Gaussian stationary stochastic processes describing the random variations $\beta_1(t)$ and $\beta_2(t)$ relative to the amplification coefficient value at the first and second channels of signal amplifier

$n_1(t)$ and $n_2(t)$ are the realizations of signal amplifier noise at the first and second channels

$s(t)$ and $s_0(t)$ are the investigated stochastic process and the reference signal, respectively

Let the reference signal be the stationary Gaussian stochastic process with the variance Var_{s_0}. The signal at the subtractor output takes the following form:

$$z(t) = \frac{1}{T}\int_0^T \left[x_1^2(t) - x_2^2(t)\right]dt. \tag{13.181}$$

The variance estimate is determined in the following form:

$$\mathrm{Var}_s^*(t) = z(t) + \mathrm{Var}_{s0}. \tag{13.182}$$

Define the mathematical expectation and the variance of the variance estimate under assumptions made earlier

$$\begin{cases} \langle \beta_1^2(t)\rangle = \langle \beta_2^2(t)\rangle = \mathrm{Var}_\beta; \\ \langle n_1^2(t)\rangle = \langle n_2^2(t)\rangle = \mathrm{Var}_n. \end{cases} \tag{13.183}$$

The mathematical expectation of variance estimate can be presented in the following form:

$$\langle \mathrm{Var}_s^* \rangle = \langle z\rangle + \mathrm{Var}_{s0} = (1+\mathrm{Var}_\beta)(\mathrm{Var}_s - \mathrm{Var}_{s0}) + \mathrm{Var}_{s0}. \tag{13.184}$$

As we can see from (13.184), the bias of variance estimate can be presented in the following form:

$$b\{\mathrm{Var}_s^*\} = \mathrm{Var}_\beta(\mathrm{Var}_s - \mathrm{Var}_{s0}). \tag{13.185}$$

If the variance of the reference stochastic process is controlled then the variance measurement process can be reduced to zero signal aspects at the output indicator. This procedure is called the *method with zero instant*. Naturally, the variance estimate is unbiased.

The variance of the variance estimate can be presented in the following form:

$$\begin{aligned} \mathrm{Var}\{\mathrm{Var}_s^*\} = {} & \frac{4\mathrm{Var}_\beta}{T}\int_0^T \left(1-\frac{\tau}{T}\right) r_\beta(\tau)[2 + r_\beta(\tau)\mathrm{Var}_\beta] \\ & \times\{(\mathrm{Var}_s + \mathrm{Var}_n)^2 + (\mathrm{Var}_{s0} + \mathrm{Var}_n)^2 + 2\{[R_s(\tau)+R_n(\tau)]^2 + [R_{s0}(\tau)+R_n(\tau)]^2\}\}d\tau \\ & + (1+\mathrm{Var}_\beta)^2\frac{4}{T}\int_0^T\left(1-\frac{\tau}{T}\right)\{[R_s(\tau)+R_n(\tau)]^2 + [R_{s0}(\tau)+R_n(\tau)]^2\}d\tau. \end{aligned} \tag{13.186}$$

We can simplify (13.186) applied to zero measurement procedure. In doing so, owing to identity of two channels of amplifier we can think that $R_s(\tau) = R_{s_0}(\tau)$. Taking into consideration the condition that $\text{Var}_\beta \ll 1$ and neglecting the terms with double frequency $2\omega_0$, we can write

$$\text{Var}\left\{\text{Var}_s^*\right\} = (1+q)^2 \frac{4\text{Var}_s^2}{T}\int_0^T\left(1-\frac{\tau}{T}\right)[r^2(\tau)+4r_\beta(\tau)\text{Var}_\beta+4r^2(\tau)r_\beta(\tau)\text{Var}_\beta]d\tau. \tag{13.187}$$

By analogy with the compensation method, (13.187) allows us to obtain the time interval of observation corresponding to the sensitivity threshold. To satisfy this condition, the following equality needs to be satisfied:

$$\frac{\text{Var}\left\{\text{Var}_s^*\right\}}{\text{Var}_s^2} = 1. \tag{13.188}$$

Comparing (13.187) with (13.162), we can see that under the considered procedure the variance of the variance estimate is twice as high as the variance of the variance estimate at the compensation method. This increase in the variance of the variance estimate is explained by the presence of the second channel, which results in an increase in the total variance of the output signal. We need to note that although there is an increase in the variance of the variance estimate for the considered procedure, the present method is a better choice compared to the compensation method if the variance of intrinsic noise of amplifier varies after compensation procedure.

Let the normalized correlation functions, as before, be described by (13.165). Then (13.187) based on the following conditions $\alpha T \gg 1$ and $\alpha \gg \gamma$ can be presented in the following form:

$$\text{Var}\left\{\text{Var}_s^*\right\} = (1+q)^2 \frac{2\text{Var}_s^2}{\alpha T}\left\{1+8\frac{\alpha}{\gamma}\text{Var}_\beta\frac{\gamma T-1+\exp\{-\gamma T\}}{\gamma T}\right\}. \tag{13.189}$$

In other words, the variance of the variance estimate in the case of the considered procedure of measurement is twice as high as the variance of the variance estimate given by (13.168) obtained using compensation method. This is true also when the random variations of the amplification coefficient are absent. In doing so, the time interval of observation corresponding to the sensitivity threshold increases twofold compared to the time interval of observation of the investigated stochastic process while using the compensation method to measure the variance.

Using the deterministic harmonic signal

$$s_0(t) = A_0\cos(\omega_0 t+\varphi_0) \tag{13.190}$$

as a reference signal, applied to the zero measurement procedure, that, $\text{Var}_{s_0} = 0.5A_0^2$, and when the random variations of the amplification coefficient are absent, the variance of the variance estimate can be presented in the following form:

$$\text{Var}\left\{\text{Var}_s^*\right\} = 2\,\text{Var}_s^2(1+2q+2q^2)\times\frac{1}{T}\int_0^T\left(1-\frac{\tau}{T}\right)r^2(\tau)d\tau. \tag{13.191}$$

As applied to the exponential normalized correlation function given by (13.165), we can transform (13.191) into the following form:

$$\mathrm{Var}\left\{\mathrm{Var}_s^*\right\} = 2\mathrm{Var}_s^2(1+2q+2q^2)\frac{2\alpha T - 1 + \exp\{-2\alpha T\}}{(2\alpha T)^2}. \tag{13.192}$$

At $\alpha T \gg 1$, we have

$$\mathrm{Var}\left\{\mathrm{Var}_s^*\right\} \approx \frac{\mathrm{Var}_s^2(1+2q+2q^2)}{\alpha T}. \tag{13.193}$$

Comparing (13.193) with (13.189) at $\mathrm{Var}_\beta = 0$, we can see that in the case of weal signals, that is, the signal-to-noise ratio is small or $q \ll 1$, with the use of the harmonic signal the variance of the variance estimate is twice as less as when the wideband stochastic process is used. Otherwise, at the high signal-to-noise ratio, $q > 1$, under the use of the harmonic signal the variance of the variance estimate can be higher compared to the case when we use the stochastic process. This phenomenon is explained by the presence of uncompensated components of the output signal formed by the terms of high order under expansion of the amplifier noise in series. In addition, the considered procedure possesses the errors caused by the nonidentity of channels and presence of statistical dependence between amplifier channels.

13.5.3 Correlation Method of Variance Measurement

While using the correlation method to measure the stochastic process variance, the stochastic process comes in at the inputs of two channels of amplifier. The intrinsic amplifier channel noise samples are independent of each other. Because of this, their mutual correlation functions are zero and the mutual correlation function of the investigated stochastic process is not equal to zero and the coincidence instants are equal to the variance of the investigated stochastic process. A flowchart of the variance measurement made using the correlation method is presented in Figure 13.10.

Let the independent channels of amplifier operate at the same frequency. Then the stochastic processes forming at the amplifier outputs can come in at the mixer inputs directly. We assume, as before, that the integrator is the ideal, that is, $h(t) = T^{-1}$. The signal at the integrator output defines the variance estimate

$$\mathrm{Var}_s^* = \frac{1}{T}\int_0^T x_1(t)x_2(t)dt, \tag{13.194}$$

where

$$\begin{cases} x_1(t) = [1+\upsilon_1(t)]\,[s(t)+n_1(t)]; \\ x_2(t) = [1+\upsilon_2(t)]\,[s(t)+n_2(t)] \end{cases} \tag{13.195}$$

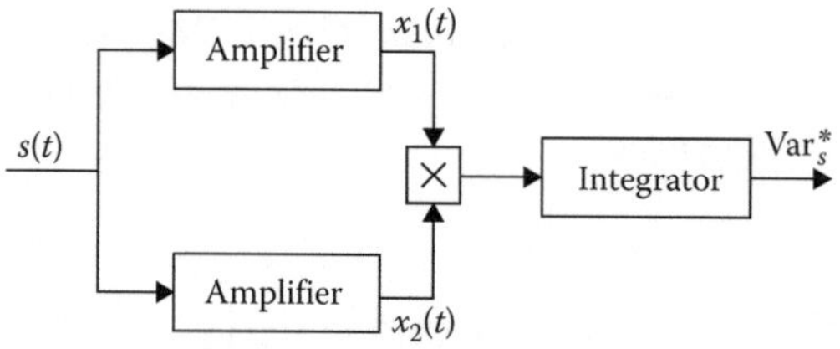

FIGURE 13.10 Correlation method of variance measurement.

are the realizations of stochastic processes at the outputs of the first and second channels, respectively. The mathematical expectation of estimate can be presented in the following form:

$$\left\langle \mathrm{Var}_s^* \right\rangle = \frac{1}{T}\int_0^T \langle s^2(t)\rangle\, dt = \mathrm{Var}_s; \tag{13.196}$$

that is, the variance estimate is unbiased under the correlation method of variance measurement and when the identical channels are independent.

The variance of the output signal or the variance of the variance estimate of the input stochastic process in the case when the input stochastic process is Gaussian takes the following form:

$$\mathrm{Var}\left\{\mathrm{Var}_s^*\right\} = \frac{2\mathrm{Var}_s^2}{T}\int_0^T \left(1-\frac{\tau}{T}\right)\left[R_{\beta_1}(\tau)+R_{\beta_2}(\tau)+R_{\beta_1}(\tau)R_{\beta_2}(\tau)\right]d\tau$$

$$+\frac{2}{T}\int_0^T \left(1-\frac{\tau}{T}\right)\left[1+R_{\beta_1}(\tau)+R_{\beta_2}(\tau)+R_{\beta_1}(\tau)R_{\beta_2}(\tau)\right]$$

$$\times\left[2R_s^2(\tau)+R_{n_1}(\tau)R_s(\tau)+R_{n_2}(\tau)R_s(\tau)+R_{n_1}(\tau)R_{n_2}(\tau)\right]d\tau, \tag{13.197}$$

where $R_{\beta 1}(\tau)$ and $R_{\beta 2}(\tau)$ are the correlation functions of random components of the amplification coefficients of the first and second channels of amplifier.

We can simplify (13.197) taking into consideration that the correlation interval of random components of amplification coefficients is longer compared to the correlation interval of the intrinsic noise $n_1(t)$ and $n_2(t)$ and the investigated stochastic process $\varsigma(t)$. At the same time, we can introduce the function $R_\beta(0) = \mathrm{Var}_\beta$ instead of the correlation function $R_\beta(\tau)$ in (13.197). Then, taking into consideration the condition accepted before, that is, $\mathrm{Var}_\beta < 1$ or $\mathrm{Var}_\beta^2 \ll 1$, identity of channels, and (13.153) through (13.155), we obtain

$$\mathrm{Var}\left\{\mathrm{Var}_s^*\right\} = \frac{2\mathrm{Var}_s^2}{T}(1+q+0.5q^2)\int_0^T \left(1-\frac{\tau}{T}\right) r^2(\tau)d\tau + \frac{4\mathrm{Var}_s^2\mathrm{Var}_\beta}{T}\int_0^T \left(1-\frac{\tau}{T}\right) r_\beta(\tau)d\tau. \tag{13.198}$$

Formula (13.198) allows us to define the time interval of observation T corresponding to the sensitivity threshold at the condition given by (13.188). Let the normalized correlation functions $r(\tau)$ and $r_\beta(\tau)$ be defined by the exponents given by (13.165) and substituting into (13.198), we obtain

$$\mathrm{Var}\left\{\mathrm{Var}_s^*\right\} = 2\mathrm{Var}_s^2(1+q+0.5q^2)\frac{2\alpha T-1+\exp\{-2\alpha T\}}{(2\alpha T)^2} + 4\mathrm{Var}_s^2\mathrm{Var}_\beta\frac{\gamma T-1+\exp\{-\gamma T\}}{(\gamma T)^2}. \tag{13.199}$$

When the random variations of the amplification coefficients are absent and the condition $T \gg \tau_{cor}$ is satisfied, the variance of the variance estimate can be presented in the following form:

$$\mathrm{Var}_0\left\{\mathrm{Var}_s^*\right\} = \mathrm{Var}_s^2\frac{1+q+0.5q^2}{\alpha T}. \tag{13.200}$$

As we can see from (13.200), it is not difficult to define the time interval of observation corresponding to the sensitivity threshold in the case of the correlation method of measurement of the stochastic process variance.

Comparing (13.198) through (13.200) and (13.162), (13.166), and (13.169), we can see that the sensitivity of the correlation method of the stochastic process variance measurement is higher compared to the compensation method sensitivity. This difference is caused by the compensation of noise components with high order while using the correlation method and by the compensation of errors caused by random variations of the amplification coefficients. However, when the channels are not identical and there is a statistical relationship between the intrinsic receiver noise and random variations of the amplification coefficients, there is an estimate bias and the variance of estimate also increases, which is undesirable.

The correlation method used for variance measurement leads to additional errors caused by the difference in the performance levels of the used mixers and the ideal mixers. As a rule, to multiply two processes, we use the following operation:

$$(a+b)^2-(a-b)^2=4ab. \tag{13.201}$$

In other words, a multiplication of two stochastic processes is reduced to quadratic transformation of sum and difference of multiplied stochastic processes and subtraction of quadratic forms. The highest error is caused by quadratic operations.

Consider briefly an effect of spurious coupling between the amplifier channels on the accuracy of measurement of the stochastic process variance. Denote the mutual correlation functions of the amplification coefficients and intrinsic noise of amplifier at the coinciding instants as

$$\begin{cases} \langle \beta_1(t)\beta_2(t)\rangle = R_{\beta_{12}}, \\ \langle n_1(t)n_2(t)\rangle = R_{n_{12}}. \end{cases} \tag{13.202}$$

As we can see from (13.195), the bias of variance estimate is defined as

$$b\{\mathrm{Var}^*\} = \mathrm{Var_n}(1+R_{\beta_{12}})+R_{\beta_{12}}\mathrm{Var}_s. \tag{13.203}$$

Naturally, the variance of estimate increases due to the relationship between the amplifier channels. However, the spurious coupling, as a rule, is very weak and can be neglected in practice.

13.5.4 Modulation Method of Variance Measurement

While using the modulation method to measure the stochastic process variance, the received realization $x(t)$ of stochastic process is modulated at the amplifier input by the deterministic signal $u(t)$ with audio frequency. Modulation, as a rule, is carried out by periodical connection of the investigated stochastic process to the amplifier input (see Figure 13.11) or amplifier input to the investigated stochastic processes and reference process (see Figure 13.12). After amplification and quadratic transformation, the stochastic processes come in at the mixer input, the second input of which is used by the deterministic signal $u_1(t)$ of the same frequency of the signal $u(t)$. The mixer output process comes in at the low-pass filter input that make smoothing or averaging of stochastic fluctuations. As a result, we obtain the variance estimate of the investigated stochastic process.

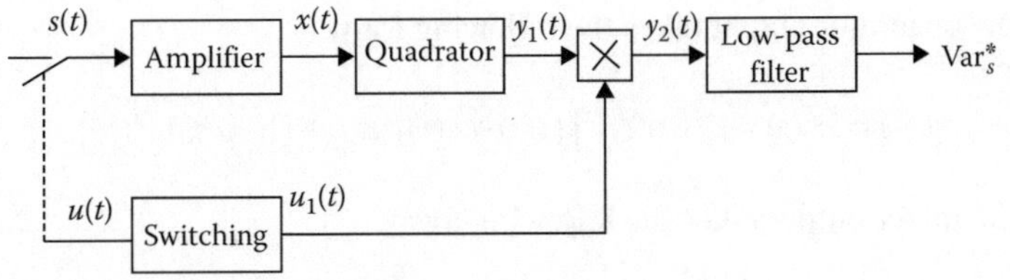

FIGURE 13.11 Modulation method of variance measurement.

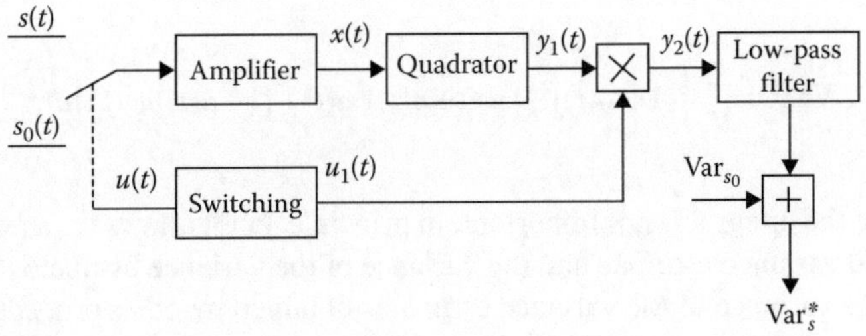

FIGURE 13.12 Modulation method of variance measurement using the reference process.

Define the characteristics of the modulation method of the stochastic process variance measurement assuming that the quadratic transformer has no inertia. The ideal integrator with the integration time T serves as the low-pass filter. Under these assumptions and conditions we can write

$$u(t) = \begin{cases} 1 \quad \text{at} \quad kT_0 \le t \le kT_0 + 0.5T_0; \\ 0 \quad \text{at} \quad kT_0 + 0.5T_0 < t < (k+1)T_0; \\ k = 0, \pm 1, \ldots; \end{cases} \tag{13.204}$$

$$u_1(t) = \begin{cases} 1 \quad \text{at} \quad kT_0 \le t \le kT_0 + 0.5T_0; \\ -1 \quad \text{at} \quad kT_0 + 0.5T_0 < t < (k+1)T_0; \\ k = 0, \pm 1, \ldots. \end{cases} \tag{13.205}$$

As we can see from (13.204), $u^2(t) = u(t)$ and the average value in time of the function $u(t)$ can be presented as $\overline{u(t)} = 0.5$. Analogously we have

$$[1 - u(t)]^2 = 1 - u(t). \tag{13.206}$$

Note that due to the orthogonality of the functions $u(t)$ and $1 - u(t)$, their product is equal to zero.

At first, we consider the modulation method of the stochastic process variance measurement presented in Figure 13.11. The total realization of stochastic process coming in at the quadrator input can be presented in the following form:

$$x(t) = [1 + \upsilon(t)]\{[s(t) + n(t)]u(t) + [1 - u(t)]n(t)\}. \tag{13.207}$$

The process at the quadrator output takes the following form:

$$y_1(t) = x^2(t) = [1+\upsilon(t)]^2\{[s(t)+n(t)]^2 u(t) + [1-u(t)]n^2(t)\}. \tag{13.208}$$

The process at the mixer output takes the following form:

$$y_2(t) = [1+\upsilon(t)]^2\{[s(t)+n(t)]^2 u(t) - [1-u(t)]n^2(t)\}. \tag{13.209}$$

The signal at the ideal integrator output (the variance estimate) takes the following form:

$$\text{Var}_s^* = \frac{2}{T}\int_0^T [1+\upsilon(t)]^2\{[s(t)+n(t)]^2 u(t) - [1-u(t)]n^2(t)\}dt. \tag{13.210}$$

The factor 2 before the integral is not important, in principle, but it allows us, as we can see later, to obtain the unbiased variance estimate and the variance of the variance estimate that are convenient to compare with the variance of the variance estimates obtained by other procedures and methods.

By averaging the random estimate Var_s^*, we obtain

$$\left\langle \text{Var}_s^* \right\rangle = \text{Var}_s(1+\text{Var}_\beta). \tag{13.211}$$

As we can see from (13.211), when the random variations of the amplification coefficients are absent $\text{Var}_\beta = 0$, the variance estimate of stochastic process using the modulation method of variance measurement is unbiased as is the case with the compensation method. To determine the variance of the variance estimate we need to present the function $u(t)$ by the Fourier series expansion

$$u(t) = \frac{1}{2} + \frac{2}{\pi}\sum_{k=1}^{\infty}\frac{\sin(2k-1)\Omega t}{2k-1}, \tag{13.212}$$

where

$$\Omega = \frac{2\pi}{T_0} \tag{13.213}$$

is the switching function frequency.

Ignoring oscillating terms and taking into consideration that $\text{Var}_\beta \ll 1$ and the correlation interval of random component of the amplification coefficients is much more the correlation interval of the investigated stochastic process, the variance of the variance estimate, can be presented in the following form:

$$\begin{aligned}\text{Var}\left\{\text{Var}_s^*\right\} = {} & \frac{32}{\pi^2 T}\sum_{k=1}^{\infty}\frac{1}{(2k-1)^2}\int_0^T\left(1-\frac{\tau}{T}\right) \\ & \times\left\{\left[R_s^2(\tau)+2R_s(\tau)R_n(\tau)+4R_n^2(\tau)\right]+2R_\beta(\tau)\left(\text{Var}_s^2+4\text{Var}_s Var_n+4\text{Var}_n^2\right)\right\} \\ & \times\cos[(2k-1)\Omega\tau]d\tau+\frac{4}{T}\int_0^T\left(1-\frac{\tau}{T}\right)\left[R_s^2(\tau)+2R_s(\tau)R_n(\tau)+2R_\beta(\tau)\text{Var}_s^2\right]d\tau\end{aligned} \tag{13.214}$$

Taking into consideration (13.153) through (13.155), (13.214) can be written in the following form:

$$\mathrm{Var}\{\mathrm{Var}_s^*\} = \frac{16\mathrm{Var}_s^2}{\pi^2 T}\sum_{k=1}^{\infty}\frac{1}{(2k-1)^2}$$

$$\times\int_0^T\left(1-\frac{\tau}{T}\right)\left[(1+2q+4q^2)r^2(\tau)+4\mathrm{Var}_\beta r_\beta(\tau)(1+4q+4q^2)\right]\cos[(2k-1)\Omega\tau]d\tau$$

$$+\frac{2\mathrm{Var}_s^2}{T}\int_0^T\left(1-\frac{\tau}{T}\right)\left[(1+2q)r^2(\tau)+4\mathrm{Var}_\beta r_\beta(\tau)\right]d\tau. \tag{13.215}$$

In practice, the correlation interval of the investigated stochastic process is much less than the modulation period T_0 and, consequently, the time interval T of observation, since in real applications, the effective spectrum bandwidth of the investigated stochastic process is more than 10^4–10^5 Hz and at the same time, the modulation frequency $f_{\mathrm{mod}} = \Omega \times (2\pi)^{-1}$ is several hundred hertz. In this case, we can assume that the functions $\cos[(2k-1)\Omega\tau]$ are not variable functions within the limits of the correlation interval, and we can think that this function can be approximated by unit, that is, $\cos[(2k-1)\Omega\tau] \approx 1$. This statement is true for the components of variance of the variance estimate caused by stochastic character of variation of the processes $s(t)$ and $n(t)$. Taking into consideration that

$$\sum_{k=1}^{\infty}(2k-1)^{-2} = \frac{\pi^2}{8}, \tag{13.216}$$

we obtain

$$\mathrm{Var}\{\mathrm{Var}_s^*\} = \frac{4\mathrm{Var}_s^2}{T}(1+2q+2q^2)\int_0^\infty r^2(\tau)d\tau + \frac{8\mathrm{Var}_s^2\mathrm{Var}_\beta}{T}$$

$$\times\left\{\int_0^T\left(1-\frac{\tau}{T}\right)r_\beta(\tau)d\tau + \frac{8(1+4q+4q^2)}{\pi^2}\sum_{k=1}^{\infty}\frac{1}{(2k-1)^2}\right.$$

$$\left.\times\int_0^T\left(1-\frac{\tau}{T}\right)r_\beta(\tau)\cos[(2k-1)\Omega\tau]d\tau\, d\tau\right\}. \tag{13.217}$$

The approximation $\cos[(2k-1)\Omega\tau] \approx 1$ is true within the limits of the correlation interval of investigated stochastic process owing to fast convergence of the series (13.216) to its limit. If, in addition to the aforementioned condition, the correlation interval of the amplification coefficients is much less than the time interval of observation, that is, $\tau_\beta \ll 1$ but $\tau_\beta > T_0$, then

$$\mathrm{Var}\{\mathrm{Var}_s^*\} \approx \frac{4\mathrm{Var}_s^2}{T}(1+2q+2q^2)\int_0^\infty r^2(\tau)d\tau + \frac{8\mathrm{Var}_s^2\mathrm{Var}_\beta}{T}\int_0^\infty r_\beta(\tau)d\tau. \tag{13.218}$$

When the random variations of the amplification coefficients are absent, based on (13.217), we can write

$$\mathrm{Var}_0\left\{\mathrm{Var}_s^*\right\} = \frac{4\mathrm{Var}_s^2}{T}(1+2q+2q^2)\int_0^\infty r^2(\tau)d\tau. \tag{13.219}$$

Comparing (13.219) and (13.162) at $\mathrm{Var}_\beta = 0$ while using the compensation method to measure variance, we can see that the relative value of the variance of the variance estimate defining the sensitivity of the modulation method is twice as high compared to the relative variation in the variance estimate obtained by the compensation method at the same conditions of measurement. Physically this phenomenon is caused by a twofold decrease in the time interval of observation of the investigated stochastic process owing to switching.

Now, consider the modulation method used for variance measurement based on a comparison between the variance Var_s of the observed stochastic process $\zeta(t)$ within the limits of amplifier bandwidth and the variance Var_{s_0} of the reference calibrated stochastic process $s_0(t)$ in accordance with the block diagram shown in Figure 13.12. In this case, the signal at the amplifier input can be presented in the following form:

$$x(t) = [1+\upsilon(t)]\left\{[s(t)+n(t)]u(t)+[s_0(t)+n(t)]\,[1-u(t)]\right\}. \tag{13.220}$$

The signal at the ideal integrator output can be presented in the following form:

$$z(t) = \frac{2}{T}\int_0^T [1+\upsilon(t)]^2\left\{[s(t)+n(t)]^2u(t)-[s_0(t)+n(t)]^2\,[1-u(t)]\right\}dt. \tag{13.221}$$

The variance estimate is given by

$$\mathrm{Var}_s^* = z+\mathrm{Var}_{s0}. \tag{13.222}$$

Averaging by realizations, we have

$$\left\langle \mathrm{Var}_s^*\right\rangle = (1+\mathrm{Var}_\beta)(\mathrm{Var}_s-\mathrm{Var}_{s0})+\mathrm{Var}_{s0}; \tag{13.223}$$

that is, the bias of variance estimate can be presented in the following form:

$$b\{\mathrm{Var}_s^*\} = \mathrm{Var}_\beta(\mathrm{Var}_s-\mathrm{Var}_{s0}). \tag{13.224}$$

When the random variations of the amplification coefficients are absent, the process at the modulation measurer output can be calibrated in values of difference between the variance of investigated stochastic process and the variance of reference stochastic process. This measurement method is called the *modulation method* of variance measurement with direct reading. When the variances of investigated and reference stochastic processes are the same, the bias of the considered variance estimate is zero. If the variance of the reference stochastic process can be controlled, the measurement is reduced to fixation of zero readings. This method is called the modulation method of measurement with zero reading. The modulation method with zero reading can be automated by a tracking device.

Determine the variation in the variance estimate of the modulation method assuming as before that $\text{Var}_\beta \ll 1$ and neglecting oscillating terms under integration and taking into consideration that $\tau_{\text{cor}} \ll \tau_\beta$. In this case, the variance of the variance estimate of the investigated stochastic process can be presented in the following form:

$$\begin{aligned}
\text{Var}\{\text{Var}_s^*\} &= \frac{32}{\pi^2 T}\sum_{k=1}^{\infty}\frac{1}{(2k-1)^2}\\
&\times\int_0^T\left(1-\frac{\tau}{T}\right)\left[R_{s0}^2(\tau)+R_s^2(\tau)+2R_s(\tau)R_n(\tau)+2R_{s0}(\tau)R_n(\tau)+4R_n^2(\tau)\right]\cos[(2k-1)\Omega\tau]d\tau\\
&+\frac{64}{\pi^2 T}\left[\text{Var}_{s0}^2+\text{Var}_s^2+2\text{Var}_{s0}\text{Var}_n+4\text{Var}_{s0}\text{Var}_n+4\text{Var}_s\text{Var}_n+4\text{Var}_n^2\right]\\
&\times\sum_{k=1}^{\infty}\frac{1}{(2k-1)^2}\int_0^T\left(1-\frac{\tau}{T}\right)R_\beta(\tau)\cos[(2k-1)\Omega\tau]d\tau\\
&+\frac{4}{T}\int_0^T\left(1-\frac{\tau}{T}\right)\left[R_s^2(\tau)+2R_s(\tau)R_n(\tau)+R_{s0}^2(\tau)+2R_{s0}(\tau)R_n(\tau)\right]d\tau\\
&+\frac{8}{T}(\text{Var}_s-\text{Var}_{s0})^2\int_0^T\left(1-\frac{\tau}{T}\right)R_\beta(\tau)d\tau.
\end{aligned}\tag{13.225}$$

As applied to the zero method of variance measurement, we have

$$R_s(\tau)=R_{s0}(\tau)=r(\tau)\text{Var}_s\cos\omega_0\tau.\tag{13.226}$$

Taking into consideration (13.226), we obtain

$$\begin{aligned}
\text{Var}\{\text{Var}_s^*\} &= \frac{32\text{Var}_s^2}{\pi^2 T}(1+2q+2q^2)\sum_{k=1}^{\infty}\frac{1}{(2k-1)^2}\int_0^T\left(1-\frac{\tau}{T}\right)r^2(\tau)\cos[(2k-1)\Omega\tau]d\tau\\
&+\frac{128\text{Var}_s^2\text{Var}_\beta}{\pi^2 T}(1+q^2)\sum_{k=1}^{\infty}\frac{1}{(2k-1)^2}\int_0^T\left(1-\frac{\tau}{T}\right)r_\beta(\tau)\cos[(2k-1)\Omega\tau]d\tau\\
&+\frac{4\text{Var}_s^2(1+2q)}{T}\int_0^T\left(1-\frac{\tau}{T}\right)r^2(\tau)d\tau.
\end{aligned}\tag{13.227}$$

In the considered case, $\tau_{cor} \ll \tau_\beta$ and $\tau_{cor} \ll T_0$, we have

$$\text{Var}\left\{\text{Var}_s^*\right\} = \frac{8\text{Var}_s^2(1+q)^2}{T}\int_0^T\left(1-\frac{\tau}{T}\right)r^2(\tau)d\tau$$

$$+\frac{256\text{Var}_s^2\text{Var}_\beta}{\pi^2 T}(1+q^2)\sum_{k=1}^{\infty}\frac{1}{(2k-1)^2}\int_0^T\left(1-\frac{\tau}{T}\right)r_\beta(\tau)\cos[(2k-1)\Omega\tau]d\tau. \quad (13.228)$$

If, in addition to the foregoing statements, the random variations of the amplification coefficients are absent or the conditions $\tau_\beta \ll T$ and $\tau_\beta \gg T_0$ are satisfied, we can write

$$\text{Var}_0\left\{\text{Var}_s^*\right\} \approx \frac{8\text{Var}_s^2(1+q)^2}{T}\int_0^T r^2(\tau)d\tau. \quad (13.229)$$

Comparing (13.229) and (13.162) at $\text{Var}_\beta = 0$ in the case of compensation method, we can see that due to the twofold decrease in the total time interval of observation of investigated stochastic process and presence of the reference stochastic process, the variance of the variance estimate under the modulation method is four times higher compared to the variance of the variance estimate measured via the compensation method at the same conditions of measurement.

When we use the deterministic harmonic signal

$$s_0(t) = A_0\cos(\omega_0 t + \varphi_0) \quad (13.230)$$

as the reference signal, we are able to obtain the variance of the variance estimate analogously. As applied to the modulation method of variance measurement with zero reading in the case of absence of the random variations of the amplification coefficients, the stochastic process variance estimate is unbiased and the variance of the variance estimate can be presented in the following form:

$$\text{Var}\left\{\text{Var}_s^*\right\} = \frac{4\text{Var}_s^2(1+2q+2q^2)}{T}\int_0^T\left(1-\frac{\tau}{T}\right)r^2(\tau)d\tau. \quad (13.231)$$

Comparison between (13.231) and (13.191) shows us that due to a twofold decrease in the total time interval of observation, the variation in the variance estimate obtained via the modulation method measurement is twice as high compared to the compensation method.

13.6 SUMMARY AND DISCUSSION

We summarize briefly the main results discussed in this chapter.

As we can see from (13.12), the variance of the optimal variance estimate is independent of the values of the normalized correlation function between the samples of observed stochastic process. This fact may lead to some results that are difficult to explain from the physical viewpoint or cannot be explained absolutely. As a matter of fact, by increasing the number of samples within the limits of the finite small time interval, we can obtain the estimate of variance with infinitely small estimate variance according to (13.12). The variance of optimal variance estimate of Gaussian stochastic process approaching zero is especially evident while moving from discrete to continuous observation of stochastic process.

The optimal variance estimate of Gaussian stochastic process based on a discrete sample is equivalent to the error given by (13.24) for the same sample size (the number of samples). This can be explained by the fact that under the optimal signal processing, the initial sample multiplies with the newly formed uncorrelated sample. However, if the normalized correlation function is unknown or known with error, then the optimal variance estimate of the stochastic process has the finite variance depending on the true value of the normalized correlation function. To simplify consideration and investigation of this problem, we need to compute the variance of the variance estimate of Gaussian stochastic process with zero mean by two samples applied to the estimate by the likelihood function maximum for the following cases: The normalized correlation function, or, as it is often called, the correlation coefficient, is completely known, unknown, and known with error.

Experimental statistical characteristics of the variance estimate obtained with sufficiently high accuracy are matched with theoretical values. The maximal relative errors of definition of the mathematical expectation estimate and the variance estimate do not exceed 1% and 2.5%, respectively. Table 13.1 presents the experimental data of the mathematical expectations and variance of the variance estimate at various values of the correlation coefficient ρ_0 between samples. Also, the theoretical data of the variation in the variance estimate based on the algorithm given by (13.24) and determined in accordance with the formula (13.40) are presented for comparison in Table 13.1. Theoretical value of the variance of the optimal variance estimate is equal to $\text{Var}\{\text{Var}_E\} = 7.756$, according to (13.5). Thus, the experimental data prove the previously discussed theoretical definition of estimates and their characteristics, at least at $N = 2$.

Difference between the transform characteristic and the square-law function can lead to high errors under definition of the stochastic process variance. Because of this, while using measurers to define the stochastic process variance, we need to pay serious attention to the transform performance. Verification of the transform performance is carried out, as a rule, by sending the harmonic signal of known amplitude at the measurer input. The methods discussed thus far that measure the variance assume an absence of limitations of instantaneous values of the investigated stochastic process. The presence of these limitations leads to additional errors while measuring the variance. The ideal integrator $h(t) = T^{-1}$ plays the role of averaging or smoothing filter.

In the majority of cases, we need to define the variance estimate by investigating a single realization of stochastic process. For measuring the variance varying in time based on a single realization of stochastic process there are similar problems as in the course of estimation of the mathematical expectation varying in time. Actually, on one hand, to decrease the variance estimate caused by the finite observation time interval, the last must be as soon as large. On the other hand, there is a need to choose an integration time as short as possible for the best definition of variance variations. Evidently, there must be a compromise.

The simplest way to define the variance of time-varying stochastic process at the instant t_0 is averaging the transformed input data of stochastic process within the limits of finite time interval. Thus, let $x(t)$ be the realization of stochastic process $\xi(t)$ with zero mathematical expectation. Measurement of variance of this stochastic process at the instant t_0 is carried out by averaging ordinate quadrature ordinates $x(t)$ within the limits of the interval about the given value of argument $[t_0 - 0.5T, t_0 + 0.5T]$. In this case, the variance estimate is defined by (13.106). Average of the variance estimate by realizations can be defined by (13.307). Thus, as in the case of time-varying mathematical expectation, the mathematical expectation of the estimate of the variance varying in time does not coincide with its true value in a general case, but it is obtained by smoothing the variance within the limits of finite time interval $[t_0 - 0.5T, t_0 + 0.5T]$. As a result of such averaging, the bias of variance of stochastic process can be presented by (13.108).

Comparing (13.187) with (13.162), we can see that in the case of comparison method, the variance of the variance estimate is twice as high compared to the variance of the variance estimate obtained via compensation method. This increase in the variance of the variance estimate is explained by the presence of the second channel because of which there is an increase in the total variance of the output signal. We need to note that although there is an increase in the variance of the variance estimate for the considered procedure, the present method is a better choice compared to the compensation method if the variance of intrinsic noise of amplifier changes after compensation procedure. This is also true when the random variations of the amplification coefficient are absent. In doing so, the time interval of observation corresponding to the sensitivity threshold increases twofold compared to the time interval of observation of the investigated stochastic process while using the compensation method to measure variance.

Comparing (13.198) through (13.200) and (13.162), (13.166), and (13.169), we can see that the sensitivity of the correlation method of the stochastic process variance measurement is higher compared to the sensitivity of the compensation method. This difference is caused by the compensation of noise components with high order while using the correlation method and by the compensation of errors caused by random variations of the amplification coefficients. However, when the channels are not identical and there is a statistical relationship between the intrinsic receiver noise and random variations of the amplification coefficients there is an estimate bias and an increase in the variance of estimate, which is undesirable.

Comparing (13.219) and (13.162) at $\text{Var}_\beta = 0$ in the case of compensation method of variance measurement, we can see that the relative value of the variance of the variance estimate defining the sensitivity of the modulation method is twice as high compared to the relative variation in the variance estimate of the compensation method at the same conditions of measurement. Physically, this phenomenon is caused by a twofold decrease in the time interval of observation of the investigated stochastic process due to switching.

When the random variations of the amplification coefficients are absent, the process at the modulation measurer output can be calibrated using the values of difference between the variance of the investigated stochastic process and the variance of reference stochastic process. This measurement method is called the modulation method of variance measurement with direct reading. When the variances of the investigated and the reference stochastic processes are the same, the bias of the considered variance estimate is zero. If the variance of the reference stochastic process can be controlled, the measurement is reduced to fixation of zero readings. This method is called the modulation method of measurement with zero reading. The modulation method with zero reading can be automated by a tracking device.

Comparing (13.229) and (13.162) at $\text{Var}_\beta = 0$ in the case of compensation method of variance measurement, we can see that owing to the twofold decrease in the total time interval of observation of the investigated stochastic process and the presence of the reference stochastic process the variance of the variance estimate while using the modulation method is four times higher compared to the variance of the variance estimate of the compensation method at the same conditions of measurement.

Comparison between (13.231) and (13.191) shows us that owing to the twofold decrease in the total time interval of observation in the variance of the variance estimate while using the modulation method is two times higher compared to the compensation method.

REFERENCES

1. Slepian, D. 1958. Some comments on the detection of Gaussian signals in Gaussian noise. *IRE Transactions on Information Theory*, 4(2): 65–68.
2. Kay, S. 2006. *Intuitive Probability and Random Processes Using MATLAB*. New York: Springer Science + Business Media, LLC.
3. Vilenkin, S.Ya. 1979. *Statistical Processing of Stochastic Functions Investigation Results*. Moscow, Russia: Energy.
4. Esepkina, N.A., Korolkov, D.V., and Yu.N. Paryisky. 1973. *Radar Telescopes and Radiometers*. Moscow, Russia: Nauka.

14 Estimation of Probability Distribution and Density Functions of Stochastic Process

14.1 MAIN ESTIMATION REGULARITIES

Experimental definition of the one-dimensional probability distribution function $F(x)$ and pdf $p(x)$ is formed in the simplest way for ergodic stochastic processes and is based on an investigation of the stochastic process within the limits of the time interval $[0, T]$. Actually, a researcher observes the realization $x(t)$ of a continuous ergodic process $\xi(t)$. An example of the realization $x(t)$ is shown in Figure 14.1a. According to the definition of ergodic stochastic process [1], the probability distribution function $F(x)$ can be defined approximately in the following form (see Figure 14.1a):

$$F^*(x) \approx \frac{1}{T}\sum_{i=1}^{N}\tau_i, \tag{14.1}$$

where $F^*(x)$ is the estimate of the probability distribution function $F(x)$ defined by the total time when the value of the realization $x(t)$ is below the level x within the limits of the observation interval $[0, T]$.

It is natural to suppose that the larger the observation interval and the total time when the value of the realization $x(t)$ is below the level x the closer the estimate $F^*(x)$ to the true value $F(x)$. Thus, in the limiting case, when the observation interval $[0, T]$ is large, the following equality

$$F(x) = P[\xi(t) \le x] = \lim_{T\to\infty}\frac{1}{T}\sum_{i=1}^{N}\tau_i \tag{14.2}$$

is satisfied.

The approximation accuracy of the defined experiment estimate $F^*(x)$ to the true probability distribution function $F(x)$ is characterized by the random variable

$$\Delta F(x) = F(x) - F^*(x). \tag{14.3}$$

Based on (14.3), we obtain the bias of the probability distribution function estimate

$$b[F^*(x)] = \langle \Delta F(x)\rangle = F(x) - \langle F^*(x)\rangle \tag{14.4}$$

and the variance of the probability distribution function estimate

$$\mathrm{Var}\{F^*(x)\} = \left\langle [F^*(x) - \langle F^*(x)\rangle]^2\right\rangle, \tag{14.5}$$

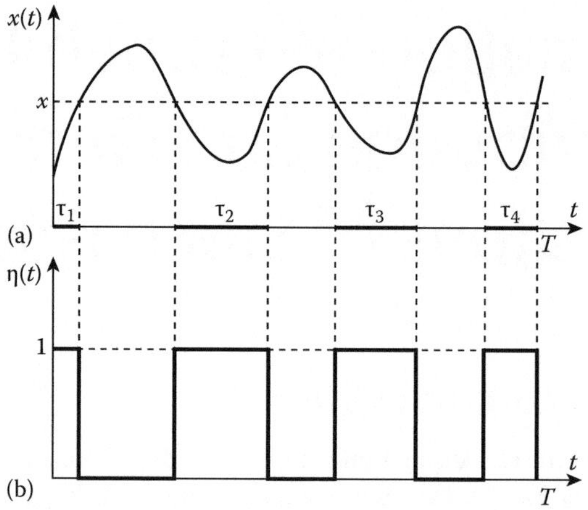

FIGURE 14.1 (a) Realization of stochastic process within the limits of the observation interval [0, *T*]; (b) Sequence of rectangular pulses formed in accordance with the nonlinear inertialess transformation (14.13).

which are the functions both of the averaging time and of the reference level.

The one-dimensional pdf of ergodic stochastic process is defined based on the following relationship:

$$p(x) = \frac{dF(x)}{dx} = \lim_{T\to\infty} \frac{1}{T\Delta x} \sum_{i=1}^{N} \tau_i'. \tag{14.6}$$

As we can see, (14.6) is the ratio between the probabilities of event that the values of stochastic process $\xi(t)$ are within the limits of the interval $[x - 0.5dx; x + 0.5dx]$ and dx, where dx is the interval length. In other words,

$$p[x(t)] = \frac{P\{(x - 0.5dx) \le x(t) \le (x + 0.5dx)\}}{dx}. \tag{14.7}$$

As we can see from (14.6), $\sum_{i=1}^{N} \tau_i'$ is the total time when the values of the realization $x(t)$ of stochastic process are within the limits of the interval $[x \pm 0.5dx]$ (see Figure 14.2a).

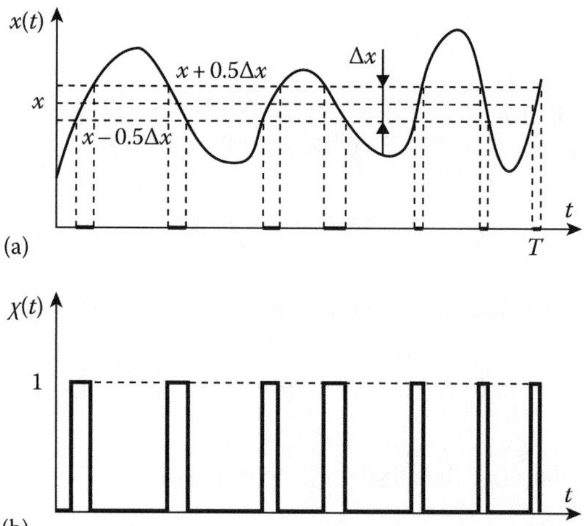

FIGURE 14.2 (a) Realization of stochastic process within the limits of the observation interval [0, *T*] with quantization by amplitude within the limits of the interval $[x \pm 0.5\Delta x]$; (b) Sequence of rectangular pulses formed in accordance with the nonlinear transformation (14.23).

In practice, a measurement of the pdf is carried out within the limits of the fixed time interval $[0, T]$. The interval Δx takes the finite value and $\Delta x \neq 0$. In this case, the pdf defined by experiment can be presented in the following form:

$$p^*(x) = \frac{1}{T\Delta x}\sum_{i=1}^{N}\tau_i' \tag{14.8}$$

and is different from the true values of the pdf $p(x)$ on the random variable defined as

$$\Delta p(x) = p(x) - p^*(x). \tag{14.9}$$

The bias and variance of the pdf estimate can be presented in the following forms:

$$b[p^*(x)] = p(x) - \langle p^*(x)\rangle, \tag{14.10}$$

$$\text{Var}\{p^*(x)\} = \left\langle [p^*(x) - \langle p^*(x)\rangle]^2 \right\rangle. \tag{14.11}$$

According to (14.8), the pdf is defined experimentally like the average value of the pdf within the limits of the interval $[x \pm 0.5dx]$. Because of this, from the viewpoint of approximation of the measured pdf to its true value at the level x, it is desirable to decrease the interval Δx. However, a decrease in the Δx at the fixed observation time interval $[0, T]$ leads to a decrease in time of the stochastic process within the limits of the interval Δx and, consequently, to an increase in the variance of the pdf estimate owing to a decrease in the statistical data sample based on which the decision about the pdf $p(x)$ magnitude is made. Therefore, there is an optimal value of the interval Δx, under which the dispersion of the pdf estimate

$$\hat{D}[p^*(x)] = b^2[p^*(x)] + \text{Var}\{p^*(x)\} \tag{14.12}$$

takes the minimal magnitude at the fixed observation time T.

There are several ways to measure the total time of stochastic process observation below the given level or within its limits. Consider these ways applied to an experimental definition of the probability distribution function estimate $F^*(x)$. The first way is the direct measurement of total time when a magnitude of the realization $x(t)$ of continuous ergodic process $\xi(t)$ is below the fixed level x. In this case, the realization $x(t)$ is transformed into the sequence of rectangular pulses $\eta(t)$ (see the Figure 14.1b) with the unit amplitude and duration τ_i equal to the time when a magnitude of the realization $x(t)$ of continuous ergodic process $\xi(t)$ is below the fixed level x. This transformation takes the following form:

$$\eta(t) = \begin{cases} 1 & \text{at} \quad x(t) \leq x, \\ 0 & \text{at} \quad x(t) > x. \end{cases} \tag{14.13}$$

Then, the area occupied by all rectangular pulses is equal by value to the total time when the realization $x(t)$ of continuous ergodic process $\xi(t)$ is below the fixed level x:

$$\sum_{i=1}^{N}\tau_i = \int_0^T \eta(t)dt. \tag{14.14}$$

As a result, the probability distribution function estimate takes the following form:

$$F^*(t) = \frac{1}{T}\int_0^T \eta(t)dt. \tag{14.15}$$

The nonlinear inertialess transformation (14.13) is carried out without any problems by corresponding limitations in amplitudes and threshold devices.

The second way to define and measure the probability distribution function is based on computation of the number of quantized by amplitude and duration of the sampled pulses. In this case, at first, the continuous realization $x(t)$ of stochastic process $\xi(t)$ is transformed by the corresponding pulse modulator into sampled sequence x_i and comes in at the input of the threshold device with the threshold level x. Then, a ratio of the number of pulses N_x that do not exceed the threshold to the total number of pulses N corresponding to the observation time interval $[0, T]$ is equal by value to the probability distribution function estimate, i.e.,

$$F^*(x) = \frac{N_x}{N}. \tag{14.16}$$

The total number of pulses is associated with the observation time accurate within a single pulse by the following relationship:

$$N = \frac{T}{T_{\text{p}}}, \tag{14.17}$$

where T_{p} is the period of pulse repetition.

The number of pulses N_x that do not exceed the threshold x can be presented in the form of summation of the pulses with unit amplitude

$$N_x = \sum_{i=1}^{N} \eta_i, \tag{14.18}$$

where

$$\eta_i = \begin{cases} 1 & \text{at} \quad x_i \le x, \\ 0 & \text{at} \quad x_i > x. \end{cases} \tag{14.19}$$

Then the probability distribution function estimate takes the following form:

$$F^*(x) = \frac{1}{N}\sum_{i=1}^{N} \eta_i. \tag{14.20}$$

The number of pulses N and N_x is defined by various analog and digital counters. The digital counters are preferable. In practice, it is convenient to employ the nonlinear transformations of the following form:

$$\eta_1(t) = 1 - \eta(t) = \begin{cases} 1 & \text{at} \quad x(t) \ge x, \\ 0 & \text{at} \quad x(t) < x. \end{cases} \tag{14.21}$$

In this case, we use the following probability distribution function:

$$F_1(x) = 1 - F(x). \tag{14.22}$$

This statement is correct with respect to the discrete stochastic process. Experimental definition of the pdf of ergodic stochastic process is carried out in an analogous way. In the case of the first method, we carry out the following nonlinear transformation (see Figure 14.2b):

$$\chi(t) = \begin{cases} 1 & \text{at} \quad x - 0.5\Delta x \le x(t) \le x + 0.5\Delta x, \\ 0 & \text{at} \quad x(t) > x(t) < x - 0.5\Delta x,\ x + 0.5\Delta x. \end{cases} \tag{14.23}$$

In the case of the second method, using a computation of pulses, we carry out the following nonlinear transformation:

$$\chi_i = \begin{cases} 1 & \text{at} \quad x - 0.5\Delta x \le x_i \le x + 0.5\Delta x, \\ 0 & \text{at} \quad x_i < x - 0.5\Delta x, \quad x_i > x + 0.5\Delta x. \end{cases} \tag{14.24}$$

Then, an experimental measurement of the stochastic process pdf is reduced to the following procedures:

$$p^*(x) = \frac{1}{T\Delta x}\int_0^T \chi(t)dt \tag{14.25}$$

and

$$p^*(x) = \frac{1}{N\Delta x}\sum_{i=1}^{N} \chi_i. \tag{14.26}$$

14.2 CHARACTERISTICS OF PROBABILITY DISTRIBUTION FUNCTION ESTIMATE

Determine the bias and variance of the probability distribution function estimate based on the direct measurement of the total time when the realization $x(t)$ of continuous ergodic process $\xi(t)$ is below the fixed level x according to (14.15). Deviation of measured value of the probability distribution function estimate $F^*(x)$ with respect to the true value $F(x)$ can be presented as follows:

$$\Delta F(x) = \frac{1}{T}\int_0^T \eta(t)dt - F(x). \tag{14.27}$$

According to (14.13), the mathematical expectation of realizations $\eta(t)$ can be presented in the following form:

$$\langle \eta(t) \rangle = P[\xi(t) \le x] = \int_{-\infty}^{x} p(x')dx' = F(x). \tag{14.28}$$

As we can see from (14.27), the probability distribution function estimate is unbiased. In (14.28), $p(x')$ is the one-dimensional pdf of the investigated stochastic process.

Define the correlation function of the probability distribution function estimate $F^*(x)$ at various levels x_1 and x_2:

$$\langle \Delta F(x_1)\Delta F(x_2)\rangle = R_F(x_1,x_2) = \frac{1}{T^2}\int_0^T\int_0^T \langle \eta(t_1)\eta(t_2)\rangle dt_1 dt_2 - F(x_1)F(x_2). \tag{14.29}$$

The instantaneous function $\langle \eta(t_1)\eta(t_2)\rangle$ of the stochastic process obtained as a result of nonlinear operation given by (14.13) is numerically equal to the probability of joint event that $\xi(t_1) \le x_1$ and $\xi(t_2) \le x_2$, i.e., a value of the two-dimensional probability distribution function $F(x_1, x_2; \tau)$ at the points x_1 and x_2:

$$F(x_1,x_2;\tau) = \langle \eta(t_1)\eta(t_2)\rangle = P[\xi(t_1) \le x_1, \xi(t_2) \le x_2] = \int_{-\infty}^{x_1}\int_{-\infty}^{x_2} p_2(x_1', x_2';\tau)dx_1'dx_2', \tag{14.30}$$

where $p_2(x_1', x_2';\tau)$ is the two-dimensional pdf of the investigated stochastic process. In the case of a stationary stochastic process, the two-dimensional pdf depends only on the absolute difference in time instants $\tau = t_2 - t_1$. For this reason, the instantaneous function given by (14.30) also depends only on the absolute difference in time instants $\tau = t_2 - t_1$. Based on this fact, the double integral in (14.29) can be transformed into a single integral. For this purpose, introduce new variables $\tau = t_2 - t_1$ and $t_1 = t$ and change the order of integration taking into consideration a parity of the two-dimensional probability distribution function $F(x_1, x_2; \tau) = F(x_1, x_2; -\tau)$. As a result, we obtain

$$\int_0^T\int_0^T \langle \eta(t_1)\eta(t_2)\rangle dt_1 dt_2 = 2T\int_0^T \left(1-\frac{\tau}{T}\right)F(x_1,x_2;\tau)d\tau. \tag{14.31}$$

Substituting (14.31) into (14.29), we obtain

$$R_F(x_1,x_2) = \frac{2}{T}\int_0^T \left(1-\frac{\tau}{T}\right)F(x_1,x_2;\tau)d\tau - F(x_1)F(x_2). \tag{14.32}$$

The variance of the probability distribution function estimate at the level x is defined by (14.32) substituting $x_1 = x_2 = x$:

$$\text{Var}\{F^*(x)\} = \frac{2}{T}\int_0^T \left(1-\frac{\tau}{T}\right)F(x,x;\tau)d\tau - F^2(x). \tag{14.33}$$

By analogous method, we can define characteristics under discrete transformation of the stochastic process. In this case, the deviation of the probability distribution function estimate with respect to its true value can be presented in the following form:

$$\Delta F(x) = \frac{1}{N}\sum_{i=1}^{N}\eta_i - F(x). \tag{14.34}$$

As we can see, the probability distribution function estimate is unbiased and the correlation function of the probability distribution function estimate at different levels x_1 and x_2 can be written in the following form:

$$R_F(x_1,x_2)=\frac{1}{N^2}\sum_{i=1,j=1}^{N}\langle\eta_i\eta_j\rangle-F(x_1)F(x_2), \tag{14.35}$$

where

$$\langle\eta_i\eta_j\rangle=F[x_1,x_2;(i-j)T_p]=\int_{-\infty}^{x_1}\int_{-\infty}^{x_2}p_2[x_1',x_2';(i-j)T_p]dx_1'dx_2'. \tag{14.36}$$

The double sum in (14.35) can be presented in the following form:

$$\sum_{i=1,j=1}^{N}\langle\eta_i\eta_j\rangle=N\langle\eta_i^2\rangle+\sum_{\substack{i=1,j=1\\ i\neq j}}^{N}\langle\eta_i\eta_j\rangle. \tag{14.37}$$

The first term in (14.37) depends on the relationship between values x_1 and x_2 and can be presented in the following form:

$$N\langle\eta_i^2\rangle=\begin{cases}NF(x_1) & \text{if } x_1\le x_2,\\ NF(x_2) & \text{if } x_2\le x_1.\end{cases} \tag{14.38}$$

Actually, at $x_1<x_2$, the probability of joint event that $\xi\le x_1$ and $\xi\le x_2$ is not equal to zero if and only if $\xi\le x_1$. In this case, the inequality $\xi<x_2$ is satisfied with the probability equal to unit. Therefore, the probability of the joint event is equal to the probability that $\xi\le x_1$. For the second term in (14.37), we can introduce a new summation index $k=i-j$ and change the order of summation. As a result, we obtain

$$\sum_{\substack{i=1,j=1\\ i\neq j}}^{N}F[x_1,x_2;(i-j)T_p]=2\sum_{k=1}^{N-1}(N-k)F[x_1,x_2;kT_p]. \tag{14.39}$$

Substituting (14.38) and (14.39) into (14.35), the correlation function of the probability distribution function estimate $F^*(x)$ takes the following form:

$$R_F(x_1,x_2)=\begin{cases}\dfrac{F(x_1)}{N}+\dfrac{2}{N}\sum\limits_{k=1}^{N-1}\left(1-\dfrac{k}{N}\right)F[x_1,x_2;kT_p]-F(x_1)F(x_2), & x_1\le x_2,\\[2ex] \dfrac{F(x_2)}{N}+\dfrac{2}{N}\sum\limits_{k=1}^{N-1}\left(1-\dfrac{k}{N}\right)F[x_1,x_2;kT_p]-F(x_1)F(x_2), & x_1\ge x_2.\end{cases} \tag{14.40}$$

The variance of the correlation function of the probability distribution function estimate $F^*(x)$ of the stochastic process is defined based on (14.40) by substituting $x_1 = x_2 = x$:

$$\mathrm{Var}\{F^*(x)\} = \frac{F(x)}{N} + \frac{2}{N}\sum_{k=1}^{N-1}\left(1-\frac{k}{N}\right)F[x,x;kT_{\mathrm{p}}] - F^2(x). \tag{14.41}$$

If samples are uncorrelated, then the correlation function given by (14.40) is simplified since at $k \neq 0$

$$F[x_1,x_2;kT_{\mathrm{p}}] = F(x_1)F(x_2). \tag{14.42}$$

As a result, we obtain

$$R_F(x_1,x_2) = \begin{cases} \dfrac{F(x_1)[1-F(x_2)]}{N}, & x_1 \leq x_2, \\ \dfrac{F(x_2)[1-F(x_1)]}{N}, & x_1 \geq x_2. \end{cases} \tag{14.43}$$

In doing so, the variance of the probability distribution function estimate $F^*(x)$ of the stochastic process takes the following form:

$$\mathrm{Var}\{F^*(x)\} = \frac{F(x)[1-F(x)]}{N}. \tag{14.44}$$

As we can see from (14.44), the variance of the probability distribution function estimate $F^*(x)$ of the stochastic process depends essentially on the level x and reaches the maximum

$$\mathrm{Var}_{\max}\{F^*(x)\} = \frac{1}{4N} \tag{14.45}$$

at the level x corresponding to the condition $F(x) = 0.5$.

In addition to the finite observation time period of the realization $x(t)$ of continuous ergodic process $\xi(t)$, the characteristics of the probability distribution function estimate $F^*(x)$ of the stochastic process depends also on the stability of the threshold level that can possess a random component characterized by the value δ, i.e., instead of the level x we use the level $x + \delta$ in practice. Then, the obtained characteristics of the probability distribution function estimate $F^*(x)$ of stochastic process are conditional, and to obtain the unconditional characteristics of the probability distribution function estimate $F^*(x)$ of stochastic process, there is a need to employ an averaging by all possible values of the random variable δ. We assume that the correlation interval of the random variable δ is much more in comparison with the observation time interval $[0, T]$. Consequently, we can think that the random variable δ does not change during measurement time of the probability distribution function estimate $F^*(x)$ of stochastic process and has the same statistical characteristics for all possible values x. Additionally, it is natural to suppose that random variations of the threshold are negligible and a difference between $F(x + \delta)$ and $F(x)$ is very small in average sense. In this case, the probability distribution function $F(x + \delta)$ can be expanded in Taylor series about a point x and can be limited by the first three terms of expansion:

$$F(x+\delta) \approx F(x) + \delta p(x) + 0.5\delta^2\frac{dp(x)}{dx}. \tag{14.46}$$

Averaging (14.46) by realizations of the random variable δ, we obtain

$$\langle F(x+\delta)\rangle \approx F(x)+E_{\delta}p(x)+0.5D_{\delta}\frac{dp(x)}{dx}, \tag{14.47}$$

where E_δ and D_δ are the mathematical expectation and dispersion of random variations of the threshold value x. Based on (14.27) and (14.34), we obtain that the random variations of the threshold level lead to the bias of the probability distribution function estimate $F^*(x)$ of stochastic process:

$$b[F^*(x)] = E_{\delta}p(x)+0.5D_{\delta}\frac{dp(x)}{dx}. \tag{14.48}$$

Naturally, the presence of random variable δ leads to an increase in the variance of the probability distribution function estimate $F^*(x)$ of stochastic process.

14.3 VARIANCE OF PROBABILITY DISTRIBUTION FUNCTION ESTIMATE

14.3.1 Gaussian Stochastic Process

The two-dimensional pdf of a Gaussian stochastic process with zero mathematical expectation is defined in the following form:

$$p_2(x_1,x_2;\tau)=\frac{1}{2\pi\sigma^2\sqrt{1-\mathcal{R}^2(\tau)}}\exp\left\{-\frac{x_1^2-2\mathcal{R}(\tau)x_1x_2+x_2^2}{2\sigma^2[1-\mathcal{R}^2(\tau)]}\right\}, \tag{14.49}$$

where

σ^2 is the variance

$\mathcal{R}(\tau)$ is the normalized correlation function of the investigated stochastic process

However, the written form of the two-dimensional pdf of Gaussian stochastic process with zero mathematical expectation in (14.49) does not allow us to obtain the formulas for the variance of the probability distribution function estimate $F^*(x)$ that are convenient for further analysis. Therefore, we present the two-dimensional pdf of Gaussian stochastic process with zero mathematical expectation in the form of a series with respect to derivatives from the Gaussian Q-function, i.e., $Q(x)$ given by (12.157), where we assume that the mathematical expectation is equal to zero ($E = 0$):

$$p_2(x_1,x_2;\tau)=\frac{1}{\sigma^2}\sum_{\nu=0}^{\infty}\left\{1-Q^{\nu+1}\left(\frac{x_1}{\sigma}\right)\right\}\left\{1-Q^{\nu+1}\left(\frac{x_2}{\sigma}\right)\right\}\frac{\mathcal{R}^{\nu}(\tau)}{\nu!}. \tag{14.50}$$

Substituting (14.50) into (14.30), we obtain

$$F(x_1,x_2;\tau)=\left\{1-Q^{\nu+1}\left(\frac{x_1}{\sigma}\right)\right\}\left\{1-Q^{\nu+1}\left(\frac{x_2}{\sigma}\right)\right\}+\sum_{\nu=0}^{\infty}\left\{1-Q^{\nu+1}\left(\frac{x_1}{\sigma}\right)\right\}\left\{1-Q^{\nu+1}\left(\frac{x_2}{\sigma}\right)\right\}\frac{\mathcal{R}^{\nu}(\tau)}{\nu!}. \tag{14.51}$$

Taking into consideration that in the case of Gaussian stochastic process

$$F(x)=1-Q\left(\frac{x}{\sigma}\right), \tag{14.52}$$

the correlation function given by (14.32) and the variance given by (14.33) of the probability distribution function estimate $F^*(x)$ take the following form:

$$R_F(x_1,x_2)=\sum_{\nu=0}^{\infty}\frac{1}{\nu!}\left\{1-Q^{\nu}\left(\frac{x_1}{\sigma}\right)\right\}\left\{1-Q^{\nu}\left(\frac{x_2}{\sigma}\right)\right\}\frac{2}{T}\int_0^T\left(1-\frac{\tau}{T}\right)\mathcal{R}^{\nu}(\tau)d\tau; \tag{14.53}$$

$$\mathrm{Var}\{F^*(x)\}=\sum_{\nu=0}^{\infty}\frac{1}{\nu!}\left\{1-Q^{\nu}\left(\frac{x}{\sigma}\right)\right\}^2\frac{2}{T}\int_0^T\left(1-\frac{\tau}{T}\right)\mathcal{R}^{\nu}(\tau)d\tau. \tag{14.54}$$

Since

$$\left|Q^{\nu}(-z)\right|=\left|Q^{\nu}(z)\right|,\quad \nu=1,2,\ldots, \tag{14.55}$$

based on (14.54), we obtain that the variance $\mathrm{Var}\{F^*(x)\}$ of the probability distribution function estimate $F^*(x)$ is the even function with respect to the threshold level x.

Table 14.1 represents the values of coefficients

$$a_{\nu}=\frac{1}{\nu!}[1-Q^{\nu}(z)]^2 \tag{14.56}$$

as a function of the normalized level

$$z=\frac{x}{\sigma}. \tag{14.57}$$

TABLE 14.1
Values of the Coefficients $a_\nu = (1/\nu!)[1 - Q^\nu(z)]^2$ as a Function of the Normalized Level $z = x/\sigma$

z	$\nu = 1$	$\nu = 2$	$\nu = 3$	$\nu = 4$	$\nu = 5$	$\nu = 6$	$\nu = 7$
0.0	0.15915	0.00000	0.02653	0.00000	0.01193	0.00000	0.00710
0.1	0.15757	0.00079	0.02574	0.00059	0.01135	0.00048	0.00662
0.2	0.15291	0.00306	0.02349	0.00223	0.00971	0.00181	0.00530
0.3	0.14543	0.00605	0.02007	0.00462	0.00738	0.00362	0.00353
0.5	0.12395	0.01549	0.01162	0.00976	0.00252	0.00679	0.00054
0.7	0.09750	0.02389	0.00423	0.01254	0.00007	0.00709	0.00025
1.0	0.05854	0.02927	0.00000	0.00976	0.00195	0.00293	0.00297
1.5	0.01678	0.01887	0.00436	0.00088	0.00413	0.00031	0.00157
2.0	0.00291	0.00583	0.00437	0.00048	0.00061	0.00131	0.00007
2.5	0.00031	0.00096	0.00141	0.00084	0.00005	0.00019	0.00035
3.0	0.00002	0.00009	0.00021	0.00026	0.00015	0.00001	0.00003
3.5	0.00000	0.00000	0.00001	0.00003	0.00004	0.00002	0.00000
4.0	0.00000	0.00000	0.00000	0.00000	0.00000	0.00000	0.00000

As we can see from Table 14.1, the a_ν defining the variance of the probability distribution function estimate $F^*(x)$ of Gaussian stochastic process decreases with an increase in the number ν. Taking into consideration that in the case of stochastic processes the factor

$$c_\nu = \frac{2}{T}\int_0^T \left(1 - \frac{\tau}{T}\right)\mathcal{R}^\nu(\tau)d\tau \tag{14.58}$$

decreases also with an increase in the number ν, in practice, we can be limited by the first 5–7 terms of series expansion.

When the observation time interval $[0, T]$ is much more than the correlation interval of stochastic process, as earlier, the coefficients c_ν can be approximated by

$$c_\nu \approx \frac{2}{T}\int_0^T \mathcal{R}^\nu(\tau)d\tau. \tag{14.59}$$

If, in addition to the condition $T \gg \tau_{\text{cor}}$ we assume that $T \to \infty$, then, according to (14.59) and (14.54) the variance of the probability distribution function estimate $F^*(x)$ of the Gaussian stochastic process tends to approach zero, which should be expected; i.e., the probability distribution function estimate $F^*(x)$ of the Gaussian stochastic process is consistent.

In the case when the observation time is lesser than the correlation interval of the stochastic process, i.e., $T < \tau_{\text{cor}}$, we have

$$\text{Var}\{F^*(x)\} \approx \sum_{\nu=1}^{\infty} a_\nu. \tag{14.60}$$

Formula (14.60) characterizes the maximal variance of the probability distribution function estimate $F^*(x)$ of Gaussian stochastic process.

As an example, consider the Gaussian stochastic process with the exponential normalized correlation function given by (12.13). Substituting (12.13) into (14.58), we obtain

$$c_\nu = \frac{2}{p^2\nu^2}[\exp(-p\nu) + p\nu - 1], \tag{14.61}$$

where p is given by (12.140). In doing so, if $p \gg 1$, then

$$c_\nu \approx \frac{2}{p\nu}. \tag{14.62}$$

The root-mean-square deviation $\sigma\{F^*(x)\}$ of measurements of the probability distribution function of Gaussian stochastic process as a function of the normalized level z for various values of p is presented in Figure 14.3. As we can see from Figure 14.3, the maximal values of $\sigma\{F^*(x)\}$ are obtained at the zero level.

The maximal value of $\sigma\{F^*(x)\}$ as a function of the parameter p given by (12.40) is shown in Figure 14.4. As we can see from Figure 14.4 and based on (14.54) and (14.61), the variance of the probability distribution function estimate $F^*(x)$ decreases linearly as a function of p starting from $p \geq 10$. This graph allows us to define the minimal required time to observe a realization of the

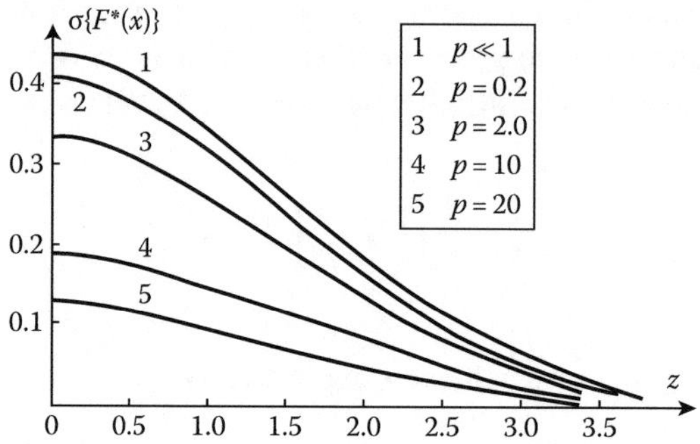

FIGURE 14.3 Root-mean-square deviations $\sigma\{F^*(x)\}$ as a function of the normalized level z. Gaussian stochastic process.

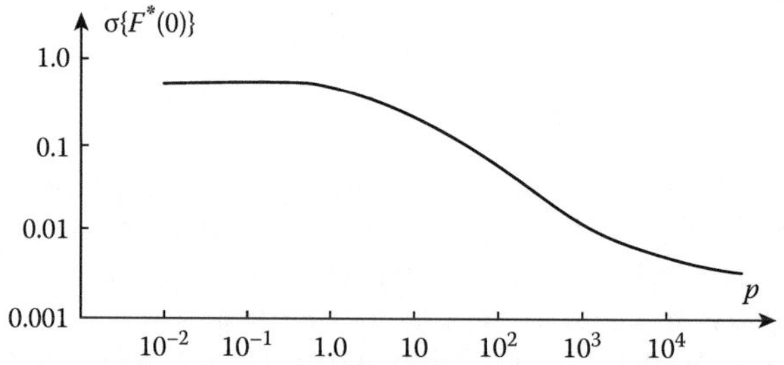

FIGURE 14.4 Maximal value of $\sigma\{F^*(0)\}$ as a function of p. Gaussian stochastic process.

Gaussian stochastic process when the maximal value of variance is not more than the previously given definite magnitude. For example, if we require that $\sigma\{F^*(x)\} \le 0.01$, the condition $p \ge 3000$ should be satisfied.

When we investigate a discrete Gaussian stochastic process, the variance of the probability distribution function estimate $F^*(x)$ is defined as

$$\operatorname{Var}\{F^*(x)\} = \frac{1}{N}\left\{1 - Q\left(\frac{x}{\sigma}\right)\right\}Q\left(\frac{x}{\sigma}\right) + \sum_{v=1}^{\infty}\frac{1}{v!}\left\{1 - Q^{v}\left(\frac{x}{\sigma}\right)\right\}^{2} c'_v, \tag{14.63}$$

where

$$c'_v = \frac{2}{N}\sum_{k=1}^{N-1}\left(1 - \frac{k}{N}\right)\mathcal{R}^{v}(kT_{\text{p}}). \tag{14.64}$$

As applied to the exponential normalized correlation function

$$\mathcal{R}^{v}(kT_{\text{p}}) = \exp\{-\alpha\,|\,kT_{\text{p}}\,|\}, \tag{14.65}$$

(14.64) can be presented in the following form [2]:

$$c'_v = \frac{2}{N} \times \frac{\exp\{-v\alpha T_{\text{p}}\}\left\{1 - \exp\{-v\alpha T_{\text{p}}\} + N^{-1}\{\exp\{-v\alpha T_{\text{p}}\} - 1\}\right\}}{\{1 - \exp\{-v\alpha T_{\text{p}}\}\}^{2}}. \tag{14.66}$$

TABLE 14.2
Normalized Correlation Function (14.68) as a Function of the Normalized Threshold Level z

	z_1						
z_2	−1.5	−1.0	−0.5	0	0.5	1.0	1.5
−1.5	1.00	0.62	0.40	0.27	0.18	0.12	0.07
−1.0	0.62	1.00	0.65	0.43	0.29	0.19	0.12
−0.5	0.40	0.65	1.00	0.67	0.45	0.29	0.18
0	0.27	0.43	0.67	1.00	0.67	0.43	0.27
0.5	0.18	0.29	0.45	0.67	1.00	0.65	0.40
1.0	0.12	0.19	0.29	0.43	0.65	1.00	0.62
1.5	0.07	0.12	0.18	0.27	0.40	0.62	1.00

At $N \gg 1$

$$c'_v \approx \frac{2}{N} \times \frac{\exp\{-v\alpha T_p\}}{1-\exp\{-v\alpha T_p\}}. \tag{14.67}$$

In the case of uncorrelated samples, we have $c'_v = 0$. As a result, (14.63) coincides with (14.44).

Thus, (14.63) demonstrates that the variance of the probability distribution function estimate $F^*(x)$ of stochastic process by correlated samples increases in comparison with the variance of the probability distribution function estimate $F^*(x)$ of stochastic process by uncorrelated samples of the same sample size N.

Table 14.2 represents the normalized correlation function

$$\rho(x_1, x_2) = \frac{R_F(x_1, x_2)}{\sqrt{\text{Var}\{F^*(x_1)\}\text{Var}\{F^*(x_2)\}}} \tag{14.68}$$

of the probability distribution function estimate $F^*(x)$ of stochastic process for various values of the threshold level z computed based on (14.43) in the case of uncorrelated samples. As we can see from Table 14.2, there is a high correlation between the probability distribution function estimations $F^*(x)$ of stochastic process.

14.3.2 Rayleigh Stochastic Process

Two-dimensional pdf of Rayleigh stochastic process $\xi(t)$ can be presented in the following form [1]:

$$p_2(x_1, x_2; \tau) = \frac{x_1 x_2}{\sigma^4[1-\mathcal{R}^2(\tau)]} \exp\left\{-\frac{x_1^2 + x_2^2}{2\sigma^2[1-\mathcal{R}^2(\tau)]}\right\} I_0\left\{\frac{x_1 x_2}{\sigma^2} \times \frac{\mathcal{R}(\tau)}{1-\mathcal{R}^2(\tau)}\right\}, \tag{14.69}$$

where

$I_0(z)$ is the zero-order Bessel function of imaginary argument

σ^2 and $\mathcal{R}(\tau)$ are the parameters of the pdf associated with the initial second moment $\xi(t)$ and the normalized correlation function $\rho(\tau)$ of Rayleigh stochastic process by (12.165)

Introduce new variables

$$y_1 = \frac{x_1^2}{2\sigma^2} \quad \text{and} \quad y_2 = \frac{x_2^2}{2\sigma^2}. \tag{14.70}$$

New top integration limits in (14.30) correspond to $z_1 = y_1$ and $z_2 = y_2$. New two-dimensional pdf can be presented by expansion in series using the orthogonal Laguerre polynomials [2]:

$$p_2(x_1, x_2; \tau) = \exp(-y_1 - y_2) \sum_{\nu=0}^{\infty} \frac{L_\nu(y_1) L_\nu(y_2) \mathcal{R}^{2\nu}(\tau)}{(\nu!)^2}, \tag{14.71}$$

where $L_\nu(y)$ is the Laguerre polynomial satisfying the following definition:

$$L_\nu(y) = \exp\{y\} \frac{d^\nu [y \exp(-y)]}{dy^\nu} = \sum_{\mu=0}^{\nu} (-1)^\mu C_\nu^\mu \frac{\nu!}{\mu!} y^\mu, \tag{14.72}$$

where

$$C_\nu^\mu = \frac{\nu!}{\mu!(\nu - \mu)!}; \tag{14.73}$$

$$L_0(y) = 1; \tag{14.74}$$

$$L_1(y) = -y + 1. \tag{14.75}$$

Substituting (14.71) into (14.30), we obtain

$$F(x_1, x_2; \tau) = [1 - \exp(-z_1)][1 - \exp(-z_2)] + \sum_{\nu=1}^{\infty} \frac{\mathcal{R}^{2\nu}(\tau)}{(\nu!)^2} \int_0^{z_1} \exp(-y_1) L_\nu(y_1) dy_1 \int_0^{z_2} \exp(-y_2) L_\nu(y_2) dy_2, \tag{14.76}$$

where

$$\begin{cases} z_1 = \dfrac{x_1^2}{2\sigma^2}; \\ z_2 = \dfrac{x_2^2}{2\sigma^2}. \end{cases} \tag{14.77}$$

Based on Ref. [2], we can write

$$\int_0^{\infty} \exp(-y) L_\nu(y) dy = 0; \tag{14.78}$$

$$\int_0^\infty \exp(-y)L_\nu(y)dy = \exp(-\beta)[L_\nu(\beta) - \nu L_{\nu-1}(\beta)]. \tag{14.79}$$

Taking into consideration (14.77) through (14.79), we can write

$$F(x_1, x_2; \tau) = [1 - \exp(-z_1)][1 - \exp(-z_2)]$$
$$+ \sum_{\nu=1}^{\infty} \frac{\mathcal{R}^{2\nu}(\tau)}{(\nu!)^2} \exp(-z_1 - z_2)[\nu L_{\nu-1}(z_1) - L_\nu(z_1)][\nu L_{\nu-1}(z_2) - L_\nu(z_2)]. \tag{14.80}$$

Substituting (14.80) into (14.32) and taking into consideration that in the case of the Rayleigh stochastic process

$$F(x) = 1 - \exp\left\{-\frac{x^2}{2\sigma^2}\right\}, \tag{14.81}$$

we obtain the correlation function of the probability distribution function estimations $F^*(x)$ of Rayleigh stochastic process at various levels:

$$R_F(x_1, x_2) = \sum_{\nu=1}^{\infty} \frac{d_\nu}{(\nu!)^2} \exp\left\{-\frac{x_1^2 + x_2^2}{2\sigma^2}\right\}\left\{\nu L_{\nu-1}\left(\frac{x_1^2}{2\sigma^2}\right) - L_\nu\left(\frac{x_1^2}{2\sigma^2}\right)\right\}\left\{\nu L_{\nu-1}\left(\frac{x_2^2}{2\sigma^2}\right) - L_\nu\left(\frac{x_2^2}{2\sigma^2}\right)\right\}, \tag{14.82}$$

where

$$d_\nu = \frac{2}{T}\int_0^T \left(1 - \frac{\tau}{T}\right)\mathcal{R}^{2\nu}(\tau)d\tau. \tag{14.83}$$

Substituting into (14.82) $x_1 = x_2 = x$, we obtain the variance of the probability distribution function estimations $F^*(x)$ of Rayleigh stochastic process:

$$\text{Var}\{F^*(x)\} = \sum_{\nu=1}^{\infty} \frac{d_\nu}{(\nu!)^2} \exp\left\{-\frac{x^2}{\sigma^2}\right\}\left\{\nu L_{\nu-1}\left(\frac{x^2}{2\sigma^2}\right) - L_\nu\left(\frac{x^2}{2\sigma^2}\right)\right\}^2. \tag{14.84}$$

When $T \gg \tau_{\text{cor}}$, where τ_{cor} is the correlation interval, in the case of stochastic process with the normalized correlation function $\mathcal{R}(\tau)$, we obtain

$$d_\nu \approx \frac{2}{T}\int_0^T \mathcal{R}^{2\nu}(\tau)d\tau. \tag{14.85}$$

If, in addition to the condition $T \gg \tau_{\text{cor}}$, we suppose that $T \to \infty$, then based on (14.85) we can conclude that $d_\nu \to 0$ and the variance of the probability distribution function estimations $F^*(x)$ of

Rayleigh stochastic process tends to approach zero and is consistent for the considered case. The variance of the probability distribution function estimation $F^*(x)$ of Rayleigh stochastic process also tends to approach zero at $z = 0$ $(x = 0)$ and $z \to \infty$ $(x \to \infty)$, as in both cases the coefficients

$$b_\nu = \frac{1}{(\nu!)^2}\exp\left\{-\frac{x^2}{\sigma^2}\right\}\left\{\nu L_{\nu-1}\left(\frac{x^2}{2\sigma^2}\right) - L_\nu\left(\frac{x^2}{2\sigma^2}\right)\right\}^2 \tag{14.86}$$

tend to approach zero for any values of the ratio T/τ_{cor}.

When $T \ll \tau_{cor}$, the variance of the probability distribution function estimations $F^*(x)$ of Rayleigh stochastic process can be presented in the following form:

$$\mathrm{Var}\{F^*(x)\} = \sum_{\nu=1}^{\infty} b_\nu. \tag{14.87}$$

Table 14.3 represents the coefficients b_ν as a function of the number ν and the normalized level z. As we can see from Table 14.3, the coefficients b_ν decrease slowly and nonmonotonically with an increase in the number ν. In practice, we can be limited by 4–5 terms of the sum given by (14.84).

As an example, we consider the Rayleigh stochastic process with the normalized correlation function $\mathcal{R}(\tau)$ given by (12.13). In this case, we can write

$$d_\nu = \frac{\exp\{-2p\nu\} + 2p\nu - 1}{2p^2\nu^2}. \tag{14.88}$$

When $p = T/\tau_{cor} \gg 1$, we have

$$d_\nu \approx \frac{1}{p\nu}. \tag{14.89}$$

TABLE 14.3
Coefficients b_ν as a Function of the Number ν and Normalized Level z

z	$\nu = 1$	$\nu = 2$	$\nu = 3$	$\nu = 4$	$\nu = 5$
0.0	0.00000	0.00000	0.00000	0.00000	0.00000
0.1	0.00010	0.00010	0.00009	0.00010	0.00009
0.2	0.00148	0.00143	0.00137	0.00131	0.00131
0.3	0.00673	0.00615	0.00560	0.00512	0.00470
0.4	0.01850	0.01575	0.01320	0.01100	0.00920
0.5	0.03750	0.02900	0.02200	0.01620	0.01190
0.7	0.09000	0.05120	0.02720	0.01295	0.00519
1.0	0.13469	0.03360	0.00360	0.00024	0.00346
1.2	0.11640	0.00912	0.01000	0.00705	0.00827
1.5	0.05620	0.00088	0.00935	0.00576	0.0083
2.0	0.00535	0.00535	0.00061	0.00060	0.00112
2.5	0.00014	0.00065	0.00020	0.00014	0.00002
3.0	0.00000	0.00002	0.00004	0.00001	0.00002
3.5	0.00000	0.00000	0.00000	0.00000	0.00000

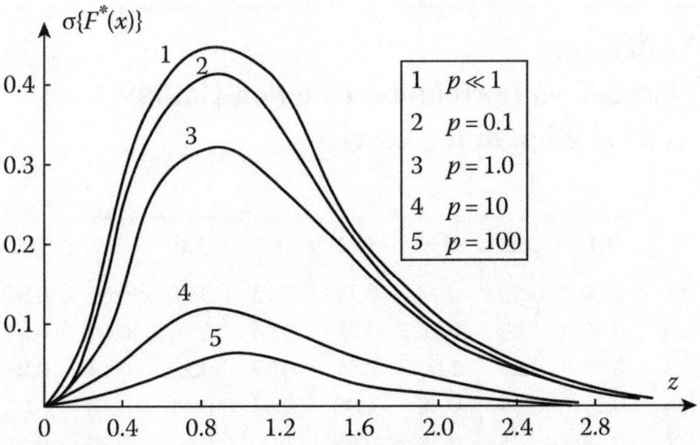

FIGURE 14.5 Maximal value of $\sigma\{F^*(x)\}$ as a function of p. Rayleigh stochastic process.

The root-mean-square deviation $\sigma\{F^*(x)\}$ of measurements of the probability distribution function of Rayleigh stochastic process as a function of the normalized level z for various values of p is presented in Figure 14.5. As we can see from Figure 14.5, the maximal value of $\sigma\{F^*(x)\} = 0.5$ is obtained about the point $z = 0.83$.

The root-mean-square deviation $\sigma\{F^*(x)\}$ at $z = 1$ as a function of the normalized observation interval p is shown in Figure 14.6. Based on Figure 14.6 and (14.89), we can conclude that starting from $p \geq 10$, in the case of Rayleigh stochastic process, the root-mean-square deviation $\sigma\{F^*(x)\}$ is inversely proportional to the ratio $p = T/\tau_{\text{cor}}$. In discrete case, when the sample size is equal to N, the variance of the probability distribution function estimations $F^*(x)$ of Rayleigh stochastic process in accordance with (14.41) and (14.80) can be presented in the following form:

$$\text{Var}\{F^*(x)\} = \frac{1}{N}\exp\left\{-\frac{x^2}{2\sigma^2}\right\}\left[1-\exp\left\{-\frac{x^2}{2\sigma^2}\right\}\right] + \sum_{\nu=1}^{\infty} d'_\nu b_\nu, \tag{14.90}$$

where

$$d'_\nu = \frac{2}{N}\sum_{k=1}^{N-1}\left(1-\frac{k}{N}\right)\mathcal{R}^{2\nu}(kT_{\text{p}}). \tag{14.91}$$

Applied to the Rayleigh stochastic process with uncorrelated samples, we obtain

$$\text{Var}\{F^*(x)\} = \frac{1}{N}\exp\left\{-\frac{x^2}{2\sigma^2}\right\}\left[1-\exp\left\{-\frac{x^2}{2\sigma^2}\right\}\right]. \tag{14.92}$$

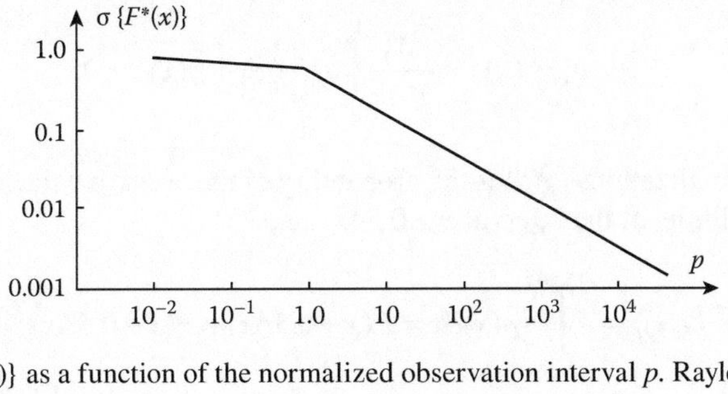

FIGURE 14.6 $\sigma\{F^*(x)\}$ as a function of the normalized observation interval p. Rayleigh stochastic process.

TABLE 14.4
Normalized Correlation Function (14.68) as a Function of the Level z

	z_2							
z_1	**0.1**	**0.2**	**0.3**	**0.5**	**0.7**	**1.0**	**1.5**	**2.0**
0.1	1.00	0.50	0.33	0.19	0.13	0.08	0.03	0.01
0.2	0.50	1.00	0.66	0.38	0.25	0.15	0.07	0.03
0.3	0.33	0.66	1.00	0.58	0.39	0.23	0.11	0.04
0.5	0.19	0.38	0.58	1.00	0.67	0.41	0.18	0.07
0.7	0.13	0.25	0.39	0.67	1.00	0.61	0.27	0.11
1.0	0.08	0.15	0.23	0.41	0.61	1.00	0.45	0.18
1.5	0.03	0.07	0.11	0.18	0.27	0.45	1.00	0.40
2.0	0.01	0.03	0.04	0.07	0.11	0.18	0.40	1.00

Table 14.4 represents the normalized correlation function given by (14.68) in the case of probability distribution function estimation $F^*(x)$ of Rayleigh stochastic process at different values of the level z under investigations by independent samples (14.67).

14.4 CHARACTERISTICS OF THE PROBABILITY DENSITY FUNCTION ESTIMATE

Determine the bias and variance of the pdf estimate of ergodic stochastic process based on an investigation of the continuous realization $x(t)$ within the limits of the observation interval $[0, T]$ in accordance with (14.25). Deviation of the measured pdf $p^*(x)$ from the true value $p(x)$ can be presented in the following form:

$$\Delta p(x) = p^*(x) - p(x) = \frac{1}{T\Delta x}\int_0^T \chi(t)dt - p(x), \tag{14.93}$$

where

$\chi(t)$ is given by (14.23)
Δx is the interval between the adjacent levels x_k and x_{k+1}

$$x = \frac{x_k + x_{k+1}}{2}. \tag{14.94}$$

In other words, the pdf is measured at the point corresponding to the middle of adjacent fixed levels.

The bias of the pdf estimate of ergodic stochastic process can be presented in the following form:

$$b\{p^*(x)\} = \frac{1}{T\Delta x}\int_0^T \langle\chi(t)\rangle dt - p(x). \tag{14.95}$$

The average value by realizations $\langle\chi(t)\rangle$ is the probability of the event that the investigated stochastic process is within the limits of the interval $x \pm 0.5\Delta x$, i.e.,

$$\langle\chi(t)\rangle = \int_{x-0.5\Delta x}^{x+0.5\Delta x} p(x)dx = F(x+0.5\Delta x) - F(x-0.5\Delta x). \tag{14.96}$$

Based on (14.95) and (14.96), we are able to define the bias of the pdf estimate of ergodic stochastic process

$$b\{p^*(x)\} = \frac{1}{\Delta x}[F(x+0.5\Delta x) - F(x-0.5\Delta x)] - p(x). \tag{14.97}$$

We can use the Taylor series expansion for the functions $F(x + 0.5\Delta x)$ and $F(x - 0.5\Delta x)$ about the point x:

$$F(x+0.5\Delta x) = \sum_{\mu=0}^{\infty} \frac{dF^{\mu}(x)}{dx^{\mu}} \frac{1}{\mu!}(0.5\Delta x)^{\mu}; \tag{14.98}$$

$$F(x-0.5\Delta x) = \sum_{\mu=0}^{\infty} \frac{dF^{\mu}(x)}{dx^{\mu}} \frac{(-1)^{\mu}}{\mu!}(0.5\Delta x)^{\mu}. \tag{14.99}$$

The bias of the pdf estimate of stochastic process can be presented in the following form:

$$b\{p^*(x)\} = \sum_{\mu=1}^{\infty} \frac{d^{2\mu+1}F(x)}{dx^{2\mu+1}} \times \frac{(0.5\Delta x)^{2\mu}}{(2\mu+1)!}. \tag{14.100}$$

In practice, we can use the following approximation:

$$b\{p^*(x)\} \approx \frac{\Delta x^2}{24} \times \frac{d^2 p(x)}{dx^2} + \frac{\Delta x^4}{1920} \times \frac{d^4 p(x)}{dx^4}. \tag{14.101}$$

As we can see from (14.97) through (14.101), the bias of the pdf estimate of stochastic process decreases proportionally to a decrease in the interval Δx between two adjacent fixed levels. It is clear from the physical viewpoint. In the limiting case, as $\Delta x \to 0$, the bias of the pdf estimate of stochastic process tends to approach zero. In other words, as $\Delta x \to 0$, the pdf estimate of stochastic process is unbiased. Actually, as $\Delta x \to 0$

$$\lim_{\Delta x \to 0} \frac{F(x+0.5\Delta x) - F(x-0.5\Delta x)}{\Delta x} - p(x) = \frac{dF(x)}{dx} - p(x) = 0. \tag{14.102}$$

With a decrease in the interval Δx, the variance of the pdf estimate of stochastic process increases since with a decrease in the interval Δx we observe a high deviation of total time values of stochastic process within the limits of the interval Δx from realization to realization.

Apply (14.100) to the Gaussian and Rayleigh stochastic processes. Since

$$\frac{d^{\nu}\{1 - Q(x/\sigma)\}}{dx^{\nu}} = \frac{1 - Q^{\nu}(z)}{\sigma^{0.5\nu}}, \tag{14.103}$$

where

$$z = \frac{x}{\sigma}, \tag{14.104}$$

the bias of the pdf estimate of Gaussian stochastic process takes the following form:

$$b\{p^*(x)\} = \frac{1}{\sigma}\sum_{\nu=1}^{\infty}\frac{\{\Delta x/2\sigma\}^{2\nu}}{(2\nu+1)!}\left\{1-Q^{2\nu+1}\left(\frac{x}{\sigma}\right)\right\}$$

$$\approx \frac{1}{\sigma}\left\{\frac{\{\Delta x/\sigma\}^2}{24}\left\{1-Q^3\left(\frac{x}{\sigma}\right)\right\}+\frac{\{\Delta x/\sigma\}^4}{1920}\left\{1-Q^5\left(\frac{x}{\sigma}\right)\right\}\right\}. \tag{14.105}$$

The minimal bias under the condition

$$\frac{\Delta x}{\sigma} < 1 \tag{14.106}$$

corresponds to the value $z \approx 1$ at which $Q^3(z) \approx 0$.

Applied to the Rayleigh stochastic process, at the first approximation the bias of the pdf estimate of the Rayleigh stochastic process (the first term in (14.101)) can be presented in the following form:

$$b\{p^*(x)\} \approx \frac{(\Delta x)^2/\sigma^2}{24\sigma}(x^2\sigma^{-2}-3)x\sigma^{-1}\exp\left\{-\frac{x^2}{2\sigma^2}\right\}. \tag{14.107}$$

The relative bias

$$\frac{|\,b\{p^*(x)\}\,|}{p(x)} \tag{14.108}$$

of the pdf estimates of Gaussian (the solid line) and Rayleigh (the dashed line) stochastic processes as a function of the normalized interval

$$\Delta z = \frac{\Delta x}{\sigma}$$

at various values of the normalized level z are shown in Figure 14.7. As we can see from Figure 14.7, the relative bias of pdf estimate does not exceed the level 0.03 at $\Delta z \le 0.5$, which is acceptable in practice.

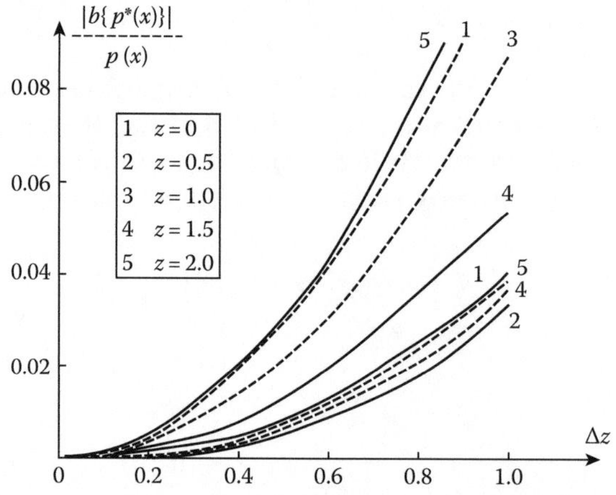

FIGURE 14.7 Relative bias of pdf estimate versus the normalized observation interval p.

Determine the variance of the pdf estimate of stochastic process. According to (14.25), we can write

$$\operatorname{Var}\{p^*(x)\} = \frac{1}{T^2(\Delta x)^2}\int_0^T\int_0^T \langle\chi(t_1)\chi(t_2)\rangle dt_1 dt_2 - \left\{\frac{1}{T\Delta x}\int_0^T \langle\chi(t)\rangle dt\right\}^2, \tag{14.109}$$

where

$$\langle\chi(t_1)\chi(t_2)\rangle = \int_{x-0.5\Delta x}^{x+0.5\Delta x}\int_{x-0.5\Delta x}^{x+0.5\Delta x} f(x_1', x_2'; t_1, t_2) dx_1' dx_2' \equiv \mathcal{P}(x \pm 0.5\Delta x; t_1, t_2) \tag{14.110}$$

is the probability of the event that the stochastic process $\xi(t)$ lies within the limits of the interval $x \pm 0.5\Delta x$ at the instants t_1 and t_2. In the case of stationary stochastic process, the following condition is satisfied:

$$\mathcal{P}(x \pm 0.5\Delta x; \tau) = \mathcal{P}(x \pm 0.5\Delta x; -\tau), \tag{14.111}$$

where $\tau = t_2 - t_1$. By introducing new variables $\tau = t_2 - t_1$ and $t = t_1$, the double integral in (14.109) is transformed into a single integral, i.e., we obtain

$$\operatorname{Var}\{p^*(x)\} = \frac{2}{T(\Delta x)^2}\int_0^T \left(1 - \frac{\tau}{T}\right)\mathcal{P}(x \pm 0.5\Delta x) d\tau - (\langle p^*(x)\rangle)^2. \tag{14.112}$$

In practice, at the first approximation, we can suppose that the one- and two-dimensional pdfs are constant within the limits of the interval $x \pm 0.5\Delta x$. Then $\langle p^*(x)\rangle \approx p(x)$ and

$$\mathcal{P}(x \pm 0.5\Delta x; \tau) \approx p_2(x, x; \tau) \times (\Delta x)^2. \tag{14.113}$$

In this case, the variance of probability density function estimate of stochastic process can be presented as

$$\operatorname{Var}\{p^*(x)\} \approx \frac{2}{T}\int_0^T \left(1 - \frac{\tau}{T}\right) p_2(x, x; \tau) d\tau - p^2(x). \tag{14.114}$$

In the case of Gaussian and Rayleigh stochastic processes, the two-dimensional pdfs are given by (14.50) and (14.71), respectively.

In the case of discrete stochastic process, in accordance with (14.26) the variance of the pdf estimate can be presented in the following form:

$$\operatorname{Var}\{p^*(x)\} = \frac{1}{N^2(\Delta x)^2}\sum_{i=1,j=1}^{N} \langle\chi_i\chi_j\rangle - \left\{\frac{1}{N\Delta x}\sum_{i=1}^{N}\langle\chi_i\rangle\right\}^2, \tag{14.115}$$

where $\langle \chi_i\, \chi_j \rangle$ is given by (14.110). When the stochastic process samples are uncorrelated, we can write

$$\langle \chi_i \chi_j \rangle = \begin{cases} \langle \chi_i \rangle & \text{at} \quad i = j, \\ \langle \chi_i \rangle \langle \chi_j \rangle & \text{at} \quad i = j. \end{cases} \tag{14.116}$$

As a result, the variance of the pdf estimate of stochastic process takes the following form:

$$\text{Var}\{p^*(x)\} = \frac{[F(x+0.5\Delta x) - F(x-0.5\Delta x)][1 - F(x+0.5\Delta x) + F(x-0.5\Delta x)]}{N(\Delta x)^2}. \tag{14.117}$$

Applying the Taylor series expansion for the function $F(x \pm 0.5\Delta x)$ about the point x and limiting by the first three terms, we obtain

$$\text{Var}\{p^*(x)\} \approx \frac{p(x)}{N\Delta x}[1 - \Delta x p(x)]. \tag{14.118}$$

For the considered Δx, we have the following restriction $p(x)\Delta x < 1$. As we can see from (14.118), with an increase in the interval Δx the variance of the pdf estimate of stochastic process is decreased, which is confirmed by physical representation.

As applied to the Gaussian stochastic process, the variance of the pdf estimate of stochastic process defined by uncorrelated samples can be presented in the following form:

$$\text{Var}\{p^*(x)\} = \frac{[Q(z-0.5\Delta z) - Q(z+0.5\Delta z)][Q(z-0.5\Delta z) - Q(z+0.5\Delta z)]^2}{N(\Delta z)\sigma^2}. \tag{14.119}$$

The dependences $\sqrt{\text{Var}\{p^*(x)\}N}\big/p(x)$ versus the normalized interval between the levels Δz and various values of the normalized level z under investigation of the Gaussian (the solid line) and Rayleigh (the dashed line) stochastic processes in the case of uncorrelated samples are shown in Figure 14.8.

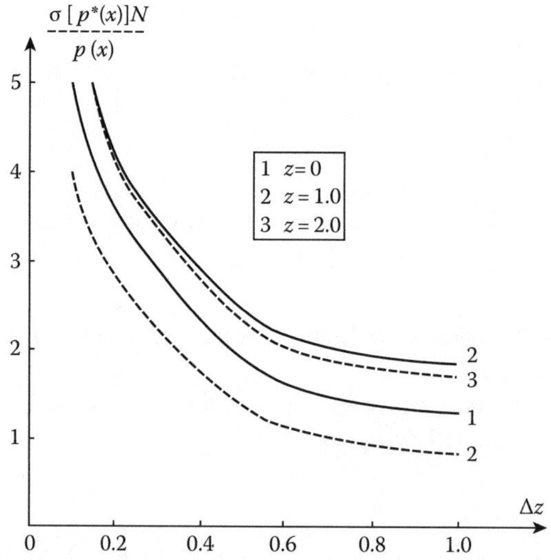

FIGURE 14.8 $\sqrt{\text{Var}\{p^*(x)\}N}\big/p(x)$ versus the normalized interval between the levels Δz.

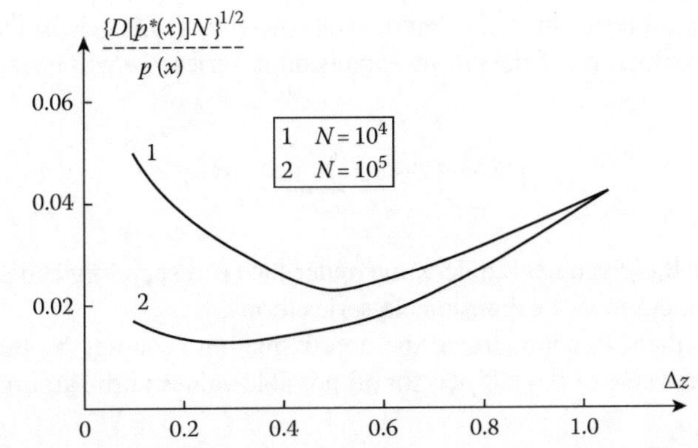

FIGURE 14.9 Normalized error of pdf measurement versus the normalized interval between the levels Δz.

As we can see from Figure 14.8, as opposed to the bias of the pdf estimate, the variance is decreased with an increase in the interval between levels. However, under experimental measurement of the pdf the interval between the levels Δx must be defined based on the condition of minimization of dispersion of the pdf estimate, namely,

$$D\{p^*(x)\} = \text{Var}\{p^*(x)\} + b^2\{p^*(x)\}. \tag{14.120}$$

The normalized resulting errors of measurement of the pdf of Gaussian stochastic process $\sqrt{D\{p^*(x)\}}/p(x)$ as a function of the normalized interval between levels Δz at $z = 0$ and $N = 10^4$ and $N = 10^5$ are shown in Figure 14.9. As we can see from Figure 14.9, for the given number of samples and at $z = 0$, the minimal magnitude of the resulting error is decreased with a decrease in the values Δz under an increase in the number of samples.

14.5 PROBABILITY DENSITY FUNCTION ESTIMATE BASED ON EXPANSION IN SERIES COEFFICIENT ESTIMATIONS

The one-dimensional stochastic process pdf can be presented in the form of series using the previously given orthogonal functions $\varphi_k(x)$ [2]:

$$p(x) = \sum_{k=1}^{\infty} c_k \varphi_k(x), \tag{14.121}$$

where the unknown coefficients are determined as

$$c_k = \int_{-\infty}^{\infty} \varphi_k(x) p(x) dx. \tag{14.122}$$

In the case of normalized orthogonal functions, we can use the following relations:

$$\int_{-\infty}^{\infty} \varphi_k(x)\varphi_l(x)dx = \delta_{kl} = \begin{cases} 1, & k = l, \\ 0, & k \neq l. \end{cases} \tag{14.123}$$

Practically, the number of terms in expansion in series given by (14.121) is limited by some value ν. Therefore, under approximation of the pdf by expansion in series, we will have forever an error

$$\varepsilon(x) = p(x) - \sum_{k=1}^{\nu} c_k \varphi_k(x) \tag{14.124}$$

that can be reduced to the given negligible value under the corresponding choice of the orthogonal functions $\varphi_k(x)$ and the number of expansions in series terms.

Further, it is convenient to characterize the approximation accuracy by the root-mean-square deviation from the true value of the pdf $p(x)$ for all possible values of the argument x

$$\varepsilon^2 = \int_{-\infty}^{\infty} \varepsilon^2(x)dx = \int_{-\infty}^{\infty} \left\{ p(x) - \sum_{k=1}^{\nu} c_k \varphi_k(x) \right\}^2 dx. \tag{14.125}$$

Taking into consideration (14.122) and (14.123), the formula (14.125) takes the following form:

$$\varepsilon^2 = \int_{-\infty}^{\infty} p^2(x)dx - \sum_{k=1}^{\nu} c_k^2. \tag{14.126}$$

As we can see from the definition, the coefficients c_k of expansion in series represent, in general, the mathematical expectations of random variables subjected to the nonlinear transformation $\varphi_k(x)$. As applied to ergodic stochastic process, the coefficients c_k can be defined by averaging the realization of stochastic process $\varphi_k[x(t)]$ transformed in time (continuously or discretely). Taking into consideration these statements, the method to measure the pdf based on generation of orthogonal functions and estimation of coefficients of expansion in series of measured pdf was supposed in Ref. [3].

The estimate of the pdf $p^*(x)$ can be presented in the following form:

$$p^*(x) = \sum_{k=1}^{\nu} c_k^* \varphi_k(x), \tag{14.127}$$

where the estimations of coefficients c_k^* in the cases of continuous and discrete methods of investigation of the realization $x(t)$ can be presented in the following form:

$$c_k^* = \frac{1}{T} \int_0^T \varphi_k[x(t)]dt; \tag{14.128}$$

$$c_k^* = \frac{1}{N} \sum_{i=1}^{N} \varphi_k(x_i), \tag{14.129}$$

where T and N are the observation time and number of samples, respectively.

A flowchart of the considered pdf measurer functioning in real time is shown in Figure 14.10. At the measurer output, the pdf estimate is formed as a function of time

$$p^*(t) = \sum_{k=1}^{\nu} c_k^* \varphi_k(t). \tag{14.130}$$

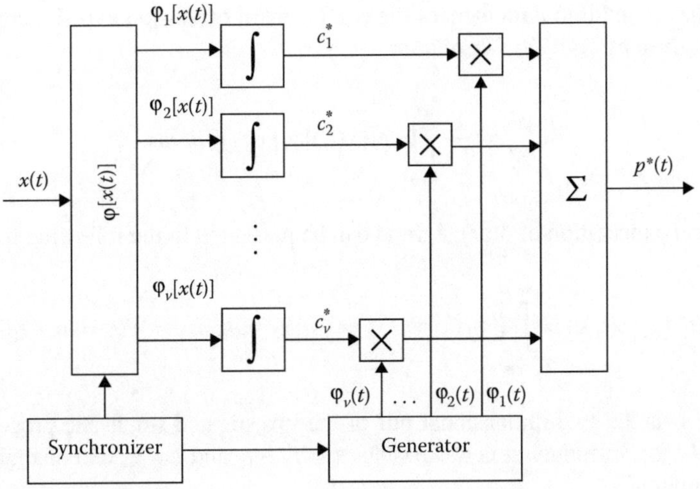

FIGURE 14.10 Measurer of pdf.

The measurer operates in the following way. The input realization $x(t)$ of the stochastic process is subjected to multichannel inertialess transformation $\varphi[x(t)]$ and is averaged after that by ideal integrators. As a result, the estimations of the expansion in series coefficients c_k^* are formed. These coefficients c_k^* come in at the mixer inputs. The orthogonal functions of time $\varphi_k(t)$ come in at the second mixer input. The estimate of the pdf $p^*(t)$ as a function of time is formed at the summator output. The orthogonal time functions $\varphi_k(t)$ coming at the mixers are delayed on the time T. This time is required to obtain the estimations of coefficients c_k^*.

Determine the characteristics of the pdf estimations. Since

$$\langle \varphi_k[x(t)] \rangle = \int_{-\infty}^{\infty} \varphi_k(x) p(x) dx, \tag{14.131}$$

we can see from (14.128) and (14.129) that the estimations of coefficients c_k^* are unbiased, i.e.,

$$\left\langle c_k^* \right\rangle = c_k. \tag{14.132}$$

The resulting dispersion of the pdf estimate can be presented in the following form:

$$D\{p^*\} = \left\langle \int_{-\infty}^{\infty} [p(x) - p^*(x)]^2 dx \right\rangle = \varepsilon^2 + \sum_{k=1}^{\nu} \left[\left\langle c_k^{*2} \right\rangle - c_k^2 \right]. \tag{14.133}$$

The second term

$$\text{Var}\{p^*\} = \sum_{k=1}^{\nu} \left[\left\langle c_k^{*2} \right\rangle - c_k^2 \right] \tag{14.134}$$

is the variance of the pdf estimate caused by random errors under estimation of the coefficients c_k^*.

Determine the second initial moment of the coefficient of expansion in series applied to the continuous method given by (14.128). We obtain

$$\left\langle c_k^{*2}\right\rangle = \frac{1}{T^2}\int_0^T\int_0^T \langle \varphi_k[x(t_1)]\varphi_k[x(t_2)]\rangle dt_1 dt_2. \tag{14.135}$$

The mathematical expectation of the integrand can be presented in the following form:

$$\langle \varphi_k[x(t_1)]\varphi_k[x(t_2)]\rangle = \int_{-\infty}^{\infty}\int_{-\infty}^{\infty} \varphi_k(x_1)\varphi_k(x_2)p_2(x_1,x_2;\tau)dx_1dx_2 = \Psi(\tau = |t_2 - t_1|), \tag{14.136}$$

where $p_2(x_1, x_2; \tau)$ is the two-dimensional pdf of the investigated stochastic process. Substituting (14.136) into (14.135), introducing new variables $\tau = t_2 - t_1$ and $t = t_1$, and changing the order of integration, we obtain

$$\left\langle c_k^{*2}\right\rangle = \frac{2}{T}\int_0^T \left(1 - \frac{\tau}{T}\right)\Psi(\tau)d\tau. \tag{14.137}$$

As applied to investigation of stochastic process realization at discrete instants

$$\left\langle c_k^{*2}\right\rangle = \frac{1}{N^2}\sum_{i=1}^{N}\sum_{j=1}^{N}\langle \varphi_k(x_i)\varphi_k(x_j)\rangle, \tag{14.138}$$

where $\langle \varphi_k(x_i)\varphi_k(x_j)\rangle$ is defined by analogous way as in (14.136).

Under analysis of stochastic process realizations at independent instants, the components

$$\Delta\varphi_k(x_i) = \varphi_k(x_i) - c_k \tag{14.139}$$

will be independent random variables. Therefore, the mathematical expectation of the square of the estimate of the expansion in series coefficients $\left\langle c_k^{*2}\right\rangle$ given by (14.138) can be presented in the following form:

$$\left\langle c_k^{*2}\right\rangle = c_k^2 + \frac{1}{N^2}\sum_{i=1}^{N}\left\langle \Delta\varphi_k^2(x_i)\right\rangle. \tag{14.140}$$

The variance of random component $\Delta\varphi_k(x_i)$ is defined by the well-known relationship

$$\mathrm{Var}\{\Delta\varphi_k\} = \left\langle \Delta\varphi_k^2(x_i)\right\rangle = \int_{-\infty}^{\infty}[\varphi_k(x) - c_k]^2 p(x)dx \tag{14.141}$$

and does not depend on the number of sample. Because of this, we can write the variance of the pdf estimate in the following form:

$$\mathrm{Var}\{p^*\} = \frac{1}{N}\sum_{k=1}^{\nu}\mathrm{Var}\{\Delta\varphi_k\}. \tag{14.142}$$

As we can see from (14.142), the variance of the pdf estimate is decreased with an increase in the number of independent samples and is increased with an increase in the number of expansion in series terms ν. Since the approximation accuracy is also increased with increasing the number of expansion in series terms, under approximation of the measured pdf by expansion in series, the number of terms must be chosen based on the condition of minimization of estimate dispersion:

$$D\{p^*\} = \int_{-\infty}^{\infty} p^2(x)dx - \sum_{k=1}^{\nu} c_k^2 + \frac{1}{N}\sum_{k=1}^{\nu} \mathrm{Var}\{\Delta\varphi_k\}. \tag{14.143}$$

In practice, to design and construct the digital or analog measurer of the pdf, the non-normalized orthogonal functions of threshold type can be chosen as the functions $\varphi_k(x)$, i.e.,

$$\varphi_k(x) = \begin{cases} 1 & \text{if } \; x_k \leq x \leq x_{k+1}, \\ 0 & \text{if } \; x < x_k, x > x_{k+1}. \end{cases} \tag{14.144}$$

In this case, the coefficients of expansion in series are defined in the following form:

$$c_k = \frac{\int_{x_k}^{x_{k+1}} \varphi_k(x)p(x)dx}{\int_{x_k}^{x_{k+1}} \varphi_k^2(x)dx} = \frac{\Delta F_k}{\Delta x_k}, \tag{14.145}$$

where

$$\Delta F_k = F(x_{k+1}) - F(x_k) \tag{14.146}$$

is the probability of the event that the realizations of stochastic process lie within the limits of the interval $\Delta x_k = x_{k+1} - x_k$ of argument x. Formula (14.145) is obtained by multiplication of the left and right parts of (14.121) on $\varphi_k(x)$, integration with respect to x within the limits from x_k to x_{k+1} and taking into consideration the condition of orthogonality. For the orthogonal functions given by (14.144) the coefficients c_k are the ordinates of the approximated pdf and look like the discrete curve shown in Figure 14.11. It is easy to see that the coefficient estimations are unbiased.

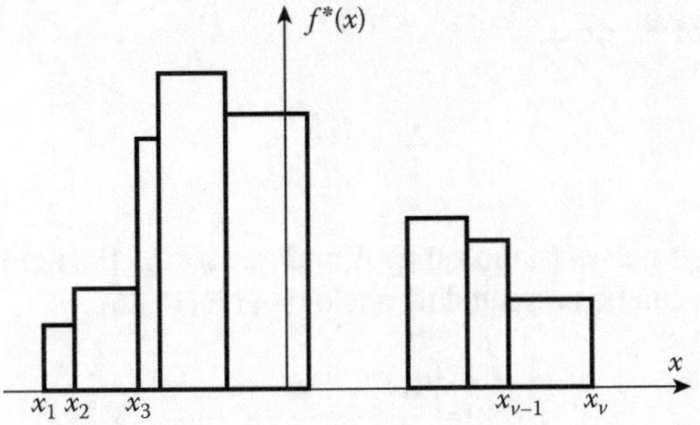

FIGURE 14.11 Histogram of distribution.

As applied to discrete sampling, in the considered case, the coefficient estimations of expansion in series can be presented in the following form:

$$c_k^* = \frac{1}{\Delta x_k} \times \frac{1}{N}\sum_{i=1}^{N} \varphi_k(x_i). \tag{14.147}$$

In the case of an independent sample of stochastic process, the variance of estimations of stochastic components $\varphi_k(x_i)$ can be presented in the following form:

$$\text{Var}\{\Delta\varphi_k\} = \int_{x_k}^{x_{k+1}} \varphi_k^2(x)p(x)dx - \left\{\int_{x_k}^{x_{k+1}} \varphi_k(x)p(x)dx\right\}^2 = \Delta F_k(1-\Delta F_k). \tag{14.148}$$

By analogous way, the variance of the pdf estimate can be presented as

$$\text{Var}\{p^*\} = \frac{1}{N}\sum_{k=1}^{v} \frac{\text{Var}\{\Delta\varphi_k\}}{\Delta x_k^2} = \frac{1}{N}\sum_{k=1}^{v} \frac{\Delta F_k(1-\Delta F_k)}{\Delta x_k^2}. \tag{14.149}$$

If the interval between samples is constant, i.e., $\Delta x_k = \Delta x = \text{const}$, then

$$\text{Var}\{p^*\} = \frac{1}{N\Delta x^2}\sum_{k=1}^{v} \Delta F_k(1-\Delta F_k) = \sum_{k=1}^{v} \text{Var}\{p^*(x_k)\}, \tag{14.150}$$

where $\text{Var}\{p^*(x_k)\}$ coincides with (14.117) for the level x_k, as expected.

14.6 MEASURERS OF PROBABILITY DISTRIBUTION AND DENSITY FUNCTIONS: DESIGN PRINCIPLES

Under practical realization of measurers of the stochastic process pdf, the multichannel amplitude analyzers are widely used. The number of channels is very high and can be more than 1024 or 4096 channels. As applied to investigation of continuous stochastic process, the input realization $x(t)$ is transformed into a sequence of pulses, the duration of which is much less than the correlation interval of stochastic process and the pulse amplitude corresponds to instantaneous value of the input realization of stochastic process. In doing so, a range of possible values of the input realization $x(t)$ of stochastic process should with high probability be within the limits of the interval $[c,d]$ defined by the multichannel analyzer. As a rule, the range of possible values $[c,d]$ is divided by v equal intervals with the width given as

$$\Delta x = \frac{d-c}{v}. \tag{14.151}$$

Giving the total number of pulses (samples) by N and measuring the number of pulses N_j at the jth channel, the pdf estimate can be presented by analogy with (14.26)

$$p_j^*(x) = \frac{N_j}{N\Delta x} = \frac{N_j}{N} \times \frac{v}{d-c}. \tag{14.152}$$

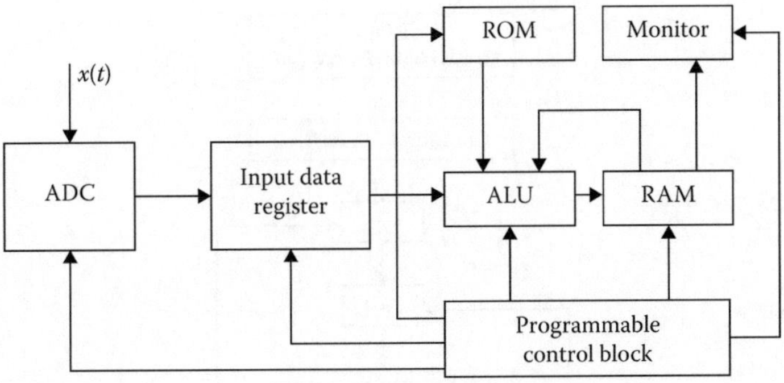

FIGURE 14.12 Measurer of pdf.

The upper bound of the *j*th channel can be presented in the following form:

$$g_j = c + j\Delta x, \quad j = 1, 2, \ldots, \nu. \tag{14.153}$$

As a result, we obtain the estimations of ordinate of the stochastic process pdf at the points

$$x_j = \frac{c + \Delta x(2j - 1)}{2} \tag{14.154}$$

corresponding to the middle between the levels g_j and g_{j-1}.

The principal flowchart of a pdf measurer is shown in Figure 14.12. The functioning principles are the following. The input realization $x(t)$ of a stochastic process comes in at the ADC input. After sampling in time and amplitude, the input signal comes in at the input of input data register and, after that, at the ALU input, in which a determination of the pdf is carried out according to algorithm. The final result, i.e., the ALU output, changes information in RAM for output data, where the values of probability density function estimates for all ν channels are stored. The value of the pdf is reproduced by a monitor. Initial data required to compute the pdf estimations are stored by ROM in the form of total number of samples N, interval between the channels (or levels) Δx, and interval $[c, d]$ of possible values of investigated stochastic process. The channel number where we can find the reading $x[i] = x_i$ can be defined as

$$j = \left\{ \frac{x[i] - c}{\Delta x} \right\} + 1, \tag{14.155}$$

where $\{\cdots\}$ means the integer number. The algorithm to compute the pdf estimate is presented in Figure 14.13. There is a need to note that all blocks of the structural block diagram can be realized by microprocessor systems.

The previously mentioned methods to measure the pdf are based on uniform division of the range of possible stochastic process values. However, this procedure, as a rule, is not optimal, since peculiarities of measured characteristic variations are not taken into consideration, which leads finally to information excess and high errors under definition of characteristics of argument values. Additionally, there is an inverse problem—to define the argument values by the given pdf values.

Consider the method of argument definition by the given characteristic values applied to the probability distribution function estimate in the case of sampled stochastic process. Division of probability distribution function domain on discrete values F_j must be done in accordance with the applied problem. The most favorable action is a uniform division from zero to unit using the

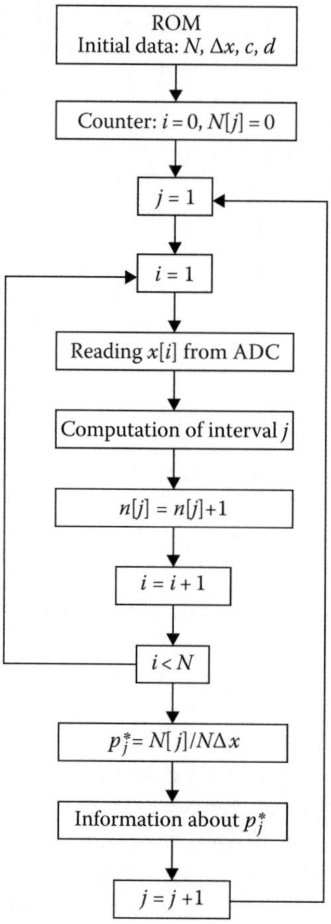

FIGURE 14.13 Algorithm of pdf computation.

equal intervals δF. The method to define an argument of probability distribution function is the following. In the case of arbitrary argument, according to the rule given by (14.20), the probability distribution function estimate $F^*(x)$ is computed and compared with the given value F_j. As a result of comparison, we obtain the signal

$$\varepsilon(x) = F_j - F^*(x), \tag{14.156}$$

based on which it is possible to measure the level x. The value x_j^* at which the signal given by (14.156) is equal to zero is the estimate of probability distribution function argument. In practice, a nonzero measure is chosen as a criterion that the estimate $F^*(x)$ is very close to F_j. The simplest way to define this measure is to compare the module of the signal given by (14.165) with the previously given value ε. In this case, the level satisfying the condition

$$|\varepsilon(x)| \leq \varepsilon. \tag{14.157}$$

can be considered as the estimate x_j^*. The probability distribution function can be given either in the form of deterministic numbers F_j within the limits of the interval [0,1] or using the values of corresponding transformations of reference stochastic process samples with the known probability distribution function [4].

Consider the measurement method when the probability distribution function is given as a set of discrete numbers F_j. In this case, the current estimate x_j^* of argument for jth value of probability distribution function can be defined using the recurrent relationship (12.342) that can be transformed and applied to the considered method of measurement in the following form:

$$x_j^*[N] = x_j^*[N-1] + \gamma[N]\{F_j - F_j^*[N]\}. \tag{14.158}$$

In doing so, the current estimate of the probability distribution function $F_j^*[N]$ at the nth step is defined as

$$F_j^*[N] = \frac{1}{N}\sum_{i=1}^{N} \eta_x[i], \tag{14.159}$$

where we use the transformation by analogy with (14.19)

$$\eta_x[i] = \begin{cases} 1 & \text{if} \quad x[i] \le x_j^*[i-1], \\ 0 & \text{if} \quad x[i] > x_j^*[i-1]. \end{cases} \tag{14.160}$$

Transformation (14.160) means that the current sample of investigated stochastic process at the ith iteration step is compared with the argument estimate obtained at the previous $(i-1)$th iteration step. Initial argument value $x_j^*[0] = a$ is chosen in arbitrary way from the range of possible values $[c, d]$. The factor $\gamma[N]$ defining the next iteration step must satisfy the condition given by (12.343). As a rule, it is inversely proportional to the number of iterations, i.e., $\gamma[N] = \kappa N^{-1}$, where $\kappa > 0$ characterizes the value of the first iteration step. The probability distribution function estimate given by (14.159), by analogous way with (12.347), can be presented in the following form:

$$F_j^*[N] = F_j^*[N-1] + N^{-1}\{\eta_x[N] - F_j^*[N-1]\}. \tag{14.161}$$

The iteration number N is defined based on the selected rule. The most widely used rule can be presented in the following form:

$$|x_j^*[N] - x_j^*[N-1]| = \alpha[N] \le \alpha, \tag{14.162}$$

where $\alpha > 0$ is the given number. When the condition (14.162) is satisfied, the iteration process is stopped. The structure of algorithm to compute the argument estimates for the given values of probability function distribution is shown in Figure 14.14.

Computing a ratio of the end differences of probability distribution function and its arguments, we can define the pdf estimate of the investigated stochastic process. As applied to the probability distribution function estimate by three of its values x_{j-1}^*, x_j^*, and x_{j+1}^*, the pdf estimate for the argument x_j^* can be presented in the following form:

$$p_j^*(x) = \frac{1}{2}\left\{\frac{F_j - F_{j-1}}{x_j^* - x_{j-1}^*} + \frac{F_{j+1} - F_j}{x_{j+1}^* - x_j^*}\right\}. \tag{14.163}$$

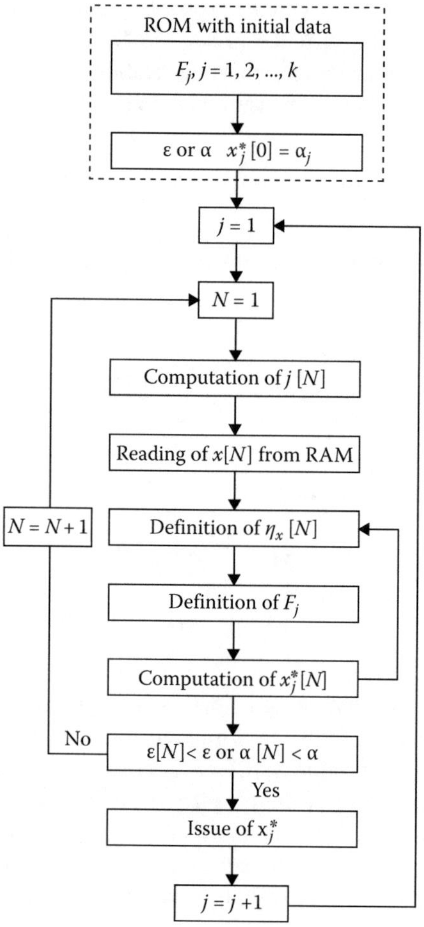

FIGURE 14.14 Algorithm of the argument estimate computation.

Under the use of reference stochastic process with the known pdf, the structure of algorithm to measure the arguments of probability distribution function is changed. Denote discrete values of realizations of investigated and reference stochastic processes as x_i and y_i, respectively. As applied to the reference stochastic process, we can theoretically define the argument value y by the given value of probability distribution function $F(y) = F_j$. If we apply the following transformation

$$\eta_{yi} = \begin{cases} 1 & \text{if} \quad y_i \le y, \\ 0 & \text{if} \quad y_i > y, \end{cases} \tag{14.164}$$

with respect to samples y_i of reference stochastic process, then the unbiased probability distribution function estimate $F^*(y)$ of random sample y_i is defined by (14.20). Comparing the probability distribution function estimate of reference stochastic process with the probability distribution function estimate $F^*(x)$ of the investigated stochastic process, we obtain the signal

$$\varepsilon(x) = F^*(y) - F^*(x) = \frac{1}{N}\sum_{i=1}^{N} z_i, \tag{14.165}$$

where $z_i = \eta_{yi} - \eta_{xi}$ takes the following values:

$$z_i = \begin{cases} 1 & \text{if} \quad x_i \le x, \ \ y_i > y, \\ 0 & \text{if} \quad \begin{cases} x_i > x, & y_i > y, \\ x_i \le x, & y_i \le y, \end{cases} \\ -1 & \text{if} \quad x_i > x, \ \ y_i \le y. \end{cases} \tag{14.166}$$

Varying the level x at which the signal $\varepsilon(x)$ given by (14.165) can satisfy (14.157), we obtain the argument estimate x_j^* for the probability distribution function F_j. The argument estimate x_j^* for the probability distribution function F_j can be found by an iteration procedure based on recurrent relationship [4]:

$$x_j^*[N] = x_j^*[N-1] + \gamma[N]z[N]. \tag{14.167}$$

Part of the flowchart of the digital measurer of the probability distribution function arguments employing the sensors of reference random numbers is presented in Figure 14.15. Operation of ROM and corresponding programmable blocks is identical to functions of the measurer of pdf (see Figure 14.12). ALU must define the end of iteration procedure according to (14.157) or (14.162). The algorithm to compute the argument estimations of probability distribution function is presented in Figure 14.16. This algorithm is different from the algorithm presented in Figure 14.14. The difference is that the arguments y_i, $j = 1, 2, \ldots, k$ corresponding to known values of the probability distribution function of reference stochastic process (RSP) are given instead of probability distribution function F_j values.

The procedure for the measurement of probability distribution function arguments allows us to define the required number N of iterations based on the given relative root-mean-square deviation of the probability distribution function estimate. The disadvantage of such a measurer is the presence of transient process under definition of argument estimate x_j^*. The considered procedures and methods of measurements of probability distribution functions and pdfs are acceptable in the case of pulse stochastic processes. In this case, the software control device and ADC must be synchronized by the investigated pulse stochastic process.

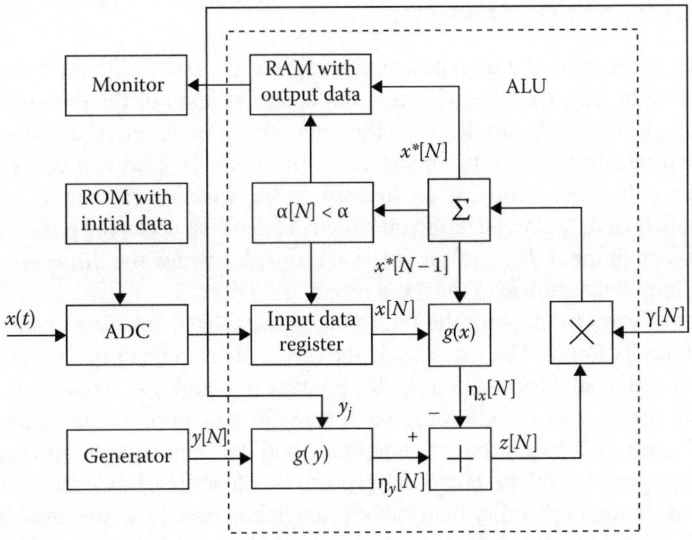

FIGURE 14.15 Measurer of probability distribution function argument.

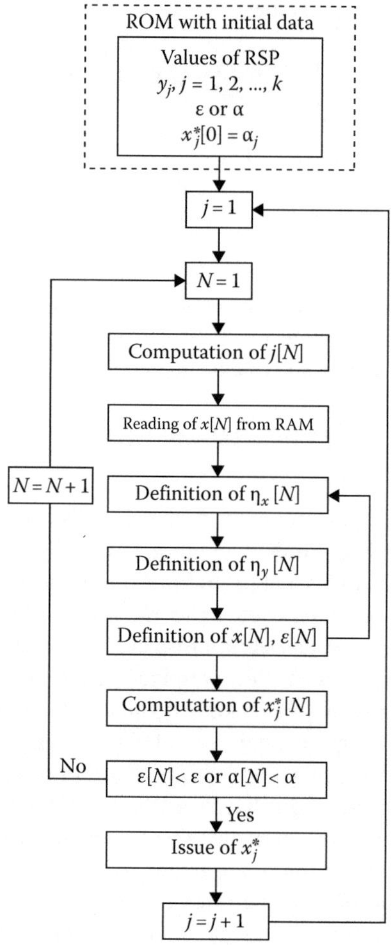

FIGURE 14.16 Algorithm of computation of the probability distribution function argument estimations.

14.7 SUMMARY AND DISCUSSION

The pdf is defined experimentally like the average value of the pdf within the limits of the interval $[x \pm 0.5dx]$. Because of this, from the viewpoint of approximation of the measured pdf to its true value at the level x, it is desirable to decrease the interval Δx. However, a decrease in the Δx at the fixed observation time interval leads to a decrease in time of the stochastic process within the limits of the interval Δx and, consequently, to an increase in the variance of the pdf estimate owing to decreasing the statistical data sample based on which the decision about the pdf magnitude is made. Therefore, there is an optimal value of the interval Δx, under which the dispersion of the pdf estimate takes the minimal magnitude at the fixed observation time.

There are several ways to measure the total time of a stochastic process observation below the given level or within its limits. The first way is the direct measurement of total time when a magnitude of the realization of continuous ergodic process is below the fixed level x. In this case, the realization of continuous ergodic process is transformed into the sequence of rectangular pulses (see the Figure 14.1b) with the unit amplitude and duration equal to the time when a magnitude of the realization of continuous ergodic process is below the fixed level. The second method to define and measure the probability distribution function is based on computation of the number of quantized by amplitude and duration of the sampled pulses. In this case, at first, the continuous realization of stochastic process is transformed by corresponding pulse modulator into sampled sequence and comes in at the input of the threshold device with the threshold level. Then, a ratio of the number of pulses that do not exceed the threshold to the total number of pulses corresponding to the observation time interval is equal by value to the probability distribution function estimate.

The root-mean-square deviation $\sigma\{F^*(x)\}$ of measurements of the probability distribution function of Gaussian stochastic process as a function of the normalized level for various values of p is presented in Figure 14.3. As we can see from Figure 14.3, the maximal values of $\sigma\{F^*(x)\}$ are obtained at the zero level. Formula (14.63) demonstrates that the variance of the probability distribution function estimate $F^*(x)$ of stochastic process by correlated samples increases in comparison with the variance of the probability distribution function estimate $F^*(x)$ of stochastic process by uncorrelated samples of the same sample size N.

The dependences $\sqrt{\mathrm{Var}\{f^*(x)\}N}\big/p(x)$ versus the normalized interval between the levels Δz and various values of the normalized level z under investigation of the Gaussian (the solid line) and Rayleigh (the dashed line) stochastic processes in the case of uncorrelated samples are shown in Figure 14.8. As we can see from Figure 14.8, as opposed to the bias of pdf estimate, the variance of the pdf estimate is decreased with an increase in the interval between levels. However, under experimental measurement of the pdf the interval between the levels Δx must be defined based on the condition of minimization of dispersion of the pdf estimate. The normalized resulting errors of measurement of the probability density function of Gaussian stochastic process $\sqrt{D\{p^*(x)\}}\big/p(x)$ as a function of the normalized interval between levels Δz at $z = 0$ and $N = 10^4$ and $N = 10^5$ are shown in Figure 14.9. As we can see from Figure 14.9, for the given number of samples and at $z = 0$, the minimal magnitude of the resulting error is decreased with a decrease the values Δz under an increase in the number of samples.

REFERENCES

1. Haykin, S. and M. Moher. 2007. *Introduction to Analog and Digital Communications*. 2nd edn. New York: John Wiley & Sons, Inc.
2. Gradshteyn, I.S. and I.M. Ryzhik. 2007. *Tables of Integrals, Series and Products*. 7th edn. London, U.K.: Academic Press.
3. Sheddon, I.N. 1951. *Fourier Transform*. New York: McGraw Hill, Inc.
4. Domaratzkiy, A.N., Ivanov, L.N., and Yu. Yurlov. 1975. *Multipurpose Statistical Analysis of Stochastic Signals*. Novosibirsk, Russia: Nauka.

15 Estimate of Stochastic Process Frequency-Time Parameters

15.1 ESTIMATE OF CORRELATION FUNCTION

Considered parameters of the stochastic process, namely, the mathematical expectation, the variance, the probability distribution and the density functions, do not describe statistic dependence between the values of stochastic process at different instants. We can use the following parameters of the stochastic process, such as the correlation function, spectral density, characteristics of spikes, central frequency of narrowband stochastic process, and others, to describe the statistic dependence subject to specific problem. Let us briefly consider the methods to measure these parameters and define the methodological errors associated, in general, with the finite time of observation and analysis applied to the ergodic stochastic processes with zero mathematical expectation.

As applied to the ergodic stochastic process with zero mathematical expectation, the correlation function can be presented in the following form:

$$R(\tau) = \lim_{T\to\infty} \frac{1}{T}\int_0^T x(t)x(t-\tau)dt. \tag{15.1}$$

Since in practice the observation time or integration limits (integration time) are finite, the correlation function estimate under observation of stochastic process realization within the limits of the finite time interval [0, T] can be presented in the following form:

$$R^*(\tau) = \frac{1}{T}\int_0^T x(t)x(t-\tau)dt. \tag{15.2}$$

As we can see from (15.2), the main operations involved in measuring the correlation function of ergodic stationary process are the realization of fixed delay τ, process multiplication, and integration or averaging of the obtained product. Flowchart of the measurer or correlator is depicted in Figure 15.1. To obtain the correlation function estimate for all possible values of τ, the delay must be variable. The flowchart shown in Figure 15.1 provides a sequential receiving of the correlation function for various values of the delay τ. To restore the correlation function within the limits of the given interval of delay values τ, the last, as a rule, varies discretely with the step $\Delta\tau = \tau_{k+1} - \tau_k$, $k = 0, 1, 2, \ldots$. If the spectral density of investigated stochastic process is limited by the maximum frequency value $f_{\max}$, then according to the sampling theorem or the Kotelnikov's theorem, in order to restore the correlation function, we need to employ the intervals equal to

$$\Delta\tau = \frac{1}{2f_{\max}}. \tag{15.3}$$

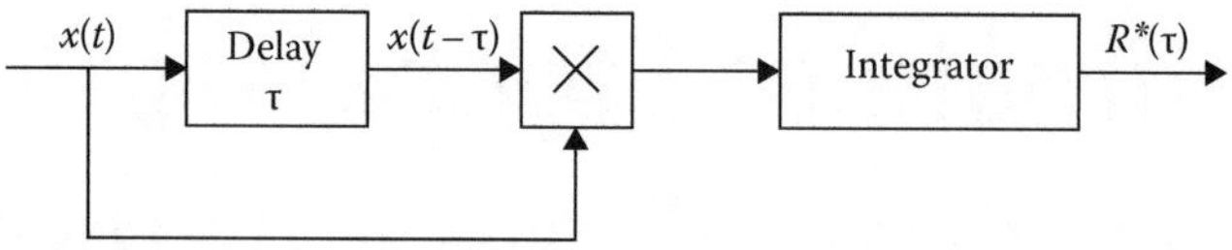

FIGURE 15.1 Correlator.

However, in practice, it is not convenient to restore the correlation function employing the sampling theorem. As a rule, there is a need to use an interpolation or smoothing of obtained discrete values. As it was proved empirically [1], it is worthwhile to define the discrete step in the following form:

$$\Delta\tau \approx \frac{1}{5-10f_{\max}}. \tag{15.4}$$

In doing so, 5 – 10 discrete estimates of correlation function correspond to each period of the highest frequency $f_{\max}$ of stochastic process spectral density.

Approximation to select the value $\Delta\tau$ based on the given error of straight-line interpolation of the correlation function estimate was obtained in Ref. [2]:

$$\Delta\tau \approx \frac{1}{\hat{f}}\sqrt{0.2\,|\Delta\mathcal{R}|}, \tag{15.5}$$

where $|\Delta\mathcal{R}|$ is the maximum allowable value of interpolation error of the normalized correlation function $\mathcal{R}(\tau)$;

$$\hat{f} = \sqrt{\frac{\int_0^\infty f^2 S(f)df}{\int_0^\infty S(f)df}} \tag{15.6}$$

is the mean-square frequency value of the spectral density $S(f)$ of the stochastic process $\xi(t)$.

Sequential measurement of the correlation function at various values of $\tau = k\Delta\tau$, $k = 0, 1, 2, \ldots, \nu$ is not acceptable forever owing to long-time analysis, in the course of which the measurement conditions can change. For this reason, the correlators operating in parallel can be employed. The flowchart of multichannel correlator is shown in Figure 15.2. The voltage proportional to the discrete value of the estimated correlation function is observed at the output of each channel of the multichannel correlator. This voltage is supplied to the channel commutator, and the correlation function estimate, as a discrete time function, is formed at the commutator output. For this purpose, the commutator output is connected with the low-pass filter input and the low-pass filter possesses the filter time constant adjusted with request speed and a priori behavior of the investigated correlation function.

In principle, a continuous variation of delay is possible too, for example, the linear variation. In this case, the additional errors of correlation function measurements will arise. These errors are caused by variations in delay during averaging procedure. The acceptable values of delay variation velocity are investigated in detail based on the given additional measurement error.

Methods of the correlation function estimate can be classified into three groups based on the principle of realization of delay and other elements of correlators: analog, digital, and analog-to-digital. In turn, the analog measurement procedures can be divided on the methods employing a representation of investigated stochastic process both as the continuous process and as the

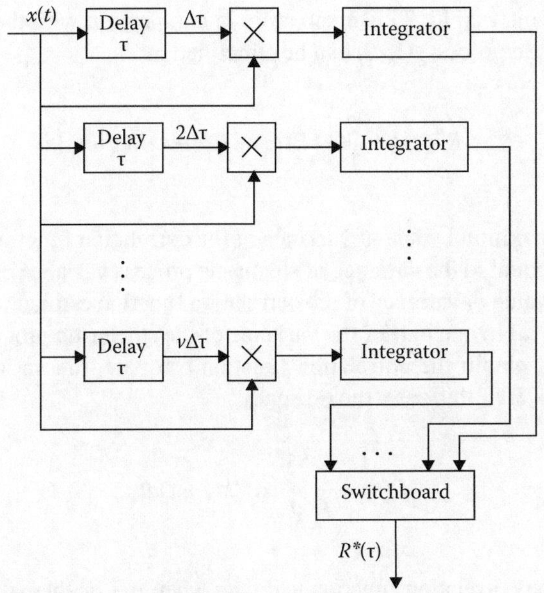

FIGURE 15.2 Multichannel correlator.

sampled process. As a rule, physical delays are used by analog methods with continuous representation of investigated stochastic process. Under discretization of investigated stochastic process in time, the physical delay line can be changed by corresponding circuits. While using digital procedures to measure the correlation function estimate, the stochastic process is sampled in time and transformed into binary number by analog-to-digital conversion. Further operations associated with the signal delay, multiplication, and integration are carried out by the shift registers, summator, and so on.

Define the statistical characteristics of the correlation function estimate (the bias and variance) given by (15.2). Averaging the correlation function estimate by an ensemble of realizations we obtain the unbiased correlation function estimate. The variance of correlation function estimate can be presented in the following form:

$$\mathrm{Var}\{R^*(\tau)\} = \frac{1}{T^2}\int_0^T\int_0^T \langle x(t_1)x(t_1-\tau)x(t_2)x(t_2-\tau)\rangle\, dt_1 dt_2 - R^2(\tau). \tag{15.7}$$

The fourth moment $\langle x(t_1)x(t_1-\tau)x(t_2)x(t_2-\tau)\rangle$ of Gaussian stochastic process can be determined as

$$\langle x(t_1)x(t_1-\tau)x(t_2)x(t_2-\tau)\rangle = R^2(\tau) + R^2(t_2-t_1) + R(t_2-t_1-\tau)R(t_2-t_1+\tau). \tag{15.8}$$

Substituting (15.8) in (15.7) and transforming the double integral into a single one by introduction of the new variable $t_2 - t_1 = \tau$, we obtain

$$\mathrm{Var}\{R^*(\tau)\} = \frac{2}{T}\int_0^T \left(1-\frac{z}{T}\right)[R^2(z) + R(z-\tau)(z+\tau)]dz. \tag{15.9}$$

If the observation time interval [0, *T*] is much more in comparison with the correlation interval of the investigated stochastic process, (15.9) can be simplified as

$$\operatorname{Var}\{R^*(\tau)\} = \frac{2}{T}\int_0^T [R^2(z) + R(z-\tau)R(z+\tau)]dz. \tag{15.10}$$

Thus, we can see that maximum value of variance of the correlation function estimate corresponds to the case $\tau = 0$ and is equal to the variance of stochastic process variance estimate given by (13.61) and (13.62). Minimum value of variance of the correlation function estimate corresponds to the case when $\tau \gg \tau_{cor}$ and is equal to one half of the variance of the stochastic process variance estimate.

In principle, we can obtain the correlation function estimate, the variance of which tends to approach to zero as $\tau \to T$. In this case, the estimate

$$\tilde{R}(\tau) = \frac{1}{T}\int_0^{T-|\tau|} x(t)x(t-\tau)dt \tag{15.11}$$

can be considered as the correlation function estimate when the mathematical expectation of stochastic process is equal to zero. This correlation function estimate is characterized by the bias

$$b[\tilde{R}(\tau)] = \frac{|\tau|}{T}R(\tau) \tag{15.12}$$

and variance

$$\operatorname{Var}\{\tilde{R}(\tau)\} = \frac{2}{T}\int_0^{T-|\tau|}\left(1 - \frac{z+|\tau|}{T}\right)[R^2(z) + R(z-\tau)R(z+\tau)]dz. \tag{15.13}$$

In practice, a realization of the estimate given by (15.11) is more difficult to achieve compared to a realization of the estimate given by (15.2) since we need to change the value of the integration limits or observation time interval simultaneously with changing in the delay τ employing the one-channel measurer or correlator. Under real conditions of correlation function measurement, the observation time interval is much longer compared to the correlation interval of stochastic process. Because of this, the formula (15.13) can be approximated by the formula (15.10). Note that the correlation function measurement process is characterized by dispersion of estimate. If the correlation function estimate is given by (15.11), in the limiting case, the dispersion of estimate is equal to the square of estimate bias. For example, in the case of exponential correlation function given by (12.13) and under the condition $T \gg \tau_{cor}$, based on (15.10) we obtain

$$\operatorname{Var}\{R^*(\tau)\} = \frac{\sigma^4}{\alpha T}[1 + (2\alpha\tau + 1)\exp\{-2\alpha\tau\}]. \tag{15.14}$$

As applied to the estimate given by (15.11) and the exponential correlation function, the variance of correlation function estimate can be presented in the following form [3]:

$$\operatorname{Var}\{\tilde{R}(\tau)\} = \frac{\sigma^4}{\alpha T}[1 + (2\alpha\tau + 1)\exp\{-2\alpha\tau\}] - \frac{\sigma^4}{2\alpha^2 T^2}[2\alpha\tau + 1(4\alpha\tau + 6\alpha^2 T^2)\exp\{-2\alpha\tau\}], \tag{15.15}$$

when the conditions $T \gg \tau_{cor}$ and $\tau \ll T$ are satisfied. Comparing (15.15) and (15.14), we can see that the variance of correlation function estimate given by (15.11) is lesser than the variance of correlation function estimate given by (15.2). When $\alpha T \gg 1$, we can discard this difference because it is defined as $(\alpha T)^{-1}$.

In addition to the ideal integrator, any linear system can be used as an integrator to obtain the correlation function estimate analogously as under the definition of estimates of the variance and the mathematical expectation of the stochastic process. In this case, the function

$$R^*(\tau) = c\int_0^{\infty} h(z)x(t-z)x(t-\tau-z)dz \tag{15.16}$$

can be considered as the correlation function estimate, where, as before, the constant c is chosen from the condition of estimate unbiasedness

$$c\int_0^{\infty} h(z)dz = 1. \tag{15.17}$$

As applied to the Gaussian stochastic process, the variance of correlation function estimate as $t \to \infty$ can be presented in the following form:

$$\mathrm{Var}\{R^*(\tau)\} = c^2\int_0^T\int_0^T h(z_1)h(z_2)[R^2(z_2-z_1)+R(z_2-z_1-\tau)R(z_2-z_1+\tau)]dz_1dz_2. \tag{15.18}$$

Introducing new variables $z_2 - z_1 = z$, $z_1 = \upsilon$ we obtain

$$\mathrm{Var}\{R^*(\tau)\} = c^2\int_0^T [R^2(z)+R(z-\tau)R(z+\tau)]dz\int_0^{T-\tau} h(z+\upsilon)h(\upsilon)d\upsilon$$
$$+\int_{-T}^0 [R^2(z)+R(z-\tau)R(z+\tau)]dz\int_0^T h(z+\upsilon)h(\upsilon)d\upsilon. \tag{15.19}$$

Introducing the variables $y = -z$, $x = \upsilon - y$ and as $T \to \infty$, we obtain

$$\mathrm{Var}\{R^*(\tau)\} = c^2\int_0^{\infty} [R^2(z)+R(z-\tau)R(z+\tau)]\, r_h(z)dz, \tag{15.20}$$

where the function $r_h(z)$ is given by (12.131) as $T - \tau \to \infty$.

Now, consider how a sampling in time of stochastic process acts on the characteristics of correlation function estimate assuming that the investigated stochastic process is a stationary process. If a stochastic process with zero mathematical expectation is observed and investigated at discrete instants, the correlation function estimate can be presented in the following form:

$$R^*(\tau) = \frac{1}{N}\sum_{i=1}^{N} x(t_i)x(t_i-\tau), \tag{15.21}$$

where N is the sample size. The correlation function estimate is unbiased, and the variance of correlation function estimate takes the following form:

$$\mathrm{Var}\{R^*(\tau)\} = \frac{1}{N^2}\sum_{i=1}^{N}\sum_{j=1}^{N}\langle x(t_i)x(t_i-\tau)x(t_j)x(t_j-\tau)\rangle - R^2(\tau). \tag{15.22}$$

As applied to the Gaussian stochastic process, the general formula, by analogy with the formula for the variance of correlation function estimate under continuous observation and analysis of stochastic process realization, is simplified and takes the following form:

$$\mathrm{Var}\{R^*(\tau)\} = \frac{2}{N}\sum_{i=1}^{N}\left(1-\frac{i}{N}\right)[R^2(iT_p) + R(iT_p-\tau)(iT_p+\tau)], \tag{15.23}$$

where we assume that the samples are taken over equal time intervals, that is, $T_p = t_i - t_{i-1}$.

If samples are pairwise independent, a definition of the variance of correlation function estimate given by (15.22) can be simplified; that is, we can use the following representation:

$$\mathrm{Var}\{R^*(\tau)\} = \frac{1}{N^2}\sum_{i=1}^{N}\langle x^2(t_i)x^2(t_i-\tau)\rangle - \frac{1}{N}R^2(\tau). \tag{15.24}$$

As applied to the Gaussian stochastic process with pairwise independent samples, we can write

$$\mathrm{Var}\{R^*(\tau)\} = \frac{\sigma^4}{N}[1+\mathcal{R}^2(\tau)], \tag{15.25}$$

where $\mathcal{R}(\tau)$ is the normalized correlation function of the observed stochastic process. As we can see from (15.25), the variance of correlation function estimate increases with an increase in the absolute magnitude of the normalized correlation function.

The obtained results can be generalized for the estimate of mutual correlation function of two mutually stationary stochastic processes, the realizations of which are $x(t)$ and $y(t)$, respectively:

$$R_{xy}^*(\tau) = \frac{1}{T}\int_0^T x(t)y(t-\tau)dt. \tag{15.26}$$

At this time, we assume that the investigated stochastic processes are characterized by zero mathematical expectations. Flowcharts to measure the mutual correlation function of stochastic processes are different from the flowcharts to measure the correlation functions presented in Figures 15.1 and 15.2 by the following: The processes $x(t)$ and $y(t-\tau)$ or $x(t-\tau)$ and $y(t)$ come in at the input of the mixer instead of the processes $x(t)$ and $x(t-\tau)$. The mathematical expectation of mutual correlation function estimate can be presented in the following form:

$$\left\langle R_{xy}^*(\tau)\right\rangle = \frac{1}{T}\int_0^T \langle x(t)y(t-\tau)\rangle\, dt = R_{xy}(\tau) \tag{15.27}$$

that means the mutual correlation function estimate is unbiased.

As applied to the Gaussian stochastic process, the variance of mutual correlation function estimate takes the following form:

$$\mathrm{Var}\{R_{xy}^{*}(\tau)\} = \frac{1}{T^2}\int_0^T\int_0^T \{R_{xx}(t_2 - t_1)R_{yy}(t_2 - t_1) + R_{xy}[\tau - (t_2 - t_1)]R_{xy}[\tau + (t_2 - t_1)]\}dt_1 dt_2. \tag{15.28}$$

As before, we introduce the variables $t_2 - t_1 = z$ and $t_1 = \upsilon$ and reduce the double integral to a single one. Thus, we obtain

$$\mathrm{Var}\{R_{xy}^{*}(\tau)\} = \frac{2}{T}\int_0^T \left(1 - \frac{z}{T}\right)[R_{xx}(z)R_{yy}(z) + R_{xy}(\tau - z)R_{xy}(\tau + z)]dz. \tag{15.29}$$

At $T \gg \tau_{\mathrm{cor}_x}$, $T \gg \tau_{\mathrm{cor}_y}$, $T \gg \tau_{\mathrm{cor}_{xy}}$, where $\tau_{\mathrm{cor}_{xy}}$ is the interval of mutual correlation of stochastic processes and is determined analogously to (12.21), the integration limits can be expanded until $[0, \infty)$ and we can neglect z/T in comparison with unit. Since the mutual correlation function can reach the maximum value at $\tau \neq 0$, the maximum variance of mutual correlation function estimate can be obtained at $\tau \neq 0$.

Now, consider the mutual correlation function between stochastic processes both at the input of linear system with the impulse response $h(t)$ and at the output of this linear system. We assume the input process is the "white" Gaussian stochastic process with zero mathematical expectation and the correlation function

$$R_{xx}(\tau) = \frac{\mathcal{N}_0}{2}\delta(\tau). \tag{15.30}$$

Denote the realization of stochastic process at the linear system input as $x(t)$. Then, a realization of stochastic process at the linear system output in stationary mode can be presented in the following form:

$$y(t) = \int_0^\infty h(\upsilon)x(t - \upsilon)d\upsilon = \int_0^\infty h(t - \upsilon)x(\upsilon)d\upsilon. \tag{15.31}$$

The mutual correlation function between $x(t - \tau)$ and $y(t)$ has the following form:

$$R_{yx}(\tau) = \langle y(t)x(t - \tau)\rangle = \int_0^\infty h(\upsilon)R_{xx}(\upsilon - \tau)d\upsilon = \begin{cases} 0.5\mathcal{N}_0 h(\tau), & \tau \geq 0, \\ 0, & \tau < 0. \end{cases} \tag{15.32}$$

In (15.32) we assume that the integration process covers the point $\upsilon = 0$.

As we can see from (15.32), the mutual correlation function of the stochastic process at the output of linear system in stationary mode, when the "white" Gaussian noise excites the input of linear system, coincides with the impulse characteristic of linear system accurate with the constant factor. Thus, the method to measure the impulse response is based on the following relationship:

$$h^{*}(\tau) = \frac{2}{\mathcal{N}_0}R_{yx}^{*}(\tau) = \frac{2}{\mathcal{N}_0}\frac{1}{T}\int_0^T y(t)x(t - \tau)dt, \tag{15.33}$$

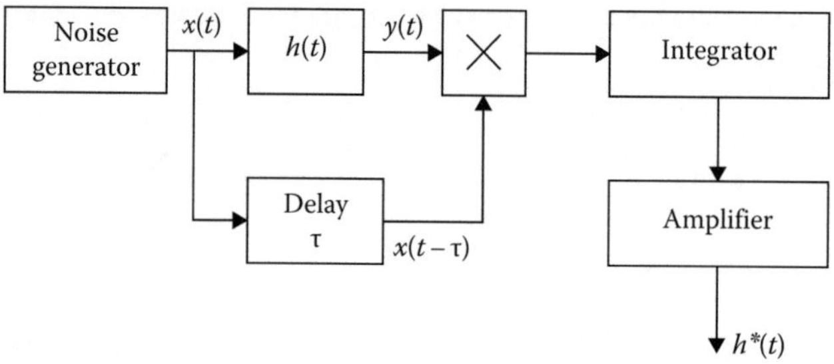

FIGURE 15.3 Measurement of linear system impulse response.

where $y(t)$ is given by (15.31). The measuring process of the impulse characteristic of linear system is shown in Figure 15.3. Operation principles are clear from Figure 15.3.

Mathematical expectation of the impulse characteristic estimate

$$\left\langle h^*(\tau)\right\rangle = \frac{2}{\mathcal{N}_0 T}\int_0^T\int_0^\infty h(\upsilon)\langle x(t-\upsilon)x(t-\tau)\rangle\, d\upsilon dt = h(\tau), \tag{15.34}$$

that is, the impulse response estimate is unbiased. The variance of impulse characteristic estimate takes the following form:

$$\mathrm{Var}\{h^*(\tau)\} = \frac{4}{\mathcal{N}_0^2 T^2}\int_0^T\int_0^T \langle y(t_1)y(t_2)x(t_1-\tau)x(t_2-\tau)\rangle\, dt_1 dt_2 - h^2(\tau). \tag{15.35}$$

Since the stochastic process at the linear system input is Gaussian, the stochastic process at the output of the linear system is Gaussian, too. If the condition $T \gg \tau_{\mathrm{cor}_y}$ is satisfied, where τ_{cor_y} is the correlation interval of the stochastic process at the output of linear system, then the stochastic process at the output of linear system is stationary. For this reason, we can use the following representation for the variance of impulse characteristic estimate

$$\mathrm{Var}\{h^*(\tau)\} = \frac{4}{\mathcal{N}_0^2 T^2}\int_0^T\int_0^T [R_{xx}(t_2-t_1)R_{yy}(t_2-t_1) + R_{yx}(t_2-t_1-\tau)R_{yx}(t_2-t_1+\tau)]dt_1 dt_2. \tag{15.36}$$

Introducing new variables $t_2 - t_1 = z$, as before, the variance of impulse characteristic estimate can be presented in the following form:

$$\mathrm{Var}\{h^*(\tau)\} = \frac{8}{\mathcal{N}_0^2 T}\int_0^T \left(1-\frac{z}{T}\right)[R_{xx}(z)R_{yy}(z) + R_{yx}(\tau-z)R_{yx}(\tau+z)]dt_1 dt_2$$

$$+\frac{2}{T}\int_0^T \left(1-\frac{z}{T}\right)h(\tau-z)h(\tau+z)dz, \tag{15.37}$$

where [3]

$$R_{yy}(z) = \frac{\mathcal{N}_0}{2}\int_0^\infty h(\upsilon)h(\upsilon+|z|)d\upsilon \tag{15.38}$$

is the correlation function of the stochastic process at the linear system output in stationary mode. While calculating the second integral in (15.37), we assume that the observation time interval $[0, T]$ is much more compared to the correlation interval of the stochastic process at the output of linear system τ_{cor_y}. For this reason, in principle, we can neglect the term zT^{-1} compared to the unit. In this case, the upper integration limits can be approximated by ∞. However, taking into consideration the fact that at $\tau < 0$ the impulse response of linear system is zero, that is, $h(\tau) = 0$, the integration limits with respect to the variable z must satisfy the following conditions:

$$\begin{cases} 0 < z < \infty, \\ \tau - z > 0, \\ \tau + z > 0. \end{cases} \tag{15.39}$$

Based on (15.39), we can find that $0 < z < \tau$. As a result, we obtain

$$\mathrm{Var}\{h^*(\tau)\} = \frac{2}{T}\left\{\int_0^\infty h^2(\upsilon)d\upsilon + \int_0^\tau h(\tau-\upsilon)h(\tau+\upsilon)d\upsilon\right\}. \tag{15.40}$$

As applied to the impulse responses of the form

$$h_1(\tau) = \frac{1}{T_1}, \quad 0 < \tau < T_1, \quad T > T_1 \tag{15.41}$$

and

$$h_2(\tau) = \alpha\exp(-\alpha\tau), \tag{15.42}$$

the variances of impulse characteristic estimates can be presented in the following form:

$$\mathrm{Var}\{h_1^*(\tau)\} = \frac{2}{TT_1}\times\begin{cases} 1+\dfrac{\tau}{T_1}, & 0 \le \tau \le 0.5T_1, \\ 2-\dfrac{\tau}{T_1}, & 0.5T_1 \le \tau \le T_1, \end{cases} \tag{15.43}$$

$$\mathrm{Var}\{h_2^*(\tau)\} = \frac{\alpha}{T}[1 + 2\alpha\tau\times\exp\{-2\alpha\tau\}]. \tag{15.44}$$

15.2 CORRELATION FUNCTION ESTIMATION BASED ON ITS EXPANSION IN SERIES

The correlation function of stationary stochastic process can be presented in the form of expansion in series with respect to earlier-given normalized orthogonal functions $\varphi_k(t)$:

$$R(\tau) = \sum_{k=0}^{\infty}\alpha_k\varphi_k(\tau), \tag{15.45}$$

where the unknown coefficients α_k are given by

$$\alpha_k = \int_{-\infty}^{\infty} \varphi_k(\tau)R(\tau)d\tau, \tag{15.46}$$

and (14.123) is true in the case of normalized orthogonal functions. The number of terms of the expansion in series (15.45) is limited by some magnitude v. Under approximation of correlation function by the expansion in series with the finite number of terms, the following error

$$\varepsilon(\tau) = R(\tau) - \sum_{k=0}^{v} \alpha_k\varphi_k(\tau) = R(\tau) - R_v(\tau) \tag{15.47}$$

exists forever. This error can be reduced to an earlier-given negligible value by the corresponding selection of the orthogonal functions $\varphi_k(t)$ and the number of terms of expansion in series.

Thus, the approximation accuracy will be characterized by the total square of approximated correlation function $R_v(\tau)$ deviation from the true correlation function $R(\tau)$ for all possible values of the argument τ

$$\varepsilon^2 = \int_{-\infty}^{\infty} \varepsilon^2(\tau)d\tau = \int_{-\infty}^{\infty} R^2(\tau)d\tau - \sum_{k=0}^{v} \alpha_k^2. \tag{15.48}$$

Formula (15.48) is based on (15.45). The original method to measure the correlation function based on its representation in the form of expansion in series

$$R_v^*(\tau) = \sum_{k=0}^{v} \alpha_k^*\varphi_k(\tau) \tag{15.49}$$

by the earlier-given orthogonal functions $\varphi_k(t)$ and measuring the weight coefficients α_k was discussed in Ref. [4]. According to (15.46), the following representation

$$\alpha_k = \lim_{T\to\infty} \int_{-\infty}^{\infty} \varphi_k(\tau)\left\{\frac{1}{T}\int_0^T x(t)x(t-\tau)dt\right\}d\tau \tag{15.50}$$

is true in the case of ergodic stochastic process with zero mathematical expectation. In line with this fact, the estimates of unknown coefficients of expansion in series can be obtained based on the following representation:

$$\alpha_k^* = \frac{1}{T}\int_0^T x(t)\left\{\int_0^{\infty} x(t-\tau)\varphi_k(\tau)d\tau\right\}dt. \tag{15.51}$$

The integral in the braces

$$y_k(t) = \int_0^{\infty} x(t-\tau)\varphi_k(\tau)d\tau \tag{15.52}$$

is the signal at the output of linear filter operating in stationary mode with impulse response given by

$$h_k(t) = \begin{cases} 0 & \text{if} \quad t < 0, \\ \varphi_k(t) & \text{if} \quad t \geq 0, \end{cases} \tag{15.53}$$

matched with the earlier-given orthogonal function $\varphi_k(t)$. As we can see from (15.51), the mathematical expectation of estimate

$$\langle \alpha_k^* \rangle = \int_0^\infty \varphi_k(\tau) \left\{ \frac{1}{T} \int_0^T \langle x(t)x(t-\tau) \rangle \, dt \right\} d\tau = \int_0^\infty \varphi_k(\tau) R(\tau) d\tau = \alpha_k \tag{15.54}$$

is matched with the true value; in other words, the estimate of coefficients of expansion in series is unbiased.

The estimate variance of expansion in series coefficients can be presented in the following form:

$$\text{Var}\{\alpha_k^*\} = \frac{1}{T^2} \int_0^T \int_0^T \langle x(t_1)x(t_2)y_k(t_1)y_k(t_2) \rangle \, dt_1 dt_2 - \left\{ \frac{1}{T} \int_0^T \langle x(t)y_k(t) \rangle \, dt \right\}^2. \tag{15.55}$$

As applied to the stationary Gaussian stochastic process, the stochastic process forming at the output of orthogonal filter will also be the stationary Gaussian stochastic process for the considered case. Because of this, we can write

$$\text{Var}\{\alpha_k^*\} = \frac{1}{T^2} \int_0^T \int_0^T [R(t_2 - t_1) R_{y_k}(t_2 - t_1) + R_{xy_k}(t_2 - t_1) R_{y_k x}(t_2 - t_1)] dt_1 dt_2, \tag{15.56}$$

where

$$R_{y_k}(\tau) = \int_0^\infty \int_0^\infty R(\tau + \nu - \kappa) \varphi_k(\kappa) \varphi_k(\nu) d\kappa \, d\nu; \tag{15.57}$$

$$R_{xy_k}(\tau) = \int_0^\infty R(\tau - \nu) \varphi_k(\nu) d\nu; \tag{15.58}$$

$$R_{y_k x}(\tau) = \int_0^\infty R(\tau + \nu) \varphi_k(\nu) d\nu. \tag{15.59}$$

Introducing new variables $t_2 - t_1 = \tau$, $t_1 = t$ and changing the order of integration analogously as shown in Section 12.4, we obtain

$$\text{Var}\{\alpha_k^*\} = \frac{2}{T} \int_0^T \left(1 - \frac{\tau}{T}\right) \int_0^T \int_0^T [R(\tau) R_{y_k}(\tau) + R_{xy_k}(\tau) R_{y_k x}(\tau)] d\tau. \tag{15.60}$$

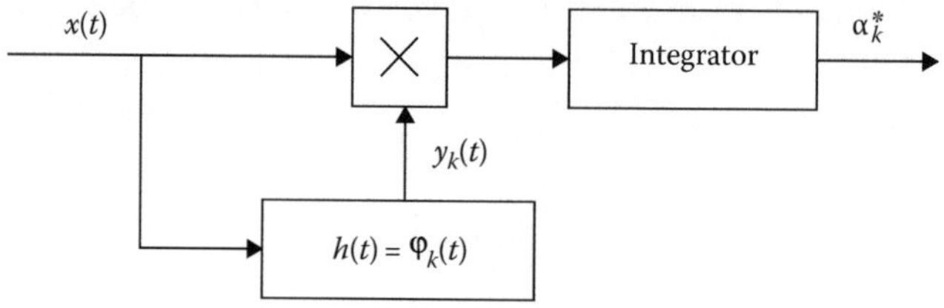

FIGURE 15.4 One-channel measurer of coefficients α_k.

Taking into consideration that

$$\left\langle(\alpha_k^*)^2\right\rangle = \text{Var}\{\alpha_k^*\} + \alpha_k^2 \tag{15.61}$$

and the condition given by (14.123), we can define the "integral" variance of correlation function estimate

$$\text{Var}\{R_\nu^*(\tau)\} = \int_{-\infty}^{\infty} \left\langle \left[R_\nu^*(\tau) - \left\langle R_\nu^*(\tau)\right\rangle\right]^2 \right\rangle d\tau = \sum_{k=0}^{\nu} \text{Var}\{\alpha_k^*\}. \tag{15.62}$$

As we can see from (15.62), the variance of correlation function estimate increases with an increase in the number of terms of expansion in series ν. Because of this, we must take into consideration choosing the number of terms under expansion in series.

The foregoing formulae allow us to present the flowchart of correlation function measurer based on the expansion of this correlation function in series and the estimate of coefficients of this expansion in series. Figure 15.4 represents the one-channel measurer of the current value of the coefficient α_k^*. Operation of measurer is clear from Figure 15.4. One of the main elements of block diagram of the correlation function measurer is the generator of orthogonal signals or functions (the orthogonal filters with the impulse response given by [15.53]). If the pulse with short duration τ_p that is much shorter than the filter constant time and amplitude τ_p^{-1} excites the input of the orthogonal filter, then a set of orthogonal functions $\varphi_k(t)$ are observed at the output of orthogonal filters.

The flowchart of correlation function measurer is shown in Figure 15.5. Operation control of the correlation function measurer is carried out by the synchronizer that stimulates action on the generator of orthogonal signals and allows us to obtain the correlation function estimate with period that is much longer compared to the orthogonal signal duration.

The functions using the orthogonal Laguerre polynomials given by (14.72) and presented in the following form

$$L_k(\alpha t) = \exp\{\alpha t\}\frac{d^k[t^k \exp\{-\alpha t\}]}{dt^k} = \sum_{\mu=0}^{k} \frac{(-\alpha t)^\mu (k!)^2}{(k-\mu)!(\mu!)^2} \tag{15.63}$$

are the simplest among a set of orthogonal functions, where α characterizes a polynomial scale in time. To satisfy (14.123), in this case, the orthogonal functions $\varphi_k(t)$ take the following form:

$$\varphi_k(t) = \frac{1}{k!}\sqrt{\alpha}\exp\{-0.5\alpha t\}L_k(\alpha t). \tag{15.64}$$

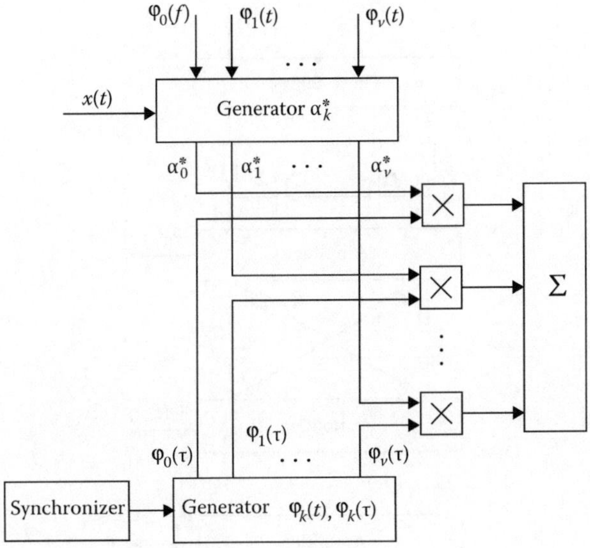

FIGURE 15.5 Measurer of correlation function.

Carrying out the Laplace transform

$$\varphi_k(p) = \int_0^\infty \exp\{-pt\}\varphi_k(t)dt, \tag{15.65}$$

we can find that the considered orthogonal functions correspond to the transfer functions

$$\varphi_k(p) = \frac{2}{\sqrt{\alpha}} \times \frac{0.5\alpha}{p+0.5\alpha}\left[\frac{p-0.5\alpha}{p+0.5\alpha}\right]^k. \tag{15.66}$$

The multistep filter based on RC elements, that is, $\alpha = 2(RC)^{-1}$, which has the transfer characteristic given by (15.66) accurate with the constant factor $2\alpha^{-0.5}$ is shown in Figure 15.6. The phase inverter is assigned to generate two signals with equal amplitudes and shifted by phase on 90° and the amplifiers compensate attenuation in filters and ensure decoupling between them.

If the stationary stochastic process is differentiable ν times, the correlation function $R(\tau)$ of this process can be approximated by expansion in series about the point $\tau = 0$:

$$R(\tau) \approx R_\nu(\tau) = \sum_{i=0}^{\nu} \left.\frac{d^{2i}R(\tau)}{d\tau^{2i}}\right|_{\tau=0} \times \frac{\tau^{2i}}{(2i)!}. \tag{15.67}$$

The approximation error of the correlation function is defined as

$$\varepsilon(\tau) = R(\tau) - R_\nu(\tau). \tag{15.68}$$

The even $2i$th derivatives of the correlation function at the point $\tau = 0$ accurate within the coefficient $(-1)^i$ are the variances of ith derivatives of the stochastic process

$$(-1)^i \left.\frac{d^{2i}R(\tau)}{d\tau^{2i}}\right|_{\tau=0} = \left\langle \left[\frac{d^i\xi(t)}{dt^i}\right]^2 \right\rangle = \text{Var}_i. \tag{15.69}$$

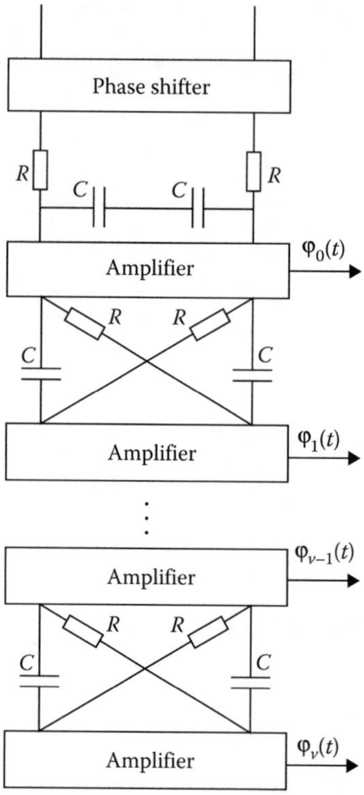

FIGURE 15.6 Multistep *RC* filter.

As applied to the ergodic stochastic process, the coefficients of expansion in series given by (15.67) can be presented in the following form:

$$\alpha_i = \left.\frac{d^{2i}R(\tau)}{d\tau^{2i}}\right|_{\tau=0} = (-1)^i \lim_{T\to\infty} \frac{1}{T}\int_0^T \left\{\frac{d^i x(t)}{dt^i}\right\}^2 dt. \tag{15.70}$$

In the case when the observation time interval is finite, the estimate α_i^* of the coefficient α_i is defined as

$$\alpha_i^* = (-1)^i \frac{1}{T}\int_0^T \left\{\frac{d^i x(t)}{dt^i}\right\}^2 dt. \tag{15.71}$$

The correlation function estimate takes the following form:

$$R^*(\tau) = \sum_{i=1}^{\nu} \frac{\alpha_i^* \tau^{2i}}{(2i)!}. \tag{15.72}$$

Flowchart of measurer based on definition of the coefficients of correlation function expansion in power series is shown in Figure 15.7. The investigated realization $x(t)$ is differentiable ν times. The obtained processes $y_i(t) = d^i x(t)/dt^i$ are squared and integrated within the limits of the observation time interval $[0, T]$ and come in at the input of calculator with corresponding signs. According to (15.72), the

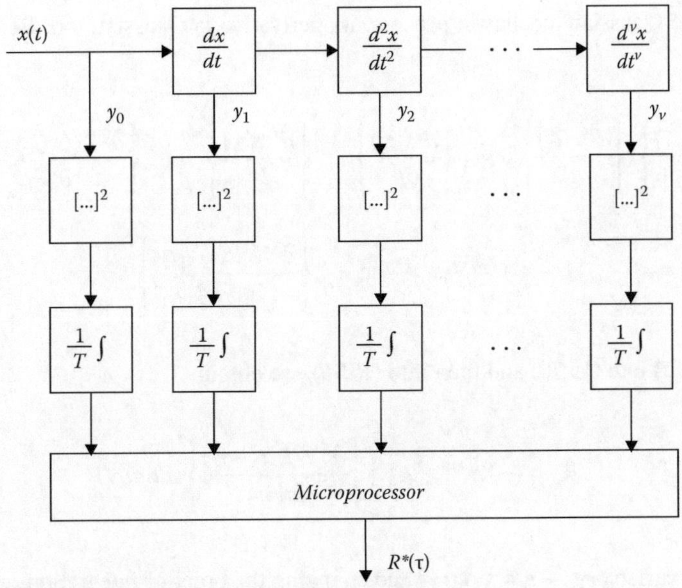

FIGURE 15.7 Correlation function measurement.

correlation function estimate is formed at the calculator output. Define the statistical characteristics of the coefficient estimate α_i^*. The mathematical expectation of the estimate α_i^* has the following form:

$$\langle\alpha_i^*\rangle = (-1)^i \frac{1}{T}\int_0^T \left\langle \left\{\frac{d^i x(t)}{dt^i}\right\}^2 \right\rangle dt. \tag{15.73}$$

We can see that

$$\left\{\frac{d^i x(t)}{dt^i}\right\}^2 = \frac{\partial^{2i} R(t_1 - t_2)}{\partial t_1^i \partial t_2^i}\bigg|_{t_1=t_2=t} = (-1)^i \frac{d^{2i} R(\tau)}{d\tau^{2i}}\bigg|_{\tau=0}. \tag{15.74}$$

An introduction of new variables $t_2 - t_1 = \tau$, $t_2 = t$ is made. Substituting (15.74) into (15.73), we see that the estimates of coefficients of expansion in series are unbiased. The correlation function of estimates is

$$R_{ip} = \langle\alpha_i^*\alpha_p^*\rangle - \langle\alpha_i^*\rangle\langle\alpha_p^*\rangle, \tag{15.75}$$

where

$$\langle\alpha_i^*\alpha_p^*\rangle = (-1)^{(i+p)} \frac{1}{T^2}\int_0^T\int_0^T \left\langle \left\{\frac{d^i x(t_1)}{dt_1^i}\right\}^2 \left\{\frac{d^p x(t_1)}{dt_2^p}\right\}^2 \right\rangle dt_1 dt_2. \tag{15.76}$$

As applied to the Gaussian stochastic process, its derivative is Gaussian too. Because of this, we can write

$$\left\langle\left\{\frac{d^i x(t_1)}{dt_1^i}\right\}^2\left\{\frac{d^p x(t_1)}{dt_2^p}\right\}^2\right\rangle=\left\langle\left\{\frac{d^i x(t_1)}{dt_1^i}\right\}^2\right\rangle\left\langle\left\{\frac{d^p x(t_1)}{dt_2^p}\right\}^2\right\rangle+2\left\{\frac{\partial^{(i+p)}\langle x(t_1)x(t_2)\rangle}{\partial t_1^i \partial t_2^p}\right\}^2$$

$$=\mathrm{Var}_i\times\mathrm{Var}_p+2\left\{\frac{\partial^{(i+p)}R(t_2-t_1)}{\partial t_1^i\partial t_2^p}\right\}^2. \tag{15.77}$$

Substituting (15.77) into (15.76) and then into (15.75), we obtain

$$R_{ip}=(-1)^{(i+p)}\frac{2}{T^2}\int_0^T\int_0^T\left\{\frac{\partial^{(i+p)}R(t_2-t_1)}{\partial t_1^i\partial t_2^p}\right\}^2 dt_1dt_2. \tag{15.78}$$

Introducing new variables $t_2 - t_1 = \tau$, $t_2 = t$ and changing the order of integration, we obtain

$$R_{ip}=\frac{4}{T}\int_0^T\left(1-\frac{\tau}{T}\right)\left\{\frac{d^{(i+p)}R(\tau)}{d\tau^{(i+p)}}\right\}^2 d\tau. \tag{15.79}$$

If the observation time interval is much longer than the correlation interval of stochastic process and its derivatives, we can write

$$R_{ip}=\frac{2}{T}\int_{-\infty}^{\infty}\left\{\frac{d^{(i+p)}R(\tau)}{d\tau^{(i+p)}}\right\}^2 d\tau. \tag{15.80}$$

As applied to the conditions, for which (15.80) is appropriate, the derivatives of the correlation function can be written using the spectral density $S(\omega)$ of the stochastic process:

$$\frac{d^{(i+p)}R(\tau)}{d\tau^{(i+p)}}=\frac{1}{2\pi}\int_{-\infty}^{\infty}(j\omega)^{(i+p)}S(\omega)\exp\{j\omega\tau\}d\omega. \tag{15.81}$$

Then

$$R_{ip}=\frac{1}{T\pi}\int_{-\infty}^{\infty}\omega^{2(i+p)}S^2(\omega)d\omega. \tag{15.82}$$

The variance of the estimate α_i^* of expansion in series coefficients can be presented in the following form:

$$\mathrm{Var}\{\alpha_i^*\}=\frac{1}{T\pi}\int_{-\infty}^{\infty}\omega^{4i}S^2(\omega)d\omega. \tag{15.83}$$

Let us define the deviation of correlation function estimate from the approximated value, namely,

$$\varepsilon(\tau) = R_v(\tau) - R_v^*(\tau). \tag{15.84}$$

Averaging $\varepsilon(\tau)$ by realizations of the investigated stochastic process, we can see that in the considered case $\langle\varepsilon(\tau)\rangle = 0$, which means the bias of correlation function estimate does not increase due to the finite observation time interval.

The variance of correlation function estimate can be presented in the following form:

$$\mathrm{Var}\{R_v^*(\tau)\} = \sum_{i=1}^{v}\sum_{p=1}^{v} \frac{\tau^{2(i+p)}}{(2i)!(2p)!}\frac{4}{T}\int_0^T \left(1-\frac{\tau}{T}\right)\left[\frac{d^{(i+p)}R(\tau)}{d\tau^{(i+p)}}\right]^2 d\tau. \tag{15.85}$$

At $T \gg \tau_{\text{cor}}$, we can write

$$\mathrm{Var}\{R_v^*(\tau)\} = \frac{1}{T\pi}\sum_{i=1}^{v}\sum_{p=1}^{v} \frac{\tau^{2(i+p)}}{(2i)!(2p)!}\int_{-\infty}^{\infty} \omega^{2(i+p)} S^2(\omega)d\omega. \tag{15.86}$$

Let the correlation function of the stochastic process be approximated by

$$R(\tau) = \sigma^2 \exp\{-\alpha^2\tau^2\}, \tag{15.87}$$

which corresponds to the spectral density defined as

$$S(\omega) = \sigma^2 \frac{\sqrt{\pi}}{\alpha} \exp\left\{-\frac{\omega^2}{4\alpha^2}\right\}. \tag{15.88}$$

Substituting (15.88) into (15.86), we obtain

$$\mathrm{Var}\{R_v^*(\tau)\} = \frac{\sigma^4\sqrt{2\pi}}{T\alpha}\sum_{i=1}^{v}\sum_{p=1}^{v} \frac{[2(i+p)-1]!!}{(2i)!(2p)!}(\alpha\tau)^{2(i+p)}, \tag{15.89}$$

where

$$[2(i+p)-1]!! = 1\times 3\times 5\times\cdots\times[2(i+p)-1]. \tag{15.90}$$

As we can see from (15.89), the variance of correlation function estimate increases with an increase in the number of terms under correlation function expansion in power series.

15.3 OPTIMAL ESTIMATION OF GAUSSIAN STOCHASTIC PROCESS CORRELATION FUNCTION PARAMETER

In some practical conditions, the correlation function of stochastic process can be measured accurately with some parameters defining a character of its behavior. In this case, a measurement of correlation function can be reduced to measurement or estimation of unknown parameters of correlation function. Because of this, we consider the optimal estimate of arbitrary correlation function parameter assuming that the investigated stochastic process $\xi(t)$ is the stationary Gaussian stochastic process observed within the limits of time interval [0, T] in the background of Gaussian stationary noise $\zeta(t)$ with known correlation function.

Thus, the following realization

$$y(t) = x(t,l_0) + n(t), \quad 0 \le t \le T \tag{15.91}$$

comes in at the measurer input, where $x(t, l_0)$ is the realization of the investigated Gaussian stochastic process with the correlation function $R_x(t_1, t_2, l)$ depending on the estimated parameter l; $n(t)$ is the realization of Gaussian noise with the correlation function $R_n(t_1, t_2)$. True value of estimated parameter of the correlation function $R_x(t_1, t_2, l)$ is l_0. Thus, we assume that the mathematical expectation of both the realization $x(t, l_0)$ and the realization $n(t)$ is equal to zero and the realizations $x(t, l_0)$ and $n(t)$ are statistically independent of each other. Based on input realization, the optimal receiver should form the likelihood ratio functional $\Lambda(l)$ or some monotone function of the likelihood ratio. The stochastic process $\eta(t)$ with realization given by (15.91) is the Gaussian stochastic process with zero mathematical expectation and the correlation function

$$R_y(t_1,t_2,l) = R_x(t_1,t_2,l) + R_n(t_1,t_2). \tag{15.92}$$

As applied to the investigated stochastic process $\eta(t)$, the likelihood functional can be presented in the following form [5]:

$$\Lambda(l) = \exp\left\{\frac{1}{2}\int_0^T\int_0^T y(t_1)y(t_2)[\vartheta_n(t_1,t_2) - \vartheta_x(t_1,t_2;l)]dt_1dt_2 - \frac{1}{2}H(l)\right\}. \tag{15.93}$$

We can write the derivative of the function $H(l)$ in the following form:

$$\frac{dH(l)}{dl} = \int_0^T\int_0^T \frac{\partial R_x(t_1,t_2,l)}{\partial l}\vartheta_x(t_1,t_2;l)dt_1dt_2 \tag{15.94}$$

and the functions $\vartheta_x(t_1, t_2; l)$ and $\vartheta_n(t_1, t_2)$ can be found from the following equations:

$$\int_0^T [R_x(t_1,t_2,l) + R_n(t_1,t_2)]\vartheta_x(t_1,t_2;l)dt = \delta(t_2 - t_1), \tag{15.95}$$

$$\int_0^T R_n(t_1,t_2)\vartheta_n(t_1,t_2)dt = \delta(t_2 - t_1). \tag{15.96}$$

Evidently, we can use the logarithmic term of the likelihood functional depending on the observed data as the signal at the receiver output

$$M_1(l) = \frac{1}{2}\int_0^T\int_0^T y(t_1)y(t_2)\vartheta(t_1,t_2;l)dt_1dt_2, \tag{15.97}$$

where

$$\vartheta(t_1,t_2;l)=\vartheta_n(t_1,t_2)-\vartheta_x(t_1,t_2;l). \tag{15.98}$$

We suppose that the correlation intervals of the stochastic processes $\xi(t)$ and $\zeta(t)$ are sufficiently small compared to the observation time interval $[0, T]$. In this case, we can use infinite limits of integration. Under this assumption, we have

$$\begin{cases}\vartheta_x(t_1,t_2;l)=\vartheta_x(t_1-t_2;l),\\ \vartheta_n(t_1,t_2)=\vartheta_n(t_1-t_2),\end{cases} \tag{15.99}$$

and

$$\vartheta(t_1,t_2;l)=\vartheta(t_1-t_2;l). \tag{15.100}$$

Introducing new variables $t_2 - t_1 = \tau$, $t_2 = t$ and changing the order of integration, we obtain

$$M_1(l)=\frac{1}{2}\left\{\int_0^T\vartheta(\tau;l)\int_0^{T-\tau}y(t+\tau)y(t)dtd\tau+\int_{-T}^0\vartheta(\tau;l)\int_0^T y(t+\tau)y(t)dtd\tau\right\}. \tag{15.101}$$

Introducing new variables $\tau=-\tau'$, $t'=t+\tau=t-\tau'$ and taking into consideration that the correlation interval of stochastic process $\eta(t)$ is shorter compared to the observation time interval $[0, T]$, we can write

$$M_1(l)=T\int_0^T R_y^*(\tau)\vartheta(\tau;l)d\tau, \tag{15.102}$$

where

$$R_y^*(\tau)=\frac{1}{T}\int_0^{T-\tau}y(t)y(t+\tau)dt\approx\frac{1}{T}\int_0^T y(t)y(t-\tau)dt \tag{15.103}$$

is the correlation function estimate of investigated input process consisting of additive mixture of the signal and noise.

Applying the foregoing statements to (15.94), we obtain

$$\frac{dH(l)}{dl}=2T\int_0^T\frac{\partial R_x(\tau;l)}{\partial l}\vartheta_x(\tau;l)d\tau. \tag{15.104}$$

Substituting (15.102) and (15.104) into (15.93), we obtain

$$\Lambda(l)=\exp\left\{T\int_0^T R_y^*(\tau)\vartheta(\tau;l)d\tau-\frac{1}{2}H(l)\right\}. \tag{15.105}$$

In the case, when the observation time interval [0, T] is much longer compared to the correlation interval of the investigated stochastic process, we can use the spectral representation of the function given by (15.100)

$$\vartheta(\tau;l) = \frac{1}{2\pi}\int_{-\infty}^{\infty}[\vartheta_n(\omega) - \vartheta_x(\omega;l)]\exp\{j\omega\tau\}d\omega. \tag{15.106}$$

Then (15.104) takes the following form:

$$\frac{dH(l)}{dl} = \frac{T}{2\pi}\int_0^T \frac{\partial S_x(\omega;l)}{\partial l}\vartheta_x(\omega;l)d\omega. \tag{15.107}$$

In (15.106) and (15.107), $\vartheta_n(\omega)$, $\vartheta_x(\omega; l)$, and $S_x(\omega; l)$ are the Fourier transforms of the corresponding functions $\vartheta_n(\tau)$, $\vartheta_x(\tau; l)$, and $R_x(\tau; l)$. Applying the Fourier transform to (15.95) and (15.96), we obtain

$$\begin{cases} R_n(\omega)\vartheta_n(\omega) = 1, \\ \vartheta_x(\omega;l)[S_n(\omega) + R_x(\omega;l)] = 1. \end{cases} \tag{15.108}$$

Taking into consideration (15.107) and (15.108), we can write

$$\vartheta(\tau;l) = \frac{1}{2\pi}\int_{-\infty}^{\infty}\frac{S_x(\omega;l)S_n^{-1}(\omega)}{S_x(\omega;l) + S_n(\omega)}\exp\{j\omega\tau\}d\omega, \tag{15.109}$$

$$H(l) = \frac{T}{2\pi}\int_{-\infty}^{\infty}\ln\left[1 + \frac{S_x(\omega;l)}{S_n(\omega)}\right]d\omega. \tag{15.110}$$

The signal at the optimal receiver output takes the following form:

$$M(l) = T\int_0^T R_y^*(\tau)\vartheta(\tau;l)d\tau - \frac{1}{2}H(l). \tag{15.111}$$

Flowchart of the optimal measurer is shown in Figure 15.8. This measurer operates in the following way. The correlation function $R_y^*(\tau)$ is defined based on the input realization of additive mixture of the signal and noise. This correlation function is integrated as the weight function with the signal $\vartheta(\tau; l)$. The signals forming at the outputs of the integrator and generator of the function $H(l)$ come

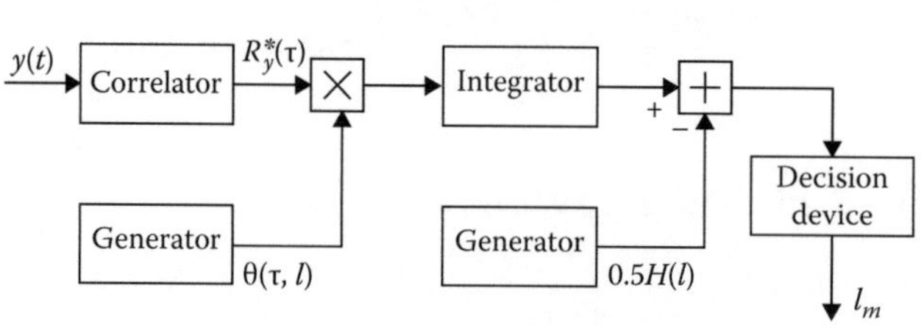

FIGURE 15.8 Optimal measurement of correlation function parameter.

in at the input of summator. Thus, the receiver output signal is formed. The decision device issues the value of the parameter l_m, under which the output signal takes the maximum value.

If the correlation function of the investigated Gaussian stochastic process has several unknown parameters $\mathbf{l} = \{l_1, l_2,\ldots, l_\mu\}$, then the likelihood ratio functional can be found by (15.93) changing the scalar parameter l on the vector parameter $\mathbf{l}$ and the function $H(\mathbf{l})$ is defined by its derivatives

$$\frac{\partial H(\mathbf{l})}{\partial l_i} = \int_0^T\int_0^T \frac{\partial R_x(t_1,t_2;\mathbf{l})}{\partial l_i} \vartheta_x(t_1,t_2;\mathbf{l})dt_1dt_2. \tag{15.112}$$

The function $\vartheta_x(t_1, t_2; \mathbf{l})$ is the solution of integral equation that is analogous to (15.95).

To obtain the estimate of maximum likelihood ratio of the correlation function parameter the optimal measurer (receiver) should define an absolute maximum l_m of logarithm of the likelihood ratio functional

$$M(l) = \frac{1}{2}\int_0^T\int_0^T y(t_1)y(t_2)[\vartheta_n(t_1,t_2) - \vartheta_x(t_1,t_2,l)]dt_1dt_2 - \frac{1}{2}H(l). \tag{15.113}$$

To define the characteristics of estimate of the maximum likelihood ratio l_m introduce the signal

$$s(l) = \langle M(l)\rangle \tag{15.114}$$

and noise

$$n(l) = M(l) - \langle M(l)\rangle \tag{15.115}$$

functions. Then (15.113) takes the following form:

$$M(l) = s(l) + n(l). \tag{15.116}$$

Prove that if the noise component is absent in (15.116), that is, $n(l) = 0$, the logarithm of likelihood ratio functional reaches its maximum at $l = l_0$, that is, when the estimated parameter takes the true value. Define the first and second derivatives of the signal function given by (15.114) at the point $l = l_0$:

$$\left.\frac{ds(l)}{dl}\right|_{l=l_0} = \left.\left\langle\frac{dM(l)}{dl}\right\rangle\right|_{l=l_0}. \tag{15.117}$$

Substituting (15.113) into (15.117) and averaging by realizations $y(t)$, we obtain

$$\left.\frac{ds(l)}{dl}\right|_{l=l_0} = \left\{-\frac{1}{2}\int_0^T\int_0^T [R_n(t_1,t_2) + R_x(t_1,t_2;l)]\frac{\partial\vartheta_x(t_1,t_2;l)}{\partial l}dt_1dt_2 - \frac{1}{2}\int_0^T\int_0^T \vartheta_x(t_1,t_2;l)\frac{\partial R_x(t_1,t_2;l)}{\partial l}dt_1dt_2\right\}_{l=l_0}$$

$$= -\frac{1}{2}\left\{\frac{d}{dt}\int_0^T\int_0^T [R_n(t_1,t_2) + R_x(t_1,t_2;l)]\vartheta_x(t_1,t_2;l)dt_1dt_2\right\}_{l=l_0}. \tag{15.118}$$

Since

$$R(t_1,t_2;l) = R_n(t_1,t_2) + R_x(t_1,t_2;l) = R(t_2,t_1;l), \tag{15.119}$$

in accordance with (15.95) we have

$$\left.\frac{ds(l)}{dl}\right|_{l=l_0} = -\frac{1}{2}\int_0^T \left\{\frac{d}{dl}\int_0^T [R_n(t_2,t_1) + R_x(t_2,t_1:l)]\vartheta_x(t_1,t_2;l)dt_1\right\}_{l=l_0} dt_2 = 0. \tag{15.120}$$

Now, let us define the second derivative of the signal component at the point $l = l_0$:

$$\left.\frac{d^2 s(l)}{dl^2}\right|_{l=l_0} = \left.\left\langle \frac{d^2 M(l)}{dl^2}\right\rangle\right|_{l=l_0}$$

$$= \frac{1}{2}\left\{\int_0^T\int_0^T \frac{\partial R_x(t_2,t_1:l)}{\partial l}\frac{\partial \vartheta_x(t_1,t_2;l)}{\partial l}dt_1 dt_2 - \frac{d^2}{dl^2}\int_0^T\int_0^T [R_n(t_2,t_1)+R_x(t_2,t_1:l)]\vartheta_x(t_1,t_2;l)dt_1 dt_2\right\}_{l=l_0}. \tag{15.121}$$

In accordance with (15.95) we can write

$$\left\{\frac{d^2}{dl^2}\int_0^T\int_0^T [R_n(t_2,t_1) + R_x(t_2,t_1:l)]\vartheta_x(t_1,t_2;l)dt_1 dt_2\right\}_{l=l_0}$$

$$= \int_0^T \left\{\frac{d^2}{dl^2}\int_0^T [R_n(t_2,t_1) + R_x(t_2,t_1:l)]\vartheta_x(t_1,t_2;l)dt_1\right\}_{l=l_0} dt_2 = 0. \tag{15.122}$$

Because of this

$$\left.\frac{d^2 s(l)}{dl^2}\right|_{l=l_0} = \frac{1}{2}\left\{\int_0^T\int_0^T \frac{\partial R_x(t_2,t_1:l)}{\partial l}\frac{\partial \vartheta_x(t_1,t_2;l)}{\partial l}dt_1 dt_2\right\}_{l=l_0}. \tag{15.123}$$

Let us prove that the condition

$$\left.\frac{d^2 s(l)}{dl^2}\right|_{l=l_0} < 0 \tag{15.124}$$

is satisfied forever. For this purpose, we define the averaged quadratic first derivative of the likelihood ratio functional logarithm at the point $l = l_0$

$$m^2 = \left\langle \left\{\left.\frac{dM(l)}{dl}\right|_{l=l_0}\right\}^2\right\rangle = \left\langle \left\{\left.\frac{dN(l)}{dl}\right|_{l=l_0}\right\}^2\right\rangle, \tag{15.125}$$

which is a positive value. Substituting (15.91) into (15.113), differentiating by l, and averaging by realizations $y(t)$, we obtain the second central moment of the first derivative of the likelihood ratio functional logarithm:

$$\frac{\partial^2}{\partial l_1 \partial l_2}\langle [M(l_1)-\langle M(l_1)\rangle][M(l_2)-\langle M(l_2)\rangle]\rangle$$

$$=\frac{1}{2}\int_0^T\int_0^T\int_0^T\int_0^T [R_n(t_1,t_3)+R_x(t_1,t_3;l_0)][R_n(t_2,t_4)+R_x(t_2,t_4;l_0)]$$

$$\times \frac{\partial \vartheta_x(t_1,t_2;l_1)}{\partial l_1}\frac{\partial \vartheta_x(t_3,t_4;l_2)}{\partial l_2}dt_1dt_2dt_3dt_4. \tag{15.126}$$

Assuming $l_2 = l_1 = l_0$ and taking into consideration that

$$\int_0^T [R_n(t_1,t)+R_x(t_1,t;l_0)]\frac{\partial \vartheta_x(t_1,t_2;l)}{\partial l_1}dt = -\int_0^T \vartheta_x(t,t_2;l)\frac{\partial R_x(t_1,t;l)}{\partial l}dt, \tag{15.127}$$

we obtain

$$m^2 = -\left\{\frac{1}{2}\int_0^T\int_0^T\int_0^T\int_0^T [R_n(t_2,t_4)+R_x(t_2,t_4;l_0)]\vartheta_x(t_1,t_2;l)\frac{\partial R_x(t_1,t_3;l)}{\partial l}\frac{\partial \vartheta_x(t_3,t_4;l)}{\partial l}dt_1dt_2dt_3dt_4\right\}_{l=l_0}$$

$$= -\left\{\frac{1}{2}\int_0^T\int_0^T \frac{\partial R_x(t_1,t_2;l)}{\partial l}\frac{\partial \vartheta_x(t_1,t_2;l)}{\partial l}dt_1dt_2\right\}_{l=l_0}. \tag{15.128}$$

In (15.128) we have implemented (15.95) again. Comparing (15.128) with (15.123), we can see that

$$\left.\frac{d^2 s(l)}{dl^2}\right|_{l=l_0} = -m^2 \tag{15.129}$$

and, consequently, (15.124) is satisfied forever.

Introduce the signal-to-noise ratio (SNR)

$$\text{SNR} = \frac{s^2(l_0)}{\langle n^2(l_0)\rangle} \tag{15.130}$$

and the normalized signal and noise functions

$$\begin{cases} \mathbf{S}(l) = \dfrac{s(l)}{s(l_0)}, \\ \mathbf{N}(l) = \dfrac{n(l)}{\sqrt{\langle n^2(l_0)\rangle}}. \end{cases} \tag{15.131}$$

Taking into consideration (15.120), (15.124), and (15.131), we can see that

$$\begin{cases} \max \mathbf{S}(l) = \mathbf{S}(l_0) = 1, \\ \langle \mathbf{N}^2(l_0) \rangle = 1. \end{cases} \tag{15.132}$$

In addition, as follows from the definition, the mathematical expectation of the noise function $n(l)$ is zero. Taking into consideration the introduced notations, we can write the logarithm of the likelihood ratio functional in the following form:

$$M(l) = s(l_0)[\mathbf{S}(l) + \varepsilon \mathbf{N}(l)], \tag{15.133}$$

where $\varepsilon = 1/\sqrt{\text{SNR}}$. Taking into consideration (15.133), the likelihood ratio equation for the estimate of correlation function parameter of Gaussian stochastic process can be presented in the following form:

$$\left\{ \frac{d\mathbf{S}(l)}{dl} + \varepsilon \frac{d\mathbf{N}(l)}{dl} \right\}_{l=l_m} = 0. \tag{15.134}$$

Usually, under the measurement of stochastic process parameters, SNR is high and, consequently, $\varepsilon \ll 1$. Then, by analogy with Ref. [5], the approximated solution of likelihood ratio equation can be searched in the form of expansion in power series

$$l_m = l_0 + \varepsilon l_1 + \varepsilon^2 l_2 + \varepsilon^3 l_3 + \cdots. \tag{15.135}$$

To define the approximations l_1, l_2, l_3 and grouping the terms with small value ε of the same power, we obtain

$$\mathbf{s}_1 + \varepsilon(l_1\mathbf{s}_2 + \mathbf{n}_1) + \varepsilon^2(l_2\mathbf{s}_2 + l_1\mathbf{n}_2 + 0.5l_1^2\mathbf{s}_3) + \varepsilon^3\left(l_3\mathbf{s}_2 + l_2\mathbf{n}_2 + 0.5l_1^2\mathbf{n}_3 + \frac{l_1^3\mathbf{s}_4}{6} + l_1 l_3 \mathbf{s}_2\right) + \cdots = 0, \tag{15.136}$$

where we use the following notations:

$$\begin{cases} \mathbf{s}_i = \left.\dfrac{d^i\mathbf{S}(l)}{dl^i}\right|_{l=l_0}, \\ \mathbf{n}_i = \left.\dfrac{d^i\mathbf{N}(l)}{dl^i}\right|_{l=l_0}. \end{cases} \tag{15.137}$$

Since the system of functions 1, x, x^2,... is linearly independent, the equality given by (15.136) is satisfied for any ε if and only if all coefficients of terms with power equal to ε are equal to zero. Zero approximation is matched with the true value of the parameter l_0 since $\mathbf{S}(l)$ reaches its absolute maximum at

$$l = l_0. \tag{15.138}$$

Equating to zero the coefficients at ε, ε^2, and ε^3, we obtain equations to define l_1, l_2, and l_3. We can write solutions of these equations in the following form:

$$\begin{cases} l_1 = -\dfrac{\mathbf{n}_1}{\mathbf{s}_2}, \\ l_2 = -\dfrac{l_1\mathbf{n}_2 + 0.5l_1^2\mathbf{s}_3}{\mathbf{s}_2}, \\ l_3 = -\dfrac{l_2\mathbf{n}_2 + 0.5l_1^2\mathbf{n}_3 + 6^{-1}l_1^3\mathbf{s}_4 + l_1l_2\mathbf{s}_3}{\mathbf{s}_2}. \end{cases} \tag{15.139}$$

Taking into consideration the first three approximations l_1, l_2, and l_3, the conditional bias and variance of maximum likelihood ratio take the following form:

$$b(l_m|l_0) = \varepsilon\langle l_1\rangle + \varepsilon^2\langle l_2\rangle + \varepsilon^3\langle l\rangle_3, \tag{15.140}$$

$$\begin{aligned} \mathrm{Var}\{l_m|l_0\} &= \varepsilon^2[\langle l_1^2\rangle - \langle l_1\rangle^2] + 2\varepsilon^3[\langle l_1l_2\rangle - \langle l_1\rangle\langle l_2\rangle] \\ &\quad + \varepsilon^4[\langle l_2^2\rangle - \langle l_2\rangle^2 + 2\langle l_1l_3\rangle - 2\langle l_1\rangle\langle l_3\rangle]. \end{aligned} \tag{15.141}$$

Averaging is carried out by all possible realizations of the total stochastic process $\eta(t)$ at the fixed value of estimated parameter l_0. The relative error of estimate bias and variance that can be defined as the ratio of the first term with small order to the first term of expansion takes the order ε^2.

We are limited by consideration of the first approximation. In doing so, the random error of a single measurement can be presented in the following form:

$$\Delta l = l_m - l_0 = \varepsilon l_1 = -\varepsilon \left. \frac{\dfrac{d\mathbf{N}(l)}{dl}}{\dfrac{d^2\mathbf{S}(l)}{dl^2}} \right|_{l=l_0} = -\left. \frac{\dfrac{dn(l)}{dl}}{\dfrac{d^2 s(l)}{dl^2}} \right|_{l=l_0}. \tag{15.142}$$

For the first approximation, the estimate of arbitrary parameter of correlation function will be unbiased, since $\langle n(l)\rangle = 0$. Taking into consideration (15.128) and (15.129), the variance of estimate can be presented in the following form:

$$\mathrm{Var}\{l_m|l_0\} = \frac{\left. \dfrac{\partial^2}{\partial l_1 \partial l_2} \langle n(l_1)n(l_2)\rangle \right|_{l=l_0}}{\left. \left[\dfrac{d^2\mathbf{S}(l)}{dl^2}\right]^2 \right|_{l=l_0}} = m^{-2}. \tag{15.143}$$

If the observation time interval is much longer than the correlation interval of the investigated stochastic process $\eta(t)$, a flowchart of optimal measurer is significantly simplified. In this case, the logarithm of the likelihood ratio functional is given by (15.111), where the signal function can be described in the following form:

$$s(l) = T\int_0^T \left\langle R_y^*(\tau)\right\rangle \vartheta(\tau;l)d\tau - \frac{1}{2}H(l). \tag{15.144}$$

The first term in (15.144) can be presented in the following form:

$$\int_0^T \langle R_y^*(\tau)\rangle \vartheta(\tau;l)d\tau = \frac{1}{2}\int_{-T}^{T}[R_n(\tau)+R_x(\tau;l_0)]\vartheta(\tau;l)d\tau \approx \frac{1}{2}\int_{-\infty}^{\infty}[R_n(\tau)+R_x(\tau;l_0)]\vartheta(\tau;l)d\tau$$
$$= \frac{1}{4\pi}\int_{-\infty}^{\infty}[S_n(\omega)+S_x(\omega;l_0)]\vartheta(\omega;l)d\omega = \frac{1}{4\pi}\int_{-\infty}^{\infty}\frac{S_x(\omega;l)[S_n(\omega)+S_x(\omega;l_0)]}{S_n(\omega)[S_n(\omega)+S_x(\omega;l)]}d\omega. \tag{15.145}$$

Substituting (15.145) into (15.114) and taking into consideration (15.110), we obtain the signal function in the following form:

$$s(l) = \frac{T}{4\pi}\int_{-\infty}^{\infty}\left\{\frac{S_x(\omega;l)[S_n(\omega)+S_x(\omega;l_0)]}{S_n(\omega)[S_n(\omega)+S_x(\omega;l)]} - \ln\left[1+\frac{S_x(\omega;l)}{S_n(\omega)}\right]\right\}d\omega. \tag{15.146}$$

The signal function reaches its maximum at $l = l_0$:

$$s(l_0) = \frac{T}{4\pi}\int_{-\infty}^{\infty}\left\{\frac{S_x(\omega;l_0)}{S_n(\omega)} - \ln\left[1+\frac{S_x(\omega;l_0)}{S_n(\omega)}\right]\right\}d\omega. \tag{15.147}$$

Analogously, we can define the variance of noise component $n(l)$ given by (15.115)

$$\langle n^2(l)\rangle = \frac{T}{4\pi}\int_{-\infty}^{\infty}\frac{S_x^2(\omega;l_0)}{S_n^2(\omega)}d\omega. \tag{15.148}$$

Consequently, SNR given by (15.130) can be presented in the following form:

$$\text{SNR} = \frac{T}{4\pi}\frac{\left\{\int_{-\infty}^{\infty}\left\{\frac{S_x(\omega;l_0)}{S_n(\omega)} - \ln\left[1+\frac{S_x(\omega;l_0)}{S_n(\omega)}\right]\right\}d\omega\right\}^2}{\int_{-\infty}^{\infty}\frac{S_x^2(\omega;l_0)}{S_n^2(\omega)}d\omega}. \tag{15.149}$$

If SNR is high, the variance of correlation function estimate is defined by (15.143), where the value m^2 given by (15.129) can be presented in the following form using the spectral density components:

$$m^2 = -T\int_0^T \frac{\partial R_x(\tau;l)}{\partial l}\times\frac{\partial \vartheta_x(\tau;l)}{\partial l}d\tau\bigg|_{l=l_0} \approx -\frac{T}{4\pi}\int_{-\infty}^{\infty}\frac{\partial S_x(\omega;l)}{\partial l}\times\frac{\partial \vartheta_x(\omega;l)}{\partial l}d\omega\bigg|_{l=l_0}$$
$$= \frac{T}{4\pi}\int_{-\infty}^{\infty}\frac{\left[\frac{\partial S_x(\omega;l)}{\partial l}\right]^2}{[S_n(\omega)+S_x(\omega;l)]^2}d\omega\Bigg|_{l=l_0}. \tag{15.150}$$

We can define the second set of approximations for the bias and variance of arbitrary parameter estimate of the correlation function of the investigated stochastic process in accordance with (15.140) and (15.141). After cumbersome mathematical transformations, we obtain that the bias and variance of estimate take the following form:

$$b(l_m|l_0) = -\frac{1}{2} J_{12}^{11}\left(J_{10}^{20}\right)^{-2}, \tag{15.151}$$

$$\mathrm{Var}\{l_m|l_0\} = \frac{2}{J_{10}^{20}} + \frac{4}{\left[J_{10}^{20}\right]^3}\left\{12J_{10}^{40} - 6J_{12}^{21} - J_{13}^{11} + \frac{1}{J_{10}^{20}}\left[\frac{7}{2}\left[J_{12}^{11}\right]^2 + 6J_{12}^{11}J_{10}^{30} - 12\left[J_{10}^{30}\right]^2\right]\right\}, \tag{15.152}$$

where

$$J_{ij}^{pq} = \frac{T}{2\pi}\int_{-\infty}^{\infty}\left\{\frac{\partial^i S_x(\omega;l)}{\partial l^i}\right\}_{l=l_0}^{q}\left\{\frac{\partial^j S_x(\omega;l)}{\partial l^j}\right\}_{l=l_0}^{p}[S_x(\omega;l_0)+S_n(\omega)]^{-(p+q)}d\omega. \tag{15.153}$$

Comparing (15.129), (15.150), and (15.153), we see that

$$m^2 = \frac{1}{2}J_{10}^{20}. \tag{15.154}$$

Based on the formulae obtained, we can define the statistical characteristics of the estimate of correlation function parameter α of the investigated stochastic process additively mixed with the white noise possessing the one-sided power spectral density $\mathcal{N}_0$

$$R_x(\tau;\alpha) = \sigma^2\exp\{-\alpha|\tau|\}. \tag{15.155}$$

To reduce mathematical transformations and computations, the bias of estimate of the correlation function parameter α is defined taking into consideration the second approximation given by (15.151) and the variance of estimate of the correlation function parameter α is defined taking into consideration the first approximation given by (15.129) (the first term in the right side of (15.152)).

The correlation function parameter α defines the effective bandwidth of spectral density of stochastic process, that is, $\Delta f_{\mathrm{ef}} = 0.25\alpha$. Thus, we can write

$$\begin{cases} S_x(\omega;\alpha) = \dfrac{2\alpha\sigma^2}{\alpha^2+\omega^2}, \\ S_n(\omega) = \dfrac{\mathcal{N}_0}{2}. \end{cases} \tag{15.156}$$

Introduce the following notations:

$$\begin{cases} q^2 = \dfrac{4\sigma^2}{\mathcal{N}_0\alpha}, \\ p = \dfrac{T}{\tau_{\mathrm{cor}}}, \\ \beta = \sqrt{1+q^2}, \end{cases} \tag{15.157}$$

q^2 is the ratio of the investigated stochastic process variance to the white noise power within the limits of effective bandwidth of the signal; p is the ratio between the observation time interval of the investigated stochastic process and its correlation interval. Using these notations, we can write

$$J_{10}^{20} = \frac{pq^4(\beta^3 - \beta^2 + 3\beta + 1)}{\alpha_0^2\beta^3(1+\beta)^3}; \tag{15.158}$$

$$J_{12}^{11} = -\frac{2pq^4(2\beta^3 - \beta^2 + 4\beta + 1)}{\alpha_0^2\beta^3(1+\beta)^4}, \tag{15.159}$$

where α_0 is the true value of estimated correlation function parameter. Substituting (15.158) and (15.159) into (15.151) and (15.152), we obtain

$$b(\alpha_m|\alpha_0) = \frac{\alpha_0(1+\beta)^2\beta^3(2\beta^3 - \beta^2 + 4\beta + 1)}{pq^4(\beta^3 - \beta^2 + 3\beta + 1)^2}; \tag{15.160}$$

$$\text{Var}\{\alpha_m|\alpha_0\} = \frac{2\alpha_0^2(1+\beta)^3\beta^3}{pq^4(\beta^3 - \beta^2 + 3\beta + 1)^2}. \tag{15.161}$$

When $q^2 \ll 1$ and $4\sigma^2 T\mathcal{N}_0^{-1} \gg 1$, in this case (15.160) and (15.161) are correct, then $\beta \approx 1$ and (15.160) and (15.161) take a simple form:

$$b(\alpha_m|\alpha_0) \approx \frac{3\alpha_0}{2pq^4} = \frac{3\mathcal{N}_0^2\alpha_0^2}{32\sigma^4 T}; \tag{15.162}$$

$$\text{Var}\{\alpha_m|\alpha_0\} \approx \frac{4\alpha_0^2}{pq^4} = \frac{\mathcal{N}_0^2\alpha_0^3}{4\sigma^4 T}. \tag{15.163}$$

If $q^2 \ll 1$ and $\beta \approx q$, (15.162) and (15.163) take the following form:

$$b(\alpha_m|\alpha_0) = \frac{\alpha_0}{pq^2}; \tag{15.164}$$

$$\text{Var}\{\alpha_m|\alpha_0\} = \frac{2\alpha_0^2}{pq}. \tag{15.165}$$

The relative shift of estimate bias $pb(\alpha_m|\alpha_0)/\alpha_0$ and relative root-mean-square deviation $\sqrt{(p\,\text{Var}\{\alpha_m|\alpha_0\})/2\alpha_0^2}$ as a function of ratio between the variance of investigated stochastic process to power noise q^2 within the limits of effective spectral bandwidth are presented in Figures 15.9 and 15.10.

Consider the second example. For this purpose, we analyze the correlation function of the narrowband stochastic process $\xi(t)$

$$R_x(\tau;\nu) = \sigma^2\rho_{en}(\tau)\cos\nu\tau, \tag{15.166}$$

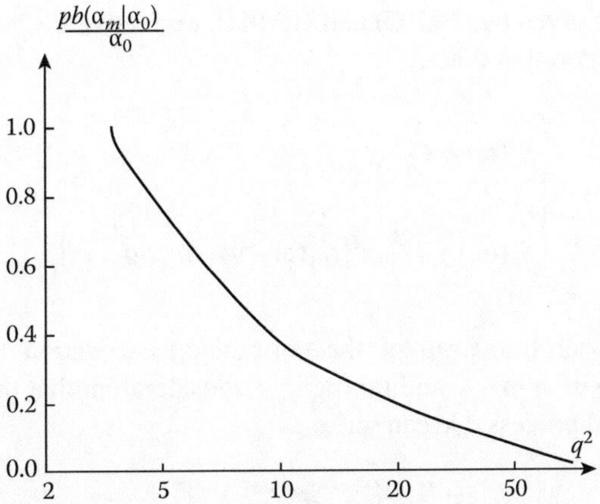

FIGURE 15.9 Relative estimate bias shift as a function of the ratio between the variance of the investigated stochastic process and power noise.

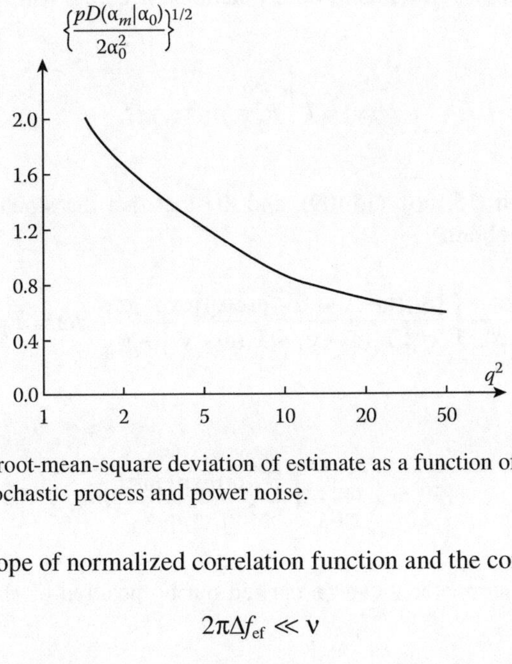

FIGURE 15.10 Relative root-mean-square deviation of estimate as a function of the ratio between the variance of the investigated stochastic process and power noise.

where $\rho_{en}(\tau)$ is the envelope of normalized correlation function and the condition

$$2\pi\Delta f_{ef} \ll \nu \tag{15.167}$$

is satisfied. We estimate the parameter ν. In the narrowband stochastic process case, the parameter ν plays a role of the central spectral density frequency. We assume that the stochastic process with the correlation function given by (15.166) is investigated in the white noise with the correlation function

$$R_n(\tau) = \frac{\mathcal{N}_0}{2}\delta(\tau), \tag{15.168}$$

where $\delta(\tau)$ is the Dirac delta function and the observation time interval $[0, T]$ is much longer than the correlation stochastic process interval.

In accordance with (15.111), the logarithm of likelihood ratio functional can be presented in the following form:

$$M(\nu) = M_1(\nu) - 0.5H(\nu), \tag{15.169}$$

where $M_1(\nu)$ and $H(\nu)$ are given by (15.102) and (15.104), respectively. Consider the second term in (15.169) taking into consideration that

$$\begin{cases} S_n(\omega) = \dfrac{\mathcal{N}_0}{2}, \\ S_x(\omega;\nu) = \dfrac{1}{2}\sigma^2[\mathcal{F}_1(\omega-\nu)+\mathcal{F}_1(\omega+\nu)], \end{cases} \tag{15.170}$$

where $\mathcal{F}_1(\omega)$ is the Fourier transform of the normalized correlation function envelope $\rho_{en}(\tau)$. Introducing new variable $\omega' = \omega - \nu$ and taking into consideration that the investigated stochastic process is the narrowband process, we can write

$$H(\nu) \approx \frac{T}{\pi}\int_{-\infty}^{\infty} \ln\left[1+\frac{\sigma^2}{\mathcal{N}_0}\mathcal{F}_1(\omega)\right]d\omega = \text{const.} \tag{15.171}$$

Consequently, the logarithm of likelihood ratio functional accurate with the constant factor coincides with the output signal

$$M(\nu) = T\int_0^T R_y^*(\tau)\vartheta(\tau;\nu)d\tau. \tag{15.172}$$

Taking into consideration (15.106), (15.109), and the fact that the stochastic process $\xi(t)$ is the narrowband process, we obtain

$$\vartheta(\tau;\nu) = \frac{\sigma^2}{\pi\mathcal{N}_0}\int_{-\infty}^{\infty}\frac{[\mathcal{F}_1(\omega-\nu)+\mathcal{F}_1(\omega+\nu)]\exp\{j\omega\tau\}}{\sigma^2[\mathcal{F}_1(\omega-\nu)+\mathcal{F}_1(\omega+\nu)]+\mathcal{N}_0}d\omega = \tilde{\mathcal{F}}_1(\tau)\cos\nu\tau, \tag{15.173}$$

where

$$\tilde{\mathcal{F}}_1(\tau) = \frac{2\sigma^2}{\pi\mathcal{N}_0}\int_{-\infty}^{\infty}\frac{\mathcal{F}_1(\omega)\exp\{j\omega\tau\}}{\sigma^2\mathcal{F}_1(\omega)+\mathcal{N}_0}d\omega. \tag{15.174}$$

Thus, the estimation of parameter ν can be carried out by position of absolute maximum of the function

$$M_1(\nu) = T\int_0^T R_y^*(\tau)\tilde{\mathcal{F}}_1(\tau)\cos\nu\tau d\tau, \tag{15.175}$$

where $R_y^*(\tau)$ is the correlation function estimate of the total process given by (15.103).

Define the bias and variance of estimate of the parameter ν limiting only by the first approximation. Under this approximation, the estimate will be unbiased and, in accordance with (15.143) and (15.150), the variance of estimate can be presented in the following form:

$$\text{Var}\{\nu_m|\nu_0\} = \frac{2\pi}{T\sigma^4\displaystyle\int_{-\infty}^{\infty}\left\{\frac{d\mathcal{F}_1(\omega)}{d\omega}\right\}^2\frac{d\omega}{[\sigma^2\mathcal{F}_1(\omega)+\mathcal{N}_0]^2}}. \tag{15.176}$$

If the normalized correlation function envelope takes the form

$$\rho_{\text{en}}(\tau) = \exp\{-\alpha \,|\, \tau \,|\}, \tag{15.177}$$

the Fourier transform can be presented as

$$\mathcal{F}_1(\omega) = \frac{2\alpha}{\alpha^2 + \omega^2}. \tag{15.178}$$

As a result, the variance of the correlation function parameter estimate takes the following form:

$$\text{Var}\{\nu_m \,|\, \nu_0\} = \alpha^2 \left(1 + \sqrt{1+q^2}\right)^3 \frac{\sqrt{1+q^2}}{q^4 p}, \tag{15.179}$$

where

$$q^2 = \frac{2\sigma^2}{\mathcal{N}_0 \alpha}. \tag{15.180}$$

If $q^2 \ll 1$ and $2\sigma^2 T \mathcal{N}_0^{-1} \gg 1$, the variance of correlation function parameter estimate is simplified

$$\text{Var}\{\nu_m \,|\, \nu_0\} = \frac{8\alpha^2}{q^4 p}. \tag{15.181}$$

If $q^2 \gg 1$, then

$$\text{Var}\{\nu_m \,|\, \nu_0\} = \frac{\alpha^2}{p}, \tag{15.182}$$

or the variance of the central frequency estimate of stochastic process spectral density is inversely proportional to the product between the correlation interval and observation time interval. Figure 15.11 presents the root-mean-square deviation $\sqrt{p\,\text{Var}\{\nu_m \,|\, \nu_0\}/\alpha^2}$ as a function of ratio between the variance of the investigated stochastic process and the power noise q^2 within the limits of effective spectral bandwidth. In doing so, we assume that for all values of q^2 the following inequality $q^2 p \gg 1$ is satisfied.

The optimal estimate of stochastic process correlation function can be found in the form of estimations of the elements R_{ij} of the correlation matrix $\mathbf{R}$ or elements C_{ij} of the inverse matrix $\mathbf{C}$. In the case of Gaussian stationary stochastic process with the multidimensional probability density function given by (12.169), the solution of likelihood ratio equation

$$\frac{\partial f_N(x_1, x_2, \ldots, x_N \,|\, \mathbf{C})}{\partial C_{ij}} = 0 \tag{15.183}$$

allows us to obtain the estimates C_{ij} and, consequently, the estimates of elements R_{ij} of correlation matrix $\mathbf{R}$.

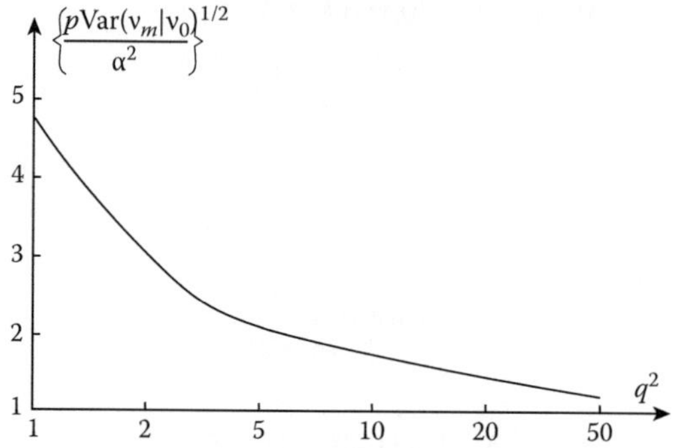

FIGURE 15.11 Relative root-mean-square deviation $\sqrt{p\,\text{Var}\{\nu_m|\nu_0\}/\alpha^2}$ as a function of the ratio between the variance of the investigated stochastic process and power noise.

15.4 CORRELATION FUNCTION ESTIMATION METHODS BASED ON OTHER PRINCIPLES

Under practical realization of analog correlators based on the estimate given by (15.2), a multiplication of two stochastic processes is most difficult to carry out. We discussed previously that for this purpose there is a need to use the circuits performing a multiplication in accordance with (13.201). The described flowchart consists of two quadrators. There are two methods [1] called the *interference* and *compensation methods* using a single quadrator. In doing so, it is assumed that the variance of investigated stochastic process is known very well. The interference method is based on the following relationship:

$$R(\tau) = \pm\left\{\frac{1}{2}\langle[x(t-\tau) \pm x(t)]^2\rangle \mp \sigma^2\right\}. \tag{15.184}$$

It is natural to use the following function to estimate the correlation function given by (15.184)

$$\tilde{R}(\tau) = \pm\left\{\frac{1}{2T}\int_0^T [x(t-\tau) \pm x(t)]^2\,dt \mp \sigma^2\right\}. \tag{15.185}$$

As we can see from (15.184) and (15.185), the estimate of correlation function is not biased.

Let us define the variance of correlation function estimate assuming that the investigated process is Gaussian. Suppose that we use the sign "+" in the square brackets in (15.184) and (15.185). Then

$$\begin{aligned}\text{Var}\{\tilde{R}(\tau)\} &= \left\langle\left\{\frac{1}{T}\int_0^T x(t)x(t-\tau)dt\right\}^2\right\rangle - R^2(\tau)\\ &+\frac{1}{T^2}\int_0^T\int_0^T \langle x(t_1)x(t_1-\tau)[x^2(t_2)+x^2(t_2-\tau)]\rangle dt_1dt_2\\ &-2\sigma^2R(\tau) + \left\langle\left\{\frac{1}{2T}\int_0^T [x^2(t)+x^2(t-\tau)]dt\right\}^2\right\rangle - \sigma^4.\end{aligned} \tag{15.186}$$

The first and second terms in the right side of (15.186) represent the variance of the correlation function estimate $R^*(\tau)$ according to (15.2). Other terms on the right side of (15.186) characterize an increase in the variance of the correlation function estimate $R^*(\tau)$ according to (15.185) compared to the estimate given by (15.2). Making mathematical transformations with the introduction of new variables, as it was done earlier, we can write

$$\operatorname{Var}\{\tilde{R}(\tau)\} = \operatorname{Var}\{R^*(\tau)\} + \frac{1}{T}\int_0^T \left(1-\frac{\tau}{T}\right)$$

$$\times\left\{2R^2(z)+R^2(z+\tau)+R^2(z-\tau)+4R(z)[R(z+\tau)+R(z-\tau)]\right\}dz. \quad (15.187)$$

As $\tau \to 0$

$$\operatorname{Var}\{\tilde{R}(0)\} \approx \frac{16}{T}\int_0^T 1-\frac{\tau}{T} R^2(z)dz, \quad (15.188)$$

and, in this case, the variance of correlation function estimate given by (15.185) exceeds by four times the variance of correlation function estimate given by (15.2).

If the condition $T \gg \tau_{\text{cor}}$ is satisfied, the variance of correlation function estimate given by (15.185) can be presented in the following form:

$$\operatorname{Var}\{\tilde{R}(\tau)\} = \frac{2}{T}\int_0^\infty [R^2(z)+R(z+\tau)R(z-\tau)]dz$$

$$+\frac{1}{T}\int_0^T \left\{2R^2(z)+R^2(z+\tau)+R^2(z-\tau)+4R(z)[R(z+\tau)+R(z-\tau)]\right\}dz. \quad (15.189)$$

At $T \gg \tau \gg \tau_{\text{cor}}$, we have

$$\operatorname{Var}\{\tilde{R}(\tau)\} \approx \frac{6}{T}\int_0^\infty R^2(z)dz. \quad (15.190)$$

Under accepted initial conditions, we obtain

$$\int_0^\infty R^2(z-\tau)dz \approx \int_{-\infty}^\infty R^2(\upsilon)d\upsilon = 2\int_0^\infty R^2(\upsilon)d\upsilon. \quad (15.191)$$

As applied to the exponential correlation function given by (12.13) and if the condition $T \gg \tau_{\text{cor}}$ is satisfied, the variance of correlation function estimate given by (15.185) can be written in the following form:

$$\operatorname{Var}\{\tilde{R}(\tau)\} = \frac{\sigma^4}{\alpha T}[3+4(1+\alpha T)\exp\{-\alpha T\}+(1+2\alpha T)\exp\{-2\alpha T\}]. \quad (15.192)$$

Using the compensation method to measure the correlation function, the function

$$\mu(\tau,\gamma) = \langle [x(t-\tau) - \gamma x(t)]^2 \rangle \tag{15.193}$$

is formed and a selection of the factor γ ensuring a minimum of the function $\mu(\tau, \gamma)$ is performed. In doing so, the factor γ becomes numerically equal to the normalized correlation function value. Thus, defining the minimum of the function $\mu(\tau, \gamma)$ based on the condition

$$\frac{d\mu(\tau,\gamma)}{d\gamma} = 0 \quad \text{if} \quad \frac{d^2\mu(\tau,\gamma)}{d\gamma^2} > 0, \tag{15.194}$$

we obtain

$$\gamma = \frac{\langle x(t)x(t-\tau)\rangle}{\langle x^2(t)\rangle} = \mathcal{R}(\tau). \tag{15.195}$$

Consequently, the compensation measurer of correlation function should generate the function of the following form:

$$\mu^*(\tau,\gamma) = \frac{1}{T}\int_0^T [x(t-\tau) - \gamma x(t)]^2 dt. \tag{15.196}$$

Minimizing the function $\mu^*(\tau, \gamma)$ given by (15.196) with respect to the parameter γ, we are able to obtain the estimate of normalized correlation function $\gamma = \mathcal{R}(\tau)$. Solving the equation

$$\frac{d\mu^*(\tau,\gamma)}{d\gamma} = 0, \tag{15.197}$$

we can see that the procedure to define the estimate of the correlation function $R^*(\tau)$ is equivalent to the estimate that can be presented in the following form:

$$\gamma^* = \tilde{\mathcal{R}}(\tau) = \frac{(1/T)\int_0^T x(t)x(t-\tau)dt}{(1/T)\int_0^T x^2(t)dt}. \tag{15.198}$$

As was shown in Ref. [1], as applied to the estimate by minimum of the function $\mu(\tau, \gamma)$ given by (15.196), the requirements of quadrator are less stringent compared to the requirements of quadrators used by the previously discussed methods of correlation function measurement.

Determine the statistical characteristics of normalized correlation function estimate of the Gaussian stochastic process. For this purpose, we present the numerator and denominator in (15.198) in the following form:

$$\frac{1}{T}\int_0^T x(t)x(t-\tau)dt = \sigma^2\mathcal{R}(\tau) + \sigma^2\Delta\mathcal{R}(\tau). \tag{15.199}$$

$$\frac{1}{T}\int_0^T x^2(t)dt = \sigma^2\left[1+\frac{\Delta\text{Var}}{\sigma^2}\right]. \tag{15.200}$$

As discussed earlier,

$$\langle\Delta\mathcal{R}(\tau)\rangle = 0, \tag{15.201}$$

$$\langle\Delta\text{Var}\rangle = 0. \tag{15.202}$$

Their variances are given by (15.9) and (13.62), respectively. Henceforth, we assume that the condition $T \gg \tau_{\text{cor}}$ is satisfied. In this case, the error of variance estimate is negligible compared to the true value of variance

$$\frac{\langle(\Delta\text{Var})^2\rangle}{\sigma^4} \ll 1. \tag{15.203}$$

Because of this, we can use the following approximation of estimate given by (15.198)

$$\tilde{\mathcal{R}}(\tau) = \frac{\mathcal{R}(\tau)+\Delta\mathcal{R}(\tau)}{1+(\Delta\text{Var}/\sigma^2)} \approx \mathcal{R}(\tau)+\Delta\mathcal{R}(\tau)\left[1-\frac{\Delta\text{Var}}{\sigma^2}+\left(\frac{\Delta\text{Var}}{\sigma^2}\right)^2\right]. \tag{15.204}$$

Under the definition of the bias and variance of estimate, a limitation is imposed by the terms containing the moments of random variables $\Delta\mathcal{R}(\tau)$ and $\Delta\text{Var}/\sigma^2$, and the order of these terms is not higher than 2. Under this approximation, the mathematical expectation of estimate of the normalized correlation function (15.204) can be presented in the following form:

$$\langle\tilde{\mathcal{R}}(\tau)\rangle = \mathcal{R}(\tau) - \frac{\langle\Delta\mathcal{R}(\tau)\Delta\text{Var}\rangle}{\sigma^2} + \mathcal{R}(\tau)\left(\frac{\Delta\text{Var}}{\sigma^2}\right)^2. \tag{15.205}$$

Thus, the estimate of the normalized correlation function given by (15.198) is characterized by the bias

$$b[\tilde{\mathcal{R}}(\tau)] = \langle\tilde{\mathcal{R}}(\tau)\rangle - \mathcal{R}(\tau) = -\frac{\langle\Delta\mathcal{R}(\tau)\Delta\text{Var}\rangle}{\sigma^2} + \mathcal{R}(\tau)\left(\frac{\Delta\text{Var}}{\sigma^2}\right)^2. \tag{15.206}$$

The product moment $\langle\Delta\mathcal{R}(\tau)\Delta\text{Var}\rangle$ can be presented in the following form:

$$\begin{aligned}\langle\Delta\mathcal{R}(\tau)\Delta\text{Var}\rangle &= \left\langle\left\{\frac{1}{\sigma^2 T}\int_0^T x(t)x(t-\tau)dt - \mathcal{R}(\tau)\right\}\times\left\{\frac{1}{T}\int_0^T x^2(t)dt-\sigma^2\right\}\right\rangle \\ &= \frac{1}{\sigma^2 T}\int_0^T\int_0^T \left\langle x(t_1)x(t_1-\tau)x^2(t_2)\right\rangle dt_1 dt_2 - \sigma^2\mathcal{R}(\tau).\end{aligned} \tag{15.207}$$

Determining the fourth product moment and making transformations and introducing new variables under the condition $T \gg \tau_{\text{cor}}$, as it was done before, we can write

$$\frac{\langle \Delta\mathcal{R}(\tau)\Delta\text{Var}\rangle}{\sigma^2} = \frac{2}{T}\int_0^{\infty} \mathcal{R}(z)[\mathcal{R}(z+\tau)+\mathcal{R}(z-\tau)]dz. \tag{15.208}$$

Taking into consideration (15.208) and the variance of variance estimate given by (13.63), we obtain the estimate bias in the following form:

$$b[\tilde{\mathcal{R}}(\tau)] = -\frac{2}{T}\int_0^{\infty} \mathcal{R}(z)[\mathcal{R}(z+\tau)+\mathcal{R}(z-\tau)]dz + \mathcal{R}(z)\frac{4}{T}\int_0^{T} \mathcal{R}^2(z)dz. \tag{15.209}$$

To define the variance of the normalized correlation function estimate

$$\text{Var}\{\tilde{\mathcal{R}}(\tau)\} = \langle \tilde{\mathcal{R}}^2(\tau)\rangle - [\langle \tilde{\mathcal{R}}(\tau)\rangle]^2 \tag{15.210}$$

we determine $\langle \tilde{\mathcal{R}}^2(\tau)\rangle$ accurate with the terms of the moments $\Delta\mathcal{R}(\tau)$ and ΔVar of the second order

$$\begin{aligned}\langle \tilde{\mathcal{R}}^2(\tau)\rangle &= \left\langle \frac{[\mathcal{R}(\tau)+\Delta\mathcal{R}(\tau)]^2}{[1+(\Delta\text{Var}/\sigma^2]^2}\right\rangle \\ &\approx \left\langle \{\mathcal{R}^2(\tau)+2\mathcal{R}(\tau)\Delta\mathcal{R}(\tau)+[\Delta\mathcal{R}(\tau)]^2\}\left\{1-2\frac{\Delta\text{Var}}{\sigma^2}+3\left(\frac{\Delta\text{Var}}{\sigma^2}\right)^2\right\}\right\rangle \\ &\approx \mathcal{R}^2(\tau)+\langle \Delta\mathcal{R}^2(\tau)\rangle + 3\mathcal{R}^2(\tau)\frac{\langle(\Delta\text{Var})^2\rangle}{\sigma^4} - 4\mathcal{R}(\tau)\frac{\langle \Delta\text{Var}\Delta\mathcal{R}(\tau)\rangle}{\sigma^2}.\end{aligned} \tag{15.211}$$

Taking into consideration (15.205) and the earlier-given moments, we obtain

$$\begin{aligned}\text{Var}\{\tilde{\mathcal{R}}(\tau)\} &= \frac{2}{T}\int_0^{\infty}[\mathcal{R}^2(z)+\mathcal{R}(z+\tau)\mathcal{R}(z-\tau)]dz + \mathcal{R}^2(\tau)\frac{4}{T}\int_0^{T}\mathcal{R}^2(z)dz - \mathcal{R}(\tau) \\ &\quad\times \frac{4}{T}\int_0^{\infty}\mathcal{R}(z)[\mathcal{R}(z+\tau)+\mathcal{R}(z-\tau)]dz.\end{aligned} \tag{15.212}$$

As we can see from (15.209) and (15.212), as $\tau \to 0$ the bias and variance of estimate given by (15.198) tend to approach zero, since at $\tau = 0$, according to (15.198), the normalized correlation function estimate is not a random variable.

As applied to the Gaussian stochastic process with the exponential correlation function given by (12.13), the bias and variance of the normalized correlation function estimate take the following form:

$$b[\tilde{\mathcal{R}}(\tau)] = -\frac{2}{T}\exp\{-\alpha\tau\}, \tag{15.213}$$

$$\operatorname{Var}\{\tilde{\mathcal{R}}(\tau)\} = \frac{1}{T\alpha}[1-(1+2\alpha\tau)\exp\{-2\alpha\tau\}]. \tag{15.214}$$

The sign or polar methods of measurements allow us to simplify essentially the experimental investigation of correlation and mutual correlation functions. Delay and multiplication of stochastic processes can be realized very simply by circuitry. The *sign methods* of correlation function measurements are based on the existence of a functional relationship between the correlation functions of the initial stochastic process $\xi(t)$ and the transformed stochastic process $\eta(t) = \operatorname{sgn}\xi(t)$. The stochastic process $\eta(t)$ is obtained by nonlinear inertialess transformation of initial stochastic process $\xi(t)$ by the ideal two-sided limiter with transformation characteristic given by (12.208).

As applied to the Gaussian stochastic process, its normalized correlation function $\mathcal{R}(\tau)$ is related to the correlation function $\rho(\tau)$ of the transformed stochastic process $\eta(t) = \operatorname{sgn}\xi(t)$ by the following relationship:

$$\mathcal{R}(\tau) = \sin[0.5\pi\rho(\tau)] = -\cos[2\pi P_+(\tau)], \tag{15.215}$$

where

$$P_+(\tau) = \int_0^{\infty}\int_0^{\infty} p_2(x_1, x_2; \tau)dx_1dx_2 \tag{15.216}$$

is the probability of coincidence between the positive signs of functions $\eta(t)$ and $\eta(t-\tau)$. The estimate of the probability $P_+(\tau)$ can be obtained as a signal averaging by time at the matching network output of positive values of the stochastic functions $\eta(t)$ and $\eta(t-\tau)$ realizations. If the stochastic process is non-Gaussian, a relationship between the correlation functions of initial and transformed by the ideal limiter stochastic processes is very complex. For this reason, the said method of correlation function measurement is restricted. The method of correlation function measurement using additional processes by analogy with the discussed method of estimation of the mathematical expectation and variance of stochastic process is widely used.

Assume that the investigated stochastic process $\xi(t)$ has zero mathematical expectation. Consider two sign functions

$$\begin{cases}\eta_1(t) = \operatorname{sgn}[\xi(t) - \mu_1(t)],\\ \eta_2(t-\tau) = \operatorname{sgn}[\xi(t-\tau) - \mu_2(t-\tau)],\end{cases} \tag{15.217}$$

where the mutual independent additional stationary stochastic processes $\mu_1(t)$ and $\mu_2(t)$ have the same uniform probability density functions given by (12.205) and the condition (12.206) is satisfied. As mentioned previously, the conditional stochastic processes $\eta_1(t|x_1)$ and $\eta_2[(t-\tau)|x_2]$ are mutually independent at the fixed values $\xi(t) = x_1$ and $\xi(t-\tau) = x_2$. Taking into consideration (12.209), we obtain

$$\langle\eta_1(t|x_1)\eta_2[(t-\tau)|x_2]\rangle = \frac{x_1x_2}{A^2}. \tag{15.218}$$

The unconditional mathematical expectation of product between two stochastic functions can be presented in the following form:

$$\langle\eta_1(t)\eta_2(t)\rangle = \frac{1}{A^2}\int_{-\infty}^{\infty}\int_{-\infty}^{\infty} x_1x_2p_2(x_1, x_2; \tau)dx_1dx_2 = \frac{R(\tau)}{A^2}. \tag{15.219}$$

Thus, the function

$$\tilde{\mathcal{R}}(\tau) = \frac{A^2}{N}\sum_{i=1}^{N} y_{1i}y_{2i}, \tag{15.220}$$

where y_{1i} and y_{2i} are the samples of stochastic sequences η_{1i} and η_{2i}, can be considered as the estimate of correlation function to be used for the investigation of stochastic process at discrete instants. In this case, the estimate will be unbiased. When additional stochastic functions are carried out to estimate the variance (see Section 13.3), the operations of product and summation in (15.220) are easily changed by operations of definition of estimate difference between the probability of polarity coincidence and noncoincidence of sampled values y_{1i} and y_{2i}. Delay operations of sign functions can be implemented by circuitry.

Determine the variance of correlation function estimate given by (15.220) assuming that the samples are pairwise independent, that is,

$$\langle y_{1i}y_{1j}\rangle = \langle y_{2i}y_{2j}\rangle = 0 \tag{15.221}$$

we obtain

$$\mathrm{Var}\{\tilde{R}(\tau)\} = \frac{A^4}{N^2}\sum_{i=1}^{N}\sum_{j=1}^{N}\langle y_{1i}y_{2i}y_{1j}y_{2j}\rangle - R^2(\tau). \tag{15.222}$$

The double sum can be presented in the form of two sums by analogy with (13.94). At this time, (13.55) is true. Define the conditional product moment $\langle(\eta_{1i}\eta_{2i}\eta_{1j}\eta_{2j}|x_{1i}, x_{2i}, x_{1j}, x_{2j})\rangle$ if $i \neq j$ under the condition

$$\begin{cases} \xi(t_i) = x_{1i}, \\ \xi(t_i - \tau) = x_{2i}, \\ \xi(t_j) = x_{1j}, \\ \xi(t_j - \tau) = x_{2j}. \end{cases} \tag{15.223}$$

Taking into consideration a statistical independence between η_1 and η_2, mutual independence between the conditional random values $\eta_1(t_i|x_{1i})$ and $\eta_2[(t_i - \tau)|x_{2i}]$, and (12.209), the conditional product moment can be written in the following form:

$$\langle(\eta_{1i}\eta_{2i}\eta_{1j}\eta_{2j}\,|\,x_{1i}, x_{2i}, x_{1j}, x_{2j})\rangle = \frac{x_{1i}x_{2i}x_{1j}x_{2j}}{A^4}. \tag{15.224}$$

Given that the random variables η_i and η_j are independent of each other, the unconditional product moment can be presented in the following form:

$$\langle\eta_{1i}\eta_{2i}\eta_{1j}\eta_{2j}\rangle = \frac{1}{A^4}\left\{\int_{-\infty}^{\infty}\int_{-\infty}^{\infty} x_1x_2p_2(x_1, x_2;\tau)dx_1dx_2\right\}^2 = \frac{R^2(\tau)}{A^4}. \tag{15.225}$$

Substituting (13.96) and (15.225) into (13.94) and then into (15.222), we obtain

$$\mathrm{Var}\{\tilde{R}(\tau)\} = \frac{A^4}{N}\left[1 - \frac{R^2(\tau)}{A^4}\right]. \tag{15.226}$$

According to (15.220), the correlation function estimate satisfies the condition given by (12.206), that is, $\sigma^2 \ll A^2$. For this reason, the variance of correlation function estimate is defined by half-intervals of possible values of additional stochastic processes. Comparing (15.226) and (15.25), we obtain

$$\frac{\mathrm{Var}\{\tilde{R}(\tau)\}}{\mathrm{Var}\{R^*(\tau)\}} = \frac{A^4}{\sigma^4} \times \frac{1-(\sigma^4/A^4)\mathcal{R}^2(\tau)}{1+\mathcal{R}^2(\tau)}. \tag{15.227}$$

As we can see from (15.227), since the condition $\sigma^2 \ll A^2$ is satisfied, the correlation function estimate given by (15.220) is worse compared to the correlation function estimate given by (15.21).

15.5 SPECTRAL DENSITY ESTIMATE OF STATIONARY STOCHASTIC PROCESS

By definition, the spectral density of stationary stochastic process is the Fourier transform of correlation function

$$S(\omega) = \int_{-\infty}^{\infty} R(\tau)\exp\{-j\omega\tau\}d\tau. \tag{15.228}$$

The inverse Fourier transform takes the following form:

$$R(\tau) = \frac{1}{2\pi}\int_{-\infty}^{\infty} S(\omega)\exp\{j\omega\tau\}d\omega. \tag{15.229}$$

As we can see from (15.229), at $\tau = 0$ we obtain the variance of stochastic process:

$$\mathrm{Var} = R(\tau = 0) = \frac{1}{2\pi}\int_{-\infty}^{\infty} S(\omega)d\omega. \tag{15.230}$$

As applied to the ergodic stochastic process with zero mathematical expectation, the correlation function is defined by (15.1). Because of this, we can rewrite (15.1) in the following form:

$$S_1(\omega) = \lim_{T\to\infty}\frac{1}{T}\int_0^T x(t)\left\{\int_{-\infty}^{\infty} x(t-\tau)\exp\{-j\omega\tau\}d\tau\right\}dt. \tag{15.231}$$

The received realization of stochastic process can be presented in the following form:

$$x(t) = \begin{cases} x(t) & \text{if } 0 \le t \le T, \\ 0 & \text{if } |t| > T. \end{cases} \tag{15.232}$$

In the case of physically realized stochastic processes, the following condition is satisfied:

$$\int_0^T x^2(t)dt = \int_{-\infty}^{\infty} x^2(t)dt < \infty \tag{15.233}$$

For the realization of the stochastic process, the Fourier transform takes the following form:

$$X(j\omega) = \int_0^T x(t)\exp\{-j\omega\tau\}dt = \int_{-\infty}^{\infty} x(t)\exp\{-j\omega\tau\}dt. \tag{15.234}$$

Introducing a new variable $z = t - \tau$, we can write

$$\int_{-\infty}^{\infty} x(t-\tau)\exp\{-j\omega\tau\}d\tau = X(-j\omega)\exp\{-j\omega\tau\}. \tag{15.235}$$

Substituting (15.235) into (15.231) and taking into consideration (15.234), we obtain

$$S_1(\omega) = \lim_{T\to\infty}\frac{1}{T}|X(j\omega)|^2. \tag{15.236}$$

Formula (15.236) is not correct for the definition of spectral density as the characteristic of stochastic process averaged in time. This phenomenon is caused by the fact that the function $T^{-1}|X(j\omega)|^2$ is the stochastic function of the frequency ω. As the stochastic function $x(t)$, this function changes randomly by its mathematical expectation and possesses the variance that does not tend to approach zero with an increase in the observation time interval. Because of this, to obtain the averaged characteristic corresponding to the definition of spectral density according to (15.228), the spectral density $S_1(\omega)$ should be averaged by a set of realizations of the investigated stochastic process and we need to consider the function

$$S(\omega) = \lim_{\substack{T\to\infty \\ N\to\infty}}\frac{1}{T}\sum_{i=1}^{N}|X_i(j\omega)|^2. \tag{15.237}$$

Consider the statistical characteristics of estimate of the function

$$S_1^*(\omega) = \frac{|X_i(j\omega)|^2}{T}, \tag{15.238}$$

where the random spectrum $X(j\omega)$ of stochastic process realization is given by (15.234).

The mathematical expectation of spectral density estimate given by (15.238) takes the following form:

$$\langle S_1^*(\omega)\rangle = \frac{\langle|X_i(j\omega)|^2\rangle}{T} = \frac{1}{T}\int_0^T\int_0^T \langle x(t_1)x(t_2)\rangle \exp\{-j\omega(t_2-t_1\}dt_1dt_2$$

$$= \frac{1}{T}\int_0^T\int_0^T R(t_2-t_1)\exp\{-j\omega(t_2-t_1)\}dt_1dt_2. \tag{15.239}$$

After introduction of new variables $\tau = t_2 - t_1$ and $t_2 = t$, the double integral in (15.239) can be transformed into a single integral, that is,

$$\left\langle S_1^*(\omega)\right\rangle = \int_{-T}^{T}\left(1-\frac{|\tau|}{T}\right)R(\tau)\exp\{-j\omega\tau\}d\tau. \tag{15.240}$$

If the condition $T \gg \tau_{\text{cor}}$ is satisfied, we can neglect the second term compared to the unit in parenthesis in (15.240), and the integration limits are propagated on $\pm\infty$. Consequently, as $T \to \infty$, we can write

$$\left\langle S_1^*(\omega)\right\rangle = S(\omega), \tag{15.241}$$

that is, as $T \to \infty$, the spectral density estimate of stochastic process is unbiased.

Determine the correlation function of spectral density estimate

$$\begin{aligned}
R(\omega_1,\omega_2) &= \left\langle S_1^*(\omega_1)S_1^*(\omega_2)\right\rangle - \left\langle S_1^*(\omega_1)\right\rangle\left\langle S_1^*(\omega_2)\right\rangle \\
&= \frac{1}{T^2}\int_0^T\int_0^T\int_0^T\int_0^T \langle x(t_1)x(t_2)x(t_3)x(t_4)\rangle \exp\{-j\omega_1 t_1 + j\omega_1 t_2 - j\omega_2 t_3 + j\omega_2 t_4\}dt_1dt_2dt_3dt_4 \\
&\quad - \frac{1}{T^2}\int_0^T\int_0^T \langle x(t_1)x(t_2)\rangle \exp\{-j\omega_1 t_1 + j\omega_1 t_2\}dt_1dt_2 \\
&\quad \times \int_0^T\int_0^T \langle x(t_3)x(t_4)\rangle \exp\{-j\omega_2 t_3 + j\omega_2 t_4\}dt_3dt_4
\end{aligned} \tag{15.242}$$

As applied to the Gaussian stochastic process, (15.242) can be reduced to

$$\begin{aligned}
R(\omega_1,\omega_2) &= \frac{1}{T^2}\int_0^T\int_0^T\int_0^T\int_0^T [R(t_1-t_3)R(t_2-t_4) + R(t_1-t_4)R(t_2-t_3)] \\
&\quad \times \exp\{-j\omega_1 t_1 + j\omega_1 t_2 - j\omega_2 t_3 + j\omega_2 t_4\}dt_1dt_2dt_3dt_4.
\end{aligned} \tag{15.243}$$

Taking into consideration (15.229) and (12.122) and as $T \to \infty$, we obtain

$$\begin{aligned}
R(\omega_1,\omega_2) &= \frac{S(\omega_1)S(\omega_2)}{T^2}\left\{\int_0^T \exp\{-j(\omega_1+\omega_2)\}tdt \int_0^T \exp\{j(\omega_1+\omega_2)\}tdt \right. \\
&\quad \left. + \int_0^T \exp\{-j(\omega_1-\omega_2)\}tdt \int_0^T \exp\{j(\omega_1-\omega_2)\}tdt \right\} \\
&= S(\omega_1)S(\omega_2)\left\{\left[\frac{2\sin\dfrac{\omega_1+\omega_2}{2}T}{(\omega_1+\omega_2)T}\right]^2 + \left[\frac{2\sin\dfrac{\omega_1-\omega_2}{2}T}{(\omega_1-\omega_2)T}\right]^2\right\}.
\end{aligned} \tag{15.244}$$

If the condition $\omega T \gg 1$ is satisfied, we can use the following approximation:

$$R(\omega_1,\omega_2) \approx S(\omega_1)S(\omega_2)\left\{\frac{\sin[0.5(\omega_1-\omega_2)T]}{0.5(\omega_1-\omega_2)T}\right\}^2. \tag{15.245}$$

If

$$\omega_1-\omega_2=\frac{2i\pi}{T}, \quad i=1,2,\ldots, \tag{15.246}$$

then

$$R(\omega_1,\omega_2)=0, \tag{15.247}$$

which means, if the frequencies are detuned on the value multiple $2\pi T^{-1}$, the spectral components $S_1^*(\omega_1)$ and $S_1^*(\omega_2)$ are uncorrelated. When $0.5(\omega_1-\omega_2)T \gg 1$, we can neglect the correlation function between the estimates of spectral density given by (15.238).

The variance of estimate of the function $S(\omega_1)$ can be defined by (15.244) using the limiting case at $\omega_1=\omega_2=\omega$:

$$\mathrm{Var}\{S_1^*(\omega)\}=S^2(\omega)\left\{\frac{\sin^2\omega T}{(\omega T)^2}+1\right\}. \tag{15.248}$$

If the condition $\omega T \gg 1$ is satisfied (as $T\to\infty$), we obtain

$$\lim_{T\to\infty}\mathrm{Var}\{S_1^*(\omega)\}=S^2(\omega). \tag{15.249}$$

Thus, according to (15.238), in spite of the fact that the estimate of spectral density ensures unbiasedness, it is not acceptable because the value of the estimate variance is larger than the squared spectral density true value.

Averaging the function $S_1^*(\omega)$ by a set of realizations is not possible, as a rule. Some indirect methods to average the function $S_1^*(\omega)$ are discussed in Refs. [3,5,6]. The first method is based on implementation of the spectral density averaged by frequency bandwidth instead of the estimate of spectral density defined at the point (the estimate of the given frequency). In doing so, the more the frequency range, within the limits of which the averaging is carried out, at $T=\text{const}$, the lesser the variance of estimate of spectral density. However, as a rule, there is an estimate bias that increases with an increase in the frequency range, within the limits of which the averaging is carried out. In general, this averaged estimate of spectral density can be presented in the following form:

$$\overline{S_2^*}(\omega)=\frac{1}{2\pi}\int_{-\infty}^{\infty}W(\omega)S_1^*(\omega-\nu)d\nu, \tag{15.250}$$

where $W(\omega)$ is the even weight function of frequency ω or as it is called in other words the *function of spectral window*. The widely used functions $W(\omega)$ can be found in Refs. [3,5,6].

As $T \to \infty$, the bias of spectral density estimate $S_2^*(\omega)$ can be presented in the following form:

$$b\{S_2^*(\omega)\} = \frac{1}{2\pi}\int_{-\infty}^{\infty} S(\omega - \nu)W(\omega)d\nu - S(\omega). \tag{15.251}$$

As applied to the narrowband spectral window $W(\omega)$, the following expansion

$$S(\omega - \nu) \approx S(\omega) - S'(\omega) + 0.5S''(\omega)\nu^2 \tag{15.252}$$

is true, where $S'(\omega)$ and $S''(\omega)$ are the derivatives with respect to the frequency ω. Because of this, we can write

$$b\{S_2^*(\omega)\} \approx \frac{S''(\omega)}{4\pi}\int_{-\infty}^{\infty} \omega^2 W(\omega)d\omega. \tag{15.253}$$

If the condition $\omega T \gg 1$ is satisfied (as $T \to \infty$), we obtain [1]

$$\mathrm{Var}\{S_2^*(\omega)\} \approx \frac{S^2(\omega)}{2\pi T}\int_{-\infty}^{\infty} W^2(\omega)d\omega. \tag{15.254}$$

As we can see from (15.254), as $T \to \infty$, $\mathrm{Var}\{S_2^*(\omega)\} \to 0$; that is, the estimate of the spectral density $S_2^*(\omega)$ is consistent.

The second method to obtain the consistent estimate of spectral density is to divide the observation time interval $[0, T]$ on N subintervals with duration $T_0 < T$ and to define the estimate $S_{1i}^*(\omega)$ for each subinterval and subsequently to determine the averaged estimate

$$S_3^*(\omega) = \frac{1}{N}\sum_{i=1}^{N} S_{1i}^*(\omega), \quad N = \frac{T}{T_0}. \tag{15.255}$$

Note that according to (15.240), at $T = \text{const}$, an increase in N (or decrease in T_0) leads to bias of the estimate $S_3^*(\omega)$. If the condition $T_0 \gg \tau_{\text{cor}}$ is satisfied, the variance of estimate $S_3^*(\omega)$ can be approximated by the following form:

$$\mathrm{Var}\{S_3^*(\omega)\} \approx \frac{S^2(\omega)}{N}. \tag{15.256}$$

As we can see from (15.256), as $T \to \infty$, the estimate given by (15.255) will be consistent.

In radar applications, sometimes it is worthwhile to obtain the current estimate $S_1^*(\omega, t)$ instead of the averaged summation given by (15.225). Subsequently, the obtained function of time is smoothed by the low-pass filter with the filter constant time $\tau_{\text{filter}} \gg T_0$. This low-pass filter is equivalent to estimate by ν uncorrelated estimations of the function $S_1^*(\omega)$, where $\nu = \tau_{\text{filter}}T_0^{-1}$. In practice, it makes sense to consider only the positive frequencies $f = \omega(2\pi)^{-1}$. Taking into consideration that the correlation function and spectral density remain even, we can write

$$G(f) = 2S(\omega = 2\pi f), \quad f > 0. \tag{15.257}$$

According to (15.228) and (15.229), the spectral density $G(f)$ and the correlation function $R(\tau)$ take the following form:

$$G(f) = 4\int_0^{\infty} R(\tau)\cos 2\pi f\tau\, d\tau, \tag{15.258}$$

$$R(\tau) = \int_0^{\infty} G(f)\cos 2\pi f\tau\, df. \tag{15.259}$$

In this case, the current estimate, by analogy with (15.238), can be presented in the following form:

$$G_1^*(f,t) = \frac{2A^2(f,t)}{T_0}, \tag{15.260}$$

where the squared current spectral density can be presented in the following form:

$$A^2(f,t) = |X(j\omega,t)|^2 = A_c^2(f,t) + A_s^2(f,t) = \left\{\int_{t-T_0}^{t} x(t)\cos 2\pi ft\, dt\right\}^2 + \left\{\int_{t-T_0}^{t} x(t)\sin 2\pi ft\, dt\right\}^2, \tag{15.261}$$

where $A_c^2(f,t)$ and $A_s^2(f,t)$ are the cosine and sine components of the current spectral density of realization $x(t)$. As a result of smoothing the estimate $G_1^*(f,t)$ by the filter with the impulse response $h(t)$, we can write the averaged estimate of spectral density in the following form:

$$G_2^*(f,t) = \int_0^{\infty} h(z)G_1^*(f,t-z)dz. \tag{15.262}$$

If

$$h(t) = \alpha_0 \exp\{-\alpha_0 t\}, \quad t > 0, \quad \tau_{\text{filter}} = \frac{1}{\alpha_0}, \tag{15.263}$$

the variance of spectral density estimate $G_2^*(f,t)$ can be approximated by

$$\text{Var}\{G_2^*(f,t)\} \approx \frac{G^2(f)}{\alpha_0 T_0}. \tag{15.264}$$

The flowchart illustrating how to define the current estimate $G_2^*(f,t)$ of spectral density is shown in Figure 15.12. The input realization of stochastic process is multiplied by the sine and cosine signals of the reference generator in quadrature channels, correspondingly. Obtained products are integrated, squared, and come in at the summator input. The current estimate $G_2^*(f,t)$ forming at the summator output comes in at the smoothing filter input. The smoothing filter possesses the impulse response $h(t)$. The smoothed estimate $G(f, t)$ of spectral density is issued at the filter output.

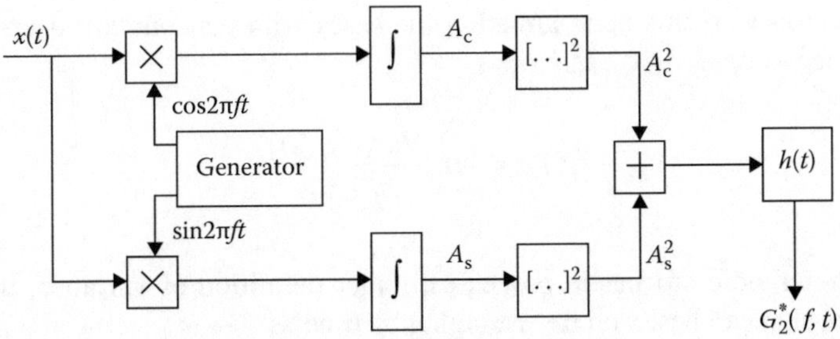

FIGURE 15.12 Definition of the current estimate of spectral density.

To measure the spectral density within the limits of whole frequency range we need to change the reference generator frequency discretely or continuously, for example, by the linear law.

In practice, the filtering method is widely used. The essence of filtering method is the following. The investigated stationary stochastic process is made to pass through the narrowband (compared to the stochastic process spectrum bandwidth) filter with the central frequency $\omega_0 = 2\pi f_0$. The ratio between the variance of stochastic process at the narrowband filter output and the bandwidth Δf of the filter is considered as the estimate of spectral density of stochastic process.

Let $h(t)$ be the impulse response of the narrowband filter. The transfer function corresponding to the impulse response $h(t)$ is $\mathcal{H}(j\omega)$. The stationary stochastic process forming at the filter output takes the form:

$$y(t,\omega_0) = \int_0^\infty h(t-\tau)x(\tau)d\tau. \tag{15.265}$$

Spectral density $\tilde{G}(f)$ at the filter output can be presented in the following form:

$$\tilde{G}(f) = G(f)\mathcal{H}^2(f), \tag{15.266}$$

where $\mathcal{H}(f)$ is the *filter transfer function module* with the maximum defined as $\mathcal{H}_{max}(f) = \mathcal{H}_{max}$. The narrowband filter bandwidth can be defined as

$$\Delta f = \int_0^\infty \frac{\mathcal{H}^2(f)}{\mathcal{H}^2_{max}} df. \tag{15.267}$$

The variance of stochastic process at the filter output in stationary mode takes the following form:

$$\mathrm{Var}\{y(t,f_0)\} = \langle y^2(t,f_0)\rangle = \int_0^\infty G(f)\mathcal{H}^2(f)df. \tag{15.268}$$

Assume that the filter transfer function module is concentrated very closely about the frequency f_0 and we can think that the spectral density is constant within the limits of the bandwidth Δf, that is, $G(f) \approx G(f_0)$. Then

$$\mathrm{Var}\{y(t,f_0)\} \approx G(f_0)\Delta f \mathcal{H}^2_{max}. \tag{15.269}$$

Naturally, the accuracy of this approximation increases with concomitant decrease in the filter bandwidth Δf, since as $\Delta f \to 0$ we can write

$$G(f_0) = \lim_{\Delta f \to 0} \frac{\operatorname{Var}\{y(t, f_0)\}}{\Delta f \mathcal{H}_{\max}^2}. \tag{15.270}$$

As applied to the ergodic stochastic processes, under definition of variance, the averaging by realization can be changed based on the averaging by time as $T \to \infty$

$$G(f_0) = \lim_{\substack{\Delta f \to 0 \\ T \to \infty}} \frac{1}{T \Delta f \mathcal{H}_{\max}^2} \int_0^T y^2(t, f_0) dt. \tag{15.271}$$

For this reason, the value

$$G^*(f_0) = \frac{1}{T \Delta f \mathcal{H}_{\max}^2} \int_0^T y^2(t, f_0) dt \tag{15.272}$$

is considered as the estimate of spectral density under designing and construction of measurers of stochastic process spectral density. The values Δf and $\mathcal{H}_{\max}^2$ are known before. Because of this, a measurement of the stochastic process spectral density is reduced to estimate of stochastic process variance at the filter output. We need to note that (15.272) envisages a correctness of the condition $T\Delta f \gg 1$, which means the observation time interval is much longer compared to the narrowband filter time constant.

Based on (15.272), we can design the flowchart of spectral density measurer shown in Figure 15.13. The spectral density value at the fixed frequency coincides accurately within the constant factor with the variance of stochastic process at the filter output with known bandwidth. Operation principles of the spectral density measurer are evident from Figure 15.13. To define the spectral density for all possible values of frequencies, we need to design the multichannel spectrum analyzer and the central frequency of narrowband filter must be changed discretely or continuously. As a rule, a shift by spectrum frequency of investigated stochastic process needs to be carried out using, for example, the linear law of frequency transformation instead of filter tuning by frequency. The structure of such measurer is depicted in Figure 15.14. The sawtooth generator controls the operation of measurer that changes the frequency of heterodyne.

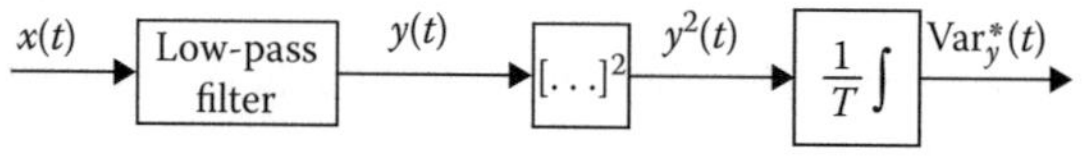

FIGURE 15.13 Measurement of spectral density.

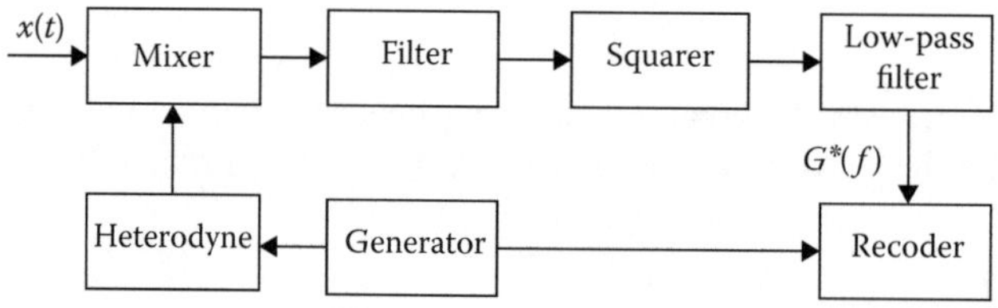

FIGURE 15.14 Measurement of spectral density by spectrum frequency shift.

Define the statistical characteristics for using the filter method to measure the spectral density of stochastic process according to (15.272). The mathematical expectation of spectral density estimate at the frequency f_0 takes the following form:

$$\langle G^*(f_0)\rangle = \frac{\langle y^2(t,f_0)\rangle}{\Delta f \mathcal{H}^2_{\max}} = \frac{1}{\Delta f \mathcal{H}^2_{\max}} \int_0^\infty G(f)\mathcal{H}^2(f)df. \tag{15.273}$$

In a general case, the estimate of spectral density will be biased, that is,

$$b\{G_0(f)\} = \langle G^*(f_0)\rangle - G(f_0). \tag{15.274}$$

The variance of spectral density estimate is defined by the variation in the variance estimate of stochastic process $y(t, f_0)$ at the filter output. If the condition $T \gg \tau_{cor}$ is satisfied for the stochastic process $y(t, f_0)$, the variance of estimate is given by (13.64), where instead of $S(\omega)$ we should understand

$$S_y(\omega) = |\mathcal{H}(j\omega)|^2 S(\omega). \tag{15.275}$$

As applied to introduced notations $G(f)$ and $\mathcal{H}(f)$, we can write

$$\mathrm{Var}\{G^*(f_0)\} = \frac{1}{T(\Delta f)^2 \mathcal{H}^4_{\max}} \int_0^\infty G^2(f)\mathcal{H}^4(f)df. \tag{15.276}$$

To define the bias and variance of spectral density estimate of stochastic process we assume that the module of transfer function is approximated by the following form:

$$\mathcal{H}(f) = \begin{cases} \mathcal{H}_{\max}, & f_0 - 0.5\Delta f \le f \le f_0 + 0.5\Delta f; \\ 0, & f_0 - 0.5\Delta f > f, \quad f_0 + 0.5\Delta f < f, \end{cases} \tag{15.277}$$

where $\Delta f = \delta f$. We apply an expansion in power series about the point $f = f_0$ for the spectral density $G(f)$ and assume that there is a limitation imposed by the first three terms of expansion in series, namely,

$$G(f) \approx G(f_0) + G'(f_0)(f - f_0) + 0.5G''(f_0)(f - f_0)^2, \tag{15.278}$$

where $G'(f_0)$ and $G''(f_0)$ are the first and second derivatives with respect to the frequency f at the point f_0. Substituting (15.278) and (15.277) into (15.273) and (15.274), we obtain

$$b\{G^*(f_0)\} \approx \frac{1}{24}(\Delta f)^2 G''(f_0). \tag{15.279}$$

Thus, the bias of spectral density estimate of stochastic process is proportional to the squared bandwidth of narrowband filter. To define the variance of estimate for the first approximation, we can assume that the condition $G(f) \approx G(f_0)$ is true within the limits of the narrowband filter bandwidth. Then, according to (15.276), we obtain

$$\mathrm{Var}\{G^*(f_0)\} \approx \frac{G^2(f_0)}{T\Delta f}. \tag{15.280}$$

The dispersion of spectral density estimate of the stochastic process takes the following form:

$$D\{G^{*}(f_0)\} \approx \frac{G^2(f_0)}{T\Delta f} + \frac{1}{576}[\Delta f G''(f_0)]^2. \tag{15.281}$$

15.6 ESTIMATE OF STOCHASTIC PROCESS SPIKE PARAMETERS

In many application problems we need to know the statistical parameters of stochastic process spike (see Figure 15.15a): the spike mean or the average number of down-up cross sections of some horizontal level M within the limits of the observation time interval $[0, T]$, the average duration of the spike, and the average interval between the spikes. In Figure 15.15a, the variables τ_i and θ_i mean the random variables of spike duration and the interval between spikes, correspondingly. To measure these parameters of spikes, the stochastic process realization $x(t)$ is transformed by the nonlinear transformer (threshold circuitry) into the pulse sequence normalized by the amplitude η_τ with duration τ_i (Figure 15.15b) or normalized by the amplitude η_θ with duration θ_i (Figure 15.15c), correspondingly:

$$\eta_\tau(t) = \begin{cases} 1 & \text{if} \quad x(t) \ge M, \\ 0 & \text{if} \quad x(t) < M; \end{cases} \tag{15.282}$$

$$\eta_\theta(t) = \begin{cases} 1 & \text{if} \quad x(t) \le M, \\ 0 & \text{if} \quad x(t) > M. \end{cases} \tag{15.283}$$

Using the pulse sequences η_τ and η_θ, we can define the aforementioned parameters of stochastic process spike. Going forward, we assume that the investigated stochastic process is ergodic, as mentioned previously, and the following condition $T \gg \tau_{cor}$ is satisfied.

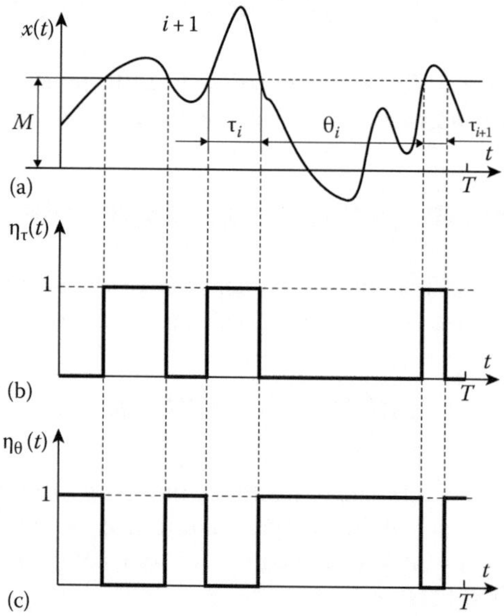

FIGURE 15.15 Transformation of stochastic process realization $x(t)$ into the pulse sequence: (a) Example of stochastic process spike; (b) Pulse sequence normalized by the amplitude η_τ with duration T_i; (c) Pulse sequence normalized by the amplitude η_θ with duration θ_i.

15.6.1 Estimation of Spike Mean

Taking into consideration the assumptions stated previously, the estimate of the spike number in the given stochastic process realization $x(t)$ within the limits of the observation time interval $[0, T]$ at the level M can be defined approximately as

$$N^* = \frac{1}{\tau_{av}} \int_0^T \eta_\tau(t)dt = \frac{1}{\theta_{av}} \int_0^T \eta_\theta(t)dt, \tag{15.284}$$

where τ_{av} and θ_{av} are the average duration of spike and the average interval between spikes within the limits of the observation time interval $[0, T]$ of the given stochastic process realization at the level M. The true values of the average duration of spikes $\bar{\tau}$ and the average interval between spikes $\bar{\theta}$ obtained as a result of averaging by a set of realizations are defined in accordance with Ref. [1] in the following form:

$$\bar{\tau} = \frac{1}{\bar{N}} \int_M^\infty p(x)dx = \frac{1}{\bar{N}}[1 - F(M)], \tag{15.285}$$

$$\bar{\theta} = \frac{1}{\bar{N}} \int_{-\infty}^M p(x)dx = \frac{F(M)}{\bar{N}}, \tag{15.286}$$

where

$F(M)$ is the probability distribution function

$\bar{N}$ is the average number of spikes per unit time at the level M defined as

$$\bar{N} = \int_0^\infty \dot{x} p_2(M, \dot{x}) d\dot{x}, \tag{15.287}$$

where $p_2(M, \dot{x})$ is the two-dimensional pdf of the stochastic process and its derivative at the same instant.

Note that $\bar{\tau} = \bar{\theta}$ corresponds to the level M_0 defined from the equality

$$F(M_0) = 1 - F(M_0) = 0.5. \tag{15.288}$$

If the condition $M \gg M_0$ is satisfied, the probability of event that on average there will be noninteger number of intervals between the stochastic process spikes θ_i within the limits of the observation time interval $[0, T]$ is high; otherwise, if the condition $M \ll M_0$ is satisfied, the probability of event that on average there will be the noninteger number of spike duration τ_i within the limits of the observation time interval $[0, T]$ is high. This phenomenon leads, on average, to more errors while measuring N^* using the only formula (15.284). Because of this, while determining the statistical characteristics of the estimate of the average number of spikes, the following relationship

$$N^* = \begin{cases} \dfrac{1}{\bar{\theta}} \displaystyle\int_0^T \eta_\theta(t)dt & \text{at} \quad M \le M_0(\bar{\tau} \ge \bar{\theta}), \\ \dfrac{1}{\bar{\tau}} \displaystyle\int_0^T \eta_\tau(t)dt & \text{at} \quad M \ge M_0(\bar{\tau} \le \bar{\theta}) \end{cases} \tag{15.289}$$

can be considered as the first approximation, where we assume that

$$\frac{\tau_{av} - \bar{\tau}}{\bar{\tau}} \ll 1 \quad \text{and} \quad \frac{\theta_{av} - \bar{\theta}}{\bar{\theta}} \ll 1.$$

For this reason, we use the approximations $\tau_{av} \approx \bar{\tau}$ and $\theta_{av} \approx \bar{\theta}$.

The mathematical expectation of the average number of stochastic process spikes can be determined in the following form:

$$\langle N^* \rangle = \begin{cases} \dfrac{1}{\bar{\theta}} \displaystyle\int_0^T \langle \eta_\theta(t) \rangle \, dt & \text{at} \quad M \le M_0 (\bar{\tau} \ge \bar{\theta}), \\ \dfrac{1}{\bar{\tau}} \displaystyle\int_0^T \langle \eta_\tau(t) \rangle \, dt & \text{at} \quad M \ge M_0 (\bar{\tau} \le \bar{\theta}). \end{cases} \tag{15.290}$$

According to (15.282), (15.283), (15.285), and (15.286), we obtain

$$\langle \eta_\theta(t) \rangle = \int_{-\infty}^{M} p(x) dx = \bar{\theta} \times \bar{N}, \tag{15.291}$$

$$\langle \eta_\tau(t) \rangle = \int_{M}^{\infty} p(x) dx = \bar{\tau} \times \bar{N}. \tag{15.292}$$

Substituting (15.291) and (15.292) into (15.290), we obtain

$$\langle N^* \rangle = \bar{N} \times T. \tag{15.293}$$

In other words, the estimate of average number of the stochastic process spikes at the level M within the limits of the observation time interval $[0, T]$ is unbiased as a first approximation.

The estimate variance of the average number of stochastic process spikes at the level M can be presented in the following form:

$$\text{Var}\{N^*\} = \begin{cases} \dfrac{1}{\bar{\theta}} \displaystyle\int_0^T\int_0^T \langle \eta_\theta(t_1)\eta_\theta(t_2) \rangle dt_1 dt_2 - [\bar{N}T]^2, & M \le M_0, \\ \dfrac{1}{\bar{\tau}} \displaystyle\int_0^T\int_0^T \langle \eta_\tau(t_1)\eta_\tau(t_2) \rangle dt_1 dt_2 - [\bar{N}T]^2, & M \ge M_0. \end{cases} \tag{15.294}$$

In the case of ergodic stochastic processes, the average values can be written in the following form:

$$\langle \eta_\theta(t_1)\eta_\theta(t_2) \rangle = R_\theta(t_1 - t_2) \tag{15.295}$$

and

$$\langle \eta_\tau(t_1)\eta_\tau(t_2)\rangle = R_\tau(t_1 - t_2). \tag{15.296}$$

As we can see from (15.295) and (15.296), the average values are equal to the probabilities of nonexceeding and exceeding the level M by the stochastic process realization $x(t)$ at the instants t_1 and t_2

$$R_\theta(t_1 - t_2) = \int_{-\infty}^{M}\int_{-\infty}^{M} p_2(x_1, x_2; t_1 - t_2)dx_1dx_2; \tag{15.297}$$

$$R_\tau(t_1 - t_2) = \int_{M}^{\infty}\int_{M}^{\infty} p_2(x_1, x_2; t_1 - t_2)dx_1dx_2. \tag{15.298}$$

Taking into consideration (15.294) through (15.298), introducing new variables $t = t_1 - t_2$, and changing the order of integration, we can write

$$\frac{\mathrm{Var}\{N^*\}}{[\bar{N}T]^2} = \begin{cases} \dfrac{2}{T\{F(M)\}^2}\displaystyle\int_0^T\left(1-\frac{\tau}{T}\right)R_\theta(t)dt - 1, & M \le M_0, \\ \dfrac{2}{T\{1-F(M)\}^2}\displaystyle\int_0^T\left(1-\frac{\tau}{T}\right)R_\tau(t)dt - 1, & M \ge M_0, \end{cases} \tag{15.299}$$

where $\mathrm{Var}\{N^*\}/[\bar{N}T]^2$ is the normalized variance of the average number of stochastic process spikes or the relative variance of the average number of stochastic process spikes.

As applied to the Gaussian and Rayleigh stochastic processes, we can present the two-dimensional probability distribution functions in the form (14.50) and (14.71). In the case of Gaussian stochastic process with zero mathematical expectation, that is, $F(M_0) = 1 - F(M_0)$ at $M_0 = 0$. Substituting (14.50) into (15.297) and (15.298), we obtain

$$R_\theta(t) = \left\{1 - Q\left[\frac{M}{\sigma}\right]\right\}^2 + \sum_{\nu=1}^{\infty}\left\{1 - Q^{(\nu)}\left[\frac{M}{\sigma}\right]\right\}^2 \frac{R^\nu(t)}{\nu!}, \tag{15.300}$$

$$R_\tau(t) = \left\{Q\left[\frac{M}{\sigma}\right]\right\}^2 + \sum_{\nu=1}^{\infty}\left\{1 - Q^{(\nu)}\left[\frac{M}{\sigma}\right]\right\}^2 \frac{R^\nu(t)}{\nu!}. \tag{15.301}$$

In accordance with (15.299), we obtain

$$\frac{\mathrm{Var}\{N^*\}}{[\bar{N}T]^2} = \begin{cases} \left\{1 - Q\left[\dfrac{M}{\sigma}\right]\right\}^{-2}\displaystyle\sum_{\nu=1}^{\infty} a_\nu c_\nu, & \dfrac{M}{\sigma} \le 0, \\ \left\{Q\left[\dfrac{M}{\sigma}\right]\right\}^{-2}\displaystyle\sum_{\nu=1}^{\infty} a_\nu c_\nu, & \dfrac{M}{\sigma} \le 0, \end{cases} \tag{15.302}$$

where a_v and c_v are defined analogously as in (14.56) and (14.58). In doing so, the values of the coefficients a_v are presented in Table 14.1 as a function of v and the normalized level $z = M\sigma^{-1}$.

Since

$$1-Q\left[-\frac{M}{\sigma}\right]=Q\left[-\frac{M}{\sigma}\right], \tag{15.303}$$

we can see from (15.302) that the normalized variance of the average number of stochastic process spikes $\text{Var}\{N^*\}/[\bar{N}T]^2$ is symmetrical with respect to the level $M/\sigma = 0$. Because of this, we can write

$$\frac{\text{Var}\{N^*\}}{[\bar{N}T]^2}=\left\{Q\left[\left|\frac{M}{\sigma}\right|\right]\right\}^{-2}\sum_{v=1}^{\infty}a_v c_v. \tag{15.304}$$

As applied to the Gaussian stochastic process with the normalized correlation function

$$\mathcal{R}(t)=\exp\{-\alpha^2 t^2\}, \tag{15.305}$$

we obtain

$$c_v=\frac{\sqrt{\pi}}{p\sqrt{v}}\left\{1-Q(p\sqrt{v})+\frac{1-\exp\{-vp\}}{p\sqrt{v}}\right\}, \tag{15.306}$$

where $p = \alpha T$. If $p \gg 1$

$$c_v=\frac{\sqrt{\pi}}{p\sqrt{v}}. \tag{15.307}$$

The normalized squared deviation of the average number of stochastic process spikes $\sqrt{\text{Var}\{N^*\}}/[\bar{N}T]$ as a function of the normalized level $|z = M\sigma^{-1}|$ for realizations of stochastic process with fixed duration, the parameter $p = \alpha T$ is shown in Figure 15.16. As we can see from

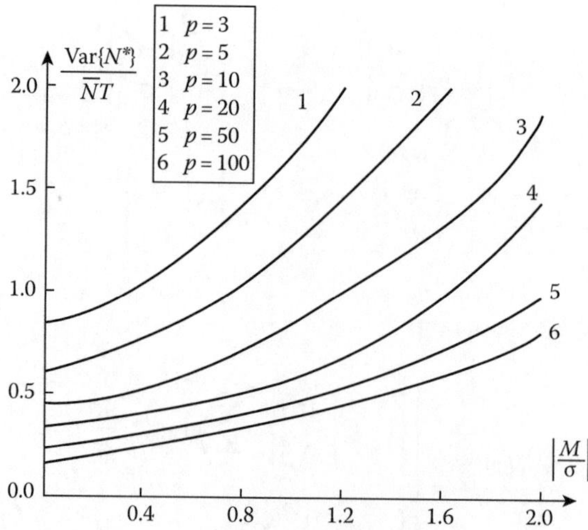

FIGURE 15.16 Normalized squared deviation of the average number of stochastic process spikes as a function of the normalized level for Gaussian realizations of a stochastic process with fixed duration.

Figure 15.16, the normalized squared deviation $\sqrt{\text{Var}\{N^*\}}/[\bar{N}T]$ of the average number of stochastic process spikes is increased with increasing in the absolute level value $|z = M\sigma^{-1}|$ and decreasing the parameter $p = \alpha T$. At $p = \alpha T \geq 10$ and $|z = M\sigma^{-1}| = 0$, we can write

$$\frac{\text{Var}\{N^*\}}{[\bar{N}T]^2} \approx \frac{1.3}{\alpha T}. \tag{15.308}$$

As applied to the Rayleigh stochastic process, we have that $\bar{\tau} = 0$ when

$$\frac{M}{\sqrt{2}\sigma} = \sqrt{\ln 2} \approx 0.83. \tag{15.309}$$

Substituting (14.71) into (15.297) and (15.298), we obtain

$$R_\theta(t) = \left[1 - \exp\left\{-\frac{M^2}{2\sigma^2}\right\}\right]^2 + \sum_{\nu=1}^{\infty} \frac{R^{2\nu}(t)}{(\nu!)^2} \exp\left\{-\frac{M^2}{\sigma^2}\right\}\left\{L_\nu\left[\frac{M^2}{2\sigma^2}\right] - \nu L_{\nu-1}\left[\frac{M^2}{2\sigma^2}\right]\right\}^2; \tag{15.310}$$

$$R_\tau(t) = \exp\left\{-\frac{M^2}{2\sigma^2}\right\} + \sum_{\nu=1}^{\infty} \frac{R^{2\nu}(t)}{(\nu!)^2} \exp\left\{-\frac{M^2}{\sigma^2}\right\}\left\{L_\nu\left[\frac{M^2}{2\sigma^2}\right] - \nu L_{\nu-1}\left[\frac{M^2}{2\sigma^2}\right]\right\}^2. \tag{15.311}$$

In accordance with (15.299), we obtain

$$\frac{\text{Var}\{N^*\}}{[\bar{N}T]^2} = \begin{cases} \exp\left\{-\dfrac{M^2}{2\sigma^2}\right\} \displaystyle\sum_{\nu=1}^{\infty} b_\nu d_\nu, & M \geq 0.83, \\ \left\{1 - \exp\left\{-\dfrac{M^2}{2\sigma^2}\right\}\right\}^{-2} \displaystyle\sum_{\nu=1}^{\infty} b_\nu d_\nu, & M < 0.83, \end{cases} \tag{15.312}$$

where b_ν and d_ν are defined analogously as in (14.86) and (14.83). The values of coefficients are given in Table 14.3. The normalized squared deviation of the average number of stochastic process spikes $\sqrt{\text{Var}\{N^*\}}/[\bar{N}T]$ as a function of the normalized level $M/\sqrt{2\sigma^2}$ at $p_1 = \sqrt{2}\alpha T = 10$ in the case of the exponential normalized correlation function given by (15.305) is shown in Figure 15.17. As we can see from Figure 15.17, the normalized squared deviation of the average number of stochastic process spikes $\sqrt{\text{Var}\{N^*\}}/[\bar{N}T]$ increases with an increase in the deviation of the normalized level $M/\sqrt{2\sigma^2}$ with respect to the median of the probability distribution function $M_0/\sqrt{2\sigma^2} = \sqrt{\ln 2}$ at the fixed $p_1 = \sqrt{2}\alpha T$. This phenomenon is explained by a decrease in the number of spikes higher or lower than the level $M_0/\sqrt{2\sigma^2} = \sqrt{\ln 2}$.

15.6.2 Estimation of Average Spike Duration and Average Interval between Spikes

Considering the stochastic process realization presented in Figure 15.15a, we can see that with higher number of spikes N within the limits of the observation time interval $[0, T]$ the estimate of

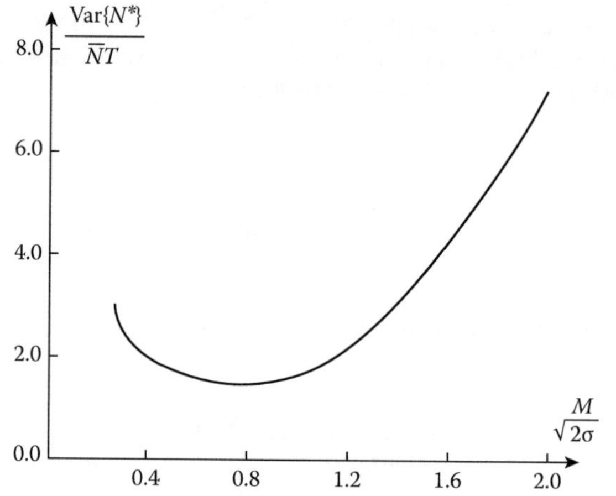

FIGURE 15.17 Normalized squared deviation of the average number of stochastic process spikes as a function of the normalized level for Rayleigh realizations of the stochastic process with fixed duration.

spike duration average τ^* and the estimate of average of the interval θ^* between the spikes can be presented as

$$\begin{cases} \tau^* = \dfrac{1}{N}\sum_{i=1}^{N}\tau_i, \\ \theta^* = \dfrac{1}{N}\sum_{i=1}^{N}\theta_i. \end{cases} \tag{15.313}$$

for the given realization. The relationships given by (15.313) can be presented in the following form:

$$\begin{cases} \tau^* = \dfrac{1}{N}\int_0^T \eta_\tau(t)dt, \\ \theta^* = \dfrac{1}{N}\int_0^T \eta_\theta(t)dt, \end{cases} \tag{15.314}$$

where $\eta_\tau(t)$ and $\eta_\theta(t)$ are given by (15.282) and (15.283).

The device to measure the spike duration average τ^* and the average of interval θ^* between the spikes can be designed based on (15.313) and (15.314). As applied to the estimate of the spike duration average τ^*, the flowchart of measurer is depicted in Figure 15.18. This measurer consists

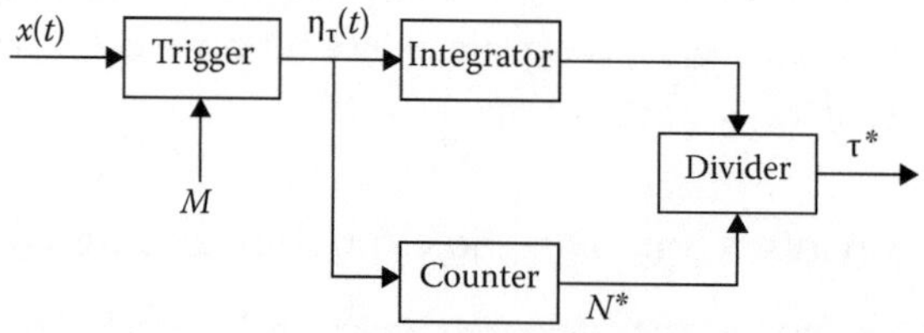

FIGURE 15.18 Measurement of spike duration and the average of interval between the spikes.

of the trigger forming at the output normalized by amplitude and shape function $\eta_\tau(t)$, the counter of spikes, the integrator, and the divisor determining the estimate of spike duration average τ^*. The threshold M is given by external source.

To define the statistical characteristics of estimates τ^* and θ^* we assume that the condition $T \gg \tau_{cor}$ is satisfied. In this case, we can approximately assume $N \approx \bar{N}T$. Then

$$\begin{cases} \tau^* = \dfrac{1}{\bar{N}T}\displaystyle\int_0^T \eta_\tau(t)dt, \\ \theta^* = \dfrac{1}{\bar{N}T}\displaystyle\int_0^T \eta_\theta(t)dt. \end{cases} \tag{15.315}$$

The mathematical expectation of estimates can be presented in the following form:

$$\begin{cases} \langle\tau^*\rangle = \dfrac{1}{\bar{N}T}\displaystyle\int_0^T \langle\eta_\tau(t)\rangle\, dt, \\ \langle\theta^*\rangle = \dfrac{1}{\bar{N}T}\displaystyle\int_0^T \langle\eta_\theta(t)\rangle\, dt. \end{cases} \tag{15.316}$$

Taking into consideration (15.291) and (15.292), we see that $\langle\tau^*\rangle = \bar{\tau}$ and $\langle\theta^*\rangle = \bar{\theta}$. In other words, the estimates of the spike duration average τ^* and the average of interval θ^* between the spikes are unbiased as a first approximation.

Determine the estimate variance of the average spike duration of stochastic process at level M:

$$\operatorname{Var}\{\tau^*\} = \langle(\tau^* - \bar{\tau})^2\rangle = \frac{1}{[\bar{N}T]^2}\int_0^T\int_0^T \langle\eta_\tau(t_1)\eta_\tau(t_2)\rangle\, dt_1dt_2 - \bar{\tau}^2. \tag{15.317}$$

The mathematical expectation

$$\langle\eta_\tau(t_1)\eta_\tau(t_2)\rangle = \mathcal{R}_\tau(t_1 - t_2) \tag{15.318}$$

is defined by (15.298). By analogy with (15.299), we can define the estimate variance of the average spike duration τ^*

$$\operatorname{Var}\{\tau^*\} = \frac{1}{\bar{N}^2} \times \frac{2}{T}\int_0^T \left(1 - \frac{\tau}{T}\right)\mathcal{R}_\tau(t)dt - \bar{\tau}^2. \tag{15.319}$$

The estimate variance of the average of interval θ^* between the spikes can be presented in the following form:

$$\operatorname{Var}\{\theta^*\} = \frac{1}{\bar{N}^2} \times \frac{2}{T}\int_0^T \left(1 - \frac{\tau}{T}\right)\mathcal{R}_\theta(t)dt - \bar{\theta}^2. \tag{15.320}$$

As applied to the Gaussian stochastic process, if the condition $T \gg \tau_{cor}$ is satisfied the normalized correlation functions $\mathcal{R}_\tau(t)$ and $\mathcal{R}_\theta(t)$ are given by (15.300) and (15.301) and the average number of stochastic process spikes at level M can be determined as

$$\bar{N}\left[\frac{M}{\sigma}\right] = \frac{1}{2\pi}\sqrt{-\frac{d^2\mathcal{R}(t)}{dt^2}\bigg|_{t=0}}\exp\left\{-\frac{M^2}{2\sigma^2}\right\}. \tag{15.321}$$

Taking into consideration (15.285) and substituting (15.319), we obtain

$$\text{Var}\{\tau^*\} = \frac{8\pi^2}{T\left(d^2\mathcal{R}(t)/dt^2\right)\big|_{t=0}}\exp\left\{\frac{M^2}{\sigma^2}\right\}\sum_{\nu=1}^{\infty}\alpha_\nu\int_0^T \mathcal{R}^\nu(t)dt, \tag{15.322}$$

where α_ν is given by (14.56) and Table 14.1. The formula (15.322) is correct for measuring the estimate variance of the average of interval θ^* between the spikes.

As applied to the normalized correlation function given by (15.305), the estimate variance of the average spike duration τ^*

$$\text{Var}\{\tau^*\} = \frac{2\pi^2\sqrt{\pi}}{p}\exp\left\{\frac{M^2}{\sigma^2}\right\}\sum_{\nu=1}^{\infty}\frac{\alpha_\nu}{\sqrt{\nu}}, \tag{15.323}$$

where $p = \alpha T$. As we can see from (15.323), in the Gaussian stochastic process and the fixed duration of the observation interval $[0, T]$ case, the estimate variance of the spike duration average τ^* and the estimate variance of the average of interval θ^* between the spikes are minimum at $M/\sigma = 0$.

As applied to the Rayleigh stochastic process, if the condition $T \gg \tau_{cor}$ is satisfied the normalized correlation functions $\mathcal{R}_\tau(t)$ and $\mathcal{R}_\theta(t)$ are given by (15.310) and (15.311) and the average number of stochastic process spikes at level M can be determined as

$$\bar{N}\left[\frac{M}{\sigma}\right] = \frac{1}{\sqrt{2\pi}}\sqrt{-\frac{d^2\mathcal{R}(t)}{dt^2}\bigg|_{t=0}}\exp\left\{\frac{M}{\sigma}\right\}\exp\left\{-\frac{M^2}{2\sigma^2}\right\}. \tag{15.324}$$

Substituting $\mathcal{R}_\tau(t)$ into (15.319) and taking into consideration (15.285) and (15.324), we obtain

$$\text{Var}\{\tau^*\} = \frac{2\pi}{T\left(d^2\mathcal{R}(t)/dt^2\right)\big|_{t=0}}\exp\left\{\frac{M^2}{\sigma^2}\right\}\frac{2\sigma^2}{M^2}\sum_{\nu=1}^{\infty}b_\nu\int_0^T \mathcal{R}^{2\nu}(t)dt. \tag{15.325}$$

It is easy to prove that (15.235) is true to define the estimate variance of the average of interval θ^* between the spikes.

15.7 MEAN-SQUARE FREQUENCY ESTIMATE OF SPECTRAL DENSITY

The mean-square frequency $\overline{f}$ given by (15.6) is widely used as a parameter characterizing the spectral density of stochastic process. The value $\overline{f}$ defines a dispersion of component of the stochastic process spectral density relative to zero frequency. As applied to low-frequency stationary stochastic processes, the mean-square frequency $\overline{f}$ characterizes the effective bandwidth of spectral density. Taking into consideration (15.259), we can present the mean-square frequency $\overline{f}$ in the following form:

$$\overline{f} = \frac{1}{2\pi}\sqrt{-\frac{\frac{d^2R(\tau)}{d\tau^2}}{R(\tau)}\Bigg|_{\tau=0}} = \frac{1}{2\pi}\sqrt{\frac{\left\langle\left[\frac{dx(t)}{dt}\right]^2\right\rangle}{\langle x^2(t)\rangle}}. \tag{15.326}$$

Here and further, we assume that the investigated stochastic process possesses zero mathematical expectation. As applied to the Gaussian stochastic process, the mean-square frequency $\overline{f}$ is matched with the average number of stochastic process spikes per unit time at the zero level (15.321).

According to (15.326), it is worthwhile to consider for the investigated stationary stochastic process the following value

$$\overline{f}^* = \frac{1}{2\pi}\sqrt{\frac{\frac{1}{T}\int_0^T\left[\frac{dx(t)}{dt}\right]^2 dt}{\frac{1}{T}\int_0^T x^2(t)dt}} \tag{15.327}$$

as the estimate of the mean-square frequency that tends to approach $\overline{f}$ as $T \to \infty$. The flowchart of measurer of the mean-square frequency of the stochastic process is shown in Figure 15.19. To define the characteristics of the mean-square frequency estimate we can use the following representation of numerator and denominator in (15.327) in the following form:

$$\frac{1}{T}\int_0^T\left[\frac{dx(t)}{dt}\right]^2 dt = \mathrm{Var}_{\dot{x}} + \Delta\mathrm{Var}_{\dot{x}}, \tag{15.328}$$

$$\frac{1}{T}\int_0^T x^2(t)dt = \mathrm{Var}_x + \Delta\mathrm{Var}_x, \tag{15.329}$$

where

Var_x and $\mathrm{Var}_{\dot{x}}$ are the mathematical expectations of variances of the stochastic process and its derivative

$\Delta\mathrm{Var}_x$ and $\Delta\mathrm{Var}_{\dot{x}}$ are the random errors of definition of the earlier-mentioned variances within the limits of the finite observation time interval $[0, T]$

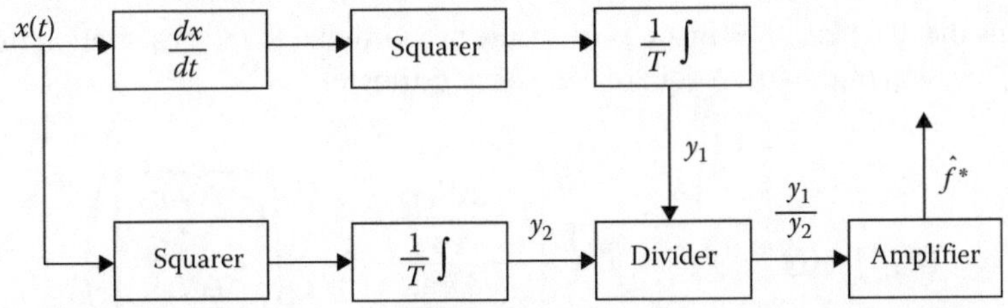

FIGURE 15.19 Measurement of the averaged squared frequency of stochastic process.

As we discussed before, the estimates of variances are unbiased and the variances of variance estimates are defined by (13.62) where we need to use the corresponding correlation function of the investigated stochastic process and its first derivative instead $\sigma^2\mathcal{R}(\tau)$.

Going forward, we assume that the condition $T \gg \tau_{\text{cor}}$ is satisfied. In this case, the errors ΔVar_x and $\Delta\text{Var}_{\dot{x}}$ will be small compared to Var_x and $\text{Var}_{\dot{x}}$. To define the bias of the mean-square frequency under previous assumptions, we can think that

$$\bar{f}^* = \frac{1}{2\pi}\sqrt{\frac{\text{Var}_{\dot{x}} + \Delta\text{Var}_{\dot{x}}}{\text{Var}_x + \Delta\text{Var}_x}} = \bar{f}\sqrt{\frac{1+\dfrac{\Delta\text{Var}_{\dot{x}}}{\text{Var}_{\dot{x}}}}{1+\dfrac{\Delta\text{Var}_x}{\text{Var}_x}}} \approx \bar{f}\left\{1+\frac{1}{2}\frac{\Delta\text{Var}_{\dot{x}}}{\text{Var}_{\dot{x}}} - \frac{1}{8}\frac{\Delta\text{Var}_{\dot{x}}^2}{\text{Var}_{\dot{x}}^2}\right\}\left\{1-\frac{1}{2}\frac{\Delta\text{Var}_x}{\text{Var}_x} + \frac{3}{8}\frac{\Delta\text{Var}_x^2}{\text{Var}_x^2}\right\}. \tag{15.330}$$

Now, we are able to obtain the relative bias of estimate

$$\frac{\langle\Delta\bar{f}\rangle}{\bar{f}} = \frac{\langle\bar{f}^* - \bar{f}\rangle}{\bar{f}} \approx -\frac{1}{8}\left\{\frac{\langle\Delta\text{Var}_{\dot{x}}^2\rangle}{\text{Var}_{\dot{x}}^2} - 3\frac{\langle\Delta\text{Var}_x^2\rangle}{\text{Var}_x^2} + 2\frac{\langle\Delta\text{Var}_{\dot{x}}\Delta\text{Var}_x\rangle}{\text{Var}_{\dot{x}}\text{Var}_x}\right\}, \tag{15.331}$$

where under the condition $T \gg \tau_{\text{cor}}$ we have

$$\langle\Delta\text{Var}_{\dot{x}}\Delta\text{Var}_x\rangle = \frac{4}{T}\int_0^T\left[\frac{dR(\tau)}{d\tau}\right]^2 d\tau. \tag{15.332}$$

As a result, the relative bias can be presented in the following form:

$$\frac{\langle\Delta\bar{f}\rangle}{\bar{f}} = \frac{1}{2T}\left\{-\int_0^\infty\left[\frac{\mathcal{R}''(\tau)}{\mathcal{R}''(0)}\right]^2 d\tau + 3\int_0^\infty \mathcal{R}^2(\tau)d\tau + 2\int_0^\infty \frac{\mathcal{R}'^2(\tau)}{\mathcal{R}''(0)}d\tau\right\}, \tag{15.333}$$

where $\mathcal{R}(\tau)$, $\mathcal{R}'(\tau)$, and $\mathcal{R}''(\tau)$ are the normalized correlation function of the investigated stochastic process and its first and second derivatives. As applied to the exponential normalized correlation function of the investigated stochastic process given by (15.305), we can write

$$\frac{\langle\Delta\bar{f}\rangle}{\bar{f}} = \frac{5\sqrt{2\pi}}{32\alpha T}, \tag{15.334}$$

that is, it means that the bias of estimate is inversely proportional to the observation time interval T.

Define a dispersion of the mean-square frequency estimate

$$D\{\bar{f}^*\} = \langle(\bar{f}^* - \bar{f})^2\rangle = \bar{f}^2\left\langle\left\{\frac{1+\dfrac{\Delta\text{Var}_{\dot{x}}}{\text{Var}_{\dot{x}}}}{1+\dfrac{\Delta\text{Var}_x}{\text{Var}_x}} + 1 - 2\sqrt{\frac{1+\dfrac{\Delta\text{Var}_{\dot{x}}}{\text{Var}_{\dot{x}}}}{1+\dfrac{\Delta\text{Var}_x}{\text{Var}_x}}}\right\}\right\rangle. \tag{15.335}$$

Using two-dimensional Taylor expansion in series for the first and third terms in (15.335) about the points

$$\frac{\Delta \mathrm{Var}_{\dot{x}}}{\mathrm{Var}_{\dot{x}}} = 0 \quad \text{and} \quad \frac{\Delta \mathrm{Var}_{x}}{\mathrm{Var}_{x}} = 0$$

and limiting by terms of the second order, we obtain

$$\frac{D\{\bar{f}^*\}}{\bar{f}^2} \approx \left\langle \left\{ \left[1 + \frac{\Delta \mathrm{Var}_{\dot{x}}}{\mathrm{Var}_{\dot{x}}}\right] \left[1 - \frac{\Delta \mathrm{Var}_{x}}{\mathrm{Var}_{x}} + \frac{\Delta \mathrm{Var}_{x}^2}{\mathrm{Var}_{x}^2}\right] + 1 - 2\left[1 + \frac{1}{2}\frac{\Delta \mathrm{Var}_{\dot{x}}}{\mathrm{Var}_{\dot{x}}} - \frac{1}{8}\frac{\Delta \mathrm{Var}_{\dot{x}}^2}{\mathrm{Var}_{\dot{x}}^2}\right]\right.\right.$$

$$\left.\left. \times \left[1 - \frac{1}{2}\frac{\Delta \mathrm{Var}_{x}}{\mathrm{Var}_{x}} + \frac{3}{8}\frac{\Delta \mathrm{Var}_{x}^2}{\mathrm{Var}_{x}^2}\right]\right\}\right\rangle. \tag{15.336}$$

Under averaging to a first approximation, we obtain

$$\frac{D\{\bar{f}^*\}}{\bar{f}^2} \approx \frac{1}{T}\left\{ \int_0^\infty \left[\frac{\mathcal{R}''(\tau)}{\mathcal{R}''(0)}\right]^2 d\tau + \int_0^\infty \mathcal{R}^2(\tau) d\tau + 2\int_0^\infty \frac{\mathcal{R}'^2(\tau)}{\mathcal{R}''(0)} d\tau \right\}. \tag{15.337}$$

As applied to the exponential normalized correlation function given by (15.305), we obtain

$$\frac{D\{\bar{f}^*\}}{\bar{f}^2} \approx \frac{3\sqrt{2\pi}}{16\alpha T} \approx \frac{0.47}{\alpha T}. \tag{15.338}$$

Comparing (15.338) with the relative variance of estimate of the average number of Gaussian stochastic process spikes for the same normalized correlation function given by (15.308), we see that according to (15.326), the average mean-square frequency estimate or the average number of spikes leads to the bias of estimate and decrease in estimate dispersion approximately in 2.8 times.

15.8 SUMMARY AND DISCUSSION

Methods of the correlation function estimate can be classified on three groups subject to a principle of realization of delay and other elements of correlators: analog, digital, and analog-to-digital. In turn, the analog measurement procedures can be divided based on the methods using a representation of the investigated stochastic process both as the continuous process and as the sampled process. As a rule, physical delays are used by analog methods with continuous representation of the investigated stochastic process. Under discretization of investigated stochastic process in time, the physical delay line can be changed by corresponding circuits. Under the use of digital procedures to measure the correlation function estimate, the stochastic process is sampled in time and transformed into binary number by analog-to-digital conversion. Further operations associated with the signal delay, multiplication, and integration are carried out by the shift registers, summator, and so on.

We can see that the maximum value of variance of the correlation function estimate corresponds to the case $\tau = 0$ and is equal to the variance of the stochastic process variance estimate given by (13.61) and (13.62). Minimum value of variance of the correlation function estimate corresponds to the case when $\tau \gg \tau_{\mathrm{cor}}$ and is equal to one-half of the variance of the stochastic process variance estimate.

The correlation function of stationary stochastic process can be presented in the form of expansion in series with respect to earlier-given normalized orthogonal functions. The variance of correlation function estimate increases with an increase in the number of terms of expansion in series ν. Because of this, we must take into consideration this fact choosing the number of terms under expansion in series.

In some practical problems, the correlation function of stochastic process can be measured accurately with some parameters defining a character of its behavior. In this case, the measurement of correlation function can be reduced to measurement or estimation of unknown parameters of correlation function. Because of this, we consider the optimal estimate of correlation function arbitrary parameter assuming that the investigated stochastic process $\xi(t)$ is the stationary Gaussian stochastic process observed within the limits of time interval $[0, T]$ in the background of Gaussian stationary noise $\zeta(t)$ with known correlation function.

The optimal estimate of stochastic process correlation function can be found in the form of estimations of the elements R_{ij} of the correlation matrix $\mathbf{R}$ or elements C_{ij} of the inverse matrix $\mathbf{C}$. In the case of Gaussian stationary stochastic process with the multidimensional pdf given by (12.169), the solution of likelihood ratio equation allows us to obtain the estimates C_{ij} and, consequently, the estimates of elements R_{ij} of the correlation matrix $\mathbf{R}$.

Based on (15.272), we can design the flowchart of spectral density measurer shown in Figure 15.13. The spectral density value at the fixed frequency coincides accurately within the constant factor with the variance of stochastic process at the filter output with known bandwidth. Operation principles of the spectral density measurer are evident from Figure 15.13. To define the spectral density for all possible values of frequencies, we need to design the multichannel spectrum analyzer and the central frequency of narrowband filter must be changed discretely or continuously. As a rule, we need to carry out a shift by the spectrum frequency of the investigated stochastic process using, for example, the linear law of frequency transformation instead of filter tuning by frequency. The structure of such measurer is depicted in Figure 15.14. The sawtooth generator controls the operation of measurer changing a frequency of heterodyne.

In many applications, we need to know the statistical parameters of stochastic process spike (see Figure 15.15a): the spike mean or the average number of down-up cross sections of some horizontal level M within the limits of the observation time interval $[0, T]$, the average duration of spike, and the average interval between spikes. To measure these parameters of spikes, the stochastic process realization $x(t)$ is transformed by the nonlinear transformer (threshold circuitry) into the pulse sequence normalized by the amplitude η_τ with duration τ_i (Figure 15.15b) or normalized by the amplitude η_θ with duration θ_i (Figure 15.15c).

The mean-square frequency $\overline{f}$ given by (15.6) is widely used as a parameter characterizing the spectral density of the stochastic process. The value $\overline{f}$ defines a dispersion of component of the stochastic process spectral density relative to zero frequency. As applied to low-frequency stationary stochastic processes, the mean-square frequency $\overline{f}$ characterizes the effective bandwidth of spectral density.

REFERENCES

1. Lunge, F. 1963. *Correlation Electronics*. Leningrad, Russia: Nauka.
2. Ball, G.A. 1968. *Instrumental Correlation Analysis of Stochastic Processes*. Moscow, Russia: Energy.
3. Kay, S.M. 1993. *Fundamentals of Statistical Signal Processing: Estimation Theory*. Upper Saddle River, NJ: Prentice Hall, Inc.
4. Lampard, D.G. 1955. New method of determining correlation functions of stationary time series. *Proceedings of the IEE*, C-102(1): 343.
5. Kay, S.M. 1998. *Fundamentals of Statistical Signal Processing: Detection Theory*. Upper Saddle River, NJ: Prentice Hall, Inc.
6. Gribanov, Yu.I. and V.L. Malkov. 1978. *Selected Estimates of Spectral Characteristics of Stationary Stochastic Processes*. Moscow, Russia: Energy.

Notation Index

AD	arithmetical device
ADC	analog-to-digital converter
AF	additional linear system of filter
ALU	arithmetic and logic unit
AOD	associative output device
ASIC	application-specific integrated circuit
ATC	address translation controller
BA	buffer accumulator
BSOR	block of subsystem operation realizations
CB	control block
CCB	central control block
CCD	central control device
CD	correlation device
CMP	central microprocessor
CRS	complex radar system
DAC	digital-to-analog converter
DDC	digital downconverter
DFT	discrete Fourier transforms
DGD	digital generalized detector
DMF	digital matched filter
DSP	digital signal processing
DTDS	documenting and transferring data subsystems
EM	external memory
FFT	fast Fourier transform
FPGA	field programmable gate array
GD	generalized detector
HDL	hardware description language
IBA	input buffer accumulator
IC	integrated circuits
IDFT	inverse discrete Fourier transform
IF	intermediate frequency
IFFT	inverse fast Fourier transform
I/O	input/output
LO	local oscillator
MMSE	minimum mean-square error
MOTR	multi-object tracking radar
MSD	more significant digit
MSG	model signal generator or local oscillator
MTI	moving target indicator
OBA	output buffer accumulator
PF	preliminary linear system or filter
PRI	pulse repetition interval
RAM	random access memory
RF	radio frequency
ROM	read-only memory

QoS	quality of service
QS	queuing system
SMP	specific microprocessors
SPCS	special-purpose computer subsystems
VLIC	very large integrated circuit
i.i.d.	independent and identically distributed
NLE	not less or equal
R	the register
SNR	signal-to-noise ratio
SR	shift register
$\mathbf{\Delta Y}_n$	the error vector of coordinate measuring
$\mathbf{\Gamma}_n$	the known $s \times h$ matrix
$\mathbf{\Pi}(r)$	the matrix of transient probabilities at the rth radar antenna scanning step
$\boldsymbol{\eta}_n$	the n-dimensional vector of target track parameter disturbance
$\boldsymbol{\eta}_{n-1}$	the $(h \times 1)$ vector of disturbance of target track parameters
$\mathbf{\Theta}$	the space of possible values of the estimated vector parameter θ_n
$\boldsymbol{\theta}(t)$	the target track parameter vector
$\hat{\boldsymbol{\theta}}(t_i)$	the estimate vector of target track parameters
$\hat{\boldsymbol{\theta}}_{\text{ex}_n}$	the $(s\times1)$ vector of extrapolated target track parameters
$\boldsymbol{\theta}_n$	the s-dimensional target track parameter vector at the nth step
$\hat{\boldsymbol{\theta}}_n$	the $(s \times 1)$ vector of estimated target track parameters
$\boldsymbol{\theta}'_{n+1}$	the deterministic (undisturbed) component of the target track parameter vector at the $n + 1$th step
$\mathbf{\Phi}_n$	the transfer $(s \times s)$ matrix of target track model
$\mathbf{\Psi}$	the error correlation matrix of linear target track parameter estimations
$\mathbf{\Psi}_\eta$	the correlation $(h \times h)$ matrix of target track random disturbances
$\mathbf{\Psi}_{\text{ex}_n}$	the correlation $(s \times s)$ matrix of extrapolation of target track parameters
$\mathbf{\Psi}_n$	the correlation $(s \times s)$ matrix of errors of target track parameter estimation
$\mathbf{F}$	the vector-operator of functioning the controlled object
$\mathbf{F}_X(k)$	the discrete Fourier transform
$\mathbf{G}_n$	the matrix filter gain
$\mathbf{g}_{\text{m}_n}$	the l-dimensional vector of the disturbed target track parameters
$\dot{\mathbf{H}}[i]$	the sequence of readings of the complex impulse response of the convolving filter
$\mathbf{H}_n$	the known $m \times s$ matrix defining a function between the observed coordinates and the estimated parameters of target track
$\mathbf{I}$	the identity matrix
$\mathbf{l} = \{l_1, l_2,\ldots, l_\mu\}$	the estimated multidimensional parameter vector
$\mathbf{P}(r-1)$	the row vector of probabilities of states at the previous $(r-1)$th radar antenna scanning step
$\mathbf{Q}$	the appropriate QoS vector
$\lvert\mathbf{R}_N\rvert$	the determinant of correlation matrix
$\mathbf{R}_N^{-1}$	the inverse correlation matrix of measurer error
$\mathbf{R}_n$	the $m \times m$ correlation error matrix of target track coordinate measurements
$\mathbf{S}_{\text{tg}}$	the effective scattering area of target
$\mathbf{U}(t)$	the control signal
$\dot{\mathbf{X}}[i]$	the sequence of complex readings of the input (convolved) function (the target return signal)
$\mathbf{X}_U(t)$	the internal parameters
$\hat{\mathbf{Y}}_{\text{ex}_n}$	the $(m \times 1)$ vector of extrapolated coordinate magnitudes
$\mathbf{Y}_n$	the $(m \times 1)$ vector of measured coordinate magnitudes

A_j^{chp}	the amplitude of chaotic pulse interference
α	the coefficient depending on a direction to the noise source
B_N	the receiver bandwidth
$b[F^*(x)]$	the bias of the probability distribution function estimate
$b(\gamma_E \mid E_0)$	the conditional estimate bias
$b\{\mathrm{Var}^*(t_0, T)\}$	the bias of variance of stochastic process
C	the threshold
D^i	the operator delaying the input data on i cycles
$D(\gamma_E \mid E_0)$	the conditional dispersion of the obtained estimate
d	the length of antenna aperture
$\det \lVert \ldots \rVert$	the determinant of matrix
$\det \boldsymbol{\Psi}_{nf}$	the determinant of the diagonal or factored correlation matrix of errors
$d(T_{sc})$	the variation of radar range coordinate within the limits of the radar antenna scanning period T_{sc}
d_r	the dynamic range of analog part of receiver
E^*	the mathematical expectation estimate of stochastic process
E_0	the true value
E_E, $\hat{E}$	the mathematical expectation estimate
E_S	the energy during scanning
E_s	the energy of the received signal
$E(t_0)$	the mathematical expectation of stochastic process at the instant t_0
$E_Z[\hat{\rho}]$	the mean of decision statistic when the input process samples are correlated with each other
$E[X]$	the mathematical expectation of random variable X
F	the repetition frequency of scanning signals
$F(x)$	the one-dimensional probability distribution function
$F^*(x)$	the estimate of the probability distribution function $F(x)$
$F[x(t) \mid E_0]$	the pdf functional of the Gaussian process
$F(x_1, x_2; \tau)$	the two-dimensional probability distribution function
$\hat{f}$	the mean-square frequency value of the spectral density $S(f)$ of stochastic process $\xi(t)$
$\overline{f_{D_{pi}}}$	the estimator of the average Doppler frequency of passive interference
f_c	the carrier frequency
f_{l-1}, f_l, f_{l+1}	the nonweighted signals at the outputs of l – 1th, lth, and l + 1th Doppler channels
f_l^{weight}	the weighted signal at the output of lth Doppler channel
f_{max}	the cutoff frequency of target return signal spectrum
f_{mod}	the modulation frequency
f_s	the sampling rate
$f(\Delta R, \Delta\beta, \Delta\varepsilon)$	the pdf of target coordinate deviation from the target designation area center
$G(nf_s)$	the Fourier transform of $x(t)$
$G^*(f_0)$	the estimate of spectral density
G_{can}	the coefficient of passive interference cancellation
G_{im}	the coefficient of improvement
G_s	the coefficient characterizing a signal passing through the passive interferences equalizer
G_{tr_0}	the amplifier coefficient of transmit antenna on the phased array axis
G_{tr}	the transmit antenna gain
G_r	the receive antenna gain
G_{r_0}	the amplifier coefficient of receive antenna on the phased array axis

$g(\beta_i, \beta_{tg})$	the function defining the radar antenna directional diagram envelope in the scanning plane
$g(x)$	the quantization step
g_{fi}	the probability of detection at least one false target pip within the gate volume V_i
g_m	the distortion parameter
$h(t)$	the impulse response of linear system
$h_{AF}(\tau)$	the AF impulse response
$h_{i_{rf}}$	the coefficients of digital recursive filter impulse response
$h_{PF}(\tau)$	the PF impulse response
$I_0(x)$	the zero order Bessel function of the first kind
i^{true}	the number of target pip considered true
K	the detection threshold
K_g	the DGD threshold
K_i^{load}	the coefficient of loading
K_{mes}	the covariance matrix of estimation errors
K_{mes}^{tr}	the covariance matrix of estimation errors of target trajectory parameters
K_{red}	the coefficient of reduction
k	the Boltzmann constant
k_{tr}^{can}	the criterion to cancel the target range tracking
k_{use}	the coefficient of energy use
L_{gate}	the number of resolved elements within the gate
$L_\nu(y)$	the Laguerre polynomial
l_{eff}	the effective length of antenna aperture normalized with respect to wavelength
M_R	the number of discrete resolution elements by radar range
$N(E)$	the normalized noise component
N_0	the power spectral density of set noise
$\overline{N}_{a_i}$	the average number of arithmetical operations required for the ith algorithm
N_b	the number of code bits
N_{bz}	the number of scanned directions within the limits of the buffer zone
N_{cell}	the number of cells within the radar coverage
$\overline{N}_D^{true}$	the average number of the true target tracks in detection process
$\overline{N}_{D_j}$	the average number of gates with the number j
N_d	the deliberate interference
N_e	the number of elementary signals
$\overline{N}_{false}$	the average number of false target pips
$\overline{N}_{ftr}$	the average number of false target range tracks
N_i	the number of scans in the radar coverage along the ith direction
$\overline{N}_j^{lock\text{-}in}$	the average number of the primary lock-in gates with the number j
$\overline{N}_j^{scan}$	the average number of jth target tracking gates for all true and false target tracking trajectories
$N_{jl}^{(k)}$	the number of sweeps during the sth scanning of the jth priority zone at the kth control cycle
N_{MTI}	the MTI number
N_{mp}	the required number of the same type microprocessors
N_{N+I}	the power spectral density of total noise and interferences
N_{pi}	the sample size of passive interferences
N_q	the number of quantization levels
N_s	the number of target return pulses in train
N_{sc}	the number of scans in each direction

N_{sp}	the number of convolution operations
N_{tg}	the number of targets within the limits of scanning area
N_{total}	the total average number of operations carried out under a single realization of digital signal processing algorithm
$\bar{N}_{\text{true}}$	the average number of true target pips
$\bar{N}_{\text{true}}^{\text{lock-in}}$	the average number of the primary lock-in gates of true target tracks
N_{φ}	the number of phase counts
$\bar{n}^{\mathcal{H}_1}$	the average number of tests at sequential analysis procedure in the direction of target presence
$\bar{n}^{\mathcal{H}_0}$	the average number of tests at sequential analysis procedure in the direction of target absence
n_0	the number of discrete elements of the model signal
$\bar{n}_2^{\mathcal{H}_1}$	the average number of tests at sequential analysis procedure in the direction of target presence at the second stage
$\bar{n}_2^{\mathcal{H}_0}$	the average number of tests at sequential analysis procedure in the direction of target absence at the second stage
n_{b}	the number of bit site
$\bar{n}_{\text{false}}^{\text{scan}}$	the average number of false target pips inside the target tracking gates under each scanning
$\bar{n}_k$	the average cycle of sequential analysis at the kth radar range resolution element
n_{sc}	the number of points of location on the target tracking
$n_{\text{tg}}^{\text{track}}$	the number of tracked targets
n_{tr}	the truncation threshold
$\bar{n}_{\text{true}}^{\text{scan}}$	the average number of true target pips coming in the target tracking gates under each scanning
P	the probability
$P\left(g_{m_{jn}} \| \{\mathbf{Y}\}_n\right)$	the a posteriori probability of the event $g_{m_{jn}} = g_{m_j}$ by data of n measurements $\{\mathbf{Y}\}_n$
$P\left(g_{m_{jn}} \| \{\mathbf{Y}\}_{n-1}\right)$	the a priori probability of the parameter g_{m_j} at the nth step
$P_{1j}^{(r-i)}$	the probability of system transition from the initial state a_1 to the state a_j for $r-j$ steps
P_{2_j}	the probability of success under the parameter clarification of the jth target
$P_{3_{ij}}$	the probability of destruction of the jth target
P_{beg}	the probability that the target pip belongs to beginning the new target track
P_{br}	the probability of breaking up in the target tracking
$P_{\text{circuit}}(t_{\text{real}})$	the probability of microprocessor subsystem circuit drop-in within the limits of t_{real}
P_{D}	the probability of detection
P_{DD}	the probability of the true target pip detection into the detected target track gates
$P_{\text{D}}^{\text{gate}}$	the probability of detection of true target pips within the limits of the target tracking gates
$P_{\text{D}}^{\text{lock-in}}$	the probability of the true target pip detection into the primary lock-in gates
P_{D}^{tr}	the probability of target trajectory detection
$P_{\text{D}_j}^{\text{false}}$	the probability of the false target pip hit into the detected target track gates with the number j
P_{dl}	the power of noise source
$P_{\text{er}}(C)$	the probability of wrong decision at sequential analysis
$P_{\text{er}}^{\text{tr}}$	the probability of error at the truncation procedure
$P_{\text{er}}(\bar{n} < n_{\text{tr}})$	the probability of wrong decision at sequential analysis procedure before truncation

P_F	the probability of false alarm
P_{F_1}	the probability of false alarm at the first stage
P_F^{tr}	the probability of target trajectory false alarm
$P_{F_j}^{lock\text{-}in}$	the probability of false target pip hit into the primary lock-in gates with the number j
$P_{F_j}^{scan}$	the probability of the false target pip hit into the jth target tracking gate
$P_{f_{tr}}$	the unconditional probability of detection of the false target range track
$P_{fail}(t_{real})$	the probability of microprocessor subsystem failure within the limits of t_{real}
$P_{failure}$	the probability of target tracking failure within the limits of scanning range
P_{ind}	the probability of correct indication
$P_{ind}(V_j)$	the probability of indication of true target pip among false target pips within the gate volume V_i
P_{miss}	the probability of miss
P_{new_i}	the probability of event that the target pip belongs to a newly detected target
P_{pi}^{out}	the power of the passive interference at the equalizer output
P_p	the transmit pulse power
P_{pi}^{in}	the power of the passive interference at the equalizer input
P'_{sc}	the probability of duplicate scanning in the "no" signal direction
P_{suc}^{dest}	the probability of target destruction
P_{ij}	the probability of the jth armed attack facility damage by the ith defended facility
P_p	the power of transmitted pulse
$P_s(\phi_i)$	the probability of target detection in the ith cell under the condition that the energy φ_i is required to do it
$\bar{P}_{scan}$	the average power of scanning signal
P_{scan}	the power of scanning pulse
P_{suc}^{dest}	the conditional probability of target destruction
P_t^{av}	the transmit power
$P_t(i)$	the a priori probability of target presence in the ith cell
$P_{t_N}^{*}(\upsilon)$	the a posteriori probability of target presence in the υ th cell determined at the instant t_N
P_{td}	the probability to lock-in the target
P_{td_j}	the probability of success under the target designation with respect to the jth target
$P_{tg}(r)$	the probability of target detection at the rth radar antenna scanning
$P_{tgp_i}^{true}$	the probability of detection of a newly obtained target pip
$P_{x_{ij}}$	the probability of deviation of the ith target pip from the extrapolated point of the jth target range track
$p\left(\mathbf{Y}_n \middle\| g_{m_{j,n-1}}\right)$	the conditional pdf of the observed magnitude of the coordinate $\mathbf{Y}_n$
$p(\boldsymbol{\theta}_n \| \{\mathbf{Y}\}_n)$	the a posteriori pdf of the current value of parameter vector $\boldsymbol{\theta}_n$ by sequence of measured data $\{\mathbf{Y}\}_n$
$p(\mathbf{Y}_n \| \boldsymbol{\theta}_n)$	the likelihood function of the last nth coordinate measuring
$p(x_1, x_2, \ldots, x_n \| l_{fix})$	the pdf of observed data sample at some fixed value of the estimated random process parameter l_{fix}
$p_2(x_1, x_2; \tau)$	the two-dimensional pdf of the observed stochastic process
$p^*(t)$	the pdf estimate
$p_{post}(\gamma)$	the a posteriori pdf
$p_{pr}(l)$	the a priori pdf
$p_{tr}(r)$	the probability of true target pip detection in the course of the rth radar antenna scanning
p_{beg}	the conditional probability of new target range track beginning

p_{conf}	the conditional probability of the confirmation of new target range track beginning
$Q(x)$	the Gaussian Q-function
Q_{BM}	the required buffer memory
Q_{ROM}	the ROM size
Q_{digit}	the RAM cell array assigned to store numerical information
Q_{in}	the RAM cell array assigned to receive external information
Q_{out}	the RAM cell array assigned to store the digital signal processing results
$Q_{routine}$	the RAM cell array assigned to store the routines
$Q_{working}$	the RAM cell array participating at the calculation process
q^2	the signal-to-interference-plus-noise ratio
q_0^2	the signal-to-noise ratio (SNR)
q_μ	the signal-to-noise ratio for the μth component (or harmonic) of the mathematical expectation
$R(t_1, t_2)$	the correlation function
R_0	the far-field region boundary of destruction area
R_{dI}	the distance between the complex radar system and the noise source
R_{eff}	the effective radius of target destruction
$R_n[k]$	the correlation function of the noise
$\hat{R}[k]$	the estimate of noise correlation function
$R_{ss^*}(\tau_0 - t)$	the autocorrelation function of expected signal $s(t, \alpha)$
$R_{td_j}^{eff}$	the effective radar range
R_0	the far-field region boundary of destruction area
$R_F(x_1, x_2)$	the correlation function of the probability distribution function estimate $F^*(x)$
$R_{ij}(\nu \mid l)$	the conditional mutual correlation function of the estimations of the parameters l_i and l_j
R_t	the distance to the target
$R_\zeta(t_1, t_2)$	the correlation function of random component of the investigated squared stochastic process with respect to its variance at the instant t
r_0	the range of the nearest to origin point on the target track
r_n	the target track parameters representing the target track coordinate
$\dot{r}_n$	the velocity of target track coordinate variation
$\ddot{r}_n$	the acceleration by the target track coordinate
r_{tg}, r_n	the target range
r_{tg_j}	the target range measure by a single reading
r_{ε_i}	the distance from the radar to the radar coverage bound under the given elevation angle value ε_i
$S(t_i, \alpha)$	the target return signal
$S(E)$	the normalized signal component
S_1	the gate with primary lock-in
S_{eff}	the effective area of the receive antenna
S_{gate}	the gate area
S_{pi}	the energy spectrum of passive interferences
S_t^{ef}	the effective target reflective surface
S_{tg}	the information density equal to the average number of target traverses of the radar coverage external border per unit time
$s(t)$	the signal
$s^*(t, \alpha)$	the expected signal model generated by a local oscillator in a receiver or a detector
sinc(x)	the sinc function

T	the absolute temperature
T_0	the absolute temperature of signal source
T_c	the time of conversion
T_{cc}	the time of conversion cycle
T_k^{total}	the assigned period of scanning for the kth zone
$T_{(l+1)_i}^{total}$	the period of the ith scanning for the l + 1th zone
T_m	the average duration of target maneuver
T_{max}	the maximal repetition period
T_{mp}^{single}	the realization time of all algorithms for a single microprocessor subsystem
T_{new}	the period of getting new information about the target
T_p	the period of pulse repetition
T_{renew}	the period of refreshment at the preliminary target tracking stage
T_{S_i}	the period of the ith zone scanning
T_s	the sampling period
T_{scan}	the radar antenna scanning period
T_{scan}^{per}	the permissible period of scanning
T_{th}	the threshold value
T_{sol}	the time to solve the problem
$t_\Sigma^{(k)}$	the time required to scan the priority zones
t_0	the observation time
t_{d_j}	the time required to detect the jth target
t_{finish}	the instant to finish the target maneuver
$\overline{t_i}^{wait}$	the average waiting time for request queue before the ith phase
t_k^*	the limiting value of average request-queuing system time for requests of the kth incoming signal flux
$\overline{t_{k_i}^{total}}$	the average time required to scan the priority zone at the ith scanning
t_{l_j}	the time of lock-in of the jth target to track
$t_{(l+1)_i}^{total}$	the time required to scan the no priority (l + 1)th zone at the ith scanning
$\overline{t}_{rc}$	the average time of target remaining within the limits of radar coverage
t_{real}	the time of a single realization of digital signal processing algorithm for the solved problem
$\overline{t}_{S_i}$	the average time to search the ith zone
t_{start}	the instant to start the target maneuver
$t_{tr_{jtd}}$	the time of the jth target tracking
t_{td_j}	the time required to transmit information about the target designation
t_{tr_j}	the tracking time of the jth target to track
$\overline{t}_{wait}$	the average waiting time for all requests
$\mathrm{tr}\,\boldsymbol{\Psi}_{n_f}$	the correlation matrix of errors
U_0	the threshold of binary quantization
V_{el}	the volume of total error ellipsoid
V_{par}	the volume of total error parallelepiped
V_A	the scanning speed of antenna beam
Var*	the variance estimate
$\mathrm{Var}(\gamma \vert l)$	the estimate variance
Var_{prior}	the variance of a priori distribution
$\mathrm{Var}\{\mathrm{Var}^*(t_0, T)\}$	the variance of variance estimate
$V_{DS}^{effective}$	the effective speed of the operation of the detector–selector
V_{eff}	the average number of operations per second under realization of the specific digital signal processing algorithm

$V_{\text{eff}\,i}$	the effective operation speed of the *i*th microprocessor
V_{rel_j}	the relative velocity of the *j*th target motion
V_{tg}	the target velocity
$\hat{V}_{X_n}^{\text{tg}}$	the estimation of target velocity by the Cartesian coordinates *X*
$\hat{V}_{x_{\max}}$	the maximal speed of the variations of the sampled and quantized target return signal
$\hat{V}_{Y_n}^{\text{tg}}$	the estimation of target velocity by the Cartesian coordinates *Y*
$W(\omega)$	the even weight function of frequency ω or the function of spectral window
w	the target return signal weight
$w[i]$	the weight "window" function
$w(t)$	the stationary white Gaussian noise
$X(t)$	the low-frequency signal (envelope)
$\dot{X}(t)$	the complex envelope of the narrowband signal
X_i	the samples of the input signal
$x(t)$	the target return signal
$x(nT_s)$	the sampled target return signal
$x_I(t)$	the in-phase component of the narrowband target return signal $x(t)$
x_i	the relative envelope amplitude
$\tilde{x}_i$	the reference noise with a priori information, a "no" signal
$x_0(t) = x(t) - E(t)$	the centered Gaussian stochastic process
$x_Q(t)$	the quadrature component of the narrowband target return signal $x(t)$
$x_\delta(t)$	the instantaneously (ideal) sampled target return signal
$\bar{Z}_{nil}^2$	the squared signal amplitude on next (*n*th) update step by the data of three Doppler zero velocity filters at *i*th resolution element by radar range at *l*th azimuth direction
$\hat{Z}_{(n-1)il}^2$	the previous estimation of squared signal amplitude at *i*th resolution element by radar range at *l*-th azimuth direction
$Z_{\text{GD}}^{\text{out}}(t)$	the process forming at the GD output
$Z_{\text{DGD}}^{\text{out}}(r_i)$	the known rank function
$Z_{\text{IQ}_{\text{nr}}}^{\text{in}}[n-i]$	the in-phase and quadrature components of signal at the filter input (the DGD output)
Δ_β	the gate sampling interval by the coordinates β
Δ_r	the gate sampling intervals by the coordinate *r*
$\Delta\beta_{\text{gate}}$	the angular size by azimuth in radian
$\Delta\hat{\varphi}_{\text{D}_{\text{pi}}}$	the equivalent Doppler shift in phase within the limits of the period *T*
ΔF_0	the target return signal frequency deviation
Δf_{Dmax}	the Doppler frequency range
Δf_{dI}	the noise bandwidth
Δ_{F}	the AF or PF bandwidth
Δf_{pi}	the bandwidth of passive interference spectrum at the level 0.5
Δf_{s}	the signal bandwidth
$\Delta f_{\text{s}}^{\text{eff}}$	the effective signal bandwidth
Δf_{tg}	the spectrum bandwidth of the fluctuated target return signal
Δx	the quantization step
$\Delta\mathscr{L}_{g\mu}$	the likelihood ratio increment at the μ th step of sequential analysis
$\lvert\Delta\mathcal{R}\rvert$	the maximum allowable value of interpolation error of the normalized correlation function $\mathcal{R}(\tau)$
$\Lambda(l)$	the likelihood ratio
$\hat{\Lambda}(l)$	the likelihood function
α	the regularization parameter
α_k	the weight coefficients

β_0	the azimuth of the nearest to origin point on the target track
β_{angle}	the angular direction of phased array axis
β_i	the value of azimuth angle under receiving the ith pulse of the pulse train
β_k	the losses caused by the queue length for each signal flux request queuing
β_Q	the losses caused by idle mode
β_n	the azimuth
$\gamma(\beta_i, \hat{\beta}_{tg})$	the discrete weight function
$\gamma = \gamma[x(t)]$	the point estimate
δ_{ij}	the Kronecker symbol
$\delta(t)$, $\delta(\tau)$, $\delta(z)$	the Dirac delta function
δf_D	the complex radar system resolution by Doppler frequency
$\delta\theta_{tg}$	the required resolution to estimate the target angular coordinate
ε_{angle}	the angular direction of phased array axis
ε_n	the elevation
$\zeta(t)$	the fluctuations of square of the investigated stochastic process realization with respect to its mathematical expectation at the same instant t
$\zeta_1(kT_s)$	the amplitude quantization Gaussian noise
$\zeta_2(kT_s)$	the amplitude sampling Gaussian noise
$\zeta_\Sigma(kT_s)$	the summary additive amplitude quantization and sampling interference
η	the figure of merit
$\eta_{\hat{x}_{N+1}}(i)$	the weight function of measured coordinate magnitudes
Θ	the space of the possible values of estimated target track parameter
θ	the direction of arrival
θ_{av}	the average interval between the spikes of stochastic process realization at the level M
θ_b	the radar antenna directional diagram beam width
γ_{rel}	the relative threshold
λ	the wavelength
μ_{ν_i}	the central moment of the νth order
ν	the dynamic range in dB for quantized sequence of samples
$\nu(t, \theta_i)$	the loss function formed by the target with the parameter θ_i
ν_{in}	the sample size of interference determined by analogous way as for ν_s
ν_s	the sample size of the target return signal equal to $N_s - 1$
$\xi(t)$	the Gaussian stochastic process
$\xi_{AF}(t)$	the narrowband Gaussian noise forming at the AF output
$\xi_{AF}^2(t) - \xi_{PF}^2(t)$	the background noise forming at the GD output
$\xi_{AF_\Sigma}(kT_s)$	the noise forming at the AF output of DGD input linear system
$\xi_{PF}(t)$	the narrowband Gaussian noise forming at the PF output
$\xi_{PF_\Sigma}(kT_s)$	the noise forming at the PF output of DGD input linear system
$\rho_{en}(\tau)$	the envelope of normalized correlation function
ρ_{ij}	the interperiod correlation coefficient
$\rho_{pi}[\cdot]$	the coefficient module of passive interference interperiod correlation
ρ_S	the density of false target pips per unit square of gate
ρ_s	the coefficient of correlation of the target return signal
ρ_{scan}	the density of newly obtained target pips per unit area of scanning
σ_0^2	the variance of receiver noise
$\hat{\sigma}_{chp}^2$	the estimation of chaotic pulse interference variance
$\sigma_{extr_j}^2$	the variance of errors under coordinate extrapolation of the jth target range track

σ_{f_D} the root-mean-square value of potential error under the measurement of the Doppler frequency
σ_g^2 the variance of target maneuver intensity for each target
σ_{hit}^2 the error variance to hit the target by missile
$\sigma_{launcher}^2$ the error variance of antitarget guidance by a launcher
$\sigma_{l_n}^2$ the variance of error of lth parameter estimation at nth filtering step
σ_m^2 the variance of the target maneuver intensity
$\sigma_{meas_i}^2$ the variance of errors of coordinate measurement of the ith target pip
σ_{mes}^2 the variance of estimation error
σ_{miss}^2 the variance of miss
$\sigma_{missile}^2$ the error variance of missile flight path
σ_n^2 the white Gaussian noise variance
σ_{out}^2 the total error variance at the digital filter output
$\sigma_{pi_{in}}^2$ the variance of the passive interference at the equalizer input
$\sigma_{pi_{out}}^2$ the variance of the passive interference at the equalizer output
$\sigma_{r_j}^2$ the variance of estimation of the target range measure by a single reading
σ_q^2 the variance of quantization errors
$\sigma_{q\Sigma}^2$ the variance of statical quantization errors
$\sigma_{S_i}^2$ the variance of target return signal amplitude
$\sigma_{t_d}^2$ the variance of error of a single reading of the time delay t_d
$\sigma_{x_m}^2$ the variance of coordinating measuring errors
$\sigma_Z^2(\hat{\rho})$ the variance of decision statistic when input process samples are correlated with each other
σ_η the mean-square deviations of true target pips from the gate center along the axes η
$\sigma_{\zeta_1}^2$ the variance of amplitude quantization errors
$\sigma_{\zeta_2}^2$ the variance of amplitude sampling errors
$\sigma_{\zeta_\Sigma}^2$ the summary additive amplitude quantization and sampling interference variance
$\sigma_{\hat{\theta}_j accept}^2$ the acceptable value of filtering error variance at the line of the release of information
σ_ξ the mean-square deviations of true target pips from the gate center along the axes ξ.
$\sigma_{\xi_{AF\Sigma}^2-\xi_{PF\Sigma}^2}^2$ the variance of the total background noise and interferences forming at the DGD output
σ_θ the root-mean-square value of potential error under measurement of the angular coordinates
σ_τ the root-mean-square value of potential error under measurement of delay
τ the pulse duration
τ_0 the duration of the sampled signal
τ_{av} the average duration of spike of stochastic process realization at the level M
τ_{clean} the time to clean memory cells
τ_{cor} the correlation interval
τ_{cycle}^{read} the cycle time at reading
τ_{cycle}^{rec} the cycle time at recording
$\overline{\tau}_{DS}$ the average request queuing time
τ_{ex} the interval of extrapolation time
$\overline{\tau}_i$ the average request queuing time during the ith phase

$\overline{\tau}_{\text{memory}}$	the average time to fill out the memory
τ_{queue}	the time of a single request queue
$\overline{\tau}_{\text{queue}}$	the average time queue
τ_{read}	the time to read information
τ_{rebuild}	the time to repair information destroyed under reading
τ_{rec}	the time to record information
τ_s	the duration of scanning signal
$\overline{\tau}_{\text{scan}}$	the average time of directional analysis, in which there are targets
$\tau_{\text{scan}}^{\text{comp}}$	the duration of compressed linear-frequency scanning signal
τ_{search}	the time to search information
$\overline{\tau}_{\text{tg}}$	the average time interval between two considered events (the target traverses)
$\Phi(x)$	the standard normal Gaussian pdf
$\Phi_0(x)$	the integral of probability
φ_0	the one-half main beam width of the radar antenna directional diagram at zero level
$\varphi_k(t)$	the orthogonal functions
$\varphi(\vartheta)$	the function characterizing the shape of radar antenna directional diagram
$\varphi(t)$	the phase modulation law of the narrowband signal
φ_{dd}	the width of radar antenna directional diagram in the searching plane at the given power P level
φ_i	the energy to scan the ith cell
χ_{DS}	the loading factor
χ_{MTI}	the MTI loading factor
Ω	the switching function frequency
ω_0	the resonance frequency
$\varpi(ij\|i_0j_0)$	the set of the weight coefficients
$\mathscr{F}(\cdot)$	the coefficient of asymptotic relative efficiency
$\mathscr{F}[z]$	the nearest integer *NLE* z
$\mathscr{F}_1(\omega)$	the Fourier transform of the normalized correlation function envelope $\rho_{\text{en}}(\tau)$
$\mathscr{H}(f)$	the filter transfer function module
$\mathscr{K}$	the generalized loss factor
$\mathscr{L}$	the total loss factor
$\mathscr{L}(x\|\theta)$	the conditional likelihood ratio
$\mathscr{L}(\boldsymbol{\theta}_N)$	the likelihood function of the vector parameter $\boldsymbol{\theta}_N$ estimated by sequent measurements $\{\mathbf{Y}_N\}$
$\mathscr{L}(\beta_{\text{tg}})$	the likelihood function of the estimated target angular coordinate β_{tg}
$\mathscr{L}(\gamma, E)$	the quadratic loss function
$\mathscr{L}(\gamma, l)$	the loss function
$\mathscr{L}_{\text{ac}}$	the coefficient taking into consideration losses under accumulation process
$\mathscr{L}_{\text{g}}$	the likelihood ratio for the generalized signal processing algorithm
$\mathscr{L}_{g_{\mu-1}}$	the accumulated likelihood ratio over $\mu - 1$ steps
$\mathscr{M}_{\text{tg}}$	the target importance
$\mathscr{N}(t_i)$	the noise
$\mathscr{N}_0$	the two-sided power spectral density of white noise
$\mathscr{N}_{\Sigma}$	the total noise power spectral density at the receiver input
$\mathscr{N}_{\text{in}}^{\text{active}}$	the average power spectral density of active interferences
$\mathbf{N}(l)$	the normalized noise function
$\mathscr{P}$	the value of classification factor
$\mathscr{R}^{\text{av}}(\cdots)$, $\mathscr{R}(\gamma)$	the average risk

$\mathcal{R}_m$	the Bayes risk
$\mathcal{R}_{\text{post}}(\gamma)$	the a posteriori risk
$\mathcal{R}(i\Delta)$	the normalized correlation function of observed stochastic process
$\Re(\tau)$	the normalized correlation function
$\mathbf{S}(l)$	the normalized signal function
$\mathbf{S}(\omega)$	the Fourier transform of mathematical expectation of the function $s(t)$ – the spectral density
$\mathbf{S}(j\omega)$	the frequency characteristic of the low-pass filter
$\mathcal{S}(\omega)$	the spectral power density

Index

D

N

O

P

Q

R

S

For Product Safety Concerns and Information please contact our EU representative GPSR@taylorandfrancis.com Taylor & Francis Verlag GmbH, Kaufingerstraße 24, 80331 München, Germany